THE ROUTLEDGE HANDBOOK OF URBANIZATION AND GLOBAL ENVIRONMENTAL CHANGE

Urban areas are major drivers of global environmental change that has become a central element of the sustainability challenge. In response, this volume takes the position that the fundamental question is not how we achieve urban sustainability, but rather how we achieve sustainability in an urbanizing world. To address this issue analytically, the volume provides a comprehensive review and synthesis of more than 10 years of research for researchers, practitioners and students interested in the effects of contemporary urbanization on the global environment as well as how global environmental change affects urban areas. Drawing on the knowledge of over 70 recognized experts from around the world, the Handbook provides a cutting-edge introduction to the field, benefiting from the huge breadth of disciplines represented.

Organized into four thematic sections, the Handbook first considers the impact that urbanization has on global change, before looking next at how global environmental change already has begun to affect urban areas and their management. The third section assesses the extent to which urban systems have adapted in response to environmental stressors, with the concluding section exploring critical new urban research and the implications for the future.

This authoritative volume is an essential resource across a range of disciplines, including environmental studies, sustainability science, urban studies, geography, climate science, planning and related disciplines.

Karen C. Seto is Associate Dean of Research and Professor of Geography and Urbanization Science at the Yale School of Forestry and Environmental Studies, Yale University, USA, and Co-Chair of the Urbanization and Global Environmental Change Project of Future Earth.

William D. Solecki is Professor of Geography at Hunter College, City University of New York, USA, and Director Emeritus of the CUNY Institute for Sustainable Cities.

Corrie A. Griffith is Executive Officer of the Urbanization and Global Environmental Change Project of Future Earth, headquartered at the Julie Ann Wrigley Global Institute of Sustainability, Arizona State University, USA.

THE ROUTLEDGE HANDBOOK OF URBANIZATION AND GLOBAL ENVIRONMENTAL CHANGE

*Edited by Karen C. Seto,
William D. Solecki and Corrie A. Griffith*

LONDON AND NEW YORK

First published 2016
by Routledge
2 Park Square, Milton Park, Abingdon, Oxon OX14 4RN

and by Routledge
711 Third Avenue, New York, NY 10017

Routledge is an imprint of the Taylor & Francis Group, an informa business

© 2016 Karen C. Seto, William D. Solecki and Corrie A. Griffith

The right of the editors to be identified as the author of the editorial material, and of the authors for their individual chapters, has been asserted by them in accordance with sections 77 and 78 of the Copyright, Designs and Patents Act 1988.

All rights reserved. No part of this book may be reprinted or reproduced or utilised in any form or by any electronic, mechanical, or other means, now known or hereafter invented, including photocopying and recording, or in any information storage or retrieval system, without permission in writing from the publishers.

Trademark notice: Product or corporate names may be trademarks or registered trademarks, and are used only for identification and explanation without intent to infringe.

British Library Cataloging in Publication Data
A catalogue record for this book is available from the British Library

Library of Congress Cataloging in Publication Data
The Routledge handbook of urbanization and global environmental change /
edited by Karen Seto, William Solecki and Corrie Griffith.—1 Edition.
    pages cm
Includes bibliographical references and index.
1. Urbanization. 2. Sustainable development. 3. Land use. I. Seto, Karen Ching-Yee, editor. II. Solecki, William, editor. III. Griffith, Corrie, editor.
HT361.R68 2016
307.76—dc23   2015027517

ISBN: 978-0-415-73226-0 (hbk)
ISBN: 978-1-315-84925-6 (ebk)

Typeset in Bembo
by Keystroke, Station Road, Codsall, Wolverhampton

Printed and bound in the United States of America by Publishers Graphics, LLC on sustainably sourced paper.

Dedicated to the memory of JoAnn Carmin (1957–2014)

JoAnn Carmin was an internationally recognized expert on environmental governance, policy and urban climate change adaptation. In the course of her career, she made significant contributions to our scientific understanding of environmental inequalities and marginalized populations. She also was a pioneer in the study of how cities adapt to climate change. Her work in this area helped establish this rapidly growing field and brought much needed policy and scientific attention to vulnerable urban populations and places. JoAnn will be remembered for her intellectual rigor, refreshing candor, compassion and passion for great food.

CONTENTS

List of figures	*xi*
List of tables	*xv*
Contributors	*xvii*
Acknowledgments	*xxiii*

Introduction   1
William D. Solecki, Karen C. Seto and Corrie A. Griffith

PART I
Pathways by which urbanization impacts environmental systems and become drivers of global environmental change   7

1   Urbanization, economic growth and sustainability   9
Michail Fragkias

2   Urbanization, food consumption and the environment   27
Shannon Murray, Samara Brock and Karen C. Seto

3   Urbanization and impacts on agricultural land in China   42
Xiangzheng Deng

4   Urban land use in the global context   50
Dagmar Haase and Nina Schwarz

5   Contemporary urbanization in India   64
Chandana Mitra, Bhartendu Pandey, Nick B. Allen and Karen C. Seto

Contents

6 Resource use for the construction and operation of built environments 77
Burak Güneralp

7 Harnessing urban water demand: lessons from North America 93
Patricia Gober and Ray Quay

8 Urbanization, energy use and greenhouse gas emissions 106
Peter J. Marcotullio

9 Suburban landscapes and lifestyles, globalization and exporting
the American dream 125
Robin M. Leichenko and William D. Solecki

10 Urbanization, habitat loss and biodiversity decline: solution
pathways to break the cycle 139
Thomas Elmqvist, Wayne C. Zipperer and Burak Güneralp

11 Urban precipitation: a global perspective 152
Chandana Mitra and J. Marshall Shepherd

12 Effects of urbanization on local and regional climate 169
CSB Grimmond, Helen C. Ward and Simone Kotthaus

13 Urban nutrient cycling 188
Lucy R. Hutyra

PART II
**Impacts of global environmental change on urban systems
and urbanization processes** **201**

14 A broader framing of ecosystem services in cities: benefits
and challenges of built, natural or hybrid system function 203
Nancy B. Grimm, Elizabeth M. Cook, Rebecca L. Hale
and David M. Iwaniec

15 Urbanization, vulnerability and risk 213
Patricia Romero-Lankao and Daniel Gnatz

16 Extreme events and their impacts on urban areas 229
Andrea Ferraz Young

17 Water supply and urban water availability 247
Stephan Pfister, Stefan Schultze and Stefanie Hellweg

Contents

18 Urban water quality 262
    Conor Murphy

19 Urban greening, human health and well-being 276
    Alexei Trundle and Darryn McEvoy

20 Food price volatility and urban food insecurity 293
    Marc J. Cohen and James L. Garrett

21 Urbanization and global environmental change: from a
    gender and equity perspective 310
    Gotelind Alber and Kate Cahoon

PART III
Urban responses to global environmental change **325**

22 Environmental justice and transitions to a sustainable urban future 327
    Christopher G. Boone and Sonja Klinsky

23 Global patterns of adaptation planning: results from a global survey 336
    Linda Shi, Eric Chu and JoAnn Carmin

24 Adaptation to climate change in rapidly growing cities 350
    Roberto Sánchez-Rodríguez

25 Spatial planning: an integrative approach to climate change
    response 364
    Shu-Li Huang and Szu-Hua Wang

26 Climate change mitigation in high-income cities 377
    Eugene Mohareb, David Bristow and Sybil Derrible

27 Climate change mitigation in medium-sized, low-income cities 406
    Shuaib Lwasa

28 Urban and peri-urban agriculture: cultivating urban climate resilience 421
    *John P. Connors, Corrie A. Griffith, Camille L. Nolasco, Bolanle Wahab
    and Frank Mugagga*

29 Integrating biodiversity and ecosystem services into urban
    planning and conservation 441
    Robert McDonald

Contents

30 The potential of the green economy and urban greening for
addressing urban environmental change 455
David Simon

31 Positive externalities in the urban boundary: the case of industrial
symbiosis 470
Marian Chertow, Junming Zhu and Valerie Moye

32 Resilient urban infrastructure for adapting to extreme
environmental disruptions 488
Rae Zimmerman

33 Soft and hard infrastructure co-production and lock-in: the
challenges for a post-carbon city 513
Stephanie Pincetl

PART IV
Urbanization, global change and sustainability: critical
emerging integrative research **525**

34 Adaptation, transformation and transition: approaches to the
sustainability challenge 527
Charles L. Redman

35 City action for global environmental change: assessment and
case study of Durban, South Africa 534
Debra Roberts

36 Climate change mitigation in rapidly developing cities 549
Alexander Aylett, Boyd Dionysius Joeman, Benoit Lefevre,
Andrés Luque-Ayala, Atiqur Rahman, Debra Roberts,
Sarah Ward and Mark Watkins

Conclusion: the road ahead for urbanization and sustainability research 561
Karen C. Seto and William D. Solecki

Index 576

FIGURES

1.1	Percent urban population and real GDP per capita, 1970–2012, world countries with available data for above years
1.2	Scatterplot of log of percentage urban population and log of per capita real GDP, World countries, 1970–2011 including linear fit
1.3	Scatterplot of logarithms of total personal income against log of total population, United States MSAs, 1992–2008, including regression fitted values
1.4	Scatterplot of logarithms of metropolitan GDP against log of total population, 275 world MSAs, 2010, including regression fitted values
1.5	Scatterplot of metropolitan CO_2 emissions per capita over population density, 275 world MSAs, 2010
1.6	Scatterplot of the natural log of metropolitan CO_2 emissions against log of GDP per capita, 275 world MSAs, 2010, including linear fit
2.1	A conceptual framework: links between urbanization, the food system and global environmental change
2.2	Trends in global meat production, 1960–2050
2.3	Urban population and food consumption at home (national scale), 2012
4.1	Percentage urban and location of urban agglomerations with at least 500,000 inhabitants, 2014
4.2	Urban and rural population as proportion of total population, by major areas, 1950–2050
4.3	Mean soil sealing per urban morphological zone (UMZ) and soil sealing per capita in European cities
4.4	Types of rural-urban gradients
5.1	a) State-wise aggregates of area under urban expansion (1998–2008), as quantified from DMSP/OLS NTLs and SPOT-VGT data; location of urban land use changes in b) northern states and c) southern states
5.2	Rate and magnitude of India's population growth
5.3	Population distribution by settlement size, 1951–2011
5.4	State-wise patterns of agricultural land loss
5.5	Urban expansion in the greater Kolkata metropolitan area from 1756 to 2000, derived from historical maps and GIS analysis

1.1   13
1.2   14
1.3   16
1.4   17
1.5   18
1.6   19
2.1   28
2.2   30
2.3   33
4.1   52
4.2   53
4.3   56
4.4   59
5.1   65
5.2   66
5.3   68
5.4   70
5.5   72

Figures

5.6	Graphical representation of changes in urban area, Kolkata city, 1756–2000	73
6.1	Global cement and steel production, 1970–2012	79
7.1	Indoor water consumption in the US	94
7.2	Monthly water production in the City of Portland on a per capita basis	95
7.3	Residential metering and per capita consumption in Canada, 1991–2009	97
7.4	Declining per capita water use in Phoenix, 1990–2013	97
8.1	A critical knowledge pathway to low-carbon, sustainable futures: integrated understanding of urbanization, urban areas and carbon	107
9.1	The rich spread out: median density of large cities (persons per hectare)	128
10.1	Endemic vertebrate species expected to be lost due to urban area expansion: (A) the 25 most threatened ecoregions are shown with red dots; (B) the majority of species loss due to urbanization will be in a small fraction of ecoregions	141
10.2	Urban extent within a distance of 50 km of PAs by geographic region *c.* 2000 and as forecasted in 2030	142
10.3	Urban extent in biodiversity hotspots *c.* 2000 and as forecasted in 2030	143
11.1	Towards a conceptualization of the urban rainfall effect	156
11.2	Maximum downwind precipitation suppression and invigoration (cm) for downwind regions northeast and southeast of circular city center for cities of different sizes	157
11.3	Urban precipitation studies conducted across developed and developing world cities	159
11.4	Monthly precipitation distribution (mm) in the urban and rural areas of Benin City	160
11.5	Cumulative rains over an eight-day period (12 to 20 April 2004) in the Sylhet region of northeastern Bangladesh	161
12.1	The effects of the urban landscape and urban emissions on key parameters and processes resulting from urbanization	170
12.2	Key spatial scales in the urban atmosphere	171
12.3	Measured wind speeds (m s^{-1}) at various heights in central London during a particularly windy period in February 2014	172
12.4	Mean daily air temperature, relative humidity, vapor pressure deficit and solar (short-wave) radiation for a two-year period from an urban (dark grey, central London, Kotthaus and Grimmond 2014a,b), suburban (medium grey, residential Swindon, Ward et al. 2013) and rural (light grey, Alice Holt deciduous forest, Wilkinson et al. 2012) site in southern England	174
12.5	Effect of surface materials on reflected short-wave radiation: (a) circle shows 80 percent field of view (15.2 m radius) for the radiation sensor; (b) median reflected short-wave radiation observed (2012) by solar azimuth angle and distance of maximum, specular reflection Rs	175
12.6	Energy balance fluxes normalized by incoming (short- and long-wave) radiation	176
12.7	Spatial variability of heat emissions (2008) at 200×200 m^2 resolution by sector for London (classes by Jenks natural breaks): (a) total, (b) building, (c) metabolism and (d) road traffic values in W m^{-2} [number of grids in each class]	177

Figures

13.1	(A) *Ailanthus altissima* tree growing within a matrix of parking lots and building in Boston, MA, USA. The circle denotes the location of an *Ailanthus* tree growing adjacent to the building; (B) with pavement completely surrounding the base of the tree; (C) this tree is thriving despite all of the surrounding soil being covered by impervious surfaces 189
13.2	(A) Simplified diagram of the active pools and fluxes within the global carbon cycle; (B) non-linear relationship between ISA and fossil fuel emissions in Massachusetts; (C) gradient in carbon exchange for biological and anthropogenic sources 192
14.1	Gradients of infrastructure solutions appropriate for cities, including some common terms used to describe them (Hulsman et al., 2011; Naylor et al., 2011) 205
15.1	Urbanization and vulnerability 222
15.2	Urban risk, its dimensions and its drivers 223
17.1	Water scarcity maps 251
17.2	Regression analysis of water use and GDP for individual countries (left) and projection of water consumption for 2030 (right) 255
17.3	Effect of new urban areas in 2030 on water consumption compared to current water consumption by agriculture 256
18.1	Key changes in runoff processes between idealized natural and urbanized conditions 263
19.1	Categorization of health benefits 278
19.2	Green roof runoff benefits 280
19.3	Linking greening and physical activity levels 281
19.4	Smoothed recurrence of land-based extreme heat in the Northern Hemisphere 1951–2011 282
19.5	Equivalent floor-area densities in three different layouts: low-rise single-story homes (left); multi-story medium-rise (middle); high-rise towers (right) 285
20.1	Food price index, 1961–2015 294
20.2	Rural and urban incidences of hunger (calorie deficiency) 295
20.3	Determinants of food, nutrition and health security 296
23.1	Percentages of cities conducting risk or vulnerability assessments by city size 342
23.2	Proportional sources of funding for climate adaptation for local governments by size 344
25.1	Spatial planning framework for integrative urban climate change strategies 365
26.1	Per capita emissions for a variety of cities in industrialized countries 378
26.2	GHG emissions from seven high-income cities 379
26.3	Annual population growth and average annual GHG reductions for eight high-income cities 379
26.4	Distribution of residential housing stock by construction age for Europe and North America 381
26.5	The relationship between Singapore's development (GDP per capita) and its GHG emissions over the last four decades (per capita energy use) 382
27.1	A generalized model of scenario planning 410
28.1	Main contributions from UPA across social, economic and ecological dimensions to support climate resilient cities 425

Figures

29.1	Ecosystem services and the urban rent gradient	451
31.1	The growth of an industrial symbiosis network	472
31.2	Industrial facilities in Kalundborg, Denmark are located around the local harbor	476
31.3	A conceptual model for determining industrial symbiosis potential in urban areas	481
31.4	Geographic distribution of industrial parks around Mysore city boundary	482
33.1	The urban metabolism of Brussels, Belgium, in the early 1970s; early comprehensive example of an urban metabolism analysis	516
35.1	Biodiversity and climate change response milestones	536
36.1	Installation of solar hot water systems on the roof of a luxury residential tower in Thane, Mumbai Metropolitan Region (India)	555
36.2	Solar hot water systems displaying the PROCEL Seal for Energy Savings in São Paulo (Brazil)	556
C.1	$PM_{2.5}$ is significantly higher in urban agglomerations compared to other regions in China	571
C.2	High degree of within-city dynamics in the well-developed urban cores of Beijing	572

TABLES

1.1	List of OECD metropolitan areas with share of metropolitan GDP over 40 percent, sample of 275 world MSAs, 2010	17
4.1	Selected landscape metrics to describe the spatial configuration and composition of cities	54
4.2	Characterization of the Urban Atlas land cover dataset as one typical example of urban land cover/use data	55
6.1	Total production of cement and steel by the top five producers, 2005, 2012, and Average Annual Growth Rate (AAGR)	79
8.1	Range in urban final energy use and urban percent of total final energy use by region, 2005	108
8.2	Range in urban GHG emissions and urban percent of total GHG emissions by region, 2000	110
11.1	Development of urban climate study over a century	154
11.2	US cities where observational and modeling studies mostly occur	159
12.1	Causes of urban warming and examples of mitigation strategies	182
13.1	Summary of key nitrogen cycle compounds and description of urban modifications	193
14.1	Examples of built (technological) ecosystem components, their functions and related services provided to urban inhabitants	205
15.1	Examples of research questions on urban vulnerability	215
17.1	Summary of the five metropolitan water supply systems	252
17.2	Resulting water stress indices based on two different approaches: the WSI of the watershed in which the city is located (no distinction of water origins) and the WSI of the watershed of external reservoirs	252
23.1	Examples of adaptation planning guidance documents in the Global North and South	338
23.2	Types of cities surveyed by population size and topography	340
23.3	Plan development, integration or both by city size	343
26.1	GHG emissions and reductions achieved from eight industrialized country cities	380

Tables

26.2	Municipal actions in established cities to mitigate GHG emissions in the energy generation/supply sector	385
26.3	Municipal actions in established cities to mitigate GHG emissions in the building sector	387
26.4	Municipal actions in established cities to mitigate GHG emissions in the transportation sector	389
26.5	Municipal actions in established cities to mitigate GHG emissions in the government services sector	391
26.6	Municipal actions in established cities to mitigate GHG emissions in land use and spatial planning	393
26.7	Sample urban (re)development projects that include measures to achieve substantial GHG emission reductions	395
28.1	Common sources of urban and peri-urban agriculture contamination	431
29.1	Most important ecosystem services for cities, classified according to the scheme of the MEA (2003)	444
31.1	Environmentally related benefits of industrial symbiosis in the UK for the period 2005–2012 as reported by Internatonal Synergies for the National Industrial Symbiosis Programme	474
31.2	Number of firms in various industries located in rural or urban areas of Mysore	483
32.1	Estimated duration of selected electric power outages and initial restoration time for selected storm events, US, 1965–2013	498
32.2	Estimated duration of selected nuclear power plant restoration times for Hurricanes Irene and Sandy, northeastern US	499
34.1	System dynamics under adaptation, transition and transformation	529

CONTRIBUTORS

Gotelind Alber is an Independent Researcher based in Germany and advisor on sustainable energy and climate change policy with a special focus on gender issues, climate justice and multi-level governance. She is on the Board of Directors of GenderCC.

Nick B. Allen is a Master's student in the Department of Urban Studies and Planning at the Massachusetts Institute of Technology, USA.

Alexander Aylett is Assistant Professor of Urban Climate Governance at the National Institute for Scientific Research in Montreal, Canada.

Christopher G. Boone is Dean of the School of Sustainability at Arizona State University, USA. He is a member of the Scientific Steering Committee of the Urbanization and Global Environmental Change Project of Future Earth.

David Bristow is a Postdoctoral Fellow at the Centre for Resilience of Critical Infrastructure at the University of Toronto, Canada.

Samara Brock is a PhD candidate at the Yale School of Forestry and Environmental Studies, Yale University, USA.

Kate Cahoon is Project Coordinator at GenderCC – Women for Climate Justice e.V., Berlin, Germany.

JoAnn Carmin[†] was Associate Professor of Environmental Policy and Planning at the Massachusetts Institute of Technology, USA.

Marian Chertow is Associate Professor of Industrial Environmental Management, Director of the Program on Solid Waste Policy, and Director of the Industrial Environmental Management Program at the Yale School of Forestry and Environmental Studies, Yale University, USA.

Contributors

Eric Chu is Assistant Professor of Urban Studies in the Department of Geography, Planning, and International Development Studies at the University of Amsterdam, the Netherlands.

Marc J. Cohen is Senior Researcher at Oxfam America and Professorial Lecturer at Johns Hopkins University, USA.

John P. Connors is a post-doctoral associate in the Frederick S. Pardee Center for the Study of the Longer-Range Future at Boston University, USA.

Elizabeth M. Cook is a postdoctoral scholar in the Institute for Environmental Science and Evolution at the Universidad Austral de Chile.

Xiangzheng Deng is Professor of Geography at the Institute of Geographic Sciences and Natural Resources, and Senior Research Fellow at the Center for Chinese Agricultural Policy, Chinese Academy of Sciences, Beijing, China. He is a member of the Scientific Steering Committee of the Urbanization and Global Environmental Change Project of Future Earth.

Sybil Derrible is Assistant Professor of Sustainable Infrastructure Systems in the Civil and Materials Engineering Department and Director of the Complex and Sustainable Urban Networks Lab at the University of Illinois at Chicago, USA.

Thomas Elmqvist is Professor in Natural Resource Management at the Stockholm Resilience Centre, Stockholm University, Sweden.

Michail Fragkias is Assistant Professor in the Department of Economics at Boise State University, USA.

James L. Garrett is Senior Research Fellow at the International Food Policy Research Institute, USA (based in Italy).

Daniel Gnatz is an environmental writer based in Boulder, Colorado, USA.

Patricia Gober is Interim Director and Research Scientist at the School of Geographical Sciences and Urban Planning at Arizona State University, USA.

Corrie A. Griffith is Executive Officer for the Urbanization and Global Environmental Change Project of Future Earth, headquartered at the Julie Ann Wrigley Global Institute of Sustainability, Arizona State University, USA.

Nancy B. Grimm is Professor in the School of Life Sciences at Arizona State University, USA, and Director of the Central Arizona–Phoenix Long-Term Ecological Research Project.

CSB (Sue) Grimmond is Professor in the Department of Meteorology at the University of Reading, UK.

Burak Güneralp is Professor in the Department of Geography at Texas A&M University, USA.

Contributors

Rebecca L. Hale is Assistant Research Professor in the Department of Biological Sciences at Idaho State University, USA.

Dagmar Haase is Professor of Landscape Ecology and Land Use Modeling at the Humboldt University, Berlin, and a Guest Scientist at the Helmholtz Centre for Environmental Research in Leipzig, Germany.

Stefanie Hellweg is Professor at the Institute of Environmental Engineering, ETH Zurich, Switzerland.

Shu-Li Huang is Distinguished Professor in the Graduate Institute of Urban Planning at National Taipei University, Taiwan. He is a member of the Scientific Steering Committee of the Urbanization and Global Environmental Change Project of Future Earth.

Lucy R. Hutyra is Assistant Professor of Earth and Environment at Boston University, USA, where she is also a Pardee Center Faculty Research Fellow.

David M. Iwaniec is Assistant Research Professor in the Julie Ann Wrigley Global Institute of Sustainability at Arizona State University, USA.

Boyd Dionysius Joeman is Acting Head of Environment, Iskandar Regional Development Authority, Malaysia.

Sonja Klinsky is Assistant Professor in the School of Sustainability at Arizona State University, USA.

Simone Kotthaus is a Postdoctoral Research Associate in the Department of Meteorology at the University of Reading, UK.

Benoit Lefevre is the Director of Energy and Climate for EMBARQ, the sustainable transport program of the World Resources Institute, Washington, DC, USA.

Robin M. Leichenko is Professor and Chair in the Department of Geography at Rutgers University and Co-Director of the Rutgers Climate Institute, USA.

Andrés Luque-Ayala is a Postdoctoral Research Associate at the Geography Department at Durham University, UK, and Coordinator of the International Network on Comparative Urban Low Carbon Transitions.

Shuaib Lwasa is Associate Professor in the Department of Geography, Geo-Informatics and Climatic Sciences at Makerere University, Uganda. He is a member of the Scientific Steering Committee of the Urbanization and Global Environmental Change Project of Future Earth.

Robert McDonald is Senior Scientist for Urban Sustainability at The Nature Conservancy in Arlington, Virginia, USA.

Contributors

Darryn McEvoy is the Program Leader of the Climate Change Adaptation Program at RMIT University in Melbourne, Australia. He is a member of the Scientific Steering Committee of the Urbanization and Global Environmental Change Project of Future Earth.

Peter J. Marcotullio is Professor of Geography at Hunter College, City University of New York (CUNY), USA, and Director of the CUNY Institute for Sustainable Cities. He is a member of the Scientific Steering Committee of the Urbanization and Global Environmental Change Project of Future Earth.

Chandana Mitra is Assistant Professor in the Department of Geosciences at Auburn University, USA.

Eugene Mohareb is Lecturer in Sustainable Technologies, School of the Built Environment, University of Reading, UK, NSERC Post-Doctoral Fellow, Centre for Sustainable Development, University of Cambridge, UK, and Policy Advisor, Pembina Institute for Appropriate Development, Canada.

Valerie Moye is Project Manager at Socrata in Washington, DC, USA.

Frank Mugagga is Senior Lecturer in the Department of Geography, Geo Informatics and Climatic Sciences at Makerere University, Uganda.

Conor Murphy is Lecturer in the Department of Geography at Maynooth University, Ireland.

Shannon Murray is a Research Associate with the Yale School of Forestry and Environmental Studies, Yale University, USA.

Camille L. Nolasco is a PhD Candidate in Earth System Science at the Earth System Science Centre, National Institute for Space Research [INPE], São José dos Campos, Brazil.

Bhartendu Pandey is a PhD student with the Yale School of Forestry and Environmental Studies, Yale University, USA.

Stephan Pfister is Senior Research Associate at the Chair of Ecological Systems Design, ETH Zurich, Switzerland.

Stephanie Pincetl is Director and Professor-in-Residence at the California Center for Sustainable Communities at the University of California Los Angeles, USA.

Ray Quay is a Research Professional with the Decision Center for a Desert City Project at Arizona State University, USA.

Atiqur Rahman is Associate Professor, Department of Geography, Jamia Millia Islamia University, New Delhi, India.

Charles L. Redman is Professor and Founding Director of the School of Sustainability at Arizona State University, USA.

Contributors

Debra Roberts founded and leads the Environmental Planning and Climate Protection Department of eThekwini Municipality, Durban, South Africa. She is a member of the Future Earth Engagement Committee.

Patricia Romero-Lankao is a Scientist III with the Climate Science and Applications Program at the National Center for Atmospheric Research in Boulder, Colorado, USA. She is a member of the Scientific Steering Committee of the Urbanization and Global Environmental Change Project of Future Earth.

Roberto Sánchez-Rodríguez is Professor in the Department of Urban and Environmental Studies at El Colegio de la Frontera Norte, Mexico, and Professor Emeritus in the Department of Environmental Sciences at the University of California, Riverside, USA. He is Co-Chair of the Urbanization and Global Environmental Change Project of Future Earth.

Stefan Schultze is a student of Environmental Engineering at Swiss Federal Institute of Technology in Zurich, Switzerland.

Nina Schwarz is a Scientist at the Helmholtz Centre for Environmental Research in Leipzig, Germany.

Karen C. Seto is the Associate Dean of Research and Professor of Geography and Urbanization Science at the Yale School of Forestry and Environmental Studies, Yale University, USA, and Co-Chair of the Urbanization and Global Environmental Change Project of Future Earth.

J. Marshall Shepherd is the Georgia Athletic Association Distinguished Professor of Geography and Atmospheric Sciences and Director of the Atmospheric Sciences Program at the University of Georgia, USA.

Linda Shi is a PhD Candidate in the Department of Urban Studies and Planning at the Massachusetts Institute of Technology, USA.

David Simon is Professor of Development Geography at Royal Holloway, University of London, UK, and Director of Mistra Urban Futures headquartered in Gothenburg, Sweden. He is a member of the Scientific Steering Committee of the Urbanization and Global Environmental Change Project of Future Earth.

William D. Solecki is Professor of Geography at Hunter College, City University of New York, USA, and Director Emeritus of the CUNY Institute for Sustainable Cities.

Alexei Trundle is a Research Officer in RMIT University's Global Cities Research Institute Climate Change Adaptation Program in Melbourne, Australia.

Bolanle Wahab is Senior Lecturer in the Department of Urban and Regional Planning, and Coordinator of the Indigenous Knowledge Program at the University of Ibadan, Nigeria.

Szu-Hua Wang is Assistant Professor in the Department of Urban Affairs and Environmental Planning at the Chinese Culture University, Taiwan.

Contributors

Helen C. Ward is a Postdoctoral Research Associate in the Department of Meteorology at the University of Reading, UK.

Sarah Ward is the Manager for Energy and Climate Change at the City of Cape Town, South Africa.

Mark Watkins is Project Coordinator for the Urbanization and Global Environmental Change Project of Future Earth, headquartered at the Julie Ann Wrigley Global Institute of Sustainability, Arizona State University, USA.

Andrea Ferraz Young is a Researcher at State University of Campinas in the Center for Applied Meteorological and Climate Research in São Paolo, Brazil.

Junming Zhu is a Postdoctoral Associate at the Yale School of Forestry and Environmental Studies, USA.

Rae Zimmerman is Professor of Planning and Public Administration at the Robert F. Wagner Graduate School of Public Service at New York University, USA, and Director of NYU-Wagner's Institute for Civil Infrastructure Systems.

Wayne C. Zipperer is a Research Forester with the U.S. Department of Agriculture Forest Service Southern Research Station, USA.

ACKNOWLEDGMENTS

This Handbook is a culmination of many years of work that would not have been possible without the effort and patience of all the contributing authors—thank you to everyone involved for your commitment to and participation in the project.

We also wish to express our gratitude to the following people who provided encouragement, contributed to and supported the ideas, design and overall development of the book:

- Mark Watkins, National Science Foundation (NSF) funded Urbanization and Global Environmental Change (UGEC) International Project Office, housed at the Julie Ann Wrigley Global Institute of Sustainability, Arizona State University, for his overall development support.
- Michael Dorsch, City University of New York (CUNY), for editorial assistance.
- Alain H. Clarke for graphics and editorial assistance.
- Cary Simmons, Kellie Stokes and Meredith Reba, Yale University, for cover images and design assistance.
- Michail Fragkias, Boise State University, for feedback on earlier drafts of the proposal.
- Andrew Mould and Sarah Gilkes, Routledge, for their support and patience.

INTRODUCTION

William D. Solecki, Karen C. Seto and Corrie A. Griffith

This Handbook is the synthesis of more than ten years of research by an international community of urbanization and global environmental scholars that have generated the core knowledge that is the basis of much of our understanding of the relationship between urbanization and global environmental change. In the Fall of 2002, the International Human Dimensions Programme on Global Environmental Change (IHDP)[1] convened a scoping meeting to discuss the idea of creating a new international science project on cities and urbanization. At that time, the science of urbanization and global change was just emerging, and IHDP recognized an opportunity to develop a new science project that would frame, catalyze and coordinate research in this area. Over the next three years, a Science Plan[2] would be written, reviewed and published, a Scientific Steering Committee (SSC) appointed and convened, an International Project Office (IPO) inaugurated and staffed, and the Project officially launched.

From its conception, the strategy of the Urbanization and Global Environmental Change (UGEC) Project was to build on the substantial knowledge already accumulated from existing but fragmented research in the social sciences, natural sciences, engineering, and the arts and humanities on the human dimensions of global environmental change in urban areas. In the early years of the Project, new networks were cultivated and existing research communities were engaged through research activities, workshops and conferences held in different regions of the world. The scholarly community grew rapidly in the ten years of the Project's lifespan, new areas of research emerged and the accompanying scientific literature grew manifold. Today, the community of scholars working at the intersection of urbanization and global environmental change extends well beyond the original international science project.

With any rapidly growing literature, it becomes an increasing challenge to identify main intellectual and methodological developments, debates and areas of consensus. Especially for researchers and practitioners now entering this field, there is a wide and deep literature from which to draw, but it may be difficult to see the main threads. Although there have been a number of articles and special issues of journals that have synthesized some aspects of the research, the literature is vast and remains largely fragmented. This Handbook aims to be a comprehensive reference and source for critical reflection on contemporary ideas and debates on the interactions and feedbacks between urbanization and global environmental change at local, regional and global scales. The main goal of the Handbook therefore is to give a 'state of the art' survey of the topics, explain how the issues are important and critically discuss the leading views in these rich and dynamic topics.

The connections between urbanization and global sustainability challenges are a key issue for the future of the planet. This volume takes the position that the fundamental question is not how do we achieve urban sustainability, but rather how do we achieve sustainability in an urbanizing world? Two dominant conceptual frames underpin this Handbook. First is that urbanization is a process that is not bounded by a city's administrative or territorial boundaries including extended metropolitan regions. Many chapters in the book use concepts of flows, linkages, and networks to describe and examine the critical relationships between local urbanization and its impacts and global scale environmental changes. A second key frame is that urban areas are in a continuous cycle of change. This volume intentionally uses the term urbanization to describe a dynamic process of change that involves simultaneous transformations of livelihoods and societies, governance and institutions, demography, economies and the biosphere. Thus, urbanization describes a condition as well as transitions between stages of urbanization.

To comprehensively address these framing elements, the Handbook is centered on four goals to:

1   Provide a comprehensive overview of the current and emerging interactions and feedbacks between urbanization and global environmental change.
2   Critically review and synthesize the conceptual approaches, empirical knowledge and understanding of urbanization and global environmental change science.
3   Share knowledge and experiences across world regions through the global scope of coverage and contributors.
4   Provide the scientific foundations to guide policy and future scientific inquiry on urbanization and global environmental change issues.

Handbook structure

To best achieve the stated goals with the context of the two framing issues, the Handbook content is divided into four parts, each covering an array of urbanization and global environmental change interactions. Part I addresses the pathways across which urbanization drives global environmental change. Part II examines the mechanisms by which global environmental change affects the urban system. Part III focuses on the interactions and responses within the urban system in response to global environmental change. Part IV of the Handbook centers on critical emerging research in the domain of urbanization, global environmental change and sustainability. Below we briefly describe each part and define their contribution to the Handbook.

Part I: Pathways by which urbanization impacts environmental systems and become drivers of global environmental change

The chapters herein evaluate core elements of the urbanization process: concentrated populations, the rapid and large-scale flow of financial and natural resource capital, and the abundance and complexity of civil society institutions. They explore how and why urbanization today differs from urban processes in the past and discusses urbanization as an outcome of decisions, actors and institutions. While population size has been the dominant indicator to describe urban areas,

Introduction

this part introduces new multi-dimensional urbanization taxonomies to describe and investigate the intersections between urbanization and global environmental change.

A central focus in Part I is contemporary urbanization trends, and how different urbanization variables lead to the development of different types of urban areas. Critical urbanization trajectories are examined including population growth, flows of ideas and materials, resource stocks and social contracts. Together, these factors create the conditions for growth of urban settlements. Key to the discussion is how these factors become drivers of urbanization and the relative role of actors and decision-makers. A specific question is: How do these processes manifest as differing patterns of urban spatial development, urban form and the built environment? Each outcome is conceptualized as having varying sets of urban resource demands that maintain and supply critical infrastructure and urban systems (e.g. transportation networks, energy demands).

Chapters in Part I also examine the links between urbanization and ecosystem services and the interactions between urbanization and climate change at different spatial scales. The main questions addressed in the chapters include: How do the conditions of urbanization translate to environmental change pressures within cities as well as connect to distant sites to foster environmental demands in those locations—i.e. via ecological footprints and urban land teleconnections? How does this process play out across different cities? Of specific interest are the key variables and system functions that alter the intensity of impact. Embedded in this discussion is the fundamental issue of how and under what conditions urbanization alters local, regional and global climates.

Part II: Impacts of global environmental change on urban systems and urbanization processes

Part II of the Handbook examines the ways in which global environmental change modulates and reshapes urbanization processes and urban areas. Global environmental change and its manifestations have a wide spectrum of effects on urban processes. The increasing frequency and magnitude of climate related extreme events in urban areas during the past decade are some of the clearest indicators of the vulnerability and shifting exposure of urban areas. There are, however, many pathways through which global environmental change affects urban systems. The impacts will depend on a number of factors, including the resilience of people and places as well as the ability of formal and informal institutions to respond. The chapters collectively attempt to address how to better grasp the pathways through which specific types of global environmental change affect local and regional urban processes and ultimately human well-being.

For example, we need improved understanding of how climate change and related global scale processes impact urban areas and create social-ecological systems, and of tight coupling between the urban and natural systems. Some chapters in Part II focus on the competition for resources and how global environmental change affects fundamental urbanization processes, such as the concentration of populations, rapid and large-scale flow of financial and natural resource capital, and the abundance and complexity of civil society institutions. Historically, cities influenced by global economic forces such as industrialization and globalization often have undergone dramatic transformations, socially, economically or environmentally. These chapters examine how global environmental change is altering the rate and scale of change in cities and may alter urbanization processes in the future. To complete an analysis of these issues, Part II also includes a discussion of the conditions that create resource competition and unstable resource supply for cities.

Part III: Urban responses to global environmental change

Part III of the Handbook focuses on urban action on climate change including climate change mitigation and adaptation. Herein it is understood that how urban systems respond to global environmental change greatly depends on the existing interactions between socio-economic and geopolitical processes and the built environment, and, in turn, urban form and function modify local environments and influence human behavior. Furthermore, built environments and their communities have different capacities to cope, adapt and respond to the impacts of global environmental change.

Thousands of cities around the world of varying sizes and levels of economic development are taking action on climate change and sustainability. Chapters focus on what motivates these responses and actions as well as the strategies that help cities become more resilient in the face of increasing challenges from global environmental change. It is important to understand how global environmental changes, combined with the politics of global environmental change (e.g. Global North versus Global South; local versus national actors) and the policies of global environmental change (e.g. climate change agreements and multinational environmental agreements), are reshaping local civic agendas, institution-building and decision-making.

Also addressed in Part III is how the international global environmental change agenda will reshape local urban dynamics, and how the response of local strategies and actions will change national and international policy agendas. The ways in which cities are building or rebuilding (including sustainable urban and infrastructure engineering) in reaction to these opportunities and challenges are diverse and take place under a variety of social, political and environmental contexts. Part III therefore highlights the interface of science with planning and policy-making within urban areas as well as best practices or lessons learned that have the potential to be scaled up or adopted across a range of cities.

Part IV: Urbanization, global change and sustainability: critical emerging integrative research

The final section of the Handbook focuses on emerging research issues and domains. The coupling of urbanization with global environmental change will lead to a multitude of impacts emerging from multifaceted interactions between sets of cities, cities with rural areas and individual city pairings. This system level perspective provides an opportunity to examine the simultaneity and the complexity of the connections complete with time lag effects, tipping points and transitions. Part IV broadens beyond the academic community to include perspectives from urban practitioners and decision-makers; a representative few of a diverse web of urban actors responding to such complex challenges at various scales and institutional levels. It provides insight into the on-the-ground realities that occur within our cities, the decisions that shape future urbanization and what is required for the strengthening of future urban sustainability research, policy and action. The concluding chapter provides a broad arc of where science advances have come in the last ten years and proposes new and emerging research questions, including: How does the confluence of urbanization and global environmental change present new ways to think about vulnerability and crisis, and how can we integrate the concepts of tipping points and transitions into urbanization and sustainability issues? Also, what is missing in our understanding of metrics, indicators and analytical frameworks, and how can we move beyond new taxonomies of urbanization alone to tightly coupled urbanization-global change or urbanization-sustainability frameworks? The volume

Introduction

closes with an argument for the need for building an urbanization science that is theory-driven and testable across multiple urban conditions and domains around the world.

A handbook for the urban century

What historically has been a local phenomenon is now global in scope, with urbanization transforming nearly every region of the world. With urbanization affecting nearly all aspects of the global environment, and with global environmental change affecting many dimensions of urban systems, the literature on urbanization and global environmental change has grown rapidly and has consequently become extremely diverse. A wide range of researchers and practitioners who bring diverse perspectives, experience and knowledge to the topic is creating new bodies of knowledge.

How did cities come to be? What roles do they play in society? How do cities shape their environment? How does environmental change shape cities? These are questions that have been topics of discussion and debate for as long as human settlements have existed. Engineers, designers and planners have shaped and constructed urban settlements for millennia. Over the past several hundred years, there have been at least two significant shifts in the consciousness of those who build and study cities. Starting with the Renaissance during the 14th through 17th centuries, and subsequently since then, those who construct and examine cities have begun to recognize the burden of history, and how and why cities thrive and succeed or decline and fail. In turn, cities of the 19th and 20th centuries were built from lessons of the past with an eye toward the present.

The perspectives of contemporary urbanists—whether they study or build urban settlements—have been further transformed by two concurrently occurring phenomena: urbanization increasingly being recognized as a global and not only local issue, and the growing threat of global environmental change on human societies. Consequently, the study and building of cities today reflects an understanding of the past, present and increasingly the future. The confluence of global urbanization and global environmental change signals the entry into a new era of profound dynamism and non-stationarity, where the past will be less and less useful for predicting the future. Yet, it is increasingly necessary to make predictions about impacts of urban decision-making over decadal and century-long timescales. In order to do this, we need to be able to identify, describe and model feedbacks and tipping points between urban and non-urban systems. To provide insight into these complex relationships, we have put together this Handbook with chapter contributions from urban specialists who reflect the range of perspectives, experience and knowledge found in the literature and in practice. We have also intentionally invited contributors at different stages in their careers. All too often, perspective volumes become retrospective. It is clear that the study of urbanization and global environmental change is still very much in its infancy and that the next generation of scholars have much to contribute to building the core knowledge base.

User guide and final points

This volume aims to be a tool for scholars, students, and practitioners to understand how the processes of urbanization and global environmental change are connected, and where might space exist to employ new practices that will help the transition to sustainability. The Handbook provides cutting-edge knowledge and information about these connections as we enter into a crucial moment in the drive toward global sustainability. With the promulgation of the United Nations sponsored Sustainable Development Goals, the United Nations

Framework Convention on Climate Change (UNFCCC) COP21 agreements and the UN Habitat III declarations, cities will be positioned at the forefront of the discussions of how to create viable, economically beneficial and equitable pathways to a more sustainable world.

In order to ensure success, these pathways must be meaningful at the local level to those who live in or carry out their lives in urban areas. The Handbook attempts to incorporate these perspectives throughout the chapters of the volume. Embedded in many of the chapters individually and the volume collectively is that knowledge and information about the processes of urbanization and global environmental change must be co-produced through integration of expert and local knowledge, and include the skills, perspectives and tools of scientists, practitioners and residents.

The Handbook illustrates how both the formal and informal sectors of urbanization are parts of a whole. Understanding how these parts interact and push forward urbanization and become agents for sustainability is critical. The everyday of contemporary urban areas is often splintered into separate social and spatial realms. As reflected in the cover images of the Handbook, it is the position of the editors and other authors that these disparate pieces not only are present but that there might be ways to see the pieces as a mosaic within each urban area and across the community of urban areas worldwide. The Handbook carries this vision of the urban mosaic forward.

Note

1 IHDP was established in 1990 by the International Social Science Council (ISSC) and in 1996, jointly sponsored by both ISSC and the International Council for Science (ICSU). IHDP, together with the World Climate Research Programme (WCRP), DIVERSITAS, and the International Geosphere-Biosphere Program (IGBP) constituted the four Global Change Programmes which aimed to frame and coordinate international natural and social science research on global change. In 2014, IHDP and DIVERSITAS were dissolved and their work and networks were transferred to a new global initiative called *Future Earth*.
2 Sánchez-Rodríguez, R., Seto, K. C., Simon, D., Solecki, W. D., Kraas, F., and Laumann, G. (2005). Science Plan Urbanization and Global Environmental Change Project (IHDP Report No. 15). Bonn, Germany: International Human Dimensions Programme on Global Environmental Change.

PART I

Pathways by which urbanization impacts environmental systems and become drivers of global environmental change

Part I explores the fundamental characteristics and dynamic processes that influence contemporary urbanization and how these drive and interact with environmental changes across regions. The chapters within this part take a critical look at a) *urbanization trends* including changing demographics, land use and land change to provide an understanding of how today's patterns and processes of global urbanization differ from those of the past (Haase and Schwartz, Chapter 4; Deng, Chapter 3; Mitra et al., Chapter 5). Fundamental to the profound global urban transition we are experiencing is how socio-economic/political and ecological sub-systems are implicated by such change at multiple scales and the roles they have in further driving and shaping worldwide urban development.

This part highlights the understanding that urbanization is not solely a local phenomenon, but has far reaching affects on flows of natural and financial resources, patterns of consumption (i.e. food, water and energy) as well as other biophysical systems (i.e. carbon, hydrological and atmospheric cycles). Thus, b) *existing linkages between urbanization and resource demand* is a second critical topic addressed within this part. The understanding of how current urbanization trends and their effect on the supply and integrity of the Earth's life support system and how to sustain these valuable resources as populations, consumption and economic activity continue to concentrate in urban areas are growing concerns amongst urban sustainability researchers. The authors herein examine the many complexities associated with urban systems as they offer insight to, for example: what effects does urbanization have on greenhouse gas (GHG) emissions and energy use including the increasing demand for water and materials associated with construction and operation of a growing urban populace (Gober and Quay, Chapter 7; Marcotullio, Chapter 8; Güneralp, Chapter 6)? What are the local and far reaching effects on the environment including urban nutrient cycles (Hutyra, Chapter 13) as well as ecosystem services and biodiversity functioning and health (Elmqvist et al., Chapter 10)?

Just as resources are often teleconnected to rural and distal places much farther away from the center of urban activities, lifestyle choices and consumption patterns, typically associated with high-income countries, are becoming a growing 'export' of the global urbanization process. Thus, an understanding of how urbanization not only influences land and resources, but also c) *urbanization influences on human and social behaviors* including changing diets and food demand (Murray et al., Chapter 2); housing preferences (Leichenko and Solecki, Chapter 9); and other lifestyle choices are significant contributions to the urbanization and global

Urbanization impacts on environmental systems

environmental change knowledge base. These tightly coupled socio-economic and ecological processes driving global scale urbanization and other global changes consequently have large implications for human well-being. As urban economies continue to grow and expand, an increasingly important question relates to whether a positive relationship exists with human well-being and sustainability (Fragkias, Chapter 1); i.e. under what conditions is urbanization 'good' for both economic growth and the environment?

The fourth topic included in Part I is d) *urbanization and climate*, which uncovers the ways in which urbanization processes alter the local, regional and global climate. Over the last few decades, research in this area has grown tremendously as a result of more and improved data and methodologies that aid our understanding of the multi-scale influences urbanization has with other bio-physical systems. The chapters addressing this fourth topic reveal that cities and urban processes modify urban surface and atmospheric interactions, creating distinct urban climates (Grimmond et al., Chapter 12) including, for example, changes in precipitation (Mitra and Shepherd, Chapter 11) and the well-known urban heat island effect.

The authors included in Part I provide a variety of disciplinary perspectives on contemporary urbanization processes, emphasizing the profound nature of these influences—so large in scale, fast in pace and extensive in geographic reach that aggregated globally, they affect all aspects of the Earth system.

1

URBANIZATION, ECONOMIC GROWTH AND SUSTAINABILITY

Michail Fragkias

Historically, the organization of societies in dense settlements is closely correlated with the rise in incomes and human well-being. Analogies for the ways that dense human settlements lead to processes of wealth generation abound. Braudel (1982) discusses towns as electric transformers that "increase tension, accelerate the rhythm of exchange and constantly recharge human life" (p. 497). Cities have also been described as 'engines' of economic growth (Lucas, 1988). More recently, cities have drawn the analogy of stars—the changing characteristics of urban agglomeration as their size increases is paralleled to the functioning of stars "which burn faster and brighter (superlinearly) with increasing mass. [. . .] although the form of cities may resemble the vasculature of [. . .] biological organisms, their primary function is as open-ended social reactors" (Bettencourt, 2013, p. 1441). Notwithstanding all analogies, the concentration of human capital, large-scale physical infrastructure and public and market institutions in cities enable increases in innovation, economic activity and efficiencies from scale (Seto et al., 2010; Puga, 2010).

Urbanization provides economic benefits through what economists call agglomeration economies. Urban areas showcase unique outcomes, emerging only when population agglomerates in a single location. The agglomeration economies made possible by the concentration of individuals (and firms) are responsible for making cities the ideal setting for innovation, job and wealth creation (Carlino et al., 2007; Knudsen et al., 2008; Puga, 2010; Rosenthal and Strange, 2004). Larger agglomerations of individuals entail a wider 'repertoire' of intellectual capabilities and include a larger stock of 'recipes' for producing economic output. These two factors facilitate the creation of new ideas (partly through the recombination of old ideas). Larger agglomerations also increase the likelihood that people will interact; the aforementioned research suggests that these interactions are eventually responsible for the generation and dissemination of new ideas and recipes. Consequently, the heterogeneity and diversity of cities becomes a source of economic growth (Quigley, 1998). This self-reinforcing process scales super-linearly with an increase of the size of these agglomerations (Bettencourt et al., 2007; Bettencourt, 2013). Urban economists have also documented the positive correlations between urban (population) size and productivity measured as average wage or value added (Melo et al., 2009). New interdisciplinary urban theory (Bettencourt et al., 2010; Glaeser and Resseger, 2010) provides novel explanations for the strong positive relationship between urban size and productivity.

Given the advantages that cities provide, it is thus not surprising that the turn of the century marked a milestone in human history: in 2008, for the first time in human history, more than half of the world's population lived in urban areas. Projections suggest that cities will continue to be the dominant type of human settlement for the foreseeable future (UN, 2014). Thus, the bidirectional positive feedback loop of urban population agglomeration and wealth is expected to continue its cycle. At the same time, the era of the Great Transformation is advancing (Kates et al., 1990). Through centuries of interventions, humans have literally changed the 'face' of the earth. Although the transformation started in ancient times through the use of fire, irrigation, afforestation and new technologies that allowed us to make steady progress in increasing our material living standards, it has accelerated through time resulting in extensive changes in the functioning of social-ecological systems (Kates et al., 1990). Today, the results of global environmental change (GEC) force us to face a sobering reality—human activity is a planetary force, and its scale has important human well-being implications. Entering a new geological era, branded the Anthropocene, scientists claim today that humanity has surpassed limits in Earth system functioning and operates today in an 'unsafe space' (Rockström et al., 2009). At the same time, the debate regarding the energetic limits to economic growth continues (Brown et al., 2011).

As agglomeration phenomena across scales continue to drive urban economic growth, the urbanization process—ranging from local neighborhood dynamics to regional clusters of cities and global urban interdependencies—has significant effects on the local and global environment. Urban agglomeration has complex interconnections with GEC (agricultural land conversion, habitat loss, urban form, added wealth generation, etc.). The study of urban growth is complicated by the fact that as a process it operates across scales and spatial boundaries. Urban agglomeration processes lead to consumption and production decisions that in turn result in GEC (e.g. climate change, biodiversity and habitat loss, air pollution), but could also result in other global socio-economic pathologies (e.g. inequalities, poverty, unrest, etc.). Urban agglomeration and economic growth can be in turn affected by GEC, resulting in a suite of issues such as altered migration patterns, shifts in infrastructure investments and changes in quality of life.

Furthermore, in light of the emergence of sustainability as a new paradigm for organizing our thinking about cities, calls have been made for a new 'science of cities' and an 'urbanization science' to address the challenges of the Anthropocene directly (Bettencourt et al., 2007; Seitzinger et al., 2012; Solecki et al., 2013). These authors suggest that a better understanding of the interconnections between urbanization, wealth and sustainability can only emerge through an interdisciplinary, systems-oriented, integrated study of cities as a process—not through a 'place-only' approach. A new urbanization science could potentially provide ideas for improved national and local policies as well as more effective institutions.

The following sections address major questions at the intersection of urbanization, economic growth, human well-being and sustainability, while providing an interdisciplinary viewpoint on the above themes, connecting distinct strands of urban research. The chapter is not a comprehensive summary of the literature, but an overview of selected themes that are ignored in the past reviews and recent findings from a multiplicity of disciplines. The first section reviews the often forgotten historical connections between urbanization, economic growth and sustainability. The second section focuses on basic facts and empirical evidence on systems of cities and their human-environment interactions globally, exploring the broader effects of these interactions at the regional, national and international levels and across time. It also examines recent data on the relationship among urban growth, national economic growth and economic development. The third section reports on future projections of the relationship between urban sustainability and economic well-being. The fourth section concludes the chapter.

Historical foundations of urbanization, economic growth and sustainability

The connection of the formation and interaction of cities with economic growth and sustainability has been the subject of many important works in the fields of economic history and urban studies (Pirenne, 1925; Jacobs, 1969, 1984; Bairoch, 1988; Hall, 1998). The roots of this scholarship go back in time, more than 200 years ago; Smith (1776) identifies a major source of the wealth of nations as the division of labor and suggests that the "division of labor must always be limited by [. . .] the extent of the market" (I.3.1). Braudel (1982) points out that historically "[w]here there is a town, there will be division of labour, and where there is any marked division of labour, there will be a town" with markets occupying a central position in dense settlements. For Braudel, it is the town that is responsible for the development and diversification of consumption; dense settlements *"generalize the market into a widespread phenomenon"* (p. 481) since town-dwellers depend on central markets for obtaining basic goods such as their food supply. It is thus the decision of people to move into cities that brings prominence to the institution of the market and the workings of commerce and capitalism, the institutions responsible for the rapid growth in standards of living and wealth accumulation.

As markets developed and scholars better understood their functioning, the concept of 'knowledge' or 'information spillovers' became central to the question of what causes economic growth and the formation, existence and flourishing of cities. The 'informational' role of cities was pinpointed early on in economic scholarship (Marshall, 1890); cities simply allow for more frequent interactions between people and firms, facilitating the exchange of ideas that lead to higher productivities. Knowledge spillover is such an important concept that Jacobs (1969) even hypothesizes that this aspect of cities is responsible for the agricultural revolution circa 8000 BCE, a thesis that has been deemed improbable by archeologists. It is worthwhile to note though that the historical exploration of the relationship of the urban and the rural reveals a 'reciprocity of perspectives' involving mutual creation, domination and exploitation (Braudel, 1982); point being, rural economies gain from proximity to urban areas. Knowledge spillovers eventually became a central idea in economic growth theory (Romer, 1986; Lucas, 1988) driving the development of modern macroeconomics and urban economics.

In addition to the functions of 'generalizing the market' and creating knowledge spillovers across interacting economic agents, the formation of cities and towns brought about the need for different forms of power, typically through legal and governmental structures, protective and coercive (Braudel, 1982). These forms of power provided a certain degree of stability for social institutions to develop and provide the foundations for the operation of markets. Smith (1776), for example, points out that "[o]rder and good government, and along with them the liberty and security of individuals, were, in this manner, established in cities at a time when the occupiers of land in the country were exposed to every sort of violence." The move of people in cities, then, leads to a nexus of conditions that favor the increase in material prosperity: the strengthening of market institutions, social institutions such as security and trust as well as the more frequent exchange of ideas that allowed the rapid rise in economic production and material living standards, but also the adverse effects of GEC.

Naturally, the success of cities was also dependent on additional factors and conditions. For example, the location of settlements was particularly important for the residents' prosperity, since the opportunity for trade in coastal areas or at close proximity to navigable waters clearly drove economic growth (Smith, 1776). It is perhaps ironic that the path dependence of these early location decisions affects significantly the prospects of economic well-being of present-day residents: coastal urban zones are expected to experience significant economic losses due to the effects of climate change. Furthermore, trade across cities has been identified as an important factor leading to increased wealth. Researchers have emphasized the importance of urban hierarchies since "[t]he town only exists as a town in relation to a form of life lower than its own," but also the role of international connections: "[t]here can be no door to the rest of the world, no international trade without towns" (Braudel, 1982, p. 481). The historical trajectories of urban areas also affect the present-day teleconnections of hierarchically structured urbanization, affecting the responses of urban areas to GEC.

A particularly interesting facet of urban life across the ages is a relatively short-lived burst of inventive activity, known as a 'golden age' for cities. Detailing a long history of golden ages of cities, Hall (1998) discusses different factors and conditions that can lead to extraordinary performance in cities, common across the ages, factors for which Marshall (2006) provides a succinct summary. An important factor is the existence of people that are 'foreign-born' or not fully integrated in the local culture; Hall discusses the metics of Ancient Athens—an equivalent of a resident alien today—who were responsible for many of the advancements experienced circa 500 BCE. Disorder is also an all too common condition for golden ages of cities—violent places faced with rapidly changing social order (e.g. Renaissance Florence) that is assisted by growing wealth. Specialization in one or more directions is another common occurrence in highly creative and prospering cities (Paris in the 1920s, New York City). The State has historically been an important player in the flourishing of cities, aiding the operation of markets. Historical examples include the funding of the arts (Paris in its Belle Époque in the 19th century), but more recently, the importance of government intervention is exhibited by the success of research and development clusters such as the one in Silicon Valley. Along similar lines, while specific individuals act as catalysts for golden eras, the group is important. Shakespeare or Picasso did not flourish in a vacuum, but were constantly challenged and supported by a community of peers (Marshall, 2006). It is important to realize that golden eras in cities last for short periods of time. According to Hall, great epochs are not sustainable (Marshall, 2006); interestingly, new theory supports this claim (Bettencourt et al., 2007).

In summary, urban research has thus established a direct link between initial phases of urbanization through the emergence of dense settlements and the realization of the importance of a widespread functioning of markets for material prosperity and the conversion of natural to manufactured capital. This argument, on the surface, establishes early urbanization *processes* as a factor for (un)economic growth, present-day GEC and decline in environmental conditions on the planet. While today's cities continue to be important in further promoting the institution of the market, the link of urbanization, economic growth and sustainability is debated. Presently, the discussion focuses on the causal chain that begins from people moving to dense settlements, leading to increases in the density of social interactions, which creates gains in material prosperity arising from mutually beneficial exchange that eventually leads to further environmental degradation threatening today's life support systems. On the other hand, the more dense exchange of ideas that occurs in urban agglomerations can produce technological innovations that lead to 'cleaner' industrial production processes and save space for nature.

Examining the link between urbanization, economic well-being and sustainability from a historical perspective raises several important questions: Why do people decide to move into dense settlements in the first place? Is people's attraction to dense settlements and migration patterns going to change with the projected GEC? Will GEC bring about new conditions for migration and city formation? Under what conditions could urbanization be 'good' for both economic growth and the environment? Will urbanization and resulting innovation through the increased chances for creative interactions lead to novel sustainability solutions? Do current trends in urbanization force us to rethink whether the scale of urbanization can lead to continued growth in innovation and environmental solutions? The remainder of this chapter attempts to shed additional light on major sustainability questions.

Facts and theory on urbanization, economic growth and sustainability

National levels of urbanization

Academic research has shown that the national level of urbanization—the percentage of a country's population living in cities—and income per capita are highly correlated. At a macro scale, economic growth seems to be a phenomenon almost synonymous to the process of urbanization. Examining data spanning from the 1960s to the present day, Henderson (2010) finds that the level of urbanization explains 57 percent of the variation in levels of income per capita. Furthermore, additional variation can be explained by the different definitions of urbanization across countries.

Figures 1.1 and 1.2 showcase the fundamental relationships between the percentage of the population that is urbanized in a country and the country's real GDP per capita. Figure 1.1 shows that a certain level of urbanization at the national level is strongly correlated with

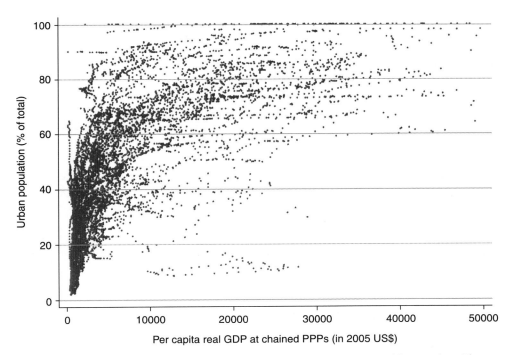

Figure 1.1 Percent urban population and real GDP per capita, 1970–2012, world countries with available data for above years

Sources: World Development Indicators for urban population data and Penn World Tables v.8 for real GDP data—WGB (2015); Feenstra et al. (forthcoming 2015).

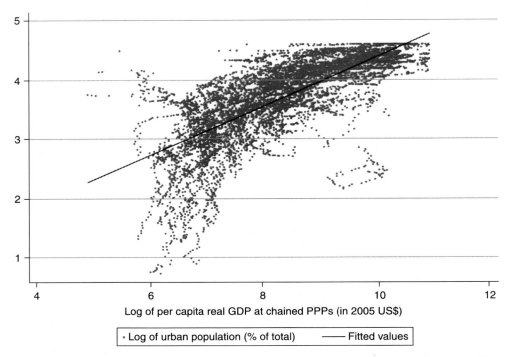

Figure 1.2 Scatterplot of log of percentage urban population and log of per capita real GDP, World countries, 1970–2011 including linear fit

Sources: World Development Indicators for urban population data and Penn World Tables v.8 for real GDP data—WGB (2015); Feenstra et al. (forthcoming 2015).

higher per capita real GDP. Very few countries make it to a per capita real GDP in the range of ten to twenty thousand USD without reaching a 50–60 percent urbanization level. Figure 1.2 plots the logarithmic transformation of the same data, showcasing the constant elasticity relationship.

Scholars suggest that the role of urbanization is more nuanced; for example, Bloom et al. (2008) claim that while levels of income across the globe are highly correlated with the proportion of a country's urban population, they find no evidence that the level of urbanization causally affects the rate of economic growth. Instead they point to the possibility of reverse directionality of causation (income levels and growth affect urbanization levels) or the existence of a confounding factor (another variable that affects both urbanization and income levels when these two variables don't affect each other). Past research corroborates these findings (Henderson, 2003). Indeed, research in the context of low- and middle-income countries reveals that while the strength of agglomeration economies differs across nations, the relationship between urbanization and wealth is defined by other factors such as infrastructure investment and local institutions on rural to urban mobility and land use (Turok and McGranahan, 2013). It is worth noting that in a few cases, high levels of urbanization in countries with non-existent or even negative economic growth exist. Examples include Argentina and Uruguay circa 2000; Lebanon; the UAE and Qatar in 2005; Iceland and Venezuela in 2010.

Others suggest that it is not urbanization level per se that affects productivity growth, but the degree of urban concentration (Henderson, 2003). One of the main messages from this

Urbanization, economic growth and sustainability

research is that countries can fine tune degrees of urban concentration for maximizing productivity growth. Furthermore, the optimal degree of urban concentration within a country depends on various factors such as level of economic development and the size of the country; the optimal degree of concentration declines as countries develop economically and their total population grows. Finally, the degree of urban concentration could be a major policy lever given that countries are currently experiencing 'sub-optimal' productivity growth given excessive or insufficient primacy levels. While most nations are flexible to act towards specific urban concentration targets in their transitions to higher urbanization levels, other nations have been fully urbanized.

Metropolitan region level

A variety of disciplines have pushed forward our understanding of urban growth dynamics, especially in relation to differential growth of economic output across space. A key feature of global economic geography is agglomeration economies—processes that drive the geographic concentration of economic activity across different scales. An agglomeration can form as clustering in neighborhoods (at the smallest scale), a city formation (at a medium scale) or a core-periphery structure of nations (at the largest scale). Various forces disperse economic activity—land rents and immobile factors (e.g. land and natural resources)—while other forces concentrate it (e.g. forward and backward production linkages and labor markets) (Krugman, 1998). While agglomeration economies have been classified in a variety of ways (Jacobs, 1969; Puga, 2010), important research revolves around the concept of dynamic externalities. Rosenthal and Strange (2004) evaluate the empirical literature of the last 30 years and find evidence of economic agglomeration forces in three broad areas: labor market pooling, input sharing and knowledge spillovers. Factors such as natural advantage, home market effects, consumption opportunities and rent-seeking also contribute to agglomeration.

Thus, switching the unit of observation to the individual city and studying national systems of cities provides a distinct viewpoint. Past research documents that urban size (measured by the population of a metropolitan area) is related to productivity, as measured by average wage or value added (Glaeser and Maré, 2001; Melo et al., 2009). Others have pointed to a positive feedback between physical urban expansion and regional economic growth (Bai et al., 2012). Such findings have led researchers to derive a production function for cities that can explain productivity patterns in urban economies (Lobo et al., 2013).

A typical result of the recent literature on the 'new science of cities' identifies a strong relationship that holds for multiple countries across the globe: urban wealth scales superlinearly (Bettencourt et al., 2007). As the size of metropolitan areas doubles, metropolitan GDP is expected to more than double—and in actuality grows by about 115 percent. Figure 1.3 shows this remarkable scaling law relationship between wealth (personal income—total earnings from wages, investment interest and other sources) and urban population for US core-based statistical areas (the collection of metropolitan areas as well as smaller urban settlements) between 1992 and 2008. Figure 1.4 reports the same relationship for the OECD Metro Explorer dataset of 275 world metropolitan statistical areas (MSAs) for the year 2010.

While the debate of the association between national urbanization rates and levels of wealth due to higher productivity levels continues, a big challenge in urban research is the lack of detailed measures of output and output growth at the urban/metropolitan scale for an array of countries. Metropolitan areas globally are primary drivers of their national economy. Urban areas dominate the global economy—they produce more than 90 percent of the world's GDP (Seto et al., 2010). This is part of their innovation function: more than

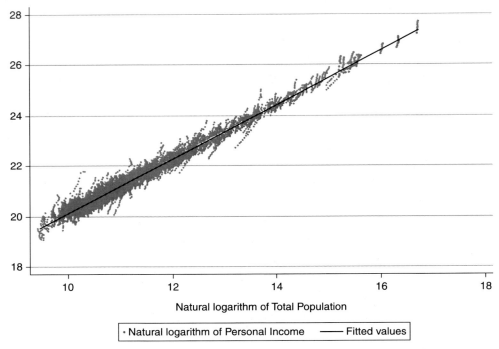

Figure 1.3 Scatterplot of logarithms of total personal income against log of total population, United States MSAs, 1992–2008, including regression fitted values

Source: Data from Fragkias et al. (2013).

81 percent of OECD patents are filed by entities in urban areas. The OECD provides socio-economic data that is available for the largest OECD metropolitan areas in 29 OECD countries—with more than 1.5 million inhabitants (OECD, 2012). That dataset reveals that most OECD metro-regions have higher GDP per capita and labor productivity levels than their national average (84.6 percent and 83.3 percent, respectively). Many OECD metro-regions also grow faster than the country of which they are part (OECD, 2006).

The degree to which various large metro-regions concentrate a significant portion of the national economic activity is remarkable (Seto et al., 2010). OECD data show that Budapest (Hungary), Seoul (S. Korea), Copenhagen (Denmark), Dublin (Ireland), Helsinki (Finland) and Randstad (the conurbation of Amsterdam, Rotterdam, The Hague and Ultrecht in Holland) produce between 40–60 percent of their national GDP. Oslo (Norway), Auckland (New Zealand), Prague (Czech Republic), London (U.K.), Stockholm (Sweden), Tokyo (Japan) and Paris (France) are examples of urban areas that are responsible for about one-third of their national GDP. GDP growth for other metropolitan areas is lagging behind the national average while at the same time their unemployment rates are higher than their national average (OECD, 2006). In countries where one large city does not dominate with a disproportionately high share of the national GDP, metro-regions, as a group, constitute the main engine for economic growth. In China, for example, 53 metro-regions, home to under 30 percent of the country's population, produced 64 percent of the country's GDP in 2004, up from 55 percent in 1998 (OECD 2006, 2012).

Urbanization, economic growth and sustainability

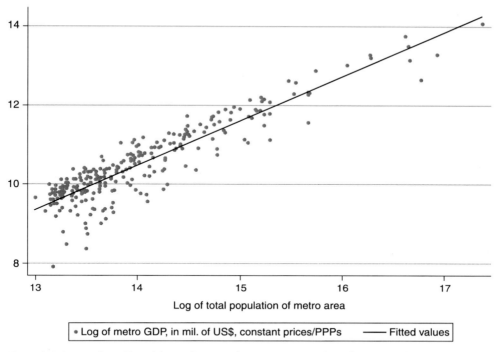

Figure 1.4 Scatterplot of logarithms of metropolitan GDP against log of total population, 275 world MSAs, 2010, including regression fitted values

Source: OECD (2012).

Urban agglomeration and sustainability

The economics of agglomeration and economic growth of cities has significant implications for issues of social and environmental sustainability, especially as it regards the demands for material resources, energy, water, land and ecosystem services and public decisions over the form and function of urban spaces and the technologies employed in major infrastructure projects.

Table 1.1 List of OECD metropolitan areas with share of metropolitan GDP over 40 percent, sample of 275 world MSAs, 2010

City name	Country	Share of metropolitan GDP over national GDP (%)
Tallinn	Estonia	59.85
Dublin	Ireland	48.27
Budapest	Hungary	47.63
Santiago	Chile	47.26
Athens	Greece	45.25
Seoul Incheon	Korea	44.56
Copenhagen	Denmark	42.85

Source: Data from OECD Metro Explorer.

A first issue regards the spatial arrangement of people and firms in urban environments—in particular, expansive urban development or urban sprawl. Expansive physical urban growth has important implications for land use, biodiversity, energy use, transportation, public infrastructure and other sustainability issues. Economic growth has been identified as one of the three main drivers of expansive urban development, the other two being a growing population and falling transportation costs (Brueckner, 2000). The relationship between wealth and demand for larger dwellings and space has been established in many societies across time; for example, Morris (2004) summarizes excavated house data from ancient Greece and finds that house sizes increased on average by more than 350 percent, from 80 m^2 to about 360 m^2 between 800 and 300 BCE—an era of economic growth for cities in Greece (Ober, 2010). Today, urban sprawl typically invokes discussions of both market and government failures—primarily through lack of appropriate pricing mechanisms, land use and zoning regulations (Anas and Rhee, 2006). This understanding of the relationship between wealth and expansive urban growth poses an interesting challenge for the capacity of cities to effectively respond to climate change by regulating emissions of greenhouse gases (Anderson et al., 1996). Figure 1.5 shows the average relationship between population density and per capita CO_2 emissions using OECD data; if cities of tomorrow across the globe do not densify, ceteris paribus, we are faced with a greater challenge for mitigation to climate change through other areas of intervention (Seto et al., 2010).

The relationship between energy use and economic growth is also important for addressing GEC challenges. Urban economic growth is a key driver of energy use and greenhouse gas

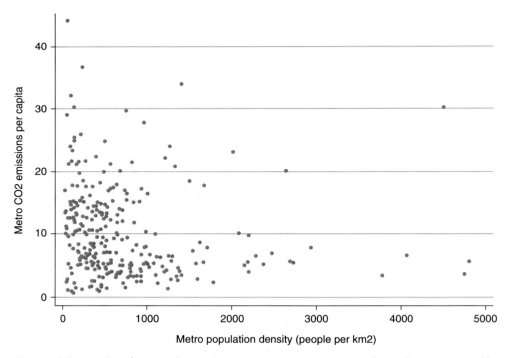

Figure 1.5 Scatterplot of metropolitan CO_2 emissions per capita over population density, 275 world MSAs, 2010

Source: Data from OECD Metro Explorer.

(GHG) emissions (Satterthwaite, 2009; Seto et al., 2014), and higher incomes are associated with a higher consumption of energy and GHG emissions (Weisz and Steinberger, 2010; Zheng et al., 2011). While there is theoretical support to the idea that incomes drive energy use (energy being a normal good), empirical research also points to the reverse causal path (Liddle and Lung, 2014). Economic theory also hypothesizes that economic growth coevolves with an increased capacity for innovation and technological change that can increase energy efficiency, leading to reductions in GHG emissions. Furthermore, other parameters influencing the degree to which economic growth drives GHG emissions include the type of economic specialization of urban activities and the energy supply mix (Kennedy et al., 2012). A positive association can be expected in the growth of both economic output and emissions of an urban area with a higher proportion of industries specialized in energy-intensive and carbon-intensive activities. However, also expected is a lack of association in the growth of economic output and emissions when an urban area has a high proportion of its industry specialized in less energy-intensive industries (Marcotullio et al., 2014).

The centrality of economic output for CO_2 emissions in the developed world context is exhibited in Figure 1.6, showing the relationship between CO_2 emissions per capita and GDP per capita in 275 metropolitan areas of OECD countries. Regression analysis reveals that the elasticity of CO_2 emissions per capita with respect to GDP per capita is 1.21. That is, as GDP per capita increases by 1 percent, CO_2 emissions per capita are expected to increase by 1.21 percent, holding population density constant; thus CO_2 emissions scale superlinearly with the

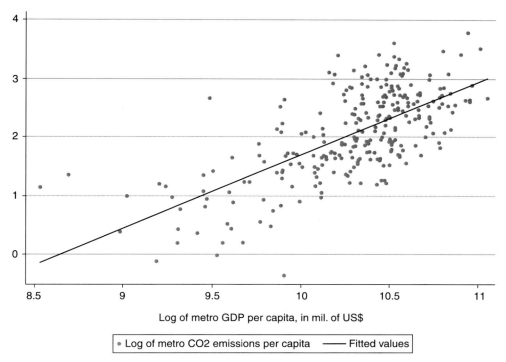

Figure 1.6 Scatterplot of the natural log of metropolitan CO_2 emissions against log of GDP per capita, 275 world MSAs, 2010, including linear fit

Source: OECD, 2012.

economic size of the metropolitan area. At the same time, an increase of population density by 100 percent (doubling population density) reduces CO_2 emissions per capita by 8 percent, holding GDP per capita constant. It seems that it will be very difficult to counter the effects of growing GDP per capita via measures of population density alone.

In practice, theoretical considerations regarding the dynamics of economic output and emissions have been tested within urban settings through the environmental Kuznets curve (EKC) hypothesis, suggesting an inverted-U shaped relationship between GDP per capita and pollution (Grossman and Krueger, 1995). Brock and Taylor (2005) distinguish three critically interacting parts in the EKC hypothesis: scale, composition and technique. Larger scale implies a proportionally higher level of economic activity and more emissions. Composition and technique are responsible for ameliorating or worsening the effects of scale. Composition refers to the (lack of) proportionality due to emissions-intensive products that the economy produces, which affects the overall level of emissions. Technique refers to recipes and technologies employed in production. Although the literature on the EKC is mixed, a significant majority of empirical studies do not support the EKC hypothesis for CO_2, suggesting that economic growth will not automatically solve the problem of emissions growth (e.g. Stern, 2004; He and Richard, 2010; Hamit-Haggar, 2012). Other studies have focused on the direction of causality between economic growth, energy consumption and carbon emissions reaching inconclusive results (Zhang and Cheng, 2009).

In an era of GEC, cities offer opportunities to mitigate climate change through their concentration of people, activities, resources, technology and institutions, but very importantly, wealth—the main focus of this chapter. Cities can be instrumental in mitigating climate change through their economic agglomeration effects (Dhakal, 2010). A stream of research in the past few years provides several examples of the benefits, but also some potential pitfalls of agglomeration. Globally, per capita urban energy consumption in industrialized countries is often lower than national averages (Dodman, 2009). In almost every US metropolitan area, carbon emissions are significantly lower for people who live in central cities than for people who live in suburbs (Glaeser and Kahn, 2010). Recent research also suggests that large urban areas in the US are not more CO_2 emissions efficient than small ones (Fragkias et al., 2013) or even, depending on the definition of city boundaries, CO_2 emissions inefficient (Oliveira et al., 2014).

Cities also offer opportunities for climate change mitigation and adaptation through their wider policy structures, institutions and governance (Bulkeley, 2010). Large wealthy cities can use urban design and integrated land use and transportation planning—a critical aspect in climate change mitigation efforts (Weisz and Steinberger, 2010; Huang et al., 2010). A recent meta-analysis shows that compact and mixed urban land use coupled with co-located high residential and employment densities can reduce energy consumption and emissions through reducing vehicle miles traveled (Ewing and Cervero, 2010). The importance of urban form has been argued in a wide array of cities across the globe (Liu and Deng, 2011; Liu et al., 2014). Policy-makers at sub-national administrative levels have been developing climate action plans and initiatives that set emission targets, usually to be realized through sector-specific GHG reductions (Lutsey and Sperling, 2008). Early findings on whether these opportunities bear fruit are at present mixed. For example, case studies in municipalities in California show that climate change planning is not causally connected to specific outcomes (Millard-Ball, 2013).

While median incomes have grown globally, urbanization is characterized by income inequality and substantial GDP per capita differences within and across metropolitan areas (OECD, 2006; UN-HABITAT, 2008; Seto et al., 2010). Economic development does not always coincide with higher levels of urbanization; this is clearly the case when more than

30 percent of urban dwellers worldwide live in slums, faced with deteriorating conditions across time (Cohen, 2006). Hundreds of millions of people live in informal settlements, with many routinely facing serious health-threatening conditions (UN, 2005; UN-HABITAT, 2003). Cities include different classes of vulnerable populations—due to lack of resources, connections to services and political voice (OECD, 2006). Inequality is thus a major challenge for social sustainability, and ignoring it fundamentally affects the prospect of sustainable urban futures.

Future prospects for cities, economic growth and sustainability

Projecting economic output into the future has been a common practice in economics and other disciplines for the last 100 years, since the inception of GDP as a measure of economic activity. Such projections are typically founded on national economic accounts and measure potential growth of national economies and supra-national regions (Colm, 1958). Various international organizations, governance bodies, think tanks, investment firms and nonprofits are in the business of producing income projections for nations and regions globally at regular intervals (TCB, 2015; IMF, 2013; IMF, 2014). The emergence of the city as an important unit of observation for economic growth leads naturally to the need for accurate measurement of economic activity at the sub-national level. With few exceptions, it is thus surprising that global projections for even fundamental measures of economic activity are nearly non-existent.

For example, the McKinsey Global Institute (MGI) reports urban economic projections for the world's largest 600 metropolitan areas (Dobbs et al., 2011). MGI projects that we will soon experience a reshuffling in the list of the world's largest 600 urban economies; urban economies of the Global South—and in particular those in Southeast Asia—will become dominant in the club of the 600 largest urban economies. The report suggests that the largest 600 cities will keep producing approximately 60 percent of the total GDP of the world. Furthermore, the new members of the largest urban economy club will be cities in low- and middle-income countries. Most of these economic powerhouses will be Chinese urban agglomerations.

Placing the above projections in a sustainability context, the implication of future wealth for urban sustainability is not evident. Early work on urban sustainability suggests that the term 'urban sustainability' may itself be an oxymoron (Rees, 1997). As researchers solidify the strong association between the process of urbanization and wealth generation, we are naturally led to the possibility of a continued vicious cycle of unsustainable city and economic growth. Current wealth projections that are agnostic with regards to the potential transformation of our market economy are problematic. The sustainability context requires an exploration of the conditions in which urbanization can be beneficial for both increased human prosperity (material or not) and the environment. The answer to this fundamental urban sustainability question is dependent, in part, on the form of our economic systems.

It is straightforward to envision a main dichotomy in possible urban futures. If future economic growth replicates the dominant patterns of past and present-day capitalism, we can expect that added wealth will lead to an added material throughput and severely negative consequences for GEC. In this scenario, we can expect more expansive growth, limited use of alternative means of transportation, continued use of carbon-intensive technologies, lack of accounting of natural capital, highly inequitable distribution of resources, etc. If a new understanding of human well-being and prosperity becomes a core human value that leads to a transformation of our market system towards a new 'green' economy, decoupling wealth from material throughput, we would be exiting the era of (un)economic growth and place limited effects on the global environment (see Simon, Chapter 30 in this volume, for a more detailed discussion of the green economy in addressing urban environmental change). This scenario could entail accounting for all forms of capital that contribute to our well-being (human, manufactured, financial and natural capital), a substantial reduction in material throughput and a stronger 'service' orientation in our capitalist system (Hawken et al., 1999).

Irrespective of the above considerations, in an era of increasingly stronger urban land teleconnections (Seto et al., 2012), we can expect that new urban wealth and the creation of a new global middle class will have environmental impacts far away from traditional boundaries of cities (see Leichenko and Solecki, Chapter 9 in this volume). Greening an urban economy in a specific locality may be a first step towards urban sustainable futures, but in a globalized planet characterized by international trade, urban consumption patterns may have deleterious effects in other parts of the globe—rural, urban and ex-urban. We simply cannot afford to ignore the increasingly fuzzier boundaries of urban processes. The agenda thus needs to center around sustainability in the context of an urbanizing world, rather than urban sustainability.

Finally, while general wealth trends cannot be disputed, one can be critical about the above projections for a lack of integration of GEC realities in the estimation of future wealth. While we expect that urban areas will continue to be attractive to migrants and internally displaced people searching for new economic opportunities, we do not know to what degree the assumed migration patterns will change due to GEC (Tacoli, 2009). Historically, environmental change brings about new conditions for migration and city formation (Barrios et al., 2006). The era of the Anthropocene is expected to amplify this trend (Lambin et al., 2001); future research needs to further address possible future patterns of internal and external migration.

Conclusions

In the last few years, we have observed the intensification of efforts for a new urban research synthesis, attempting to connect seemingly disparate topics or themes and integrate disciplinary knowledge (Solecki et al., 2013). In this effort, cities and their material wealth can now be viewed jointly through the lenses of complexity science, geography, urban economics, regional science and sociology. They are typically examined as interlinked systems of networks and flows, within and across regions and administrative boundaries. Emergent properties of these systems (such as scaling and power laws) are progressively being better understood by the development of micro-foundations—theories of interactions of agents in an economy—and generative processes common to past and present urban life across the world. New and bigger datasets are increasingly available for the testing of hypotheses that are more relevant for the challenges of this century. Scientists and practitioners are faced with the task of jointly formulating, building and utilizing better theories and models about cities.

This chapter explores a central theme in this new 'science of cities' or 'urbanization science': the essential connection between the urbanization process, the generation of wealth and the path towards sustainability. The chapter argues that, in the past, cities were very important elements for the emergence and generalization of the market and material prosperity and that today markets are critical for the growth of cities. That is, the city helped create an institution, the market, that is now a defining factor in the future prosperity of urban dwellers and a city's long-term success—both directly, through the generation of wealth and indirectly, through the effects of GEC. It may be the case that in the near future, this interaction

Urbanization, economic growth and sustainability

will change directionality again and the city of the 21st century will generalize a new form of market and a new idea of prosperity, addressing the current major concerns about the unsustainability of our economic, social and ecological systems. Notwithstanding, the bi-directional, symbiotic relationship between cities and prosperity (through the market mechanism) is important for good policymaking and urban governance and a defining factor towards sustainable urban futures.

New research must further examine the theoretical and policy relevance of the relationship between the concentration of population in urban areas, the wealth that is generated in urban areas and GEC. Several topics are great candidates for our next research steps: (i) exploring the implications of the new lenses and knowledge produced within a new urbanization science for our capacity to respond to climate change; (ii) better understanding of the impact of urban growth on energy use; (iii) clarifying the role of innovation—through urbanization—in counteracting the environmental impacts of the very same processes of urbanization; (iv) envisioning sustainable urban futures that allow for a better management of biodiversity-rich landscapes; and, (v) better addressing low-probability, high-impact events or tipping points in urban ecosystems. We can only hope that the new urban research synthesis can offer novel perspectives to stakeholders and the wider GEC community.

Key chapter messages

- The relationship between urbanization, economic growth and sustainability is a central theme in urbanization science.
- The link between urbanization and economic growth has been shifting throughout history, with important implications for sustainability.
- Examining the still evolving interactions between urbanization and wealth generation is critical for sustainability in the 21st century.

Key research questions and/or forward looking research agenda items

- Under what conditions is urbanization 'good' for both economic growth and the environment?
- How can we most effectively capture the full effect of wealth generated in cities, especially exploring the idea of urban teleconnections?
- How can we better inform urban economic projections by integrating our current understanding of future GEC impacts?

Note

The author wishes to thank the editors for helpful comments and suggestions; all errors, omissions and choices of literature to include or exclude belong to the author alone.

References

Anas, A., and Rhee, H.-J. (2006). Curbing excess sprawl with congestion tolls and urban boundaries. *Regional Science and Urban Economics, 36*(4), 510–541. doi:10.1016/j.regsciurbeco.2006.03.003

Anderson, W. P., Kanaroglou, P. S., and Miller, E. J. (1996). Urban form, energy and the environment: a review of issues, evidence and policy. *Urban Studies, 33*(1), 7–35. doi:10.1080/00420989650012095

Bai, X., Chen, J., and Shi, P. (2012). Landscape urbanization and economic growth in China: positive feedbacks and sustainability dilemmas. *Environmental Science & Technology, 46*(1), 132–139. doi:10.1021/es202329f

Bairoch, P. (1988). *Cities and economic development: from the dawn of history to the present* (Translation of: De Jericho a Mexico: villes et economie dans l'histoire, Gallimard, Paris, 1985.). Chicago, IL: University of Chicago Press.

Barrios, S., Bertinelli, L., and Strobl, E. (2006). Climatic change and rural–urban migration: the case of sub-Saharan Africa. *Journal of Urban Economics*, *60*(3), 357–371. doi:10.1016/j.jue.2006.04.005

Bettencourt, L. M. A. (2013). The origins of scaling in cities. *Science*, *340*(6139), 1438–1441. doi:10.1126/science.1235823

Bettencourt, L. M. A., Lobo, J., Helbing, D., Kuhnert, C., and West, G. B. (2007). Growth, innovation, scaling, and the pace of life in cities. *Proceedings of the National Academy of Sciences*, *104*(17), 7301–7306. http://doi.org/10.1073/pnas.0610172104

Bettencourt, L. M. A., Lobo, J., Strumsky, D., and West, G. B. (2010). Urban scaling and its deviations: revealing the structure of wealth, innovation and crime across cities. *PLoS ONE*, *5*(11), e13541. doi:10.1371/journal.pone.0013541

Bloom, D. E., Canning, D., and Fink, G. (2008). Urbanization and the wealth of nations. *Science*, *319*(5864), 772–775. doi:10.1126/science.1153057

Braudel, F. (1982). *Civilization and capitalism, 15th–18th century* (1st U.S. ed., Vol. 1). New York: Harper & Row.

Brock, W. A., and Taylor, M. S. (2005). Chapter 28: Economic growth and the environment: a review of theory and empirics. In *Handbook of Economic Growth* (Vol. 1, pp. 1749–1821). Elsevier. Retrieved from http://linkinghub.elsevier.com/retrieve/pii/S1574068405010282

Brown, J. H., Burnside, W. R., Davidson, A. D., DeLong, J. P., Dunn, W. C., Hamilton, M. J., et al. (2011). Energetic limits to economic growth. *BioScience*, *61*(1), 19–26. doi:10.1525/bio.2011.61.1.7

Brueckner, J. K. (2000). Urban sprawl: diagnosis and remedies. *International Regional Science Review*, *23*(2), 160–171. doi:10.1177/016001700761012710

Bulkeley, H. (2010). Cities and the governing of climate change. *Annual Review of Environment and Resources*, *35*(1), 229–253. doi:10.1146/annurev-environ-072809-101747

Carlino, G. A., Chatterjee, S., and Hunt, R. M. (2007). Urban density and the rate of invention. *Journal of Urban Economics*, *61*(3), 389–419. doi:10.1016/j.jue.2006.08.003

Cohen, B. (2006). Urbanization in developing countries: current trends, future projections, and key challenges for sustainability. *Technology in Society*, *28*(1–2), 63–80. doi:10.1016/j.techsoc.2005.10.005

Colm, G. (1958). Economic projections: tools of economic analysis and decision making. *American Economic Review*, *48*(2), 178–187.

Dhakal, S. (2010). GHG emissions from urbanization and opportunities for urban carbon mitigation. *Current Opinion in Environmental Sustainability*, *2*(4), 277–283. doi:10.1016/j.cosust.2010.05.007

Dobbs, R., Smit, S., Remes, J., Manyika, J., Roxburgh, C., and Restrepo, A. (2011). *Urban world: mapping the economic power of cities*. New York: McKinsey Global Institute.

Dodman, D. (2009). Blaming cities for climate change? An analysis of urban greenhouse gas emissions inventories. *Environment and Urbanization*, *21*(1), 185–201. doi:10.1177/0956247809103016

Ewing, R., and Cervero, R. (2010). Travel and the built environment: a meta-analysis. *Journal of the American Planning Association*, *76*(3), 265–294. doi:10.1080/01944361003766766

Feenstra, R. C., Inklaar, R., and Timmer, M. P. (forthcoming, 2015). The next generation of the Penn World Table. *American Economic Review*.

Fragkias, M., Lobo, J., Strumsky, D., and Seto, K. C. (2013). Does size matter? Scaling of CO_2 emissions and US urban areas. *PLoS ONE*, *8*, e64727.

Glaeser, E. L., and Maré, D. C. (2001). Cities and skills. *Journal of Labor Economics*, *19*(2), 316–342. doi:10.1086/319563

Glaeser, E. L., and Kahn, M. E. (2010). The greenness of cities: carbon dioxide emissions and urban development. *Journal of Urban Economics*, *67*(3), 404–418. doi:10.1016/j.jue.2009.11.006

Glaeser, E. L., and Resseger, M. G. (2010). The complementarity between cities and skills. *Journal of Regional Science*, *50*(1), 221–244. doi:10.1111/j.1467-9787.2009.00635.x

Grossman, G. M., and Krueger, A. B. (1995). Economic growth and the environment. *The Quarterly Journal of Economics*, *110*(2), 353–377. doi:10.2307/2118443

Hall, P. (1998). *Cities in civilization*. New York: Pantheon Books.

Hamit-Haggar, M. (2012). Greenhouse gas emissions, energy consumption and economic growth: a panel cointegration analysis from Canadian industrial sector perspective. *Energy Economics*, *34*(1), 358–364. doi:10.1016/j.eneco.2011.06.005

Hawken, P., Lovins, H., and Lovins, A. (1999). *Natural capitalism: creating the next industrial revolution* (1st ed.). Boston: Little, Brown and Co.

He, J., and Richard, P. (2010). Environmental Kuznets curve for CO2 in Canada. *Ecological Economics, 69*(5), 1083–1093. doi:10.1016/j.ecolecon.2009.11.030

Henderson, J. V. (2010). Cities and development. *Journal of Regional Science, 50*(1), 515–540. doi:10.1111/j.1467-9787.2009.00636.x

Henderson, V. (2003). The urbanization process and economic growth: the so-what question. *Journal of Economic Growth, 8*(1), 47–71.

Huang, S.-L., Yeh, C.-T., and Chang, L.-F. (2010). The transition to an urbanizing world and the demand for natural resources. *Current Opinion in Environmental Sustainability, 2*(3), 136–143. doi:10.1016/j.cosust.2010.06.004

IMF. (2013). *World Economic Outlook (WEO): coping with high debt and sluggish growth.* Washington, D.C.: International Monetary Fund (IMF).

IMF. (2014). *World Economic Outlook (WEO): legacies, clouds, uncertainties.* Washington, D.C.: International Monetary Fund (IMF).

Jacobs, J. (1969). *The economy of cities.* New York: Vintage.

Jacobs, J. (1984). *Cities and the wealth of nations.* New York: Vintage.

Kates, R. W., Turner II, B. L., and Clark, W. C. (1990). The great transformation. In B. L. Turner II, W. C. Clark, R. W. Kates, J. F. Richards, J. T. Mathews, and W. B. Meyer (Eds.), *The Earth as transformed by human action: global and regional changes in the biosphere over the past 300 years* (pp. 1–18). Cambridge, U.K.: Cambridge University Press.

Kennedy, C., Demoullin, S., and Mohareb, E. (2012). Cities reducing their greenhouse gas emissions. *Energy Policy, 49*, 774–777. doi:10.1016/j.enpol.2012.07.030

Knudsen, B., Florida, R., Stolarick, K., and Gates, G. (2008). Density and creativity in U.S. regions. *Annals of the Association of American Geographers, 98*(2), 461–478. doi:10.1080/00045600701851150

Krugman, P. (1998). What's new about the new economic geography? *Oxford Review of Economic Policy, 14*(2), 7–17. doi:10.1093/oxrep/14.2.7

Lambin, E. F., Turner, B. L., Geist, H. J., Agbola, S. B., Angelsen, A., Bruce, J. W., . . . Xu, J. (2001). The causes of land-use and land-cover change: moving beyond the myths. *Global Environmental Change, 11*(4), 261–269. doi:10.1016/S0959-3780(01)00007-3

Liddle, B., and Lung, S. (2014). Might electricity consumption cause urbanization instead? Evidence from heterogeneous panel long-run causality tests. *Global Environmental Change, 24*, 42–51. doi:10.1016/j.gloenvcha.2013.11.013

Liu, J., and Deng, X. (2011). Impacts and mitigation of climate change on Chinese cities. *Current Opinion in Environmental Sustainability, 3*(3), 188–192. doi:10.1016/j.cosust.2010.12.010

Liu, Y., Song, Y., and Song, X. (2014). An empirical study on the relationship between urban compactness and CO2 efficiency in China. *Habitat International, 41*, 92–98. doi:10.1016/j.habitatint.2013.07.005

Lobo, J., Bettencourt, L. M. A., Strumsky, D., and West, G. B. (2013). Urban scaling and the production function for cities. *PLoS ONE, 8*(3), e58407. doi:10.1371/journal.pone.0058407

Lucas, R. (1988). On the mechanics of economic development. *Journal of Monetary Economics, 22*, 3–42.

Lutsey, N., and Sperling, D. (2008). America's bottom-up climate change mitigation policy. *Energy Policy, 36*(2), 673–685. doi:10.1016/j.enpol.2007.10.018

Marcotullio, P. J., Hughes, S., Sarzynski, A., Pincetl, S., Sanchez Peña, L., Romero-Lankao, P., . . . Seto, K. C. (2014). Urbanization and the carbon cycle: contributions from social science. *Earth's Future, 2*(10), 496–514. http://doi.org/10.1002/2014EF000257

Marshall, A. (1890). *Principles of economics.* London: Macmillan and Co.

Marshall, A. (2006). The golden flame flickers most brightly in cities: review of cities and civilization. *Metropolis Magazine.*

Melo, P. C., Graham, D. J., and Noland, R. B. (2009). A meta-analysis of estimates of urban agglomeration economies. *Regional Science and Urban Economics, 39*(3), 332–342. doi:10.1016/j.regsciurbeco.2008.12.002

Millard-Ball, A. (2013). The limits to planning: causal impacts of city climate action plans. *Journal of Planning Education and Research, 33*(1), 5–19. doi:10.1177/0739456X12449742

Morris, I. (2004). Economic growth in Ancient Greece. *Journal of Institutional and Theoretical Economics, 160*, 709–742.

Ober, J. (2010). Wealthy hellas. *Transactions of the American Philological Association, 140*(2), 241–286.

OECD. (2006). *Competitive cities in the global economy.* Paris: OECD.

OECD. (2012). *Redefining urban: a new way to measure metropolitan areas.* Paris: OECD Publishing.

Oliveira, E. A., Jr, J. S. A., and Makse, H. A. (2014). Large cities are less green. *Scientific Reports, 4.* doi:10.1038/srep04235

Pirenne, H. (1925). *Medieval cities: their origins and the revival of trade* (Reprint). Princeton, NJ: Princeton University Press.

Puga, D. (2010). The magnitude and causes of agglomeration economies. *Journal of Regional Science, 50*(1), 203–219. doi:10.1111/j.1467-9787.2009.00657.x

Quigley, J. M. (1998). Urban diversity and economic growth. *Journal of Economic Perspectives, 12*(2), 127–138. doi:10.1257/jep.12.2.127

Rees, W. E. (1997). Is "sustainable city" an oxymoron? *Local Environment, 2*(3), 303–310. doi:10.1080/13549839708725535

Rockström, J., Steffen, W., Noone, K., Persson, Å., Chapin, F. S., Lambin, E. F., . . . Foley, J. A. (2009). A safe operating space for humanity. *Nature, 461*(7263), 472–475. doi:10.1038/461472a

Romer, P. (1986). Increasing returns and long run growth. *Journal of Political Economy, 94*, 1002–1037.

Rosenthal, S. S., and Strange, W. C. (2004). Chapter 49: Evidence on the nature and sources of agglomeration economies. In *Handbook of Regional and Urban Economics* (Vol. 4, pp. 2119–2171). Elsevier. Retrieved from http://linkinghub.elsevier.com/retrieve/pii/S1574008004800063

Satterthwaite, D. (2009). The implications of population growth and urbanization for climate change. *Environment and Urbanization, 21*(2), 545–567. doi:10.1177/0956247809344361

Seitzinger, S. P., Svedin, U., Crumley, C. L., Steffen, W., Abdullah, S. A., Alfsen, C., . . . Sugar, L. (2012). Planetary stewardship in an urbanizing world: beyond city limits. *AMBIO, 41*(8), 787–794. doi:10.1007/s13280-012-0353-7

Seto, K. C., Sánchez-Rodríguez, R., and Fragkias, M. (2010). The new geography of contemporary urbanization and the environment. *Annual Review of Environment and Resources, 35*(1), 167–194. doi:10.1146/annurev-environ-100809-125336

Seto, K. C., Reenberg, A., Boone, C. G., Fragkias, M., Haase, D., Langanke, T., . . . Simon, D. (2012). Urban land teleconnections and sustainability. *Proceedings of the National Academy of Sciences, 109*(20), 7687–7692. doi:10.1073/pnas.1117622109

Seto, K. C., Dhakal, S., Blanco, H., Delgado, G. C., Dewar, D., Huang, L., . . . Ramaswami, A. (2014). Chapter 12: Human settlements, infrastructure, and spatial planning. In *IPCC Fifth Assessment Report: Mitigation of Climate Change.* Berlin, Germany: Intergovernmental Panel on Climate Change (IPCC).

Smith, A. (1776). *An inquiry into the nature and causes of the wealth of nations.* (E. Cannan, Ed., 1994). New York: Modern Library.

Solecki, W. D., Seto, K. C., and Marcotullio, P. (2013). It's time for an urbanization science. *Environment: Science & Policy for Sustainable Development, 55*(1), 12–17.

Stern, D. I. (2004). The rise and fall of the environmental Kuznets curve. *World Development, 32*(8), 1419–1439. doi:10.1016/j.worlddev.2004.03.004

Tacoli, C. (2009). Crisis or adaptation? Migration and climate change in a context of high mobility. *Environment and Urbanization, 21*(2), 513–525. doi:10.1177/0956247809342182

TCB. (2015). *Global Economic Outlook (GEO).* New York, NY: The Conference Board, Inc. (TCB). Retrieved from www.conference-board.org/data/globaloutlook.cfm

Turok, I., and McGranahan, G. (2013). Urbanization and economic growth: the arguments and evidence for Africa and Asia. *Environment and Urbanization, 25*(2), 465–482. doi:10.1177/0956247813490908

UN. (2005). *A home in the city.* London: Earthscan.

UN. (2014). *World urbanization prospects, the 2014 revision.* New York: United Nations, Department of Economic and Social Affairs (UN-DESA), Population Division.

UN-HABITAT. (2003). *The challenge of slums: global report on human settlements.* London: Earthscan.

UN-HABITAT. (2008). *State of the world's cities: harmonious cities.* London: Earthscan/UN-HABITAT.

WBG. (2015). *World development indicators 2015.* Washington D.C.: International Bank for Reconstruction and Development/The World Bank. Retrieved from doi:10.1596/978-1-4648-0440-3

Weisz, H., and Steinberger, J. K. (2010). Reducing energy and material flows in cities. *Current Opinion in Environmental Sustainability, 2*(3), 185–192. doi:10.1016/j.cosust.2010.05.010

Zhang, X.-P., and Cheng, X.-M. (2009). Energy consumption, carbon emissions, and economic growth in China. *Ecological Economics, 68*(10), 2706–2712. doi:10.1016/j.ecolecon.2009.05.011

Zheng, S., Wang, R., Glaeser, E. L., and Kahn, M. E. (2011). The greenness of China: household carbon dioxide emissions and urban development. *Journal of Economic Geography, 11*(5), 761–792. doi:10.1093/jeg/lbq031

2

URBANIZATION, FOOD CONSUMPTION AND THE ENVIRONMENT

Shannon Murray, Samara Brock and Karen C. Seto

Recent projections suggest that the global urban population will reach between 5.6 and 7.1 billion by the middle of the 21st century—up from 3.5 billion today (UN DESA, 2014). Urbanization in low- and middle-income countries over the last century has been a major factor in shaping the current global food system, from large-scale food production to distribution networks that stretch across continents. The 'urban transition' that is underway in much of the world will be paralleled by new types of urban developments, new innovations in urban design as well as radical changes in lifestyles, preferences and diet. Simultaneously, prime agricultural land is being lost to urban expansion and environmental degradation, and farmers are experiencing greater competition for land, water and energy (Godfray et al., 2010; Foley et al., 2005). Given these patterns, what are the implications of contemporary urbanization for sustainability of the global food system?

The first part of this chapter identifies some of the key trends in the current food system that are related to food consumption demands in urban areas. For example, many of the changing demands on global agriculture can be linked to the transition from rural to urban life and associated changes in gender roles including an increasing proportion of women in the workforce. Moreover, technological innovations such as packaging have enabled urban lifestyles and contribute to changes in the global supply chain and distribution network of the food system. As rising incomes and infrastructure development expand the availability of food products, a 'nutrition transition' is occurring in low- and middle-income countries that was once only found in 'Westernized' countries (Popkin, 1993). In addition to its linkages with a rise in non-communicable diseases (NCDs), this diet shift also contributes to environmental impacts of increased agricultural activity such as soil erosion, nutrient losses, reduction in biodiversity and water scarcity. In many ways, the global supply chain that supports the distribution of food to cities reinforces this system of agriculture through complex processes of storage and transportation. Production and distribution systems that feed major cities contribute to environmental degradation in many cases and are, in turn, vulnerable to natural disasters, infrastructure issues and global climate change.

In this chapter, the authors argue that urbanization creates changes in the food system, which in turn impacts the global environment. The ways in which food is produced, distributed and consumed influence patterns of urbanization and create positive feedback within this system. Thus, urban consumption shapes the structure and functioning of food systems, which

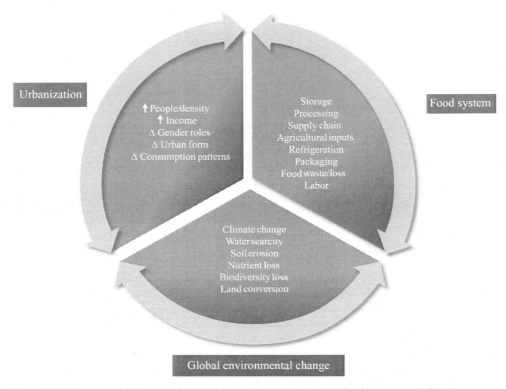

Figure 2.1 A conceptual framework: links between urbanization, the food system and global environmental change

shapes the built environment and urban populations. Additionally, aspects of global environmental change, such as climate change and water scarcity, impact processes in both cities and the food system. The interactions discussed here are complex and evolving. This conceptual framework (see Figure 2.1) is therefore intended to highlight components of these systems that interact across multiple spatial scales including specific patterns associated with urbanization (i.e. rising incomes and changing gender roles). The connections between these key trends and system components are discussed further in this chapter. While much of the popular media and scientific literature focus on the negative impacts of urbanization on food and agriculture, there are many positive effects of urbanization on food systems. The second part of this chapter highlights areas where changes can be made across the supply chain as well as urban planning mechanisms and food policies that cities across the world are using to support the development of a more sustainable food system. Based on analysis of these challenges and opportunities, the chapter closes with recommendations for further research.

Changing diets

The nutrition transition

Dietary patterns are changing worldwide, driven by a combination of factors including high rates of urbanization, rising incomes and greater availability of foods such as cheap vegetable oils and fats (Drewnowski and Popkin, 1997). This trend, described as the 'nutrition transition,' is characterized by shifts from a grain-based diet high in fiber and unrefined carbohydrates to a diet featuring more animal products, vegetable oils, sweeteners and processed carbohydrates (Abrahams et al., 2011; Popkin and Gordon-Larsen, 2004). The nutrition transition was initially an urban issue. Based on data from the 1970s to the 1990s, researchers observed that people in cities were transitioning away from cereal-based diets at far lower income levels than had been previously possible, as people at the very lowest income levels had increased access to vegetable oils and refined sweeteners (Drewnowski and Popkin, 1997). Dietary changes linked with the nutrition transition correspond with a sharp increase in the incidence of NCDs, particularly obesity, heart disease and diabetes (Kearney, 2010). As discussed further below, the system of industrial agriculture that currently supports this consumer demand is associated with intensive energy use, agrochemical inputs, land degradation and loss of natural resources (i.e. biodiversity, freshwater, soil) on a global scale (Foley et al., 2005; Godfray et al., 2010). Figure 2.2 highlights recent patterns in global livestock production including environmental impacts, which have been driven largely by urban demand for meat.

The nutrition transition in low- and middle-income countries is well underway and is proceeding at a much faster rate than was first observed in high-income economies. Dietary shifts that unfolded over four decades in North America took place on a much larger scale in many low- and middle-income countries over the course of a single decade (Ng et al., 2014). However, it is possible that some of the current public health challenges and environmental impacts of the nutrition transition can be attenuated, reversed or averted. The anticipated changes in household income and demographics for the fastest growing urban areas of the world are likely to differ from those studied in the nutrition transitions in North America, China and South and Central America due to differences in culture, historical and contemporary contexts and current conditions of the global food system. The most pronounced population growth rates are expected to be in cities in sub-Saharan Africa, where the effects of the nutrition transition are even less certain (de Ruiter et al., 2014). For example, the anticipated lack of cultural acceptability of increased meat consumption in fast-growing, predominantly Muslim areas of sub-Saharan Africa suggests that the nutrition transition patterns observed in many Asian and Latin American countries may not be globally inevitable (Alexandratos and Bruinsma, 2012).

Diet and environmental impacts

Agricultural production has met the consumption demands of a rapidly growing urban population by using land, water and energy inputs more intensively (Satterthwaite et al., 2010), with global agricultural yields doubling since the 'Green Revolution' of the 1960s (Tilman et al., 2001). In the past, this growth in production capacity largely depended on bringing new land under cultivation. However, the availability of fertile agricultural land is declining, and significant environmental impacts are generated when land is converted for food production. Globally, an estimated 25 percent of greenhouse gas (GHG) emissions result from land clearing, crop production and fertilizer use (Tilman et al., 2011). Much of the world's prime arable lands are already under cultivation, with stagnating yields predicted for key crops in certain regions due to changes in growing season temperature and precipitation (Lobell and Field, 2007). Therefore, most of the expansion of agriculture is likely to be on less productive, or marginal, land where the cost of improving degraded or unproductive land is significant. Such cultivation of marginal lands is often supported by increased use of agrochemical inputs (i.e. pesticides, herbicides and fertilizer) and irrigation to maintain crop

Box 2.1 Changing diets: urban meat consumption

The environmental impacts of changing diets are clearly evident in the livestock sector. Growth of large-scale livestock operations has been largely driven by rising incomes and urban demand for meat, in addition to improvements in infrastructure such as cold chains that allow for greater distribution on a global scale (Thornton, 2010). As Figure 2.2 illustrates, beef production has more than doubled since the 1960s and the global cattle population is projected to increase from 1.5 billion to 2.6 billion between 2000 and 2050, leading to expanded crop production of feed stocks such as soybean and corn (Thornton, 2010). Industrial poultry and swine production is currently growing at an even greater rate than cattle production.

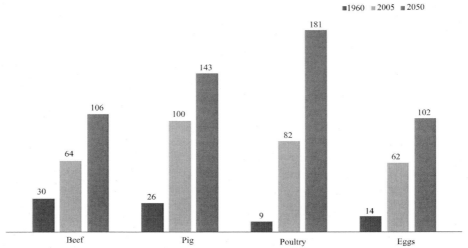

Figure 2.2 Trends in global meat production, 1960–2050 (in million tonnes)
Source: Alexandratos and Bruinsma (2012).

Consumption of animal products is identified as an important driver of land conversion, greenhouse gas emissions and pressure on water resources. GHG emissions from agriculture are linked with climate change (Smith et al., 2008), and animal production alone accounts for roughly 14 percent of total anthropogenic GHG emissions (CO_2e) (Pelletier and Tyedmers, 2010). In high-income countries, approximately 70 percent of cereal crops are consumed by animals, and the livestock sector accounts for over 8 percent of global human water use, primarily for irrigation of feed crops (Pretty, 2008; Steinfeld and FAO, 2006).

---

yields. If current production trends continue, global pesticide use in 2050 will be an estimated 2.7 times greater than levels used in the early 20th century (Tilman et al., 2001), leading to greater environmental degradation and long-term production costs as farmers increasingly rely on external inputs to limit pests and build soil nutrients. Approximately 17 percent of global croplands now depend on irrigation to maintain production (Thenkabail et al., 2009), leading to concerns about the future quantity and quality of water available under conflicting resource demands from agriculture and urban areas. Water scarcity is predicted to be a regional concern in countries such as China and India, where demands on water for irrigation and urban consumption are already impacting natural groundwater recharge and limiting the availability of water supplies (Tilman et al., 2002; see Pfister et al., Chapter 17 in this volume, for a discussion of urban water supply and availability). These irrigation systems are used in part to support the production of high-value horticultural crops in arid regions and in low- and middle-income countries such as Kenya, for export to cities and wealthy nations (Omiti et al., 2008).

To a certain degree, the availability of prime agricultural lands is also threatened by expansion of the built environment as rising urban populations drive the development of new urban infrastructure. In many parts of the world, cities have historically developed in or near fertile regions such that the direct expansion of urban areas creates a loss of arable land (Deng, Chapter 3 in this volume). Global urban land cover is likely to triple between 2000 and 2030 to accommodate the growth in urban population (Angel et al., 2011; Seto et al., 2011; Seto et al., 2012). A recent study shows that approximately 8 percent of the total global crop yield and 4.5 percent of global cultivated lands will be lost to urban expansion by 2030 (Reitsma et al., in review). While the total area of agricultural land impacted by direct expansion of urban areas is relatively small on a global scale, the loss of this arable land in rapidly urbanizing areas may create instability in regional food systems. Thus, the combination of growth in the built environment with consumer demand for a greater quantity and diversity of food products can significantly impact agricultural areas.

Continued growth in the food production sector is often cited as a necessity in order to feed a rapidly urbanizing population, with recent studies estimating that food production will need to increase by 70 to 100 percent in order to feed an expected nine billion people in the 21st century (Godfray et al., 2010). Researchers have drawn attention to multiple production solutions—including improving crop yields on land currently under cultivation (closing the yield gap) and selecting regionally adapted crop varieties that may provide additional resilience under current climate change projections (Foley et al., 2011; Lobell et al., 2008). In addition, the food supply currently produces an estimated 3,000 (FAO, 2014) to 3,900 (Kummu et al., 2012) calories per person per day, representing an excess of calories in the food supply beyond scientific recommendations for individual consumption (Lichtenstein et al., 2006). Improvements in efficiency and equity in food distribution may therefore be of critical importance in meeting global consumption demands.

Distribution and supply chain networks

Food consumption demand in cities has been a major driver in developing the complex supply chain that is responsible for much of the processing, storage and transportation of agricultural products. Improvements in cold chain technology have spurred shipment of perishable food around the world (Rees, 2013), while also enabling high product turnover and increasing the variety of food offered year round through modern retail formats (i.e. hypermarkets, supermarkets and convenience stores).

The rise of supermarkets

The growth and purchasing power of urban populations continues to shape production and distribution trends within the global food supply chain. Although local production may provide a supplementary food source, much of the food consumed in cities is still produced in other countries and travels many miles before reaching a retail market. In high-income countries, urban residents rely heavily on supermarkets or corner stores to access food on a daily basis (Walker et al., 2010), and much of the food available in these retail formats is processed and packaged. As discussed further in this chapter, the economies of scale and long supply chains required to support this system contribute significantly to the environmental impacts of food waste throughout the food system. While supermarkets have long been viewed as shopping destinations in major cities and wealthy countries, the share of modern retail formats (such as hypermarkets, supermarkets and convenience stores) in the food retail sector is currently rising in many low- and middle-income countries (Reardon et al., 2003).

The retail power of supermarkets in these regions began with Latin America in the 1990s, followed by more rapid growth in Southeast Asia five to seven years later; the most recent wave of expansion is happening in both Africa and Eastern Europe (Traill, 2006). However, supermarket development is occurring through a markedly different process in these countries, achieving a lower market share of the food retail sector when compared to patterns previously observed in North America and Western Europe. Despite the fact that supermarkets controlled an estimated 48 percent share of the Chinese urban food retail market in 2001 (Reardon et al., 2003), multiple surveys in the same decade find that 60 to 70 percent of urban food consumers purchase most of their fruit and vegetables at wet markets (Zhang and Pan, 2013). Shopping patterns in these cities may also contribute to consumption of more fresh produce, differences in time spent preparing and cooking meals and urban linkages with regional producers through traditional markets. Surveys in Hong Kong show that, on average, consumers shop at wet markets 5.8 times per week, in addition to less frequent supermarket trips for nonperishable and packaged foods (Goldman et al., 1999). Similar patterns have been observed in Southeast Asia and Latin America (Reardon et al., 2003).

The link between urbanization and the growth of supermarkets has several important implications for how food is produced and transported. Supplying food to large markets and chain stores favors large-scale agriculture and transnational corporations, leading to considerable changes in the distribution and marketing of food (Kennedy et al., 2004). Within the global food supply chain, vertically integrated companies have the ability to source large volumes of food while directing quality standards for produce sourced from large-scale farms, often increasing food waste during harvest and processing.

Risk and resilience in food supply chains

Urban areas depend on complex global food supply chains, and the food found on grocery store shelves has often traveled long distances and arrived 'just in time' for consumer retail. Food supplies in urban areas are thus particularly vulnerable to natural disasters and disruptions in the global supply chain (Satterthwaite et al., 2010). Over the last decade, natural disasters have disrupted global transportation systems and the distribution of food multiple times. The eruption of Iceland's Mount Eyjafjallajökull in 2010 caused airport closures throughout Europe for approximately one week and disrupted food supplies. While European cities were able to adapt by sourcing food through multiple supply chains, this isolated natural disaster created economic impacts for Kenyan farmers, who produce high-value produce for export markets along with other sub-Saharan countries (Justus, 2014). In the United States, Hurricane Katrina and Superstorm Sandy affected travel and delivery of food for days due to large-scale road closures (Wilkie et al., 2012).

While these storms did not lead to widespread food shortages, their impact on urban provisioning infrastructure highlights food supply risk in the current distribution model. Extreme weather events and fuel shortages have drawn attention to the fragility of food distribution networks that supply major cities such as New York City and London; recent estimates predict that such urban areas store only a three-day food supply (Simms, 2010). As climate change leads to changes in the frequency, intensity and duration of extreme weather events (IPCC, 2013), these vulnerabilities in the food supply could be critical.

Several studies have shown that cultivated areas in close proximity to cities, and even within cities, serve as buffers against potential food shortages that may come with disruption of long-distance trade networks and thus increase food security. This suggests that urbanization's direct expansion on global croplands must be taken into consideration in developing strategies for food security at both national and global levels. In addition, as discussed later in this chapter, many cities are instituting policies to support regional food systems in an effort to preserve valuable local farmland and bolster regional food security.

Urban food consumption trends

Urbanization affects not only what people eat, but also *how* and *where* they eat. An essential characteristic of urban life is an increase in activities away from home, be it for employment, shopping, recreation or transit. Urban residents purchase and eat more of their foods away from home on a daily basis. At a national scale, as Figure 2.3 illustrates, the relative amount of food consumed at home decreases as the percentage of the urban population rises.

Although the growth of urban populations is just one contributing factor, this pattern has important impacts on both the global food system and public health of urban residents. Foods consumed away from home often have higher calories per meal as well as less dairy, fresh fruits and vegetables (Diliberti et al., 2004; Taveras et al., 2005; Poti and Popkin, 2011). In

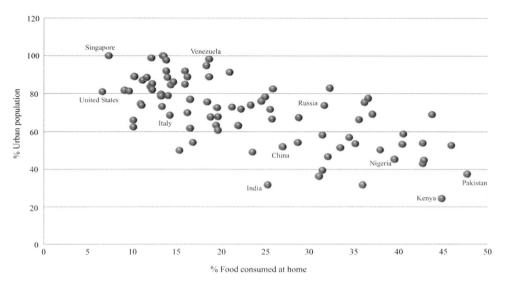

Figure 2.3 Urban population and food consumption at home (national scale), 2012

Source: Calculated from World Bank (2014).

general, eating meals away from home results in consumption of larger portions than meals consumed at home (French et al., 2001). Thus, the growing trend in consumption of foods away from home is increasing total caloric intake and overall quantity of food consumed while simultaneously supplanting nutritious foods with more processed options. As previously discussed, the nutrition transition (including consumption of processed foods) has been linked with multiple environmental externalities of large-scale food production, including increases in land, water and energy use (Satterthwaite et al., 2010).

Rising incomes and changes in urban employment patterns also influence consumption, particularly as greater numbers of women work outside of the home and people work longer hours (Satterthwaite et al., 2010). Combined with factors such as access to large quantities of food in modern retail formats, global improvements in the cold chain and household refrigerator use have increased the amount of food purchased at the consumer level in cities. Refrigerator ownership (and size, particularly in the United States) grew alongside rising incomes in both North America and Europe during the 20th century, and a similar pattern is occurring today in Asia. As of 2005, only 12 percent of rural families in China owned refrigerators, compared to 80 percent in urban areas (Rees, 2013). In addition, studies show that household demand for food away from home increases along with the income of the household member who is primarily responsible for food preparation (Stewart, 2011). These changes in disposable income and the reduction in time spent on cooking have created growing demand for convenience. As studies of consumer spending on food in low- and middle-income countries reveal, variation in these patterns may exist between rapidly urbanizing areas and cities in high-income countries. However, global trends such as the nutrition transition and lifestyle changes such as food consumption outside of the home, have been supported by continued urban demand and technological investments. Innovations in food packaging have further enabled long-distance transport, storage, distribution, preservation, containment, protection and convenience of pre-prepared meals (Robertson, 2012). Although packaged, prepared and processed foods have in part supported urban economic growth and preferences, their use has also contributed to significant waste across various sectors of the global supply chain.

Food loss and waste

Although food waste is difficult to quantify, recent studies estimate that 30 to 50 percent of the food produced globally is lost during the processes of harvesting, storage, packaging, distribution and consumption (Godfray et al., 2010; Parfitt et al., 2010). Food waste generates global impacts beyond the direct loss of calories and easily visible loss of biomass. Globally, roughly 24 percent of the freshwater resources used in crop production and 23 percent of synthetic fertilizer use can be attributed to losses in the food supply chain (Kummu et al., 2012). The environmental cost of wasted food is therefore magnified because of the natural resources that are used as inputs in food production.

Urban demand for processed food and preferences for visual qualities can significantly contribute to food waste. As consumers purchase a growing proportion of their food in supermarkets, expectations about the aesthetics and standardization of quality lead to food losses during both production and retail. Farmers are less likely to harvest crops that show insect damage or asymmetrical growth forms because of supermarket standards, leaving edible food to rot in fields (Parfitt et al., 2010). Moreover, routine retail strategies of overstocking display shelves, disposing of blemished or bruised items, and using promotions that encourage consumers to buy more food contribute to food waste in supermarkets.

Food losses occur throughout all sectors of the food supply chain, although there are distinct differences in food waste trends between regions and income levels. In high-income countries, a significant proportion of food waste is linked with consumer behavior and therefore may be relatively easy to prevent. Such waste stems from confusion over 'best before' and 'use by' dates, uncertainty about the edible portion of foods, lack of knowledge regarding food storage and proper food preparation and over-supply of prepared foods. The large size of household refrigerators, particularly in North America, contributes to consumer purchase of larger quantities of food, which may spoil before it can be eaten (Freidberg, 2009).

In contrast, much of the food waste in low- and middle-income countries is historically related to post-harvest practices and distribution, including lack of proper storage and cooling techniques, long travel times between points of production and consumption and insufficient cold chain infrastructure throughout the entire supply chain (Parfitt et al., 2010; Kummu et al., 2012). Improvements to distribution infrastructure, continued expansion of cold chain storage and greater communication between supply chain sectors present an important potential for improvement. The issue of food waste occurs at a global scale, and while private farms in the processing and distribution sectors have greater control over food losses at these points, cities may exert some control over losses at the retail, consumption and post-consumption stages of the supply chain.

Supporting sustainable food systems: the role of cities

Urbanization and consumption demand in cities are important drivers of several key trends in the global food system. Diet shifts attributed to the nutrition transition, the rise of supermarkets and the complex supply chains that support them and urban preferences for convenience generate economic and ecological impacts related to how food is produced. In turn, these processes contribute to food waste and resource losses across different sectors of the supply chain. The next section identifies several key ways that cities are striving to create a more sustainable food system.

Regional food systems: culture, knowledge and practice

Beyond the direct impacts caused by expansion of the built environment, urbanization is also contributing to the loss of accumulated and traditional knowledge shared between generations of farmers including production practices based on seasonal variations in precipitation and temperature and locally adapted seed varieties (Kremen et al., 2012). This practical knowledge supports greater food security and sustained food production on a long-term scale, while also providing farmers with the knowledge base to adapt to changing environmental conditions. Transmission of this knowledge within the farming community is being challenged as younger generations move to urban areas, driven largely by economic opportunities in cities. As urban populations become disconnected—both physically and culturally—from the process of food production, basic skills such as cultivation of crops, food preparation and cooking are often lost. However, urban populations in many areas of the world are currently experiencing a renewed interest in learning how to produce, preserve and cook their own food. This has resulted in innovative direct marketing and urban agriculture systems and in some cases a return to rural areas by a new generation of people interested in farming. For example, urban gardens are now common in many cities around the world. Cityfarm, in Hong Kong (http://cityfarm.hk), covers an area of 10,000 square feet and attracts urban dwellers eager to produce their own vegetables, partly in response to growing concern about tainted foods from mainland China. Urban agricultural activities can have benefits that extend well beyond food provision, for example, in the large-scale effort to establish coastal agroforestry in Kesbewa, Sri Lanka. This food forest was designed in part to protect the highly urbanized area inland from flooding, which is anticipated to worsen with climate change (Kekulandala et al., 2012).

Urban and peri-urban agriculture

Growing food in and around cities is often cited as a key strategy for building resilience while improving local food economies. Food production in urban areas, or 'urban and peri-urban agriculture' (UPA), includes both the intra-urban activities within the built environment of the city core and the peri-urban agricultural activities occurring on the periphery of the city (FAO, 2011). Production occurs in cities around the world, largely specializing in high-value, highly perishable foods (Orsini et al., 2013). As further discussed in Griffith et al. (Chapter 28, this volume), UPA has the potential to offer a wide range of social, environmental and economic benefits. Due to a range of financial, policy and cultural barriers including high opportunity costs for urban land, restrictive policies concerned with health and sanitation and cultural norms, UPA remains a poorly understood part of urban development (De Zeeuw and Dubbeling, 2009).

Due to space constraints in urban areas, UPA cannot provide a significant portion of a city's diet. Feasibility studies for large cities in the high-income world including New York, London and Vancouver, indicate that the theoretical output of all intra-urban growing areas would supply less than one percent of the city's total macronutrient demands (Mendes et al., 2008; Peters et al., 2007). Despite this, UPA can still make important contributions to the sustainability of urban food systems. For low-income urbanites whose food costs represent the majority of total household expenses, UPA can make a significant contribution to both household food security and income (Gallaher et al., 2013).

In the past decade, several municipal governments have enacted legislation supporting UPA, including legalizing many previously prohibited practices such as keeping chickens and honeybees, constructing greenhouses and garden structures, marketing produce, recycling food wastes and also offering tax incentives to make urban land available for food production (Ackerman, 2011; Mendes et al., 2008; Parfitt et al., 2010). Policy support for UPA varies at the municipal level, and further research is needed to support development of food production at this scale.

Local and regional policy

Municipal policies have also been introduced that are aimed at larger food system issues of urban food access and the viability of nearby rural areas (Deelstra and Girardet, 2000; Satterthwaite et al., 2010). Access-oriented policies that have been enacted by city governments include supports for farmers markets and distribution points, healthy meal options for mobile vendors, improvements in school meal programs and strategies for food assistance recipients to purchase more healthy options. These policies specifically target access in under-served urban areas, often described as 'food deserts' (Allen, 2010). Examples of such policies include New York City's incentives for corner stores to stock fresh produce as part of the 'Healthy Bodegas' program, Washington DC's 'SmartKarts' initiative that offers a streamlined licensing process for food carts with healthy options and the city of Chicago's 'Food Desert Action,' which brings a mobile market to under-served neighborhoods in a retrofitted bus (Quinn and The New York City Council, 2011; The Harvard Law School Food Law and Policy Clinic and The Community Food Security Coalition, 2012).

Municipal policies that aim to support rural viability include efforts to increase demand for local foods through institutional (i.e. hospitals and schools) food purchasing policies, direct farmland preservation efforts including land trusts and agricultural land reserves and indirect farmland preservation through the creation of urban containment boundaries (Condon et al., 2010; Zasada, 2011). For example, the City of Vancouver spearheaded the creation of a 'green zone' in partnership with other regional municipal governments that provides both a form of regional farmland protection and an urban growth containment function (Mendes et al., 2008). These broad, system-oriented policies indicate that municipal governments increasingly see themselves as playing a significant role in creating sustainable food systems (Morgan, 2009; Sonnino, 2009).

Final thoughts

The growth and purchasing power of urban populations continues to shape production and distribution trends within the global food supply chain, with broad impacts on the global environment. Within this complex system, many cities are now actively working to shape food policy at a local and regional scale. Norms can change rapidly in cities, and urban areas have the potential to create change in the food system as centers of innovation. Understanding how urbanization affects food consumption, distribution, production and waste will be an important aspect of rethinking food sustainability in an urbanizing era.

Key messages

- As dietary patterns shift worldwide in a trend known as the nutrition transition, consumption of foods such as animal products, vegetable oils, sweeteners and processed carbohydrates are on the rise. When combined with the physical expansion of the built environment, these urbanization patterns are driving changes in water and energy use, land conversion and ecological impacts of agriculture at a global scale.
- Urban purchasing power and technological innovations play an important role in creating a complex, but vulnerable global supply chain. Cities are now emerging as food policy actors, with a particular focus on food systems at the local and regional scale.
- Changes in *how* and *where* food is consumed are linked with urban income and time constraints. Increased consumption of processed and packaged food has impacts for the global supply chain including generation of food waste and production practices used to supply large quantities of standardized food to urban areas.
- Patterns that have characterized urbanization and food system development of the past 100 years do not necessarily dictate future trends. With significant changes in urban (and rural) lifestyles in Asia and Africa in the coming decades, many more changes in the food system are expected.

Future research directions

- Access to food retail is an important determinant of the foods that people purchase and consume. That is, land use and urban planning are powerful tools that can increase availability and accessibility of markets that provide nutritious food for future urban populations. How can existing cities support more sustainable distribution and consumption patterns? Which features of the built environment could be encouraged in cities that have yet to be developed?

- What is the role of efficiency improvements in food distribution and waste in meeting consumption demands of an urbanizing world?
- How can changes in food consumption norms within cities generate broader impacts on the food system?
- How can cities develop networks to learn from each other and catalyze change at larger scales beyond a single city?

References

Abrahams, Z., Mchiza, Z., and Steyn, N. P. (2011). Diet and mortality rates in Sub-Saharan Africa: Stages in the nutrition transition. *BMC Public Health, 11*(1), 801.

Ackerman, K. (2011). The potential for urban agriculture in New York City: Growing capacity, food security, and green infrastructure. *New York: Urban Design Lab at the Earth Institute Columbia University.*

Alexandratos, N., and Bruinsma, J. (2012). World Agriculture: Towards 2030/2050. The 2012 revision. ESA Working Paper No. 12-03. Food and Agriculture Organization of the United Nations (FAO). Rome, Italy. Available at: www.fao.org/docrep/016/ap106e/ap106e.pdf.

Allen, P. (2010). Realizing justice in local food systems. *Cambridge Journal of Regions, Economy and Society, 3*(2), 295–308.

Angel, S., Parent, J., Civco, D. L., Blei, A., and Potere, D. (2011). The dimensions of global urban expansion: Estimates and projections for all countries, 2000–2050. *Progress in Planning, 75*(2), 53–107.

Condon, P. M., Mullinix, K., Fallick, A., and Harcourt, M. (2010). Agriculture on the edge: Strategies to abate urban encroachment onto agricultural lands by promoting viable human-scale agriculture as an integral element of urbanization. *International Journal of Agricultural Sustainability, 8*(1–2), 104–115.

Deelstra, T., and Girardet, H. (2000). Urban agriculture and sustainable cities. Leusden, The Netherlands: Resource Centres on Urban Agriculture and Food Security, 43–65.

de Ruiter, H., Kastner, T., and Nonhebel, S. (2014). European dietary patterns and their associated land use: Variation between and within countries. *Food Policy, 44*, 158–166.

De Zeeuw, H., and Dubbeling, M. (2009). Cities, food and agriculture: Challenges and the way forward. Working Paper No 3. RUAF Working Paper Series. Leusden, The Netherlands: Resource Centres on Urban Agriculture and Food Security.

Diliberti, N., Bordi, P. L., Conklin, M. T., Roe, L. S., and Rolls, B. J. (2004). Increased portion size leads to increased energy intake in a restaurant meal. *Obesity Research, 12*(3), 562–568.

Drewnowski, A., and Popkin, B. M. (1997). The nutrition transition: New trends in the global diet. *Nutrition Reviews, 55*(2), 31–43.

Foley, J. A., DeFries, R., Asner, G. P., Barford, C., Bonan, G., Carpenter, S. R., and Snyder, P. K. (2005). Global consequences of land use. *Science, 309*(5734), 570–574.

Foley, J. A., Ramankutty, N., Brauman, K. A., Cassidy, E. S., Gerber, J. S., Johnston, M., . . . and Zaks, D. P. (2011). Solutions for a cultivated planet. *Nature, 478*(7369), 337–342.

Food and Agriculture Organization of the United Nations (FAO) (2011). The place of urban and peri-urban agriculture (UPA) in national food security programmes. Available at: www.fao.org/docrep/014/i2177e/i2177e00.pdf.

Food and Agriculture Organization of the United Nations (FAO) (2014). FAOSTAT online statistics division. Rome, Italy. Available at: http://faostat.fao.org/.

Freidberg, S. (2009). *Fresh: a perishable history.* Cambridge, Mass.: Belknap Press of Harvard University Press.

French, S. A., Story, M., and Jeffery, R. W. (2001). Environmental influences on eating and physical activity. *Annual Review of Public Health, 22*(1), 309–335.

Gallaher, C. M., Kerr, J. M., Njenga, M., Karanja, N. K., and WinklerPrins, A. M. (2013). Urban agriculture, social capital, and food security in the Kibera slums of Nairobi, Kenya. *Agriculture and Human Values, 30*(3), 389–404.

Godfray, H. C. J., Beddington, J. R., Crute, I. R., Haddad, L., Lawrence, D., Muir, J. F., and Toulmin, C. (2010). Food security: The challenge of feeding 9 billion people. *Science, 327*(5967), 812–818.

Goldman, A., Krider, R., and Ramaswami, S. (1999). The persistent competitive advantage of traditional food retailers in Asia: Wet markets' continued dominance in Hong Kong. *Journal of Macromarketing*, *19*(2), 126–139.

IPCC. (2013). Summary for policymakers. In: *Climate Change 2013: The Physical Science Basis. Contribution of Working Group I to the Fifth Assessment Report of the Intergovernmental Panel on Climate Change* [T. F. Stocker, D. Qin, G.-K. Plattner, M. Tignor, S. K. Allen, J. Boschung, A. Nauels, Y. Xia, V. Bex and P. M. Midgley (eds.)]. Cambridge, UK and New York: Cambridge University Press, pp. 1–30, doi:10.1017/CBO9781107415324.004

Justus, F. K. (2014). Coupled effects on Kenyan horticulture following the 2008/2009 post-election violence and the 2010 volcanic eruption of Eyjafjallajökull. *Natural Hazards*, 1–14.

Kearney, J. (2010). Food consumption trends and drivers. *Philosophical Transactions of the Royal Society B: Biological Sciences*, *365*(1554), 2793–2807.

Kekulandala, B., Nandana A., and Hettige, V. (2012). Land use mapping for promoting urban and periurban agriculture and forestry in Kesbewa urban council area. Janathakshan Gurantee Ltd, Colombo, Sri Lanka. Report prepared for RUAF Foundation and UN-Habitat.

Kennedy, G., Nantel, G., and Shetty, P. (2004). Globalization of food systems in developing countries: A synthesis of country case studies. *Globalization of Food Systems in Developing Countries: Impact on Food Security and Nutrition*, 1–25.

Kremen, C., Iles, A., and Bacon, C. (2012). Diversified farming systems: An agroecological, systems-based alternative to modern industrial agriculture. *Ecology and Society*, *17*(4), 44.

Kummu, M., De Moel, H., Porkka, M., Siebert, S., Varis, O., and Ward, P. J. (2012). Lost food, wasted resources: Global food supply chain losses and their impacts on freshwater, cropland, and fertiliser use. *Science of the Total Environment*, *438*, 477–489.

Lichtenstein, A. H., Appel, L. J., Brands, M., Carnethon, M., Daniels, S., Franch, H. A., . . . and Wylie-Rosett, J. (2006). Diet and lifestyle recommendations revision 2006: A scientific statement from the American Heart Association nutrition committee. *Circulation*, *114*(1), 82–96.

Lobell, D. B., and Field, C. B. (2007). Global scale climate–crop yield relationships and the impacts of recent warming. *Environmental Research Letters*, *2*(1), 014002.

Lobell, D. B., Burke, M. B., Tebaldi, C., Mastrandrea, M. D., Falcon, W. P., and Naylor, R. L. (2008). Prioritizing climate change adaptation needs for food security in 2030. *Science*, *319*(5863), 607–610.

Mendes, W., Balmer, K., Kaethler, T., and Rhoads, A. (2008). Using land inventories to plan for urban agriculture: Experiences from Portland and Vancouver. *Journal of the American Planning Association*, *74*(4), 435–449.

Morgan, K. (2009). Feeding the city: The challenge of urban food planning. *International Planning Studies*, *14*(4), 341–348.

Ng, M., Fleming, T., Robinson, M., Thomson, B., Graetz, N., Margono, C., . . . and Gupta, R. (2014). Global, regional, and national prevalence of overweight and obesity in children and adults during 1980–2013: A systematic analysis for the Global Burden of Disease Study 2013. *The Lancet*, *384*(9945), 766–781.

Omiti, J., Otieno, D., Nyanamba, T., and McCullough, E. B. (2008). The transition from maize production systems to high-value agriculture in Kenya. In: *The Transformation of Agri-Food Systems. Globalization, Supply Chains and Smallholder Farmers* [E. B. McCullough, P. L. Pingali and K. G. Stamoulus (eds.)]. London: FAO and Earthscan, pp. 235–257.

Orsini, F., Kahane, R., Nono-Womdim, R., and Gianquinto, G. (2013). Urban agriculture in the developing world: A review. *Agronomy for Sustainable Development*, *33*(4), 695–720.

Parfitt, J., Barthel, M., and Macnaughton, S. (2010). Food waste within food supply chains: Quantification and potential for change to 2050. *Philosophical Transactions of the Royal Society B: Biological Sciences*, *365*(1554), 3065–3081.

Pelletier, N., and Tyedmers, P. (2010). Forecasting potential global environmental costs of livestock production 2000–2050. *Proceedings of the National Academy of Sciences*, *107*(43), 18371–18374.

Peters, C. J., Wilkins, J. L., and Fick, G. W. (2007). Testing a complete-diet model for estimating the land resource requirements of food consumption and agricultural carrying capacity: The New York State example. *Renewable Agriculture and Food Systems*, *22*(02), 145–153.

Popkin, B. M. (1993). Nutritional patterns and transitions. *Population and Development Review*, 138–157.

Popkin, B. M., and Gordon-Larsen, P. (2004). The nutrition transition: Worldwide obesity dynamics and their determinants. *International Journal of Obesity*, *28*, S2–S9.

Poti, J. M., and Popkin, B. M. (2011). Trends in energy intake among US children by eating location and food source, 1977–2006. *Journal of the American Dietetic Association, 111*(8), 1156–1164.

Pretty, J. (2008). Agricultural sustainability: Concepts, principles and evidence. *Philosophical Transactions of the Royal Society B: Biological Sciences, 363*(1491), 447–465.

Quinn, C., and The New York City Council. (2011). FoodWorks: One year later. New York: The New York City Council. Available at: council.nyc.gov/downloads/pdf/foodworks1.pdf.

Reardon, T., Timmer, C. P., Barrett, C. B., and Berdegué, J. (2003). The rise of supermarkets in Africa, Asia, and Latin America. *American Journal of Agricultural Economics, 85*(5), 1140–1146.

Rees, J. (2013). *Refrigeration Nation: A History of Ice, Appliances, and Enterprise in America.* Baltimore, MD: JHU Press.

Reitsma, F., Barthel, S., Guneralp, B., and K. Seto. (in review). Impacts of future urban expansion on global food production. *Environmental Research Letters.*

Robertson, G. L. (2012). *Food packaging: Principles and practice.* Boca Raton, FL: CRC Press.

Satterthwaite, D., McGranahan, G., and Tacoli, C. (2010). Urbanization and its implications for food and farming. *Philosophical Transactions of the Royal Society B: Biological Sciences, 365*(1554), 2809–2820.

Seto, K. C., Fragkias, M., Güneralp, B., and Reilly, M. K. (2011). A meta-analysis of global urban land expansion. *PloS One, 6*(8), e23777.

Seto, K. C., Güneralp, B., and Hutyra, L. R. (2012). Global forecasts of urban expansion to 2030 and direct impacts on biodiversity and carbon pools. *Proceedings of the National Academy of Sciences, 109*(40), 16083–16088.

Simms, A. (2010, January 11). Nine meals from anarchy. *The Guardian.* Retrieved from www.theguardian.com.

Smith, P., Martino, D., Cai, Z., Gwary, D., Janzen, H., Kumar, P., . . . and Smith, J. (2008). Greenhouse gas mitigation in agriculture. *Philosophical Transactions of the Royal Society B: Biological Sciences, 363*(1492), 789–813.

Sonnino, R. (2009). Feeding the city: Towards a new research and planning agenda. *International Planning Studies, 14*(4), 425–435.

Steinfeld, H., and Food and Agriculture Organization of the United Nations. (2006). *Livestock's long shadow: Environmental issues and options.* Rome: Food and Agriculture Organization of the United Nations.

Stewart, H. (2011). Food away from home. In *The Oxford Handbook of Food Consumption and Policy* [J. Lusk, J. Roosen, and J. Shogren (eds.)]. Oxford, UK: Oxford University Press, pp. 647–666.

Taveras, E. M., Berkey, C. S., Rifas-Shiman, S. L., Ludwig, D. S., Rockett, H. R., Field, A. E., and Gillman, M. W. (2005). Association of consumption of fried food away from home with body mass index and diet quality in older children and adolescents. *Pediatrics, 116*(4), e518–e524.

The Harvard Law School Food Law and Policy Clinic, and The Community Food Security Coalition. (2012). Good laws, good food: Putting local food policy to work for our communities. Available at: http://blogs.law.harvard.edu/foodpolicyinitiative/files/2011/09/FINAL-LOCAL-TOOLKIT2.pdf.

Thenkabail, P. S., Biradar, C. M., Noojipady, P., Dheeravath, V., Li, Y., Velpuri, M., . . . and Dutta, R. (2009). Global irrigated area map (GIAM), derived from remote sensing, for the end of the last millennium. *International Journal of Remote Sensing, 30*(14), 3679–3733.

Thornton, P. K. (2010). Livestock production: Recent trends, future prospects. *Philosophical Transactions of the Royal Society B: Biological Sciences, 365*(1554), 2853–2867.

Tilman, D., Fargione, J., Wolff, B., D'Antonio, C., Dobson, A., Howarth, R., . . . and Swackhamer, D. (2001). Forecasting agriculturally driven global environmental change. *Science, 292*(5515), 281–284.

Tilman, D., Balzer, C., Hill, J., and Befort, B. L. (2011). Global food demand and the sustainable intensification of agriculture. *Proceedings of the National Academy of Sciences, 108*(50), 20260–20264.

Tilman, D., Cassman, K. G., Matson, P. A., Naylor, R., and Polasky, S. (2002). Agricultural sustainability and intensive production practices. *Nature, 418*(6898), 671–677.

Traill, W. B. (2006). The rapid rise of supermarkets? *Development Policy Review, 24*(2), 163–174.

UN DESA. (2014). Population density and urbanization. Available at: http://unstats.un.org/unsd/demographic/sconcerns/densurb/densurbmethods.htm.

Walker, R. E., Keane, C. R., and Burke, J. G. (2010). Disparities and access to healthy food in the United States: A review of food deserts literature. *Health & Place, 16*(5), 876–884.

Wilkie, C., Chun, J., and Hines, A. (2012, October 31). Hurricane Sandy disrupts food distribution, 'thousands of trucks' in limbo. *The Huffington Post*. Retrieved from www.huffingtonpost.com.

World Bank. (2014). World Bank open data. Available at: http://data.worldbank.org/.

Zasada, I. (2011). Multifunctional peri-urban agriculture—A review of societal demands and the provision of goods and services by farming. *Land Use Policy*, *28*(4), 639–648.

Zhang, Q. F., and Pan, Z. (2013). The transformation of urban vegetable retail in China: Wet markets, supermarkets and informal markets in Shanghai. *Journal of Contemporary Asia*, *43*(3), 497–518.

3
URBANIZATION AND IMPACTS ON AGRICULTURAL LAND IN CHINA

Xiangzheng Deng

Urbanization is the most dramatic human form of environmental transformation, resulting in widespread changes to the structure and functioning of ecosystems including those of agricultural systems (Seto et al., 2012). Urbanization has several impacts on agricultural systems. As discussed elsewhere in this volume (see Murray et al., Chapter 2 in this volume), urbanization has direct impacts on the conversion and loss of agricultural land. Urbanization also exerts indirect impacts on agricultural systems. Changes in urban diet put additional pressure on agricultural systems and rural–urban migration, driven by better economic and employment opportunities leads to agricultural land abandonment.

Urban settlements have existed for hundreds of years. Beginning with the Neolithic revolution, the transition from hunter-gatherers to agriculture and eventually sedentary societies began more than 10,000 years ago. Advances in food cultivation including irrigation systems and food storage allowed for an increase in food supply, which enabled larger and larger populations. The modern urbanization process initially flourished during the 18th century due to the Industrial Revolution. As humanity became more technologically advanced, economies became more dependent on natural resources, which in turn catalyzed intensive land conversion for urban development.

Urban expansion and associated land conversion are key components of the urbanization process. Currently, the rate of urban land expansion is slower in developed countries such as North America and Europe, in part because they experienced their large-scale urban expansion during the 19th and 20th centuries (Seto et al., 2011). In contrast, many low- to middle-income countries such as India and China are now undergoing large-scale urban expansion in conjunction with economic development and industrialization (Naab et al., 2013).

Urbanization tends to result in conversion of agricultural lands in part because of relatively high levels of rural population densities and proximity to urban centers. Thus, as urban areas grow, they tend to expand into adjacent areas, which are often agricultural areas. Throughout the world, there are many examples where the area of cultivated land decreases as urbanization and industrialization increase. For example, in Asia, cultivated land in Japan has been declining 1 percent annually from the 1970s to the 2000s due to economic development and unprofitable agricultural production (Saizen et al., 2006); South Korea also experienced declines in agricultural areas during the 1970s (Kim and Pauleit, 2007). In many countries of Europe, agricultural land decreased slightly between 1975 and 1995 (Antrop, 2000) and in the United

States, agricultural land declined at a rate of 0.1–0.3 percent annually due to economic development and despite environmental conservation efforts (Imhoff et al., 2004; Jantz et al., 2005).

This chapter explores agricultural land loss due to urbanization as a pathway through which urbanization impacts environmental systems and becomes a driver of regional environmental change. The case of China is examined in order to better understand how current urbanization trends affect the supply and integrity of agricultural systems. Additionally, the opportunities for sustaining agricultural land are discussed and guidelines for future policy-making making regarding sustainable land development are provided.

The effects of urbanization on agricultural land

Agriculture contributes to urbanization in the form of food provision, raw material, labor, capital and land. Urbanization also plays an important role in the development of agriculture by providing markets for edibles and other agricultural products as well as technological advances. Since the late 1970s, China has experienced rapid urban transformation, resulting in large-scale expansion of the urban landscape. Studies using high resolution satellite imagery show that the urban extent of China grew by nearly 25 percent during the 1990s (Liu et al., 2005). Throughout the country, the expansion of urban land cover is growing faster than the growth of urban population (Seto et al., 2011).

One result of urban expansion is the loss of agricultural land throughout the country, but especially in the coastal and central provinces. At the same time, the rising demand for agricultural products and the loss of agricultural land results in cultivating land previously not used for agriculture. The cultivation of new lands is especially noteworthy in the northern and border provinces of the country (Deng et al., 2006). Although figures for site-specific agricultural productivity are unavailable, some studies find that yields from these new agricultural lands are lower than the converted land (Yan et al., 2009). In fact, the productivity of the agricultural land converted for urban use was found to be 80 percent higher than marginal lands converted for agricultural use (Yan et al., 2009). This is in part due to the differences in inherent geographic features that affect agricultural productivity including soil quality, water availability, slope, aspect and drainage. Much of the lands that are brought into cultivation are less suited for agricultural production than the original farmlands that were lost to urban expansion. In fact, much of the lower quality cultivated land is converted from grassland (Deng et al., 2006).

In China, urban land expansion is correlated with agricultural land loss, and GDP growth associated with industrial activity has negative impacts on agricultural land (Jiang et al., 2012). Throughout the country, high quality cultivated lands are converted to industrial, commercial and residential uses. This will change the spatial patterns of agricultural land use. In the inland provinces, areas of 'intensive farming' are likely to occur, with significant environmental consequences. In order to compensate for lower agricultural suitability, farmers will need to increase inputs such as fertilizers and/or apply alternate management practices. Thus, it can be argued that land use throughout a country—and not just land immediately adjacent to urban areas—is continually shaped by demand-driven economic growth from within urban areas (Deng et al., 2010).

Spatial patterns of agricultural land loss

Studies of agricultural land loss in China have largely focused on one of two scales: national or city-region. The majority of the research on urban expansion is devoted to the study of the growth of individual cities or metropolitan regions (He et al., 2008; Long et al., 2007). Among the studies at the national scale (Tan et al., 2005; Liu et al., 2003, 2005; Deng et al., 2008, 2010), few examine spatial variations in the causes or patterns of agricultural land loss. Results from Jiang et al. (2012, 2013) are referred to here as well as unpublished results from those studies to explore the spatial patterns of agricultural land loss.

Agricultural land conversion is most prevalent in low- and middle-income provinces. For example, the Beijing–Tianjin–Hebei metropolitan area expanded by 71 percent between 1990 and 2000. Seventy-four percent of the urban expansion occurred on arable land (Tan et al., 2005). Why does urban expansion occur largely on agricultural land? One explanation is the relative differences in land rent between urban areas and cultivated land (Seto and Kaufmann, 2003; Jiang et al., 2013). Relatively higher urban land rent stimulates more cultivated land conversion, which in turn can result in a further rise of the urban land rent of a region. New urban development typically occurs near or adjacent to existing urban areas. Due to the differences in land rent, farmland near existing urban areas is the most at risk of conversion.

Additionally, agricultural land is at risk for conversion due to the differences in wages. When off-farm (non-agricultural) wages increase, the opportunity costs of farming increase, and may result in labor scarcity in the agricultural sector. This in turn can lead to farmland abandonment and a higher risk of the conversion of farmland into non-agricultural uses (Jiang et al., 2013). Across the country, small cities tend to experience high rates of agricultural land conversion due to urbanization, as urban land is highly correlated with arable land in terms of the spatial distribution (Tan et al., 2005). Moreover, the cumulative effects of succession and dominance factors related to land change have made land increasingly scarce for peri-urban farmers (Naab et al., 2013).

In China, the total conversion of cultivated land to other types of land uses led to a total net loss of 347.56 million tons of grain or about 0.1 percent of the total production potential in 1988–2008; but a decrease of 499.28 million tons of grain (about 65.4 percent of the total production potential) was due to the conversion of cultivated land to built-up area.

Furthermore, rapid urban growth coupled with inadequate land and building regulation has led to inefficient urban land use and fragmented landscapes, with dire consequences for environmental quality. An example of this is the city of Ordos in Inner Mongolia, a typical resource-based medium-sized city, which entered into a period of rapid urbanization in the 1990s. The population increased from 812,000 in 1990 to more than one million in 2007 (Dong et al., 2012). Urbanization has had profound effects on the local ecology, society and economy. The landscape has become fragmented, with different patch types exhibiting distinctive spatial characteristics. Ordos has become a symbol of unsuccessful urbanization: it sits largely empty. In the new district of Kangbashi, prime farmland was developed into housing and high-rises that are vacant or incomplete. It is an example of wasted farmland and under-utilized urban land.

From a national perspective, agricultural land in some regions has been 'sacrificed' in order to support urbanization and environmental protection elsewhere. For instance, some cultivated lands in Inner Mongolia have been converted to grasslands or woodlands in order to provide some level of protection for Beijing from sand storms. Other national projects such as the Payments for Environmental Services and Grain for Green Program (also known as the Sloping Land Conversion Program), have set aside land in anticipation for supporting strategic urbanization. One large drawback, however, is that economists have provided only weak evidence that these programs benefit the poor, or that farmers have been able to obtain gainful employment from off-farm activities (Uchida et al., 2007).

Additionally, the accelerated urbanization following population increases has greatly affected agricultural land through rural–urban migration and the transformation of rural settlements into towns and cities. Cities can more easily provide cheap, plentiful labor for emerging factories and offer abounding job opportunities, attracting rural residents. Rural–urban migration and urbanization polices have further promoted additional agricultural land loss, which has been most significant in recent decades. To some extent, urbanization promotes higher average income in cities, but lower average income in rural areas. The large amounts of cultivated land left by agricultural laborers, the high rates of rural unemployment, low income from farming, combined with uncertain and unfavorable climate conditions and natural hazards are drivers of rural–urban migration.

Additionally, urban remittances to rural family members are significantly changing the structure of agricultural households. In China, statistics show that around 250 million rural residents have moved to cities for employment (China Development Research Foundation, 2012). This means that roughly half of all Chinese farmers are either partially or entirely separated from agricultural production, and instead derive their incomes mainly from working in urban or off-farm conditions (Démurger and Li, 2013). The majority of these agricultural workers are middle-aged with technical farming expertise. While they leave for the cities, their children and aging parents remaine in rural areas, further creating new rural sociologies and household dynamics. The prevalence of farmers working in off-farm employment further increases the rural–urban income gap and encourages even more agricultural workers to migrate to cities, creating a vicious cycle of agricultural abandonment (Feng et al., 2010).

Debates on the impacts of urbanization on agricultural land

Large-scale urban conversion of agricultural land consumes natural resources, often negatively impacts the environment and causes many serious socio-economic problems. Rural–urban migration may lead to the scarcity of infrastructure for urban housing, stimulating urban expansion on to the peripheries of agricultural land. Abandoned cultivated land can be transferred to another farmer, but more often these lands are left unused. In order to make up for the reduction of cultivated land and the loss of agricultural production, large amounts of grassland and unused land are often converted into cultivated land, which creates another threat to natural environments. In this sense, urbanization near cultivated land tends to result in a net decrease of cultivated land area.

In developed countries like the United States and Canada, the loss of agricultural labors can be balanced by industrial farming. However, this is difficult in countries with lower levels of economic development where farming technologies may be too expensive or unavailable. This is particularly true for remote and hilly areas in developing countries, where rates of agricultural mechanization are low and considerable labor and expertise are necessary in order to operate the available machinery on small-scale farms (Siciliano, 2012). There is also debate over whether the abandoned agricultural land is cultivated by the remaining farmers in rural areas. In reality, the flight of farmers to urban areas tends to be of large magnitude, and generally, land tenure is usually transferred among farmers without permission from local government. In some countries such as China, property is not individually owned. In turn, economic growth and associated urbanization along with low farm commodity prices often means that it is far more profitable in the long-term for farmers to rent or even abandon their land than to continue farming. The key challenge of urbanization with respect to agriculture is the large-scale conversion of prime agricultural land. Urbanization often occurs where there are good natural conditions for farming.

Although urbanization often results in the loss of agricultural land, there is also potential in some places for urbanization to increase land use efficiency. Urbanization can increase the accessibility of agricultural land and therefore reduce the time from farm to market. The reduced cost of transportation and close linkages with the markets can increase agricultural productivity by making inputs cheaper, agricultural technology more accessible and farm-gate prices of agricultural commodities higher. If households are refocused into intensive agriculture and off-farm employment, the pressure on cultivated land may actually be reduced. Worldwide, average urban incomes are higher than average rural incomes, and the price index in urban areas is also higher than that in many rural areas. Thus, from an economic perspective, new urban land results in higher economic values than other types of land. The intensive land use in urban areas leads to higher densities of built-up land. Under these circumstances, converting other types of land into built-up land may benefit economic development.

Overall, urbanization is an important process for industrialization and economic development. However, it should not continue to the extent that it diminishes the functions of agricultural land. This reverts back to the debate about their relationship: in order to assess the economic and ecological value of urbanization and agricultural land, future development modes must examine tradeoffs.

Sustaining agricultural land

Rapid urbanization is expected to continue so long as it is considered an engine of economic growth. To formulate policy that promotes economic growth and urban development while minimizing environmental impacts, policy-makers must understand the factors that drive urban expansion. Without a clear sense of the conditions that cause urban encroachment on farmland, policies are likely to be ineffectual.

Loss of agricultural land through urban expansion is not the only concern for policy-makers. As discussed in this chapter, migration from rural communities is a major threat to the sustainability of agricultural land. In order to preserve and sustain agricultural production, we need a better understanding of the socio-economic-political causes of farmland abandonment. A few studies have found that increasing opportunity cost of farming—that is, the increasing economic opportunities of off-farm employment—is one of the main factors why farmers abandon farmland (Xie et al., 2014). Abandonment of farmland in turn leads to a reduction in farmland labor, which further increases the cost of farming. The lack of technical on-farm expertise also threatens farmland sustainability.

Another issue with respect to sustaining agricultural land is that of agricultural land quality. Although the total amount of agricultural land in China remains somewhat stable, it is clear that the quality of agricultural land that is converted to urban uses is much higher than the quality of marginal land that is brought into agricultural production. Thus, it will be critical for policy-makers to develop strategies to not only maintain total agricultural land, but rather to keep highly productive land in production. However, this is challenging because national-scale land use assessments mainly keep track of the total amount of farmland, not quality. The government has established that China needs a minimum of 1.8 billion mu (120 million hectares) of arable land to maintain food security (XinhuaNet, 2007). Policy-makers call this the 'red line' for food security that cannot be breached (Xinhua News Online, 2010). This functional and psychological 'red line' uses farmland quantity as a proxy for quality. In order to truly sustain agricultural land, policy-makers must develop new accounting measures that explicitly include quality.

However, the challenge of sustaining quality agricultural land in China is not only a matter of mitigating agricultural land loss. Sustaining agricultural land will require mitigating pollution on agricultural land. It is estimated that more than three million hectares of farmland in China are contaminated with pollution (Kong, 2014). The Ministry of Environmental Protection confirms that more than 16 percent of Chinese soil exceeds national standards for pollutants and that an estimated 20 percent of agricultural soils are polluted (Duggan, 2014). Metal contamination of soils and plants can be traced throughout the food chain. In some regions, rice is grown in soils contaminated with cadmium and zinc, commonly found in heavy industries, metals mining, smelting and untreated wastewater (Wang et al., 2003), posing risks to food security and ultimately human health.

In addition to soil contamination, water pollution and water scarcity are already posing significant risks to agricultural sustainability. In northern and northwestern China, two of the country's main wheat producing regions, severe water shortage is a major problem for sustainable agricultural production. By some accounts, the region has less than half the water per person as Egypt (Varis and Vakkilainen, 2001). Water infrastructure in many agricultural regions is outdated and inefficient, leading to overuse of a limited resource. Water pollution in China is due to industrial discharges as well as agricultural activities. For many years, increasing agricultural output in China was dependent on increasing agrochemical inputs. If current trends of fertilizer and pesticide use are not reversed, the long-term capacity of sustaining agricultural production will be compromised. Zhang Weili, a prominent agricultural expert with the Chinese Academy of Sciences, argues that water pollution caused by intensive use of agricultural inputs is one of the most significant challenges to sustainable development (Outlook, 2010). Thus, China's agriculture faces two critical issues with regard to water: there is simply not very much of it and what it has is highly polluted.

Integrated urban–rural planning

For China, it is clear that a new path forward is necessary in order to harmonize urbanization with agricultural land preservation. The key to sustaining urbanization and agricultural land lies in the coordination of their relationship and the dynamic balance of policies for development of urban and rural areas. For instance, more job opportunities and improved infrastructure would incentivize rural living and agricultural livelihoods. In coastal China, land is too scarce to support local development needs. However, local governments are implementing a national policy that aims to offset increases in urban land with decreases in rural built-up land. Centralized efforts to develop human settlements is the principal model, but it is causing land displacement while building new countrysides in coastal China (Long et al., 2009). Integrated urban–rural development indicates that urban functions are being spread over larger and larger geographic areas so that the traditional distinction between urban and rural is becoming increasingly blurred.

Integrated urban–rural planning will also require urban and national leaders to explicitly acknowledge the links between sustainable urban development and sustainable agricultural areas. Peri-urbanization and peri-urban communities are increasingly the norm in China, referring to spatially and structurally dynamic transition zones where land use, populations and economic activities are neither fully urban nor rural. Peri-urban households may have some family members living in rural areas, but not employed in agricultural activities and other family members living in urban areas but engaged in agricultural activities. At the same time, urban remittances to rural areas are changing the character of what it means to be an agricultural household. In other words, rural and urban livelihoods and landscapes are increasingly intertwined in China, and policy-makers need to acknowledge these new realities and conditions and adjust development strategies accordingly so that development in one area does not come at the expense of the other.

Key messages and concluding thoughts

- As China continues to urbanize, it is highly likely that prime farmland will continue to be abandoned or converted.
- Urbanization in China affects agricultural land not only through direct land conversion, but also through rural to urban migration and the reduction in farm laborers.
- Rural to urban migration is creating new household dynamics in rural farming areas in China.
- Integrated rural–urban planning and development is key for the long-term sustainability of both urban and agricultural areas.

References

Antrop, M. (2000). Changing patterns in the urbanized countryside of Western Europe. *Landscape Ecology*, 15(3), 257–270.

China Development Research Foundation. (2012). China Development Report 2011/2012: Changes of Demographic Conditions and Resulting Policy Adjustment. Online resource (in Chinese). Retrieved from: www.cdrf.org.cn/plus/view.php?aid=739

Démurger, S., and Li, S. (2013). Migration, remittances, and rural employment patterns: evidence from China. *Research in Labor Economics*, 37, 31–63.

Deng, X., Huang, J., Rozelle, S., and Uchida, E. (2006). Cultivated land conversion and potential agricultural productivity in China. *Land Use Policy*, 23(4), 372–384.

Deng, X., Huang, J., Rozelle, S., and Uchida, E. (2008). Growth, population and industrialization, and urban land expansion of China. *Journal of Urban Economics*, 63(1), 96–115.

Deng, X., Huang, J., Rozelle, S., and Uchida, E. (2010). Economic growth and the expansion of urban land in China. *Urban Studies*, 47(4), 813–843.

Dong, N., Han, X. G., and Wu, J. G. (2012). Changes in the spatiotemporal pattern of urbanization in Erdos of Inner Mongolia and related driving forces. *Chinese Journal of Applied Ecology*, 4, 035.

Duggan, J. (2014, April 18). One fifth of China's farmland polluted. *The Guardian*. Retrieved from: www.theguardian.com/environment/chinas-choice/2014/apr/18/china-one-fifth-farmland-soil-pollution

Feng, S., Heerink, N., Ruben, R., and Qu, F. (2010). Land rental market, off-farm employment and agricultural production in Southeast China: a plot-level case study. *China Economic Review*, 21(4), 598–606.

He, C., Okada, N., Zhang, Q., Shi, P., and Li, J. (2008). Modelling dynamic urban expansion processes incorporating a potential model with cellular automata. *Landscape and Urban Planning*, 86(1), 79–91.

Imhoff, M. L., Bounoua, L., DeFries, R., Lawrence, W. T., Stutzer, D., Tucker, C. J., and Ricketts, T. (2004). The consequences of urban land transformation on net primary productivity in the United States. *Remote Sensing of Environment*, 89(4), 434–443.

Jantz, P., Goetz, S., and Jantz, C. (2005). Urbanization and the loss of resource lands in the Chesapeake Bay watershed. *Environmental Management*, 36(6), 808–825.

Jiang, L., Deng, X., and Seto, K. C. (2012). Multi-level modeling of urban expansion and cultivated land conversion for urban hotspot counties in China. *Landscape and Urban Planning*, 108(2–4), 131–139.

Jiang, L., Deng, X., and Seto, K. C. (2013). The impact of urban expansion on agricultural land use intensity in China. *Land Use Policy*, 35, 33–39.

Kim, K. H., and Pauleit, S. (2007). Landscape character, biodiversity and land use planning: the case of Kwangju City Region, South Korea. *Land Use Policy*, 24(1), 264–274.

Kong, X.-B. (2014, February 6). China must protect high-quality arable land. *Nature*, 506, 7. Retrieved from: www.nature.com/polopoly_fs/1.14646!/menu/main/topColumns/topLeftColumn/pdf/506007a.pdf

Liu, J., Liu, M., Zhuang, D., Zhang, Z., and Deng, X. (2003). Study on spatial pattern of land-use change in China during 1995–2000. *Science in China Series D: Earth Sciences*, 46(4), 373–384.

Liu, J., Zhan, J., and Deng, X. (2005). Spatio-temporal patterns and driving forces of urban land expansion in China during the economic reform era. *AMBIO: A Journal of the Human Environment*, 34(6), 450–455.

Long, H., Tang, G., Li, X., Helling, G. H. (2007). Socio-economic driving forces of land-use change in Kunshan, the Yangtze River Delta economic area of China. *Journal of Environmental Management*, 83(3), 351–364.

Long, H., Liu, Y., Wu, X., and Dong, G. (2009). Spatio-temporal dynamic patterns of farmland and rural settlements in Su–Xi–Chang region: implications for building a new countryside in coastal China. *Land Use Policy*, 26, 322–333.

Naab, F. Z., Dinye, R. D., and Kasanga, R. K. (2013). Urbanisation and its impact on agricultural lands in growing cities in developing countries: a case study of Tamale in Ghana. *Modern Social Science Journal*, 2, 256–287.

Outlook *(News Weekly)*. (2010, September 19). China's cultivated land quality: a case of alarm.

Saizen, I., Mizuno, K., and Kobayashi, S. (2006). Effects of land-use master plans in the metropolitan fringe of Japan. *Landscape and Urban Planning*, 78(4), 411–421.

Seto, K. C., and Kaufmann, R. K. (2003). Modeling the drivers of urban land use change in the Pearl River Delta, China: integrating remote sensing with socioeconomic data. *Land Economics*, 79(1), 106–121.

Seto, K. C., Fragkias, M., Güneralp, B., and Reilly, M. K. (2011). A meta-analysis of global urban land expansion. *PloS One*, 6(8), e23777.

Seto, K. C., Reenberg, A., Boone, C. G., Fragkias, M., Haase, D., Langanke, T., . . . and Simon, D. (2012). Urban land teleconnections and sustainability. *Proceedings of the National Academy of Sciences*, 109(20), 7687–7692.

Siciliano, G., (2012). Urbanization strategies, rural development and land use changes in China: a multiple-level integrated assessment. *Land Use Policy*, 29(1), 165–178.

Tan, M., Li, X., Xie, H., and Lu, C. (2005). Urban land expansion and arable land loss in China—a case study of Beijing–Tianjin–Hebei region. *Land Use Policy*, 22(3), 187–196.

Uchida, E., Xu, J., Xu, Z., and Rozelle, S. (2007). Are the poor benefiting from China's land conservation program? *Environment and Development Economics*, 12, 593–620.

Varis, O., and Vakkilainen, P. (2001). China's 8 challenges to water resources management in the first quarter of the 21st Century. *Geomorphology*, 41(2–3), 93–104.

Wang, Q.–R., Cui, Y.–S., Liu, X.–M., Dong Y.-T., and Christie, P. (2003). Soil contamination and plant uptake of heavy metals at polluted sites in China. *Journal of Environmental Science and Health, Part A: Toxic/Hazardous Substances and Environmental Engineering*, 38(5), 823–838.

Xie, H., Wang, P., and Yao, G. (2014). Exploring the dynamic mechanisms of farmland abandonment based on a spatially explicit economic model for environmental sustainability: a case study in Jiangxi Province, China. *Sustainability*, 6, 1260–1282.

XinhuaNet. (2007, March 5). Wen Jiabao emphasises that the 1.8 billion mu basic farmland must be defended. Retrieved from: http://news.xinhuanet.com/misc/2007-03/05/content_5801278.htm

Xinhua News Online (2010, December 12). The red line of 1.8 billion mu of arable land is the lifeline of China s food security. Retrieved from: www.chinanews.com/cj/2010/12-12/2716084.shtml

Yan, H. M., Liu, J. Y., Huang, H. Q., Tao, B., and Cao, M. (2009). Assessing the consequence of land use change on agricultural productivity in China. *Global and Planetary Change*, 67(1–2), 13–19. doi: 10.1016/j.gloplacha.2008.12.012

4

URBAN LAND USE IN THE GLOBAL CONTEXT

Dagmar Haase and Nina Schwarz

As the global share of people who live in urban areas continues to grow, understanding urban dynamics is increasingly important with respect to land use, socio-demographics, human health and global environmental impacts. There are ongoing discussions about how to define a city including comparatively straightforward accounts such as using the administrative unit with a defined number of inhabitants (McIntyre et al., 2000). Cities and urban regions can be delineated in many other ways including by population density, built-up area and commuting distance. In addition to discussions about city definitions, there are other discourses about the need for new concepts to characterize the 'urban'. Recently, the concept of *urbanity* was proposed, defined by "the magnitude and qualities of livelihoods, lifestyles, connectivity, and place that create urban-ness of intertwined human experiences and land configurations" (Boone et al., 2014, p. 314). The concept of urbanity comes out of a growing consensus that the classic urban versus rural classification to categorize land is insufficient for research, analysis or practice. Importantly, the concept of urbanity is a continuum which can be applied outside the administrative boundaries of cities and therefore can extend to multiple dimensions including livelihoods, land uses and economies. Urbanity can also be used to understand how land changes in non-urban areas are connected to underlying urbanization dynamics. In this way, urbanity is closely tied to another new conceptual framework of urban land teleconnections, which seeks to link land change to underlying urbanization dynamics (Seto et al., 2012).

The purpose of this chapter is to examine and discuss urban land use in the context of global land use. Here, urban land use is considered in terms of form, size and shape across regions at different scales and across different continents. Urban land use differs from non-urban land use in many ways, but for the purposes of this chapter and this handbook, the most salient distinguishing features are its high degree of impervious cover, the high share of artificial material and the high degree of built-up space (Haase and Nuissl, 2010). It is these three characteristics of urban land use that largely define and shape the relationships between urbanization and global environmental change. The increase in impervious surface transforms the biosphere, alters local and regional surface energy balances, and results in the modification of temperature and precipitation (Kalnay and Cai, 2003; Seto and Shepherd, 2009). The materials used in built-up environments (e.g. concrete, asphalt, steel) absorb and trap heat, which not only influence local and regional climate, but also reduce local air quality by altering atmospheric chemistry and aerosol composition (Stone Jr., 2008). The form, size, and shape of the built-up area and urban form affect many dimensions of urban areas, urban livelihoods, economies and ultimately a city's character. The spatial form of an urban area or a city—such as its density or compactness—is an important factor for both quality of life and environmental impacts (Schwarz, 2010).

Urbanization: a globally heterogeneous and dynamic process

Many different studies have used remote sensing to map and characterize urban land use. Satellite-based efforts do not agree on the size and pattern of global urban land use, with estimates ranging from 0.2 to 2.4 percent of terrestrial land surface depending on the sensor used (Potere and Schneider, 2007). Urban land use is not equally distributed across the world. Urban expansion over the last 30 years has been greatest along coastlines and low lying elevation zones (Seto et al., 2011). The majority of the world's cities are growing with a population growth rate of ≥1 percent per year (Oswalt and Rieniets, 2006), but an increasing portion is stable, with zero growth, and some are even declining, with a growth rate of ≤1 percent per year. In the future, urbanization will continue and the percentage of people living in urban areas will grow; the most recent forecasts from the United Nations expects two out of three inhabitants in 2050 to live in urban areas (UN DESA, 2014). Most of this urban growth will take place in Asia and West Africa with population growth rates of 3–5 percent per year (UN DESA, 2014). However, global data also show that urban population growth in the developing world is expected to fall from 3–5 percent to an annual 1.8 percent in 2030 (UN Habitat, 2008). By 2050, urban dwellers will likely account for 86 percent of the population in the Global North and 67 percent in the Global South. Urban land use is also highly multifunctional, complex and dynamic. There is a lot of variation over short distances in both vertical and horizontal dimensions. A single urban location, whether it is a pixel or patch, can have many different functions. For example, a single high-rise building can serve as housing, retail and commercial space. Studies using remote sensing show that urban land use exhibits much variation in the horizontal dimension, leading to what remote sensing scientists call 'high texture,' or variability over short distances.

Urbanization is a demographic process of an increasing number of people living in urban areas. Urbanization is also a spatial, land use process that predominantly results in the physical growth of built-up areas, whether it is horizontal or vertical (UN Habitat, 2010). By 2050, 65 percent of the population in developing countries and nearly 90 percent of the population in developed countries will be in urban areas (UN DESA, 2014). In many parts of the world, the physical expansion of urban areas has been growing faster than the urban population (Angel et al., 2011b), suggesting declining densities. However, there are some exceptions. In some East Asian cities, notably those in China, there has been a significant increase over the last decade in urban built-up densities (Frolking et al., 2013).

From the perspective of land use, urbanization takes up very little space, relatively. Compared with other major land uses such as agriculture or forestry, urban areas occupy little of the Earth's surface. Even in the future, with the unprecedented rates of urbanization, urban areas are not expected to significantly increase in their global footprint. A forecast of urban land expansion predicts that urban land will triple by 2030 from 2000 (Seto et al., 2012), but urban areas will remain less than 10 percent of Earth's land surface. Thus, the concentration of urban populations and urban activities will continue to occur on a small area of the world's land surface. However, this relatively small area should not be confounded with small impact; despite the remarkably small area, urbanization has a disproportionate impact on the global environment.

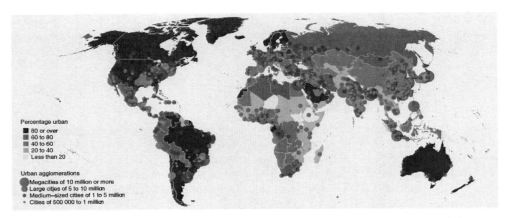

Figure 4.1 Percentage urban and location of urban agglomerations with at least 500,000 inhabitants, 2014

Source: UN DESA (2014).
A detailed, colored version of this figure can be found on the book's website, www.routledge.com/9780415732260

In the future, almost all population growth will occur in urban areas, from natural population growth or migration from rural to urban areas (Bell et al., 2010). However, urbanization, as such, is a relatively new phenomenon in terms of its global spread. In 1800, only 3 percent of the world's population lived in cities, but this figure rose to 47 percent by the end of the 20th century. In 1950, there were 83 cities with populations that exceeded one million; by 2010, this number had risen to more than 460. If this trend continues, the world's urban population will double in less than 40 years (Brinkhoff, 2012). Cities also can decline and lose population (Rieniets, 2009). Today, there are approximately 350–400 shrinking cities worldwide; most of them are situated in the early industrialized Western world, namely Europe and the US, but also Japan (Haase, 2013; see Figure 4.1). Urban shrinkage is by no means a new phenomenon, but globalization and communication have made it more visible. Shrinkage also existed at early urbanization stages when early urban agglomerations were declining and abandoned such as Rome—the first megacity on the planet, or large cities of the Egyptian, Mexican or Incan Empires.

Geographically, cities are mostly located in coastal and riverine areas. Current 'urban' hotspots are located on the East Coast of the US, Southeast China, Western Europe, Japan, West Africa, and parts of the South American Atlantic coast. Nearly all of the ten largest megacities are situated in the Global South such as Shanghai, Karachi, Beijing, Lagos, Istanbul, Guangzhou, Mumbai, Moscow, Dhaka and Cairo (UN Habitat, 2008). There are also large areas of the planet that are able to support concentrated populations that have sufficient water supply, energy supply, affordable housing, available food, access to health, education and other infrastructure, but which are not larger cities or urban agglomerations. These places are typically near or in high mountain zones and the inner tropics of Northern Asia (including the declining oil and mining cities of northeast Russia), and parts of Northern Europe.

In terms of their spatial land use patterns and morphological properties, cities differ enormously across the globe. The pre-medieval and medieval cities in Europe are typical examples of compact cities with mid-rise houses. Regions with younger urbanization such

Urban land use in the global context

as North America, Latin America and Asia tend to develop less compact cities as a whole, but with a pronounced dense Central Business District (CBD) embedded into less dense counter-urbanized surroundings. Mid-twentieth century motorization led to what is referred to as 'urban sprawl,' describing the expansion of settlements away from central cities into previously rural areas. Sprawl has also affected compact European cities after 1970. Most of the large megacities of the Global South face enormous poverty, which concentrates in informal or squatter settlements in both inner and outer parts of the city (Angel et al., 2011a). Despite growth, Angel et al. (2011b) reveal that most cities across the globe had the tendency to de-densify in terms of the number of persons per hectare from 1990 to 2000.

Today, the most urbanized regions worldwide include North America (82 percent), Latin America and the Caribbean (80 percent), followed by Europe (73 percent). In contrast, large parts of Africa and Asia (e.g. India) remain mostly rural, with 40–50 percent of their respective populations living in urban areas. Despite the current degree of urbanization or the growth rate of their cities, all regions are expected to continue urbanizing over the coming decades (see Figure 4.2). Regions that are less urbanized (e.g. Africa and Asia) are currently urbanizing faster than those with an already high share of urban population. Thus, most of the world's future urban growth will occur in the Global South, namely in Asia and Africa (Seto et al., 2011).

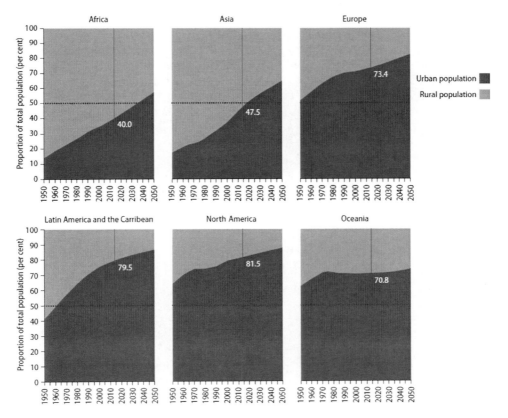

Figure 4.2 Urban and rural population as proportion of total population, by major areas, 1950–2050
Source: UN DESA (2014).

The spatial configuration of cities

The spatial configuration of cities across the globe is very heterogeneous. The physical structure of cities can also be quite diverse, ranging from skyscraper areas for business or housing found in cities of the US, Japan, Latin America or the new/upcoming Chinese cities to medieval compact mid-rise cities of Europe to large low-rise informal settlements that make up considerable parts of almost all megacities worldwide in Asia, Latin America or Africa (Angel et al., 2010; Angel et al., 2011a). Basic properties of the physical space (e.g. built-up density, building height, amount of open and green space, or the number of streets and width) differ considerably between these types of cities and consequently impact the living conditions for urban dwellers (UN Habitat, 2008).

Apart from this more visual three-dimensional (3D) interpretation of the physical land cover and surface of cities, the spatial two-dimensional (2D) configuration of urban areas is very helpful in quantitatively describing and comparing cities of different shapes. The configuration of cities is mostly analyzed in a 2D way by looking at the urban region from a bird's perspective. Differentiating land covers gives the opportunity to describe spatial patterns such as compact patterns or strip development along major transport routes. Landscape metrics, which were originally developed in landscape ecology, can be used to describe those spatial patterns (see Table 4.1). The basic spatial unit is a so-called 'patch' which encompasses all quasi-homogeneous neighboring entities (e.g. cells) of the same land cover type. Often, built-up land is contrasted with all other land covers in urban studies. However, the approach also allows for more detailed land cover classifications (see the next subsection). Furthermore, socio-economic or ecological indicators are valuable for describing spatial patterns in cities such as population density, the distribution of population, jobs or infrastructure within districts, even noise, particle or heat emissions (Weber et al., 2014a; 2014b).

Cities consist of very diverse built-up structures. Therefore, differentiating urban land covers as described in the next subsection is a necessary tool to understand urban patterns and processes.

Table 4.1 Selected landscape metrics to describe the spatial configuration and composition of cities

Landscape metric	Measurement	Interpretation
Edge density	Sum of all built-up edge lengths divided by total built-up area	Edge density indicates compact patches versus ragged built-up patches
Share of built-up area	Percentage of built-up area of total land area	The share of built-up area indicates the relationship between built-up areas and the study region
Patch size standard deviation	Deviation from mean in patch size for built-up patches	The higher the patch size standard deviation, the larger are the differences in patch size between the individual built-up patches
Number of patches	Absolute number of built-up patches	The higher the number of patches, the more scattered are the built-up patches within the region

Source: Based on Schwarz (2010).

Disentangling urban land cover

Up to now, global land cover datasets such as the MODIS IGBP land cover include only one category of 'built-up' or 'urban' land cover. However, it is obvious that land cover varies significantly within urban regions. They do not only consist of agricultural or forested land and water surfaces, but mostly of built-up areas with various building types, construction heights and densities. Decreasing the pixel size, e.g. to 500m (Schneider et al., 2009; 2010) or even 300m (European Space Agency Climate Change Initiative land cover), is one approach to provide a more detailed global dataset based upon remote sensing data. Land cover classifications for smaller geographical extents sometimes also include several categories of built-up land (see the example of the European data base 'Urban Atlas' in Table 4.2). The same holds for land cover and land use maps derived for individual urban regions. Recently, remote sensing data were used to estimate 3D structures of urban regions (Frolking et al., 2013). However, the most detailed view of urban 3D structure can be provided by cadastral data, which, if available at all, can even provide the exact shapes of individual buildings.

Urban land use is often characterized by high impervious cover, but not all urban land use types are almost completely or semi-impervious. Compact cities such as those found in Europe typically appear more sealed than scattered or polycentric cities (Westerink et al., 2012). On average, European cities are sealed up to 50 percent, with only a minority showing higher percentage of impervious cover (see Figure 4.3). The other 50 percent is forest, park and agricultural land. Cities with a CBD found in North American, Latin American or

Table 4.2 Characterization of the Urban Atlas land cover dataset as one typical example of urban land cover/use data

Spatial resolution	Polygons, 1:10.000
Spatial extent	305 urban regions in Europe
Temporal coverage	approx. 2006
Minimal mapping unit	0.25 hectares
Urban land cover classes	Continuous urban fabric
	Discontinuous dense urban fabric
	Discontinuous medium density urban fabric
	Discontinuous low-density urban fabric
	Discontinuous very low-density urban fabric
	Isolated structures
	Industrial, commercial, public, military and private units
	Fast transit roads and associated land
	Other roads and associated land
	Railways and associated land
	Port areas
	Airports
	Construction sites
	Green urban areas
	Sports and leisure facilities
	Land removed former industry areas
Non-urban land cover classes	Mineral extraction and dump sites
	Land without current use
	Agricultural, semi-natural and wetland areas
	Forest
	Water

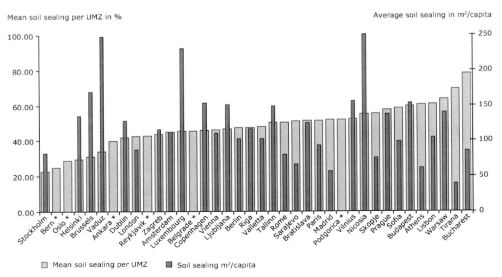

Figure 4.3 Mean soil sealing per urban morphological zone (UMZ) and soil sealing per capita in European cities

Source: EEA (2009).

Note: Cities with a high degree of soil sealing per capita, such as Brussels or Vaduz, appear less dense than those with a low percentage of soil sealing per inhabitant, such as Madrid or Tirana. Whereas the former exhibit urban sprawl, the latter are compact cities.

Chinese cities, for example, show a dense and almost completely sealed small center exhibiting tall construction heights as compared to large urban agglomerations of low-density, low-rise housing, which is often less than 40 percent sealed. Slums and other informal settlements in cities often appear unsealed, which makes them extremely vulnerable against heavy rainfall, subsequent mass movement and landslides. The highest sealing rates (up to 90 percent) are characteristic of industrial or commercial areas, followed by inner-urban high-density housing and transportation sites. Next to highly surfaced land many urban green and blue land use types are found in cities and urban agglomerations such as parks, urban gardens, lakes or forests, which by definition are almost completely unsealed (Haase and Nuissl, 2010).

Brown- and grey-fields in cities (abandoned industrial or commercial land) represent another type of land use that has been converted from being highly sealed to, in many cases, semi-open and partly to succession space. Surface sealing in cities, on the one hand, provides an important precondition of housing and transport, as it permits mobility in a very basic sense. In the developing world, many cities do not have sealed roads (that is, they are dirt or unpaved roads), but traffic density is nonetheless high (Angel et al., 2010). Soil sealing is detrimental to the environment and to ecosystems in many different ways. For example, soil sealing impairs ecosystem services such as reducing the regulation of floods and matter flows, limiting water supply, the provisioning of recreational space, and biodiversity or natural aesthetic values (Nuissl et al., 2009).

Density as significant feature of urban land use

Whereas the aforementioned discussion focused on the physical characteristics of urban land use, here another important characteristic of urban land use is discussed: urban density. In its most basic form, urban density refers to the relationship between two features, one of which concerns urban land. There are dozens of different ways to define and calculate urban density. Some examples include but are not limited to: the number of households per unit area such as square kilometer or city block; population per unit land area; or number of buildings per unit area. Other ways to calculate density in cities include vehicle traffic intensity or density; road density; or commercial or restaurant densities. All of these densities can be calculated in a number of different ways ranging from simple indicator per area measures via accessibility measures up to dissimilarity indices or activity models.

Population density is a frequently used indicator to link urban land use with urban dwellers and their quality of life (see Fragkias, Chapter 1 in this volume). It is either calculated as gross or net density: gross density is computed as population number per overall land area, including non-residential areas such as roads and other infrastructure, green spaces, etc. Net density refers only to land directly related to residential purposes such as dwelling units. Cities strongly vary in population density. Lowest gross densities are found in North America and Oceania (between 700 and 1,200 inhabitants per km^2), while population densities are much higher in less developed countries, with Dhaka, Bangladesh, having the highest population density of 44,500 inhabitants per km^2. This is mostly related to informal settlements or slum-like settlements within the city (Demographia World Urban Areas, 2013).

High population densities can either occur due to high floor-area ratios (which is the total floor area of a building divided by the plot size, resulting in large values for high-rise buildings in the CBD), small living space per person or both. Therefore, significant contrasts exist even between neighboring countries (e.g. India and China) regarding the way urbanization is taking place. Built-up densities are much higher in China than in India, even though urban population densities are not so dissimilar (Frolking et al., 2013). Population density has impacts on energy consumption due to transportation, water and electricity needs. Furthermore, population density can influence the space available for urban green spaces, recreation, and agricultural uses and thus the opportunities of the urban population to engage in urban agriculture, urban gardening and the like. Accordingly, compact, high-density development is often the target of urban planning such as for the 'smart growth' movement as this type of urbanization saves space for open land. Is there an optimal urban density? Is the compact, dense city model more sustainable than dispersed patterns of urbanization, which often have more open and green space per unit of area? The Intergovernmental Panel on Climate Change Fifth Assessment Report concludes that increasing urban density is a necessary, but not a sufficient condition to reduce urban greenhouse gas emissions (Seto et al., 2014). Thus, it is unlikely that there is an 'optimal density.' Rather our idea of acceptable density will depend in part on context and local conditions.

The rural–urban gradient

Most of the theories and conceptual models of cities and of urbanization generally support the assumption (even if they are not explicitly based on this exact assumption) that the density of urban structures, along with the amount of impervious cover, increases with decreasing distance to the city center (McDonnell et al., 1993). Furthermore, many theories imply that land use changes along an urban-to-rural gradient over distance and over time are the result of deconcentration or accumulation and de-densification or abandonment processes.

The urban-to-rural gradient lies between the poles of 'the urban' and 'the rural.' Rural land uses are predominantly vegetated and connected to agricultural production such as farming and pasture, sometimes also to forest (Haase and Nuissl, 2010). Smaller settlements are spread within the rural landscape. Urban areas, by contrast, are densely built-up spaces with a high degree of impervious cover, as previously mentioned. They serve industrial production, services and commerce, cultural facilities and housing as well as the provision of related transport and logistics facilities. On the border between the two, almost everywhere, rural land uses become more dense and less open, whereas urban land uses become more scattered and less impervious (Antrop, 2004; Haase and Nuissl, 2010). This transition between the two space types rarely corresponds to administrative boundaries. Of course, the gradient is differently shaped in different urban regions across the globe; for example, we find large and dense skyscraper settlements for housing next to small rural houses in Chinese conurbations (Seto et al., 2011), and many rural elements (houses, production units, behavior) in urban areas of India (Boone et al., 2014).

The edge between urban and rural areas does not maintain a stable structure over time, rather it transforms. The conversion of enormous amounts of arable or natural land into land with impervious cover used for industrial and residential purposes has substantially altered the character and the environmental properties of the urban-to-rural gradient (Antrop, 2004). Likewise, the 'central' segment of the urban-to-rural gradient (i.e. the urban fringe) has not only experienced an outward expansion and become more blurred, but characteristic land uses have changed. It has become the *Zwischenstadt*—something in between 'the urban' and 'the rural' (Sieverts, 2003), which also applies to features found today along the rural–urban gradient in developing countries, e.g. blurred low-rise housing structures, degraded or cut forest remnants, large sealed uncultivated wastelands, etc. In Central Europe and the US, this *Zwischenstadt* was usually characterized by one-story houses and gardens in preindustrial times; in the heyday of industrialization it was dominated by industrial facilities, multi-story housing and, in some areas, mining sites; nowadays, it is mainly detached houses, industrial, commercial and retail sites that dominate the urban-to-rural interface (Meeus and Gulinck, 2008). However, it is even more significant that this interface is now often hardly perceptible; in many places, the difference between urban, peri-urban and pure rural land use patterns has virtually disappeared due to the urbanization of 'the rural' (EEA, 2006).

Figure 4.4 shows that next to land cover and use change, the rural–urban gradient implies a range of effects for the natural environment: natural ecosystem functions and ecosystem services supply decline due to soil sealing and soil deterioration (upper two graphs and lower left graph for the city of Leipzig, and the lower middle graph for Helsinki) whereas human demand on food (see lower right graph, again for Leipzig) increases as a function of population number and density.

Modeling future urban land use change

Numerous models aim to explain the internal differentiation of the urban region and its land use change. Classic theoretical models draw on the economic competition between different land uses (von Thünen, 1826/1966; Alonso, 1964) or between social groups (Burgess, 1925; Hoyt, 1939; Harris and Ullmann, 1945). More recent computational models of urban land cover and land use change such as system dynamics models, cellular automata models or agent-based approaches, employ the (changing) distribution of population or households in the urban region/along the rural–urban gradient due to changing preferences of the population as the key variable (Lauf et al., 2012, Haase et al., 2010). Conceptual urbanization models state a sequence of three, or alternatively four phases, of urban development: urbanization; suburbanization; de-surbanization; and re-urbanization (van den Berg et al., 1982; Champion, 2001a; 2001b). Other models on the dynamics and

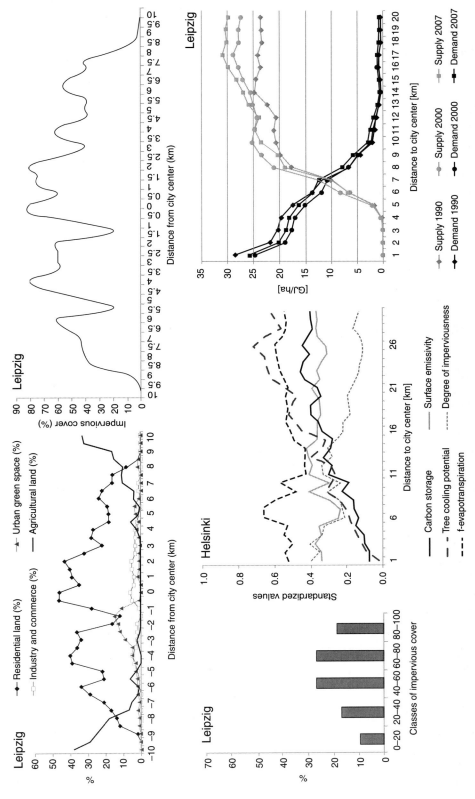

Figure 4.4 Types of rural-urban gradients: land cover and land use as well as soil sealing of the city of Leipzig, Germany (upper left and right), classes of impervious cover for Leipzig, Germany (lower left), ecosystem services supply of Helsinki, Finland (lower middle), and food demand-supply-function for Leipzig (lower right)

transformation of urban development are based on chaos theory (Wilson, 1976), the theorem of fractal development (Batty and Longley, 1989; White and Engelen, 1993; Batty, 2008) or self-organization (Portugali 2000).

Many simulation models of urban land use are based on the underlying assumption that land use change is a linear, or at least a quasi-linear, process. Only more recent modeling techniques such as agent-based modeling, allow for the inclusion of complex interactions between the different actors in urban regions and thus permit emergence (Schwarz et al., 2010). Although population trends have usually been well-documented for decades or even centuries there are only rare examples of studies that provide empirical evidence of long-term land use change and the respective land use patterns or gradients. Such a long-term observation of urban trajectories, however, is required in order to analyze the driving forces of land use change along the urban-to-rural gradient.

Concluding thoughts: where is urbanization headed?

Where is the trajectory of urbanization headed and what will an urban planet look like at the end of the 21st century? By then, nearly the entire human race will live in urban areas thus 'the urban' and human settlements will be synonymous. At that point, the rest of the world that is not urban will function as service areas—for food production and energy supply, carbon sequestration and tourism—for a growing number of cities (Haase, 2014). Humanity will live in a teleconnected urban world (Seto et al., 2012) that no longer physically touches the surroundings or rural hinterland, as discussed. This implies that areas providing ecosystem services and resource supply, production (rural) and consumption (urban) will be increasingly separated from each other and, what is more, that the rural–urban gradient might change from a real and factual one to a more virtual, teleconnected one. In addition, future urbanization will lead to higher and more densely built spaces experienced already in Asian megacities.

In terms of their footprint, the urban resident of these dense high-rise cities will live more sustainably than a rural resident due to the high population density (despite a potential decrease of density due to increasing wealth), the effectiveness of the transportation system and the short distances. Thus, future urbanization will 'save open space', as cities will increasingly inhabit vertical and multi-story spaces. Space saving is yet another urban space service and needs to be valued as such. Moreover, the urban areas in the 21st century will be more connected by wireless communication (Boone et al., 2014), which will save energy due to reduced physical transport and respective carbon emissions. In terms of urban nature, green spaces, along with lakes/ponds, might be spread over different vertical levels, starting at the ground level, but also covering roofs and walls in diverse forms (Haase, 2014). This chapter provides the following key messages:

- Urbanization is one of the land use megatrends of the 21st century. Although urban land use takes up a small percentage of the Earth's surface, its impacts are disproportionate to its size. Many of the impacts of urban land use and urbanization will be through teleconnections, rather than on-site impacts (Seto and Reenberg, 2014).
- Urban land use, including configuration, density and spatial form, varies significantly across regions around the world and even within countries. How urban areas are configured, designed and networked has important implications for resource use including material demand for buildings and other urban structures, energy use for mobility, heating and cooling, and quality of life for their residents.

Urban land use in the global context

- The sealing of urban land, especially soils and greenspaces, is one of the major ways in which urban land use affects the environment, especially ecosystem services. Surface sealing, however, is one of the key ways to increase urban mobility, and hence there is a tradeoff between transport efficiency and accessibility versus ecosystem services.
- The 'urban' is everywhere, especially if teleconnected spaces, economies, and livelihoods are considered. Increasingly, few places on Earth will be untouched by urbanization, urban economies, and urban demands for goods and services. The authors hypothesize that areas of production will be increasingly in rural places and places of consumption will be increasingly in urban areas, and that going forward, places of production and places of consumption will continue to be separate from each other.

Going forward the following research questions regarding urban land should be explored:

- What is the importance of urban form for the future?
- How will global land use be influenced by telework, energy prices after peak oil, or the aging of urban populations?
- How can global sprawl abate in favor of the compact, dense city?
- What is the relationship between the height and spatial configuration of the built-up space and environmental and human well-being?
- What is the impact of urban shrinkage on land use and urban sustainability?

References

Alonso, W. (1964). *Location and Land Use: Toward a General Theory of Land Rent*, Harvard University Press, Cambridge, MA.

Angel, S., Parent, J., Civco, D. L., Blei, A. M. (2010). 'The atlas of urban expansion.' Available: www.lincolninst.edu/subcenters/atlas-urban-expansion.

Angel, S., Parent, J., Civco, D. L., and Blei, A. M. (2011a). *Making room for a planet of cities.* Policy Focus Report/Code PF027, Lincoln Institute of Land Policy.

Angel, S., Parent, J., Civco, D. L., Blei, A., and Potere, D. (2011b). *The dimensions of global urban expansion: Estimates and projections for all countries, 2000–2050.* Progress in Planning 75, 53–107. Doi: 10.1016/j.progress.2011.04.001

Antrop, M. (2004). 'Landscape change and the urbanization process in Europe,' *Landsc Urb Plan*, vol. 67, pp. 9–26.

Batty, M. (2008). 'The size, scale, and shape of cities,' *Science*, vol. 319, pp. 769–771.

Batty, M., and Longley, P. (1989). 'Urban growth and form: scaling, fractal geometry, and diffusion-limited aggregation,' *Environment Planning A*, vol. 21, pp. 1447–1472.

Bell, S., Alves, S., Silveirinha de Oliveira, E., and Zuin, A. (2010). *Migration and Land Use Change in Europe: A Review* Living Reviews in Landscape Research, vol. 4.

Boone, C., Redman, C. L., Blanco, H., Haase, D., Koch, J., Lwasa, S., Nagendra, H., Pauleit, S., Pickett, S. T. A., Seto, K. C., and Yokohari, M. (2014). 'Reconceptualizing urban land use.' In: K. Seto and A. Reenberg, eds. *Rethinking Global Land Use in an Urban Era.* Strüngmann Forum Reports, vol. 14, Julia Lupp, series editor. MIT Press, Cambridge, MA, pp. 313–330.

Brinkhoff, T. (2012). *The Principal Agglomerations of the World.* http://www.citypopulation.de.

Burgess, E. W. (1925). 'The growth of the city: an introduction to a research project.' In: R. E. Park, E. W. Burgess, and R. McKennzie, eds. *The City*, University of Chicago Press, Chicago, pp. 47–62.

Champion, T. (2001a). 'Urbanization, suburbanization, counterurbaniszation and reurbanization.' In: R. Paddison, ed., *Handbook of Urban Studies*, Sage, London, pp. 143–161.

Champion, T. (2001b). 'A changing demographic regime and evolving polycentric urban regions: Consequences for the size, composition and distribution of city populations.' *Urban Studies*, vol. 38(4), pp. 657–677.

Demographia World Urban Areas. (2013). 9th annual edition. Available: www.demographia.com/db-worldua.pdf.

European Environmental Agency (EEA). (2006). Urban sprawl in Europe, the ignored challenge. R. Uhel, ed., Copenhagen.

European Environmental Agency (EEA). (2009). [Fast track service precursor on land monitoring – degree of soil sealing]. *Raster data set of built-up and non built-up areas including continuous degree of soil sealing ranging from 0–100% in aggregated spatial resolution (100 x 100 m and 20 x 20m)*. Retrieved from:www.eea.europa.eu/data-and-maps/data/eea-fast-track-service-precursor-on-land-monitoring-degree-of-soil-sealing#tab-interactive-maps-produced.

Frolking S., Milliman, T., Seto, K. C., and Friedl, M. A. (2013). 'A global fingerprint of macro-scale changes in urban structure from 1999 to 2009.' *Environmental Research Letters*, vol. 8, 024004. Doi: 10.1088/1748-9326/8/2/024004

Haase, D. (2013). 'Shrinking cities, biodiversity and ecosystem services.' In: T. Elmqvist, M. Fragkias, B. Güneralp, et al., eds. *Global Urbanisation, Biodiversity and Ecosystem Services: Challenges and Opportunities*. Springer Dordrecht Heidelberg, New and York London, pp. 253–274.

Haase, D. (2014). 'The nature of urban land use and why it is a special case.' In: K. Seto, and A. Reenberg, eds., *Rethinking Global Land Use in an Urban Era*. Strüngmann Forum Reports, vol. 14, MIT Press, Cambridge, MA, pp. 305–312.

Haase, D., and Nuissl, H. (2010). 'The urban-to-rural gradient of land use change and impervious cover: a long-term trajectory for the city of Leipzig,' *Land Use Science*, vol. 5(2), pp. 123–142.

Haase, D., Lautenbach, S., and Seppelt, R. (2010). 'Applying social science concepts: modelling and simulating residential mobility in a shrinking city,' *Env Mod Softw*, vol. 25, pp. 1225–1240.

Harris, C. D., and Ullmann, L. E. (1945). 'The nature of cities', *Annals of the American Academy Political and Social Science*, vol. 242, pp. 7–17.

Hoyt, H. (1939). *The Structure and Growth of Residential Neighbourhoods in American Cities*. Federal Housing Administration, Washington, DC.

Kalnay, E., and Cai, M. (2003). 'Impact of urbanization and land-use change on climate,' *Nature*, 423, pp. 528–531.

Lauf, S., Haase, D., Kleinschmidt, B., Hostert, P., and Lakes, T. (2012). 'Uncovering land use dynamics driven by human decision-making. A combined model approach using cellular automata and system dynamics,' *Env Mod Softw*, vol. 27–28, pp. 71–82.

McDonnell, M. J., Pickett, S. T. A., and Pouyat, R. V. (1993). 'The application of the ecological gradient paradigm to the study of urban effects.' In: M. J. McDonnell and S. T. A. Pickett, eds., *Humans as Components of Ecosystems*. Springer, New York, pp. 175–189.

McIntyre, N. E., Knowles-Yánez, K., and Hope, D. (2000). 'Urban ecology as an interdisciplinary field: differences in the use of "urban" between the social and natural sciences,' *Urban Ecosystems*, vol. 4, pp. 5–24.

Meeus, S. J., and Gulinck, H. (2008). 'Semi-urban areas in landscape research: a review,' *Living Reviews in Landscape Research*, vol. 2(3), p. 45.

Nuissl, H., Haase, D., Wittmer, H., and Lanzendorf, M. (2009). 'Environmental impact assessment of urban land use transitions – a context-sensitive approach,' *Land Use Policy*, vol. 26(2), pp. 414–424.

Oswalt, P., and Rieniets, T., eds. (2006). *Atlas of Shrinking Cities*. Hatje, Ostfildern.

Portugali, J. (2000). *Self-Organisation and the City*. Springer, Berlin.

Potere, D., and Schneider, A. (2007). 'A critical look at representations of urban areas in global maps,' *GeoJournal*, vol. 69, 55.

Rieniets, T. (2009). 'Shrinking cities: causes and effects of urban population losses in the twentieth century,' *Nature and Culture*, vol. 4, pp. 231–254.

Schneider, A., Friedl, M. A., and Potere, D. (2009). 'A new map of global urban extent from MODIS data,' *Environmental Research Letters*, vol. 4, article 044003.

Schneider, A., Friedl, M. A., Potere, D. (2010). 'Mapping global urban areas using MODIS 500-m data: new methods and datasets based on "urban ecoregions",' *Remote Sensing of Environment*, vol. 114, pp. 1733–1746.

Schwarz, N. (2010). 'Urban form revisited—selecting indicators for characterising European cities,' *Landscape and Urban Planning*, vol. 96, pp. 29–47.

Schwarz, N., Haase, D., and Seppelt, R. (2010). 'Omnipresent sprawl? A review of urban simulation models with respect to urban shrinkage,' *Environment and Planning B*, vol. 37, pp. 265–283.

Seto, K. C., and Shepherd, J. M. (2009). 'Global urban land-use trends and climate impacts,' *Current Opinion in Environmental Sustainability*, vol. 1(1), pp. 89-95.

Seto, K., and Reenberg, A., eds. (2014) *Rethinking Global Land Use in an Urban Era*. Strüngmann Forum Reports, vol. 14, Julia Lupp, series editor. MIT Press, Cambridge, MA.

Seto, K. C., Fragkias, M., Güneralp, B., and Reilly, M. K. (2011). 'A meta-analysis of global urban land expansion,' *Plos One*, vol. 6(8): e23777. doi:10.1371/journal.pone.0023777.

Seto, K. C., Reenberg, A., Boone, C. C., Fragkias, M., Haase, D., Langanke, T., Marcotullio, P., Munroe, D. K., Olah, B., and Simon, D. (2012). 'Teleconnections and sustainability: new conceptualizations of global urbanization and land change,' *PNAS*. Retrieved from: www.pnas.org/cgi/doi/10.1073/pnas.1117622109.

Seto, K. C., Dhakal, S., Bigio, A., Blanco, H., Delgado, G. C., Dewar, D., Huang, L., Inaba, A., Kansal, A., Lwasa, S., Mueller, D. B., Murakami, J., Nagendra, H., and Ramaswami, A. (2014). 'Human settlements, infrastructure, and spatial planning.' In: O. Edenhofer, R. Pichs-Madruga, Y. Sokona, E. Farahani, S. Kadner, K. Seyboth, A. Adler, I. Baun, S. Brunner, P. Eickemeier, B. Kriemann, J. Savolainen, S. Schlömer, C. von Stechow, T. Zwickel and J. C. Minx, eds. *Climate Change 2014: Mitigation of Climate Change. Contribution of Working Group III to the IPCC Fifth Assessment Report of the Intergovernmental Panel on Climate Change*. Cambridge University Press, Cambridge, UK and New York, NY, USA.

Sieverts, T. (2003). *Cities Without Cities: An Investigation of the Zwischenstadt*. Spon Press, London.

Stone Jr., B. (2008). 'Urban sprawl and air quality in large US cities,' *J. Environ. Manage*, vol. 86, pp. 688–698.

UN DESA. (2014). World urbanization prospects: the 2014 revision, highlights (ST/ESA/SER.A/352). United Nations, Department of Economic and Social Affairs, Population Division.

UN Habitat. (2008). State of the world's cities 2008/2009. Harmonious cities. London, UK and Sterling, VA, USA : Earthscan and on behalf of the United Nations Human Settlements Programme (UN-HABITAT).

UN Habitat. (2010). 2010/11 state of the world's cities report, "bridging the urban divide". Nairobi, Kenya: United Nations Human Settlements Programme (UN-HABITAT).

van den Berg, L., Drewett, R., Klaassens, L. H., Rossi, A., and Vijverberg, C. H. T. (1982). Urban Europe, Vol. I. A Study of Growth and Decline, Pergamon Press, Oxford.

von Thünen, J. H. (1826/1966). *The Isolated State: An English Edition of Der Isolierte Staat*, ed. P. Hall. Pergamon Press, Oxford and New York.

Weber, N., Haase, D., and Franck, U. (2014a). 'Zooming into the urban heat island: How do urban built and green structures influence earth surface temperatures in the city?,' *Science of the Total Environment*, vol. 496, pp. 289–298.

Weber, N., Haase, D., and Franck, U. (2014b). 'Assessing traffic-induced noise and air pollution in urban structures using the concept of landscape metrics,' *Lands Urb Plan*, vol. 125, pp. 105–116.

Westerink, J., Haase, D., Bauer, A., Ravetz, J., Jarrige, F., and Aalbers, C. (2012). 'Expressions of the compact city paradigm in peri-urban planning across European city regions – how do planners deal with sustainability trade-offs?,' *Europ Plan Stud*, vol. 25, pp. 1–25.

White, R., and Engelen, G. (1993). 'Cellular automata and fractal urban form: a cellular modeling approach to the evolution of urban land-use patterns,' *Environment Planning A*, vol. 25, pp. 1175–1199.

Wilson, A. G. (1976). 'Catastrophe theory and urban modelling: an application to modal choice,' *Environment Planning A*, vol. 8, pp. 351–356.

5

CONTEMPORARY URBANIZATION IN INDIA

Chandana Mitra, Bhartendu Pandey, Nick B. Allen and Karen C. Seto

Soon to be the world's most populous country, India is also assuming the mantle of this century's largest urban transition. Between 2015 and 2050, India's urban population is projected to increase by 404 million, 30 percent more than China's projected urban growth and nearly twice that of any other country over the same period (UN DESA 2014). This rapid growth, which has been a well-documented characteristic of India's largest cities since the country's independence in 1947, now prevails in India's smaller urban settlements as well.

In the post-independence era, India's urban geography has been reshaped most significantly by the 1991 economic crisis. In its wake, the central government liberalized its policies of state-led industrialization and rural development to promote market reforms, abolish the 'License Raj,' and foster international trade and foreign investment. In this post-liberalization period, economic growth has differed from the past most substantially in three ways. First, the private sector has achieved higher growth rates than the public sector (Nath 2007). Second, the growth of service industries began to dominate the manufacturing sector, which has precipitated a gradual shift away from labor-intensive production toward skill- and finance-intensive sectors (Desmet et al. 2013; Kochhar et al. 2006). Third, private and foreign investors have successfully sought deregulated zones of capital investment. These trends, especially the second and third, have re-ordered the regional characteristics of urban growth and accelerated the expansion of peripheral urban land across the subcontinent.

This chapter reviews three of the country-scale effects of the post-liberalization urban phase: patterns of urban expansion, shifts in urban population structure and country-scale losses of agricultural land. It also considers how these covalent processes manifest in Kolkata, a metropolitan area remarkably reshaped by India's political and economic inflections.

Urban land use change patterns

Like studies of global-urban land expansion in other regions (Angel et al. 2011; Seto et al. 2011), most India-specific studies indicate that the rate of urban expansion is significantly greater than the population growth rate (Sudhira et al. 2004; Farooq and Ahmad 2008; Jat, Garg, and Khare 2008; Haase and Schwartz, Chapter 4 in this volume). However, unlike other rapidly urbanizing regions such as China, Indian cities lack significant vertical growth, a pattern that has been attributed to India's strict municipal regulations on building height and the informalized

Contemporary urbanization in India

process of land acquisition (Frolking et al. 2013). Indian urbanization is also characterized by uncontrollable growth through a dynamic process of informality. Roy (2003) describes informality, which is prevalent in all urban areas of India, as a lack of any regulation or law according to which the ownership, use and purpose of land can be fixed or mapped.

At the national scale, the expansion of urban areas has been geographically imbalanced. Some of the regional disparities can be ascribed to state interventions in planning and development coupled with atomized growth engines in each state. Satellite image analysis shows that urban expansion is largely concentrated in southern states; net increases in urban area from 1998 to 2008 were greatest in Tamil Nadu (48 percent) and Andhra Pradesh (33 percent). In both states, rapid growth in the information technology and biotechnology sectors (Vaidyanathan 2008) has resulted in high levels of urban land change. In addition to service-sector activities that predominate in large metropolitan areas like Chennai and Hyderabad, the two states' small- and medium-sized cities have experienced growth in their manufacturing industries. Although the total area of urban land use change in Karnataka is less than in Tamil Nadu and Andhra Pradesh, tertiary sector growth is no different. Among northern states, Punjab has the highest urban land use change. In contrast to growth in the service sector in the southern states, Punjab's economic growth is induced by highly efficient transportation infrastructure and industrialization. Punjab, the 'bread basket of India,' has recently emerged as a manufacturing hub in northern India.

Satellite image analysis provides evidence of significant variation in urban land use change across the country but very little evidence of urban land expansion in low-income states such as Orissa, Jharkhand and Bihar. Differences in regional patterns of urban growth can be observed between northern and southern India (Figure 5.1). There is aggregated growth in

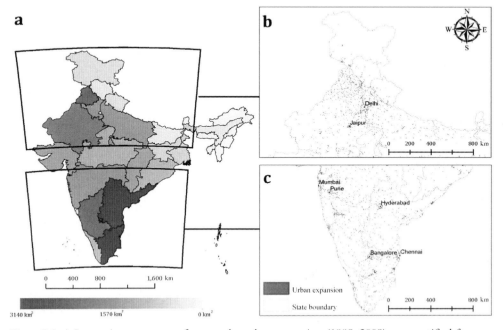

Figure 5.1 a) State-wise aggregates of area under urban expansion (1998–2008), as quantified from DMSP/OLS NTLs and SPOT-VGT data; location of urban land use changes in b) northern states and c) southern states

the south around large metropolitan cities such as Mumbai, Pune, Hyderabad, Bangalore and Chennai, whereas in northern India patterns of urban land use change are dispersed, especially along the Delhi–Jaipur and Delhi–Amritsar corridors. The prevalence of manufacturing units along these northern corridors could be a reason for the spatial dispersion of urban land use change patterns.

Trends in urban population growth

Though studies of urban land change reveal continuing, accelerating patterns of expansion across the subcontinent, the growth of India's urban population is experiencing significant shifts away from its 20th century patterns. Foremost among these changes are a declining rate of urban growth and the rise of small- and medium-sized cites as the primary poles of growth.

Country-level growth by rate and magnitude

Reports on India's urban transition often suggest the present pace of population growth is unmatched to any other country. The World Bank (2013) deems India's current demographic shift "the largest rural-urban migration of this century." The Government of India's Planning Commission (2011) states that the "speed of urbanization presents unprecedented managerial and policy challenges" for India's city planners. Although these reports convey the great magnitude of India's urban growth and frame the scope of the tasks facing planners, investors and policy-makers in the next two decades, they have also contributed to a false impression that India's urban population growth has come to an apex. In fact, challenging rates of urban population growth have been sustained through much of the 20th century, and the great increase of urban population will continue well into the 21st century.

The 'rapid urbanization' of India's post-liberalization growth is often counterposed to its growth before 1991, but Indian cities have sustained and absorbed a high level of population growth for more than 60 years. This rate of transformation peaked between 1970 and 1990 (Figure 5.2). In the 1970s alone, the population of Indian cities increased by almost

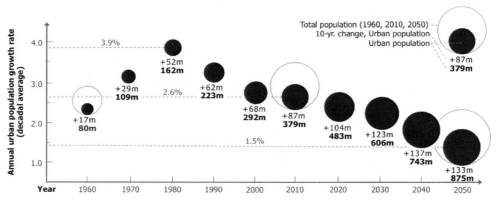

Figure 5.2 Rate and magnitude of India's population growth

Source: Adapted from Pradhan (2013); Data: UN DESA (2014) World Urbanization Prospects.

50 percent; a rate unmatched in India's modern history. Furthermore, India's 2010 population is forecast to increase 27 percent by 2020 (UN DESA 2014).

The secular decline in India's forecasted urban growth rate should not be mistaken for abatement; in absolute terms, India's urban growth will continue over most of the current century. The 230 million people expected to enter India's urban areas between 2010 and 2030 will have significant effects on city scale, economic growth and infrastructure. A larger increase of 280 million is expected in the following two decades. Year-over-year urban population increase is not expected to level off until 2045, meaning that India's substantial urban growth will likely proceed through the end of the century (UN DESA 2014). While the acceleration of India's urban population transition was the primary strain on municipal services over the past century, it is the scale that will make India notable in the century ahead. With the sole exception of China, India's urban population increase will be greater than that of any other society in history. As its subcontinental population grows to exceed that of East Asia, the planet's most widely settled and human-impacted region will also be among the most intensively urbanized.

Demographic trends in settlement size

Although aggregate data is useful for analyzing the long-term urban growth process, country-level data cannot usefully explain the economic and social processes responsible for changing India's urban landscape. While retaining the India-wide frame of this chapter, we turn to an analysis of settlements by size in order to better situate social processes in demographic change.

India's 'metropolitan cities' of greater than one million persons receive a great deal of attention because of their economic and political primacy. Greater Mumbai alone accounts for 40 percent of India's foreign trade and tax revenue (The Economist 2007; Government of Maharashtra 2006). These cities draw nearly all funding for urban infrastructure and technological development through India's urban investment schemes (Véron 2010). Although they will dominate India's economy for the foreseeable future, their pull as population magnets has waned considerably in the past three decades. The annual rate of growth of India's largest metropolitan areas—Mumbai, Delhi, Kolkata and Chennai—has dropped by half since the 1980s (Kundu 2011; UN DESA 2014). Today, the fastest-growing population centers are India's mid- and small-sized cities; those with a population of one million residents or fewer (Figure 5.3). By 2025, the majority (55 percent) of India's urban population are projected to live in and around urban centers smaller than one million people, more than twice the share forecast to live in agglomerations of greater than ten million (21 percent) (UN DESA 2014).

Despite the disproportionate growth of small- and medium-sized urban areas, these regions have been routinely overlooked by urbanization researchers (Montgomery 2008; Denis and Marius-Gnanou 2011; Suri 2011). Most social understanding of them is necessarily schematic because comparative, ground-level studies of these urban places are lacking. Compared to the increasingly international orientation of large metropolitan cities, these relatively smaller urban areas are economically and commercially oriented toward their surrounding rural villages and hinterlands (Satterthwaite and Tacoli 2003). In addition, their operational and economic management is heavily controlled by state agencies, which have concomitantly low levels of local governance capacity and democratic participation in municipal issues. This is likely due to neglect by central and state government investment (Véron 2010). District-level analysis suggests that direct foreign investment is nearly absent outside large cities (Mukim and Nunnenkamp 2012).

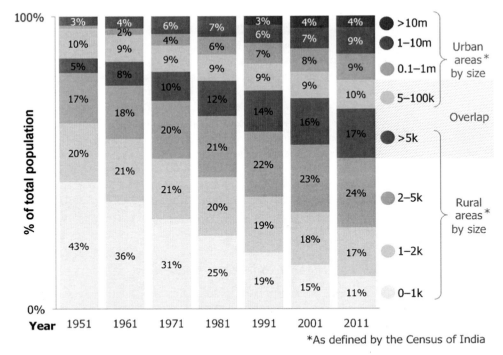

Figure 5.3 Population distribution by settlement size, 1951–2011

Source: Adapted from IIHS (2011); Data: Government of India (2011).

Settlements of 5,000–20,000 people constitute the fast-growing category of Indian settlements (Figure 5.3), but are perhaps least well understood by urban researchers. Some of this oversight is a result of India's atypical rural–urban classification scheme. The Census of India's tripartite criteria for urban places—at least 5,000 people, 75 percent non-agricultural male employment and a density of 400 people/km—has obscured the high growth of small urban places when measured by population concentration alone (Denis and Marius-Gnanou 2011). Distributional politics further compound the problem: lacking a small-city urban investment scheme, the majority of these settlements choose a rural administrative designation that allows them to draw from rural-focused investment schemes and avoid higher municipal tax burdens (Bhaghat 2005; Planning Commission 2012). One major consequence of these political tactics is that intercensal urban population growth is badly distorted by classification. In fact, reclassification of large rural villages as urban areas accounted for nearly 30 percent of all officially recorded urban population growth in the 2011 Census including more than two-thirds of urban growth in four states (Pradhan 2013).

The prevailing growth and demographic significance of small, municipality-sized settlements—whether 'rural' or 'urban' classified—is now widely acknowledged, but existing publications have little to say about why these places are leading India's urban growth (Bhagat 2005, 2011; Pradhan 2013; Denis and Marius-Gnanou 2011). Out-migration from villages and hamlets to these small urban centers suggests they are stepping-stones in India's urban transformation, and demographic studies have noted the 'push' factors—especially rural income stagnation and a decline in agricultural jobs—that induce rural–urban migration

(Nijman 2012). Yet the continued concentration of education, employment opportunities and foreign capital in larger urban centers indicate possible undiscovered reasons for the shift away from large city growth.

Agricultural land loss to urbanization

Indian agriculture is sensitive to climate change (Kavi Kumar and Parikh 2001), and population growth in India has stimulated land-intensive cropping practices (Mishra 2002) known to alter ecosystem properties (Matson et al. 1997) and regional meteorology (Douglas et al. 2009). Climate change sensitivity and agricultural land intensification together raise concerns about agricultural sustainability. Limited land resources in India—half of which are already degraded (Varughese et al. 2009)—further compound the problem. Amidst these challenges to India's agricultural economy, the scale of urban land conversion presents additional challenges in meeting the country's growing demand for agricultural products (see Deng, Chapter 3 in this volume for a comparison of agricultural land loss in China).

Economic growth and urbanization require infrastructure for housing, industry, transportation and other needs, resulting in the expansion of urban land. In the vast majority of cases, urban expansion occurs on prime agricultural land. This is primarily for two reasons: (1) the relative abundance of agricultural land in areas contiguous to cities and the tendency for urban expansion to occur outward from existing urban areas, and (2) the legislative instruments or land use policies that facilitate urban conversion of agricultural land. Urbanization can generate social and economic benefits including employment opportunities, higher income levels and access to better amenities, all of which generate higher living standards. However, urban conversion of agricultural lands has a number of negative externalities. First is the loss of farmers' livelihoods coupled with inadequate rehabilitation and resettlement provisions. In India, agricultural land acquisition for development, in several instances, has entailed inadequate compensation to the farmers (Reddy and Reddy 2007). This is primarily because land acquisition was previously administered under an archaic legal instrument, the Land Acquisition Act of 1894, which did not offer equitable compensation to farmers. In fact, land acquisition efforts, especially when they involve a large geographical area, have been contentious and have resulted in public protests and even riots in several instances. A new legislative instrument, the Right to Fair Compensation and Transparency in Land Acquisition, Rehabilitation and Resettlement Act of 2013, was designed to offer more adequate compensation and rehabilitation to agricultural landholders. However, it is not clear how this act would balance the need for swift land conversion, given India's impending urban and economic growth and equitable compensation offered to farmers. Second, urban conversion of agricultural land has direct impacts on local agricultural systems and land productivity (Fazal 2000; 2001). The loss of agricultural land in the outlying areas of an expanding urban center alters the distribution and supply chain for agricultural commodities. A regionalized agricultural system tends to produce higher-value agricultural commodities, which has environmental repercussions in the form of declining land productivity due to intensified cropping practices. In addition, agricultural land abandonment due to real estate speculation and urban development pressure further disrupts land use structure and composition in outlying areas. A case study from India shows that agricultural land cultivation is abandoned ten years before the actual development (Reddy and Reddy 2007).

Land systems are dynamic. In India, the rate of change of land use in the fringe areas of rapidly growing cities is especially high, and monitoring these frequent land use changes is a major concern (Pandey et al. 2013). The Indian Agricultural Census, which forms part of the FAO's World Agricultural Census Program, is the major source of land use change data. The area under a given category is estimated through the land records maintained by the village accountants. The agricultural census follows a nine-fold classification system which includes the category 'area under non-agricultural uses' that monitors agricultural land conversion to 'non-agricultural uses.' Because 'non-agricultural uses' might also subsume fallowing and permanent abandonment, the category does not explicitly monitor agricultural land conversion to urban uses. While contemporary urban expansion in India is unprecedented in both rate and magnitude, the agricultural census does not account for these urbanization-induced land use changes in a timely or precise manner. However, scholars have identified remote sensing and geographic information system (GIS) techniques which permit rapid and reliable assessment of land use change (Farooq and Ahmad 2008; Fazal 2000, 2001; Jat et al. 2008; Sudhira et al. 2004).

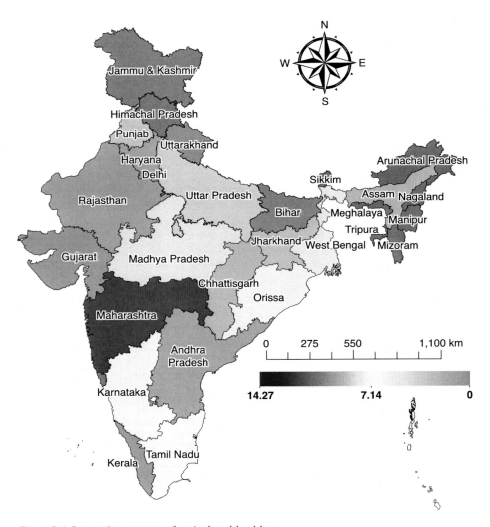

Figure 5.4 State-wise patterns of agricultural land loss

Pandey and Seto (2015) used time-series satellite data from 2000 to 2011 to monitor agriculture-to-urban land conversion across India. Their results show that the total agricultural land loss due to urban growth was roughly 0.7 million hectares from 2000–2011. In addition to urban growth, agricultural land is also abandoned or fallowed; therefore, this estimate is a very conservative measure of total agricultural land loss. These results show that agricultural land loss is concentrated mostly in industrialized states.

Estimated losses of 14 percent occurred in Maharashtra, and an additional 20 percent were concentrated in the adjacent states of Andhra Pradesh and Chhattisgarh. Agricultural land loss in northern India was also predominant in Uttar Pradesh, which subsumes one of the most fertile regions of the Indo-Gangetic plain. These results are not surprising as these states have recently experienced rapid economic growth coupled with high rates of urbanization. Our results also show a positive relationship between agricultural land loss and the agricultural suitability of land, which suggests greater urban expansion in fertile agricultural land than in areas unsuitable for cultivation (Pandey and Seto 2015). Though fallowing, abandonment and other tallied losses of agricultural land do not prohibit future farming, agricultural-to-urban land conversion is a largely irreversible land use change. In view of the need to preserve agricultural land, rational land use policies are required such that urban expansion on prime agricultural land could be mitigated.

Case study of urban expansion

Here we provide an in-depth analysis of urban change using the example of Kolkata, the capital city of West Bengal in eastern India with a population of over 14 million (Government of India 2011). Kolkata is no unique city in India; it exemplifies a typical large city, which traditionally has an older colonial section along with new built-up areas. In Kolkata, agricultural lands and wetlands have been converted to built-up area in the recent past due to population pressure.

Kolkata city

Kolkata was founded as a garrison town, becoming the East India Company's major port of trade. The city then evolved into a provincial city and eventually became the seat of the British crown in India. It is now the third-largest city in India, a status which has established a level of importance for the city and stimulated significant growth in size over the years (Kundu and Nag 1996; Mitra et al. 2012). Kolkata has a more urbanized core, known as the Kolkata Municipal Corporation (KMC), set within a larger urban agglomeration, known as the Kolkata Metropolitan Area (KMA) (Dasgupta et al. 2012).

Urban growth still continues, often in an unplanned manner. This unplanned growth scenario is true for almost all cities in India (Roy 2009). Using historical maps and GIS analysis, Mitra et al. (2012) find that the city has consistently grown since 1756, with the majority of the population growth occurring during the Indian independence phase between 1940 and 1950 (Figure 5.5). Indeed, the greatest amount of change in urban extent occurred between India's independence in 1947 and 1990. In this period, Kolkata exhibited a 'balloon bursting effect,' where rapid urban expansion occurred in all directions from the core. Figure 5.6 is a graphical representation comparing the total area, total area change and area change per year up to the year 2000 for the city.

Although there are no reliable statistics on this influx, by the time of the 1951 census, only one-third (33.2 percent) of the inhabitants of Kolkata were recorded as having been born in

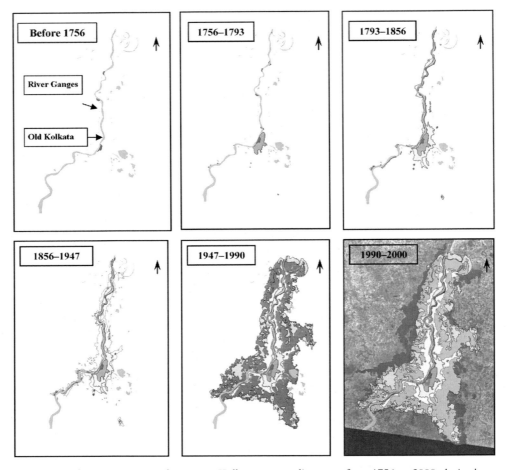

Figure 5.5 Urban expansion in the greater Kolkata metropolitan area from 1756 to 2000, derived from historical maps and GIS analysis

the city, with everyone else classified as an immigrant. The population from neighboring villages was 12.3 percent, 26.6 percent from other Indian states, and more than a quarter of the population (26.9 percent) from East Pakistan (present-day Bangladesh), a result of the communal troubles that raged between 1907 and the 1947 partition (Dutta 2003).

The partition of India and Pakistan in 1947, a result of India's independence, contributed to the influx of millions of refugees across the present Indian border. Before independence, Bangladesh and the Indian state of West Bengal were a unified political entity. They were partitioned on the basis of religion (i.e. Hindus and Muslims), leading to an exodus across the borders. An estimated four million people moved from Bangladesh to West Bengal between 1946 and 1971 (Chatterjee 1990). In the decades following, a steady stream of Bangladeshi immigrants continued to arrive in Kolkata. Most settled on whichever plot of land was available without use restrictions and planning.

Another factor contributing to the urbanization of India was the influx of people from the surrounding non-urban and less developed areas (Roy 2003). Kolkata is an important city compared with other cities in the eastern part of India. It has become the target city for

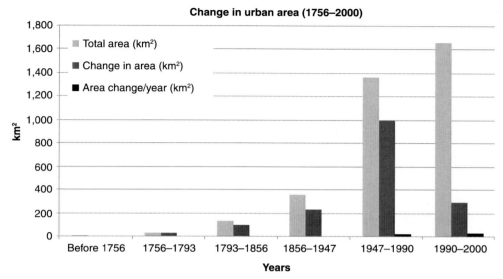

Figure 5.6 Graphical representation of changes in urban area, Kolkata city, 1756–2000
Source: Adapted from Mitra, Shepherd and Jordan (2012)

rural–urban migrants of east India and a source of financial advancement for many people. Population growth and the drive for economic development led to unplanned and sporadic expansion of this mega-city. Most rural–urban migrants came from three states: Bihar (60.6 percent), Orissa (8.4 percent) and Uttar Pradesh (15.6 percent). All of these migrants were attracted by Kolkata, the only large city within hundreds of miles (Lubell 1974).

This case study not only brings forth the dynamics of land use change over centuries, but also emphasizes the pressure the city's infrastructure and available amenities experienced under uncontrollable growth. This is true for all major Indian cities where in-migration has impacted the shape, structure and environment. As discussed earlier in the chapter, the mid- and small-sized cities of India will soon feel the effects of steady migration from surrounding rural areas.

Concluding remarks and key research questions

Indian cities receive much attention for their severe inequalities, overburdened physical infrastructure and extraordinary population pressures (Chatterjee et al. 2012). Single-city urbanization studies provide a view of the political and distributive problems facing India's urban residents and have become the framing devices of India-wide planning interventions and urban investment schemes. However, country-level studies of the effects of rapid urbanization on land use, demographic transition and urban structure suggest we should reconsider the scalar impact of India's urbanization process. The results of previous single-city urbanization and country-level studies suggest at least three definitive trends in India's urban growth paradigm:

- increasing rates of urban land expansion, well in excess of urban population growth;
- increasing areas of agricultural land lost to urban land conversion, with the largest losses concentrated in India's most fertile regions;

- a transition from population growth concentrated in several mega-cities to growth distributed across many small- and medium-sized cities (5,000–1,000,000 residents).

This dispersive phase of urban development will require broader and more demanding policy responses. Although mega-cities like Kolkata are now relatively used to rapid population growth, urban land area continues to expand at accelerating rates, resulting in a new set of planning and managerial challenges.

These trends will dramatically alter the spatial and social structure of India's urban constellations, in addition to increasingly whole agrarian systems and bio-regions, the impacts of which are rarely considered by city-centric planning and land management schemes. Mountain ranges that contain significant biodiversity including the Western Ghats and the Himalayas, are particularly threatened by contiguous areas of urban expansion (Seto et al. 2012). This urban expansion is also likely to have significant climate effects. Irrespective of global climate change, cities alter their local climate particularly by reducing rainfall and increasing nighttime temperatures (WMO 1996). Furthermore, as the structure and function of new urban land continues to evolve, mitigating emissions and adapting to global climate change will become increasingly challenging (IPCC 2014).

As India enters a phase of large-scale urbanization, there is still much that is unknown about its processes and patterns; its challenges and future hurdles:

- How are patterns of urban expansion affected by policy changes such as measures for land reform?
- How are rural livelihoods being reshaped by urbanization and wage labor migration?
- How is urbanization changing patterns of socio-economic inequalities?
- How is urbanization impacting the environment and health conditions?
- Will Indian cities, large and small, merge to form urban archipelagoes or corridors, multiplying the effects of urbanization?
- What are the spatial patterns of urban land cover change and how do they vary across the country, over time, and for different economic development levels?

References

Angel, S., J. Parent, D.L. Civco, A. Blei, and D. Potere. 2011. "The Dimensions of Global Urban Expansion: Estimates and Projections for All Countries, 2000–2050." *Progress in Planning* 75 (2): 53–107. doi:10.1016/j.progress.2011.04.001.

Bhagat, R.B. 2005. "Rural–Urban Classification and Municipal Governance in India." *Singapore Journal of Tropical Geography* 26 (1): 61–73. doi:10.1111/j.0129-7619.2005.00204.x.

———. 2011. "Emerging Pattern of Urbanisation in India." *Economic and Political Weekly* 46 (34): 10–12.

Chatterjee, I., G. Pomeroy, and A.K. Dutt. 2012. "Cities of South Asia." In S.D. Brunn, M. Hays-Mitchell, and D.J. Ziegler (eds.), *Cities of the World: World Regional Urban Development*. 5th edition. Lanham, MD: Rowman and Littlefield, pp. 381–423.

Chatterjee, M. 1990. "Town Planning in Calcutta: Past, Present and Future." In S. Chaudhuri (ed.), *Calcutta: The Living City*. Calcutta: Oxford University Press, pp. 137–147.

Dasgupta, S., A. Gosain, S. Rao, S. Roy, and M. Sarraf. 2012. "A Megacity in a Changing Climate: The Case of Kolkata." *Climate Change: Earth and Environmental Science*. 1–20. doi: 10.1007/s10584-012-0516-3.

Denis, E., and K. Marius-Gnanou. 2011. "Toward a Better Appraisal of Urbanization in India." *Cybergeo: European Journal of Geography*, November. doi:10.4000/cybergeo.24798.

Desmet, K., E. Ghani, S. O'Connell, and E. Rossi-Hansberg. 2013. "The Spatial Development of India." *Journal of Regional Science* 55 (1). doi:10.1111/jors.12100.

Douglas, E.M., A. Beltrán-Przekurat, D. Niyogi, R.A. Pielke Sr., and C.J.Vörösmarty. 2009. "The Impact of Agricultural Intensification and Irrigation on Land–Atmosphere Interactions and Indian Monsoon Precipitation—A Mesoscale Modeling Perspective." *Changes in Land Use and Water Use and Their Consequences on Climate, Including Biogeochemical Cycles* 67 (1–2): 117–128. doi:10.1016/j.gloplacha.2008.12.007.

Dutta, K. 2003. *Calcutta: A Cultural and Literary History*. Northampton, Massachusetts: Interlink Books.

The Economist. 2007. "A Cul-de-Sac of Poverty." *The Economist*, May 3. www.economist.com/node/9070705.

Farooq, S., and S. Ahmad. 2008. "Urban Sprawl Development around Aligarh City: A Study Aided by Satellite Remote Sensing and GIS." *Journal of the Indian Society of Remote Sensing* 36 (1): 77–88.

Fazal, S. 2000. "Urban Expansion and Loss of Agricultural Land – A GIS Based Study of Saharanpur City, India." *Environment and Urbanization* 12 (2): 133–149.

————. 2001. "The Need for Preserving Farmland: A Case Study from a Predominantly Agrarian Economy (India)." *Landscape and Urban Planning* 55 (1): 1–13. doi:10.1016/S0169-2046(00)00134-1.

Frolking, S., T. Milliman, K.C. Seto, and M.A. Friedl. 2013. "A Global Fingerprint of Macro-Scale Changes in Urban Structure from 1999 to 2009." *Environmental Research Letters* 8 (2): 024004. doi:10.1088/1748-9326/8/2/024004.

Government of India. 2011. Census of India. http://censusindia.gov.in/. Last accessed on December 15, 2014.

Government of Maharashtra. 2006. "Economic Profile." In *Greater Mumbai City Development Plan*. www.mcgm.gov.in/irj/portal/anonymous?NavigationTarget=navurl://095e1c7b9486b1423b881dce8b106978.

India Institute for Human Settlements (IIHS). 2011. *Urban India 2011: Evidence*. Bangalore: India Institute for Human Settlements.

IPCC. 2014. "Summary for Policymakers. " In: C.B. Field, V.R. Barros, D.J. Dokken, K.J. Mach, M.D. Mastrandrea, T.E. Bilir, M. Chatterjee, K.L. Ebi, Y.O. Estrada, R.C. Genova, B. Girma, E.S. Kissel, A.N. Levy, S. MacCracken, P.R. Mastrandrea, and L.L. White (eds.), *Climate Change 2014: Impacts, Adaptation, and Vulnerability. Part A: Global and Sectoral Aspects. Contribution of Working Group II to the Fifth Assessment Report of the Intergovernmental Panel on Climate Change*, Cambridge, UK and New York, NY, USA: Cambridge University Press, pp. 1–32.

Jat, M.K., P.K. Garg, and D. Khare. 2008. "Monitoring and Modelling of Urban Sprawl Using Remote Sensing and GIS Techniques." *International Journal of Applied Earth Observation and Geoinformation* 10 (1): 26–43. doi:10.1016/j.jag.2007.04.002.

Kochhar, K., U. Kumar, R. Rajan, A. Subramanian, and I. Tokatlidis. 2006. "India's Pattern of Development: What Happened, What Follows? " *Journal of Monetary Economics* 53 (5): 981–1019.

Kundu, A. 2011. "Method in Madness: Urban Data from 2011 Census." *Economic and Political Weekly* 46 (40): 13–16.

Kundu, A.K., and P. Nag. 1996. *Kolkata: Atlas of the City of Calcutta and Its Environs*. 2nd ed. National Atlas and Thematic Mapping Organisation. Ministry of Science and Technology, Government of India.

Kavi Kumar, K.S., and J. Parikh. 2001. "Indian Agriculture and Climate Sensitivity." *Global Environmental Change* 11 (2): 147–154. doi:10.1016/S0959-3780(01)00004-8.

Lubell, H. 1974. *Urban Development and Employment: The Prospects for Calcutta*. Washington, D.C.: International Labor Office.

Matson, P.A., W.J. Parton, A.G. Power, and M.J. Swift. 1997. "Agricultural Intensification and Ecosystem Properties." *Science* 277 (5325): 504–509.

Mishra, V. 2002. "Population Growth and Intensification of Land Use in India." *International Journal of Population Geography* 8 (5): 365–383.

Mitra, C., J.M. Shepherd, and T. Jordan (2012). "On the Relationship between the pre-Monsoonal Rainfall Climatology and Urban Land Cover Dynamics in Kolkata City, India." *International Journal of Climatology* 32 (9): 1443–1454. doi: 10.1002/joc.2366.

Montgomery, M.R. 2008. "The Urban Transformation of the Developing World." *Science* 319 (5864): 761–764. doi:10.1126/science.1153012.

Mukim, M., and P. Nunnenkamp. 2012. "The Location Choices of Foreign Investors: A District-Level Analysis in India." *The World Economy* 35 (7): 886–918. doi:10.1111/j.1467-9701.2011.01393.x.

Nath, V. 2007. *Urbanization, Urban Development and Metropolitan Cities in India*. Delhi: Concept Publishing Company.

Nijman, J. 2012. "India's Urban Challenge." *Eurasian Geography and Economics* 53 (1): 7–20. doi:10.2747/1539-7216.53.1.7.

Pandey, B., and K.C. Seto. 2015. "Urbanization and Agricultural Land Loss in India: Comparing Satellite Estimates with Census Data." *Journal of Environmental Management* 148: 53–66. doi: 10.1016/j.jenvman.2014.05.014.

Pandey, B., P.K. Joshi, and K.C. Seto. 2013. "Monitoring Urbanization Dynamics in India Using DMSP/OLS Night Time Lights and SPOT-VGT Data." *International Journal of Applied Earth Observation and Geoinformation* 23 (1): 49–61.

Planning Commission. 2011. *Faster, Sustainable, and More Inclusive Growth: An Approach to the Twelfth Five Year Plan*. New Delhi: Government of India. http://12thplan.gov.in/12fyp_docs/17.pdf.

———. 2012. *Report of the Steering Committee on Urbanisation*. New Delhi: Government of India. http://planningcommission.gov.in/aboutus/committee/strgrp12/strrep_urban0401.pdf.

Pradhan, K.C. 2013. "Unacknowledged Urbanisation: The New Census Towns of India." Delhi: Center for Policy Research. http://mpra.ub.uni-muenchen.de/41035/.

Reddy, V.R., and B.R. Reddy. 2007. "Land Alienation and Local Communities: Case Studies in Hyderabad-Secunderabad." *Economic and Political Weekly* 42 (31): 3233–3240. doi:10.2307/4419873.

Roy, A. 2003. *City Requiem, Calcutta: Gender and The Politics of Poverty*. Minneapolis: University of Minneapolis Press.

———. 2009. "Why India Cannot Plan Its Cities: Informality, Insurgence and the Idiom of Urbanization." *Planning Theory* 8 (1): 76–87. doi:10.1177/1473095208099299.

Satterthwaite, D., and C. Tacoli. 2003. *The Urban Part of Rural Development: The Role of Small and Intermediate Urban Centres in Rural and Regional Development and Poverty Reduction*. London: International Institute for Environment and Devlopment.

Seto, K.C., M. Fragkias, B. Güneralp, and M.K. Reilly. 2011. "A Meta-Analysis of Global Urban Land Expansion." *PLoS ONE* 6.8: e23777. doi:10.1371/journal.pone.0023777.

Seto, K.C., B. Güneralp, and L.R. Hutyra. 2012. "Global Forecasts of Urban Expansion to 2030 and Direct Impacts on Biodiversity and Carbon Pools." *Proceedings of the National Academy of Sciences* 109 (40): 16083–16088. doi:10.1073/pnas.1211658109.

Sudhira, H.S., T.V. Ramachandra, and K.S. Jagadish. 2004. "Urban Sprawl: Metrics, Dynamics and Modelling Using GIS." *International Journal of Applied Earth Observation and Geoinformation* 5 (1): 29–39. doi:10.1016/j.jag.2003.08.002.

Suri, S. 2011. "Decentralizing Urbanization: Harnessing the Potential of Small Cities in India." Thesis, Massachusetts Institute of Technology. http://hdl.handle.net/1721.1/65747.

Vaidyanathan, G. (2008). "Technology Parks in A Developing Country: The Case of India." *The Journal of Technology Transfer* 33 (3), 285–299.

Varughese, G.C., K.V. Lakshmi, A. Kumar, and N. Rana. 2009. "State of Environment Report: India, 2009." Delhi: Ministry of Environment and Forests, Government of India. www.moef.nic.in/downloads/home/home-SoE-Report-2009.pdf.

Véron, R. 2010. "Small Cities, Neoliberal Governance and Sustainable Development in the Global South: A Conceptual Framework and Research Agenda." *Sustainability* 2 (9): 2833–2848. doi:10.3390/su2092833.

UN DESA. 2014. *World Urbanization Prospects: The 2013 Revision*. United Nations Department of Economic and Social Affairs, Population Division.

World Bank. (2013). "India Country Overview 2013." www.worldbank.org/en/country/india/overview.

World Meteorological Organization (WMO). 1996. Climate and Urban Development. No. 844. Geneva: WMO.

6

RESOURCE USE FOR THE CONSTRUCTION AND OPERATION OF BUILT ENVIRONMENTS

Burak Güneralp

The ongoing urbanization transition around the world will have significant impacts on energy use, greenhouse gas (GHG) emissions and ultimately sustainability. The expansion of urban areas in the form of new factories, residential and business districts, and infrastructure requires various kinds of building materials, particularly, such as concrete, steel, brick and wood. The demand for natural resources for the constituents of the built environment is not limited to the requirements for construction of these structures, but also includes energy demand for their maintenance and operation. The demands of the cities' inhabitants to fulfill their preferred lifestyles are also important and interact with the demands to operate the built environment.

Urbanization is one of the major drivers of resource consumption and resulting GHG emissions (Levine and Aden 2008; Zhou et al. 2008; Vennemo et al. 2009). The land, material and energy requirements of the construction and operation of the built environment in different parts of the world can vary significantly depending on the differences in building techniques, specific construction materials as well as cultural preferences. Environmental factors such as regional climate can also play a significant role in this respect. The efficiency in material and energy use associated with the construction and operation of buildings have significantly improved in many places around the world (Worrell et al. 1995; OECD/IEA 2007; Liu et al. 2009), although significant differences among countries remain (Maruyama and Eckelman 2009; Oda et al. 2012). These improvements have been well documented and widely presented as a remedy to reduce various environmental impacts from the construction and operation of the built environment. However, less attention has been paid to how these efficiency gains compare to the increase in the magnitude of the resource demands and resulting GHG emissions due particularly to rapid urbanization in low- and middle-income countries. This chapter explores material and energy use and intensity, and associated GHG emission trends and patterns generated from constructing and operating the urban built environment.

Construction materials

The construction and operation of the built environment requires significant amounts of resources (in terms of both raw materials and energy). In particular, buildings are estimated to account for about one-sixth of the world's freshwater withdrawals, one-quarter of the global wood harvest and two-fifths of global material and energy flows (Godfried et al. 1998). In the United States, the construction and operation of the built environment accounts directly or indirectly for over half of the total energy consumption (Horvath 2004). Concrete, steel, brick and wood are the four most common building materials; the first three are also the most energy-intensive (Venkatarama Reddy and Jagadish 2003; Chaturvedi and Ochsendorf 2004). Unlike cement, steel and timber, most basic raw materials such as clay, sand and gravel required for construction are available within relatively short distances of cities (Price et al. 2005; Singh and Asgher 2005; Habert et al. 2010). Although several statistics on the amount of steel, concrete and wood used in the construction of buildings are available, the amount of other materials such as plastics and various kinds of chemicals are much less documented (Horvath 2004).

Mining activities for construction materials as well as for fossil fuels as a source for energy are a major driver of land change and other environmental impacts in rural places (Brechin, 2002), but they can also be a source for alleviating rural poverty (Bridge, 2004). Although land change due to mining has significant impacts on location as well as downstream, these have received relatively less attention than the direct land change due to expansion of urban land (Singer et al. 2013). This is partly due to the paucity of information needed to link the processes that drive urban expansion to the specific sources of the needed construction materials and energy engendered by these processes. This is needed for a more complete accounting of the effects of the construction of the built environment. In addition to these essentially technology-centered efforts, there is also a dimension of mining that has more to do with public policy, political economy and socio-cultural dynamics that needs to be taken into consideration in examining impacts of mining activities (Bridge 2004).

As urban areas expand, larger amounts of raw materials and energy are required to construct the built environment. Some of these construction materials may initially be available locally. This is especially the case for such construction materials as sand (Price et al. 2005; Habert et al. 2010). However, over time, the demand for materials may exceed the quantity or quality of the local supply, and these materials may be drawn from sources that are increasingly farther away. Of all the construction materials, cement (Steinberger et al. 2010; Peters et al. 2011), steel (Wang et al. 2007; Wang et al. 2008) and timber (Meyfroidt et al. 2010) are among those to travel the longest distances to reach their final destination. In the case of timber, however, illegal trade makes accurate characterization of the flow of this resource especially challenging (Katsigris et al. 2004).

Urban areas accumulate enormous amounts of resources in their built environment—be it in the form of buildings or other infrastructure components (e.g. wiring, piping). A vision for the future sees these urban areas as mines, where obsolete buildings serve as sources for the resources to be used in the construction of new ones or for other purposes. 'Urban mining' is increasingly becoming a reality, as existing infrastructure in mature cities become obsolete and thus a source for metals (Munro 1984; Yamasue et al. 2009; Wallsten et al. 2013). In this respect, de-urbanizing areas especially in North America and Europe, i.e. 'shrinking cities' (Oswalt and Rieniets 2006; Blanco et al. 2009), are notable, as those buildings and infrastructure that are no longer in use contain significant amounts of construction materials.

Concrete

Concrete is a construction material that is comprised of three basic elements: cement, aggregates such as gravel and limestone, and water. Cement is one of the most energy and

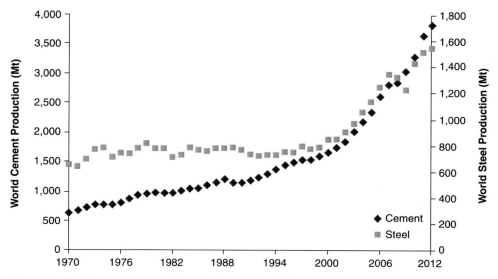

Figure 6.1 Global cement and steel production, 1970–2012
Source: USGS (2014a;b).

carbon intensive construction materials (Worrell et al. 2001; van Oss and Padovani 2003). The production of cement emits CO_2 through the combustion of fossil fuels and through the calcination of limestone in cement kilns, making it one of the largest GHG emitting industries (van Oss and Padovani 2003). While emissions of other pollutants such as nitrogen oxides (NOx), sulfur oxides (SOx), dust and certain organic compounds may also be significant, CO_2 emissions from cement production receive the most attention (Worrell et al. 2001).

The global cement production reached 3.8 billion tonnes in 2012 (Figure 6.1). In terms of production, China is ranked first with 2.1 billion tonnes, followed by India (250 million tonnes) and the United States (74 million tonnes) (USGS 2013). As the largest cement producer, China accounted for 63 percent of world production in 2012; the second-largest producer, India, accounted for only 7 percent. The consumption of cement increased from 41 Mt in 1950 to over 110 Mt in 2000 in the United States with even larger rates of increase

Table 6.1 Total production of cement and steel by the top five producers, 2005, 2012, and Average Annual Growth Rate (AAGR)

Cement (Mt)				Steel (Mt)			
Country	2005	2012	AAGR	Country	2005	2012	AAGR
China	1,040	2,150	11%	China	349	720	11%
India	145	250	8%	Australia	113	108	−1%
U.S.	101	74	−4%	Brazil	95	91	−1%
Brazil	37	70	10%	India	46	76	8%
Iran	33	65	10%	Russia	66	76	2%
World	2,310	3,400	6%	World	1,130	1,500	4%

Source: Adapted from Table 10.1 in Fischedick et al. (2014).

for other ingredients of concrete such as crushed stone, sand and gravel (Horvath 2004), but decreased to about 80 Mt by 2012 (USGS 2013). The United States accounts for the shortfall (about 6 Mt) by imports mainly from its two neighbors, but also overseas from the Republic of Korea and China. The global demand for cement is forecasted to continue increasing, particularly fuelled by urbanization in low- and middle-income countries (Schneider et al. 2011).

Cement production in China used to be particularly inefficient, exacerbating the country's CO_2 emissions problem (Soule et al. 2002; Levine and Aden 2008). Despite recent improvements, it remains less efficient than those in most of the world (OECD/IEA 2007). Although India is the second-largest producer and second-largest consumer of cement (ICR 2013; WBCSD/IEA 2013), the per capita consumption of cement was about 200 kg per capita in 2012, less than half the global average of 520 kg per capita (UN 2013; USGS 2013). For comparison, as of 2011, China's per capita cement consumption was 1,540 kg per capita (DBS Vickers Limited 2013). The International Energy Agency (IEA) projects that China's cement production will reach between 3,700 Mt and 4,400 Mt by 2050. It is notable that the production already reached the lower range of this projection in 2012 (IEA 2009) (Table 6.1). The IEA study further projects that the cement consumption in China will peak between 2015 and 2030, after which, demand from other low- and middle-income countries will become more pronounced. In particular, the annual cement production in India is forecasted to increase to anywhere between 800 and 1,350 Mt by 2050 (WBCSD/IEA 2013).

Tracking the flows of raw materials to be used in the construction of the built environment is challenging although the impacts extend well beyond the construction site (Güneralp et al. 2013). This is vividly apparent in the case of mining sites. For example, quarrying for limestone, an ingredient of cement, may decimate karstic landscapes, some of which may have cultural and ecological importance (Vermeulen and Whitten 1999). Recently, a number of 'best practices' have been established to address these impacts in relation to the siting and processing of active quarries as well as rehabilitation of those no longer in use (Misra et al. 2002; Neri and Sánchez 2010). Tepetzil (pumice) mining in Perote Valley, Mexico, is one of the few documented cases of linkages between specific urban places and specific extraction sites. Pumice is an ingredient in light-weight concrete blocks. The demand for pumice for construction in the city of Xalapa, the capital of Veracruz State, led to large-scale extraction of this raw material across the nearby Perote Valley (Fry 2011). Over time, however, these sites started to meet the demand from more distant metropolitan areas such as Mexico City and Monterrey, respectively, 300 and 1,000 km away. Increased prices for aggregates and declining returns to agriculture are among the several factors that compel farmers to convert their lands to mines in the region. Stripped of their top soil, most of these lands lose their capacity to support agriculture. Many farmers whose farms lose their productivity take off-farm jobs in nearby cities and thus perpetuate a reinforcing feedback loop where the demand for more housing and hence for more construction materials drive further mining activities.

Steel

Steel is even more energy-intensive to produce than concrete. Producing a ton of steel requires about ten to 14 times more energy than producing a ton of concrete (Chaturvedi and Ochsendorf 2004). However, in general, less steel is required to carry the same load than concrete. Worldwide, the production of steel lagged behind that of cement since the early 1970s, resulting in a leveling off of steel per capita, while cement per capita continued to increase. As of 2000, about 271 kg of cement and 139 kg of steel were required per person in the world annually. By 2012, however, per person cement and steel production increased to, respectively, 464 kg per person and 205 kg per person. The global steel production is estimated to have been about 1.6 Mt in 2012. This is more than twice the global production in 1995, the year after which steel production started to climb (Figure 6.1). In particular, China's contribution to global steel production gradually rose from 30 percent in 2005 to 48 percent in 2012 (Table 6.1). In the near future, steel use in buildings in the country may substantially increase or decrease depending on the evolution of the urban form (e.g. more high-rise buildings that require more steel), technical innovations and the overall steel production capacity of the country (Hu et al. 2010).

Almost half of the global steel production is used in construction (Wang et al. 2007). In 2004, the total amount of steel produced for construction was nearly 24 Mt (about 47 percent of the country's total steel production that year) in the United States (AISI 2005, in Pauliuk et al. 2013b). The same amounts were 3,319 t for the UK (41 percent of the country total) for the year 2000 (Dahlström et al. 2004) and about 10 Mt per year (47 percent of the country total) for India in the late 1990s (SSER not dated, in Pauliuk et al. 2013b). Overall, per capita in-use steel stocks range between six and 16 tons in the industrialized countries (about 14 tons in the United States) as of 2008, and many of these countries show saturation or signs of saturation around 10±2 tons for construction (slightly under 10.5 tons in the United States). Pauliuk et al. (2013a) forecast that the timing of saturation will vary across the world regions based on their development patterns. For example, while China is expected to reach saturation by 2050, this is not likely to happen for India until 2150. On the other hand, saturation is expected as early as 2020 for North America and the high-income countries in Asia. Another study posits that the construction sector will account for 64 percent of the projected increase in global steel demand of 2,350 Mt by 2025 (de Carvalho and Daniel 2014). Demand from residential and infrastructure projects in emerging economies will be responsible for the bulk of this increase.

Brick

Although there are no reliable time-series data on global brick production, it is estimated that around 300,000 brick kilns worldwide produce about 1.5 trillion bricks per year (Baum 2012). China accounts for over two-thirds of all the brick production in Asia, which makes up the largest share of the global production.

In Asia and Africa, bricks have been traditional building materials that were typically produced in brick kilns scattered around towns and cities. Concomitant with the rapid urbanization in these parts of the world in recent decades, demand for bricks also increased, which led to further concentration of brick production near the places where they are ultimately used (Singh and Asgher 2005; Co et al. 2009; Ahmad et al. 2012). For example, the area surrounding the city of Aligarh, India, has experienced extensive loss of farmland through quarrying of fertile top soil to make bricks since the early 1980s (Singh and Asgher 2005). Santhosh et al. (2012) report similar dynamics at Kochi city in Central Kerala, India. In a set of processes that mirror the pumice extraction dynamics in the Perote Valley, as the demand for bricks grows, many farmers lease their land to brick-makers. At the end of the lease period, the lands become degraded, unsuitable for agriculture. Most farmers who leased their land end up abandoning farming, some becoming brick-makers themselves, in a process that facilitates further land changes around the city.

As its production climbed, brick-making has become a significant source of air pollution around growing cities in many low- to middle-income countries (Schmidt 2013). In addition to its adverse impacts on human health, the pollutants emitted to the air from brick kilns can also harm nearby vegetation. Poorly regulated brick kilns in Peshawar, Pakistan, emit large amounts of fluoride that damage nearby orchards, causing significant harm to the livelihoods of farmers settled around the city (Ahmad et al. 2012). Similar problems have been reported elsewhere in Vietnam (Co et al. 2009), Kathmandu Valley, Nepal (Haack and Khatiwada 2007), and Khartoum, Sudan (Abdalla et al. 2012). Over a thousand brick kilns in the outskirts of Dhaka Metropolitan Area, Bangladesh, are responsible for 30 percent of the city's fine particulate air pollution (Begum et al. 2011). In all cases, the brick-making facilities around these rapidly growing cities proliferated as a result of the availability of fertile alluvial soils in the immediate hinterland of the cities, quick financial returns from brick-making over farming and non-existent or poorly enforced regulations.

Wood

Illegal trade makes accurate characterization of the flow of timber and other wood products particularly challenging (Brack 2003). This is especially critical in the case of China, which increasingly satisfies its demand for wood products through imports, 70 percent of which come from the Asia-Pacific region (Katsigris et al. 2004). As of 2003, over 70 percent of the timber consumption in China was due to construction and housing (Sun et al. 2005).

From a life-cycle perspective, wood is less energy-intensive than either concrete or steel. A study comparing the environmental costs of wood with concrete and steel in housing construction finds that houses that are built mainly from wood appear to have the lowest embodied energy levels of the three materials; 232 GJ (Gigajoules) of embodied energy in a wood house compared to 396 GJ and 553 GJ in a concrete house and in a steel house, respectively (Glover et al. 2002). Although there are significant uncertainties in estimates, there seems to be a general consensus that concrete-frame buildings are more energy-intensive than wood-frame buildings (Lenzen and Treloar 2002).

Operation of the built environment

The energy demands for the operation of buildings, particularly space heating, cooling and lighting, constitute a significant proportion of total energy use in many parts of the world. For example, in the United States, energy consumption from buildings (residential and commercial combined) increased 3.8 times since 1950 to about 38 QBtu (quadrillion BTU) in 2012, over 90 percent of which came from fossil fuel sources (EIA 2013). From 1980 to 2009, the percentage of households with air conditioning (AC) units increased from 56 to 86 percent and virtually all households had space heating. The household energy consumption for space heating and AC decreased from 7.2 to 5.2 QBtu from 1978 to 2005, which means that per capita energy consumption decreased from 32.1 MBtu (Million BTU) per capita to 17.4 MBtu per capita over the same time period (EIA 2013; UN 2013).

In China, the world's second-largest energy user next to the United States, the total residential energy consumption increased from 92 Mtce (Million tonnes of carbon equivalent) in 1980 to 385 Mtce in 2011 nearly 60 percent of which was urban residential consumption (CED 2013). Likewise, the total commercial energy consumption increased from 4.93 Mtce to 89.7 Mtce between these two years. Consequently, the share of fossil fuels as a source for residential and commercial energy demand dropped, respectively, from 98.8 and 95 percent in 1980 to 88.1 and 87.4 percent in 2011. In the meantime, energy efficiency in the country's residential and commercial sectors increased from 29.1 percent in 1980 to 66.2 percent in

2000 (CED 2013). In contrast, the number of AC units per 100 households increased from 0.08 in 1985 to 122 in 2011.

Around the world, the changes in incomes and urbanization levels as well as technological improvements and climatic changes are expected to bring large changes in energy demand for operation of the built environment. Buildings accounted for about a third of total global final energy use in 2010 (Lucon et al. 2014). Isaac and van Vuuren (2009) forecast that by the end of the century, the energy demand for cooling will increase by 72 percent while that for heating will decrease by 34 percent. The warming trends projected for much of the planet are responsible for these changes, especially in regions such as South Asia where cooling energy demand is forecasted to increase by 50 percent due to climate change. If current practices persist, global building final energy use could double or even triple by the mid-21st century. The median of the forecasts of the integrated model scenarios point to an approximate 75 percent increase in final energy use by 2050 compared to 2010 (Lucon et al. 2014). However, rather than doubling or even tripling, final energy use may virtually remain constant or even decline by mid-century, if existing cost-effective best practices and technologies are broadly adopted. Deep retrofits on existing buildings worldwide resulted in energy savings of 50 to 90 percent (Lucon et al. 2014). Moreover, in many parts of the world, new buildings could be 70 percent more efficient than existing ones (OECD/IEA 2006). There is also empirical evidence that behavior, lifestyle and cultural practices can have an effect on buildings' energy use comparable to that from technologies and architecture. Such social factors may account for differences of up to three to five times in the energy use, all other things being equal (Lucon et al. 2014).

Energy use in the operation of the built environment is significantly affected by the composition and efficiency of the urban infrastructure (Boyle et al. 2010). For example, compact urban areas can increase the feasibility of such energy-efficient options as district heating and combined heat and power. In addition, there is evidence that efficiency in material and energy requirements to operate the built environment increases with increasing size of urban areas (Bettencourt et al. 2007). Up to three billion more middle-class consumers, most of them in low- and middle-income countries, will emerge in the next 20 years. This will lead to dramatic expansion in the urban infrastructure in these countries (MGI 2011). What types of urban planning and development are encouraged will greatly influence the types and amounts of infrastructure being built (Kennedy and Corfee-Morlot 2013).

A more holistic, landscape-oriented approach that integrates waste, water, transport, food and energy flows is important to attain a robust infrastructure framework (Belanger 2009). Differences are present in the implementation of green or integrated infrastructure initiatives depending on the specific sustainability objectives adopted by cities in different geographical contexts (Mattingly and Winarso 2002; Marcucci and Jordan 2013; Mell 2013). For example, although concepts from green infrastructure serve as guidelines in conservation planning in both the United States and Europe (Gill et al. 2007; Mell 2010; Marcucci and Jordan 2013), there is a significant difference between the two. While engineering-oriented solutions and economic considerations still linger in the former (Newell et al. 2013), in the latter, the approach is more holistic encompassing social, economic and environmental concerns concurrently (Marcucci and Jordan 2013; Mell, 2013).

Addressing many of the issues revolving around the integrated infrastructure initiatives requires an interdisciplinary perspective including a robust understanding of social, cultural and institutional dimensions of the problem, in addition to those more technological in nature (Mattingly and Winarso 2002; Wilbanks 2003; Burch 2011; Marcucci and Jordan 2013; Mell 2013). Bringing in infrastructure capacity and cost considerations to land suitability assessments early on in the development planning process is critical in integrated planning of land use and infrastructure. This allows determining in advance whether there are locations with spare infrastructure capacity, hence that are more feasible to develop. Ensuring social equity or social inclusiveness emerges as a critical component in this respect, not only for low- and middle-income country cities (Lebel et al. 2007; Ness 2008; Jordán and Infante 2012), but also those in high-income countries (Lindsey et al. 2001; Mell 2013; Newell et al. 2013).

Placing resource use efficiency in a broader context

Efficiency gains in energy and material use can substantially reduce resource demands and associated environmental impacts at the scale of an individual building (Gustavsson and Sathre 2006; Fernández 2007). These gains are typically achieved through technological improvements and innovative features in the design of the building, but ignore the spatial configuration of urban land use. However, if one takes the metropolitan region as the spatial scale of interest beyond the individual building, the sheer scale and the spatial configuration of the built environment emerges as a key consideration in resource use (Marshall 2008; Kohler et al. 2009). In their case study of a rapidly urbanizing region in South China, Güneralp and Seto (2012) analyze how improved efficiencies in construction at the building scale compare to the rate of urban expansion at the metropolitan scale. They find that, in spite of the efficiency gains in building construction and operation, both energy consumption and CO_2 emissions continued to rise during the study period.

Land use planning and urban form can markedly influence energy consumption and resulting GHG emissions (Lucon et al. 2014). Compact urban form tends to reduce energy consumption due to several factors including lower per capita floor area, reduced building surface to volume ratio, increased shading, and more opportunities for district heating and cooling systems (Ürge-Vorsatz et al. 2012). Although greater compactness may exacerbate urban heat island effect in warmer climates, the overall impact of increased compactness appears to be reducing GHG emissions due to lower energy demand from buildings (Wheeler 2003; Ewing and Rong 2008; Sovacool and Brown 2010). Changes in urban form are typically influenced by the energy infrastructure. Therefore, improving the infrastructure that supplies energy to buildings can result in large emissions reductions in addition to those that can be attained through demand reductions (Lucon et al. 2014).

Increasingly, urban practitioners are implementing efficiency at these two scales. Leadership in Energy and Environmental Design (LEED) certification developed by the US Green Building Council (USGBC) is perhaps the best known example encouraging energy and material use efficiency at the building scale; on the other hand, its lesser known counterpart, LEED for Neighborhood Development Rating System (LEED-ND), takes the neighborhood or urban region as the focal point in determining the overall efficiency of resource use in the construction and operation of the built environment (United States Green Building Council 2010; Talen et al. 2013). Also, Building Research Establishment Environmental Assessment Methodology (BREEAM), originated in the UK in 1990, is the world's oldest and most widely used method of certification to assess the sustainability of buildings (BRE 2014). Like LEED, BREEAM also aims to raise awareness about sustainable, more energy- and materials-efficient solutions in the construction and operation of buildings.

Initiatives such as the LEED, BREEAM, and most studies on urban form and energy use have so far primarily focused on cities in industrialized countries (Norman et al. 2006; LADWP 2007; Ewing and Rong 2008; Marshall 2008). The situation, however, is changing with the increasing number of publications and reports on low- and middle-income country cities. In addition, many large cities have efficiency programs and building standards in place to promote efficient use of energy in buildings (Sovacool and Brown 2010). Although low- and middle-income country cities generally have relatively high population densities compared to their counterparts in high-income countries, there is evidence that these density levels will likely decline into the future (Angel et al. 2011). If the urbanization process results in such lower density development, this can have significant implications for the overall demand for resources to construct and operate the built environment.

Significant uncertainty also exists regarding the extent to which efficiency gains will materialize in the construction sector due to the regionally varying influence of finances, institutional and regulatory structures, and resource availability (OECD/IEA 2007; WBCSD/IEA 2010). In general, however, new international initiatives that can spur systemic eco-innovations in key areas such as cement and steel production are of critical importance to improve resource productivity (Fischedick et al. 2014). On the other hand, high-income countries have most of their built environment along with their infrastructure already in place, which are—at varying levels—inefficient in their energy use. Therefore, their challenge is to retrofit their existing built environment through employing new technologies coupled with limited changes in the spatial configuration to attain improved efficiencies in energy use for the operation of the built environment.

Buildings and other infrastructure, once constructed, can remain in service for decades. This phenomenon, called lock-in, can significantly limit how quickly the emissions in the use phase of buildings can be reduced (Davis et al. 2010). Moreover, Davis et al. (2010) estimate that the cumulative emissions that would result from a continued expansion of fossil fuel-based infrastructure would range from about 3,000 to 7,400 Gt (Gigton) of CO_2 by the end of the 21st century. This implies that devices and infrastructure that did not yet exist in 2010 will be the major cause of atmospheric concentrations reaching levels greater than 600 ppm (parts per million) (Davis et al., 2010). Likewise, Müller et al. (2013) estimate that the GHG emissions of the global infrastructure would increase almost four-fold to about 470 Gt CO_2-eq by 2050. This is about half to one-third of the cap on cumulative emissions that is not to be exceeded during the 2000–2050 period to prevent average global temperature rise to 2°C above pre-industrial levels (Meinshausen et al. 2009).

Other studies also indicate that, without attaining the highest efficiency levels, global building energy use will continue to rise (Ürge-Vorsatz et al. 2012). These studies estimate that by 2050, due to lock-in, the global building heating and cooling final energy use could reach almost 80 percent of what it was in 2005. This increase in energy use due to lock-in varies notably across regions; for example, the increase in heating and cooling energy use during the same time period is estimated to be 200 percent in Southeast Asia (including India) (Ürge-Vorsatz et al. 2012). There are serious challenges in front of achieving technologically feasible efficiencies. Nevertheless, there are also potential opportunities such as policies, regulations and innovative financing mechanisms (Lucon et al. 2014). The nature of these challenges and opportunities, however, vary by location, type of building and socio-cultural factors.

Gaps in knowledge

In the first three decades of the 21st century, the global urban land cover is expected to increase by over 200 percent whereas the global urban population is expected to grow by over 70 percent (Fragkias et al. 2013). Importantly, most of these increases will take place in low- and middle-income countries. Given the likely scale of global urbanization in the near future, the level of needed efficiency gains will be much greater than those attained in the past to offset and reverse the increasing trends in resource consumption from the construction and operation of the built environment. Thus, a comprehensive resolution to resource demands and accompanying environmental impacts of urbanization requires paying equal attention to the spatial scale (and form) of urban development and to efficiency gains. Importantly, scenarios and forecasts of resource demands for the construction and operation of the built environment should take into account the physical make-up of settlements, which can have significant influence on these resource demands. An important future research venue is evaluating how urban form influences the building material demand and energy demand for heating and cooling at the metropolitan scale. There are currently no comprehensive data available at the relevant spatial and temporal scales to carry out such an evaluation, at least for rapidly urbanizing countries such as China and India. Innovative use of data from various remote sensing sensors (Frolking et al. 2013) as well as novel geospatial approaches (Tanikawa and Hashimoto 2009) can be put to use to track building stocks at urban scales.

Systems-oriented and analytic approaches that take the spatiality of urban areas into consideration are needed to tackle cumulative resource requirements for the construction and operation of the built environment (Güneralp and Seto 2012). Such approaches are also effective in more completely accounting for the various environmental and social impacts throughout the life cycles of the materials and energy used in the construction and operation of the built environment (Brattebo et al. 2009; Lifset and Eckelman 2013; Yücel 2013). To trace the flow of resources from where they are mined or extracted to their destinations require innovative and complementary use of existing methodological approaches (Güneralp et al. 2013). Examples of such analytical methods that can track the flows and accumulations of materials and information include spatially explicit consequential lifecycle assessment (LCA) (Guggemos and Horvath 2005), commodity chains analysis and material flow analysis (MFA) (Hu et al. 2010). For a further discussion of the mapping of energy, water and resource flows see Pincetl (Chapter 33 in this volume). In addition, entropy-based approaches for measuring the accumulations and flows of materials and energy (Bejan 1996; Ayres et al. 1998) take into account not only the quantity of materials and energy, but also their quality (Gößling-Reisemann 2008). Notwithstanding their promise as more comprehensive measures of resource consumption, these methods have not yet seen wide-spread acceptance beyond a relatively limited academic circle.

Another significant gap is quantification of specific land use impacts of raw material extraction for building materials. There are methodological advances in this direction; yet, practical application is still some distance away (Habert et al. 2010). The studies on the flow of cement, steel, and timber highlight a knowledge gap in information on destinations and origins of material flows to construct the built environment. Tracking the flow of resources and energy used for construction from their sources to demand centers is still extremely challenging, not because the data do not exist, but because it is fragmented among countless companies and firms involved in their manufacture and trade; these data are also typically regarded as trade secrets.

Perhaps a bigger challenge beyond tracking the movement of physical quantities of resources and materials is bringing in social, institutional, and cultural factors that ultimately drive and influence the flows of these resources from where they remained for millennia to their points of use in buildings (Fry 2013). Nonetheless, more detailed representations of various drivers of urbanization, and of changes in consumption patterns with income growth and urbanization are urgently needed. Improving the representation of feedbacks among the components of the social, ecological and infrastructural systems of settlements are critical to identify policy levers for climate adaptation and mitigation (Ramaswami et al. 2012; Fry 2013; Lifset and Eckelman 2013). The quantification of relative influences of these multiple factors on resource demands would be an important step towards reaching full accounts of the impacts and implications of resource use in construction and operation of the built environment.

Main messages/summary

- Expansion of urban areas in the form of new factories, residential and business districts, and infrastructure will have significant impacts on energy use, GHG emissions and ultimately sustainability.
- A comprehensive resolution to resource demands and accompanying environmental impacts of urbanization requires paying equal attention to the spatial scale (and form) of urban development and to efficiency gains.
- Significant uncertainty exists regarding the extent to which efficiency gains will materialize in the construction sector due to the regionally varying influence of finances, institutional and regulatory structures, and resource availability.

Needs for further research

- Systems-oriented and analytic approaches that take the spatiality of urban areas into consideration are needed to tackle cumulative resource requirements for the construction and operation of the built environment.
- Innovative approaches are needed to quantify specific land use impacts of raw material extraction for building materials.
- An important future research avenue is evaluating how urban form influences the building material demand and energy demand for heating and cooling at the metropolitan scale.

References

Abdalla, I., S.B. Abdalla, K. El-Siddig, D. Möller and A. Buerkert (2012) Effects of red brick production on land use, household income, and greenhouse gas emissions in Khartoum, Sudan. *Journal of Agriculture and Rural Development in the Tropics and Subtropics* 113(1): 51–60.

Ahmad, M.N., L.J.L. Van Den Berg, H.U. Shah, T. Masood, P. Büker, L. Emberson and M. Ashmore (2012) Hydrogen fluoride damage to vegetation from peri-urban brick kilns in Asia: A growing but unrecognised problem? *Environmental Pollution* 162: 319–324.

AISI (American Iron and Steel Institute) (2005) Annual statistical report of the American Iron and Steel Institute. American Iron and Steel Institute, Washington, DC: s.n.

Angel, S., J. Parent, D.L. Civco, A. Blei and D. Potere (2011) The dimensions of global urban expansion: Estimates and projections for all countries, 2000–2050. *Progress in Planning* 75(2): 53–107.

Ayres, R.U., L.W. Ayres and K. Martínas (1998) Exergy, waste accounting, and life-cycle analysis. *Energy* 23(5): 355–363.

Baum, E. (2012) Present status of brick production in Asia. Instituto Nacional de Ecología (INE) Proceedings of the Workshop on public policies to mitigate environmental impact of artesanal brick production. September 4–6, 2012. Retrieved November 3, 2014, from www.ine.gob.mx/cenica-memorias/1111-taller-ladrilleras-2012-eng.

Begum, B.A., S.K. Biswas and P.K. Hopke (2011) Key issues in controlling air pollutants in Dhaka, Bangladesh. *Atmospheric Environment* 45(40): 7705–7713.

Bejan, A. (1996) Entropy generation minimization: The new thermodynamics of finite-size devices and finite-time processes. *Journal of Applied Physics* 79(3): 1191–1218.

Belanger, P. (2009) Landscape as infrastructure. *Landscape Journal* 28(1): 79–95.

Bettencourt, L.M.A., J. Lobo, D. Helbing, C. Kuhnert and G.B. West (2007) Growth, innovation, scaling, and the pace of life in cities. *Proceedings of the National Academy of Sciences* 104(17): 7301–7306.

Blanco, H., M. Alberti, R. Olshansky, S. Chang, S.M. Wheeler, J. Randolph, J.B. London, J.B. Hollander, K.M. Pallagst, T. Schwarz, F.J. Popper, S. Parnell, E. Pieterse and V. Watson (2009) Shaken, shrinking, hot, impoverished and informal: Emerging research agendas in planning. *Progress in Planning* 72(4): 195–250.

Boyle, C., G. Mudd, J.R. Mihelcic, P. Anastas, T. Collins, P. Culligan, M. Edwards, J. Gabe, P. Gallagher, S. Handy, J.J. Kao, S. Krumdieck, L.D. Lyles, I. Mason, R. McDowall, A. Pearce, C. Riedy, J. Russell, J.L. Schnoor, M. Trotz, R. Venables, J.B. Zimmerman, V. Fuchs, S. Miller, S. Page and K. Reeder-Emery (2010) Delivering sustainable infrastructure that supports the urban built environment. *Environmental Science and Technology* 44(13): 4836–4840.

Brack, D. (2003) Illegal logging and the illegal trade in forest and timber products. *International Forestry Review* 5(3): 195–198.

Brattebo, H., H. Bergsdal, N.H. Sandberg, J. Hammervold and D.B. Müller (2009) Exploring built environment stock metabolism and sustainability by systems analysis approaches. *Building Research & Information* 37(5): 569–582.

BRE (2014) What is BREEAM?, Building Research Establishment (BRE). Retrieved September 12, 2014, from www.breeam.org/about.jsp?id=66.

Brechin, G. (2002) *Imperial San Francisco: Urban Power, Earthly Ruin.* Berkeley/Los Angeles: University of California Press, 402 pp.

Bridge, G. (2004) Contested terrain: Mining and the environment. *Annual Review of Environment and Resources* 29: 205–259.

Burch, S. (2011) Sustainable development paths: Investigating the roots of local policy responses to climate change. *Sustainable Development* 19(3): 176–188.

CED (2013) *China Energy Databook v. 8.0.* Berkeley, CA, Lawrence Berkeley National Laboratory.

Chaturvedi, S. and J. Ochsendorf (2004) Global environmental impacts due to cement and steel. *Structural Engineering International* 14: 198–200.

Co, H.X., N.T. Dung, H.A. Le, D.D. An, K. van Chinh and N.T.K. Oanh (2009) Integrated management strategies for brick kiln emission reduction in Vietnam: A case study. *International Journal of Environmental Studies* 66(1): 113–124.

Dahlström, K., P. Ekins, J. He, J. Davis and R. Clift (2004) Iron, Steel and Aluminium in the UK: Material Flows and their Economic Dimensions. Executive Summary Report, April. Centre for Environmental Strategy, University of Surrey, Guildford/Policy Studies Institute, London.

Davis, S.J., K. Caldeira and H.D. Matthews (2010) Future CO_2 emissions and climate change from existing energy infrastructure. *Science* 329(5997): 1330–1333.

DBS Vickers Limited (2013) China/Hong Kong Industry Focus: China Cement Sector. Hong Kong, 43 pp.

de Carvalho, A. and L. Daniel (2014) Outlook for the steel market. *Sherpa group meeting.* Retrieved 27 October 2014, from http://ec.europa.eu/transparency/regexpert/index.cfm?do=groupDetail. groupDetailDoc&id=12706&no=3.

EIA (U.S. Energy Information Administration) (2013) Annual Energy Review 2012. Washington, DC: Office of Energy Statistics. U.S. Department of Energy.

Ewing, R. and F. Rong (2008) The impact of urban form on U.S. residential energy use. *Housing Policy Debate* 19(1): 1–30.

Fernández, J.E. (2007) Resource consumption of new urban construction in China. *Journal of Industrial Ecology* 11(2): 99–115.

Fischedick, M., J. Roy, A. Abdel-Aziz, A. Acquaye, J.M. Allwood, J.-P. Ceron, Y. Geng, H. Kheshgi, A. Lanza, D. Perczyk, L. Price, E. Santalla, C. Sheinbaum, and K. Tanaka (2014) Chapter 10: Industry. In *Climate Change 2014: Mitigation of Climate Change. Contribution of Working Group III to the Fifth Assessment Report of the Intergovernmental Panel on Climate Change.* Edited by: O. Edenhofer, R. Pichs-Madruga, Y. Sokona, E. Farahani, S. Kadner, K. Seyboth, A. Adler, I. Baum, S. Brunner, P. Eickemeier, B. Kriemann, J. Savolainen, S. Schlömer, C. von Stechow, T. Zwickel and J.C. Minx, Cambridge, UK and New York, NY, USA: Cambridge University Press.

Fragkias, M., B. Güneralp, K.C. Seto and J. Goodness (2013) A synthesis of global urbanization projections. In *Urbanization, Biodiversity and Ecosystem Services: Challenges and Opportunities.* Edited by: T. Elmqvist, M. Fragkias, J. Goodness, B. Güneralp et al., Netherlands: Springer, 409–435 pp.

Frolking, S., T. Milliman, K.C. Seto and M.A. Friedl (2013) A global fingerprint of macro-scale changes in urban structure from 1999 to 2009. *Environmental Research Letters* 8. doi: 10.1088/1748-9326/8/2/024004

Fry, M. (2011) From crops to concrete: Urbanization, deagriculturalization, and construction material mining in central Mexico. *Annals of the Association of American Geographers* 101(6): 1285–1306.

――――― (2013) Cement, carbon dioxide, and the 'necessity' narrative: A case study of Mexico. *Geoforum* 49(0): 127–138.

Gill, S.E., J.F. Handley, A.R. Ennos and S. Pauleit (2007) Adapting cities for climate change: The role of the green infrastructure. *Built Environment* 33(1): 115–133.

Glover, J., D.O. White and T.A.G. Langrish (2002) Wood versus concrete and steel in house construction: A life cycle assessment. *Journal of Forestry* 100(8): 34–41.

Godfried, A., A.R. Pearce and C.J. Kibert (1998) Sustainable construction in the United States of America: A perspective to the year 2010. CIB-W82 Report, Georgia Institute of Technology, Atlanta, GA.

Gößling-Reisemann, S. (2008) What is resource consumption and how can it be measured? Theoretical considerations. *Journal of Industrial Ecology* 12(1): 10–25.

Guggemos, A.A. and A. Horvath (2005) Comparison of environmental effects of steel- and concrete-framed buildings. *Journal of Infrastructure Systems* 11(2): 93–101.

Güneralp, B. and K.C. Seto (2012) Can gains in efficiency offset the resource demands and CO_2 emissions from constructing and operating the built environment? *Applied Geography* 32: 40–50.

Güneralp, B., K.C. Seto and M. Ramachandran (2013) Evidence of urban land teleconnections and impacts on hinterlands. *Current Opinion in Environmental Sustainability* 5(5): 445–451.

Gustavsson, L. and R. Sathre (2006) Variability in energy and carbon dioxide balances of wood and concrete building materials. *Building and Environment* 41(7): 940–951.

Haack, B.N. and G. Khatiwada (2007) Rice and bricks: Environmental issues and mapping of the unusual crop rotation pattern in the Kathmandu Valley, Nepal. *Environmental Management* 39(6): 774–782.

Habert, G., Y. Bouzidi, C. Chen and A. Jullien (2010) Development of a depletion indicator for natural resources used in concrete. *Resources, Conservation and Recycling* 54(6): 364–376.

Horvath, A. (2004) Construction materials and the environment. *Annual Review of Environment & Resources* 29(1): 181–204.

Hu, M., S. Pauliuk, T. Wang, G. Huppes, E. van der Voet and D.B. Müller (2010) Iron and steel in Chinese residential buildings: A dynamic analysis. *Resources, Conservation and Recycling* 54(9): 591–600.

ICR (2013) The Global Cement Report: World Overview: 10th Edition. London: Tradeship Publications Ltd.

IEA (2009) Technology Roadmap: Cement. International Energy Agency and World Business Council for Sustainable Development, 36 pp.

Isaac, M. and D.P. van Vuuren (2009) Modeling global residential sector energy demand for heating and air conditioning in the context of climate change. *Energy Policy* 37(2): 507–521.

Jordán, R. and B. Infante (2012) A strategic planning approach for developing eco-efficient and socially inclusive urban infrastructure. *Local Environment* 17(5): 533–544.

Katsigris, E., G.Q. Bull, A. White, C. Barr, K. Barney, Y. Bun, F. Kahrl, T. King, A. Lankin, A. Lebedev, P. Shearman, A. Sheingauz, Y. Su and H. Weyerhaeuser (2004) The China forest products trade: Overview of Asia-Pacific supplying countries, impacts and implications. *International Forestry Review* 6(3–4): 237–253.

Kennedy, C. and J. Corfee-Morlot (2013) Past performance and future needs for low carbon climate resilient infrastructure: An investment perspective. *Energy Policy* 59: 773–783.

Kohler, N., P. Steadman and U. Hassler (2009) Research on the building stock and its applications. *Building Research & Information (Special Issue: Research on Building Stocks)* 37(5): 449–454.

LADWP (2007) GREEN LA: An Action Plan to lead the Nation in Fighting Global Warming. Los Angeles Department of Water and Power pp.

Lebel, L., P. Garden, M.R.N. Banaticla, R.D. Lasco, A. Contreras, A.P. Mitra, C. Sharma, H.T. Nguyen, G.L. Ooi and A. Sari (2007) Integrating carbon management into the development strategies of urbanizing regions in Asia: Implications of urban function, form, and role. *Journal of Industrial Ecology* 11(2): 61–81.

Lenzen, M. and G. Treloar (2002) Embodied energy in buildings: Wood versus concrete—reply to Börjesson and Gustavsson. *Energy Policy* 30(3): 249–255.

Levine, M.D. and N.T. Aden (2008) Global carbon emissions in the coming decades: The case of China. *Annual Review of Environment and Resources* 33(1): 19–38.

Lifset, R. and M. Eckelman (2013) Material efficiency in a multi-material world. *Philosophical Transactions of the Royal Society A: Mathematical, Physical and Engineering Sciences* 371(1986). doi: 10.1098/rsta.2012.0002

Lindsey, G., M. Maraj and S. Kuan (2001) Access, equity and urban greenways: An exploratory investigation. *Professional Geographer* 53(3): 332–346.

Liu, X., G.J.D. Hewings and S. Wang (2009) Evaluation on the impacts of the implementation of civil building energy efficiency standards on Chinese economic system and environment. *Energy and Buildings* 41(10): 1084–1090.

Lucon O., D. Ürge-Vorsatz, A. Zain Ahmed, H. Akbari, P. Bertoldi, L.F. Cabeza, N. Eyre, A. Gadgil, L.D.D. Harvey, Y. Jiang, E. Liphoto, S. Mirasgedis, S. Murakami, J. Parikh, C. Pyke, and M.V. Vilariño (2014) Chapter 9: Buildings. In *Climate Change 2014: Mitigation of Climate Change. Contribution of Working Group III to the Fifth Assessment Report of the Intergovernmental Panel on Climate Change*. Edited by: O. Edenhofer, R. Pichs-Madruga, Y. Sokona, E. Farahani, S. Kadner, K. Seyboth, A. Adler, I. Baum, S. Brunner, P. Eickemeier, B. Kriemann, J. Savolainen, S. Schlömer, C. von Stechow, T. Zwickel and J.C. Minx. Cambridge, United Kingdom and New York, NY, USA: Cambridge University Press.

Marcucci, D.J. and L.M. Jordan (2013) Benefits and challenges of linking green infrastructure and highway planning in the United States. *Environmental Management* 51(1): 182–197.

Marshall, J.D. (2008) Energy-efficient urban form. *Environmental Science & Technology* 42(9): 3133–3137.

Maruyama, N. and M.J. Eckelman (2009) Long-term trends of electric efficiencies in electricity generation in developing countries. *Energy Policy* 37(5): 1678–1686.

Mattingly, M. and H. Winarso (2002) Spatial planning in the programming of urban investments: The experience of Indonesia's Integrated Urban Infrastructure Development Programme. *International Development Planning Review* 24(2): 109–125.

Meinshausen, M., N. Meinshausen, W. Hare, S.C.B. Raper, K. Frieler, R. Knutti, D.J. Frame and M.R. Allen (2009) Greenhouse-gas emission targets for limiting global warming to 2°C. *Nature* 458(7242): 1158–1162.

Mell, I.C. (2010) Green infrastructure: Concepts, perceptions and its use in spatial planning. School of Architecture, Planning and Landscape, Newcastle University, 291 pp.

———— (2013) Can you tell a green field from a cold steel rail? Examining the "green" of Green Infrastructure development. *Local Environment* 18(2): 152–166.

Meyfroidt, P., T.K. Rudel and E.F. Lambin (2010) Forest transitions, trade, and the global displacement of land use. *Proceedings of the National Academy of Sciences* 107(49): 20917–20922.

MGI (2011) Resource revolution: Meeting the world's energy, material, food, and water needs. Summary. McKinsey Global Institute, M. Kinsey&Company, 20 pp.

Misra, K.K., P.K. Mittal, S. Kalra, J. Janet Barber and E. Hamilton-Smith (2002) Management of land use, landscape and biodiversity. Substudy 11. World Business Council for Sustainable Development. Retrieved November 22, 2013, from www.wbcsd.org/web/projects/cement/tf5/final_report11.pdf.

Müller, D.B., G. Liu, A.N. Løvik, R. Modaresi, S. Pauliuk, F.S. Steinhoff and H. Brattebø (2013) Carbon emissions of infrastructure development. *Environmental Science and Technology* 47(20): 11739–11746.

Munro, R.R. (1984) Urban mining – recycling concrete and asphalt. *Mining Engineering*, Society of Mining Engineers of AIME.

Neri, A.C. and L.E. Sánchez (2010) A procedure to evaluate environmental rehabilitation in limestone quarries. *Journal of Environmental Management* 91(11): 2225–2237.

Ness, D. (2008) Sustainable urban infrastructure in China: Towards a Factor 10 improvement in resource productivity through integrated infrastructure systems. *International Journal of Sustainable Development and World Ecology* 15(4): 288–301.

Newell, J.P., M. Seymour, T. Yee, J. Renteria, T. Longcore, J.R. Wolch and A. Shishkovsky (2013) Green Alley Programs: Planning for a sustainable urban infrastructure? *Cities* 31: 144–155.

Norman, J., H.L. MacLean and C.A. Kennedy (2006) Comparing high and low residential density: Life-cycle analysis of energy use and greenhouse gas emissions. *Journal of Urban Planning and Development* 132(1): 10–21.

Oda, J., K. Akimoto, T. Tomoda, M. Nagashima, K. Wada and F. Sano (2012) International comparisons of energy efficiency in power, steel, and cement industries. *Energy Policy* 44(0): 118–129.

OECD/IEA (2006) Energy technology perspectives: Scenarios and strategies to 2050. OECD and International Energy Agency, I. E. Agency, 479 pp.

———— (2007) Tracking industrial energy efficiency and CO_2 emissions. International Energy Agency, S. Media, France, 321 pp.

Oswalt, P. and T. Rieniets (2006) Atlas of shrinking cities. , Ostfildern, Germany: Hatje Cantz Publishers, 160 pp.

Pauliuk, S., R.L. Milford, D.B. Müller and J.M. Allwood (2013a) The steel scrap age. *Environmental Science and Technology* 47(7): 3448–3454.

Pauliuk, S., T. Wang and D.B. Müller (2013b) Steel all over the world: Estimating in-use stocks of iron for 200 countries. *Resources, Conservation and Recycling* 71(0): 22–30.

Peters, G.P., J.C. Minx, C.L. Weber and O. Edenhofer (2011) Growth in emission transfers via international trade from 1990 to 2008. *Proceedings of the National Academy of Sciences* 108(21): 8903–8908.

Price, O., D. Milne and C. Tynan (2005) Poor recovery of woody vegetation on sand and gravel mines in the Darwin region of the Northern Territory. *Ecological Management and Restoration* 6(2): 118–123.

Ramaswami, A., C. Weible, D. Main, T. Heikkila, S. Siddiki, A. Duvall, A. Pattison and M. Bernard (2012) A social-ecological-infrastructural systems framework for interdisciplinary study of sustainable city systems: An integrative curriculum across seven major disciplines. *Journal of Industrial Ecology* 16(6): 801–813.

Santhosh, V., D. Padmalal, B. Baijulal and K. Maya (2012) Brick and tile clay mining from the paddy lands of Central Kerala (southwest coast of India) and emerging environmental issues. *Environmental Earth Sciences* 68(7): 2111–2121.

Schmidt, C.W. (2013) Modernizing artisanal brick kilns: A global need. *Environmental Health Perspectives* 121(8): A242–A249.

Schneider, M., M. Romer, M. Tschudin and H. Bolio (2011) Sustainable cement production—present and future. *Cement and Concrete Research* 41(7): 642–650.

Singer, M.B., R. Aalto, L.A. James, N.E. Kilham, J.L. Higson and S. Ghoshal (2013) Enduring legacy of a toxic fan via episodic redistribution of California gold mining debris. *Proceedings of the National Academy of Sciences* 110(46): 18436–18441.

Singh, A.L. and M.S. Asgher (2005) Impact of brick kilns on land use/landcover changes around Aligarh city, India. *Habitat International* 29(3): 591–602.

Soule, M.H., J.S. Logan and T.A. Stewart. (2002) Toward a sustainable cement industry: Trends, challenges, and opportunities in China's cement industry. Battelle study for the World Business Council for Sustainable Development. Columbus, OH: Battelle.

Sovacool, B.K. and M.A. Brown (2010) Twelve metropolitan carbon footprints: A preliminary comparative global assessment. *Energy Policy* 38(9): 4856–4869.

SSER (not dated) Spark Steel & Economy Research. Spark Steel & Economy Research. Retrieved November 23, 2013, from www.steelscenario.com/research.

Steinberger, J.K., F. Krausmann and N. Eisenmenger (2010) Global patterns of materials use: A socioeconomic and geophysical analysis. *Ecological Economics* 69(5): 1148–1158.

Sun, X., L. Wang and Z. Gu (2005) China and forest trade in the Asia-Pacific region: Implications for forests and livelihoods. DFID (Department for International Development), 14 pp.

Talen, E., E. Allen, A. Bosse, J. Ahmann, J. Koschinsky, E. Wentz and L. Anselin (2013) LEED-ND as an urban metric. *Landscape and Urban Planning* 119: 20–34.

Tanikawa, H. and S. Hashimoto (2009) Urban stock over time: Spatial material stock analysis using 4d-GIS. *Building Research & Information* 37(5): 483–502.

UN (2013) World population prospects: The 2012 revision, key findings and advance tables. Working Paper No. ESA/P/WP.227. United Nations Department of Economic and Social Affairs/Population Division, New York, New York pp.

United States Green Building Council (2010) Leadership in Energy and Environmental Design (LEED). Retrieved June 24, 2010, from www.usgbc.org/DisplayPage.aspx?CMSPageID=1988.

Ürge-Vorsatz, D., N. Eyre, P. Graham, D. Harvey, E. Hertwich, Y. Jiang, C. Kornevall, M. Majumdar, J.E. McMahon, S. Mirasgedis, S. Murakami and A. Novikova (2012) Chapter 10: Energy End-Use: Buildings. In *Global Energy Assessment: Towards a Sustainable Future*. Cambridge, UK and New York, NY, USA: Cambridge University Press and the International Institute for Applied Systems Analysis, Laxenburg, Austria, 649–760 pp.

USGS (2013) Mineral commodity summaries. United States Geological Survey. Retrieved November 22, 2013, from http://minerals.usgs.gov/minerals/pubs/mcs/.

———— (2014a) Mineral commodity summaries: Cement. Retrieved October 19, 2014, from http://minerals.usgs.gov/minerals/pubs/commodity/cement/index.html.

———— (2014b) Mineral commodity summaries: Iron and steel. Retrieved October 19, 2014, from http://minerals.usgs.gov/minerals/pubs/commodity/iron_&_steel/.

van Oss, H.G. and A.C. Padovani (2003) Cement manufacture and the environment part II: Environmental challenges and opportunities. *Journal of Industrial Ecology* 7(1): 93–126.

Venkatarama Reddy, B.V. and K.S. Jagadish (2003) Embodied energy of common and alternative building materials and technologies. *Energy and Buildings* 35(2): 129–137.

Vennemo, H., K. Aunan, H. Lindhjem and H.M. Seip (2009) Environmental pollution in China: Status and trends. *Review of Environmental Economics and Policy* 3(2): 209–230.

Vermeulen, J. and T. Whitten (1999) Biodiversity and cultural property in the management of limestone resources: Lessons from East Asia. Washington, DC: The International Bank for Reconstruction and Development/The World Bank.

Wallsten, B., A. Carlsson, P. Frändegård, J. Krook and S. Svanström (2013) To prospect an urban mine – Assessing the metal recovery potential of infrastructure "cold spots" in Norrköping, Sweden. *Journal of Cleaner Production* 55(0): 103–111.

Wang, T., D.B. Müller and T.E. Graedel (2007) Forging the anthropogenic iron cycle. *Environmental Science & Technology* 41(14): 5120–5129.

Wang, T., J. Mao, J. Johnson, B. Reck and T. Graedel (2008) Anthropogenic metal cycles in China. *Journal of Material Cycles and Waste Management* 10(2): 188–197.

WBCSD/IEA (2010) Cement technology roadmap 2009: Carbon emissions reductions up to 2050. World Business Council for Sustainable Development (WBCSD) and International Energy Agency (IEA), Corlet, 31 pp.

———— (2013) Technology roadmap low-carbon technology for the Indian cement industry. World Business Council for Sustainable Development (WBCSD) and International Energy Agency (IEA), IEA, India, 53 pp.

Wheeler, S.M. (2003) The evolution of urban form in Portland and Toronto: Implications for sustainability planning. *Local Environment* 8(3): 317–336.

Wilbanks, T.J. (2003) Integrating climate change and sustainable development in a place-based context. *Climate Policy* 3(SUPPL 1): S147–S154.

Worrell, E., R. Smit, G. Phylipsen, K. Blok, F. van der Vleuten and J. Jansen (1995) International comparison of energy efficiency improvement in the cement industry. Proceedings of American Council for an Energy-Efficient Economy (ACEEE), 123–134 pp.

Worrell, E., L. Price, N. Martin, C. Hendriks and L.O. Meida (2001) Carbon dioxide emissions from the global cement industry. *Annual Review of Energy and the Environment* 26(1): 303–329.

Yamasue, E., R. Minamino, T. Numata, K. Nakajima, S. Murakami, I. Daigo, S. Hashimoto, H. Okumura and K.N. Ishihara (2009) Novel evaluation method of elemental recyclability from urban mine – Concept of urban ore TMR. *Materials Transactions* 50(6): 1536–1540.

Yücel, G. (2013) Extent of inertia caused by the existing building stock against an energy transition in the Netherlands. *Energy and Buildings* 56: 134–145.

Zhou, N., M. Nishida and W. Gao (2008) Current status and future scenarios of residential building energy consumption in China. Lawrence Berkeley National Laboratory, Berkeley, CA, 16 pp.

7

HARNESSING URBAN WATER DEMAND

Lessons from North America

Patricia Gober and Ray Quay

Urbanization stresses water resources by disrupting the natural flow of Earth's freshwater systems, concentrating demand and separating modern industrial societies from the natural lakes and rivers that provide water to homes and businesses. Globally, water for urban or domestic uses averages only 10 percent of all withdrawals from nature (agriculture is by far the largest sector for consumptive water use), but is exceedingly important for human welfare (World Water Assessment Programme, 2009). McDonald et al. (2011) estimate that 150 million people now live in cities with persistent water shortage; this number is projected to rise to as much as one billion in 2050. More effective management of urban water demand is vital for climate adaptation, sustainable development, environmental protection and global food security.

The traditional paradigm of urban water resource management in North America and worldwide has emphasized supply—how to secure, store and distribute adequate water resources to meet perceived new demands (Gleick, 2000; Gober, 2013). Focus was on developing physical infrastructure, securing new sources of supply and manipulating the hydrological landscape to meet the needs of population growth and changing lifestyles. Rapid urbanization, more cross-sector competition for water and growing concerns about the environmental and economic costs of major water supply projects (e.g. dams, reservoirs, canals, aqueducts) shifted attention in the 1990s to demand-side management to improve the efficiency and timing of urban water use. Mechanisms for demand management included education, technology, conservation, economic incentives, reuse and regulatory mechanisms (Maas, 2003). Increasing recognition of the interactions between human and biophysical dimensions of global environmental change has, in several important ways, married the more traditional perspectives on supply and demand. A plethora of assessments have addressed the way climate change and population growth will affect water scarcity (O'Hara and Georgakakos, 2008; Vörösmarty et al., 2010). Growing interest in the climate sensitivity of water demand has further blurred the line between supply and demand approaches (Cashman et al., 2010; Christian-Smith et al., 2012; Kiefer et al., 2013). The global change community redirected attention from the drivers of change per se to the governance, legal and institutional frameworks for managing water under conditions of rapid and uncertain societal and environmental change. Also significant is today's increasing attention to the social processes by which scientific knowledge about changing water systems are linked to decision-makers.

These processes have been roundly criticized for failing to effectively communicate new knowledge about changing environmental systems to water managers and representatives from government and civil society (Pahl-Wostl, 2006; Pahl-Wostl and Borowski, 2007; Borowski and Hare, 2007; National Research Council, 2010; Dilling and Lemos, 2011).

This chapter focuses on urban water demand—its determinants, linkages to other urban resource systems and sensitivity to manipulation by policy and regulatory frameworks. It relies heavily on evidence from North America where recent trends show both the positive impacts of policy and regulation in reducing per household water demand and the need for a broader, more holistic perspective of urban water management. After a brief review of determinants of urban water demand, the chapter describes recent trends and highlights the linkages between water, land and energy and the need for better coordination across these sectors. Attention then turns to the importance of outdoor water use—why it is hard to measure, why it is important for sustainable development and why it will be difficult to reduce, given current lifestyles and cultural preferences. Conclusions address the relevance of this knowledge for water management in low- and middle-income countries and point to important future research directions for managing water in a time of rapid change and high uncertainty.

Determinants of urban water demand

Urban water demand in North America is affected by a wide range of demographic, economic, behavioral, cultural and policy factors. Included in demand is water for both indoor and outdoor purposes. Major uses of indoor water are for showering and bathing, clothes washing and toilet flushing. Thus, these are the areas where technology, policy and behavioral change have the greatest potential impacts (Figure 7.1). The variation in indoor use has been related to home size (a surrogate for household income and number of toilets), age of household members (small children add less to faucet use than teens and adults) and household size (Mayer et al., 1999). The relationship between household size and water use is not linear, as there are economies of scale for use with increased size, particularly for laundry and dishwashing (Coomes et al., 2010; Howe and Linaweaver, 1967). Wentz and Gober (2007) find the presence of strong neighborhood effects on urban water demand, with nearby districts exhibiting similar use patterns, controlling for socio-demographic status and housing characteristics.

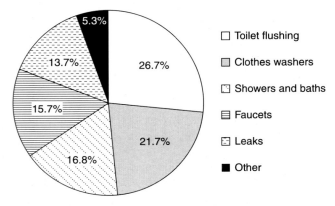

Figure 7.1 Indoor water consumption in the US

Data source: Mayer et al. (1999).

Harnessing urban water demand

Outdoor water is used to irrigate lawns and gardens, support swimming pools and spas, wash cars and maintain outdoor water features such as misters and fountains. Use varies with the size and type of landscaping, presence of pools and climate conditions (DeOreo et al., 2011). Customers in hot, dry climates tend to use more outdoor water than those in wetter and cooler ones (Coomes et al., 2010). Estimates of outdoor use have included regional measures of precipitation during the growing season, minutes of sunshine, wind speed and annual rainfall on household demand (Arbués et al., 2003). Within cities, water use has been shown to vary with urban heat island effects (Guhathakurta and Gober, 2007, 2010). The build-up of heat in impervious city surfaces and its slow release during the nighttime increases water use, especially in neighborhoods adjacent to the city center. Urban heat island impacts on water use suggest that cities facing climate change may need to balance goals of water conservation and temperature amelioration (Gober et al., 2010; Gober et al., 2012).

Heavy outdoor water use can be problematic for urban infrastructure (e.g. pipes, treatment plants and pumping stations) because it is highly peaked during the warm summer months. The monthly production capacity of Portland, Oregon reflects its summer-dry Mediterranean climate (Figure 7.2). Water systems are planned and designed, in large part, for these peak demands. Heavy summertime water use can strain the production capacity and lead to periodic shortages in heavily populated and fully allocated water basins. Hill and Polsky (2007) studied suburbs in the Boston, Massachusetts area and find water shortage conditions during summer, despite precipitation patterns that conform to historical averages. They conclude that 'infrastructure drought' is in fact 'suburban drought' due to high water demands in low-density suburbs. Preferences for artificially irrigated lawns and suburban drought conditions are not limited to North America, however. Del Moral Ituarte and Giansante (2000), and Kallis and Coccossis (2003) link water crisis and drought in Southern Europe to the growth of low-density suburbs. In a study of outdoor water use in 700 California

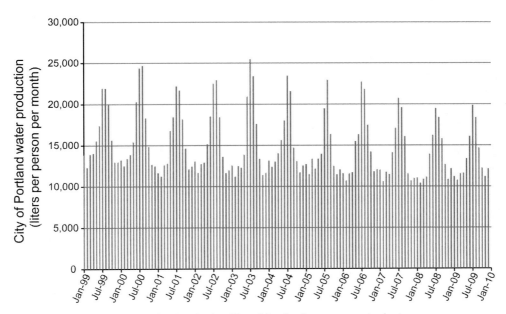

Figure 7.2 Monthly water production in the City of Portland on a per capita basis
Data source: Portland Water Bureau; obtained through public records request.

households, DeOreo et al. (2011) show that excess use of irrigated water occurs on only 18 percent of irrigated lots. From a policy perspective, this means there is significant potential for conservation in outdoor use by changing the behaviors of the relatively few households that are over-irrigating. To the extent that the affluent few in low- and middle-income countries do, in fact, have reliable water supplies and do seek to reproduce urban lifestyles of the industrialized world, conservation efforts can be focused on moderating their behaviors and conservation practices.

The effect of pricing on water demand is a much debated and complex issue in the field of water economics because of its obvious potential for policy intervention and the relevance of pricing structures for efficiency, peak use, equity, public health, financial stability and public acceptability (Arbués et al., 2003). Economic theory suggests that rising prices will reduce demand, but water prices are so low in most North American cities and across the world that they represent a very small portion of total household income. Most consumers thus are unaware of either the rate structure or the magnitude of their personal water use. Outdoor water use is generally assumed to be more price-sensitive than indoor use because it is more discretionary in nature than indoor water uses such as for drinking, cooking, cleaning and sanitation. Summer use is more price-sensitive than winter use for a similar reason (Arbués et al., 2003).

Municipal rate structures are designed to collect revenues and to communicate the value of water to users. Theoretically, rates are a function of (1) the utility's operations and maintenance costs, (2) costs to procure and develop additional water supplies to meet growing demands and (3) the social and environmental 'opportunity costs' of losing other benefits of water in its natural state (e.g. impacts to fisheries, recreational opportunities and watershed health) (Western Resource Advocates, 2013). The European Environmental Agency (2012) criticizes European utilities for setting artificially low prices that do not reflect environmental costs and do not therefore lead to the most efficient use of the resource.

Metering increases the transparency of consumer behavior and is thought to reduce use. It is a key mechanism of water governance and a prerequisite for volumetric pricing, in other words, charging households on the basis of how much water they use. To reflect the importance of transparency in effective urban demand, the number of metered properties in England and Wales rose from 3 percent in 1992–1993 to 40 percent in 2010 (European Environmental Agency, 2012; Ofwat, 2013). Similar large-scale efforts in Canada are thought to have reduced residential water use on a per capita basis (Figure 7.3). Metered households in Canada show reductions in water use, with the greatest savings occurring during the peak-use summer months. In 2009, metered households on volume-based water pricing schemes used 39 percent less water per person than unmetered households on flat-rate water pricing schedules (Environment Canada, 2013). Metering is, by no means, a universal characteristic of water systems in low- and middle-income countries. Most households are not metered, and those that are often produce inaccurate information about actual use (Sharma and Vairavamoorthy, 2009).

Trends in urban water demand

Per capita and per household water consumption has declined in many North American cities during the past several decades (Frost, 2013). Water demand in Seattle (Washington state) declined 36 percent between 1990 and 2009 (Flory, 2013). Per capita water use in Phoenix (Arizona) fell from 541 liters (143 gallons) per capita per day in 1990 to 401 liters (106 gallons) in 2013 (Figure 7.4). In a recent study of residential water consumption in 11 urban

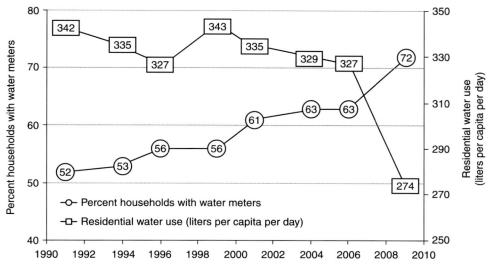

Figure 7.3 Residential metering and per capita consumption in Canada, 1991–2009

Data source: Environment Canada (2013).

water utilities in North America, Coomes et al. (2010) find a slow, but steady, 20-year decline in water use per residential customer. This trend appears to be policy-related (Vickers and Bracciano, 2014). To comply with the USA's Energy Policy Act of 1992, major plumbing manufacturers produced low-volume toilets, urinals, showerheads and faucets. Over the past several decades, these products have slowly appeared in new and remodeled homes. A recent study of per household water use in North America concludes that the steady decline in per household usage is primarily the result of the penetration of water efficient fixtures and appliances into the market (Coomes et al., 2010).

Significant, however, is that these technological and regulatory changes have occurred without obvious lifestyle changes from urban consumers. If anything, higher incomes have

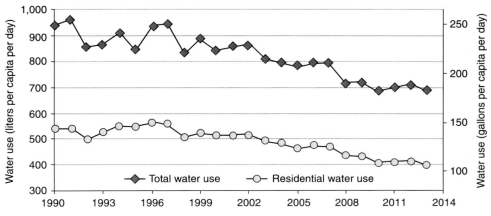

Figure 7.4 Declining per capita water use in Phoenix, 1990–2013

Data source: City of Phoenix Water Services Department; obtained through public records request.

led to larger homes with more fixtures and more outdoor water use, and less sensitivity to price. The installation of lawn irrigation systems, for example, can easily lead to as much water use outdoors as indoors. Coomes et al. (2010) find that 52 percent of Louisville (Kentucky) households consistently water outside landscapes, 4.9 percent have a spa or hot tub filled year-round and 8 percent have a pool. Subsequent statistical analysis shows the substantial impact of landscape watering, pools and spas on household water demand. Garden watering adds 11,356 liters (3,000 gallons) annually, a spa adds 20,441 liters (5,400 gallons) and a pool adds 89,714 liters (23,700 gallons).

Declining overall demand has had unintended consequences for water quality (see Murphy, Chapter 18 in this volume, for a detailed discussion of water quality) and utility revenues. Urban water systems are designed to maintain the quality of water based on anticipated flow rates and pressures. When the operational characteristics change, either the systems characteristics must be restored or water quality must be modified. Problems have also occurred in achieving the required capacity for fire flows, in other words, water to maintain sufficient pressure for fire trucks to pump water from a hydrant, in case of a fire emergency. In some commercial and high-density residential settings, fire flow demands can now exceed domestic water demand and have thus become the size criteria for pipes and other infrastructure.

Increased conservation and efficiency also has led to declining revenues with little institutional and political capacity to raise rates. Frost (2013) notes significant pushback against higher rates among customers, elected officials and utility directors.

Links to land and energy

The view of cities as complex and dynamic environmental systems has raised awareness of critical linkages among water, energy, land and the need for better coordination across these sectors for climate adaptation and sustainable urban development (Vörösmarty et al., 2013). Bates (2011) raises the problem of a 'governance gap' between land use planning and water management, where water management has traditionally been seen as subservient to land planning (the supply-side management problem mentioned earlier). Gober et al. (2013) ask 'why land planners and water managers don't talk to one another and why they should' as a way of calling attention to the fact that unsustainable land use practices have led to unsustainable water demand in many North American cities. Domene and Saurí (2006) make this connection in Barcelona, noting its transformation from a traditionally compact Mediterranean city into a diffuse urbanized region dominated by low-density housing at the urban fringe. This transformation has been accompanied by increasing residential water demand and periodic shortage conditions. Analysis of individual household surveys matched to water records reveals that detached homes on the urban fringe used water at ten times the per-household rate of high-density settings in the urban core. They conclude that water conservation programs focused on pricing, taxation and cutbacks during periodic drought conditions miss the deeper structural cause of water insecurity in Barcelona, which is the cultural preference for irrigated lawns and gardens embedded in modern suburban land use patterns (Domene and Saurí, 2006).

The link between energy and water (the so-called energy-water nexus) also has emerged as a critical element of sustainable urban development (Stockholm International Water Institute, 2014). The consequences of the nexus vary between cities in high-income countries, where most urban residents are served by piped municipal networks that distribute water around the clock, and those in low- and middle-income countries, where both water and power supplies are unreliable and inadequate. Many residents of the latter often lack in-house access to potable water and electricity connections (Sharma and Vairavamoorthy, 2009).

Malik (2002) describes individual coping strategies for dealing with shortages and uncertainties in water and energy in urban India. Faced with restricted and unreliable water deliveries with inadequate pressure, many households augment supply with storage or use of tube wells to draw groundwater for domestic consumption, thereby increasing energy use and reliance on unreliable power delivery systems. Inadequate pressure encourages many residents not living on the ground floor to install booster pumps on the main supply lines. Households thus use energy to compensate for water supply uncertainties. Better coordination between the energy and water sectors tops the list of policy remedies for these wasteful practices, along with improved metering of consumption, eliminating illegal use of both resources, charging rates that cover the cost of operation and maintenance, and creative financing for underfunded water and energy systems. Investments to reduce these inefficiencies produce shared benefits for both water and power sectors.

Positive feedbacks between energy and water present significant challenges for continued urban growth. As much as one quarter to one half of the electricity used by US cities is consumed at municipal water and wastewater treatment plants (Rogers, 2012). Future sources of water supply including desalination and water reuse, have large embedded energy content. In a study of future energy and water needs for the Greater Mekong Subregion in Asia, Rogers (2012) finds that the region will be sorely stressed to meet the electricity demands of the rapidly growing urban water sector.

Outdoor water demand

Recent trends in North America suggest that federal policy has been effective in stimulating technological change favoring more efficient indoor water appliances and gradual reductions in per household water demand (Vickers and Bracciano, 2014; Coomes et al., 2010). These reductions have buffered the effects of population growth and climate uncertainties. Further reductions in demand will, however, need to be focused on outdoor use, as the efficiency gains in indoor use over the past 20 years are unlikely to be replicated over the next 20 years. One way to reduce outdoor demand is through water reuse. Indirect use (e.g. treated to a higher standard for municipal outdoor use) of effluent has grown in popularity in Florida, California and Arizona. In California some utilities are blending reclaimed water with groundwater. In the Phoenix, Arizona region as much as 82 percent of the region's effluent is reused for irrigation or power plant cooling (Middel et al., 2013).

Arid-region cities in the US have developed policies to require or entice homeowners to reduce or eliminate the size of their lawns. Grass front yards are now banned in Las Vegas and street boulevards have been replaced with artificial turf. The Las Vegas Valley Water District (LVVWD) pays up to 1.50 USD per square foot (16.15 USD per square meter) to replace grass (LVVWD, 2013) and has converted more than 160 million square feet (15 million square meters) of grass to less water-hungry landscape treatments (Lovett, 2013). The City of Austin (Texas) has mandatory watering restrictions year-round, limiting hose-end watering and automatic irrigation systems to one day per week after sunset (Austin Water, 2013). Mesa (Arizona) has used incentives to convert 250,000 square feet (23,000 square meters) of residential lawn to desert vegetation. Los Angeles has used a rebate program to reduce lawn coverage by one million square feet (93,000 square meters) and replaced public spaces with drought-tolerant desert plants. The net result is to make landscapes less climate-sensitive, reduce outdoor water use and peak demand and curtail energy use. These efforts have, however, been accompanied by pushback from some residents who eschew the loss of green spaces in their urban neighborhoods (Lovett, 2013).

Further reductions in outdoor water use challenge urban water managers to look beyond these traditional technological and policy solutions to consider the deeper reasons for high outdoor water demand. They are both cultural and behavioral. In *Lawn People*, Robbins (2007) observes that lawns are the fastest growing landscape in the US, requiring more time and care than any other managed landscape. Modern lawns are monoculture landscapes that are heavily dependent on chemicals and fertilizers and are now deeply ingrained in the North American psyche and economy. Cultural preferences for an idealized golfcourse-like appearance lead to heavy application of chemicals and widespread use of artificial watering systems, even in humid regions. Robbins attributes the preference for lawns to tribal practices that encourage people to behave in ways that are accepted by the larger group. High-maintenance, emerald-green lawns are not merely expressions of personal preferences; they are reflections of a larger collective identity. Messages about the value of lawns have been marketed by large seed, fertilizer and equipment companies and reinforced through global supply chains of these products.

Psychologists have begun to explore the reasons why urban residents choose lawns and other greenery for their outdoor gardens. Neel et al. (2014) assert that there are self-presentation

---

Box 7.1 Decision Center for a Desert City—Exploring water complexity and uncertainty in an arid environment

Contributed by: Ray Quay, Dave White and David Sampson, Arizona State University

The Decision Center for a Desert City (DCDC), a center with the Julie Ann Wrigley Global Institute of Sustainability at Arizona State University, is a National Science Foundation (NSF) center funded under its Decision Making Under Uncertainty (DMUU) program since 2004. The goals of NSF's DMUU are to support research, education and outreach that increase basic understanding of decision-making processes and of the information needed by decision-makers; to develop tools to support decision-makers and increase their ability to make sound decisions; and to facilitate interaction among researchers and decision-makers.

DCDC is structured as a boundary organization between academic science and practice, focusing on decision-making under uncertainty for issues related to water resources, regional growth and climate change. DCDC organizes and funds both academic research and extensive stakeholder engagement. Its philosophy is that better research and practice occur when researchers and stakeholders collaborate on key science and policy questions.

Much of the research and engagement is related to the application of an anticipatory governance approach to decision-making under uncertainty. One of the primary tools for such efforts is exploratory scenario analysis including an urban water and demand model for central Arizona— WaterSim—that is spatially explicit to the 32 water utilities in the Phoenix Active Management Area. The model includes a wide range of water resource demand and supply management strategies including demand reduction, effluent reuse, water banking, water transfers and new supply development (such as desalination). WaterSim includes simulations of the Colorado River and Salt River storage and rights allocation as well as the region's aquifer system so that various external factors such as drought and climate change can be modeled.

There are four primary clients using WaterSim: university water policy researchers, secondary and university educators, the general public and water managers.

aspects to outdoor water use. Self-presentation is a mechanism by which people seek to control how others see them. Experiments asked university students and a community sample from a southwestern city to rate/judge the socio-demographic and personal characteristics of a person who preferred certain landscapes, based on landscape descriptions and photographs. Results show that low water use landscapes are associated with lower sexual attractiveness and lower family orientation. Prototypical xeric landscapes are sometimes associated with lower home-owner status in some of the studies and a generally more negative overall evaluation. Even when people's attitudes toward water consumption reflect a desire to be water efficient, their percep-tions of their own water use may not be aligned with their actual water use. In 2007 the City of Phoenix and the Arizona Municipal Water Users Association conducted a telephone survey and linked survey results to customer water records. When comparing their water usage to others, only 5–15 percent of respondents perceive their water use as above average. When com-pared to actual water bills, people who reported reducing their water use did not use any less water than people who reported that they had not reduced their use. Many participants perceive that their indoor water use greatly exceeded their outdoor use even though their summer water use was significantly higher than their winter water use (BBC Research and Consulting, 2007).

Discussion and conclusions

This chapter outlined the determinants and trends in urban water demand, primarily from a North American perspective. Per capita water demand has declined during the past several decades with credible support for the fact that it is in response to federal regulations in the US to mandate water-efficient appliances in new and remodeled homes. As indoor water use transitions toward minimum levels for human survival, emphasis in urban water demand management will shift to outdoor uses. Reducing outdoor use will be challenged, however, by the cultural and psychological underpinnings of high outdoor water use. Higher incomes have allowed many urban residents to achieve lifestyle preferences for high water use outdoor water features. These lifestyles, along with the low-density nature of suburban development, will require new strategies for urban demand management and better coordination of the water and land sectors.

Urban water demand management is beyond the purview of the traditional water sector. Concern for sustainable development, climate adaptation and global environmental change has called attention to links between water, land and energy and for the need to manage these resources in a coordinated way, paying attention to co-benefits, positive feedbacks and unintended consequences.

Trends and relationships identified in this review may or may not be relevant to cities in low- and middle-income countries as the urban demand situation is quite different. These cities experience much higher rates of population growth and thus water demand. They have water supply systems that are incapable of providing round-the-clock services to many urban residents. Leakages are common and inefficiencies are substantial. Retrofitting these cities with water-saving devices may not be feasible in the short-term, as a majority of the population lack flushing toilets, washing machines and dishwashers. Water is underpriced and political considerations continue to favor high-visibility infrastructure development over more technological and behavioral adjustments (Sharma and Vairavamoorthy, 2009).

Although it is not realistic to simply transfer the technology and knowledge of high-income countries to the water agencies and planning departments of cities in low- and middle-income parts of the world, there are larger lessons to be learned from the North American experience over the past two decades. First, neither supply- nor demand-only solutions will produce sustainable water use in the years to come. Water systems are linked to energy and land, and the three resources should be planned and managed conjunctively for sustainable development. Second, managing urban water demand will require deeper knowledge of why people use water the way they do, both inside and outside. As Robbins and others have shown, there are larger, structural forces beyond low prices that dictate high water use. Water demand management will work differently in a place where water is seen as an expression of human identity and a human right, rather than an economic commodity. And third, special attention should be directed to the unintended consequences, especially the financial ones, of policies to reduce demand. Decades of supply-oriented approaches to water development have produced a physical and institutional infrastructure that will need to be transformed to meet the water demands of these rapidly growing and environmentally complex cities.

Content messages

- Trends in North America show a widespread and consistent decline in residential water demand coincident with the development of high efficiency appliances and federal regulations that they be mandated in new housing and renovations. As indoor consumption declines, increasing attention will shift to outdoor demand management.
- There is a paucity of knowledge about the determinants and trends in outdoor water use. Outdoor demand is, however, climate sensitive thus making North American cities more vulnerable to climate change and its uncertainties, as residential use patterns shift from indoors to outdoors. Outdoor demand may not be easily manipulated by technical solutions, as it is strongly linked to lifestyle choices and land development patterns.
- Water is not a standalone resource in the urban environment but a key factor linking biophysical and human processes in complex and dynamic urban systems. Understanding these linkages, and their economic and social implications, is an important prerequisite for policy and decision-making.
- There may be unintended consequences associated with improved water efficiencies and reductions in per household water use. Although these methods can buffer city water supplies in the face of population growth and climate uncertainties, they may trigger the need for different types of infrastructure investment and new revenue sources for urban water utilities. In cities of low- and middle-income countries, the energy-water nexus can lead to positive feedbacks wherein unreliability in the water sector triggers stress on the energy sector, which, in turn, exacerbates stress in water systems.
- The North American experience may or may not be relevant to cities of low- and middle-income countries where scarcity is commonplace, population growth is very high, supply is intermittent and unreliable and there are high losses to the distribution system. High-efficiency appliances may not work in systems with unreliable water pressure. The issue of discretionary outdoor demand is less relevant to cities struggling to supply potable drinking water and sanitation. Nonetheless, water is part of an interconnected system of resources. Governance is at the heart of adaptation and sustainable development. And, the potential for unintended consequences and positive feedbacks loom large for integrated water-energy-land management.

New research directions

- Urban water use should be modeled as a critical component of complex and dynamic water systems, with an eye toward understanding positive feedbacks, critical thresholds and unintended consequences. Resulting models should be used to examine the impact of various policy decisions in water, land and energy in the face of alternative future climate and population conditions.

- Methods to measure urban water demand are deeply flawed in both high-income and low- and middle-income cities, limiting the capacity to make informed decisions. New efforts should be made to estimate demand at very small scales in urban distribution systems and monitor response to policy and climate conditions.
- New knowledge is needed about the science-policy interface in water resource management. Specific topics include the social learning, social capital and best practices for communication.
- Urban water demand analysis needs to expand from its current emphasis on economics, technology and education to include psychologists, cultural theorists and urban planners.

Acknowledgment

This material is based upon work supported by the National Science Foundation under Grant SES-0951366, Decision Center for a Desert City II: Urban Climate Adaptation. Any opinions, findings and conclusions or recommendations expressed in this material are those of the authors and do not necessarily reflect the views of the National Science Foundation.

References

Arbués, F., Ángeles García-Valiñas, M., and Martínez-Espiñeira, R. (2003). Estimation of residential water demand: A state-of-the-art review. *Journal of Socio-Economics, 32*(1), 81–102.

Austin Water. (2013). *Stage 2 watering restrictions.* Retrieved from www.austintexas.gov/department/stage-2-watering-restrictions.

Bates, S. (2011). *Bridging the governance gap: Strategies to integrate water and land use planning* (2nd ed.). Missoula, MT: Center for Natural Resources and Environmental Policy, University of Montana.

BBC Research and Consulting. (2007). *Water conservation awareness, attitudes, and behaviors.* Phoenix, AZ: Arizona Municipal Water Users Association.

Borowski, I., and Hare, M. (2007). Exploring the gap between water managers and researchers: Difficulties of model-based tools to support practical water management. *Water Resources Management, 21*(7), 1049–1074.

Cashman, A., Nurse, L., and John, C. (2010). Climate change in the Caribbean: The water management implications. *The Journal of Environment & Development, 19*(1), 42–67.

Christian-Smith, J., Heberger, M., and Allen, L. (2012). *Urban water demand in California to 2100: Incorporating climate change.* San Francisco: Pacific Institute. Retrieved from http://pacinst.org/wp-content/uploads/sites/21/2014/04/2100-urban-water-efficiency.pdf.

Coomes, P., Rockaway, T., Rivard, J., and Kornstein, B. (2010). *North American residential water usage trends since 1992.* Denver, CO: Water Research Foundation.

Del Moral Ituarte, L., and Giansante, C. (2000). Constraints to drought contingency planning in Spain: The hyrdraulic paradigm and the case of Seville. *Journal of Contingencies and Crisis Management, 8*(2), 93–102.

DeOreo, W. B., Mayer, P. W., Martien, L., Hayden, M., Funk, A., Kramer-Duffield, M., and Davis, R. (2011). *California single-family water use efficiency study.* Boulder, CO: Aquacraft Water Engineering and Management.

Dilling, L., and Lemos, M. C. (2011). Creating usable science: Opportunities and constraints for climate knowledge use and their implications for science policy. *Global Environmental Change, 21*, 680–689.

Domene, E., and Saurí, D. (2006). Ubanisation and water consumption: Influencing factors in the metropolitan region of Barcelona. *Urban Studies, 43*(9), 1605–1623.

Environment Canada. (2013). *Residential water use indicator.* Retrieved from www.ec.gc.ca/indicateurs-indicators/default.asp?lang=en&n=7E808512-1.

European Environmental Agency. (2012). *Towards efficient use of water resources in Europe* (EEA Report No 1/2012). Retrieved from www.eea.europa.eu/publications/towards-efficient-use-of-water.

Flory, B. (2013). *Why are we here?* [PowerPoint slides]. Presented at the Urban Water Demand Roundtable: Bringing Together the Best in Current Research and Applications. Retrieved from http://dcdc.asu.edu/docs/dcdc/website/documents/2_BruceFlory_WhyAreWeHere-Flory130418.pdf.

Frost, D. (2013). The water demand revolution. *Planning, 79*(7), 12–17.

Gleick, P. H. (2000). The changing water paradigm: A look at twenty-first century water resources development. *International Water Resources Association, 25*(1), 127–138.

Gober, P. (2013). Getting outside the water box: The need for new approaches to water planning and policy. *Water Resources Management, 27*(4), 955–957. doi:10.1007/s11269-012-0222-y.

Gober, P., Brazel, A. J., Quay, R., Myint, S., Grossman-Clarke, S., Miller, A., and Rossi, S. (2010). Using watered landscapes to manipulate urban heat island effects: How much water will it take to cool Phoenix? *Journal of the American Planning Association, 76*(1), 109–121.

Gober, P., Middel, A., Brazel, A., Myint, S., Chang, H., Duh, J., and House-Peters, L. (2012). Tradeoffs between water conservation and temperature amelioration in Phoenix and Portland: Implications for urban sustainability. *Urban Geography, 33*(7), 1030–1054.

Gober, P., Larson, K. L., Quay, R., Polsky, C., Chang, H., and Shandas, V. (2013). Why land planners and water managers don't talk to one another and why they should! *Society & Natural Resources, 26*(3), 356–364. doi:10.1080/08941920.2012.713448.

Guhathakurta, S., and Gober, P. (2007). The impact of the Phoenix urban heat island on residential water use. *Journal of the American Planning Association, 73*(3), 317–329.

Guhathakurta, S., and Gober, P. (2010). Residential land use, the urban heat island, and water use in Phoenix: A path analysis. *Journal of Planning Education and Research, 30*(1), 40–51. doi:10.1177/0739456X10374187.

Hill, T., and Polsky, C. (2007). Development and drought in suburbia: A mixed methods rapid assessment of vulnerability to drought in rainy Massachusetts. *Environmental Hazards, 7*(4), 291–301.

Howe, C. W., and Linaweaver, F. P., Jr. (1967). The impact of price on residential water demand and its relationship to system design and price structure. *Water Resources Research, 3*(1), 13–32.

Kallis, G., and Coccossis, H. (2003). Managing water for Athens: From the hydraulic to the rational growth paradigm. *European Planning Studies, 11*(3), 245–262.

Kiefer, J. C., Clayton, J. M., Dziegieleski, B., and Henderson, J. (2013). *Analysis of changes in water use under regional climate change scenarios.* Denver, CO: Water Research Foundation. Retrieved from www.waterrf.org/Pages/Projects.aspx?PID=4263.

Lovett, I. (2013, August 11). Arid Southwest cities' plea: Lose the lawn. *The New York Times.* Retrieved from www.nytimes.com/2013/08/12/us/to-save-water-parched-southwest-cities-ask-homeowners-to-lose-their-lawns.html?pagewanted=all&_r=0.

LVVWD (Las Vegas Valley Water District). (2013). *Rebates and services.* Retrieved from www.lvvwd.com/conservation/ws_rebates.html.

Maas, T. (2003). *What the experts think: Understanding urban water management in Canada.* Victoria, BC: The POLIS Project on Ecological Governance, University of Victoria. Retrieved from www.poliswaterproject.org/publication/26.

McDonald, R. I., Green, P., Balk, D., Fekete, B. M., Revenga, C., Todd, M., and Montgomery, M. (2011). Urban growth, climate change, and freshwater availability. *Proceedings of the National Academy of Sciences USA, 108*(15), 6312–6317.

Malik, R. P. S. (2002). Water-energy nexus in resource-pool economies: The Indian experience. *World Resources Development, 18*(1), 47–58.

Mayer, P. W., DeOreo, W. B., Opitz, E. M., Kiefer, J. C., Davis, W. Y., Dziegielewski, B., and Nelson, J. O. (1999). *Residential end uses of water.* Denver, CO: American Water Works Association Research Foundation.

Middel, A., Quay, R., and White, D. D. (2013). *Water reuse in central Arizona.* Decision Center for a Desert City Technical Report 13-01. Tempe, AZ: Arizona State University. Retrieved from http://dcdc.asu.edu/publications/technical-papers/.

National Research Council. (2010). *Informing an effective response to climate change.* Washington, DC: The National Academies Press.

Neel, R., Sadalla, E., Ledlow, S., Berlin, A., and Neufeld, S. (2014). The social symbolism of water-conserving landscaping. *Journal of Environmental Psychology, 40*, 49–56. doi:10.1016/j.jenvp.2014.04.003.

Harnessing urban water demand

Ofwat. (2013). *Water meters – your questions answered: Information for household customers.* Retrieved from www.ofwat.gov.uk/mediacentre/leaflets/prs_lft_101117meters.pdf.

O'Hara, J. K., and Georgakakos, K. P. (2008). Quantifying the urban water supply impacts of climate change. *Water Resources Management, 22*(10), 1477–1497. doi:10.1007/s11269-008-9238-8.

Pahl-Wostl, C. (2006). The importance of social learning in restoring the multifunctionality of rivers and floodplains. *Ecology and Society, 11*(1), 10. Retrieved from www.ecologyandsociety.org/vol11/iss1/art10/.

Pahl-Wostl, C., and Borowski, I. (2007). Special issue: Methods for participatory water resources management – Preface. *Water Resources Management, 21*(7), 1047–1048.

Robbins, P. (2007). *Lawn people: How grasses, weeds, and chemicals make us who we are.* Philadelphia, PA: Temple University Press.

Rogers, P. (2012). Water-energy nexus: Sustainable urbanization in the Greater Mekong Subregion. Conference proceedings, *GMS 2020 International Conference: Balancing Economic Growth and Environmental Sustainability,* 153–162. Retrieved from www.gms-eoc.org/uploads/wysiwyg/events/GMS2020-Final-Proceedings/1. Session-Decade of Development Growth and Impacts.pdf.

Sharma, S. K., and Vairavamoorthy, K. (2009). Urban water demand management: Prospects and challenges for the developing countries. *Water and Environment Journal, 23*(3), 210–218.

Stockholm International Water Institute. (2014). World water week in Stockholm: Energy and water, 2014 programme. Retrieved from http://programme.worldwaterweek.org/.

Vickers, A., and Bracciano, D. (2014). Low-volume plumbing fixtures achieve water savings. *Opflow, 40*(7), 8–9. Retrieved from www.awwa.org/publications/opflow/abstract/articleid/46049779.aspx.

Vörösmarty, C. J., McIntyre, P. B., Gessner, M. O., Dudgeon, D., Prusevich, A., Green, P., Glidden, S., Bunn, S. E., Sullivan, C. A., Liermann, C. R., and Davies, P. M. (2010). Global threats to human water security and river biodiversity. *Nature, 467*(7315), 555–561.

Vörösmarty, C. J., Pahl-Wostl, C., and Bhaduri, A. (2013). Water in the anthropocene: New perspectives for global sustainability. *Current Opinion in Environmental Sustainability, 5*(6), 535–538.

Wentz, E., and Gober, P. (2007). Determinants of small-area water consumption for the city of Phoenix, Arizona. *Water Resources Management, 21*(11), 1849–1863.

Western Resource Advocates. (2013). *Water rate structure.* Retrieved from www.westernresourceadvocates.org/water/rates.php.

World Water Assessment Programme. (2009). *The United Nations World Water Development Report 3: Water in a changing world.* Paris: UNESCO, and London: EarthScan.

8

URBANIZATION, ENERGY USE AND GREENHOUSE GAS EMISSIONS

Peter J. Marcotullio

This chapter reviews research on urbanization, energy use and subsequent greenhouse gas (GHG) emissions. The central question is: How do urban processes and transformations manifest as energy consumption and GHG emissions? In order to address this question, the chapter presents a multi-dimensional perspective on urbanization, as it relates to varying aspects of energy demands and subsequent emissions. Urbanization is defined as a process in which socio-economic (and institutional structure), urban land and infrastructure and natural system change occurs simultaneously. From this perspective a framework on which to review various studies emerges (see Figure 8.1).

Alternatively, research on urbanization has traditionally focused on the concentration of population in growing numbers of ever larger dense settlements (Tisdale 1942). While population number and concentration is important, it is not the only aspect of urbanization that influences energy use and subsequent GHG emissions. For example, among other influences, an aging population (O'Neill et al. 2010), increases in economic wealth (Ciccone and Hall 1996), changes in urban form (Newman and Kenworthy 1999) and conversion or loss of natural systems within cities (Pataki et al. 2006) also effect energy demand and emissions levels. In order to include these and other associations, this chapter suggests that urbanization is a multi-dimensional process that alters socio-economic, built form and geographic or natural sub-systems, and that each of these spatially co-located sub-systems includes a number of elements that not only interact within constituent sub-systems, but link across sub-systems (Romero-Lankao et al. 2014).

This broader definition of urbanization allows for a more comprehensive survey of the impacts of the process as well as empirical challenges in understanding the urbanization process. This chapter demonstrates that over the past decades a large and growing body of literature has developed from a variety of disciplines identifying the contribution of a number of different urbanization influences on energy use and GHG emissions. At the same time, however, defining urbanization more broadly opens up the opportunity for multiple boundary ambiguities, which result in difficulties comparing findings and interpreting results (Bader and Bleischwitz 2009). Moreover, incorporating outcomes from larger development processes in the definition begs the question of whether urbanization indeed can be defined coherently. Arguably, however, those development processes unfolding in urban areas are unique and different from the larger scale processes in both form and outcome. Demonstrating this uniqueness, however, remains

Urbanization, energy use and GHG emissions

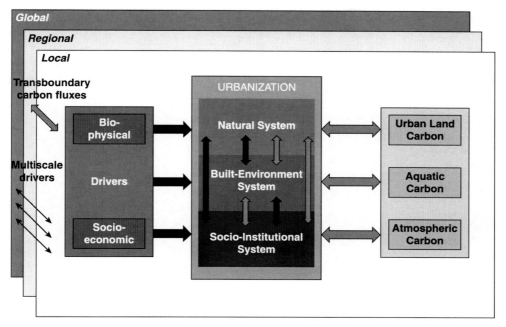

Figure 8.1 A critical knowledge pathway to low-carbon, sustainable futures: integrated understanding of urbanization, urban areas and carbon
Source: Romero-Lankao et al (2014).
Note: This figure presents urbanization as defined by changes in three core sub-systems including natural, built environment and socio-economic and institutional. These sub-systems interact and have both shared and unique drivers and effects on the carbon cycle. Finally, the figure attempts to incorporate the notion of scale as these interactions can be viewed differently over multiple scales.

difficult given data constraints (Grubler et al. 2012). Despite the challenges of using a more comprehensive definition of urbanization, the multi-dimensional definition helps to identify and organize a larger more comprehensive body of literature related to urbanization, energy use and GHG emissions. Researchers argue that findings from these more comprehensive perspectives can more directly influence policy for mitigation and adaptation than from studies using a single uni-dimensional definition (de Sherbinin et al. 2007).

The chapter first presents a review of the magnitudes of global and regional urban energy use and GHG emissions levels followed by a synthesis of factors influencing energy and GHG emissions. The chapter's conclusions identify policy implications and research gaps.

The magnitude of energy consumption and GHG emissions from urbanization

Few studies estimate the relative urban and rural shares of global GHG emissions (Dhakal 2010; Seto et al. 2014). This is largely due to the lack of urban scale energy and GHG emissions data. However, two estimates of the global picture on urban energy use have emerged recently (IEA 2008; Global Energy Assessment 2012).[1] While these studies examine different methods to estimate urban shares (bottom up and top down), both examine urban energy-related sources and calculate carbon dioxide (CO_2) emissions. The shared findings of

Peter J. Marcotullio

Table 8.1 Range in urban final energy use and urban percent of total final energy use by region, 2005

	Final energy use EJ	Percent
North America	51–63	69–87
Pacific OECD	11–16	59–92
Western Europe	31–41	64–83
Eastern Europe	4–6	51–72
Former USSR	14–20	54–78
Sub-Saharan Africa	5–10	35–71
Latin America	16–18	77–89
North Africa & Middle East	10–16	58–86
China and Central Pacific Asia	19–31	40–65
Pacific Asia	10–16	51–77
South Asia	5–10	29–51
World	**176–246**	**56–78**

Source: Grubler, et al. (2012).

these studies put the global urban final energy use share at approximately three-quarters of global final energy use (Grubler et al. 2012).[2]

Results of the Grubler et al. (2012) study are broken down by the 11 Global Energy Assessment (GEA) regions and for the global total (Table 8.1). While the central estimate for the global total is approximately 76 percent, this research also provides ranges of estimates. These results strongly suggest that urban energy use dominates global energy use and should, therefore, be a priority focus for energy sustainability challenges. At the same time, however, the results of the GEA analysis also note that in per capita terms, urban areas have a lower than national average of direct final energy use.

Marcotullio et al. (2013) identify lower average per capita GHG emissions levels in urban areas of high-income countries as compared to national averages. Higher per capita emissions levels in areas outside urban cores are confirmed by case studies of individual cities such as Paris (Barles 2009). Other studies that examine the differences in per capita energy use and GHG emissions between urban and non-urban areas also find higher levels in the outer, low-density suburbs. For example, in Toronto, VandeWeghe and Kennedy (2008) find that beyond the transit-intensive central core, private auto emissions surpass the emissions from building operations. Moreover, the top ten census tracts with high GHG emissions are located in lower-density suburbs, and their higher emissions are largely due to private automobile use. Lenzen et al. (2004) demonstrate significant differences in lifestyles between inner and outer areas of Sydney, with findings of different energy use characteristics including higher per capita energy and emissions in the outer, low-density suburbs than in the core urban area. In the US, Jones and Kammen (2014) highlight important differences between the urban core, suburban and metropolitan consumption-based, zip code-level carbon footprints. Minx et al. (2013) also find similar results for the UK. The main reason for the lower per capita energy use in urban areas is economic efficiencies (in terms of transportation) and differences in economic structure (with high-energy-consuming industries located outside of urban centers).

Alternatively, in low- and middle-income countries, most of the energy consumption technologies are concentrated in urban areas along with industrial activities. Marcotullio et al. (2013) note that in Asia, the average per capita emissions levels in cities is higher than national averages. These results confirm the findings in Grubler et al. (2012) for energy use patterns in some industrializing countries as well as differences in CO_2 emissions in several Chinese cities and national averages (Dhakal 2009). This is also due to the economic structure of cities in Asia, which are the manufacturing hubs of the country—if not the world—and the fact that energy use in non-urban areas in Asia tend to be low. In other words, the higher per capita emissions of urban dwellers in Asia is a reflection of both economic structural trends as wells low levels of development in rural areas.

There are also estimates of urban GHG emissions levels at the global scale. The IEA (2008) study suggests that the urban share of total energy use is approximately 71 percent of global CO_2 emissions. These and other high figures for the urban share of CO_2 emissions have stirred controversy, as results seem to place 'blame' or responsibility on urban areas for these emissions. For example, Satterthwaite (2008) and Dodman (2009) suggest that previous calculations overestimate or misconstrue the urban share of total GHGs. They suggest that the real culprit is high-income lifestyles, not dense settlement, pointing out the emissions reductions per capita achieved by many cities. In order to fully appreciate the emissions from cities, a consumption-based approach, or a method in which the emissions from a good or service are allocated to the individuals that consume them is needed. This approach identifies where the high consumption lifestyles are located. Most urban GHG inventories, on the other hand, are production-based, or identify the allocation of the emissions based upon where the good or service is produced. Satterthwaite (2008) suggests that his global down-scaled estimates of the urban share of total GHG emissions could be as low as 30–40 percent of total emissions.

A recent estimate of urban GHG emissions levels that examines a number of different sources (agricultural, industrial, energy service, transportation, residential and waste management) and several different GHG compounds (CO_2, CH_4, N_2O and SF_6) suggest that Satterthwaite (2008) and Dodman (2009) may be correct in that urban areas are found to be responsible for somewhere between 37 and 49 percent of total GHG emissions (Table 8.2) (Marcotullio et al. 2013). On the other hand, if the source is limited to strictly energy-related CO_2 emissions, results from this study suggest that urban areas are responsible for 44–72 percent of total energy-related CO_2 emissions, or with an upper range close to the previous IEA (2008) and GEA (2012) results. Hence, there is convergence of results for energy-related CO_2 emissions levels.

Urban energy and GHG emissions scaling effects

Given the massive urbanization that is predicted to occur over the next few decades, researchers have explored the role of urban population in GHG emissions levels (Seto et al. 2014). Will larger cities bring lower GHG emissions, as increasing size portends efficiency gains? Scaling effects describe a variable's proportionality of impact on energy and GHG emissions with increasing size. A more than proportional effect means that with increasing variable size the impact on energy and emissions increases. Scaling effects are described in detail by Bettencourt et al. (2007), who find that urban areas with larger population size have proportionally smaller energy infrastructures than smaller cities. This suggests increasing efficiencies are gained in infrastructure provision with increasing population size. Studies find, however, that in Germany for electricity consumption and for CO_2 emissions in the

Table 8.2 Range in urban GHG emissions and urban percent of total GHG emissions by region, 2000

Region	GHG emissions	
	(million tonnes)	*Percent*
North America	3,688–5,450	49–73
Pacific OECD	1,045–1,283	51–62
Western Europe	2,072–2,573	44–55
Eastern Europe	312–549	29–50
Former USSR	1,540–1,746	44–50
Sub-Saharan Africa	174–299	15–26
Latin America	635–751	25–29
North Africa and Middle East	736–840	39–44
China and Central Pacific Asia	1,312–1,781	23–31
Pacific Asia	851–1,043	40–49
South Asia	433–612	18–25
World	**12,798–16,926**	**37–49**

Source: Marcotullio et al. (2013) reallocated to GEA regions.

US, that urban population size has near proportional impacts (Fragkias et al. 2013; Bettencourt et al. 2007). On the other hand, Marcotullio et al. (2012; 2013) find super-linear population effects on urban GHG emissions, meaning that increases in urban population have a disproportionately larger impact on emissions in low- to high-income countries, although the effect is slightly smaller in high-income nations. In a study of 225 cities across both Annex 1 and non-Annex 1 countries, Grubler and Schulz (2013) find non-uniform scaling for urban final energy use within three different sets of urban areas. They find evidence of positive agglomeration economies of bigger cities (sub-linear effects) with respect to energy use. At the same time, however, these elasticities increase with the size/rank of the city in the sample to produce super-linear effects. Slopes decrease with the three groupings by size of their energy consumption (>0.5 EJ, 0.03–0.5 EJ, < 0.03 EJ) from −0.46 to −1.6 to −6.1. That is, with large energy consuming cities, a decrease in one unit of size accounts for a less than one unit decrease in energy use. In the middle and lower groupings, a decrease in one unit of size corresponds to larger than one unit of energy use. Alternatively, if one unit of size were added to the city, then there would be a larger than one unit increase in energy use. Certainly, the impact on energy and GHG emissions of increasing population size of urban areas needs further exploration, as the reason for these relationships is not entirely clear.

Drivers of energy consumption and GHG emissions from urbanization

As the understanding of urbanization, urban energy use and urban GHG emissions is critical to mitigation efforts, there has been an explosion of studies examining drivers of change. Accordingly, a model of urbanization influences can be divided into three general categories including 1) socio-economic (and institutional) characteristics; 2) geography and natural conditions; and, 3) the built urban form. Within each of these general categories the details of urban energy and GHG emissions determinants are briefly reviewed in the following

Urbanization, energy use and GHG emissions

sections (for recent detailed reviews see Chester et al. 2014; Hutyra et al. 2014; Marcotullio et al. 2014b; Seto et al. 2014).[3]

Socio-economic characteristics

Socio-economic factors covered in this chapter that are correlated to urban energy use and GHG emissions can be divided into economic and demographic sub-categories. In general, while there is strong evidence of relationships between these sets of variables and energy use and GHG emissions, the directionality and mechanisms of effects and linkages between various socio-institutional and other factors are not well understood (Marcotullio et al. 2014a).

Economic factors

Economic factors that correlate with urban energy use and GHG emissions can be broken down into income or general wealth influences, urban economic structure and urban economic function characteristics. All these categories overlap, but research on each focuses on a distinct area of economic influence. Perhaps the most important economic relationship is that of wealth or income, and energy use and GHG emissions. Researchers have long considered this relationship important at the national level (Kraft and Kraft 1978). Studies in this area find a positive correlation between income and energy use at the national level, but results of decades of research have produced diverse and inconclusive directional results (for reviews see Payne 2010; Ozturk 2010). No consensus has emerged from tests of four contending hypotheses (no causality; income growth causes increases in energy consumption; energy consumption causes income growth; bi-directionality between energy consumption and economic growth).

At the national level, researchers also explore correlations between the share of urban population, economic income and energy use, and GHG emissions. Studies show a positive relationship between the speed of economic growth and urbanization and energy demand. Poumanyvong and Kaneko (2010) find that the relationship varies across different levels of income. They find that nations with low income experience decreasing energy use with urbanization, which may be due to fuel switching in low-income societies, and that urbanization in medium- and high-income nations is associated with increasing energy use.

At the urban scale, researchers also hypothesize a positive correlation between urban income and energy, and GHG emissions (Satterthwaite 2009; Weisz and Steinberger 2010; Kahn 2009). Cross-national comparisons of urban GHG emissions and local or regional gross domestic product (GDP) demonstrate that income is a significant positive covariate with GHG emissions levels (Hoornweg et al. 2011; Kennedy et al. 2011; Marcotullio et al. 2013). Schulz (2010b) examines different cities in Africa, Latin America, OECD and non-OECD nations and finds a positive relationship between urban energy use and urban regional production. Housing size, automobile use and heating and industrial fuel use are associated with income, suggesting that larger sizes for homes, modification of the levels of thermal comfort required by city residents and mobile fuel use increase with income (Kennedy et al. 2009; Weisz and Steinberger 2010). At the household level, a number of studies from different countries (the Netherlands, India, Brazil, Denmark, Japan and Australia) note positive correlations between income and energy use (Vringer and Blok 1995; Wier et al. 2001; Pachauri and Spreng 2002; Cohen et al. 2005; Lenzen et al. 2006; Dey et al. 2007).

Analysts also suggest that urban economic structure is important for urban energy use and GHG emissions. For example, cities that are dependent on energy intensive industries are likely to contribute higher total and per capita GHG emissions than those whose economic base is in the service sector (Hoornweg et al. 2011). This relationship can be further strengthened if the energy supply mix is carbon intensive (Parikh and Shukla 1995; Sugar et al, 2012). Dhakal (2009) demonstrates that trends in urban GHG emissions in China, amongst a select group of urban areas, can be divided into three energy intensive pathways: high, medium and low. Those in the high category have energy intensive industries. Those Chinese urban areas with lower energy intensive pathways have high service sector contributions to their economies, which arguably creates lower direct emissions. In their study of Asian urban GHG emissions, Marcotullio et al. (2012) also point out that the highest per capita GHG emitters in the Asian region are typically those with industrial, or energy-producing economies.

Cities also assume different functional niches within larger regional and global urban systems (Lo and Yeung 1998; Sassen 2006), which may affect GHG emissions levels. These functional roles, defined by the hierarchical space within the New International Division of Labor (NIDL), have been associated with the variations in environmental impacts (Marcotullio 2003). Analysts have identified industrial urban areas that have specific carbon signatures, which are typically higher than those from command and control cities. For example, GHG emissions from Shanghai and Beijing are higher than that of Tokyo (Dhakal and Imura 2004). Schulz (2010a) demonstrates the significance of Singapore's energy use association with international trade in oil products. Moreover, with trade, urban carbon footprints experience 'leakage,' where through urban teleconnections (Seto et al. 2012b), contributions are off-loaded to other areas. For example, in 2004, international trade has resulted in leaking 30 percent of total US household CO_2 impact outside the US (Weber and Matthews 2008).

Demographic factors

Researchers find that demographic factors have a significant impact on energy use and GHG emissions in rapidly growing cities. All else equal, demographic factors have a smaller impact in larger cities of high-income countries, where growth is slow. Demographic factors include urban population size and share, aging and household size.

Urban energy and GHG emissions researchers traditionally examine the urban population as either the size of share or total population. Studies demonstrate that total energy use and demand and total GHG emissions for countries increase with urbanization (Jones 1991; York 2007; York et al. 2003). Arguably, at the national scale, urbanization may lead to either higher or lower emissions/energy consumption through industrialization and the shift from agricultural and primary commodity producing to manufacturing and goods producing. Those that find higher energy use and GHG emissions argue that urbanization has three positive energy use and GHG emissions effects: 1) the mechanization of agriculture, as it becomes more capital intensive; 2) transportation demand increases, as food for people must travel away from agricultural areas and into urban areas; and, 3) industry and manufacturing use more energy per unit of output and per worker than traditional agriculture (Jones 1991).

Liddle (2013b) further adds two more reasons why the increasing share of urban population may increase energy use and GHG emissions: 4) urbanization associates with economic growth and, therefore, greater energy use; and, 5) urbanization is a proxy of the amount of people with access to clean, energy sources (electricity) and this is associated with more energy consumption (see also Pauchari 2004; Pauchari and Jiang 2008). Finally, urbanization in warm climates may lead to higher energy use and GHG emissions as a result of higher energy demand (from air cooling) due to heat island effects.

Alternatively, those who argue that urbanization decreases energy use and GHG emissions suggest that dense settlement affords energy efficiencies by encouraging multi-dwelling living, public transit, walking and cycling, and winter energy use reductions from warming temperatures due to urban heat island effects. Whether national shares of urban population measure the effects of these efficiencies, however, is an open question, as urbanization (share of population in urban areas), *per se*, is not an indicator of densities that create these effects (Liddle and Lung 2010). That is, the measure of the share of population that lives in dense settlements may not directly associate with multi-dwelling living, public transit, walking, etc. Some studies find an insignificant influence or small elasticity for urbanization and energy use (Liddle and Lung 2010; Jorgenson and Clark 2012).

Studies that examine urban population at the urban scale demonstrate a positive relationship between size and energy use or GHG emissions. This is not surprising, as a larger number of individuals translates into higher total energy use and total GHG emissions, all else equal. For example, urban areas with the largest population size contribute the largest GHG emissions (Hoornweg et al. 2011; Marcotullio et al. 2013).

Some of the most interesting research in this area links urban population density to energy use and GHG emissions. The arguments suggest higher density urban areas should have lower emissions per capita than more sprawled urban areas, due to savings on transportation. The now classic studies of Newman and Kenworthy (1989), which demonstrate a negative relationship between density and transportation fuel use, are supported by more recent research on transportation energy consumption (Liddle 2013a), electricity consumption in buildings (Lariviere and Lafrance 1999) and urban GHG emissions (Marcotullio et al. 2012; Marcotullio et al. 2014b; Marcotullio et al. 2013). Despite these general results, however, scholars find the urban population density–GHG emissions relationship far from straightforward. Population density is only one way to measure compact/dense settlements (Ewing and Cervero 2001). Recent work reviewing the large literature on the relationship between vehicle miles traveled and urban development finds that destination accessibility is more important in explaining miles traveled in cars (and hence carbon emissions) than the combination of density, design and diversity of land use indicators (Ewing and Cervero 2010). These relationships are further complicated by age-structure (see also below). For example, Liddle (2011) finds in a macro-level cross-country study that young adult (20–34) transport energy consumption is intensive, but for other age groups there is a negative relationship on energy consumption for this sector. That is, nations with a larger share of population over age 65 have lower carbon emissions from road transport. In middle- and low-income countries, the population density–GHG emissions relationship is also complicated by energy access. Many densely populated cities have large slum populations that lack access to electricity and modern fuels and are, therefore, likely to use less energy than more affluent populations with access to electricity and modern fuels (Jorgenson et al. 2010).

Aging is an important demographic characteristic associated with energy use. Demographers predict that the future population will be older, resulting in a doubling of those older than 60 years of age between now and 2050 (Cohen 2003). During different life stages people have different economic activity levels, are heads of different sized households, and use power supply and transportation services differently. These shifts are captured in the household life-cycle stages, which affects patterns of purchases (Wilkes 1995) among other activities. Liddle (2013b) reports that age-structure is insignificant when age categories are not disaggregated and match appropriate lifecycle behavior categories. However, when age cohorts are disaggregated appropriately, researchers find significant association. For example, Liddle and Lung (2010) find a positive elasticity for young adults (aged 20–34) and a negative elasticity of older adults (aged 35–64) with GHG emissions from transport, residential energy and electricity consumption. Using predictive techniques, Dalton et al. (2008) and O'Neill et al. (2010) find that population aging may reduce emissions in most countries, as it will slow economic growth. O'Neill et al. (2010) examine the direct effects (through consumption of energy, food, transport, etc.) and indirect effects (through aging effects on economic growth, such as labor supply, i.e. lower labor productivity) in scenario modeling. They find that aging can have a significant influence on emissions in particular regions, separate from the effect of changes in population size. For example, they find aging can reduce emissions in the long-term (100 years) by up to 20 percent, particularly in industrialized regions.

Finally, the demographic factor of household size may affect energy use and GHG emissions. Current studies suggest that household size is decreasing globally (Liu et al. 2003). Worldwide, average household size declined from 3.6 to 2.7 between 1950 and 1990, and this trend is occurring in countries across the development spectrum, although at different rates (Mackellar et al. 1995). In terms of energy and GHG emissions, larger households may be associated with lower energy use, as households reap efficiency benefits. This notion is supported by macro-level, cross-country analyses. For example, Liddle (2004) finds that larger households correlate with lower levels of per capita road energy use in OCED countries. Cole and Neumayer (2004) find that larger households correlate with lower aggregate carbon emissions levels in low- to high-income countries. In household studies, size is associated with lower energy use in India (Pachauri 2004), Australia (Lenzen et al. 2004), the US (Weber and Matthews 2008) and Denmark and Brazil (Lenzen et al. 2006). In Japan, in contrast, larger household size correlates with slightly larger energy use (Lenzen et al. 2006). In these studies, researchers typically find that urban populations have smaller household sizes than rural populations.

Geography and natural conditions

As Hutyra et al. (2014) point out, studies of the role of urban vegetation and soils in the larger carbon cycle are relatively new and demonstrate that urban ecology can be a significant influence on urban GHG emissions in both long and short time scales (see also Hutyra Chapter 13 in this volume, for a detailed discussion of nutrient cycling in urban areas). It is generally understood that urbanization accompanies altering land cover and modification of vegetation which affect carbon storage (Imhoff et al. 2004) and carbon flows (Grimm et al. 2008; Pataki et al. 2006). Urban areas can sequester carbon (McPherson et al. 2005; Pouyat et al. 2006) and some cities are now increasing their tree cover, in part, as a climate mitigation strategy. Ultimately, however, the potential sinks for carbon in urban vegetation and soils are severely limited (Nowak and Crane 2002).

Local weather and climate can have significant impacts on energy use and, therefore, GHG emissions (Neumayer 2002). For urban areas, Toronto has cold winters and fairly warm summers, which could be linked to higher per capita energy use than London, Sydney or Hong Kong, for example (Kennedy et al. 2007). Akbari et al. (1992) find that peak urban electricity demand in five American cities (Los Angeles, CA; Washington, DC; Phoenix, AZ; Tucson, AZ; and Colorado Springs, CO) rises by 2–4 percent for each 1°C rise in daily maximum temperature above a threshold of 15–20°C. Thus, the additional air-conditioning use caused by this urban air temperature increase is responsible for 3–8 percent of urban peak electricity demand.

Research on the impact of weather and climate on energy use in buildings has a substantial history (Eto 1988). In Europe, researchers find that the energy consumption per unit of heated floor area of buildings increases with heating degree-days (HDD), or a climate measure of the demand for space heating (Balaras et al. 2005). Due to advances in building technologies, however, some researchers find that energy use intensity no longer correlates with climate (Mohareb et al. 2011). This research finds that for a sample of 57 high performance buildings around the world that have been designated 'green,' there is no significant relationship between HDD and energy demand.

The geographic location of an urban area, be it coastal, next to river or sea, or mountainous, may play a role in energy consumption and hence GHG emissions. Using US data, Glaeser and Kahn (2010) find that places with moderate temperatures, like coastal California, have significantly lower emissions than places in warmer climates. While there has been a significant body of literature on the effects of climate on urban energy use and GHG emissions, there has yet to be a significant body of studies on the effect of the type of ecosystem on energy consumption and GHG emissions.

Built urban form

Urbanization accompanies the growth of physical infrastructure including the development of roads, bridges, buildings, commercial and governmental plants, ports and airports, etc. Building technologies are cement, metal (aluminum, iron and steel) and rock, gravel and sand intensive (Shen, Chang et al. 2005). Studies of material flows into urban areas demonstrate large material inputs for urbanization (Decker et al. 2000; Kennedy et al. 2007). Cement and heavy metal industries can account for a significant share of national GHG emissions. Aluminum, cement, iron and steel, and paper industries account for approximately 60 percent of China's CO_2 emissions from the industrial sector (Rock et al. 2013). In general, given the projected increase in urban area within the near future (Seto et al. 2012a) and the global expansion of energy infrastructure used to support city growth, urbanization is a key driver of emissions across multiple scales (Davis et al. 2010). For example, some researchers have estimated that material substitution, efficient use of materials and recycling within the steel, cement and power generation industries can halve global GHG emissions by 2050 (Akashi et al. 2014) (see also Güneralp, Chapter 6 in this volume).

Notwithstanding the emissions embedded in buildings, and the energy and emissions due to building, urban researchers suggest the built environment can influence energy use and GHG emissions. This category is critical to understanding the unique characteristics of urban energy use and GHG emissions as compared to non-urban patterns. The built environment's impact on energy use and GHG emissions can be divided into three general categories: buildings and infrastructure, quality and type, and urban form (Seto et al. 2014; Grubler et al. 2012), which are reviewed below.

Buildings and infrastructure

The contribution of inputs to infrastructure (cement and steel) is estimated at approximately 16 percent of global carbon emissions in 2006 (Allwood et al. 2010) and, therefore, is an important factor to address. Using the experience of high-income countries as an example, Müller et al. (2013) calculate that the built-up infrastructure, as of 2008, embodies approximately 122 Gt CO_2 emissions, but may be as much as 350 Gt CO_2 emissions by 2050. Despite the importance of buildings and infrastructure to energy use and GHG emissions, a review of engineered systems suggests that researchers have a partial and skewed view of engineered systems, leaving much unknown about this field (Chester et al. 2014). For example, traditionally applied scientists and planners have viewed infrastructure as isolated (transportation, buildings, water and electricity) systems; findings largely focus on mature systems in high-income countries; and most studies ignore physical or institutional 'lock-in' (see Mohareb et al., Chapter 26 in this volume, for a discussion of mitigation challenges and opportunities in high-income cities).

There are some fundamental aspects of buildings and infrastructure that influence energy use and GHG emissions including design, building age and building mix. The design and thermal integrity (e.g. insulation levels) of buildings are essential for the amount of energy intensity (energy/m²) needed for heating and cooling. Research on building energy use intensity demonstrates variation among energy demand. For example, Fissore et al. (2011) find that total carbon emissions vary by a factor of four in the Minneapolis-Saint Paul metropolitan area in Minnesota, US. Building managers can also deploy various technologies to lower energy use and hence GHG emissions. For example, 'cool roofing' (reduction of internal building temperatures) includes a wide array of highly solar reflective and high infrared emittance materials (Santamouris et al. 2011). A literature review of cool roofs for residential and commercial buildings suggests results of varied energy savings, from 2–44 percent and average about 20 percent (Haberl and Cho 2004). Akbari and Konopacki (2005) calculate energy savings from increasing roof solar reflectance from a typical dark roof of 0.1 to a cool-colored roof of 0.4, for various climatic conditions worldwide. The estimated savings range from approximately 250 kWh per year for mild climates to over 1,000 kWh per year for very hot climates.

Building age is one determinant of building energy use, and hence urban GHG emissions. Larger and older buildings may be less energy efficient than newer, smaller ones and, therefore, contribute more GHG emissions. In a recent study in Melbourne, analysts find that larger buildings have more emissions than smaller ones, but smaller buildings have higher per capita emissions, and per square meter (MJ/m²) aggregate older buildings consume more energy than aggregate newer buildings (Wilkinson and Reed 2006).

Building mix is also an important characteristic of the built environment. Different building types have different energy demands. For example, Howard et al. (2012) estimate the total fuel intensities of different buildings based upon ten different functions in New York City. They calculate total fuel intensities that vary from 80.6 kWh/m² for office buildings in Manhattan to 547 kWh/m² for stores throughout the city. Norman et al. (2006) use a life-cycle analysis approach to assess residential energy use and GHG emissions, contrasting 'typical,' inner-urban, high-density and outer-urban, low-density residential developments in Toronto. They find that the energy embodied in the buildings themselves is 1.5 times higher in low-density areas than in high-density areas on a per capita basis.

Besides buildings, there are few studies of energy use and GHG emissions associated with other infrastructures, such as water supply distribution and use, and waste in cities. Typically, water studies use lifecycle assessment procedures to generate estimates of GHG emissions (Sahely and Kennedy 2007). While models to estimate GHG emissions from waste management exist (EPA 2013), how urbanization processes affect waste generation and management is largely unknown (Chester et al. 2014).

Urban form

Besides density, mentioned earlier, urban spatial patterns, which include urban form, can also influence GHG emissions. The sustainable city model, which is a low-carbon model, is often portrayed as a compact city. Despite calls for 'compact cities,' Grubler et al (2012) state that it is by no means clear that there is an ideal urban form and morphology that can maximize energy performance and satisfy all other sustainability criteria. Urban size and metropolitan form, however, may provide insights into energy use and GHG emissions. In a study of five urban areas within the Pearl River Delta, researchers find that urban land size and fragmentation of urban land use patterns is positively correlated with energy consumption, and that the dominance of the largest urban patch is negatively correlated with energy consumption (Chen et al. 2011). More of these types of studies are needed in order to fully understand the role of urban form on energy use and GHG emissions.

Conclusions: policy implications and research priorities

As demonstrated, there are a plethora of research papers addressing aspects of urbanization and energy and GHG emissions. Much of this work has been performed over the last 15 years, making this a nascent field of inquiry. This review highlights two questions for policy-makers and researchers: 1) What are the policy implications of these studies; and 2) What are the current research priorities? Each of these questions is addressed in this final section.

What policy implications can be drawn from the research?

Three general and related policy implications are clear:

- Given the importance of higher density, large population cities in the urbanized and rapidly urbanizing world, mitigation efforts may well be most successfully applied to these cities.

These cities are the sources of most urban emissions and areas of highest concentration of energy use and GHG emissions. For example, both Marcotullio et al. (2013) and Hoornweg et al. (2011) identify the importance of large cities in creating GHG emissions.

- Long-term policy may be well placed to focus on the infrastructural systems that are being developed now all over the urban world, but, particularly, in the cities of the rapidly industrializing world, where much of this infrastructure is slated to be deployed.

Creating systems that are integrated and able to both lower emissions levels and at the same time address vulnerabilities are critical. For example, Davis et al. (2010) conclude that the most threatening emissions sources include infrastructure yet to be built and that CO_2-emitting infrastructure will expand unless extraordinary efforts are undertaken to develop alternatives.

- A number of socio-institutional, built environment and urban ecology practices can be addressed that affect urban GHG emissions.

Reviews identify priority areas among these. For example, Seto et al. (2014) suggest that economic geography (trade, industry structure), income (consumption) and technologies (efficient energy end users such as buildings, vehicles, appliances, etc.) are of high importance; while infrastructure and urban form are of medium to high importance, other socio-demographic drivers are of medium importance, and fuel substitutions, urban renewables and urban ecology (afforestation) are of low importance in terms of impact on GHG emissions. At the same time, however, for urban governance policy leverage the relationships are reversed, such that the drivers of lowest impact are those of highest leverage; while for governance systems the drivers of highest impact are those of lowest leverage (see also Grubler et al. 2012). Seto et al. (2014) conclude, therefore, that pursuing policies of different scales may be most effective at reducing urban–related emissions.

What are the research priorities?

Research gaps are identified in a number of recent studies (Chester et al. 2014; Grubler et al. 2012; Marcotullio et al. 2014a; Seto et al. 2014; Hutyra et al. 2014). Five general areas are prioritized and briefly explained below:

- Develop and use more comprehensive models of urbanization processes, drivers and linkages to energy use and GHG emissions.

Recent studies call for a multi-dimensional conceptualization of urbanization as a set of processes, while recognizing the need for more careful explanations of the mechanisms embedded within these processes including how the individual dimensions of urbanization interact and co-evolve with energy use and carbon systems. As Chester et al. (2014) point out, there remains a weak understanding of urbanization as a process and how the deployment of infrastructure and use of technologies affects GHG emissions in different stages of urban growth, and for different types of growth (e.g. infill development and suburban sprawl) in low-, middle- and high-income nations.

Developing such frameworks requires that scientists prioritize theory development and primary data collection (of various kinds) that elucidate the relationships, mechanisms and interactive effects that underlie the relationship between the social, biophysical and management processes of urbanization and energy use. The looming question here is: How do we integrate the research of engineers, planners, biophysical and social scientists while recognizing the complexity and interdependence of the built environment, social and ecological systems at multiple scales associated with urbanization?

- More and better data at the urban scale.

As Hutrya et al. (2014) extensively document, the current uncertainties and paucity of data regarding urban carbon fluxes and biogeochemical and socio-economic processes that control those fluxes hamper our understanding. However, there is also a dearth of data on socio-institutional and built forms. Together this lack of data has hindered valid cross comparisons that make it possible to understand processes and drivers of urbanization, energy use and GHG emissions. More and better data about the world's cities and their energy use and GHG emissions, particularly the industrializing world, are needed.

- More research on the effect of policy and technologies on GHG emissions including lock-in.

Chester et al. (2014) argue that there remains a need for comprehensive frameworks that link infrastructure policy and planning for low GHG emissions futures in urbanization processes. Marcotullio et al. (2014a) further point out that there have been too few studies comparing governance systems and their effect on GHG emissions. Given this lacuna, there remains an insignificant understanding of what works from a policy perspective to reduce the energy use and GHG emissions associated with urbanization, how best to interface with decision-makers and how to diffuse lessons from city to city.

- Urbanization, energy use, GHG emissions and environmental justice.

Several reviews point out the limited understanding of major stakeholders, what interests are at play and how they mobilize (or not) to influence the energy use and GHG emissions associated with urbanization. Most studies of urbanization and energy use ignore the social and physical diversity that exists within cities and how these differences generate consequences for GHG emissions. While environmental justice research shows greater environmental impacts on ethnic minorities and the poor in many cities, it is less clear how group-based inequalities may impact the carbon cycle. As social groups do not settle in the same spatial patterns across regions, scholars need a better understanding of how location decisions of residences and businesses impact access to and use of energy and subsequent GHG emissions.

- Formal research programs.

The significant amount of research on this subject is largely driven by the interests of individual scientists or small research collaborations, rather than by a formal research priority with international and national science agencies. Hutyra et al. (2014) point out that there has not yet been the equivalent of an urban study of comparable scale to the North American Carbon Program's Mid-Continental Intensive experiment, not to mention an effort that attempts to integrate the natural sciences, social sciences and engineering factors relevant to carbon cycling. The Intergovernmental Panel on Climate Change (IPCC) has only recently included urban chapters in the mitigation (2007, 2014) and adaptation (2014) working groups. More formal funding for research in this area is critical for developing large-scale, long-term projects that will be able to address some of the research demands above.

Notes

1 The Grubler et al. (2012) analysis is based upon spatially explicit urban data described in Grubler et al. (2007).
2 The IEA (2008) study results suggest that the urban share of primary energy use is approximately 67 percent of total primary energy use. According to Grubler et al. (2012), the IEA calculations for this category of energy are approximately the same as the 76 percent share of final energy use found in the IIASA/GEA 2012 study.
3 In an attempt to save space, this review does not include several influential determinants of energy use and GHG emissions levels including governance, planning and policy (which are covered in several chapters in Part III of this volume), behavioral determinants (which are covered in Chapter 1, Fragkias and Chapter 30, Simon of this volume) and transportation.

References

Akashi, O., T. Hanaoka, T. Masui, and M. Kainuma. 2014. Halving global GHG emissions by 2050 without depending on nuclear and CCS. *Climatic Change* 123 (3):611–622.
Akbari, H., and S. Konopacki. 2005. Calculating energy-saving potentials of heat-island reduction strategies. *Energy Policy* 33:721–756.
Akbari, H., S. Davis, S. Dorsano, J. Huang, and S. Winnett. 1992. Cooling our Communities, A Guidebook on Tree Planting and Light-Colored Surfacing. Washington, DC: US EPA.

Allwood, J. M., J. M. Cullen, and R. L. Milford. 2010. Options for achieving a 50% cut in industrial carbon emissions by 2050. *Environmental Science and Technology* 44 (6):1888–1894.

Bader, N., and R. Bleischwitz. 2009. Measuring urban greenhouse gas emissions: The challenge of comparability. *Survey and Perspectives Integrating Environment & Society* 2 (3):7–21.

Balaras, C. A., K. Droutsa, E. Dascalaki, and S. Kontoyiannidis. 2005. Heating energy consumption and resulting environmental impact of European apartment buildings. *Energy and Buildings* 37 (5):429–442.

Barles, S. 2009. Urban metabolism of Paris and its region. *Journal of Industrial Ecology* 13 (6):898–913.

Bettencourt, L. M. A., J. Lobo, D. Helbing, C. Kuhnert, and G. B. West. 2007. Growth, innovation, scaling, and the pace of life in cities. *Proceedings of the National Academy of Sciences of the United States of America* 104 (17):7301–7306.

Chen, Y., X. Li, Y. Zheng, Y. Guan, and X. Liu. 2011. Estimating the relationship between urban forms and energy consumption: A case study in the Pearl River Delta, 2005–2008. *Landscape and Urban Planning* 102 (1):33–42.

Chester, M. V., J. Sperling, E. Stokes, B. Allenby, K. Kockelman, C. Kennedy, L. A. Baker, J. Keirstead, and C. T. Hendrickson. 2014. Positioning infrastructure and technologies for low-carbon urbanization. *Earth's Future* 2 (10):533–547.

Ciccone, A., and R. E. Hall. 1996. Productivity and the density of economic activity. *American Economic Review* 86 (1):54–70.

Cohen, C., M. Lenzen, and R. Schaeffer. 2005. Energy requirements of households in Brazil. *Energy Policy* 33 (4):555–562.

Cohen, J. E. 2003. Human population: The next half century. *Science* 302 (5648):1172–1175.

Cole, M. A., and E. Neumayer. 2004. Examining the impact of demographic factors on air pollution. *Population and Environment* 26: 5–21.

Dalton, M., B. O'Neill, A. Prskawetz, L. Jiang, and J. Pitkin. 2008. Population aging and future carbon emissions in the United States. *Energy Economics* 30:642–675.

Davis, S. J., K. Caldeira, and H. D. Matthews. 2010. Future CO2 emissions and climate change from existing energy infrastructure. *Science* 329 (5997):1330–1333.

de Sherbinin, A., D. Carr, S. Cassels, and L. Jiang. 2007. Population and environment. *Annual Review of Environment and Resources* 32:345–373.

Decker, E. H., S. Elliott, F. A. Smith, D. R. Blake, and F. S. Rowland. 2000. Energy and material flow through the urban ecosystem. *Annual Review of Energy and Environment* 25:685–740.

Dey, C., C. Berger, B. Foran, M. Foran, R. Joske, M. Lenzen, and R. Wood. 2007. Household environmental pressure from consumption: An Australian environmental atlas. In *Water, Wind, Art and Debate: How Environmental Concerns Impact on Disciplinary Research*, ed. G. Birch, 280–315. Sydney: Sydney University Press.

Dhakal, S. 2009. Urban energy use and carbon emissions from cities in China and policy implications. *Energy Policy* 37 (11):4208–4219.

———. 2010. GHG emission from urbanization and opportunities for urban carbon mitigation. *Current Opinion in Environmental Sustainability* 2 (4):277–283.

Dhakal, S., and H. Imura. 2004. Urban Energy Use and Greenhouse Gas Emissions in Asian Mega-cities, Policies for a Sustainable Future. Tokyo: Institute for Global Environmental Strategies.

Dodman, D. 2009. Blaming cities for climate change? An analysis of urban greenhouse gas emissions inventories. *Environment and Urbanization* 21 (1):185–201.

EPA. 2013. Waste Reduction Model. Washington, DC: United States Environmental Protection Agency (see www.epa.gov/WARM).

Eto, J. H. 1988. On using degree-days to account for the effects of weather on annual energy use in office buildings. *Energy and Buildings* 12 (2):113–127.

Ewing, R., and R. Cervero. 2001. Travel and the built environment. *Transportation Research Record* 1780:87–114.

———. 2010. Travel and the built environment. *Journal of the American Planning Association* 76 (3):265–294.

Fissore, C., L. A. Baker, S. E. Hobbie, J. Y. King, J. P. McFadden, K. C. Nelson, and I. Jakobsdottir. 2011. Carbon, nitrogen, and phosphorus fluxes in household ecosystems in the Minneapolis-Saint Paul, Minnesota, urban region. *Ecological Applications* 21 (3):619–639.

Fragkias, M., J. Lobo, D. Strumsky, and K. C. Seto. 2013. Does size matter? Scaling of CO2 emission and US urban areas. *PLoS ONE* 8 (6):e64727. doi:10.1371/journal.pone.0064727.

Glaeser, E. L., and M. E. Kahn. 2010. The greenness of cities: Carbon dioxide emissions and urban development. *Journal of Urban Economics* 67 (3):404–418.

Global Energy Assessment ed. 2012. *Global Energy Assessment – Toward a Sustainable Future*. Cambridge, UK and New York, NY: Cambridge University Press.

Grimm, N. B., S. H. Faeth, N. E. Golubiewski, C. L. Redman, J. Wu, X. Bai, and J. M. Briggs. 2008. Global change and the ecology of cities. *Science* 319 (5864):756–760.

Grubler, A., and N. Schulz. 2013. Urban energy use. In *Energizing Sustainable Cities: Assessing Urban Energy*, ed. A. Grubler and D. Fisk, 57–70. Oxford, UK and New York, USA: Routledge.

Grubler, A., B. O'Neill, K. Riahi, V. Chirkov, A. Goujon, P. Kolp, I. Prommer, S. Scherbov, and E. Slentoe. 2007. Regional, national, and spatially explicit scenarios of demographic and economic change based on SRES. *Technological Forecasting and Social Change* 74 (7):980–1029.

Grubler, A., X. Bai, T. Buettner, S. Dhakal, D. J. Fisk, T. Ichinose, J. E. Keirstead, G. Sammer, D. Satterthwaite, N. B. Schulz, N. Shah, J. Steinberger, and H. Weisz. 2012. Chapter 18: Urban Energy Systems. In *Global Energy Assessment – Toward a Sustainable Future*, ed. Global Energy Assessment, 1307–1400. Cambridge, UK and New York, NY: Cambridge University Press.

Haberl, J. S., and S. Cho. 2004. Literature Review of Uncertainty of Analysis Methods (Cool Roofs). Texas A&M University System: Report to the Texas Comission on Environmental Quality.

Hoornweg, D., L. Sugar, C. Lorena, and T. Gomez. 2011. Cities and greenhouse gas emissions: Moving forward. *Environment and Urbanization* 23 (1):207–227.

Howard, B., L. Parshall, J. Thompson, S. Hammer, J. Dickinson, and V. Modi. 2012. Spatial distribution of urban building energy consumption by end use. *Energy and Buildings* 45:141–151.

Hutyra, L. R., R. Duren, K. R. Gurney, N. Grimm, E. A. Kort, E. Larson, and G. Shrestha. 2014. Urbanization and the carbon cycle: Current capabilities and research outlook from the natural sciences perspective. *Earth's Future* 2 (10):473–495.

IEA. 2008. *World Energy Outlook 2008*. Paris: OECD/IEA.

Imhoff, M. L., L. Bounoua, R. DeFries, W. T. Lawrence, D. Stutzerd, C. J. Tucker, and T. Ricketts. 2004. The consequences of urban land transformation on net primary productivity in the United States. *Remote Sensing of Environment* 89 (4):434–443.

Jones, C., and D. M. Kammen. 2014. Spatial distribution of U.S. household carbon footprints reveals suburbanization undermines greenhouse gas benefits of urban population density. *Environmental Science & Technology* 48 (2):895–902.

Jones, D. W. 1991. How urbanization affects energy use in developing countries. *Energy Policy* 19 (7):621–630.

Jorgenson, A. K., and B. Clark. 2012. Are the economy and environment decoupling? A comparative international study, 1960–2005. *American Journal of Sociology* 118 (1):1–44.

Jorgenson, A. K., J. Rice, and B. Clark. 2010. Cities, slums, and energy consumption in less developed countries, 1990 to 2005. *Organization & Environment* 23 (2):189–204.

Kahn, M. E. 2009. Urban growth and climate change. *Annual Review of Resource Economics* 1:333–349.

Kennedy, C., J. Cuddihy, and J. Engel-Yan. 2007. The changing metabolism of cities. *Journal of Industrial Ecology* 11 (2):43–59.

Kennedy, C., J. Steinberger, B. Gason, Y. Hansen, T. Hillman, M. Havranck, D. Pataki, A. Phdungsilp, A. Ramaswami, and G. V. Mendez. 2009. Greenhouse gas emissions from global cities. *Environmental Science & Technology* 43 (19):7297–7302.

Kennedy, C., A. Ramaswami, S. Carney, and S. Dhakal. 2011. Greenhouse gas emission baselines for global cities and metropolitan regions. In *Cities and Climate Change: Responding to an Urgent Agenda*, eds. D. Hoornweg, M. Freire, M. J. Lee, P. Bhada-Tata, and B. Yuen, 15–54. Washington, D.C.: World Bank.

Kraft, J., and A. Kraft. 1978. On the relationship between energy and GNP. *Journal of Energy and Development* 3 (2):401–403.

Lariviere, I., and G. Lafrance. 1999. Modelling the electricity consumption of cities: Effect of urban density. *Energy Economics* 21 (1):53–66.

Lenzen, M., C. Dey, and B. Foran. 2004. Energy requirements of Sydney households. *Ecological Economics* 49 (3):375–399.

Lenzen, M., M. Wier, C. Cohen, H. Hayami, S. Pachauri, and R. Schaeffer. 2006. A comparative multivariate analysis of household energy requirements in Australia, Brazil, Denmark, India and Japan. *Energy* 31 (2–3):181–207.

Liddle, B. 2004. Demographic dynamics and per capita environmental impact: Using panel regressions and household decompositions to examien population and transport. *Population and Environment* 26 (1):23–39.

————. 2011. Consumption-driven environmental impact and age-structuer change in OECD countries: A cointegration-STIRPAT analysis. *Demographic Research* 24:749–770.

————. 2013a. The energy economic growth, urbanization nexus across development: Evidence from heterogeneous panel estaimtes robust to cross-sectional dependence. *The Energy Journal* 34 (2):223–244.

————. 2013b. Impact of population, age structure and urbanization on carbon emissions/energy consumption: Evidence from macro-level, cross-country analyses. *Population and Environment* DOI 10.1007/s11111-013-0198-4.

Liddle, B., and S. Lung. 2010. Age-structure, urbanization and climate change in developed countries: Revisting STIRPAT for disaggreagated population and consumption-related environmental impacts. *Population and Environment* 31:317–343.

Liu, J., G. C. Daily, P. R. Ehrlich, and G. W. Luck. 2003. Effects of household dynamics on resource consumption and biodiversity. *Nature* 421 (6922):530–533.

Lo, F.-c., and Y.-m. Yeung eds. 1998. *Globalization and the World of Large Cities.* Tokyo: United Nations University.

Mackellar, L., W. Lutz, C. Prinz, and A. Goujon. 1995. Population, households and CO2 emissions. *Population and Development Review* 21 (4):849–865.

McPherson, E. G., J. R. Simpson, P. F. Peper, S. E. Maco, and Q. Xiao. 2005. Municipal forest benefits and costs in five US cities. *Journal of Forestry* 103 (8):411–416.

Marcotullio, P. J. 2003. Globalization, urban form and environmental conditions in Asia Pacific cities. *Urban Studies* 40 (2):219–248.

Marcotullio, P. J., A. Sarzynski, J. Albrecht, and N. Schulz. 2012. The geography of urban greenhouse gas emissions in Asia: A regional analysis. *Global Environmental Change* 22 (4):944–958.

Marcotullio, P. J., A. Sarzynski, J. Albrecht, N. Schulz, and J. Garcia. 2013. The geography of global urban greenhouse gas emissions: An exploratory analysis. *Climatic Change* 121 (4):621–634.

Marcotullio, P. J., S. Hughes, A. Sarzynski, S. Pincetl, L. Sanchez-Pena, P. Romero-Lankao, D. Runfola, and K. C. Seto. 2014a. Urbanization and the carbon cycle: Contributions from social science. *Earth's Future* 2 (10):496–514.

Marcotullio, P. J., A. Sarzynski, J. Albrecht, and N. Schulz. 2014b. A top-down regional assessment of urban greenhouse gas emissions in Europe. *Ambio: A Journal of the Human Environment* 43 (7):957–968.

Minx, J., G. Baiocchi, T. Weidmann, J. Barrett, F. Creutzig, K. Feng, M. Forster, P.-P. Pichler, H. Weisz, and K. Hubacek. 2013. Carbon footprints of cities and other human settlements in the UK. *Environmental Research Letters* 8 (3):DOI: 10.1088/1748-9326/8/3/035039, 10 pp.

Mohareb, E. A., C. A. Kennedy, L. D. D. Harvey, and K. D. Pressnail. 2011. Decoupling of building energy use and climate. *Energy and Buildings* 43 (10):2961–2963.

Müller, D. B., G. Liu, A. N. Løvik, R. Modaresi, F. S. Steinhoff, S. Pauliuk, and H. Brattebø. 2013. Carbon emissions of infrastructure development. *Environmental Science and Technology* 47 (20):11739–11746.

Neumayer, E. 2002. Can natural factors explain any cross-country differences in carbon dioxide emissions? *Energy Policy* 30 (1):7–12.

Newman, P., and J. Kenworthy. 1989. Gasoline consumption and cities: A comparison of U.S. cities with a global survey. *Journal of American Planning Association* 55 (1):24–37.

————. 1999. *Sustainability and Cities.* Washington DC: Island Press.

Norman, J., H. L. MacLean, and C. A. Kennedy. 2006. Comparing high and low residential density: Life-cycle analysis of energy use and greenhouse gas emissions. *Journal of Urban Planning and Development* 132:10–21.

Nowak, D. J., and D. E. Crane. 2002. Carbon storage and sequestration by urban trees in the USA. *Environmental Pollution* 116 (3):381–389.

O'Neill, B. C., M. Dalton, R. Fuchs, L. Jiang, S. Pachauri, and K. Zigova. 2010. Global demographic trends and future carbon emissions. *Proceedings of the National Academy of Science* 107 (41):17521–17526.

Ozturk, I. 2010. A literature survey on energy-growth nexus. *Energy Policy* 38 (3):340–349.

Pachauri, S. 2004. An analysis of cross-sectional variations in total household energy requirements in India using micro survey data. *Energy Policy* 32(15):1723–1735.

Pachauri, S., and D. Spreng. 2002. Direct and indirect energy requirements of households in India. *Energy Policy* 30 (6):511–523.

Pachauri, S., and L. Jiang. 2008. The household energy transition in India and China. *Energy Policy* 36 (11):4022–4035.

Parikh, J., and V. Shukla. 1995. Urbanization, energy use and greenhouse effects in economic development. *Global Environmental Change* 5 (2):87–103.

Pataki, D. E., R. J. Alig, A. S. Fung, N. E. Golubiewski, C. A. Kennedy, E. G. McPherson, D. J. Nowak, R.V. Pouyat, and P. Romero-Lankao. 2006. Urban ecosystems and the North American carbon cycle. *Global Change Biology* 12 (11):2092–2102.

Payne, J. E. 2010. Survey of the international evidence on the causal relationship between energy consumption and growth. *Journal of Economic Studies* 37 (1):53–95.

Poumanyvong, P., and S. Kaneko. 2010. Does urbanization lead to less energy use and lower CO_2 emissions? A cross-country analysis. *Ecological Economics* 70 (2):434–444.

Pouyat, R.V., I. D.Yesilonis, and D.J. Nowak. 2006. Carbon storage by urban soils in the United States. *J. Environ. Qual.* 35:1566–1575.

Rock, M. T., M. Toman, Y. Cui, K. Jiang, Y. Song, and Y. Wang. 2013. Technological Learning, Energy Efficiency, and CO2 Emissions in China's Energy Intensive Industries. In *Policy Research Working Paper 6492*. Washington, DC: World Bank.

Romero-Lankao, P., K. R. Gurney, K. C. Seto, M. Chester, R. M. Duren, S. Hughes, L. R. Hutyra, P. Marcotullio, L. Baker, N. B. Grimm, C. Kennedy, E. Larson, S. Pincetl, D. Runfola, L. Sanchez, G. Shrestha, J. Feddema, A. Sarzynski, J. Sperling, and E. Stokes. 2014. A critical knowledge pathway to low-carbon, sustainable futures: Integrated understanding of urbanization, urban areas, and carbon. *Earth's Future* 2 (10):515–532.

Sahely, H. R., and C. Kennedy. 2007. Water use model for quantifying environmental and economic sustainability indicators. *Journal of Water Resources Planning and Management* 133 (6):550–559.

Santamouris, M., A. Synnefa, and T. Karlessi. 2011. Using advanced cool materials in the urban built environment to mitigate heat islands and improve thermal comfort conditions. *Solar Energy* 85 (12):3085–3102.

Sassen, S. 2006. *Cities in a World Economy*. Thousand Oaks, CA: Sage Publications.

Satterthwaite, D. 2008. Cities' contribution to global warming: Notes on the allocation of greenhouse gas emissions. *Environment and Urbanization* 20 (2):539–549.

———. 2009. The implications of population growth and urbanization for climate change. *Environment and Urbanization* 21 (2):545–567.

Schulz, N. 2010a. Delving into the carbon footprint of Singapore – comparing direct and indirect greenhouse gas emissions of a small and open economic system. *Energy Policy* 38 (9):4848–4855.

———. 2010b. Urban energy consumption database and estimations of urban energy intensities. In *GEA KM 18 Working Paper*. Vienna: IIASA.

Seto, K. C., B. Guneralp, and L. Hutyra. 2012a. Global forecasts of urban expansion to 2030 and direct impacts on biodiversity and carbon pools. *Proceedings of the National Academy of Sciences of the United States of America* 109 (40):552–563.

Seto, K. C., A. Reenberg, C. G. Boone, M. Fragkias, D. Haase, T. Langanke, P. Marcotullio, Darla K. Munroe, B. Olah, and D. Simon. 2012b. Urban land teleconnections and sustainability. *Proceedings of the National Academy of Science* 109 (18):doi: 10.1073/pnas.1117622109.

Seto, K. C., S. Dhakal, A. Bigio, H. Blanco, G. C. Delgado, D. Dewar, L. Huang, A. Inaba, A. Kansal, S. Lwasa, J. McMahon, D. Mueller, J. Murakami, H. Nagendra, and A. Ramaswami. 2014. Human settlements, infrastructure, and spatial planning. In *Climate Change 2014: Mitigation of Climate Change. Working Group III Contribution to the Fifth Assessment Report of the Intergovenmental Panel on Climate Change*, eds. O. Edenhofer, R. Pichs-Madruga, Y. Sokona, E. Farahani, S. Kadner, K. Seyboth, A. Adler, I. Baum, S. Brunner, P. Eickemeier, B. Kriemann, J. Savolainen, S. Schlömer, C. von Stechow, T. Zwickel, and J.C. Minx. Cambridge, UK and New York, NY, USA: Cambridge University Press.

Shen, L., S. Cheng, A. J. Gunson, and H. Wan. 2005. Urbanization, sustainability and the utilization of energy and mineral resources in China. *Cities* 22 (4):287–302.

Sugar, L., C. Kennedy, and E. Leman. 2012. Greenhouse gas emisssion from Chinese cities. *Journal of Industrial Ecology* 16 (4):DOI: 10.1111/j.1530-9290.2012.00481.x.

Tisdale, H. 1942. The process of urbanization. *Social Forces* 20 (3):311–316.

VandeWeghe, J. R., and C. Kennedy. 2008. A spatial analysis of residential greenhouse gas emissions in the Toronto census metropolitan area. *Journal of Industrial Ecology* 11 (2):133–144.

Vringer, K., and K. Blok. 1995. The direct and indirect energy requirements of households in the Netherlands. *Energy Policy* 23 (10):893–910.

Weber, C. L., and H. S. Matthews. 2008. Quantifying the global and distributional aspects of American household carbon footprint. *Ecol. Econ.* 66 (2–3):379–391.

Weisz, H., and J. K. Steinberger. 2010. Reducing energy and material flows in cities. *Current Opinion in Environmental Sustainability* 2 (3):185–192.

Wier, M., M. Lenzen, J. Munksgaard, and S. Smed. 2001. Effects of household consumption patterns on CO2 requirements. *Economic Systems Research* 13 (3):259–274.

Wilkes, R. E. 1995. Household life-cycle stages, transitions and product expenditures. *Journal of Consumer Research, Inc* 22 (1):27–42.

Wilkinson, S. J., and R. Reed. 2006. Office building characteristics and links with carbon emissions. *Structural Survey* 24 (3):240–251.

York, R. 2007. Demographic trends and energy consumption in European Union nations, 1960–2025. *Social Science Research* 36 (3):855–872.

York, R., E. A. Rosa, and T. Dietz. 2003. STIRPAT, IPAT and ImPACT: Analytic tools for unpacking the driving forces of environmental impacts. *Ecological Economics* 46 (3):351–365.

9

SUBURBAN LANDSCAPES AND LIFESTYLES, GLOBALIZATION AND EXPORTING THE AMERICAN DREAM

Robin M. Leichenko and William D. Solecki

Three of the great transformations of the 21st century include ever tightening economic globalization, increasing urbanization of low- and middle-income countries and the acceleration of human-induced global environmental change (GEC). Current trends suggest that increasingly interdependent economic development and associated communication linkages will bring the economies and cultures of the world into ever closer contact: that 60 percent of the population in low- and middle-income countries could live in urbanized areas by 2030 (above the current estimate of 50 percent); and that there could be as much as a 6°C degree increase in global temperature by the year 2100. Central to the inter-related processes of globalization, urbanization and environmental change are shifts in patterns of both production and consumption. While there have been numerous analyses of the connections among globalization, urbanization and changing patterns of industrial production, there is also widespread recognition of the need to examine consumption-related aspects of globalization in cities and the environmental and social consequences of that consumption (Dicken, 2011; Seto et al., 2010; Leichenko and Solecki, 2008; McCarthy, 2007; Sklair, 2002; Molotch, 2003; Bridge, 2002; Princen et al., 2002; Wilk, 2002).

This chapter examines how economic, political and cultural aspects of globalization are interacting with processes of urbanization throughout the world to create new landscapes of suburban housing consumption.[1] Although globalization forces are not ubiquitous drivers of urban change in low- and middle-income countries (McGee, 2002) and the impacts of globalization are highly varied among cities (Shatkin, 2007), global processes represent key factors that interact with regional and local level conditions and lead to transformation of housing consumption in many cities. A central proposition of the chapter is that globalization of housing consumption preferences and practices of a growing, middle-income segment of urban residents within low- and middle-income countries is leading to patterns of urban resource use that are akin to those associated with suburbanization and suburban sprawl found in higher income countries, particularly the United States and Canada. In effect, there is an export of the American Dream—the ideal of homeownership of a single-family home in a suburban area. This trend toward suburbanization within low- and middle-income countries raises critical questions about social and environmental sustainability. In this context, suburbanization is a phenomenon that allows a relatively small, but increasingly wealthy, middle class to spatially segregate itself from the vast majority of poorer residents. It may be thus understood as a spatial manifestation of the rising income inequalities associated with economic globalization. Suburbanization of low- and middle-income country cities can be expected to have dramatic impacts on the local urban and global environment. Local impacts, which have been well documented in wealthier country cities, include increased water demand and fossil fuel consumption, increased air and water pollution, and loss of agricultural lands and natural habitats. Globally, suburbanization on a mass scale will exacerbate ongoing processes of environmental change, as residents of low- and middle-income country cities increasingly emulate the highly resource-consumptive, energy intensive and ultimately unsustainable lifestyles currently practiced by suburbanites in the United States and elsewhere.

The following section discusses processes of urbanization in low- and middle-income countries and the influence of globalization on these processes. It also focuses on the changing housing structure and residential patterns of these countries' cities. The second section identifies globalization-related drivers of housing preferences. In particular, it explores linkages among globalization and consumption patterns, the definition of worldwide consumer preferences and the emergence of a 'globalized' lifestyle amongst the emerging middle classes. The third section considers several policy-related drivers of urban landscape change and their relative role in defining housing choice options for the emerging middle class. The concluding section explores implications of these inter-related processes for questions of spatial and environmental equity.

Urbanization and housing trends

According to the United Nations (2014) report, 54 percent of people globally reside in urban areas. The rate of urbanization is especially rapid in low- and middle-income countries, where urban growth rates are more than 2.5 percent per year, a rate that is more than four times the rate of urban population growth in high-income countries (United Nations, 2012). United Nations projections state that the vast majority of population growth during the next three decades will be in urban areas in low- and middle-income countries (see other chapters in this volume for discussion on global urbanization trends).

Similar to the historical pattern in the high-income countries, individuals are leaving the rural countryside in low- and middle-income countries for the cities in pursuit of a better quality of life. Importantly, this better life sought by rural in-migrants is at least partially defined and heightened by the potential of gaining access to consumer goods that are associated with globalization such as cell phones, televisions, automobiles and so forth. Another significant draw for many migrants is the possibility of employment in foreign direct investment (FDI)-related manufacturing and assembly jobs (Dicken, 2011). Globalization processes also play other roles in pushing people out of rural areas. Poor economic conditions in rural areas—particularly for small-scale farmers—are exacerbated by the globalization of agricultural markets, which has increased competition and dramatically lowered the prices paid to farmers of many raw agricultural products such as maize, coffee beans and oilseeds (Watts and Goodman, 1997; Sklair, 2002). Driven by these and other economic factors, people leave the job-poor rural areas for cities in search of economic security.

In addition to population growth, another important demographic component of housing demand in cities is the change in average household size. The average household size in

Suburban landscapes and lifestyles

low- and middle-income countries declined by almost 14 percent during 1970 and 2000 (from over 5.1 individuals per household to 4.4 per household) (Keilman, 2003). As a result of this decline, the number of households has increased faster, in percentage terms, than the population in most of these country's cities over the past three decades. This reduction in household size results from a series of factors including a decline in birth rates, personal choice, decline of extended-family communal living as well as increased affluence and the growth of a middle class in many cities. While the level of affluence in some low- and middle-income country cities is often quite low in comparison to high-income country standards, the wealth of many current households in these cities is substantially greater than that of previous generations.

As a result of these demographic trends, there has been pronounced expansion in the spatial extent of cities in recent years (Seto et al., 2012; see also Haase and Schwartz, Chapter 4 in this volume, for a discussion of urbanization and land use). For example, in China, the rate of growth of urban land area between 1990 and 1995 was more than four times faster than the urban population growth rate (Zhang, 2000). In Beijing, the built-up area of the city increased from 335km^2 in 1978 to 488km^2 in 1998, an increase of 45.6 percent (Gu and Shen, 2003). The growth in the spatial extent of these cities is partly the result of attempts to house the massive influx of low skilled, migrant workers (UN-Habitat, 2003), who frequently reside in squatter settlements and temporary housing often found on marginal lands (e.g. hazard prone) or on lands at the urban periphery. In cases such as Caracas (Venezuela) or Mumbai (India), such areas provide housing for up to half of the city's population (UN-Habitat, 2003). In Cairo and many other cities, urban growth has rapidly expanded to overwhelm the existing infrastructure (e.g. roads, water supply and waste treatment) and created tremendous environmental degradation and pollution (El Araby, 2002).

In addition to providing housing for the poor, land at the edges of many cities is also being converted into property for commercial and industrial facilities and for middle- and high-income residential housing (UN-Habitat, 2010). Most obvious of these changes are the large-scale, capital-intensive developments that have started to appear in locations distant from the old central business and administrative cores (Marcuse and van Kempen, 2000). Although the dominant trend is not single-family homes, but multi-family condos and high-rises, an important and rapidly growing component of this change is the development of low-density, American-style suburban housing, which is for the emerging middle and upper classes (see Figure 9.1) (Wang et al., 2014; Feng, 2014; D'Arcy and Veroude, 2014; Leichenko and Solecki, 2008; Aguilar and Ward, 2003; Firman, 2004; Gu and Shen, 2003; Bunnell et al., 2002; Zhang, 2000). This process of change can be associated with the construction of consumption landscapes.

The term consumption landscapes, as used here, refers to the transformation of an urban area by the influences of globalization, causing changes in the physical, economic and social characteristics. Consumption landscapes are typically constructed in sites undergoing rapid urbanization. These landscapes emerge from the interactions among urbanization, globalization, changing cultural values and consumption patterns. The interaction of these processes help set in motion a sequence of steps that can cause increases in energy and resource use among urban residents in least developed countries. Secondary issues, such as increased urban resident demands for water resources, fossil fuels for energy, air and water pollution, and the degradation of natural and agricultural lands, also result from increased globalization.

Consumption landscape development is evident in urban Asia in a variety of national contexts. While most of the associated construction involves multiple-family structures and

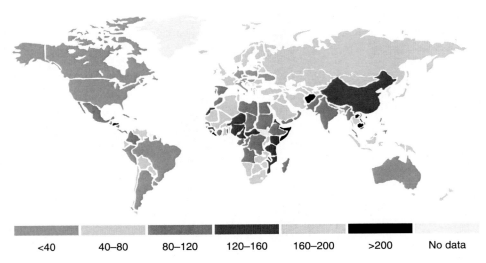

Figure 9.1 The rich spread out: median density of large cities (persons per hectare)
Source: Shlomo Angel, Lincoln Institute of Land Policy.

in many cases multi-story apartment complexes, this type of landscape is often associated with the traditional suburban ideals of lower-density, suburban-style lifestyles found in the United States, Australia and some European countries. Within India, for example, outside of some major cities, households that are wealthy tend to build 'islands of affluence' (Shirgaokar, 2012) in suburban areas with private infrastructure systems, creating gated communities. Suburbanization, therefore, often leads to poorer households being forced to live at the furthest outskirts of major cities or in other less than desirable conditions (e.g. more vulnerable to risks). According to Graham and Marvin (2001), many cities in India have undergone 'splintering urbanism,' where the movement of affluent families to the suburbs from cities is marginalizing poorer families. New suburban expansions also frequently 'leapfrog' to the fringes of Indian cities (Brueckner and Sridhar, 2012). Movement to the suburbs prompts households to shift from using forms of informal and often lower carbon-emitting transport (i.e. bicycle or other shared modes) and to purchase individual vehicles, such as mopeds, scooters or motorcycles (Shirgaokar, 2012). This condition adds to India's already growing greenhouse gas (GHG) emissions and, presently, there are few if any policies in place to curb vehicle ownership (e.g. as through regulations or increasing ownership costs) (Shirgaokar, 2012). Brueckner and Sridhar (2012) document that an increase in building height limits in Indian cities is contributing to the growth of suburban communities; households are being built higher in the distant suburbs in order to accommodate more multiple families.

The process of suburbanization is taking place to some extent in China as well. Residential suburban development has been encouraged despite a lack of sufficient public amenities and services, especially adequate transportation infrastructure. In the case of the city of Tianjin, large amounts of residential areas along the city's edge are increasingly in the form of suburbs. This has been a result of the 1996 Master Plan of Tianjin, under which residential areas were built between inner existing suburbs and the urban core to decrease population pressures on Tianjin (Liu, 2014). Yao and Wang (2014) find, over the past decade across China, that suburban sprawl is transforming into suburb-like entities around megacities (e.g. Beijing).

In urban China overall, the highest population growth rates have been around the periphery of cities. This process has been driven by a variety of factors. Most new development outside of major cities since 2008 has been along subway lines, reflecting reliance on mass transit over cars, especially for poorer households (Yao and Wang, 2014). Within the inner-city of some urban areas, such as Shanghai, inner-city demolition (in order to expand downtown commercial areas) has forced households to resettle away from the city core. This has caused an increase in jobs being further away from the city center. Around Shanghai, industrial suburbanization also has been taking place, as households settle close to industrial parks on the edge of the city in order to live closer to workplaces and reduce costs of living. This form of suburbanization has prompted migrant workers, in particular, to live in industrial suburbs (Singh, 2014). As the market economy of China continues to expand, globalization continues to play a significant role in suburban development. This interconnected process results in numerous issues including longer commute time and distance, deteriorating jobs-to-housing balance, traffic congestion, increased GHGs and air pollutant emissions (Kwan et al., 2014).

The metropolitan growth of Mexico City also documents this general trend. Globalization driven growth has promoted a decentralization of population in the region as well as the emergence of an integrated peri-urban zone that rings the older suburbs and central city. In central Mexico City, the population growth has started to slow and even become negative in some inner-city districts. With this have come three broad demographic shifts. These shifts include the flight of mostly middle-class populations to other parts of the region; further decline in inner-city population, compensated in part by an increasing density of the existing built-up area, particularly within the squatter settlements; and an inflow of new migrants both to the urban periphery, and even more so, into the peri-urban areas (Aguilar and Ward, 2003).

A few other examples of this phenomenon include Kuala Lumpur (KL), where the more highly skilled and educated residents constitute the largest single component of migrants leaving the city to live in the outer KL metropolitan area (Bunnell et al., 2002). Furthermore, recent residential planning efforts in KL call for creating attractive living environments for the middle and upper classes outside of the central city (Bunnell et al., 2002). The Jakarta Metropolitan Region is another example. During the 1990s, 25 large-scale subdivision projects (locally known as 'new towns') ranging in size from 500 to 6,000 hectares were built on the outskirts of the city (Firman, 2004). Most of these new towns are made up of low-density, single-family houses, exclusively for middle- and upper-income groups (Leaf, 1993, as discussed in Firman, 2004). The physical design of the developments very much resembles the design of residential areas in wealthier countries, specifically the United States. Many are planned and designed by expatriate architects, urban planners and property specialists who have limited knowledge of local architecture and city planning, even though they are hired by local developers (Dick and Rimmer, 1998).

It should be recognized that the extent to which these processes are emerging varies widely as a result of the varying levels of economic growth and transformation. Although these conditions are most prevalent in cities undergoing rapid globalization-associated changes (e.g. industrialization, political decentralization), such as those in East Asia and Southeast Asia (Marcotullio, 2003; Marcotullio, Chapter 8 in this volume), smaller low- and middle-income cities also are experiencing demographic and urban development shifts. In all, the urban landscape of these cities in the past decade or so has increasingly started to take on some non-traditional elements and functions (Marcuse and Van Kempen, 2000). Importantly, new suburban landscapes are spatially segregated from the central city and slum areas, and portend lifestyles that are much more resource use intensive than has been the case in the past. Some of the key globalization-related drivers creating these new consumption landscapes are detailed in the next sections.

Globalization and consumption: forces of change

Within cities of low- and middle-income countries, globalization is profoundly influencing both the cultural and economic life of residents. Globalization of mass media, the proliferation of advertising by firms and the widespread availability of Western-style consumer goods have influenced consumption preferences and brought new consumption choices (James, 2000; Sklair, 2002). This spread of Western-style material aspirations and the growing availability of products to meet these aspirations represent key points of intersection between processes of globalization and consumption in many cities.

Much of the attention to consumption aspects of globalization focuses on the growing availability of North American and Western European consumer goods such as international name brand fast food, clothing, cosmetics and carbonated beverages (Sklair, 2002). The growing globalization of such commodities has accelerated in recent years as Western companies have attempted to expand international markets in the face of increasing domestic competition and saturation of developed country markets (Schlosser, 2001). Often somewhat unseen in the face of these, at times, spectacular transformations has been the widespread adoption of Western and/or modern technologies. These devices, many of which are vital for the mass and social media transmission of knowledge about consumer goods, include the increased use of televisions (an estimated two-thirds or more of the world's population regularly watches television), computers, cell phones, satellite dishes as well as other consumer products such as air conditioners and washing machines. Even more radical changes in consumption patterns are taking place with respect to modes of transportation (Chinese Academy of Engineering and National Research Council, 2003). Many city residents now look forward to the transition from bicycle to gas-powered scooter, to motorcycle and possibly to automobile. These trends have already transformed the daily life in cities such as Bangkok, Beijing, KL and Mexico City, which now experience tremendous traffic congestion (Marcotullio, 2001; Weiner, 2002). Individually these products represent time-savings and conveniences for urban residents. Increasingly, however, these devices also are being packaged together to represent a new urban lifestyle and attitude toward material advancement (Sklair, 2002). This package of a new and better life is also increasingly linked to the ultimate expression of one's consumer aspirations—one's residence, particularly its form and location. At present, achieving this 'better life' is a reality for only a small, albeit growing, percentage of city residents (typically less than 10 percent) (Wang, 2001). It is important to recognize, however, that this vision, which Sklair (2002) terms the 'ideology of consumption,' is received not only by emerging middle-income urban inhabitants, but also by poorer urban residents and by inhabitants of many remote rural areas as well, who receive these images via Western television programming and other forms of mass and social media.

The linkages between advertisers and consumers are well illustrated in the case of Shanghai. Starting in the 1990s, a close inter-relationship developed between local advertisers, real estate developers and individual consumers in the city. After the housing shortages of the 1980s and the marketization reforms of the early 1990s, residents soon recognized that home ownership (typically associated with apartment ownership) had become a necessity for those desiring an improved quality of housing (Davis, 2002). Homes increasingly were seen as spaces where residents could assert their aspirations for individuality, personal indulgence and social distinctiveness, and homeowners felt free to use their homes to display their economic success and cultural refinement. Wide-circulation home decorating magazines and billboards encouraged consumers to leave behind the older simple housing and interior decorating styles, and strive to furnish their residences in richer cultural taste. Developers also responded to these trends by increasingly seeking customers through the promise of exclusive lifestyles rather than simply the prospect of home ownership (Fraser, 2000).

These globalization-related shifts in consumer aspirations reflect important forces of change in most low- and middle-income country cities, even though present consumption patterns remain highly uneven within particular markets and cultures (Djursaa and Krang, 1998; Dicken, 2011). Two housing surveys (one in Singapore and another in Guangzhou, China) illustrate the current discrepancies between residents' aspirations for new housing versus acquisition ability. Both surveys asked residents about their housing and mobility preferences (i.e. where they would like to move and where are they moving). The Singapore survey asked two central questions: whether it was the 'dream' of public housing residents to own a private house and what type of house people wanted to move to if they could. The results show that while a large percentage of residents had interest in owning private property in a relative low-density area, only 38 percent of respondents would allow themselves to 'dream' of such a house (Wong and Yap, 2003). Most others would not because of the high and unapproachable cost.

The Guangzhou survey focused on residential choice and mobility (Li and Siu, 2001). The results suggest that there was a general movement of residents toward the urban periphery, but that even under conditions of market transition (the survey was conducted in 1996) most of the residents were still not able to assert their housing preferences. Instead, the *Danweis* (work units) and municipal government were still seen as dominant actors in the definition of housing options for residents, and, as a result, much of the suburbanization in the region was not voluntary. Li and Siu (2001) conclude, nonetheless, that income segmentation and the rise of middle- and higher-income classes were beginning to emerge as forces of change within the housing market.

The above examples suggest that while globalization of consumption is, indeed, influencing housing preferences, housing consumption decisions in low- and middle-income countries are not simply a function of innate individual preferences and cultural changes. Rather, these decisions are influenced by both efforts to sell images of Western lifestyles *and* by international, national and local policy contexts. The imagery and marketing help create the preferences and the policies foster the conditions of capital accumulation (at both a societal and household level) that enable the actualization of these preferences. The next section explores these policies in more detail.

Global neoliberalism in cities

Concurrent with changing housing consumption patterns, there have been dramatic changes in the institutional and policy context for urban development (Seto et al., 2010). Often lumped together under the rubric of 'neoliberalism' or 'neoliberalization,' these transformations—all of which are associated with globalization—have initiated a suite of urban landscape changes (Peck and Tickell, 2002). The neoliberal drivers associated with housing changes may be grouped into three broad categories: 1) liberalization of investment and trade policies; 2) decentralization (devolution) of power and regulatory reforms; and, 3) the emergence of private real estate markets and financial institutions to support these markets. Each of these processes are, in turn, associated with the rise of new modes of capital accumulation, which help to channel and shape the emerging patterns of urban development. For example, liberalization of investment and trade policies have brought FDI and associated export-oriented industrial facilities and mega-projects (such as the Pudong area in Shanghai) to many cities. These investments, in turn, result in local capital accumulation that can be put back into the local economy to promote consumption driven economic sectors (e.g. housing development).

These investments also help to create a 'globalized' capital accumulating class within the city, a class that typically emerges out of the ranks of local political and economic cliques that historically are present in cities along with former ex-patriots (see Jones, 1998, and Zhang, 2002, for examples from Santiago and Shanghai, respectively). These individuals, together with executives from transnational corporations, globalizing state and inter-state bureaucrats, globalizing professionals (e.g. engineers, builders), and merchants and media representatives, facilitate new modes of capital accumulation and form local growth coalitions. For groups such as these, which Sklair (2002) refers to as the 'transnational capitalist class,' the creation and development of consumption driven activities become an attractive complement to production driven modes of capital accumulation.

Jakarta provides a good example of how suburban housing demand and an associated private property market may be created through the deliberate efforts of local growth coalitions that are attempting to capture and sustain the benefits of transnational capital flows (Douglass, 2001). In metropolitan Jakarta, corporate development companies play a central role in the development of urban fringe, new town construction. During the time period in which these new towns were constructed, from the late 1980s until the mid-1990s, there was easy local access to investment funds for housing development because prosperous inter-national and national financial markets brought forward a large amount of excess capital into the city (Dijkgraaf, 2000, and Leisch, 2002, as cited in Firman, 2004). As described by Firman (2004), demand for new towns in Jakarta has been in large part created by developers who were able to foster an image of these areas as symbols of 'modernism' and Western lifestyles. Underlying these 'sales efforts' is the common assumption held by developers, potential buyers and segments of the larger society that the lifestyles of Western urbanization (i.e. suburbanization) should naturally emerge as the result of economic development and industrialization.

A second key facet of neoliberalism is devolution of political power. The implementation of neoliberal political and economic policies has significant impacts on low- and middle-income countries at a range of spatial scales (Peck and Tickell, 2002). These reforms are typically instituted in response to directives from international lending or trade organizations (e.g. International Monetary Fund, World Bank or World Trade Organization) to debtor nations as a component of loan negotiations. Central governments of low- and middle-income countries are encouraged by these lending organizations to use decentralization and privatization as mechanisms to balance current accounts. The implication of these reforms for many cities is a shift of power from the central state to the local state (Mitchell, 2000). In the case of the greater Caracas metropolitan region, for example, political decentralization has been characterized by rapid development and a surge in private real estate investment in the *municipios* ringing the capital district. Specific consequences of this decentralization and municipal fragmentation include an exacerbation of long-standing problems of metropolitan infrastructure management, diseconomies in urban socio-spatial structures (e.g. a shift in real estate investment away from the central city) and ultimately, greater spatial inequalities including the development of exclusive suburban-style communities for wealthy residents in the rapidly low- and middle-income edge *municipios* (Mitchell, 2000).

Suburban landscapes and lifestyles

A third, related, change associated with economic and political liberalization is the development of market driven, local real estate economies. This process typically involves connections between the national government and the local state including, for example, national policy changes which promote the marketization of the local real estate economy and a shift toward increasing private control (ownership) of property or the creation of extended property leases. The policy transformations in China over the past 25 years provide an excellent example of this trend. Reforms started in the late 1980s have led to the current situation where property renters can sell their leases in a private real estate market. The reforms allowed the Chinese government to establish a thriving private real estate market while allowing it to maintain ownership of the land. Coupled with these reforms came an increase in the length of time leases could be held, which in most cases now exceed 50 years, and in some situations approach 100 years. A cumulative impact of all the changes has been the rapid rise of a real estate entrepreneurial class in Chinese cities (Logan, 2002).

The resolution of potential land use conflicts in cities may also result in decisions that are both economically efficient and equitable (i.e. all coalitions have an equal opportunity to profit), yet inefficient in terms of land consumption. As seen in the United States setting of metropolitan Miami, the resolution of land use conflicts may lead to patterns of residential growth and urban development that are patchy and uneven, as competing growth coalitions stake out different pieces of the urban landscape (Solecki, 2001). A similar phenomenon may be seen in Jakarta, where large-scale developments have been constructed by a wide range of different interest groups, resulting in new towns that are physically separated from each other and poorly linked to the existing infrastructure (Dijkgraaf, 2000 in Firman, 2004).

Understanding the implications of neoliberalism in cities requires the recognition that urban development patterns are influenced by each city's history and that significant variation exists in the extent to which these processes of neoliberalization have developed (Shatkin, 2007; Marcotullio, 2003). Each city's historical geography may be understood as a lens through which neoliberalizing processes intersect with local agents, institutions and the existing urban landscape (Massey and Allen, 1984). Shanghai, for example, has been attempting to re-assert its position as a global finance center, a role it lost during the early post-revolutionary period. In order to achieve this goal, a series of large-scale development efforts are ongoing throughout the city and metropolitan region. These efforts, initiated through the actions of the central Chinese government, the local municipal government and business leaders, include the massive Pudong project, a new international airport in newly built-up area and new, central city commercial office complexes made possible by the razing of huge swaths of older, low-rise residences (Shih, 2003; Wu, 2002; Olds, 2001).

Conclusion: directions for further research on GEC, suburbanization and inequality

The emergence of suburban consumption landscapes in low- and middle-income country cities signals a convergence among large-scale processes of urbanization, globalization and environmental change. This chapter suggests that key drivers of consumption landscapes include demographics (e.g. population growth, rural to urban migration, per capita income increases and disparities), the ideology of North American and increasingly many parts of Western European consumption, and globalization-related policy changes. These drivers, in turn, affect the decisions of both local real estate developers and housing consumers, with dramatic implications for both local and global environments. Two of the central concerns raised in many analyses of the impacts of globalization include the problems of accelerated environmental degradation and income polarization (Dicken, 2011; Leichenko and O'Brien, 2008; O'Brien and Leichenko, 2000, 2003; Sklair, 2002; Deardorff and Stern, 2000; Sakakibara, 2000). Both of these concerns, which suggest that globalization in its present trajectory is neither environmentally nor socially sustainable, are relevant for understanding new forms of suburbanization. The spread of suburban development and the shift to higher consumption lifestyles will have serious negative implications for the environment through effects on land use conversion, water and energy resource demands, and air and water pollution levels. Currently, only few studies have tracked the environmental impacts of increased global urbanization, in general, and suburbanization, in particular (e.g. Seto et al., 2010).

Another sustainability facet of suburbanization is that, like globalization, suburbanization enables consumers to separate themselves from the environmental and social implications of their actions (Princen, 2002). In suburbia, middle- and high-income residents distance themselves from local environmental and social problems (e.g. congestion and crime), and from other groups they deem as undesirable (e.g. the urban poor) (UN-Habitat, 2010; Lake, 2002). Several studies have illustrated that globalization has intensified spatial segregation and social fragmentation in low- and middle-income country cities (Shatkin 2007; de Queiroz Ribeiro and Telles, 2000; Chakravorty, 2000).

While low- and middle-income country cities that adopt a variant of the American Dream may potentially repeat the environmental mistakes of other cities, particularly those in the United States, this convergence in development patterns also presents many potential lessons and opportunities for two-way knowledge transfers (Seto et al., 2010). In making such connections, it will be important to note the differences between suburbanization in US cities versus other countries. The American Dream in the United States came about as a result of a confluence of forces set within a specific historical context. Individual consumption and the promotion of home ownership and suburban-style development in the United States were defined during the emergence of a new automobile-based economy and a corresponding set of federal initiatives (e.g. tax incentives, highway and bridge construction, and home mortgage subsidy measures), coupled with local land use policies (e.g. new low-density residential zoning) (Jackson, 1985). The diverse policy contexts of cities of low- and middle-income countries suggest that while the patterns of spatial development may resemble the American Dream, the processes through which this development occurs will be distinct. Understanding new processes of suburbanization, their differences with the US model, and implications for questions of environmental equity and justice remain a critical challenge for further research.

Main messages

- The global urbanization process has brought forward a great diversity of new landscapes of varying urban density including moderate to relatively lower- density non-urban core, suburban-style consumption landscapes.
- The drivers of consumption landscape formation include metropolitan-scale income disparities, neoliberal policy expansion, governance constraints and integration of Western-style (mostly North American and Western Europe) lifestyle choices.
- Suburban consumption landscape construction has important implications for the future conditions of environmental equity, vulnerability and exposure to environmental risks and hazards, and patterns of overall resource use including GHG emissions.

Future research/practice

Several key research questions and issues regarding the future practice of urbanization remain:

- Research into the feedback effects of increasing metropolitan growth, inequity and suburban consumption landscapes is needed. Does acceptance of the American Dream lifestyle in cities also serve as a way of maintaining and furthering both social inequalities and environmental injustices?
- What are the specific pathway connections between growing social and spatial inequality in low- and middle-income countries and urban environmental degradation?
- What are the aggregate environmental effects of a rise in suburban housing across low- and middle-income country cities (especially with respect to GHG emissions)? What are valid scenarios of future environmental impacts associated with widespread suburbanization?
- How do suburban lifestyle choices (i.e. the type of house, furnishings, appliances and mode of transportation) affect individuals' attitudes and decisions about resource consumption?

Note

1 This chapter is drawn heavily from an article that appeared in *Regional Studies* (Leichenko and Solecki, 2005). The chapter includes new material and references, and associated restructuring of the discussion and analysis.

References

Aguilar A.G. and Ward P.M. (2003) Globalization, regional development, and mega-city expansion in Latin America: Analyzing Mexico City's peri-urban hinterland, *Cities 20*, 3–21.

Bridge G. (2002) Grounding globalization: The prospects and perils of linking economic processes of globalization to environmental outcomes, *Economic Geography 78*, 361–386.

Brueckner J.K. and Sridhar K. S. (2012) Measuring welfare gains from relaxation of land-use restrictions: The case of India's building-height limits, *Regional Science and Urban Economics 42*, no. 6, 1061–1067.

Bunnell T., Barter P.A. and Morshidi S. (2002) Kuala Lumpur Metropolitan Region: A globalizing city-region, *Cities 19*, 357–370.

Chakravorty S. (2000) From colonial city to globalizing city? The far-from complete spatial transformation of Calcutta, in: Marcuse P. and Van Kempen R. (Eds) *Globalizing Cities: A New Spatial Order?* pp. 56–77. Blackwell Publishers, Oxford.

Chinese Academy of Engineering and National Research Council. (2003) *Personal Cars and China*, The National Academies Press, Washington, D.C.

D'arcy P. and Veroude A. (2014). Housing Trends in China and India. RBA Bulletin, 63–68. www.rba. gov.au/publications/bulletin/2014/mar/pdf/bu-0314-7.pdf

Davis D. (2002) When a house becomes his home, in: Link P., Madseb R.P. and Pickowicz P.G. (Eds) *Popular China: Unofficial Culture in a Globalizing Society*, pp. 231–250. Rowman & Littlefield, New York.

de Queiroz Ribeiro L.C. and Telles E.E. (2000) Rio de Janiero: Emerging dualization in a historically unequal city, in: Marcuse P. and Van Kempen R. (Eds) *Globalizing Cities: A New Spatial Order?* pp. 78–94. Blackwell Publishers, Oxford.

Deardorff A. and Stern R. (Eds) (2000) *Social Dimensions of U.S. Trade Policy*, The University of Michigan Press, Ann Arbor, MI.

Dick H. and Rimmer P.J. (1998) Beyond the Third World city: The new urban geography of South-East Asia, *Urban Studies 38*, 449–466.

Dicken P. (2011) *Global Shift: Mapping the Changing Contours of the World Economy, Sixth Edition, Guilford*. Guilford, New York.

Dijkgraaf C. (2000) The urban building sector in Indonesia: Before and after the crisis of 1997. Paper presented at the workshop on 'Indonesian Town Revisited,' The University of Leiden, December 6–8.

Djursaa M. and Krang S.U. (1998) Central and peripheral consumption contexts: The uneven globalization of consumer behavior, *International Business Review* 7, 23–38.

Douglass M. (2001) Urban and regional policy after the era of naive globalism, in: Kumssa A. and McGee T.G. (Eds) *Globalization and the New Regional Development: New Regional Development Paradigms; Vol. 1*, pp. 33–56. Greenwood Press, Westport, CT.

El Araby M. (2002) Urban growth and environmental degradation, *Cities 19*, 389–400.

Feng L. (2014) Urban residential area transitions in China: Patterns, trends and determinations. Proceedings of the ICE – Urban Design and Planning.

Firman T. (2004) New town development in Jakarta metropolitan region: A perspective of spatial segregation, *Habitat International 28*, 349–368.

Fraser D. (2000) Inventing oasis: Luxury housing advertisements and reconfiguring domestic space in Shanghai, in: Davis D. (Ed.) *The Consumer Revolution in Urban China*, pp. 25–53. University of California Press, Berkeley.

Graham S. and Marvin S. (2001) *Splintering Urbanism: Networked Infrastructures, Technological Mobilities and the Urban Condition*, Routledge, London.

Gu C. and Shen J. (2003) Transformation of urban socio-spatial structure in socialist market economies: The case of Beijing, *Habitat International 27*, 107–122.

Jackson K. (1985) *Crabgrass Frontier: The Suburbanization of the United States*, Oxford University Press, New York.

James J. (2000) *Consumption, Globalization, and Development*, St. Martin's Press, New York.

Jones A. (1998) Re-theorising the core: A 'globalized' business elite in Santiago, Chile, *Political Geography 17*, 295–318.

Keilman N. (2003) The threat of small households, *Nature 421*, 489–490.

Kwan M.P., Chai Y. and Tana. (2014). Reflections on the similarities and differences between Chinese and US cities. *Asian Geographer*, 1–8.

Lake R. (2002) Exclusionary environmentalism, local self-sufficiency, and dilemmas of scale in urban environmental politics. Prepared for presentation at the Annual Meeting of the Association of American Geographers, Los Angeles, CA, March 19–24.

Leaf M. (1993) Land rights for residential development in Jakarta: The colonial roots of contemporary urban dualism, *International Journal of Urban and Regional Research 17*, 477–491.

Leichenko R.M. and O'Brien K. (2008) *Environmental Change and Globalization: Double Exposures*, Oxford University Press, New York.

Leichenko R.M. and Solecki W.D. (2005) Exporting the American Dream: The globalization of suburban consumption landscapes, *Regional Studies 39*, no. 2, 241–253.

Leichenko R.M. and Solecki W.D. (2008) Consumption, inequity, and environmental justice: The making of new metropolitan landscapes in low and middle income countries, *Society and Natural Resources 21*, 611–624.

Leisch H. (2002) Gated communities in Indonesia, *Cities 19*, 341–350.

Li S. and Siu Y. (2001) Residential mobility and urban restructuring under market transition: A study of Guangzhou, China, *Professional Geographer 53*, 219–229.

Liu Z. (2014). *Applying a Spatio-Temporal Approach to the Study of Urban Social Landscapes in Tianjin, China* (Doctoral dissertation, University of Ottawa).

Logan J.R. (Ed.) (2002) *The New Chinese City: Globalization and Market Reform*, Blackwell, Malden, MA.

McCarthy J. (2007) Rural geography: Globalizing the countryside, *Progress in Human Geography 32*, 129–137.

McGee T. (2002) Reconstructing the Southeast Asian City in an era of volatile globalization, in: Bunnell T., Drummond L. and Ho K.C. (Eds) *Critical Reflections on Cities in Southeast Asia*, 31–53. Times Academic Press, Singapore.

Marcotullio P.J. (2001) The compact city, environmental transition theory and Asia-Pacific urban sustainable development. Paper presented at International Workshop, New Approaches to Land Management for Sustainable Urban Regions, University of Tokyo, Tokyo, Japan, October.

Marcotullio P.J. (2003) Globalization, urban form and environmental conditions in Asia-Pacific cities, *Urban Studies 40*, 219–247.

Marcuse P. and Van Kempen R. (Eds) (2000) *Globalizing Cities, A New Spatial Order?* Blackwell, Malden, MA.

Massey D. and Allen J. (Eds) (1984) *Geography Matters*, Cambridge University Press, New York.

Mitchell J. (2000) Political decentralization, municipal fragmentation, and the geography of real estate investment in Caracas, Venezuela, *Urban Geography 21*, 148–169.

Molotch H. (2003) *Where Stuff Comes From: How Toasters, Toilets, Cars, Computers, and Many Other Things Came to Be as They Are*, Routledge, New York.

O'Brien K.L. and Leichenko R.M. (2000) Double exposure: Assessing the impacts of climate change within the context of economic globalization, *Global Environmental Change 10*, 221–232.

O'Brien K.L. and Leichenko R.M. (2003) Winners and losers in the context of global change, *Annals of the Association of American Geographers 93*, 99–113.

Olds K. (2001) *Globalization and Urban Change: Capital, Culture, and Pacific Rim Mega-Projects*, Oxford University Press, Oxford, New York.

Peck J. and Tickell A. (2002) Neoliberalizing space, *Antipode 34*, 380–404.

Princen T. (2002) Distancing: Consumption and the severing of feedback, in: Princen T., Maniates M. and Conca K. (Eds) *Confronting Consumption*, pp. 103–131. MIT Press, Cambridge.

Princen T., Maniates M. and Conca K. (Eds) (2002) *Confronting Consumption*, MIT Press, Cambridge.

Sakakibara E. (2000) The end of progressivism: A search for new goals, in: O'Meara P., Mehlinger H. and Krain M. (Eds) *Globalization and the Challenges of a New Century: A Reader*, pp. 71–78. Indiana University Press, Bloomington.

Schlosser E. (2001) *Fast Food Nation: The Dark Side of the All-American Meal*, Houghton Mifflin, New York.

Seto K., Sánchez-Rodríguez R. and Fragkias M. (2010) The new geography of contemporary urbanization and the environment, *Annual Review of Environment and Resources 35*, 167–194.

Seto K., Güneralp B. and Hutyra L. (2012) Global forecasts of urban expansion to 2030 and direct impacts on biodiversity and carbon pools, *PNAS 109*, 16083–16088.

Shatkin G. (2007) Global cities of the South: Emerging perspectives on growth and inequality, *Cities 24*, 1–15

Shih M. (2003). Foreign capital and community displacement in Shanghai: The role of local government. Prepared for seminar on Geography of Globalization: Confronting Consumption. Department of Geography, Rutgers University.

Shirgaokar M. (2012) The rapid rise of middle-class vehicle ownership in Mumbai (Dissertation, University of California, Berkeley, Berkeley, CA).

Singh R.B. (Ed.) (2014) *Urban Development Challenges, Risks and Resilience in Asian Mega Cities*, Springer, Chicago.

Sklair L. (2002) *Globalization, Capitalism and Its Alternatives, Third Edition*, Oxford University Press, New York.

Solecki W.D. (2001) The role of global-to-local linkages in land use/land cover change in South Florida, *Ecological Economics 37*, 339–356.

UN-Habitat (2003) *The Challenge of Slums: Global Report on Human Settlements 2003*, Earthscan, London.

UN-Habitat (2010) *State of the World's Cities 2010/2011 – Cities for All: Bridging the Urban Divide*, Earthscan, London.

United Nations (2012) *World Urbanization Prospects, The 2011 Revision*, United Nations Secretariat, Population Divisions, Department of Economic and Social Affairs.

United Nations (2014) *World Urbanization Prospects, The 2014 Revision*, Highlights, UN, Department of Economic and Social Affairs, Population Division (2014). (ST/ESA/SER.A/352).

Wang Y.P. (2001) Urban house reform and finance in China: A case study of Beijing, *Urban Affairs Review 36*, 620–645.

Wang M.Y., Kee P. and Gao J. (Eds). (2014). *Transforming Chinese Cities*, Routledge, New York.

Watts M. and Goodman D. (1997) *Globalising Food: Agrarian Questions and Global Restructuring*, Routledge, London.

Weiner T. (2002) It's a first-class gridlock, but no easier to unlock, *New York Times, 151*, pA4, May 20.

Wilk R. (2002) Consumption, human needs, and global environmental change, *Global Environmental Change 12*, 5–13.

Wong T. and Yap A. (2003) From universal public housing to meeting the increasing aspiration for private housing in Singapore, *Habitat International 27*, 361–380.

Wu F. (2002) The transformation of urban space in Chinese transitional economy: With special reference to Shanghai, in: Logan J.R. (Ed.) *The New Chinese City: Globalization and Market Reform*, pp. 154–166, Blackwell, Oxford.

Yao Y-L and Wang S. (2014) Commuting tools and residential location of suburbanization: Evidence from Beijing, *Urban, Planning and Transport Research: An Open Access Journal* 2(1), 274–288, DOI: 10.1080/21650020.2014.920697

Zhang T. (2000) Land market forces and government's role in sprawl, *Cities* 17, 123–135.

Zhang T. (2002) Urban development and a socialist pro-growth coalition in Shanghai, *Urban Affairs Review* 37, 475–499.

10

URBANIZATION, HABITAT LOSS AND BIODIVERSITY DECLINE

Solution pathways to break the cycle

Thomas Elmqvist, Wayne C. Zipperer and Burak Güneralp

The interactions between urbanization with biodiversity and ecosystem services that take place defy simple generalizations. There is increasing evidence for the negative impacts of urbanization on biodiversity, most directly in the form of habitat loss and fragmentation. Recent forecasts suggest that the amount of urban land near protected areas is expected to increase, on average, by more than three times between 2000 and 2030 (from 450,000 km^2 *c.* 2000) around the world. During the same time period, the urban land in biodiversity hotspots, areas with high concentrations of endemic species, will increase by about four times on average. However, there is also ample evidence pointing to opportunities to shape urbanization strategies in a way to reconcile urban development and biodiversity conservation strategies (Elmqvist et al. 2013). While gaps in knowledge and practice remain, an increasing number of studies scrutinize the interactions of urbanization with biodiversity and ecosystem services at local, regional and global scales.

Urbanization and biodiversity

Urbanization and biodiversity interact in multifaceted and complex ways (McKinney 2002). Both the size and spatial configuration of urban areas matter for biodiversity (Alberti 2005; Tratalos et al. 2007). While some urban areas have high local species richness, this is typically at the cost of native species (McKinney 2002, 2006). Urban expansion may lead to habitat fragmentation, potentially resulting in genetic or demographic isolation of native species (Ricketts 2001). A major impact of the expansion of urban areas on native species is on their dispersal through changes in habitat configuration and connectivity (Bierwagen 2007). Urbanization is also a major threat to endemic species due to increased incidence of colonization by introduced species (McKinney 2006, 2008).

Urbanization impacts biodiversity and ecosystem services both directly and indirectly. Direct impacts primarily consist of habitat loss and degradation, altered disturbance regimes, modified soils and other physical transformations caused by the expansion of urban areas. Indirect impacts include changes in water and nutrient availability, increases in abiotic stressors such as air pollution, increases in competition from non-native species, and changes in herbivory and predation rates (Pickett and Cadenasso 2009).

Urban expansion and landcover changes

The most obvious direct impact of urbanization on biodiversity is landcover change due to the growth of urban areas. Although urban areas cover less than 3 percent of the Earth's surface, the location and spatial pattern of urban areas have significant impacts on biodiversity (Müller et al. 2013). Worldwide, urban expansion occurs faster in low-elevation coastal zones, which are biodiversity-rich, than elsewhere. Although urban land occupies less than 1 percent of land in the majority of terrestrial ecoregions, it most heavily impacts ecoregions along coasts and on islands, affecting about 10 percent of terrestrial vertebrates found therein. Likewise, more than 25 percent of all endangered or critically endangered species will be affected, directly or indirectly, by urban expansion by 2030 (Güneralp and Seto 2013; McDonald et al. 2013).

In a simple exercise, McDonald et al. (2014), assuming a linear species–area curve and using the expected amount of urban growth (and hence habitat loss) between 2000 and 2030, predict the expected number of endemic vertebrate species that might be lost due to urbanization (see Figure 10.1). They find that urban growth in 10 percent of all ecoregions would account for almost 80 percent of the expected loss in species. This implies that safeguarding species from urbanization in a relatively small number of ecoregions could have a disproportionately large benefit in terms of avoiding biodiversity loss. For example, cities have historically been concentrated along coastlines, on some islands and on major river systems, all of which are often areas of high species richness and endemism.

More than 25 percent of the world's terrestrial protected areas are within 50 km of a city (McDonald et al. 2009). This close proximity has multiple effects, positive and negative, not only on the protected area, but also on the neighboring human population. Negative effects include feral pets, vandalism, illegal dumping, poaching of animals and plants, land squatting and introduction of invasive species. Positive outcomes could include increased potential for recreational activities, eco-tourism and nature-based education thus potentially contributing to increased environmental awareness among residents and visitors. Establishing management practices such as biodiversity corridors in urbanizing regions is desirable, but will require coordinated efforts among administrative bodies within and among nations as well as local residents. The identification and implementation of such corridors may have additional significance considering the migration of species in response to shifts in their ranges with climate change (Forman 2008).

By 2030, the urban lands near protected areas (PAs) are forecasted to increase substantially in almost all world regions (see Figure 10.2; McDonald et al. 2008; Güneralp and Seto 2013). Even in the highly urbanized US, urban land near protected areas may grow by almost 70 percent by the middle of this century (Martinuzzi et al. 2015). By 2030, China will most likely have more urban land within 50 km of their respective PAs than North America or Western Europe. The largest proportional change during the same time period, however, will likely be in Mid-latitudinal Africa, where urban land near PAs is forecasted to increase about 20 times.

Of 34 identified biodiversity hotspots (Mittermeier et al. 2004; Myers et al. 2000), the Mediterranean hotspot contains the most urban land, hugging the coastlines of three continents with different geographic, cultural, social and economic characteristics. In the Mediterranean, for a hotspot that is already diminished and severely fragmented, even relatively modest decreases in habitat can cause pressure on rare species to rise disproportionately (Tilman et al. 1994). By 2030, forecasts suggest that the Mediterranean Basin may become the only hotspot containing more than 100,000 km^2 of urban land (see Figure 10.3; Güneralp and Seto 2013). On the other hand, four biodiversity hotpots that were relatively undisturbed

Urbanization, habitat and biodiversity decline

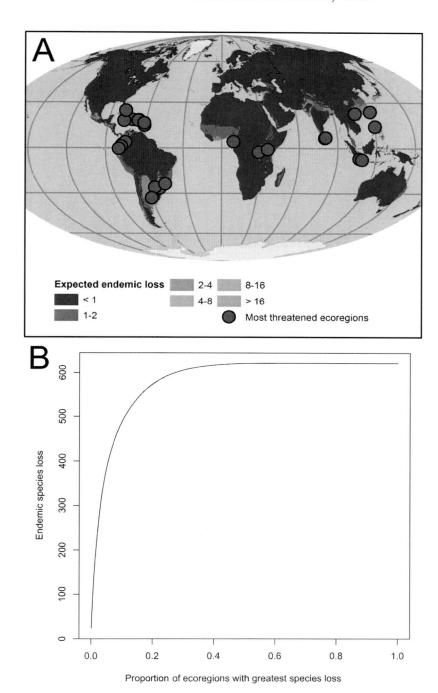

Figure 10.1 Endemic vertebrate species expected to be lost due to urban area expansion: (A) the 25 most threatened ecoregions are shown with dots; (B) the majority of species loss due to urbanization will be in a small fraction of ecoregions

Source: Reproduced under CC license from McDonald et al. (2014) with permission of the Solutions journal (2014).

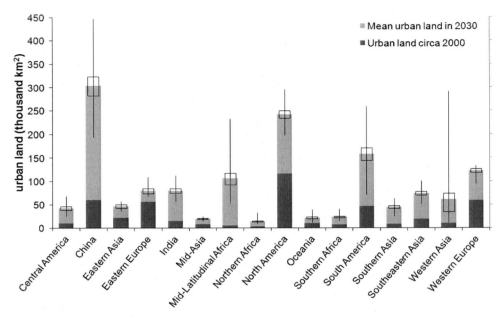

Figure 10.2 Urban extent within a distance of 50 km of PAs by geographic region *c.* 2000 and as forecasted in 2030

Source: Modified from Güneralp and Seto 2013, p. 5. Published under Creative Commons Attribution 3.0 Unported (CC-BY) license with kind permission of Environmental Research Letters 2013. All rights reserved.

by urban land change by 2000 are forecasted to experience the highest rates of increase—over ten times—in urban land cover by 2030. These hotspots are: Eastern Afromontane; Guinean Forests of West Africa; Western Ghats and Sri Lanka; and Madagascar and the Indian Ocean Islands (see Figure 10.3).

Biodiversity hotspots can span national borders creating jurisdictional challenges and issues for management and planning. These challenges and issues cannot solely be met by local level solutions and will require policy responses at a broader scale such as national and international levels. Appropriate strategies with sufficient breadth to protect biodiversity and ecosystem functioning in these multi-jurisdictional hotspots will need to be assessed and implemented through trans-border and regional cooperation among the countries involved (Chettri et al. 2007). Urban expansion will significantly impact freshwater biodiversity on a global scale. Direct and indirect impacts will be most critical in places where there is a confluence of large urban water demands relative to water availability and high freshwater endemism. For instance, Western Ghats of India is projected to have a population of 81 million people with insufficient access to water by 2050. The region also possesses 293 fish species, 29 percent of which are endemic (McDonald et al. 2008). As water resources in the region become limiting, potential species extinction could be substantial.

Urbanization and species patterns

Since cities represent a complex, interlinked system shaped by the dynamic interactions between ecological and social systems, preserving and managing urban biodiversity means going well beyond the traditional conservation approaches of protecting and restoring what are often considered 'natural ecosystems.' Indeed, there is an imperative to infuse or mimic such 'natural' elements in designing urban spaces. Although the basic ecological patterns and processes (e.g. predation, decomposition) are the same in cities and more natural areas, urban ecosystems possess features that distinguish them from other, non-urban ecosystems (Niemelä 1999). Such ecological features include the extreme patchiness of urban ecosystems, prevalence of introduced species and the high degree of disturbances in urban settings. Which species occur in any given urban area depends upon four factors: 1) site availability; 2) species availability (native and non-native); 3) species performance (how well a species does in urban landscape); and 4) site history (Pickett et al. 1987; Williams et al. 2009; Müller et al. 2013). Habitat loss and degradation and the introduction of non-native invasive species may lead not only to the loss of 'sensitive' species dependent on larger, more natural blocks of habitat, but also to the establishment of 'cosmopolitan' species, i.e. generalists that are present in most cities around the world. The net result is sometimes called 'biotic homogenization' (McKinney 2006). The flora and fauna of the world's cities indeed become more similar and homogeneous over time, but there is evidence that the proportion of native species remains high in spite of this (Pickett et al. 2011).

A recent global analysis of urban plant and bird diversity finds that urban areas filter out or exclude, on average, about one-third of native species existing in the surrounding region (Aronson et al. 2014). While this loss of diversity is problematic, it is worth noting that two-thirds of the native plant and bird species continue to occur in urban areas that are not designed with biodiversity protection in mind (although their population sizes and distribution ranges may be impacted by urbanization). In some cases, urban areas may host cultural and biodiversity-rich green spaces that serve as remnants of biodiversity of the broader landscape

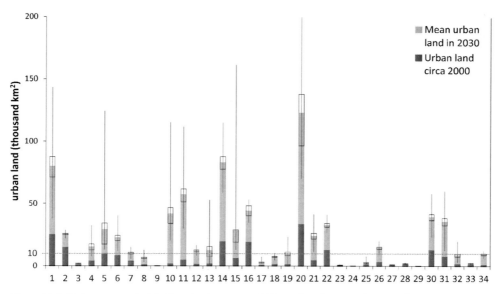

Figure 10.3 Urban extent in biodiversity hotspots *c.* 2000 and as forecasted in 2030[1]

Source: Reproduced from Güneralp and Seto 2013, Figure S4, p. 6 of supplementary data. Published under Creative Commons Attribution 3.0 Unported (CC-BY) license with kind permission of Environmental Research Letters 2013. All rights reserved.

and region, especially if the surrounding landscapes have been simplified through agriculture or forestry (Barthel et al. 2005). For instance, native species richness declines and non-native species richness increases as one moves from the rural fringe to the urban core with approximately 30–50 percent of the plant species in the urban core being non-native (Dunn and Heneghan 2011). Similarly, under some conditions of low to moderate levels of urban development (i.e. suburbanization), species richness may actually increase (McKinney 2002). The increased number of species in suburbanizing landscapes results from high habitat heterogeneity, high number of introduced species, socio-economic factors and altered disturbance regimes (see Kowarik 2011). Another species pattern observed in urban landscapes is that species tend to be non-native invasive and native generalists, which are tolerant to the urban conditions. However, the literature also provides evidence that contradicts these generalities. For example, Hope et al. (2003) report that species richness in Phoenix, Arizona, US, a city located in a desert environment, increases with urbanization because of human influences such as irrigation and ornamental landscaping.

Urbanization and ecosystem services

Urban areas affect many ecosystem services on scales ranging from local to global. One of the most critical services on a regional to global scale is the provision of freshwater (McDonald et al. 2013). Urban areas depend on freshwater availability for residential, industrial and commercial purposes; yet, they also affect the quality and amount of freshwater available to them. Water availability is likely to be a serious problem in most cities in semiarid and arid climates (Güneralp et al., 2015a; see Pfister et al., Chapter 17 in this volume). More than a fifth of urban dwellers, some 523 million, live in climates that would at least be classified as semiarid. Moreover, currently 150 million people live in cities with perennial water shortages, defined as having less than 100 L/person/day of sustainable surface and groundwater flow within their urban extent. By 2050, this number will reach almost a billion people due to population growth. Furthermore, climate change is projected to cause water shortages for an additional 100 million urbanites (McDonald et al. 2011).

Urbanization also affects regulatory hydrological services (Gómez-Baggethun and Barton 2013), which are usually defined as public goods. Consider an expanding city where new residential areas are replacing forests. This increases the impermeable surface area, which leads to increased volumes of surface water runoff, and thus increases the vulnerability to flooding of downstream communities. Depending on the rain event, vegetation reduces surface runoff following precipitation events by increasing infiltration. Urban landscapes with 50–90 percent impervious cover can lose 40–83 percent of rainfall to surface runoff compared to 13 percent in forested landscapes (Bonan 2002). Urban areas thus both depend on upstream natural habitats for regulating water flows and impact provision of this ecosystem service to downstream communities.

Ecosystem services provided by the urban forest and especially parks within cities are vital for human health and well-being, yet these are not always adequately considered during urban planning processes. These services include recreational value, aesthetic benefits and benefits to human physical and mental health (Gómez-Baggethun and Barton 2013; McDonald, Chapter 29 in this volume). Because city environments may be stressful for inhabitants, the recreational aspects of urban ecosystems are among the highest valued ecosystem services in cities. While these cultural ecosystem services are very important to the well-being of urban dwellers, they are often ignored by urban planning authorities; moreover, unplanned urban developments do not usually set aside any land for natural habitat to provide these services.

Although land is a scarce resource in a city and most will be used for development, the role of urban planning is to create an urban form that provides for the well-being of residents, which requires that some natural habitats are set aside as protected areas.

Urbanization, biodiversity, ecosystem services and governance

Minimizing habitat and biodiversity loss and limiting degradation of ecosystem services also require cities to integrate ecological knowledge into their urban planning practices (Niemelä 1999; Puppim de Oliveira et al. 2011). Specifically, urban planning practices need to become more attuned to conservation of biodiversity and preservation of ecosystem services that are of critical importance for the inhabitants of the urban areas (McDonald et al. 2014; Puppim de Oliveira et al. 2011). In this respect, the dissemination of information and connection of science to practitioners is an important aspect of formulating sound urbanization strategies that explicitly acknowledge and consider conservation of biodiversity (Güneralp and Seto 2013). However, one of the critical prerequisites to ensure this integration is that urban planners are equipped with the requisite institutional capacity (Sandström et al. 2006; Blicharska et al. 2011).

Novel ecosystems and communities composed of both native and non-native species may give us insights into how future ecosystems in urban landscapes may function. Novel plant and animal communities are continuously assembled in urban areas, either on abandoned land or with active manipulation and management. These communities can play an important role in the generation and maintenance of ecosystem services including water, fuel and food as well as recreation within the urban areas. Biodiversity-conscious urban design, therefore, has the potential to support a larger proportion of functional biodiversity within urban landscapes as well as to maintain the density, structure and distribution of the plant and animal communities (Pickett et al. 2013).

Many biodiversity hotspots threatened by urban growth are located in developing countries (countries forecasted to have the greatest increase in population), and may have fewer financial resources to devote to land protection than cities in developed countries. Moreover, since the attention of municipal governments in developing countries is often understandably focused on regulatory services such as providing clean drinking water and sanitation to their burgeoning residents, biodiversity protection may not be seen as a municipal priority. However, globally there is substantial interest by many people in preventing massive biodiversity loss in these biodiversity hotspots that face continuing urbanization. This spatial disconnect between those making the decisions in cities in biodiversity hotspots and those who care about the biodiversity losses elsewhere might only be overcome by a global effort to protect these biodiversity hotspots from further urban encroachment. This effort must include focusing conservation funding from organizations and governments in the developed world in these hotspots (McDonald et al. 2014).

In their study on potential direct impacts of urban expansion on biodiversity conservation in China, Güneralp et al. (2015b) discuss in detail the obstacles in incorporating biodiversity and ecosystem services into land use planning in the country. Examining historical patterns of urban population growth and expansion, and using forecasts from the Intergovernmental Panel on Climate Change (IPCC) on gross domestic product and projections by the United Nations on urban population growth, they predict that by 2030, urban land in China will reach over 400,000 km^2, which corresponds to a fourfold increase in urban land over 30 years. Such growth in urban areas will increase pressure on the already stressed protected areas and the biodiversity hotspots in the country. This poses a formidable challenge to the country's goal of biodiversity conservation, and calls for more effective land use planning and regulation, in particular at the regional and provincial level. China's entire planning system encompasses various government agencies that formulate and approve land use plans. Under this system, the land and fiscal policy reforms in the 1980s and 1990s created an institutional environment in which local and municipal governments came to increasingly rely on land leasing to developers as a key source of revenue. This resulted in rapid expansion of urban areas which, despite further reforms to stem the trend, continues almost unabated.

To integrate the goal of well-functioning cities with that of well-functioning ecosystems, China needs to incorporate ecological considerations in the regional and provincial-level plans and decisions and effectively regulate development decisions at the municipal level. However, it will be challenging to overcome the entrenched governance practices and special interests (Güneralp et al. 2015b). For example, although China's recently unveiled urbanization plan recognizes the over-reliance of local governments on land leasing as a source of revenue (China State Council 2014), it is uncertain if the plan, despite its lofty goals, will lead to any significant changes in planning practices.

Of the actions proposed to ameliorate urban effects on biodiversity, setting aside large parcels of native habitats in those parts of biodiversity hotspots facing urbanization pressure may provide the best opportunity for regional floral and faunal species to persist in the face of climate change and a surrounding urbanizing landscape. These protected areas would need to be large enough to contain the spectrum of natural disturbances as well as native habitats. With land conservation, a number of landscape designs are possible. For instance, one design for large parcels would make these areas composed of multiple-utilization zones (Noss and Harris 1986). The interior zone would be road-free and managed to conserve native flora and fauna. By comparison, the perimeter would serve as a buffer that is used for multiple benefits and linked to other areas. An example would be the Tijuca Forest in Rio de Janeiro, Brazil (Herzog 2013). Large parcels can, to some extent, buffer local climatic changes and contain more individuals of a single species thus enhancing its genetic breadth. Even these large areas, however, will not be immune to human intrusions; so, natural resource managers must also continually adapt to changing circumstances.

The global challenge that urbanization poses for ecosystem services and biodiversity demands a global response. McDonald et al. (2014) present three potential solutions to address this challenge: 1) treating ecosystem services as an urban utility; 2) a global effort to protect those biodiversity hotspots under urbanization pressure; and 3) international coordination for urban sustainability. In several cities around the world, these solutions are being experimented with at present with varying levels of success. Importantly, any solutions to reconcile ongoing urbanization and conservation would require policies that work in harmony across scales, from local to regional to global (McDonald et al. 2014). In particular, establishing effective biodiversity conservation strategies in regions that are expected to undergo significant urban expansion require coordinated efforts among multiple cities, provinces and even countries. Such coordination, however, has been hard to achieve even among conservation bodies under existing regional and global governance mechanisms (Larigauderie and Mooney 2010). The recently formed Intergovernmental Platform on Biodiversity and Ecosystem Services (IPBES; www.ipbes.net) aims to remedy this lack of coordination by, among others, conducting periodic sub-regional, regional and global assessments on the state of the planet's biodiversity, its ecosystems and the essential services they provide to society (Larigauderie and Mooney 2010). Established in April 2012, the IPBES will act as an independent intergovernmental body, much like the IPCC; it will be open to all member countries of the United Nations. Clearly, the impacts of urbanization on biodiversity are critical enough to be included in these assessments (McDonald et al. 2014). In this vein, the Cities and Biodiversity Outlook (CBO) that is endorsed by the Convention on Biological Diversity (CBD) is the first ever comprehensive assessment of the interaction of cities and biodiversity and ecosystem services (Elmqvist et al. 2013).

Gaps in knowledge and practice

There is no scarcity of research questions related to urbanization and its relationship to biodiversity and ecosystem services. Alongside challenges of understanding and forecasting patterns of land use change and urbanization, there are also gaps in knowledge regarding many aspects of biodiversity and ecosystem services such as connections between various ecosystem processes across spatial and temporal scales (Colding 2007). The interactions between urban and rural regions (Larondelle and Haase 2013) and feedback mechanisms among ecosystem processes within and near cities are still poorly understood, as is the impact of urbanization on values, norms and institutions related to the consumption and/or sustainable use of biodiversity and ecosystem services. Furthermore, climate change is a major driver of change that likely will affect future urban biodiversity and ecosystem services.

The need for urbanization strategies that consider biodiversity conservation is especially acute for those in developing countries where most urban expansion near protected areas and in biodiversity hotspots are expected (Niemelä 1999; Puppim de Oliveira et al. 2011; see also Roberts, Chapter 35 in this volume, for a discussion of city action and biodiversity planning in Durban, South Africa). There are two challenges to overcome in this respect: first is building a firm theoretical foundation on which to ground the research on the relationship between urbanization and biodiversity; the second is building communication channels between the researchers and the stakeholders including citizens, community organizations, planners and government representatives alike. Building such channels of information between science and practitioners for communication of concerns and insights will be an important tool for formulating more robust urbanization strategies in regards to biodiversity.

With respect to ecosystem services, little is known about the negotiated interactions that lead to trade-offs and synergies in the demand for particular bundles of ecosystem services accessible to different socio-economic or livelihood groups in urban environments (but see Colding et al. 2006; Andersson et al. 2007). These interactions play a crucial role in shaping outcomes of equity, particularly for the urban poor as well as for traditional livelihood users, such as fishers and livestock grazers in peri-urban areas (D'Souza and Nagendra 2011). Considering the multitude of services that ecosystems provide as well as the demand for these services, interdisciplinary approaches are needed to better safeguard, and benefit from, these services.

A better understanding is required of the supply, needs and management of urban ecosystem services in large regions in South Asia, Africa and Latin America, which are developing rapidly and face some of the greatest threats to protected areas and biodiversity hotspots in the future (Güneralp and Seto 2013). However, this does not necessarily mean that local knowledge is non-existent. Most likely traditional ecological knowledge, at the local level, is being used every day in more informal management decisions pertaining to ecosystem services. Indeed, this is known to be the case in many places in Asia and Africa. For instance, comparisons of residential gardens in different continents indicate that most plant species in home gardens in Europe and North America are chosen for their ornamental value, while in contrast, a large proportion of species in gardens in India are chosen for their medicinal, food or cultural properties (Jaganmohan et al. 2012). Local knowledge and practices could be mobilized in multiple ways through, for example, citizen science initiatives, and thus could support more formal governance and management of urban ecosystem services.

Ecosystem service science still lacks a robust theoretical framework that allows for consideration of social and cultural values of urban ecosystems on an equal basis with monetary values in decision-making processes. Developing such a framework involves synthesizing the large, but scattered, body of literature that has dealt with non-monetary values of the environment, and articulating this research into ecosystem services concepts, methods and classifications (Luck et al. 2012; Gómez-Baggethun and Barton 2013). For example, while much attention has been focused on provisioning and regulating ecosystem services provided by urban ecosystems, cultural services have been poorly researched (e.g. Daniel et al. 2012). Particularly across Asia and Africa, many sacred conceptualizations of nature persist in cities (e.g. protection of sacred keystone species such as *Ficus religiosa* across cities in India). There are also numerous equity and environmental justice issues related to cultural ecosystem services, but these are often poorly documented (D'Souza and Nagendra 2011). In addition, to better capture the value of biodiversity and ecosystems in reducing urban vulnerability to shocks and disturbances, there is a particular need of new valuation techniques that utilize a resilience and inclusive wealth perspective (Gómez-Baggethun et al. 2013). The insurance value of an ecosystem is closely related to its resilience and self-organizing capacity, and the extent to which it may continue to provide flows of ecosystem services benefits with stability over a range of variable environmental conditions.

Concluding remarks

Main messages from the chapter include:

- Notwithstanding uncertainties inevitable in any study on future trends, it is increasingly clear that urbanization will continue to affect biodiversity and ecosystem services around the world.
- Most of the effects on biodiversity and ecosystem services will take place in the developing world with limited means to address each and every challenge urbanization presents.
- As the world continues to urbanize, there is an increasing need for urban decision-makers and citizens to adopt policies and practices to integrate nature into daily lives; after all, cities may very well offer the key to a globally sustainable future underpinned by nature-based solutions and ecosystem-based adaptation.

Identified needs for future research and practice include:

- interdisciplinary approaches that examine the trade-offs and synergies of urban ecosystems and the interactions of user demand as it pertains to different stakeholders groups;
- mechanisms through which biodiversity and ecosystem-based solutions can be integrated into urban development and climate change adaptation decision-making and planning;
- cross scale understandings of ecosystem processes across rural–urban gradients and the resulting interactions and feedbacks from both current and future climate change and land use changes;
- more quantitative and qualitative studies on ecosystem services within developing regions, where both rapid urbanization and projected loss in protected areas and hotspots are expected to occur; and,

Urbanization, habitat and biodiversity decline

- the development of more robust theoretical frameworks that integrate the understudied social and cultural values of urban ecosystems with monetary values in decision-making processes.

Note

1 (1) Atlantic Forest, (2) California Floristic Province, (3) Cape Floristic Region, (4) Caribbean Islands, (5) Caucasus, (6) Cerrado, (7) Chilean Winter Rainfall and Valdivian Forests, (8) Coastal Forests of Eastern Africa, (9) East Melanesian Islands, (10) Eastern Afromontane, (11) Guinean Forests of West Africa, (12) Himalaya, (13) Horn of Africa, (14) Indo-Burma, (15) Irano-Anatolian, (16) Japan, (17) Madagascar and the Indian Ocean Islands, (18) Madrean Pine-Oak Woodlands, (19) Maputaland-Pondoland-Albany, (20) Mediterranean Basin, (21) Mesoamerica, (22) Mountains of Central Asia, (23) Mountains of Southwest China, (24) New Caledonia, (25) New Zealand, (26) Philippines, (27) Polynesia-Micronesia, (28) Southwest Australia, (29) Succulent Karoo, (30) Sundaland, (31) Tropical Andes, (32) Tumbes-Choco-Magdalena, (33) Wallacea, (34) Western Ghats and Sri Lanka.

References

Alberti, M. 2005. The effects of urban patterns on ecosystem function. *International Regional Science Review* 28: 168–192.

Andersson, E., S. Barthel and K. Ahrné. 2007. Measuring social-ecological dynamics behind the generation of ecosystem services. *Ecological Applications* 17(5): 1267–1278.

Aronson, M., F.A. La Sorte, C.H. Nilon, M. Katti, M.A. Goddard, C.A. Lepczyk, P.S. Warren, N.S.G. Williams, S. Cilliers, B. Clarkson, C. Dobbs, R. Dolan, M. Hedblom, S. Klotz, J.L. Kooijmans, I. Kühn, I. MacGregor-Fors, M. McDonnell, U. Mörtberg, P. Pyšek, S. Siebert, J. Sushinsky, P. Werner and M. Winter. 2014. A global analysis of the impacts of urbanization on bird and plant diversity reveals key anthropogenic drivers. *Proc. R. Soc. B* 281(1780): 1471–2954.

Barthel, S., J. Colding, T. Elmqvist and C. Folke. 2005. History and local management of a biodiversity-rich, urban cultural landscape. *Ecology and Society* 10(2): 10. [online] URL: www.ecologyandsociety.org/vol10/iss2/art10/

Bierwagen, B.G. 2007. Connectivity in urbanizing landscapes: The importance of habitat configuration, urban area size, and dispersal. *Urban Ecosystems* 10: 29–42.

Blicharska, M., P. Angelstam, H. Antonson, M. Elbakidze and R. Axelsson. 2011. Road, forestry and regional planners' work for biodiversity conservation and public participation: A case study in Poland's hotspot regions. *Journal of Environmental Planning and Management* 54(10): 1373–1395.

Bonan, G. 2002. *Ecological climatology: Concepts and applications*. Cambridge: Cambridge University Press.

Chettri, N., R. Thapa and B. Shakya. 2007. Participatory conservation planning in Kangchenjunga transboundary biodiversity conservation landscape. *Tropical Ecology* 48(2):163–176.

China State Council. 2014. Guojia xinxing chengzhenhua guihua (2014–2020) [New national urbanization plan (2014–2020)]. Retrieved April 17, 2015, from www.gov.cn/gongbao/content/2014/content_2644805.htm

Colding, J. 2007. 'Ecological land-use complementation' for building resilience in urban ecosystems. *Landscape and Urban Planning* 81(1–2): 46–55.

Colding, J., J. Lundberg and C. Folke. 2006. Incorporating green-area user groups in urban ecosystem management. *Ambio* 35(5): 237–244.

Daniel, T.C., A. Muhar, A. Arnberger, O. Aznar, J.W. Boyd, K.M.A. Chan, R. Costanza, T. Elmqvist, C.G. Flint, P.H. Gobster, A. Grêt-Regamey, R. Lave, S. Muhar, M. Penker, R.G. Ribe, T. Schauppenlehner, T. Sikor, I. Soloviy, M. Spierenburg, K. Taczanowska, J. Tam and A. Von Der Dunk. 2012. Contributions of cultural services to the ecosystem services agenda. *Proceedings of the National Academy of Sciences of the United States of America* 109(23): 8812–8819.

D'Souza, R. and H. Nagendra. 2011. Changes in public commons as a consequence of urbanization: The Agara lake in Bangalore, India. *Environmental Management* 47(5): 840–850.

Dunn, C. and L. Heneghan. 2011. Composition and diversity of urban vegetation. In *Urban ecology: Patterns, processes and applications*. Edited by: J. Niemelä, J.H. Breuste, G. Guntenspergen, N.E. McIntyre et al. Oxford: Oxford University Press, pp. 103–134.

Elmqvist, T. et al. 2013. *Urbanization, biodiversity, and ecosystem services: Challenges and opportunities*. New York: Springer.

Forman, R.T.T. 2008. *Urban regions: Ecology and planning beyond the city*. New York: Cambridge University Press.

Gómez-Baggethun, E. and D.N. Barton. 2013. Classifying and valuing ecosystem services for urban planning. *Ecological Economics* 86: 235–245.

Gómez-Baggethun, E., Å. Gren, D.N. Barton, J. Langemeyer, T. McPhearson, P. O'Farrell, E. Andersson, Z. Hamstead and P. Kremer. 2013. Urban ecosystem services. In *Urbanization, biodiversity, and ecosystem services: Challenges and opportunities*. Edited by: T. Elmqvist, M. Fragkias, J. Goodness, B. Güneralp, P.J. Marcotullio, R.I. McDonald, S. Parnell, M. Schewenius, S.M.K. Seto and C. Wilkinson. New York: Springer, pp. 175–251.

Güneralp, B. and K.C. Seto. 2013. Futures of global urban expansion: uncertainties and implications for biodiversity conservation. *Environmental Research Letters* 8(1): 1–10.

Güneralp, B., İ. Güneralp and Y. Liu (2015a) Changing global patterns of urban exposure to flood and drought hazards. *Global Environmental Change* 31: 217–225.

Güneralp, B., A.S. Perlstein and K.C. Seto. 2015b. Balancing urban growth and ecological conservation: A challenge for planning and governance in China. *Ambio* (published online first).

Herzog, L.A. 2013. Barra da Tijuca: The political economy of a global suburb in Rio de Janeiro, Brazil. *Latin American Perspectives* 40(2): 118–134.

Hope, D., C. Gries, W. Zhu, W.F. Fagan, C.L. Redman, N.B. Grimm, A.L. Nelson, C. Martin and A. Kinzig. 2003. Socioeconomics drive urban plant diversity. *Proceedings of the National Academy of Sciences of the United States of America* 100(15): 8788–8792.

Jaganmohan, M., L.S. Vailshery, D. Gopal and H. Nagendra. 2012. Plant diversity and distribution in urban domestic gardens and apartments in Bangalore, India. *Urban Ecosystems* 15(4): 911–925.

Kowarik, I. 2011. Novel urban ecosystems, biodiversity, and conservation. *Environmental Pollution* 159(8–9): 1974–1983.

Larigauderie, A. and H.A. Mooney. 2010. The intergovernmental science-policy platform on biodiversity and ecosystem services: Moving a step closer to an IPCC-like mechanism for biodiversity. *Current Opinion in Environmental Sustainability* 2(1–2): 9–14.

Larondelle, N. and D. Haase. 2013. Urban ecosystem services assessment along a rural-urban gradient: A cross-analysis of European cities. *Ecological Indicators* 29: 179–190.

Luck, G.W., K.M.A. Chan, U. Eser, E. Gómez-Baggethun, B. Matzdorf, B. Norton and M.B. Potschin. 2012. Ethical considerations in on-ground applications of the ecosystem services concept. *BioScience* 62(12): 1020–1029.

McDonald, R.I., P. Kareiva and R.T.T. Forman. 2008. The implications of current and future urbanization for global protected areas and biodiversity conservation. *Biological Conservation* 141(6): 1695–1703.

McDonald, R.I., R.T.T. Forman, P. Kareiva, R. Neugarten, D. Salzer and J. Fisher. 2009. Urban effects, distance, and protected areas in an urbanizing world. *Landscape and Urban Planning* 93(1): 63–75.

McDonald, R.I., P. Green, D. Balk, B.M. Fekete, C. Revenga, M. Todd and M. Montgomery. 2011. Urban growth, climate change, and freshwater availability. *Proceedings of the National Academy of Sciences* 108(15): 6312–6317.

McDonald, R.I., P. Marcotullio and B. Güneralp. 2013. Urbanization and trends in biodiversity and ecosystem services. In *Urbanization, biodiversity, and ecosystem services: Challenges and opportunities*. Edited by: T. Elmqvist, M. Fragkias, J. Goodness, B. Güneralp, P.J. Marcotullio, R.I. McDonald, S. Parnell, M. Schewenius, S.M.K. Seto and C. Wilkinson. New York: Springer, pp. 31–52.

McDonald, R., B. Güneralp, W. Zipperer and P.J. Marcotullio. 2014. The future of global urbanization and the environment. *Solutions* 5(6): 60–69.

McKinney, M.L. 2002. Urbanization, biodiversity, and conservation. *BioScience* 52(10): 883–890.

McKinney, M.L. 2006. Urbanization as a major cause of biotic homogenization. *Biological Conservation* 127(3): 247–260.

McKinney, M.L. 2008. Effects of urbanization on species richness: A review of plants and animals. *Urban Ecosystems* 11(2): 161–176.

Martinuzzi, S., V.C. Radeloff, L.N. Joppa, C.M. Hamilton, D.P. Helmers, A.J. Plantinga and D.J. Lewis. 2015. Scenarios of future land use change around United States' protected areas. *Biological Conservation* 184: 446–455.

Urbanization, habitat and biodiversity decline

Mittermeier, R.A., P. Robles-Gil, M. Hoffmann, J.D. Pilgrim, T.B. Brooks, C.G. Mittermeier, J.L. Lamoreux and G.A.B. Fonseca. 2004. *Hotspots revisited: Earth's biologically richest and most endangered ecoregions.* Mexico City: CEMEX.

Müller, N., M. Ignatieva, C. Nilon, P. Werner, and W. Zipperer. 2013. Patterns and trends in urban biodiversity and landscape design. In *Urbanization, biodiversity, and ecosystem services: Challenges and opportunities.* Edited by: T. Elmqvist, M. Fragkias, J. Goodness, B. Güneralp, P.J. Marcotullio, R.I. McDonald, S. Parnell, M. Schewenius, S.M.K. Seto and C. Wilkinson. New York: Springer, pp. 123–174.

Myers, N., R.A. Mittermeier, C.G. Mittermeier, G.A.B. da Fonseca and J. Kent. 2000. Biodiversity hotspots for conservation priorities. *Nature* 403(6772): 853–858.

Niemelä, J. 1999. Ecology and urban planning. *Biodiversity and Conservation* 8(1): 119–131.

Noss, R. and L. Harris. 1986. Nodes, networks, and MUMs: Preserving diversity at all scales. *Environmental Management* 10: 299–309.

Pickett, S.T. and M.L. Cadenasso. 2009. Altered resources, disturbance, and heterogeneity: a framework for comparing urban and non-urban soils. *Urban Ecosystems* 12(1): 23–44.

Pickett, S.T.A., S.L. Collins and J.J. Armesto. 1987. A hierarchical consideration of causes and mechanisms of succession. *Vegetatio* 69(1–3): 109–114.

Pickett, S.T.A., M.L. Cadenasso, J.M. Grove, C.G. Boone, P.M. Groffman, E. Irwin, S.S. Kaushal, V. Marshall, B.P. McGrath, C.H. Nilon, R.V. Pouyat, K. Szlavecz, A. Troy and P. Warren. 2011. Urban ecological systems: Scientific foundations and a decade of progress. *Journal of Environmental Management* 92(3): 331–362.

Pickett, S.T.A., C.G. Boone, B.P. McGrath, M.L. Cadenasso, D.L. Childers, L.A. Ogden, M. McHale and J.M. Grove. 2013. Ecological science and transformation to the sustainable city. *Cities* 32: S10–S20.

Puppim de Oliveira, J.A., O. Balaban, C.N.H. Doll, R. Moreno-Peñaranda, A. Gasparatos, D. Iossifova and A. Suwa. 2011. Cities and biodiversity: Perspectives and governance challenges for implementing the convention on biological diversity (CBD) at the city level. *Biological Conservation* 144(5): 1302–1313.

Ricketts, T.H. 2001. The matrix matters: effective isolation in fragmented landscapes. *American Naturalist* 158: 87–99

Sandström, U.G., P. Angelstam and A. Khakee. 2006. Urban comprehensive planning – Identifying barriers for the maintenance of functional habitat networks. *Landscape and Urban Planning* 75(1–2): 43–57.

Tilman, D., R.M. May, C.L. Lehman and M.A. Nowak. 1994. Habitat destruction and the extinction debt. *Nature* 371(6492): 65–66.

Tratalos J., R.A. Fuller, P.H. Warren, R.G. Davies and K.J. Gaston. 2007. Urban form, biodiversity potential and ecosystem services. *Landscape and Urban Planning* 83: 308–317.

Williams, N.S.G., M.W. Schwartz, P.A. Vesk, M.A. McCarthy, A.K. Hahs, S.E. Clements, R.T. Corlett, R.P. Duncan, B.A. Norton, K. Thompson and M.J. McDonnell. 2009. A conceptual framework for predicting the effects of urban environments on floras. *Journal of Ecology* 97: 4–9.

11

URBAN PRECIPITATION

A global perspective

Chandana Mitra and J. Marshall Shepherd

Urbanization directly impacts the natural environment through shifts in water cycles and microclimates. The hydroclimate (i.e. clouds, precipitation, land surface hydrology and associated flooding) is sensitive to and affected by urbanization and can have critical impacts on the fabric of society. Observational and modeling studies show compelling evidence that precipitation patterns, convective storms and flood activity in and around major urban areas have been modified in recent times. Increasingly, urban-hydroclimatic influences must now transition from an era of exploratory analysis to an integrated understanding offering actionable information. An in-depth understanding of water resources, water security, flood hazards and anomalies in the urban water cycle is vital to planning and policy-making for the future when global climate change and local urban land-atmosphere interactions will have coupled impacts on the environment.

This chapter will place urban-precipitation interactions within a contemporary context and offer a perspective that is often under-represented in the scholarly literature. Herein, urban precipitation is explored on the global scale with particular emphasis on both tropical and temperate regions. Concepts of risk and vulnerability within the context of urban precipitation juxtaposed with increasing complexity in the coupled human-natural system are explored. The chapter will not only highlight other reviews of urban precipitation studies in past years, but also elucidate the importance of conducting more well-designed studies in fast urbanizing, developing world cities.

Urban precipitation and societies at risk

Urban regions are increasingly at risk from episodic extreme weather events and long-term climate change (Shepherd et al. 2013). The emergence of rapidly urbanizing societies within the context of weather-climate extremes leads to unique, significant and far-reaching risks. The understanding of these risks requires the assessment of a confluence of factors including (1) *Hazards* such as heat and cold, hurricanes, tornadoes, droughts and floods (the risks from such hazards are further exacerbated in metropolitan regions by the urban form via land uses that affect albedo, heat capacity and infiltration, channelization of air movement, primary and secondary pollutants and removal of protective landscapes such as wetlands and green space); (2) *Exposure* of infrastructure and people leading to individual or systemic harm;

(3) *Vulnerability* of urban settings in terms of having greater sensitivity or exposure to the hazard; and (4) *Resilience* or capacity of the urban space to mitigate the hazard impacts.

Within such a risk framework, the understanding that urban environments can initiate, modify or enhance precipitation is quite relevant, as precipitation is associated with flooding, lightning and other weather hazards. It is, therefore, not surprising that scholars examine how urban environments interact with precipitation. Indeed, urban climatology offers a wealth of knowledge about heat islands, energy budgets, pollution and dynamics (see Grimmond et al., Chapter 12 in this volume). However, urban precipitation studies are a relatively late addition to this body of literature and have been episodic in productivity and knowledge content.

A short history of urban climate scholarship

More than 40 years ago, Landsberg (1970) asserted that the most pronounced and locally derived, but far-reaching, effects of human activities on the microclimate occur in cities. Urban areas have a number of characteristics or processes that affect land-surface and geophysical parameters (Mahmood et al. 2014; Grimmond et al., Chapter 12 in this volume) including rougher and uneven surfaces relative to rural areas due to manmade structures; greater atmospheric concentrations of aerosols and pollutants because of automobiles and industries; warmer surfaces due to the heat-storage capacities of concrete and asphalt; and the presence of more anthropogenic heat sources. The understanding that urban areas modify the surrounding environment has been accepted for over a century but the mechanisms through which this occurs have only begun to be understood and analyzed in the past few decades. The heat gain due to the storage capacity of urban built structures, reduction in local evapotranspiration, and anthropogenic heat altogether alter the spatio-temporal pattern of temperature and leads to the well-known urban heat island (UHI) phenomenon. According to Mills (2014), prior to the 1970s, the majority of urban climate research was observational and descriptive, whereas since this time, more importance has been given to physical dynamics (see Table 11.1).

One of the main pathways through which cities affect climate is the alteration of local and regional precipitation patterns. Horton (1921), for example, first observed thunderstorms developing over cities and Schmaus (1927) presents evidence of urban-enhanced precipitation by reporting an 11 percent increase in both light rainfall and the number of convective showers over the city of Munich (Garstang et al. 1975). Kratzer (1937, 1956), Landsberg (1956), Stout (1962) and Changnon (1962) all contribute to the further advancement of knowledge regarding urban influences on rainfall. The 'La Porte Anomaly' (Stout 1962; Changnon 1968), 'Eight Cities Study' (Huff and Changnon 1973) and Metropolitan Meteorological Experiment (METROMEX) undertaken in the St. Louis, MO area between 1971 to 1975 (Diab 1978; Braham 1981; Braham et al. 1981; Changnon 1981) confirm that urbanization modifies and impacts the hydrological cycle over cities (Lowry 1998). Many of these studies represent coordinated efforts with field campaigns, radar technology and modeling.

Resurgence and new methodologies

After the METROMEX final report (Changnon 1981) urban precipitation research slowed and focused mainly on tropical and subtropical regions, but with some notable exceptions (Lowry 1998) including two published bibliographies by Oke (1987) and Jauregui (1993) as well as three studies including Pittsburg, PA (Rosenberger and Suckling 1989), Moscow (Stulov 1993) and Mexico City (Jauregui and Romales 1996). Urban precipitation research

Table 11.1 Development of urban climate study over a century

Time period	Approach
1900s	Observation and description of urban effects using conventional meteorological equipment (thermometers, hygrometers, etc.)
1960s	Move toward measurement of 'process' variables-radiation, sensible and latent heat exchanges
	The use of statistical methods to summarize and generalize results
1970s	Application of conventional (micro-) meteorological theory to urban climates
	Use of energy budget as a framework to explain the urban effect
	Observation of process variables: radiation, estimated fluxes
	Use of computer modeling techniques
	More rigorous definition of urban 'surface', urban scales and observing urban effects
1980s	Adoption of an experimental approach: select common urban forms (streets become canyons)
	Use of scaled-physical models and direct measurement of fluxes
1990s	Relationships between real urban forms and climate effect
	Urban field projects examined by research teams
	Generalizations based on a range of settlements
2000s	Development of realistic urban climate models
	Employment of novel techniques for examining urban climate (Doppler/ polarimetric, radar, satellites, lidar)
2010s	Emergence of urban aggregate frameworks or urban climate archipelago frameworks
	Actionable policy and planning based on urban climate information
	Quantifying risk and vulnerability from urban hazards

Source: Adapted from Mills (2014).

re-emerged after 2000. Souch and Grimmond (2006) credit Shepherd et al. (2002) with this renewed interest. That work, though flawed in some aspects of the methodology, is the first to use satellite-derived rainfall to investigate what is referred to as the 'Urban Rainfall Effect' or URE (Shepherd et al. 2010a). Since 2000 there has been a gradual increase in urban precipitation studies globally. There is now significant and mounting evidence in the literature supporting the claim that urban environments modify local and regional precipitation. Studies continue to analyze and connect various influences of urban areas on the atmosphere and the mechanisms of the dynamic-microphysical processes.

Identifying the research gaps

In his review, Lowry (1998) makes recommendations for conducting more meaningful and results-oriented urban precipitation studies. He suggests more a) designed experiments; b) replication of the experiments in several urban areas; c) use of spatially small and temporally short experiment scenarios; and d) disaggregation of climatic data in smaller sample size and separation of large-scale weather systems. Methods and data availability have since continued to improve, but there is still uncertainty and scientific debate about whether urban environments increase rainfall, decrease rainfall or have no effect on rainfall (Shepherd 2013).

Urban precipitation: a global perspective

Box 11.1   Urban precipitation climatology in review – a concise history

Until recently, urban precipitation climatology review papers typically focused on the dynamic nature of urban influences on rainfall, with an emphasis on observational and numerical techniques. The first noteworthy review paper 'The structure of heat islands' (Garstang et al. 1975) focuses mainly on UHI land interactions, but gives a very good analysis of moisture fields, clouds and precipitation. Building on these efforts, the Lowry (1998) review, 'Urban effects on precipitation amount' includes an historical perspective and a critical evaluation of methods as well as an overview of the status of urban precipitation climatology and recommendations concerning future research.

'A review of current investigations of urban-induced rainfall and recommendations for the future' presents a concise review of urban precipitation studies (1990–2005) and observational and modeling studies, discussing the role of aerosols in modifying urban land-atmospheric dynamics (Shepherd 2005). Following this review, Collier (2006) assesses the urban impacts on weather while aiming at deciphering the complexities of urban land-atmospheric interactions; turbulent structure and energy transport in the urban boundary layer; the UHI; convective initiation by urban areas and suppression of convection by urban aerosols.

The most recent review papers, both on urban impacts (Shepherd 2013; Han et al. 2014), emphasize the need to understand urban rainfall effects; the mechanisms leading to urban-induced rainfall and causes behind modifications and anomalies in urban precipitation.

Enhancement or reduction?

The majority of the literature provides compelling evidence that there are variations in intensity and spatial distribution, also enhancement of rainfall due to urban impacts (Vukovich and Dunn 1978; Bornstein and Lin 2000; Shepherd et al. 2002; Changnon 2003; Dixon and Mote 2003; Mitra et al. 2012; Ashley et al. 2012; Schmid and Niyogi 2013; Haberlie et al. 2015). However, some studies indicate that precipitation can be reduced by urbanization (e.g. Guo et al. 2006; Zhang et al. 2007, 2009; Kaufmann et al. 2007; Trusilova et al. 2008). Guo et al. (2006) and Zhang et al. (2007, 2009) reveal that total accumulated precipitation (lower), evaporation (lower), surface temperatures (higher), sensible heat fluxes (larger) and boundary layer (deeper) in the Beijing region are modified by land surface conditions. Kaufmann et al. (2007), using econometric and statistical modeling, find reduced local precipitation in the Pearl River Delta as a result of urban growth, while Trusilova et al. (2008) find statistically significant increases in winter and summer decreases in precipitation in their urban simulations as compared to pre-urban simulations.

Most observational and numerical modeling studies have been deficient in addressing the co-evolving impacts of urban land cover and aerosols. Going forward, studies must address the synergistic affects. For example, Rosenfeld et al. (2008) demonstrate that a polluted cloud can reduce or enhance precipitation depending on certain microphysical or dynamic processes within it. Hence, it is reasonable to ask what role forcing from urban land cover and buildings might have on the evolution of the process. Only controlled modeling experiments and well-designed observational studies with both processes in mind can resolve such questions. Additionally, the emergence of dual-polarization radar (Thompson et al. 2014) datasets will advance our ability to qualitatively and quantitatively evaluate the microphysical stages of precipitation. Future studies in urban precipitation will need to leverage this new technology.

Towards a better understanding of urban effects on precipitation

Four possible mechanisms are identified in the literature through which urban environments impact precipitation or convection enhancement or reduction including one or a combination of the following (Shepherd, 2005, pp. 5–6):

1) enhanced convergence due to increased surface roughness in the urban environment (e.g., Changnon et al. 1981; Bornstein and Lin 2000; Thielen et al. 2000; Shem and Shepherd 2009); 2) destabilization due to UHI-thermal perturbation of the boundary layer and resulting downstream translation of the UHI circulation or UHI-generated convective clouds (e.g., Shepherd and Burian 2003; Rosenfeld and Woodley, 2003); 3) enhanced aerosols in the urban environment for cloud condensation nuclei (CCN) sources (e.g., Mölders and Olson 2004, Jin et al. 2005; van den Heever and Cotton 2007, Rosenfeld et al. 2008); or, 4) bifurcating or diverting of precipitating systems by the urban canopy or related processes (e.g. Bornstein and LeRoy 1990; Niyogi et al. 2011).

Other hypotheses contend that urban areas serve as moisture sources needed for convective development (e.g. Dixon and Mote 2003). Changnon (1992) and Huff and Changnon (1973) show that the population size of an urban area influences the horizontal extent and magnitude of urban enhanced precipitation.

One proposed schematic for ascertaining the relative roles of each process in urban rainfall was originally created by Robert Bornstein in the early 2010s, but was recently augmented by Shepherd (2013) with additional considerations (see Figure 11.1). Despite its utility in conceptualizing the URE, uncertainty remains about what conditions will lead to initiation,

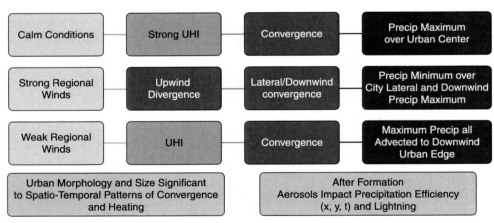

Figure 11.1 Towards a conceptualization of the urban rainfall effect

Source: Original figure courtesy of R. Bornstein as adapted and presented in Shepherd (2013).

modification or enhancement of precipitation. Other questions center on what size, shape or morphological attributes are most likely to modify precipitation processes. With possible anthropogenic aerosol cycles evident, more emphasis is also needed on possible anthropogenic impacts on weekly rainfall cycles (e.g. Bell et al. 2008; Haberlie et al. 2015). Shepherd and Mote (2011) further suggest the need to pivot research to investigate urban effects on snowfall, freezing rain, snow cover and snowmelt.

In order to understand locations of isolated convective initiation, Haberlie et al. (2015) recently developed and applied a technique to Atlanta, GA, over a 17-year period (warm season only). They find that initiation events are more likely to occur in the urban region rather than its rural counterpart. They also note a tendency for more convective initiations over the city during the weekdays rather than weekends. Moreover, the physical size of a city also has precipitation modification potential, as shown in a recent study by Schmid and Niyogi (2013), which draws upon research from Changnon (1992). Using numerical modeling, they place circular urban areas with varying radii (5 to 40 km) to determine whether a thunderstorm is modified due to urban effects (see Figure 11.2). Their findings show that storms, which pass directly over the city, produce less precipitation than with no city present. Whereas the storms, which did not pass over the urban land surface, did encounter the downwind aerosol field, were invigorated and produced more precipitation than with no city present. Shepherd et al. (2010b) also find that model simulations produce vastly different precipitation fields for the same atmospheric forcing in Houston, TX, when urban growth models are used to project land cover up to 2025.

Since urban areas are growing and becoming greater than the sum of their parts, it is imperative to understand what the aggregate effects of urban agglomerations like the United States Northeast Megalopolis, emerging Char-Lanta region (including the cities of Charlotte,

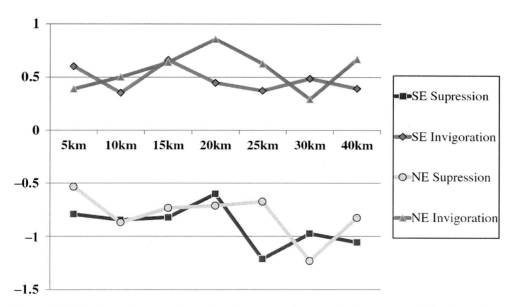

Figure 11.2 Maximum downwind precipitation suppression and invigoration (cm) for downwind regions northeast and southeast of circular city center for cities of different sizes

Source: Adapted from Schmid and Niyogi (2013).

Atlanta and Greenville) in southeastern US and chains of cities in Japan are on hydroclimate. To do this, the concept of urban climate archipelagoes (UCAs) (Shepherd et al. 2013) could be a useful new framework for thinking about urban impacts on climate, however, many more questions than answers remain for future researchers to consider. For example, there is still no formal definition of UCAs and their relative influences on the various components of the climate system reported by Seto and Shepherd (2009). Future questions will also emerge about appropriate methods to study the UCAs and what policy implications they may hold.

A strand of research is increasingly considering urban vulnerability to precipitation variability manifest in increased urban flooding events; lack of urban planning and water resources management; and associated public health impacts (Shepherd 2013). These issues will be aggravated in the near future when the pressure on the existing infrastructure will increase due to the constant population inflow to cities. A report by the World Bank (Jha et al. 2012) on cities and related flooding events emphasizes that urban flooding is becoming more dangerous and economically damaging simply because of the huge population residing in cities. Tremendous amount of investment and effort must be made to keep up with the growing pressure, and measures for prevention and risk mitigation must be implemented that benefit the urban socio-environmental system.

The importance of developing world studies

One of the aims of this chapter is to expand on existing reviews, highlighting the need to include more studies from developing countries. Because the majority of urban population growth in the 21st century will take place in the developing world, there is an urgent need to understand how the urban built environment will affect local precipitation. According to a UN report, between 2007 and 2025, the urban areas of the world are expected to gain 1.3 billion people including 261 million in China and 197 million in India, which together account for 35 percent of the total growth (UN 2007). It is well known that cities on a global scale are directly and indirectly responsible for global change in the atmosphere, hydrosphere, geosphere and biosphere (Mills 2007) and are examples of a complex, coupled human–natural system. Emphasis should be given to cities of the developing world in the tropics and sub-tropics (Hodder 1968; Mendelsohn and Dinar 1999; Mirza 2003) where urbanization is happening over very short time scales, often without much planning or restrictions of land use (Roth 2007). Climate-related environmental problems in (large) tropical urban agglomerations include (1) poor dispersion of air pollutants (generally low wind speeds and a lack of ventilation); (2) high levels of heat stress, which decreases productivity, reduces human comfort and increases mortality due to heat-related illnesses; and, (3) space cooling needs, which increase energy usage and, in turn, may exacerbate climate change (Roth 2007). Oppressive weather, poverty and unplanned infrastructure with increasing population pressure in cities will only aggravate the situation in the developing urban world.

As stated previously, a gradual shift took place in urban precipitation research beyond 1981 towards tropical and sub-tropical cities, but these often include insufficient and scattered empirical and modeling studies in these areas. Roth (2007) emphasizes that past urban climate research (which includes urban precipitation studies) has primarily focused on North American and European cities located in mid-latitudes of the Northern Hemisphere. Figure 11.3 shows the cities where urban precipitation studies have been conducted worldwide. It is evident from the map that the majority of URE studies occur in developed cities. In the US, observational as well as modeling studies have been done repeatedly on some of the cities (see Table 11.2). This list is not fully exhaustive as there are many more studies available for

Urban precipitation: a global perspective

Table 11.2 US cities where observational and modeling studies mostly occur

Phoenix	Balling and Brazel 1987; Selover 1997; Diem and Brown 2003; Shepherd 2006
Atlanta	Dixon and Mote 2003; Bornstein and Lin 2000; Craig and Bornstein 2002; Shepherd et al. 2002; Diem and Mote 2005; Mote et al. 2007; Shem and Shepherd 2009; Bentley et al. 2010; Haberlie et al. 2015
New York City	Bornstein and LeRoy 1990; Ntelekos et al. 2009
Oklahoma City	Niyogi et al. 2006; Hand and Shepherd 2009
St. Louis	Huff and Changnon 1973, 1986; Changnon et al. 1977, 1991; Ackerman et al. 1978; Changnon 1981; Westcott 1995; Rozoff et al. 2003; van den Heever and Cotton 2007
Chicago	Changnon 1980; Changnon and Westcott 2002
Houston	Huff and Changnon 1973; Orville et al. 2001; Steiger et al. 2002; Shepherd and Burian 2003; Burian and Shepherd 2005; Li et al. 2008; Carrió et al. 2010; Shepherd et al. 2010a; Carrió and Cotton 2011
Fairbanks	Mölders and Olson 2004
San Juan, Puerto Rico	Comarazamy et al. 2006; Torres-Valcárcel et al. 2014

these cities and others within the US. For example, Han et al. (2014) lists cities in Europe, Japan, Oceania and other developed countries.

Studies in the developing world are limited in number, many of which focus on cities in China (Chow and Chang 1984; Guo et al. 2006; Zhang et al. 2009; Li et al. 2011; Miao et al. 2011; Yang et al. 2012; Shao et al. 2013; Wan et al. 2013 and others). This begs the

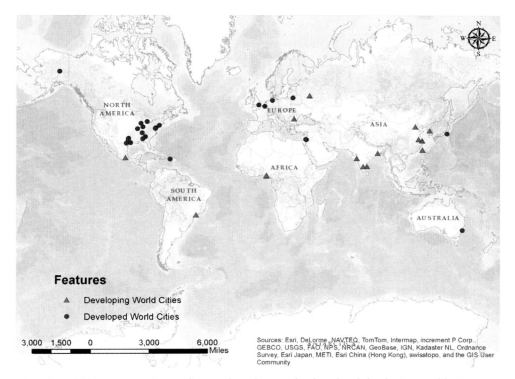

Figure 11.3 Urban precipitation studies conducted across developed and developing world cities

question as to whether or not to consider China at par with other developing countries, as China has experienced impressive economic performance since the 1978 economic reforms. Given that GDP grew 9.3 percent annually during 1979–93 (per capita net income of farmers increased by 239 percent and urban household per capita income increased by 152 percent) and the incidence of absolute poverty declined dramatically in rural areas (from 33 percent in 1978 to 12 percent in 1990) (Woo 1996), China's emerging economy differentiates it from many other developing countries. Even in the case of emissions (Zhang et al. 2007; Shiyi 2009) and urban infrastructure (Zhu and Lin 2004; Li and Yao 2009), China is further advanced relatively speaking.

For argument sake, if China is omitted, the number of urban precipitation studies done on developing world cities is clearly diminished. Some noteworthy studies are Tel Aviv (Goldreich and Manes 1979; Halfon et al. 2009); Riyadh (Shepherd 2006); Chennai (Simpson 2006); Mumbai (Khemani and Murty 1973); Kolkata (Litta and Mohanty 2008; Tyagi et al. 2011; Mitra et al. 2012); Mumbai, Bangalore and Chennai (Goswami et al. 2010); and many other urban areas in India (De and Rao 2004; Sen Roy 2008; Kishtawal et al. 2009); Sao Paulo (Farias et al. 2009; Pinto et al. 2013); Warri, (Efe 2005) and Benin (Efe and Eyefia 2014). These studies represent a mixture of observational and modeling studies, but the majority is observational. All of these studies show increasing trends in rainfall with the exception of De and Rao (2004) and Sen Roy (2008), which reveal decreasing trends in the monsoon and pre-monsoon rainfall regimes in a few Indian cities. These results reflect the dynamic and complex nature of urban areas where impacts may vary based on city size (both population and spatial extent), thermodynamic inputs, surface roughness and aerosol forcing (Shepherd 2005; Schmid and Niyogi 2013). A recent observational study shows remarkable results for Benin City, Nigeria (Efe and Eyefia 2014)—a 43 percent increase in precipitation in urban areas when compared to rural areas (see Figure 11.4). The precipitation of Benin City is also strong and heavy between 2,311 mm and 2,541 mm, which will eventually have implications on the hydrological cycle, infrastructure and socio-economic activity. This highlights the importance of conducting more planned studies on developing world cities where unprecedented and significant impacts of urbanization on precipitation may be found.

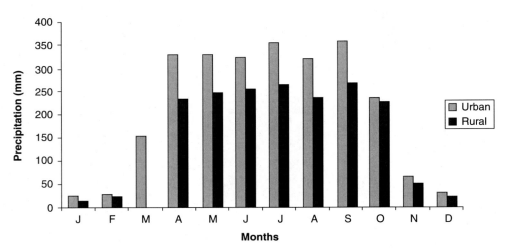

Figure 11.4 Monthly precipitation distribution (mm) in the urban and rural areas of Benin City

Source: Adapted from Efe and Eyefia (2014).

Urban precipitation: a global perspective

Historically, the majority of urban precipitation studies were conducted in European and North American cities. In the last decade or so, there have been more studies in developing countries and emerging economies, most notably a rise in studies focused on cities in India and China. In developing countries, the unplanned nature of urbanization, lower standard of living and large populace below the poverty line will not only impact the overall environment and climatic pattern, but also food production, water supply, coastal settlements, forest ecosystems, health, energy security and other sectors of society (Sathaye et al. 2006). Thus, much more research is needed to understand the interactions and feedbacks that are specific to the context of developing cities. However, obstacles exist. For example, conducting robust and in-depth urban precipitation studies is often challenging given the lack of reliable and long-term gauge, low data densities (Torres-Valcárcel et al. 2014), cost of meteorological instruments (Mills 2014) and radar data. Most meteorological data have large gaps and are also not freely available for research. These impediments would require considerable international collaboration, both financially and knowledge-wise, if they are to be overcome. Ideally, such a program would put in place both an observational and educational infrastructure to conduct some basic research (Mills 2014). Recent developments in satellite and hydroclimate data are noteworthy to mention here. These readily available datasets can be used to monitor

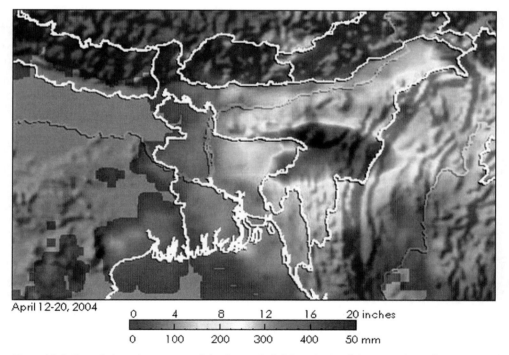

Figure 11.5 Cumulative rains over an eight-day period (12 to 20 April 2004) in the Sylhet region of northeastern Bangladesh

Source: Image courtesy of NASA.
Note: Heavy flooding was associated with the event. Data is the TRMM-based, near-real time Multi-satellite Precipitation Analysis (MPA).
A detailed, colored version of this figure can be found on the book's website, www.routledge.com/9780415732260

anomalies in the precipitation patterns over cities in any part of the world. A few examples are Tropical Rainfall Measuring Mission (TRMM; http://trmm.gsfc.nasa.gov/) (NASA 2015, April 9); GPCC (https://climatedataguide.ucar.edu/climate-data/gpcc-global-precipitation-climatology-centre) (NCAR 2014, May 5); and Aphrodite (https://climatedataguide.ucar.edu/climate-data/aphrodite-asian-precipitation-highly-resolved-observational-data-integration-towards) (NCAR 2014, August 20). Scientists working within developing world cities should utilize these datasets besides their own observational data to validate and improve the outcomes of their research.

Concluding remarks and future research directions

Complex urban land-atmospheric interactions are not easy to grasp and more research must be done to understand the impact of morphology, shape, aerosols and microscale temperature patterns on urban precipitation variability. Water scarcity and insecurity are issues that may arise due to anomalies in the urban precipitation pattern. Research must support and empower authorities to take initiative towards finding solutions to deter the many unwanted impacts of rapid and unplanned urbanization, characteristic of developing countries, on the environment. Further analysis is required to unfold the complex dynamics of the urban environment in order to build more climate change resilient cities for the future. In terms of urban rainfall, much is clear:

- There is now conclusive evidence that urban environments initiate, modify or enhance precipitating weather systems.
- New techniques like dual polarimetric radar, advanced satellite remote sensing (especially NASA's Global Precipitation Measurement Mission—GPM), and regionalized, urbanized coupled land-surface-atmospheric modeling systems are now available to evaluate aspects of the 'urban rainfall problem': relative contribution of aerosols; role of urban shape, size and morphology; impacts on frozen precipitation; signatures in weekly precipitation cycles; and, significance of aggregate urban regions or urban climate archipelagoes.
- Knowledge from urban rainfall studies need to be 'actionable'. The National Academy of Sciences (2012) report on Urban Meteorology provides some pathways and examples to be considered.
- Andersen and Marshall Shepherd (2013) remind us that flood risks are increasing. The combination of increasingly intense rainstorms (NCA 2014), more impervious surfaces (Shepherd et al. 2013; Gustafson et al. 2014) and possible interactions among them point to amplified risk. For example, extreme hydroclimatic events and disasters such as the Bangladeshi floods shown in Figure 11.5 will be further exacerbated by urban land cover and impervious surfaces.

The bi-directional interactions between precipitation processes and urbanization is clear from research, but for a more sustainable and dependable future the following recommendations for research and practice include:

- More geographically balanced studies should be done representing physical, economic, cultural and climatic aspects of developing world cities.
- The relative and synergistic roles of land cover, aerosols, urban morphology and scale on cloud and precipitation microphysics must be improved. For example, the distribution and concentration of aerosols in urban environments have been shown to enhance or

suppress rainfall (Shepherd 2013). Developing world cities, specifically Chinese cities, are significant sources of aerosols, so assessment is needed.

- Emerging remote sensing datasets for land cover, precipitation and aerosols should be implemented to fill the data gap for developing world cities. Satellite or instrument data like Global Precipitation Measurement (GPM) Mission; Moderate Resolution Imaging Spectroradiometer (MODIS) aboard Terra and Aqua; Landsat; Suomi NPP; Sentinel; GOES-R; Indian Remote Sensing (IRS); Cloud Aerosol Lidar and Infrared Pathfinder Satellite Observer (CALIPSO) as well as many others can be used to conduct in-depth research on a local scale focusing on cities.
- An accurate land cover database is required for developing world cities, in particular, in order to improve the resolution of climate models to capture the urban footprint and its interactions. Stone et al. (2010) show that climate change is very much an effect of greenhouse gases and urban processes, respectively.
- Urban precipitation studies and methodologies must be replicated in developing world cities.
- New analyses are needed to quantify how urban precipitation effects impact vulnerability and resilience frameworks relative to floods, landslides and storms.
- The Urban Climate Archipelago or other aggregate frameworks for research should be explored.

References

Ackerman, B., Changnon, S.A., Dzurisin, G., Gatz, D., Grosh, S., Hilberg, F., . . . and Semonin, R. (1978). *Summary of METROMEX, Volume 2: Causes of Precipitation Anomalies.* Bulletin 63. Urbana, Illinois: Illinois State Water Survey.

Andersen, T.K., and Marshall Shepherd, J. (2013). Floods in a changing climate. *Geography Compass*, 7(2), 95–115.

Ashley, W.S., Bentley, M.L., and Stallins, J.A. (2012). Urban-induced thunderstorm modification in the Southeast United States. *Climatic Change*, 1–18.

Balling, R.C., and Brazel, S.W. (1987). Diurnal variations in Arizona monsoon precipitation frequencies. *Monthly Weather Review*, *115*, 342–346.

Bell, T.L., Rosenfeld, D., Kim, K.M., Yoo, J.M., Lee, M.I., and Hahnenberger, M. (2008). Midweek increase in US summer rain and storm heights suggests air pollution invigorates rainstorms. *Journal of Geophysical Research*, *113*, 1–22.

Bentley, M.L., Ashley, W.S., and Stallins, J.A. (2010). Climatological radar delineation of urban convection for Atlanta, Georgia. *International Journal of Climatology*, *30*, 1589–1594.

Bornstein, R., and LeRoy, M. (1990, June). *Urban Barrier Effects on Convective and Frontal Thunderstorms.* Fourth AMS Conference on Mesoscale Processes, '90. Boulder, CO.

Bornstein, R., and Lin, Q. (2000). Urban heat islands and summertime convective thunderstorms in Atlanta: Three case studies. *Atmospheric Environment*, *34*, 507–516.

Braham, R.R. (1981). Urban precipitation processes. *Metromex: A Review and Summary, American Meteorological Society*, *40*, 75–116.

Braham, R.R., Semonin, R.G., Auer, A.H., Changnon, S.A., and Hales, J.M. (1981). Summary of urban effects on clouds and rain. In *METROMEX: A Review and Summary*. Meteorological Monograph 18. American Meteorological Society: Boston, MA, 141–152.

Burian, S.J., and Shepherd, J.M. (2005). Effect of urbanization on the diurnal rainfall pattern in Houston. *Hydrological Processes*, *19*(5), 1089–1103.

Carrió, G.G., and Cotton, W.R. (2011). Urban growth and aerosol effects on convection over Houston. Part II: Dependence of aerosol effects on instability. *Atmospheric Research*, *102*(1), 167–174.

Carrió, G.G., Cotton, W.R., and Cheng, W.Y.Y. (2010). Urban growth and aerosol effects on convection over Houston: Part I: The August 2000 case. *Atmospheric Research*, *96*(4), 560–574.

Changnon, S.A. (1962). Areal frequencies of hail and thunderstorm days in Illinois. *Monthly Weather Review*, *90*, 519–524.

Changnon, S.A. (1968). The La Port weather anomaly—Fact or fiction? *Bulletin of the American Meteorological Society, 49*, 4–11.

Changnon Jr, S.A. (1980). Evidence of urban and lake influences on precipitation in the Chicago area. *Journal of Applied Meteorology, 19*(10), 1137–1159.

Changnon, S.A. (1981). Midwestern cloud, sunshine and temperature trends since 1901: Possible evidence of jet contrail effects. *Journal of Applied Meteorology, 20*, 496–508.

Changnon, S.A. (1992). Inadvertent weather modification in urban areas: Lessons for global climate change. *Bulletin of the American Meteorological Society, 73*, 619–627.

Changnon, S.A. (2003). Urban modification of freezing-rain events. *Journal of Applied Meteorology, 42*(6), 863–870.

Changnon, S.A., and Westcott, N.E. (2002). Heavy rainstorms in Chicago: Increasing frequency, altered impacts, and future implications. *Journal of American Water Resources Association, 38*(5), 1467–1475.

Changnon, S.A., Huff, F.A., Schickedanz, P.T., and Vogel, J.L. (1977). *Summary of METROMEX, Vol. I: Weather Anomalies and Impacts.* Bulletin 62. Urbana, Illinois: Illinois State Water Survey.

Changnon, S.A., Semonin, R., Auer, A.H., Braham, R.R., and Hales, J. (1981). METROMEX: A Review and Summary. *American Meteorological Society Monograph, 18*, 81.

Changnon, S.A., Shealy, R., and Scott, R. (1991). Precipitating changes in fall, winter and spring caused by St. Louis. *Journal of Applied Meteorology, 30*(1), 126–134.

Chow, S.D., and Chang, C. (1984). Shanghai urban influences on humidity and precipitation distribution. *GeoJournal, 8*(2), 201–204.

Collier, C.G. (2006). The impact of urban areas on weather. *Quarterly Journal of the Royal Meteorological Society, 132*(614), 1–25.

Comarazamy, D.E., González, J.E., Tepley, C.A., Raizada, S., and Pandya, R.V.R. (2006). Effects of atmospheric particle concentration on cloud mirophysics over Arecibo. *Journal of Geophysical Research, 11*, 1–10.

Craig, K., and Bornstein, R. (2002). MM5 simulation of urban induced convective precipitation over Atlanta. Preprints. *Fourth Conference on the Urban Environment.* Norfolk, VA: American Meteorological Society.

De, U.S. and Rao, G.S.P. (2004). Urban climate trends – The Indian scenario. *Journal of Indian Geophysics, 8*, 199–203.

Diab, R.D. (1978). Urban effects on precipitation: A review. *South African Journal of Science, 74*(3), 87–91.

Diem, J.E., and Brown, D.P. (2003). Anthropogenic impacts on summer precipitation in central Arizona, U.S.A. *The Professional Geographer, 55*, 343–355.

Diem, J.E., and Mote, T.L. (2005). Interepochal changes in summer precipitation in the southeastern United States: Evidence of possible urban effects near Atlanta, Georgia. *Journal of Applied Meteorology, 44*, 717–730.

Dixon, P.G., and Mote, T.L. (2003). Patterns and causes of Atlanta's urban heat island-initiated precipitation. *Journal of Applied Meteorology, 42*, 1273–1284.

Efe, S.I. (2005). *Urban Effects on Precipitation Amount, Distribution and Rainwater Quality in Warri Metropolis* (Doctoral dissertation). Retrieved from Dept. of Geography and Regional Planning, Delta State University. Abraka, Delta State, Nigeria.

Efe, S.I., and Eyefia, A.O. (2014). Urban effects on the precipitation of Benin, Nigeria. *American Journal of Climate Change, 3*(1), 8–21.

Farias, W.R.G., Pinto, O., Naccarato, K.P., and Pinto, I.R.C.A. (2009). Anomalous lightning activity over the metropolitan region of São Paulo due to urban effects. *Atmospheric Research, 91*(2), 485–490.

Garstang, M., Tyson, P.D., and Emmit, G.D. (1975). The structure of heat islands. *Reviews of Geophysics and Space Physics, 13*(1), 139–165.

Goldreich, Y., and Manes, M.A. (1979). Urban effects on precipitation patterns in the greater Tel-Aviv area. *Archiv für Meteorologie, Geophysik und Bioklimatologie, Serie B, 27*(2–3), 213–224.

Goswami, P., Shivappa, H., and Goud, B.S. (2010). Impact of urbanization on tropical mesoscale events: Investigation of three heavy rainfall events. *Meteorologische Zeitschrift, 19*(4), 385–397.

Guo, X., Fu, D., and Wang, J. (2006). Mesoscale convective precipitation system modified by urbanization in Beijing City. *Atmospheric Research, 82*(1), 112–126.

Gustafson, S., Heynen, N., Rice, J.L., Gragson, T., Shepherd, J.M., and Strother, C. (2014). Megapolitan political ecology and urban metabolism in southern Appalachia. *The Professional Geographer, 66*(4), 664–675.

Urban precipitation: a global perspective

Haberlie, A.M., Ashley, W.S., and Pingel, T.J. (2015). The effect of urbanisation on the climatology of thunderstorm initiation. *Quarterly Journal of the Royal Meteorological Society*. DOI: 10.1002/qj.2499

Halfon, N., Levin, Z., and Alpert, P. (2009). Temporal rainfall fluctuations and their possible link to urban and air pollution effects. *Environmental Research Letters*, 4(2), 1–12.

Han, J.Y., Baik, J.J., and Lee, H. (2014). Urban impacts on precipitation. *Asia-Pacific Journal of Atmospheric Sciences*, 50(1), 17–30.

Hand, L.M., and Shepherd, J.M. (2009). An investigation of warm-season spatial rainfall variability in Oklahoma City. Possible linkages to urbanization and prevailing wind. *Journal of Applied Meteorology and Climatology*, 48(2), 251–269.

Hodder, B.W. (1968). *Economic Development in the Tropics*. London: Methuen and Co. Ltd.

Horton, R.E. (1921). Thunderstorm breeding spots. *Monthly Weather Review*, 49, 193–194.

Huff, F.A. and Changnon, S.A. (1973). Precipitation modification by major urban areas. *Bulletin of the American Meteorological Society*, 54(12), 1220–1232.

Huff, F.A., and Changnon Jr, S.A. (1986). Potential urban effects on precipitation in the winter and transition seasons at St. Louis, Missouri. *Journal of Applied Meteorology*, 25, 1887–1907.

Jauregui, E. (1993). Mexico City's urban heat island revisited (Die Wärmeinsel von Mexico City—ein Rückblick). *ErdKundle*, 185–195.

Jauregui, E., and Romales, E. (1996). Urban effects on convective precipitation in Mexico City. *Atmospheric Environment*, 30(20), 3383–3389.

Jha, A.K., Bloch, R., and Lamond, J. (2012). *Cities and Flooding: A Guide to Integrated Urban Flood Risk Management for the 21st Century*. Washington, DC: The World Bank.

Jin, M., Shepherd, J.M., and King, M.D. (2005). Urban aerosols and their variations with clouds and rainfall: A case study for New York and Houston. *Journal of Geophysical Research*, 110, 1–12.

Kaufmann, R.K., Seto, K.C., Schneider, A., Liu, Z., Zhou, L., and Wang, W. (2007). Climate response to rapid urban growth: Evidence of a human-induced precipitation deficit. *Journal of Climate*, 20(10), 2299–2306.

Khemani, L.T. and Murty, R. BhV. R. (1973). Rainfall variations in an urban industrial region. *Journal of Applied Meteorology*, 12, 187–194.

Kishtawal, C., Niyogi, D., Tiwari, M., Pielke, R., and Shepherd, M. (2009). Urbanization signature in the observed heavy rainfall climatology over India. *International Journal of Climatology*, 30, 1908–1916.

Kratzer, P.A. (1937). *Das Stadtklima*. 2nd ed., F. Vieweg uE Sohne, Braunschweig, F. Vieweg uE Sohne. (Trans. by the US Air Force, Cambridge Research Laboratories, Bedford, Massachusetts).

Kratzer, P.A. (1956). *Das Stadtklima*. F. Vieweg uE Sohne, Braunschweig.

Landsberg, H.E. (1956). *The Climate of Towns. Man's Role in Changing the Face of the Earth*. Chicago, IL: Univ. of Chicago Press, pp. 584–606.

Landsberg, H.E. (1970). Man-made climatic changes. *Science*, 170, 1265–1274.

Li, B., and Yao, R. (2009). Urbanisation and its impact on building energy consumption and efficiency in China. *Renewable Energy*, 34(9), 1994–1998.

Li, G., Want, Y., and Zhang, R. (2008). Implementation of a two-moment bulk microphysics scheme to the WRF model to investigate aerosol–cloud interaction. *Journal of Geophysical Research*, 113, 1–21.

Li, J., Song, C., Cao, L., Zhu, F., Meng, X., and Wu, J. (2011). Impacts of landscape structure on surface urban heat islands: A case study of Shanghai, China. *Remote Sensing of Environment*, 115(12), 3249–3263.

Litta, A.J., and Mohanty, U.C. (2008). Simulation of a severe thunderstorm event during the field experiment of STORM programme 2006, using WRF-NMM model. *Current Science*, 95(2), 204–215.

Lowry, W.P. (1998). Urban effects on precipitation amount. *Progress in Physical Geography*, 22(4), 477–520.

Mahmood, R., Pielke, R.A., Hubbard, K.G., Niyogi, D., Dirmeyer, P.A., McAlpine, C., . . . and Fall, S. (2014). Land cover changes and their biogeophysical effects on climate. *International Journal of Climatology*, 34(4), 929–953.

Mendelsohn, R., and Dinar, A. (1999). Climate change, agriculture, and developing countries: Does adaptation matter? *The World Bank Research Observer*, 14(2), 277–293.

Miao, S., Chen, F., Li, Q., and Fan, S. (2011). Impacts of urban processes and urbanization on summer precipitation: A case study of heavy rainfall in Beijing on 1 August 2006. *Journal of Applied Meteorology and Climatology*, 50(4), 806–825.

Mills, G. (2007). Cities as agents of global change. *International Journal of Climatology*, *27*(14), 1849–1857.

Mills, G. (2014). Urban climatology: History, status and prospects. *Urban Climate*, *10*, 479–489.

Mirza, M.M.Q. (2003). Climate change and extreme weather events: Can developing countries adapt? *Climate Policy*, *3*(3), 233–248.

Mitra, C., Shepherd, J.M., and Jordan, T. (2012). On the relationship between the premonsoonal rainfall climatology and urban land cover dynamics in Kolkata city, India. *International Journal of Climatology*, *32*(9), 1443–1454.

Mölders, N. and Olson, M.A. (2004). Impact of urban effects on precipitation in high latitudes. *Journal of Hydrometeorology*, *5*(3), 409–429.

Mote, T.L., Lacke, M.C., and Shepherd, J.M. (2007). Radar signatures of the urban effect on precipitation distribution: A case study for Atlanta, Georgia. *Geophysical Research Letters*, *34* (20), L20710 (4pp).

NASA. (2015, 9 April) *TRMM Tropical Rainfall Measuring Mission*. Retrieved from http://trmm.gsfc.nasa.gov

National Academy of Sciences. (2012). *Urban Meteorology: Forecasting, Monitoring, and Meeting Users' Needs*. Washington, DC: The National Academies Press.

National Center for Atmospheric Research (NCAR). (2014, 5 May). *The Climate Data Guide: APHRODITE: Asian Precipitation – Highly-Resolved Observational Data Integration Towards Evaluation of Water Resources*. Retrieved from https://climatedataguide.ucar.edu/climate-data/aphrodite-asian-precipitation-highly-resolved-observational-data-integration-towards

National Center for Atmospheric Research (NCAR). (2014, 20 August) *The Climate Data Guide: GPCC: Global Precipitation Climatology Centre. 2014. NCAR Climate Data Guide*. Retrieved from https://climatedataguide.ucar.edu/climate-data/gpcc-global-precipitation-climatology-centre

National Climate Assessment (NCA). (2014). *National Climate Assessment*. Retrieved from http://nca2014.globalchange.gov/report

Niyogi, D., Holt, T., Zhong, S., Pyle, P.C., and Basara, J. (2006). Urban and land surface effects on the 30 July 2003 mesoscale convective system event observed in the Southern Great Plains. *Journal of Geophysical Research*, *111*, 1–20.

Niyogi, D., Pyle, P., Lei, M., Arya, S.P., Kishtawal, C.M., Shepherd, M., . . . and Wolfe, B. (2011). Urban modification of thunderstorms: An observational storm climatology and model case study for the Indianapolis urban region. *Journal of Applied Meteorology and Climatology*, *50*(5), 1129–1144.

Ntelekos, A.A., Smith, J.A., Donner, L., Fast, J.D., Gustafson, W.I., Chapman, E.G., and Krajewski, W.F. (2009). The effects of aerosols on intense convective precipitation in the northeastern United States. *Quarterly Journal of the Royal Meteorological Society*, *135*(643), 1367–1391.

Oke, T.R. (1987). *Boundary Layer Climates*. London: Routledge.

Orville, R.E., Huffines, G., Nielsen-Gammon, J., Zhang, R., Ely, B., Steiger, S., . . . and Read, W. (2001). Enhancement of cloud-to-ground lightning over Houston, Texas. *Geophysical Research Letters*, *28*(13), 2597–2600.

Pinto, O., Pinto, I.R.C.A., and Ferro, M.A.S. (2013). A study of the long-term variability of thunderstorm days in southeast Brazil. *Journal of Geophysical Research: Atmospheres*, *118*(11), 5231–5246.

Rosenberger, M.S., and Suckling, P.W. (1989). Precipitation climatology in the Pittsburgh urban area during late spring and summer. *Southeastern Geographer*, *29*(2), 75–91.

Rosenfeld, D., and Woodley, W.L. (2003). Spaceborne inferences of cloud microstructure and precipitation processes: synthesis, insights, and implications. In Tao, Wei-Kuo (Ed.) *Cloud Systems, Hurricanes, and the Tropical Rainfall Measuring Mission (TRMM). A Tribute to Joanne Simpson*. Boston: American Meteorological Society, pp. 59–80.

Rosenfeld, D., Lohmann, U., Raga, G.B., O'Dowd, C.D., Kulmala, M., Fuzzi, S., and Andreae, M.O. (2008). Flood or drought: How do aerosols affect precipitation? *Science*, *321*, 1309–1313.

Roth, M. (2007). Review of urban climate research in (sub) tropical regions. *International Journal of Climatology*, *27*(14), 1859–1873.

Rozoff, C.M., Cotton, W.R., and Adegoke, J.O. (2003). Simulation of St. Louis, Missouri, land use impacts on thunderstorms. *Journal of Applied Meteorology*, *42*, 716–738.

Sathaye, J., Shukla, P.R., and Ravindranath, N.H. (2006). Climate change, sustainable development and India: Global and national concerns. *Current Science*, *90*(3), 314–325.

Schmaus, M. (1927). *Die psychologische Trinitätslehre des hl. Augustinus*. Aschendorff.

Schmid, P.E., and Niyogi, D. (2013). Impact of city size on precipitation – Modifying potential. *Geophysical Research Letters*, *40*(19), 5263–5267.

Selover, N. (1997). Proceedings from the AAG Annual Meeting '97: *Precipitation Patterns Around an Urban Desert Environment-Topographic or Urban Influences?* Fort Worth, TX: AAG.

Sen Roy, S. (2008). A spatial analysis of extreme hourly precipitation patterns in India. *International Journal of Climatology, 29,* 345–355.

Seto, K.C., and Shepherd, J.M. (2009). Global urban land-use trends and climate impacts. *Current Opinion in Environmental Sustainability, 1*(1), 89–95.

Shao, H., Song, J., and Ma, H. (2013) Sensitivity of the East Asian summer monsoon circulation and precipitation to an idealized large-scale urban expansion. *Journal of Meteorological Society of Japan, 91,* 163–177.

Shem, W., and Shepherd, M. (2009). On the impact of urbanization on summertime thunderstorms in Atlanta: Two numerical model case studies. *Atmospheric Research, 92*(2), 172–189.

Shepherd, J.M. (2005). A review of current investigations of urban-induced rainfall and recommendations for the future. *Earth Interactions, 9,* 1–27.

Shepherd, J.M. (2006). Evidence of urban-induced precipitation variability in arid climate regimes. *Journal of Arid Environment, 67,* 607–628.

Shepherd, J. (2013). *Impacts of Urbanization on PDI Precipitation and Storms: Physical Insights and Vulnerabilities. Climate Vulnerability: Understanding and Addressing Threats to Essential Resources.* Elsevier Inc., Academic Press, pp. 109–125.

Shepherd, J.M., and Burian, S.J. (2003). Detection of urban-induced rainfall anomalies in a major coastal city. *Earth Interactions, 7,* 1– 17.

Shepherd, J.M., and Mote, T.L. (2011, 6 September). *Can Cities Create Their Own Snowfall? What Observations are Required to Find Out?* Retrieved from http://earthzine.org/2011/09/06/can-cities-create-their-own-snowfall-what-observations-are-required-to-find-out/

Shepherd, J.M., Pierce, H., and Negri, A.J. (2002). Rainfall modification by major urban areas: Observations from spaceborne rain radar on the TRMM satellite. *Journal of Applied Meteorology, 41*(7), 689–701.

Shepherd, J.M., Carter, M., Manyin, M., Messen, D., and Burian, S. (2010a). The impact of urbanization on current and future coastal precipitation: A case study for Houston. *Environment and Planning B: Planning and Design, 37,* 284–304.

Shepherd, J.M., Carter, W.M., Manyin, M., Messen, D., and Burian, S. (2010b). The impact of urbanization on current and future coastal convection: A case study for Houston. *Environment and Planning, 37,* 284–304.

Shepherd, J.M., Anderson, T., Bounoua, L., Horst, A., Mitra, C., and Strother, C. (2013, 5 November). *Urban Climate Archipelagos: A New Framework for Urban-Climate Interactions.* Retrieved from www.earthzine.org/2013/11/29/urban-climate-archipelagos-a-new-framework-for-urban-impacts-on-climate/

Shiyi, C. (2009). Energy consumption, CO2 emission and sustainable development in Chinese industry. *Economic Research Journal, 4,* 1–5.

Simpson, M.D. (2006). *Role of Urban Land Use on Mesoscale Circulations and Precipitation.* (Doctoral dissertation). Retrieved from NCSU Libraries.

Souch, C., and Grimmond, S. (2006). Applied climatology: Urban climate. *Progress in Physical Geography, 30*(2), 270–279.

Steiger, S.M., Orville, R.E., and Huffines, G. (2002). Cloud-to-ground lightning characteristics over Houston, Texas: 1989–2000. *Journal of Geophysical Research, 107* (D11), 4117, http://dx.doi.org/10.1029/2001JD001142.

Stone, B., Hess J.J., and Frumkin, H. (2010). Urban form and extreme heat events: Are sprawling cities more vulnerable to climate change than compact cities. *Environmental Health Perspectives, 118*(10), 1425–1428.

Stout, G.E. (1962). *Some Observations of Cloud Initiation in Industrial Areas. In Air Over Cities.* Technical Report A62-5. Washington, DC: US Public Health Service.

Stulov, E.A. (1993). Urban effects on summer precipitation in Moscow. *Russian Meteorology and Hydrology, 11,* 34–41.

Thielen, J., Wobrock, W., Gadian, A., Mestayer, P.G., and Creutin, J.D. (2000). The possible influence of urban surfaces on rainfall development: A sensitivity study in 2D in the meso-scale. *Atmospheric Research, 54*(1), 15–39.

Thompson, E.J., Rutledge, S.A., Dolan, B., Chandrasekar, V., and Cheong, B.L. (2014). A dual-polarization radar hydrometeor classification algorithm for winter precipitation. *Journal of Atmospheric and Oceanic Technology, 31*(7), 1457–1481.

Torres-Valcárcel, Á., Harbor, J., González-Avilés, C., and Torres-Valcárcel, A. (2014). Impacts of urban development on precipitation in the tropical maritime climate of Puerto Rico. *Climate, 2*(2), 47–77.

Trusilova, K., Jung, M., Churkina, G., Karstens, U., Heimann, M., and Claussen, M. (2008). Urbanization impacts on the climate in Europe: Numerical experiments by the PSU-NCAR Mesoscale Model (MM5). *Journal of Applied Meteorology and Climatology, 47*(5), 1442–1455.

Tyagi, B., Krishna, V.N., and Satyanarayana, A.N.V. (2011). Study of thermodynamic indices in forecasting pre-monsoon thunderstorms over Kolkata during STORM pilot phase 2006–2008. *Natural Hazards, 56*(3), 681–698.

UN. (2007). *World Urbanization Prospects: The 2007 Revision.* New York: United Nations.

van den Heever, S.C., and Cotton, W.R. (2007). Urban aerosol impacts on downwind convective storms. *Journal of Applied Meteorology and Climatology, 46*, 828–850.

Vukovich, F.M., and Dunn, J.W. (1978). Theoretical study of St. Louis heat island—Some parameter variations. *Journal of Applied Meteorology, 17*, 1585–1594.

Wan, H., Zhong, Z., Yang, X., and Li, X. (2013). Impact of city belt in Yangtze River Delta in China on a precipitation process in summer: A case study. *Atmospheric Research, 125*, 63–75.

Westcott, N.E. (1995). Summertime cloud-to-ground lightning activity around major midwestern urban areas. *Journal of Applied Meteorology, 34*(7), 1633–1642.

Woo, W.T. (1996). *Chinese Economic Growth: Sources and Prospects.* Department of Economics, University of California, Davis.

Yang, F., Tan, J., Shi, Z.B., Cai, Y., He, K., Ma, Y., . . . and Chen, G. (2012). Five-year record of atmospheric precipitation chemistry in urban Beijing, China. *Atmospheric Chemistry and Physics, 12*(4), 2025–2035.

Zhang, C., Miao, S., Li, Q., and Chen, F. (2007). Incorporation of offline-resolution land use information of Beijing into numerical weather model and its assessing experiments on a summer severe rainfall. *Chinese Journal of Geophysics, 50*(5), 1172–1182.

Zhang, C.L., Chen, F., Miao, S.G., Li, Q.C., Xia, X.A., and Xuan, C.Y. (2009). Impacts of urban expansion and future green planting on summer precipitation in the Beijing metropolitan area. *Journal of Geophysical Research, 114*, 1–26.

Zhu, Y., and Lin, B. (2004). Sustainable housing and urban construction in China. *Energy and Buildings, 36*(12), 1287–1297.

12

EFFECTS OF URBANIZATION ON LOCAL AND REGIONAL CLIMATE

CSB Grimmond, Helen C. Ward and Simone Kotthaus

Urbanization has profound effects on climate. The materials and morphology of the urban surface, along with emissions from domestic, commercial and transport activities, result in changes in local climate often greater in magnitude than projected global scale climate change. Cities are commonly 2–3°C warmer than their surrounding environments, with the greatest differences at night and in winter. Such urban climate effects increase the vulnerability of residents to future environmental change, making cities prime sites for climate mitigation and adaptation. This chapter describes the major processes influencing urban climates at a range of scales, and illustrates these with data from London, one of Europe's most densely populated cities and home to over 8.2 million people.

Cities compared to rural landscapes

Major surface and atmospheric changes are associated with the construction and functioning of cities. Materials used for buildings, roads and other infrastructure, along with the three-dimensional form of the urban landscape, alter the radiative, thermal, hydrologic and aerodynamic properties of the surface, the airflow, and the energy and water exchanges. Direct anthropogenic emissions of heat, carbon dioxide and other pollutants also modify the urban climate (see Figure 12.1). Urban development typically replaces natural or agricultural vegetation with surfaces that are impermeable and inflexible to the movement of water and air. Buildings, often arranged in relatively regular rows or grids, channel and accelerate airflow, and their sharp edges can introduce highly turbulent wakes and vortices (Kastner-Klein et al. 2004; Xie et al. 2005). Networks of pipes, channels and canals replace or supplement natural drainage systems, further modifying exchanges of water and heat.

Building material properties, architectural styles and the layout of cities differ considerably (e.g. Herold et al. 2004; Kotthaus et al. 2014). Materials used for a certain type of facet (e.g. roofing materials such as asphalt, ceramic tiles, thatch, slate, steel or copper) have a wide range of radiative (albedo, emissivity) and conductive (thermal admittance, heat capacity) properties, which affect heat retention and loss. Together, these have direct implications for energy uptake and release by the buildings, indoor and outdoor temperatures, and thus heating and cooling demands (Kikegawa et al. 2003) and carbon emissions. Although vegetation cover in many cities is small, 'greenspace', i.e. parks, gardens, shrubs, trees and even green roofs and walls, have

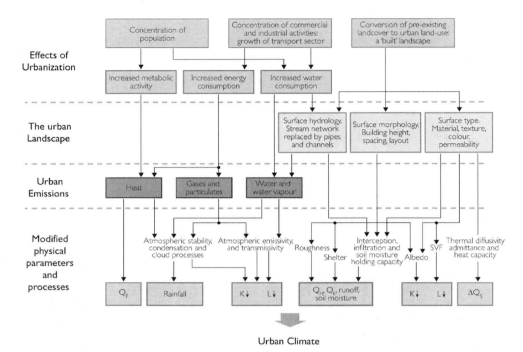

Figure 12.1 The effects of the urban landscape and urban emissions on key parameters and processes resulting from urbanization. (Symbols are defined in the text.)

Source: Cleugh and Grimmond (2012).

a particularly important influence on heat, water and carbon exchanges, and thus on temperature and humidity. Some suburbs of Western cities meet the definition of a forest (Oke 1989) and irrigation enables urban greening in arid regions (Shashua-Bar et al. 2011). The juxtaposition of built and vegetated elements is fundamental to the creation of urban climates, and it is critical to represent this adequately in numerical models (Grimmond et al. 2011).

Scales of urban climate effects

Climate variables, whether air temperature, humidity or wind speed, vary significantly both within and between cities. Air temperatures, for example, differ from one side of a street to the other (sunlit versus shaded), from a park to an industrial neighborhood or from one suburb to another. These differences also change through time. For any application, identifying the relevant scale of interest is absolutely critical: micro-, local or meso-scale (see Figure 12.2).

The fundamental elements at the *micro-scale* ($<10^2$ m) are individual buildings and trees. As buildings are typically sharp-edged, rigid and impermeable, they substantially alter the wind-flow and turbulence. The large thermal capacity of buildings enhances daytime heat storage and subsequently night-time release. In combination, building shapes and materials control how radiation is absorbed, reflected and (sometimes) transmitted. While building exteriors are designed to limit moisture exchanges (notably to restrict water entry), heat exchanges occur by conduction through all the exterior walls and by convection through gaps, vents and chimneys. How the individual buildings are operated, either by inhabitants or businesses, affects the timing and magnitude of these exhaust heat and gas emissions.

Urbanization effects on local and regional climate

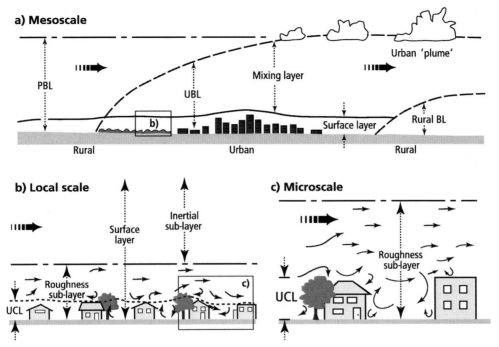

Figure 12.2 Key spatial scales in the urban atmosphere
Source: Adapted from Oke (1997).

Multiple urban canyons (two rows of buildings with a transport corridor, e.g. a street or alley way) create *city block*, *local-* or *neighborhood-scale* (10^2–10^4 m) areas within the city (see Figure 12.2). The building heights (H) and widths (W) of the spaces in between allow these areas to be characterized with respect to differences in radiative, convective heat and momentum exchanges. Cities are aerodynamically rough and therefore reduce overall wind speeds and ventilation in the urban canopy, with important implications for air quality, temperature and human exposure. Micro-scale features are blended in mean turbulent flow above the roughness sub-layer (see Figure 12.2) at the local scale. With increasing spatial extent (land use zones across several kilometers), the urban surface starts to have an influence that extends beyond the immediate group of buildings into the overlying boundary layer. Examples of local-scale units include a residential area dominated by similar style housing, an industrial area dominated by warehouses or the central business district of a city with high-rise buildings.

At the largest urban spatial scale, the *meso-scale* (10^3–10^5 m), the urban boundary layer (see Figure 12.2) reflects the influence of all the different neighborhoods, i.e. the entire city. Thermal and chemical characteristics are transported downwind, impacting a larger region extending beyond the city. The regional setting of a city, for example, if it is next to a large water body (whether lakes—Chicago, Geneva; or oceans—Los Angeles, Vancouver, Shanghai, Tokyo) or in a mountain-valley system (e.g. Salt Lake City, Phoenix, Mexico City, Innsbruck), influences how it interacts with and modifies regional thermal circulations and can cause re-circulation of air pollutants into the city and/or the surrounding areas. Such complex wind-flow regimes can generate multiple convergence zones, which modify precipitation and

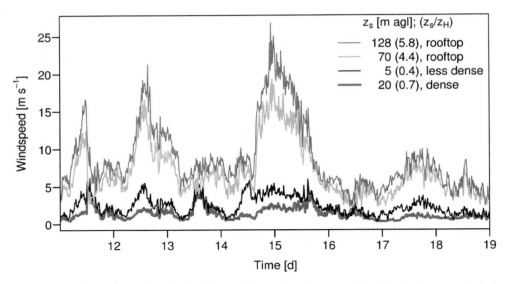

Figure 12.3 Measured wind speeds (m s^{-1}) at various heights in central London during a particularly windy period in February 2014. The four sensors are locatad at height, z_S, above ground level (agl). Two are below roof level, that is below the mean height of the buidlings (z_H) (i.e. $z_S/z_H < 1$) or within an urban canyon. The sensor in the more dense area has a lower wind speed despite being higher above the ground. The other two sensors are located well above the mean building height ($z_S/z_H > 1$). The mean height is determind here from the building heights within 1 km of the site of the anemometer.

cloud patterns (e.g. Yoshikado and Kondo 1989). These are embedded in the larger weather patterns of expected fronts and high/low pressure systems, the latitudinal variations in solar radiation and the moderation from maritime/continental influences. To illustrate effects at different scales, wind speed measurements from various settings and heights in London during a particularly windy period are shown in Figure 12.3. Within the urban canopy, the wind is significantly reduced but, as in all areas, increases with height. Such strong winds can cause significant damage to infrastructure and loss of life.

Urban energy exchanges

Modification of energy, water, momentum and carbon exchanges between the urban surface and atmosphere ultimately create distinct urban climates. Each of these exchanges can be written as a statement of conservation (of energy, mass, etc.) for a typical volume at the various scales described above (see Figure 12.2). For energy this is defined as (Oke 1987):

$$Q^\star + Q_F = Q_E + Q_H + \Delta Q_S \tag{1}$$

with inputs of net all-wave radiation ($Q^\star$) and anthropogenic heat (Q_F). The outputs are the energy to evaporate and transpire water through evapotranspiration (latent heat flux Q_E), and to heat the air (sensible heat flux Q_H) and urban volume (net storage heat flux ΔQ_S). Both biophysical and human factors regulate these exchanges. While Equation 1 is applicable at any spatial scale in the urban environment, it does assume that the net lateral transport of energy due to advection, is negligible at the scale under consideration (see Pigeon et al. 2007 for discussion of this issue).

Observations in cities, key to understanding these exchange processes, remain fewer in number than in other environments. In the last decade, a range of cities with contrasting climatic and topographic settings, varying population density, socio-economic status and urban design have been studied (see Lietzke et al. 2015). From these, the variability of flux exchanges both within and between cities and controlling factors have become evident (Loridan and Grimmond 2012). The key elements of these are described below.

Urban radiation balance

The incoming short-wave radiation (diffuse plus direct beam radiation) is reduced in urban areas, varying from minor reductions (~ 5 percent) in those cities with low aerosol concentrations to much larger reductions (up to ~ 30 percent) in cities that have high levels of particulate pollution (e.g. Rome and Athens) (Mallet et al. 2013). To illustrate urban effects on incoming solar radiation, data are presented for three sites in southern England (70 to 100 km apart at similar latitude): central London (Kotthaus and Grimmond 2014a,b); a suburban neighborhood in Swindon (Ward et al. 2013); and a woodland site (Alice Holt) (Wilkinson et al. 2012). Full details of the sites, the instrumentation and comparison of the data are presented by Ward et al. (2015). Experiencing broadly similar synoptic conditions, differences between the sites reflect local anthropogenic or natural controls. Observations of incoming solar radiation at the three sites (see Figure 12.4) illustrate general agreement in seasonal (winter–summer) and synoptic patterns (cloudy–clear), but also with a reduction of solar radiation receipt in central London, the most urbanized of the sites with the poorest air quality.

Very high concentrations of greenhouse gases (especially carbon dioxide, CO_2) and aerosols occur in urban areas where they are released from numerous sources (Crawford and Christen 2014). They increase atmospheric absorption leading to elevated temperatures and often increased absolute humidity levels. Thus, the urban atmosphere experiences an enhanced greenhouse effect, over and above that occurring at the global scale resulting in greater incoming long-wave radiation. Low urban albedo (e.g. total reflectance of 0.08 to 0.20 out of 1.00) for suburban and urban landscapes (Taha 1997) is attributed to the radiative characteristics of urban construction materials and radiative trapping within street canyons. The net effect is greater short-wave radiation absorption compared to many non-urban landscapes receiving the same global radiation. Similarly, the reduced sky view factor impedes long-wave radiation losses, so net long-wave radiation receipt is typically increased. Local scale changes in building materials and urban design can be used to modify the radiation balance (Taha 1997). However, generally, spatial differences in the net all-wave radiation are surprisingly small (Schmid et al. 1991; Arnfield 2003; Lietzke et al. 2015).

The micro-scale radiation regime of an individual element (e.g. person or tree) or facet (e.g. wall, roof or pavement) is highly variable in space and time and closely related to its surroundings; e.g. short wave reflectance of urban impervious materials may range from 0.05 to 0.54 (Kotthaus et al. 2014). The urban form (or morphology) creates shadows and reduces sky view factors; thus, tall buildings can reduce the albedo because of radiative trapping, while if clad with highly reflective, non-Lambertian materials such as glass and metal, which are often used in modern architecture, they can also generate strong specular reflections and glare (Shih and Huang 2001a,b). As one example, the façade of a new skyscraper in the City of London reflected sufficient radiation to create surface temperatures warm enough to melt plastic parts of cars parked nearby (Guardian 2013). The effects of highly reflective urban

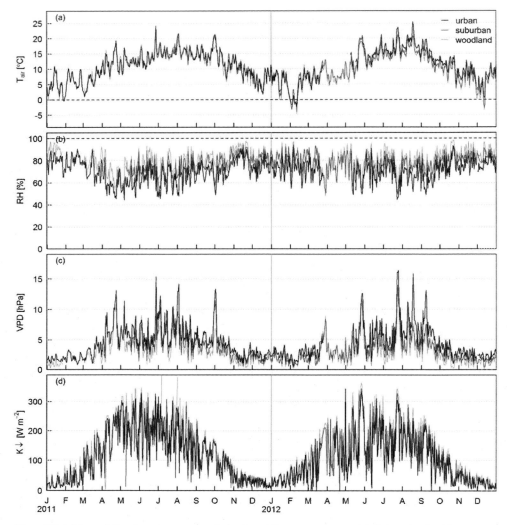

Figure 12.4 Mean daily air temperature, relative humidity, vapor pressure deficit and solar (short-wave) radiation for a two-year period from an urban (dark grey, central London, Kotthaus and Grimmond 2014a,b), suburban (medium grey, residential Swindon, Ward et al. 2013) and rural (light grey, Alice Holt deciduous forest, Wilkinson et al. 2012) site in southern England

Note: All sites experience broadly the same synoptic conditions; differences represent urban/non-urban effects. Gaps in the data are due to missing observations.

surface materials are illustrated with data of outgoing radiation at a rooftop site in London (see Figure 12.5). This shows the clear impact of a roof window. Radiation levels are crucial for human comfort (Erell et al. 2014).

Urban heat storage and anthropogenic fluxes

The change in heat storage, ΔQ_S, in all elements of the urban canopy layer (air, biomass, soil and built) is a dominant term in the urban energy balance (Equation 1). This flux is, however,

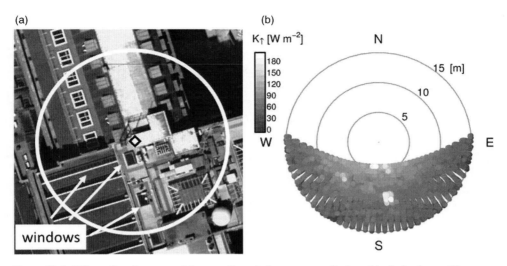

Figure 12.5 Effect of surface materials on reflected short-wave radiation: (a) circle shows 80 percent field of view (15.2 m radius) for the radiation sensor; (b) median reflected short-wave radiation observed (2012) by solar azimuth angle and distance of maximum, specular reflection Rs

Source: Aerial photo (NERC ARSF 2008).
Note: Full details of analysis and definition of Rs in Kotthaus and Grimmond (2014b).

very difficult to measure directly (Offerle et al. 2005). It can be approximated by the sum of the heat fluxes conducted through the solid-air interfaces (i.e. walls, roofs, pavement and roads, trees, lawns and gardens) (Arnfield 2003) or estimated as a residual in the energy balance when all other terms are measured and accounted for (see Roberts et al. 2006 for discussion). By day, the flux consumes between 20 and 30 percent of the net all-wave radiation in suburban land use, and up to half of the net radiation in heavily urbanized sites such as found in Mexico City (Oke et al. 1999), London (Kotthaus and Grimmond 2014a) and industrial areas of Vancouver (Grimmond and Oke 1999). By night, the net loss of radiation from the urban canopy is typically balanced by the heat storage term (and Q_F) and can lead to positive sensible heat flux exchanges (with implications for nocturnal atmospheric stability). From estimates using the residual approach, ΔQ_S for cities in North America, Africa and Europe are presented in Figure 12.6 (grey squares). The data are normalized by the incoming (short-wave and long-wave) fluxes so relative patterns are evident (otherwise controls of latitude and synoptic conditions on the total amount of energy would dominate). Over the course of the day, ΔQ_S is out of phase with the incoming radiation, leading to a non-linear, hysteresis $\Delta Q_S / Q^*$ relation (evident in Figure 12.6, particularly notable for Miami).

Anthropogenic heat flux (Q_F) from stationary and mobile sources can be important in the urban energy balance. Q_F is difficult to measure directly so most estimates are derived from energy use statistics or surrogates, such as traffic numbers; alternatively Q_F can be estimated as a residual in the urban energy balance (see review by Sailor 2011). While the average diurnal range of Q_F is estimated to be between 0.7–3.6 W m^{-2} for all urban areas globally (Allen et al. 2011), a value that is relatively small, this flux is highly scale dependent. It is much larger for small areas with dense built infrastructure in the center of major cities. For example, Ichinose et al. (1999) document fluxes > 1,000 W m^{-2} in Tokyo. Using a top-down approach for Greater London (Iamarino et al. 2012), the relative impacts of building emissions (Q_{Fb}),

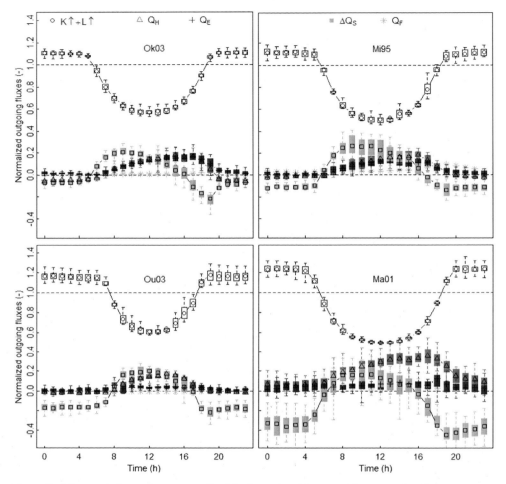

Figure 12.6 Energy balance fluxes normalized by incoming (short- and long-wave) radiation

Source: Loridan and Grimmond (2012)

Note: Mean diurnal pattern of the ratios (symbols) with box plots for data by Loridan and Grimmond (2012). Full descriptions of each site (codes below) and the appropriate source references see Loridan and Grimmond (2012). Codes: Ok03 - Oklahoma City, US; Mi95 - Miami, US; Ou03 - Ouagadougou, Burkina Faso; Ma01 - Marseille, France.

road traffic (Q_{Ft}) and metabolism (Q_{Fm}) are clear (see Figure 12.7). The combined overall annual average flux is 10.9 W m^{-2}, but in the City of London values reach 210 W m^{-2} (annual average). The commercial/service sector, which covers only 2.5 percent of London's area, has annual average fluxes > 50 W m^{-2}. If shorter time periods are considered, the peak values can be much larger. Thus, anthropogenic heat fluxes can be large relative to radiative inputs, especially in winter. The timing of human activities (such as commuting, heating of buildings, cooking of meals, etc.) and their variations between work and non-workdays all have a large impact on Q_F and its diurnal pattern and therefore on the input of additional energy to the urban system.

Urbanization effects on local and regional climate

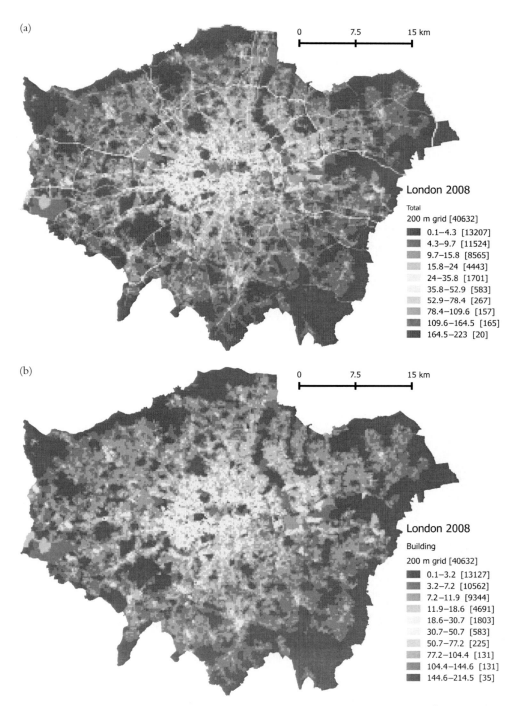

Figure 12.7 Spatial variability of anthropogenic heat emissions (2008) at 200×200 m² resolution by sector for London (classes by Jenks natural breaks): (a) total, (b) building

(continued)

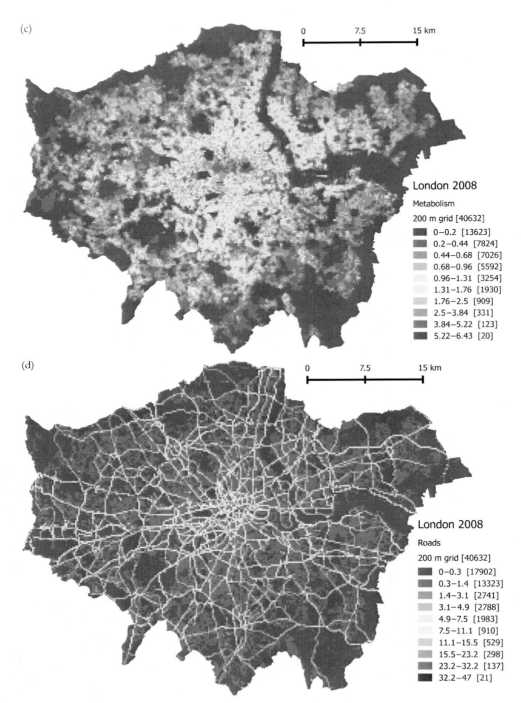

Figure 12.7 (continued) (c) metabolism and (d) road traffic values in W m^{-2} [number of grids in each class]

Source: Modified from Iamarino et al. (2012).
A detailed, colored version of this figure can be found on the book's website, www.routledge.com/9780415732260

Turbulent latent and sensible heat fluxes

Urban influences on Q_E and Q_H fluxes are large including enhancement through the aerodynamically rough urban canopy and strong dependence on surface water availability (e.g. from precipitation, irrigation). The Bowen ratio ($\beta = Q_H/Q_E$) provides a key indicator, which varies with vegetation (type, amount) and water availability. More energy is partitioned into sensible heat fluxes (Q_H) in dense urban areas if vegetated areas are small, while evaporation usually becomes more important with an increasing proportion of green surface cover (e.g. suburban areas). Typically large urban storage heat fluxes in the morning (partially because of the increased heating of the vertical surfaces due to absorption of incoming solar radiation) and the evaporation of surface water (dew, nocturnal irrigation), if present (Grimmond and Oke 1991), result in a time lag of the sensible heat flux in urban areas compared to natural surfaces (Grimmond and Oke 1999, Grimmond and Oke 2002). In dense urban areas, the turbulent sensible heat flux generally remains positive (see Figure 12.6), i.e. an energy source to the atmosphere throughout the day and night (Grimmond and Oke 2002, Kotthaus and Grimmond 2014a). This heating, which results in a persistently strong net upward motion of warm air, has an important influence on the atmospheric boundary layer in cities with implications for air quality, larger-scale meteorological processes and health (Bohnenstengel et al. 2014).

The most variable flux within and between cities is the latent heat flux: ranging from almost zero to values greater than the surrounding rural landscapes (Grimmond et al. 1993) when the study area is well watered with abundant vegetation (e.g. urban gardens, Oke 1979, or parks, Spronken-Smith et al. 2000). As evapotranspiration impacts energy, water and carbon balances, this flux can be manipulated to passively mitigate urban heating and cooling demand, and, thereby, energy consumption (e.g. Mitchell et al. 2008) (see further examples below).

Urban water balance

The urban water balance is expressed in mass (kg or as mm) exchanges per unit time (Grimmond and Oke 1991):

$$p + I + F = E + \Delta r + \Delta W \qquad (2)$$

where the inputs are precipitation p (including dew, fog, hail, rain and snow), piped water supply (I) (released indoors and outdoors), and water released with combustion (F). The output terms are evapotranspiration (E), runoff (Δr) (waste and storm water) and changes in stored water (ΔW) in the urban canopy (air, biomass, soil and built components) (see Lietzke et al. 2015 for a more extended discussion of the urban water balance). When expressed this way the water balance is assumed to relate to a volume with vertical dimensions from within the soil up to the urban canopy airspace. The lateral dimensions can apply to a single household, a neighborhood (equivalent to the local scale defined above), or a catchment defined by topography and/or the pipe supply network.

All terms in Equation 2 are impacted by urbanization, with direct implications for the urban climate. The dominant changes occur because of enhanced water supply (notably) and increased runoff associated with the greater impervious cover and piped networks (Grimmond et al. 1986). While reduced vegetation cover may lower evapotranspiration, a more secure and enhanced water supply can sustain large rates of evapotranspiration. In fact E can often be the largest output term in the urban water balance, exceeding the runoff (e.g. Grimmond et al. 1986). In the absence of water restrictions, a plentiful water supply can generate higher evaporation rates in suburban areas than in rural landscapes (e.g. Sacramento, US, presented by Grimmond et al. 1993). Urban climate and hydrology are linked directly by the evapotranspiration term, the mass flux (E) in the water balance and the energy used in the phase change of liquid water to vapor (Q_E) in the energy balance (i.e. $Q_E = L_v E$ where L_v is the latent heat of vaporization). Thus, the urban climate can be modulated by manipulating the water balance, for example by sustaining evapotranspiration through importing water (I) and/or harnessing and reusing runoff (Δr). Managing the urban water balance is a key potential strategy to mitigate excessive urban heating (Steeneveld et al. 2011). For a detailed discussion of urbanization and climate as it relates to precipitation see Mitra and Shepherd, Chapter 11 in this volume.

Urban carbon balance

Vegetation's assimilation of the greenhouse gas CO_2 during photosynthesis requires photosynthetically active radiation (PAR) and releases water vapor. Thus, there are close links between the urban carbon, energy and water balances. The control volume can again be used to define a carbon balance:

$$F_{CP} + F_{CF} = F_{CR} + F_{CM} + S_C \tag{3}$$

where the inputs of CO_2 are assimilated through photosynthesis by vegetation, F_{CP} and the import of fossil fuels F_{CF}. The CO_2 emissions are from heterotrophic (animals including people) and autotrophic (plants, algae) respiration (F_{CR}) and human activities (F_{CM}). The latter includes transport, space heating/cooling, other combustion of fuel (e.g. for cooking) and industrial emissions. The size of the net change in storage (ΔS_C) depends on both the temporal and spatial scales considered and whether a gain or loss of CO_2 occurs within the urban volume of interest. As for energy and water, the control volume, although relatively straightforward in theory, requires careful definition of the boundaries in practice.

As urban areas significantly perturb the regional and global carbon budget, understanding the drivers is critical to addressing global climate change. The ability to measure directly the net CO_2 exchanges in urban landscapes is important in assessing progress on reducing carbon emissions (see the overview by Christen 2014). With both anthropogenic and biogenic controls, urban carbon exchanges have a complex dependence on surface cover (Velasco and Roth 2010; Grimmond and Christen 2012; Järvi et al. 2012; Nordbo et al. 2012; Ward et al. 2015). In most urban environments, the biogenic sources are much smaller than the anthropogenic ones. However, when plants are actively growing (and thus sequestering CO_2) there may be a balance with respiration (i.e. $F_{CP} \cong F_{CR}$) or even a net uptake of carbon (see, for example, the summer months in Baltimore, Crawford et al. 2011). The urban climate notably the urban heat island, solar radiation (in terms of the amount of diffuse versus direct radiation), irrigation, enhanced levels of CO_2 and atmospheric deposition of nitrogen affects the biological processes of photosynthesis and respiration in urban canopies. The phenology of urban vegetation (i.e. time of leaf growth and leaf fall), which is different to rural surroundings (Gazal et al. 2008; Neil et al. 2010) and the enhanced size of urban vegetation because of elevated CO_2 concentrations (documented in Baltimore by George et al. 2009), affect net CO_2 sequestration.

Urbanization effects on local and regional climate

Cities are clearly significant net sources of carbon dioxide, with emissions closely related to the fraction of the surface covered by impervious materials (buildings and roads) (Grimmond and Christen 2012). Although urban greenspace does sequester some of the CO_2 emissions from vehicles and other urban sources, recent observations (e.g. Coutts et al. 2007; Crawford et al. 2011; Ward et al. 2015) indicate this sequestration is insufficient to offset the emissions on an annual basis. Diurnal and seasonal variations in net CO_2 fluxes are dominated by the cycle of human activities and emissions from vehicles and space heating. Net emissions are larger during weekdays and during cold periods (such as the winter months when additional heating is required). The annual emissions of CO_2 for Tokyo, Mexico City, Melbourne and Copenhagen (cited by Velasco and Roth 2010) vary between 8 and 15 kg CO_2 m^{-2} y^{-1} (i.e. 80 and 150 t CO_2 ha^{-1} y^{-1}). These emissions are about ten times the CO_2 sequestered by a mature and productive forest (Falge et al. 2002; Law et al. 2002). The interconnection of water, heat and carbon exchanges is well illustrated by the study of Pataki et al. (2009), who conclude that the greatest benefits from urban tree-planting programs come not from carbon sequestration but rather from the modification of the urban energy balance, notably shading at the micro-scale, evaporation at the local scale and the impacts these have on human comfort and energy demand.

Resulting effects and implications of urban climates

The most widely recognized and studied urban climate phenomenon, the urban heat island (UHI), refers to the frequent observation that urban areas are warmer than their surrounding regions. Canyon air temperatures tend to exhibit the greatest difference two to three hours after sunset, although daytime heat islands are also observed (evident in the temperature profiles for London, Swindon and Alice Holt forest shown in Figure 12.4). The strength of the canopy air temperature UHI (defined as ΔT_{U-R}, where U and R refer to urban and rural, respectively) can be as large as ~10°C on individual nights (Oke 1981). Typically, the peak is within the city core and declines with distance towards the urban/rural boundary. The greatest intra-urban temperature differences tend to be associated with clear skies and low wind speeds. The clear skies allow maximum solar radiation receipt during the day, thus enhancing heating of vertical surfaces and roofs. Cities usually have higher building densities in the center, so warmer temperatures tend to be found in these locations. The maxima can though be displaced downwind, with the location changing seasonally depending on winds (e.g. sea breeze circulations in Tokyo, Honjo et al. 2012). The locations of parks or other open areas can also create complex temperature patterns. However, the cooling effect of parks does not typically extend long distances downwind (Eliasson and Upmanis 2000; Nagatani et al. 2008). In extremely cold climates, the maximum urban warming is associated with intense anthropogenic heat use (e.g. Hinkel et al. 2003; Malevich and Klink 2011).

From the enormous number of studies of the UHI (e.g. within Santamouris 2007), it is evident that great care must be taken in the methods of observation and analysis (Stewart 2011). There are, in fact, many types of UHIs (Voogt 2004): near-surface urban canopy layer heat islands, subterranean urban heating effects (e.g. Allen et al. 2003), boundary layer heat islands (e.g. Bohnenstengel et al. 2011) and radiative surface heat islands (e.g. Jin et al. 2005). Remote sensing-based studies of the UHI are numerous, but determination of the surface temperature remains challenging because of limited view angles, coarse spatial resolution (mixed land cover in pixels), difficulty in obtaining the surface emissivity and atmospheric corrections (Roth et al. 1989; Voogt and Oke 2003; Dousset and Gourmelon 2003; Kotthaus et al. 2014). Understanding the causes of the UHI allows insight into strategies for mitigation with broader implications for management of energy resources. The mitigation strategies are applicable at scales ranging from individual buildings to neighborhoods (see examples in Table 12.1), and can be used in new developments, or to retrofit or refurbish existing building stock. Approaches taking advantage of material properties (e.g. Saadatian et al. 2013; Santamouris 2013) can be used on existing buildings (e.g. application of specialized coatings, exchanging roofing material) so are often more cost and time effective than completely new construction. However, new developments can alter the orientations or separations of buildings, the variability of building heights or land cover proportions and patterns. Such modifications can have environmental benefits at multiple scales, from individual buildings

Table 12.1 Causes of urban warming and examples of mitigation strategies

Urban heat island causes	Mitigation strategy*
Increased surface area	
Large vertical faces	
Reduced sky view factor	
Increased absorption of short-wave (solar) radiation	High reflection building and road materials, high reflection paints for vehicles
Decreased long-wave (terrestrial) radiation loss	Spacing of buildings
Decreased total turbulent heat transport	Variability of building heights
Reduced wind speeds	
Surface materials – thermal characteristics	
Higher heat capacities	Reduce surface temperatures (changing albedo and emissivity)
Higher conductivities	Improved roof insulation
Increased surface heat storage	
Surface materials – moisture characteristics	
Large proportion of impervious surface cover	Porous pavement
Shed water more rapidly – changes the hydrograph	Neighborhood detention ponds and wetlands which collect stormwater
Increased runoff with a more rapid peak	Increase greenspace fraction
Decreased evapotranspiration (latent heat flux, Q_E)	Green roofs, green walls
Additional supply of energy – anthropogenic heat flux – Q_F	
Electricity and combustion of fossil fuels: heating and cooling systems, machinery, vehicles	Reduced solar loading internally, reduce need for active cooling (shades on windows, change materials)
3-D geometry of buildings – canyon geometry	District heating and cooling systems
	Combined heat and power systems
	High reflection paint on vehicles and buildings to reduce temperature
Air pollution	
Human activities lead to ejection of pollutants and dust into the atmosphere	District heating and cooling systems
Increased long-wave radiation from the sky	Combined heat and power or cogeneration systems
Greater absorption and re-emission ('greenhouse effect')	

Source: Adapted from Grimmond (2007).
* Effects may vary by time of year and by cities at different latitudes.

to neighborhoods, and may even help to reduce CO_2 emissions from power generation elsewhere.

Alternative strategies to cool urban areas include adding water detention ponds and wetlands that enhance evaporation rates during the day and create more open areas to enhance cooling at night given their larger sky view factors. These have other benefits too, such as reducing peak urban runoff and reducing the need for construction of large infrastructure to protect against flash floods and/or manage the release of untreated water downstream. They can also improve water quality and provide social, cultural and psychological amenities with benefits from the creation of 'natural' space. New residential developments can employ water-sensitive urban design that involves the use of greywater to irrigate residential vegetation (Coutts et al. 2013) reducing the demand for water to be diverted into a city for irrigation purposes. The enhanced warmth of urban areas, the reduced vegetation cover and the presence of engineered structures designed to remove water rapidly from cities all have influences on a city's atmospheric moisture. The data from the urban-forest transect in southern England (see Figure 12.4) illustrate this clearly. The synoptic influences are common to each site, but superimposed are the urban effects related to both moisture and temperature. Higher vapor pressure deficits and lower relative humidities are evident in central London. In winter, if conditions are dry, water release from combustion-related sources can be important in creating an urban moisture excess (e.g. Holmer and Eliasson 1999; Kuttler et al. 2007).

Concluding comments

Urban areas cover only a small fraction (< 5 percent) of the Earth's surface (Schneider et al. 2009). Their moisture, thermal and kinematic plumes remain distinct only a few kilometers downwind, although aerosol and gaseous releases are distributed much more widely. Many of the processes instrumental in creating urban climates are the same drivers of global anthropogenic climate change: regional-scale land use changes; increased energy use; and increased emissions of climatically relevant atmospheric constituents. Thus, cities aid in the understanding of anthropogenic climate change and are also critical agents in moderating its effects. Understanding how urban surface properties affect radiation, energy, water and carbon exchanges can support mitigation and adaptation efforts centered around deliberate manipulation of surface climates, for example, through building design and materials or the use of vegetation. Cities and the drivers of urbanization are central to global environmental research. Urban populations and areas will continue to grow in size and number. Existing urban areas will experience redevelopment and refurbishment. The decisions made about how this will occur will impact not only those living within the buildings, neighborhoods and cities, but in combination will have global implications and consequences.

Key messages

- Urbanization has a profound effect on climate. The materials and morphology of the urban surface, along with emissions from domestic, commercial and transport activities, result in changes in local climate often greater in magnitude than projected global scale climate change.
- Urban climates have been studied for a long period; a rich literature with data, concepts, models and theories exists to inform current work.
- Ultimately, urban climate effects are due to changes in surface-atmosphere energy, water and carbon exchanges; these can best be understood through the frameworks of energy, water and carbon budgets.

- Understanding the scale-dependent nature of urban climates is fundamental to understanding, representative measurement programs and modelling of current and future cities for any application.
- Cities are critical agents in moderating future global climate change; the materials used in their construction, building and neighborhood design and residents' behavior all have profound effects on climate within and beyond a city's boundaries.

Future research needs

- More attention needs to be directed to understanding urban atmospheric dynamics in tropical cities where a large and ever increasing fraction of the world's population lives.
- Attention also needs to be directed to towns and medium-sized cities as well as megacities, with specific attention to the scaling of effects and processes across the full range of urban environments.
- New methods need to be developed to measure water, energy and carbon exchanges in dense urban settings (increasingly characteristic of South America and Asian cities), where current methods and theory are challenged by very tall buildings.
- High-resolution numerical models need to be further developed to resolve and link canyon, neighborhood and city scale conditions.
- Integrated studies are needed of energy, water and carbon exchanges to better understand interdependences and the implications of interventions to mitigate urban climate effects.

Acknowledgments

The writing of this chapter was supported by NERC TRUC (NE/L008971/1, G8MUREFU3FP-2201-075) and EU emBRACE (283201). We would like to thank James Morison and Matt Wilkinson (Forestry Commission) for permission to use the Alice Holt data shown here, Jon Evans (NERC-CEH) who was instrumental in the Swindon project, and Will Morrison who helped with aspects of the London plots. We would like to thank all those involved in the observations (especially Alex Bjorkegren), those who provided sites and permission to install instruments, and the agencies that have funded the data collection (NERC, EUf7, KCL).

References

Allen A, D Milenic, P Sikora 2003: Shallow gravel aquifers and the urban 'heat island' effect: A source of low enthalpy geothermal energy, *Geothermics*, 32, 569–578.
Allen L, F Lindberg, CSB Grimmond 2011: Global to city scale model for anthropogenic heat flux, *Int. J. Climat.*, 31, 1990–2005 doi:10.1002/joc.2210
Arnfield AJ 2003: Two decades of urban climate research: A review of turbulence, exchanges of energy and water, and the Urban Heat Island, *Int. J. Climatol.*, 23, 1–26.
Bohnenstengel SI, S Evans, PA Clark, SE Belcher 2011: Simulations of the London urban heat island, *Q.J.R. Meteorol. Soc.*, 137, 1625–1640 doi:10.1002/qj.855
Bohnenstengel S, SE Belcher, A Aiken, JD Allan, G Allen, A Bacak et al. 2014: Meteorology, air quality, and health in London: The ClearfLo project Bull, *Amer. Meteorol. Soc.* http://dx.doi.org/10.1175/BAMS-D-12-00245.1
Christen A 2014: Atmospheric measurement techniques to quantify greenhouse gas emissions from cities, *Urban Climate* doi: 10.1016/j.uclim.2014.04.006
Cleugh HA and CSB Grimmond 2012: Urban climates and global climate change. In Henderson-Sellers A and Mc Guffie K (eds) *The Future of the World's Climate*. Oxford: Elsevier, 47–76 + references, DOI:10.1016/B978-0-12-386917-3.00003-8
Coutts AM, J Beringer, NJ Tapper 2007: Characteristics influencing the variability of urban CO_2 fluxes in Melbourne, Australia, *Atmos. Environ.*, 41, 51–62.

Coutts, AM, NJ Tapper, J Beringer, M Loughnan, M Demuzere 2013: Watering our cities: The capacity for Water Sensitive Urban Design to support urban cooling and improve human thermal comfort in the Australian context, *Prog. Phys. Geog.*, 37, 2–28 doi:10.1177/0309133312461032

Crawford B, CSB Grimmond, A Christen 2011: Five years of carbon dioxide fluxes measurements in a highly vegetated suburban area, *Atmos. Env.*, 45, 896–905 doi:10.1016/j.atmosenv.2010.11.017

Crawford B and A Christen 2014: Spatial source attribution of measured urban eddy covariance carbon dioxide fluxes, *Theor. Appl. Climatol.* http://dx.doi.org/10.1007/s00704-014-1124-0

Dousset B and F Gourmelon 2003: Satellite multi-sensor data analysis of urban surface temperatures and landcover, *ISPRS Journal of Photogrammetry and Remote Sensing*, 58, 43–54 http://dx.doi.org/10.1016/S0924-2716(03)00016-9

Eliasson I and H Upmanis 2000: Nocturnal airflow from urban parks-implications for city ventilation, *Theoretical and Applied Climatology*, 66, 95–107. 10.1007/s007040070035.

Erell E, D Pearlmutter, D Boneh, PB Kutiel 2014: Effect of high-albedo materials on pedestrian heat stress in urban street canyons, *Urban Climate*, 10, Part 2, 367–386 doi: http://dx.doi.org/10.1016/j.uclim.2013.10.005

Falge E, D Baldocchi, J Tenhunen, M Aubinet, P Bakwin et al. 2002: Seasonality of ecosystem respiration and gross primary production as derived from FLUXNET measurements, *Agric. For. Meteorol.*, 113DI 10.1016/S0168-1923(02)00102-8

Gazal R, MA White, R Gillies, E Rodemaker, E Sparrow, L Gordon 2008: GLOBE students, teachers, and scientists demonstrate variable differences between urban and rural leaf phenology, *Glob. Change Biol.*, 14, 1568–1580 doi:10.1111/j.1365-2486.2008.01602.x

George K, LH Ziska, JA Bunce, B Quebedeaux, JL Hom, J Wolf, JR Teasdale 2009: Macroclimate associated with urbanization increases the rate of secondary succession from fallow soil, *Oecologia*, 159, 637–647.

Grimmond CSB 2007: Urbanization and global environmental change: Local effects of urban warming, *Geogr. J.*, 173, 83–88. DOI:10.1111/j.1475-4959.2007.232_3.x

Grimmond CSB and TR Oke 1991: An evapotranspiration-interception model for urban areas, *Water Resour. Res.*, 27, 1739–1755.

Grimmond CSB and TR Oke 1999: Heat storage in urban areas: Observations and evaluation of a simple model, *J. Appl. Meteorol.*, 38, 922–940.

Grimmond CSB and TR Oke 2002: Turbulent fluxes in urban areas: Observations and a local-scale urban meteorological parameterization scheme (LUMPS), *J. Appl. Meteorol.*, 41, 792–810.

Grimmond CSB, TR Oke, DG Steyn 1986: Urban water balance I: A model for daily totals, *Water Resour. Res.*, 22, 1397–1403.

Grimmond CSB, TR Oke, HA Cleugh 1993: The role of "rural" in comparisons of observed suburban and rural flux differences, *IAHS Pub.*, 212, 165–174.

Grimmond CSB, M Blackett, MJ Best, J-J Balk, SE Belcher, J Beringer et al. 2011: Initial results from phase 2 of the International Urban Energy Balance Comparison Project, *Int. J. Climatol.*, 31, 244–272 doi:10.1002/joc.2227

Grimmond S and A Christen 2012: Flux measurements in urban ecosystems, *FluxLetter*, 5(1), 1–8, 32–35 http://fluxnet.ornl.gov/sites/default/files/FluxLetter_Vol5_no1.pdf

Guardian The 2013: Walkie Talkie architect "didn't realise it was going to be so hot". www.theguardian.com/artanddesign/2013/sep/06/walkie-talkie-architect-predicted-reflection-sun-rays

Herold M, DA Roberts, ME Gardner, PE Dennison 2004: Spectrometry for urban area remote sensing: Development and analysis of a spectral library from 350 to 2400 nm, *Remote Sens. Environ.*, 91, 304–319.

Hinkel KM, FE Nelson, AE Klene, JH Bell 2003: The urban heat island in winter at Barrow, *Alaska. Int. J. Climatol.*, 23, 1889–1905 doi:10.1002/joc.971

Holmer B and I Eliasson 1999: Urban-rural vapour pressure differences and their role in the development of urban heat islands, *Int. J. Climatol.*, 19, 989–1009 doi:10.1002/(SICI)1097-0088

Honjo T et al. 2012: Daily movement of heat island in Kanto area, 6th Japanese-German Meeting on Urban Climatology, Hiroshima Institute of Technology, Japan.

Iamarino M, S Beevers, CSB Grimmond 2012: High resolution (space, time) anthropogenic heat emissions: London 1970–2025, *Int. J. Climatol.*, 32, 1754–1767 doi:10.1002/joc.2390

Ichinose T, K Shimodozono, K Hanaki 1999: Impact of anthropogenic heat on urban climate in Tokyo, *Atmospheric Environment*, 33(24), 3897–3909.

Järvi L, A Nordbo, H Junninen, A Riikonen, J Moilanen, E Nikinmaa, T Vesala 2012: Seasonal and annual variation of carbon dioxide surface fluxes in Helsinki, Finland, in 2006–2010, *Atmospheric Chemistry and Physics*, 12, 8475–8489 doi:10.5194/acp-12-8475-2012

Jin M, RE Dickinson, DL Zhang 2005: The footprint of urban areas on global climate as characterized by MODIS, *J. Climate*, 18, 1551–1565.

Kastner-Klein P, R Berkowicz, R Britter 2004: The influence of street architecture on flow and dispersion in street canyons, *Meteorology & Atmospheric Physics*, 87, 121–131 doi:10.1007/s00703-003-0065-4

Kikegawa Y, Y Genchi, H Yoshikado, H Kondo 2003: Development of a numerical simulation system toward comprehensive assessments of urban warming countermeasures including their impacts upon the urban buildings' energy-demands, *Appl. Energ.*, 76, 449–466 doi:10.1016/S0306-2619(03)00009-6

Kotthaus S and CSB Grimmond 2014a: Energy exchange in a dense urban environment – Part I: Temporal variability of long-term observations in central London, *Urban Climate*, 10, 261–280 doi:10.1016/j.uclim.2013.10.002

Kotthaus S and CSB Grimmond 2014b: Energy exchange in a dense urban environment – Part II: Impact of spatial heterogeneity of the surface, *Urban Climate*, 10, 281–307 doi:10.1016/j.uclim.2013.10.001

Kotthaus S, TEL Smith, MJ Wooster, CSB Grimmond 2014: Derivation of an urban materials spectral library through emittance and reflectance spectroscopy, *ISPRS Journal of Photogrammetry and Remote Sensing*, 94, 194–212 doi:10.1016/j.isprsjprs.2014.05.005

Kuttler W, S Weber, J Schonnefeld, A Hesselschwerdt 2007: Urban/rural atmospheric water vapour pressure differences and urban moisture excess in Krefeld, Germany, *Int. J. Climatol.*, 27, 2005–2015 doi:10.1002/joc.1558

Law BE, E Falge, L Gu, DD Baldocchi, P Bakwin, P Berbigier et al. 2002: Environmental controls over carbon dioxide and water vapor exchange of terrestrial vegetation, *Agric. For. Meteorol.*, 113, 97–120 doi:10.1016/S0168-1923(02)00104-1

Lietzke B, R Vogt, DT Young, CSB Grimmond 2015: Physical fluxes in urban environment (Chapter 4). In Chrysoulakis N, Castro E, Moors E (eds) *Understanding Urban Metabolism: A Tool for Urban Planning*. Oxon: Routledge, 29–44, www.routledge.com/books/details/9780415835114/

Loridan T and CSB Grimmond 2012: Characterization of energy flux partitioning in urban environments: Links with surface seasonal properties, *J. Appl. Meteorol. Climatol.*, 51, 219–241 doi:10.1175/JAMC-D-11-038.1

Malevich SB and K Klink 2011: Relationships between snow and the wintertime Minneapolis Urban Heat Island, *J. Appl. Meteor. Climatol.*, 50, 1884–1894 doi:http://dx.doi.org/10.1175/JAMC-D-11-05.1

Mallet M, O Dubovik, P Nabat, F Dulac, R Kahn, J Sciare et al. 2013: Absorption properties of Mediterranean aerosols obtained from multi-year ground-based remote sensing observations, *Atmos. Chem. Phys.*, 13, 9195–9210 doi:10.5194/acp-13-9195-2013

Mitchell VG, HA Cleugh, CSB Grimmond, J Xu 2008: Linking urban water balance and energy balance models to analyse urban design options, *Hydrol. Process.*, 22, 2891–2900 doi:10.1002/hyp.6868

Nagatani Y, K Umeki, T Honjo, H Sugawara, K Narita, T Mikami 2008: Analysis of movement of cooled air in Shinjuku Gyoen, Japanese progress in climatology, *Journal of Agricultural Meteorology*, 64(4), 281–288.

Neil KL, L Landrum, J Wu 2010: Effects of urbanization on flowering phenology in the metropolitan Phoenix region of USA: Findings from herbarium records, *J. Arid Environ.*, 74, 440–444.

Nordbo A, L Järvi, S Haapanala, CR Wood, T Vesala 2012: Fraction of natural area as main predictor of net CO2 emissions from cities, *Geophysical Research Letters*, 39, L20802 doi:10.1029/2012GL053087

Offerle B, CSB Grimmond, K Fortuniak 2005: Heat storage and anthropogenic heat flux in relation to the energy balance of a central European city centre, *Int. J. Climatol.*, 25, 1405–1419.

Oke TR 1979: Advectively-assisted evapotranspiration from irrigated urban vegetation, *Boundary Layer Meteorol*, 17, 167–173.

Oke TR 1981: Canyon geometry and the nocturnal urban heat island: Comparison of scale model and field observations, *J. Climatol.*, 1, 237–254.

Oke TR 1987: *Boundary Layer Climates*, Oxon: Routledge, 435 pp.

Oke TR 1989: The micrometeorology of the urban forest, *Philos. T. Roy. Soc. B*, 324, 335–349.

Oke TR 1997: Urban environments. In WG Bailey et al. (eds) *The Surface Climates of Canada*. Montréal: McGill-Queen's University Press, 303–327.

Oke TR, RA Spronken-Smith, E Jáuregui, CSB Grimmond 1999: The energy balance of central Mexico City during the dry season, *Atmos. Environ.*, 33, 3919–3930.

Pataki DE, PC Emmi, CB Forster, JI Mills, ER Pardyjak, TR Peterson et al. 2009: An integrated approach to improving fossil fuel emissions scenarios with urban ecosystem studies, *Ecol. Complex.*, 6, 1–14 doi:10.1016/j.ecocom.2008.09.003

Pigeon G, A Lemonsu, CSB Grimmond, P Durand, O Thouron, V. Masson 2007: Divergence of turbulent fluxes in the surface layer: Case of a coastal city, *Bound.-Lay. Meteorol.*, 124, 269–290.

Roberts SM, TR Oke, CSB Grimmond, JA Voogt 2006: Comparison of four methods to estimate urban heat storage, *J. Appl. Meteorol. Clim.*, 45, 1766–1781.

Roth M, TR Oke, WJ Emery 1989: Satellite-derived urban heat islands from three coastal cities and the utilization of such data in urban climatology, *Int. J. Remote Sensing*, 10, 1699–1720.

Saadatian O, K Sopian, E Salleh, CH Lim, S Riffat, E Saadatian et al. 2013: A review of energy aspects of green roofs, *Renew. Sustain. Energy Rev.*, 23, 155–168 doi:10.1016/j.rser.2013.02.022

Sailor DJ 2011: A review of methods for estimating anthropogenic heat and moisture emissions in the urban environment, *Int. J. Climatol.*, 31, 189–199 doi:10.1002/joc.2106

Santamouris M 2007: Heat island research in Europe: The state of the art, *Adv. Building Energy Research*, 1, 123–150.

Santamouris M 2013: Using cool pavements as a mitigation strategy to fight urban heat island: A review of the actual developments, *Renew. Sustain. Energy Rev.*, 26, 224–240.

Schmid HP, HA Cleugh, CSB Grimmond, TR Oke 1991: Spatial variability of energy fluxes in suburban terrain, *Boundary Layer Meteorology*, 54, 249–276.

Schneider A, MA Friedl, D Potere 2009: A new map of global urban extent from MODIS data, *Environmental Research Letters*, 4, article 044003 doi:10.1088/1748-9326/4/4/044003

Shashua-Bar L, D Pearlmutter, E Erell 2011: The influence of trees and grass on outdoor thermal comfort in a hot-arid environment, *Int. J. Climatol.*, 31, 1498–1506 doi:10.1002/joc.2177

Shih NJ and YS Huang 2001a: An analysis and simulation of curtain wall reflection glare, *Build. Env.*, 36, 619–626.

Shih NJ and YS Huang 2001b: A study of reflection glare in Taipei, *Build. Res. Inform.*, 29, 30–39.

Spronken-Smith RA, TR Oke, WP Lowry 2000: Advection and the surface energy balance across an irrigated urban park, *Int. J. Climatol.*, 20, 1033–1047.

Steeneveld GJ, S Koopmans, BG Heusinkveld, LWA van Hove, AAM Holtslag 2011: Quantifying urban heat island effects and human comfort for cities of variable size and urban morphology in the Netherlands, *Journal of Geophysical Research: Atmospheres (1984–2012)* 116, D20129 (14pp).

Stewart ID 2011: A systematic review and scientific critique of methodology in modern urban heat island literature, *Int. J. Climatol.*, 31, 200–217 doi:10.1002/joc.2141

Taha H 1997: Urban climates and heat islands: Albedo, evapotranspiration, and anthropogenic heat, *Energy and Buildings*, 25, 99–103.

Velasco E and M Roth 2010: Cities as net sources of CO2: Review of atmospheric CO2 exchange in urban environments measured by eddy covariance technique, *Geography Compass*, 4, 1238–1259 doi:10.1111/j.1749-8198.2010.00384.x

Voogt JA 2004: Urban heat islands: Hotter cities. www.actionbioscience.org/environment/voogt.html (last accessed February 23, 2014).

Voogt JA and TR Oke 2003: Thermal remote sensing of urban areas, *Remote Sensing of Environment*, 86, 370–384.

Ward HC, JG Evans, CSB Grimmond 2013: Multi-season eddy covariance observations of energy, water and carbon fluxes over a suburban area in Swindon, UK, *Atmos. Chem. Phys.*, 13, 4645–4666 doi:10.5194/acp-13-4645-2013

Ward HC, S Kotthaus, CSB Grimmond, A Bjorkegren, M Wilkinson, WTJ Morrison, JG Evans, JIL Morison, M Iamarino 2015: Effects of urban density on carbon dioxide exchanges: Observations of dense urban, suburban and woodland areas of southern England, *Environmental Pollution*, 198: 186–200 doi: 10.1016/j.envpol.2014.12.031

Wilkinson M, EL Eaton, MSJ Broadmeadow, JIL Morison 2012: Inter-annual variation of carbon uptake by a plantation oak woodland in south-eastern England, *Biogeosciences*, 9, 5373–5389 doi:10.5194/bg-9-5373-2012

Xie XM, Z Huang, JS Wang 2005: Impact of building configuration on air quality in street canyon, *Atmos. Environ.*, 39, 4519–4530 doi:10.1016/j.atmosenv.2005.03.043

Yoshikado H and H Kondo 1989: Inland penetration of the sea breeze over the suburban area of Tokyo, *Boundary-Layer Meteorology*, 48, 389–407. http://dx.doi.org/10.1007/BF00123061

13

URBAN NUTRIENT CYCLING

Lucy R. Hutyra

Nutrients are chemical elements that organisms require for life. Biogeochemistry is the study of the cycling of nutrients through the biosphere, hydrosphere, lithosphere and atmosphere. All living organisms require specific combinations of nutrients to grow, survive and reproduce. Plant leaves, for example, have an average carbon-to-nitrogen-to-phosphorus ratio of 800:30:1 (Elser et al. 2000), whereas microbes have an average ratio of 60:7:1 (Cleveland and Liptzin 2007). The specific ratios of elements in living tissue couple the cycling of these nutrients (Sterner and Elser 2002). If the availability of one element declines, it can become a limiting nutrient on overall productivity. Similarly, if one element becomes too abundant, it can cause toxicity. Six elements—Hydrogen (H), Carbon (C), Nitrogen (N), Oxygen (O), Phosphorus (P) and Sulfur (S)—constitute 95 percent of the mass of the biosphere (Schlesinger and Bernhardt 2013). Of these key elements, the C, N and P cycles have been most dramatically modified by humans and urbanization.

This chapter synthesizes current understanding of how urbanization influences the cycling of several key nutrients and highlights areas for additional research. Urban areas cover less than 5 percent of global land, but have a disproportionately large impact on global biogeochemical cycles due to the concentration of people and activities. Further, cities can often respond quickly with policy action towards environmental sustainability. Despite the biogeochemical importance of urban areas, there is a paucity of direct measurements of urban biogeochemistry due to complexities imposed by urban heterogeneity in both historic and current uses. To develop effective environmental policies, an understanding of how human activity and the physical nature of the built urban environment interact to modify biogeochemical pools and fluxes is imperative.

Urbanization and nutrient cycling

The cycling of nutrients is largely controlled by the interactions of organisms with the environment, but human-induced changes to the biosphere, particularly from urbanization, have radically altered the size of nutrient pools and the rates of nutrient fluxes on a global scale. The combustion of fossil fuels and land use changes have led to a 25 percent increase in atmospheric CO_2 during the last 50 years (Keeling et al. 2001). Industrial release of chlorofluorocarbons has changed stratospheric ozone concentrations and ultraviolet radiation, resulting in changes to vegetation that have altered the C, N and P cycles (Molina and Rowland 1974; Sinha and Hader 2002). Vast quantities of N and P fertilizers are used agriculturally to feed the global population, impacting stream and coastal water quality (Carpenter et al. 1998). The climate, physical environment of cities, and the planet as a whole are changing through warming (IPCC 2013), rainfall distributional shifts (Easterling et al. 2000; Tebaldi et al. 2006), species biogeographic shifts (Parmesan 2006) and land use changes (Vitousek et al. 1997; Foley et al. 2005).

Nutrient pools reflect the stock of elements present in a given form and within a given reservoir, like C in soils, N in leaves or P in rocks. Fluxes reflect the rate of movement and transformation of nutrients between reservoirs. For example, the combustion of fossil fuels results in C and N fluxes to the atmosphere. Both the pools and fluxes of nutrients vary in space and time as a function of the climate, land cover and intensity of human influences. Across urban-to-rural gradients, the pools of C, N and P vary with the rates of fixation and mineral weathering, but also due to the rates of fertilizer use, patterns of development and paving, variations in temperature and moisture and long-distance imports, among a multitude of other possible drivers (Kaye et al. 2006; Pickett et al. 2011).

Urban-to-rural nutrient gradients begin to emerge as a result of the initial land clearing for urban land use with the removal of vegetation and loss of organic soil layers, but evolve as the heterogeneous urban landscape evolves. For example, the entombing of soils through paving eliminates direct atmospheric inputs of organic matter causing depletion of soil nutrient pools (Raciti et al. 2012a), but leakage from sewer pipes can cause local hotspots for inputs. Figure 13.1 shows an *Ailanthus altissima* (Tree of Heaven) growing in a parking lot and under a building. With the exception of small street tree pavement cutouts (typically ~1 m^2), the nearest patch of unpaved soil is over 350 m away. Nonetheless, this tree is thriving, while the adjacent buildings have plumbing issues caused by encroachment of the tree's roots.

Figure 13.1 (A) *Ailanthus altissima* tree growing within a matrix of parking lots and building in Boston, MA, US. The circle denotes the location of an *Ailanthus* tree growing adjacent to the building; (B) with pavement completely surrounding the base of the tree; (C) this tree is thriving despite all of the surrounding soil being covered by impervious surfaces

Hotspots for nutrient availability and depletion are very likely present across urban areas, but few systematic measurements have been carried out due to the challenges of systematic sampling under pavement and private property (Raciti et al. 2012a).

Urbanization, as both a land use change process and a demographic phenomenon, affects both global and local rates of biogeochemical cycling. Globally, the vast majority of energy consumption occurs in urban areas, wherein over 70 percent of energy-related CO_2 emissions occur (Energy Information Administration (EIA) 2013). Locally, urban ecosystems experience a modified growing environment and biogeochemistry including dramatically enhanced inorganic N inputs (Rao et al. 2014), increased ambient CO_2 (Idso et al. 2001) and tropospheric O_3 (Jacob 2000), higher temperatures (Oke 1982) and lengthened growing seasons (Zhang et al. 2004), all of which impact the rates of nutrient cycling and in many respects foreshadow large-scale global environmental changes. Further, people within urban areas can directly and dramatically modify the local environment through actions like the application of synthetic fertilizers to promote the growth of residential landscaping and the active removal and export of nutrients through the collection of lawn clippings or fall leaf litter.

The carbon cycle

Carbon (C) is the central currency of life on Earth; the presence of C defines a molecule as organic. The Earth contains approximately 32×10^{23} g of C, with the vast majority stored within the lithosphere and not actively cycled (Marty 2012). The largest fluxes of C are to and from the atmosphere (Figure 13.2A). CO_2 is the second most important greenhouse gas (after water vapor) in the atmosphere and is critical for making the Earth's climate habitable. In 2011, CO_2 emissions from fossil fuel combustion and land use change were estimated to be 9.5 ± 0.5 and 0.9 ± 0.5 PgC yr^{-1}, respectively, resulting in a net increase of 3.6 ± 0.2 PgC within the atmosphere (Le Quéré et al. 2013). In 2014, the annual mean atmospheric CO_2 mixing ratios in Mauna Loa, HI, the longest running atmospheric CO_2 observatory, exceeded 400 ppm; atmospheric CO_2 has been increasing at a mean annual rate of 1.48 ± 0.11 ppm yr^{-1} since direct measurements began in 1959 (Keeling et al. 2001, http://scrippsco2.ucsd.edu).

The impact of urban areas on C budgets is especially profound (Hutyra et al. 2014). Annual urban CO_2 emissions are more than double the net terrestrial or ocean C sinks (Le Quéré et al. 2013). Expansion of urban areas in the coming decades is expected to continue to outpace urban population growth, a trend that has been observed over the last two decades in most parts of the world (Angel et al. 2011), making urban land use change and the associated impacts on regional C dynamics a critical component of the global C cycle (Brown et al. 2005; Churkina et al. 2010; Hutyra et al. 2014). Anthropogenic CO_2 emissions have been found to increase non-linearly, as the amount of development and impervious surface areas (ISA) increases (Figure 13.2B). In addition to direct urban CO_2 emissions, urban expansion and the consumptive patterns of urban dwellers impact the rates of global land cover change due to demand for resources like food, wood and fuel (DeFries et al. 2010). Urban development preferences have been evolving in the US and globally towards urban forms associated with higher C emissions. For example, between 1970 and 2012, the average size of single-family homes in the US increased from 139 m^2 (1,500 ft^2) to 215 m^2 (2,315 ft^2), central air conditioning has become much more prevalent, and garage parking space is increasing (Emrath 2013).

While these changes in urban form reflect a changing affluence, they are also associated with larger building footprints and increased energy demand and land consumption. However, there are challenges in the spatial attribution of emissions, since large emissions sources associated with energy generation are often located far from the point of consumption. Thus, C accounting is prone to double counting and leakage (Kennedy et al. 2007; Cannell et al. 1999). Recent efforts have attempted to produce separate production- and consumption-based estimates of C emissions (e.g. Peters et al. 2012); the net influence of urban areas on emissions differs dramatically with the accounting construct.

CO_2 is emitted from and taken up by a wide range of stationary sources like vegetation and buildings as well as mobile sources like people, cars and flowing waterways. While domes of CO_2 enhancement have been consistently observed around urban areas (Idso et al. 2001; Grimmond 2006; George et al. 2007; Coutts et al. 2007), data on urban C dynamics are very sparse given the importance of urban areas and the heterogeneity in urban form across the globe (Hutyra et al. 2014). It is very likely that the spatial and temporal patterns in C exchange vary with development, bioclimatic conditions and human preferences, but there is a paucity of directly measured observations to robustly quantify these relationships.

If we consider Boston, MA and Phoenix, AZ, US as two contrasting case studies where direct observations have been made, we can see the complex interplay between biological and anthropogenic fluxes across space and time. In temperate Boston, Briber et al. (2013) observe that as one moves across a 100 km gradient from rural forests to low-density urban and finally high-density urban development, the magnitude of biological fluxes decreases as anthropogenic emissions increases (Figure 13.2C). This finding is in contrast to patterns observed in arid Phoenix, where both vegetation C density and anthropogenic emissions increase within the urban area due to residential preferences for lawns and trees and humans watering their landscaping (Day et al. 2002; Jenerette et al. 2007).

The trajectories of land cover change and time since development also strongly influence nutrient pools and fluxes (Lewis et al. 2006). Urban development in Boston largely occurred on land previously forested or used for agriculture. In contrast, Phoenix's urbanization occurred on desert and former agricultural lands. In these examples, the modifications to the soil and vegetation C pools will likely differ as a function of the legacy of the previous land use and the intensity of the current usage. The C pools and fluxes will also evolve over time, as vegetation establishes and grows. Pouyat et al. (2003) offer the *urban convergence hypothesis* suggesting that human fertilizer applications and management lead to a convergence in urban ecosystems towards similar soil conditions, irrespective of the pre-development conditions.

The nutrient pools within urban soils and vegetation can be significant (Hutyra et al. 2011; Pouyat et al. 2006). A combination of field measurements and remote sensing analysis reveals that nearly half of the Boston Metropolitan Statistical Area (home to 4.6 million people) is actually vegetated, with approximately 75 percent of the C density found in nearby undeveloped areas (Raciti et al. 2012b). These findings, highlighting the presence and importance of urban vegetation, are in contrast to urban metabolism studies that have largely ignored the role of vegetation within urban C budgets (e.g. Kennedy 2012) and remote sensing products that assume urban ecosystem productivity is zero (e.g. MODIS products). Urban areas are heterogeneous, but the vegetated components of the system can be associated with significant C pools and fluxes.

While urban areas are clearly an important component in the global C cycle due to anthropogenic emissions, it is only in recent years that the C cycle science community has become keenly interested. Historically, fossil fuel CO_2 emissions were considered the best-quantified component of the US and global C budgets. In fact, emissions were typically treated as standards within atmospheric inversion studies. While emissions inventories are fairly accurate at national and annual scales, the uncertainties become very large when

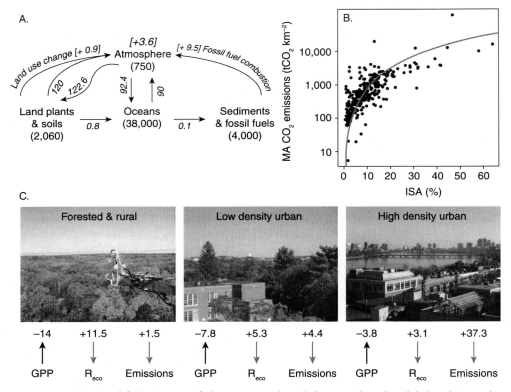

Figure 13.2 (A) Simplified diagram of the active pools and fluxes within the global carbon cycle; (B) non-linear relationship between ISA and fossil fuel emissions in Massachusetts; (C) gradient in carbon exchange for biological and anthropogenic sources

Photo credits from left to right are: Trevor Keenan (http://news.harvard.edu/gazette/story/2013/07/efficiency-in-the-forest/); Alison Dunn; and Lucy Hutyra.
Note: For (A), major pools (Gt C; black color) and fluxes (Gt C yr^{-1}, grey color) are based on estimates from the IPCC (2007), with the values for fossil fuel emissions, land cover change and net atmospheric accumulation from Le Quéré et al. (2013). Details supporting the estimation of gross primary productivity (GPP), ecosystem respiration (R$_{eco}$) and fossil fuel emissions are specified in Briber et al. (2013).

attempts are made to resolve emissions at smaller spatial and temporal scales. For example, a recent study by Gately et al. (2013) finds that annual transportation-related emissions varies by nearly 40 percent across state-level inventories and the trends in annual emissions are inconsistent. CO$_2$ emissions reduction goals are actively being set at city, state and regional scales, but the uncertainties associated with emissions inventories may exceed emissions reduction targets (Gregg et al. 2008; Peylin et al. 2009; NRC 2010).

In order for emissions reduction initiatives to succeed, better methods for monitoring, reporting and verifying urban C fluxes are required. In particular, the existing network of surface CO$_2$ observation sites intentionally avoids cities, in order to capture the background CO$_2$ conditions without the influence of point source emissions. Data on urban vegetation and soils dynamics are likewise very sparse. There is urgent need for improved observation networks and models to reliably quantify C budgets in and around cities. In 2014, the orbital carbon observatory (OCO-2) satellite was launched to quantify changes in the global C cycle, but the primary emphasis continues to be vegetation dominated ecosystems, rather than the

Urban nutrient cycling

places where most people live and emissions occur. Hutyra et al. (2014) provide an extensive review of the current understanding of the urban C cycle and offer a research agenda to advance capacity to monitor and verify urban changes in the C cycle.

The nitrogen cycle

Of all the nutrient cycles, nitrogen (N) has perhaps been modified most significantly through human activities (Vitousek et al. 1997). Since 1970, the global population increased by 78 percent and our creation of reactive nitrogen (N that is biologically or chemically available for reaction) increased by 120 percent (Galloway et al. 2008). The discovery of the Haber-Bosch process allows for the efficient production of ammonia fertilizers from N_2 and H_2 that supports rapid increases in both agricultural yields and population growth.

Nitrogen is an integral part of the DNA, proteins and the enzymes required by living tissues; and is also an important component of the photosynthetic machinery in plants, which reduces CO_2 in the atmosphere to glucose through photosynthesis. Nitrogen is the most abundant component of our atmosphere. However, over 99 percent of atmospheric N is in the form of N_2 and is chemically inert. When the triple bond of N_2 is broken, N is converted into reactive forms that actively cycle through plants, soil, water and air (Table 13.1). Conversion of N_2 to biologically available forms is referred to as *fixation*. Bacteria can reduce atmospheric N_2 to ammonia (NH_3) that can then be assimilated as organic nitrogen by plants or animals. That organic N can eventually be mineralized to ammonium (NH_4^+), which can then be further assimilated by other groups of organisms. Bacteria may also use NH_4^+ and oxidize it to nitrite (NO_2^-) and then nitrate (NO_3^-) through a process referred to as *nitrification*.

Nitrate is highly mobile in soils and easily assimilated by plants. When O_2 is limiting in soils, bacteria may use NO_3^- as an alternate oxidant to convert organic carbon to CO_2,

Table 13.1 Summary of key nitrogen cycle compounds and description of urban modifications

Compound(s)	Role in the N cycle	Urban sources and influences
Nitrogen gas (N_2)	Comprises 78 percent of the atmosphere; produced through denitrification	None, chemically inert
Ammonia ($NH3$), Ammonium (NH_4^+), Nitrite (NH_2^-) and Nitrate (NH_2^-)	Biologically available forms of N, from bacteria and fertilizers	Product of fossil fuel combustion; deposition concentrated along roadways Volatilized and leached from fertilizer applications and pet excrement
Nitrogen oxides (NO_x) (NO and NO_2)	Products of fossil fuel combustion in the presence of N and lightning	Fossil fuel combustion and industrial processes; both are concentrated in urban areas Contributes to smog, acid rain and O_3
Nitrous oxide (N_2O)	Product of denitrification; key sources are the oceans, soils (especially agricultural soils), cattle feed lots and sewage Main sink is stratospheric destruction	Lawns can be significant N_2O sources Industry, transportation and sewage are all sources of N_2O N_2O is 300x more powerful than CO_2 as a greenhouse gas
Nitric acid (HNO_3)	Product of reactions with nitrogen oxides in the presence of OH or O_3	Component of acid rain

returning the nitrogen to the atmosphere as N_2. This process of *denitrification* converts NO_3^- to N_2 and N_2O. An alternative pathway for N fixation is through high temperature oxidation of N_2 to NO through either lightning or the combustion of fossil fuels in the presence of N_2. NO in the atmosphere can be further oxidized to HNO_3, a key component of acid rain. NO and NO_2, collectively referred to as NO_x, are highly reactive in the atmosphere and central in the generation of smog, acid rain and tropospheric O_3. While the N cycle, as described above, does not depend on human or urban processes directly, the reaction rates and chemical abundances differ significantly within urban areas due to the concentration of cars, industrial activities and fertilizer inputs (Table 13.1).

Biologically available [reactive] N is a limiting element in most temperate ecosystems and is tightly cycled (LeBauer and Treseder 2008). In relatively undisturbed terrestrial ecosystems, N cycling is dominated by transfers between plants and soils with major inputs and outputs largely controlled via biological fixation and denitrification (Schlesinger and Bernhardt 2013). While agricultural fertilizer production and application is the most significant driver of changes to the N cycle globally (Galloway et al. 2008), the cycling of N is also significantly altered within and around urban areas (Driscoll et al. 2003; Kaye et al. 2006; Table 13.1).

In urban areas, wet and dry atmospheric deposition, human and animal waste and fertilizer usage can significantly enhance the inputs to the ecosystem and potentially result in increased reactive N losses from the system. For example, Rao et al. (2014) find that atmospheric deposition of reactive N within the City of Boston is 12.3 ± 1.5 Kg N ha^{-1} yr^{-1}, more than double the background deposition rates outside the city. Excrement from pets has been found to add another 15 kg ha^{-1} yr^{-1} in densely populated areas (Baker et al. 2001). Taken as a whole, urban areas concentrate key sources of reactive N including transportation (Fenn et al. 2003; Davidson et al. 2010), food and energy consumption (Galloway et al. 2008; Driscoll et al. 2003), land use change (Raciti et al. 2011) and lawn management (Groffman and Pouyat 2009), making urban areas a major direct and indirect global source of reactive N.

Enhanced N inputs can result in increased rates of nitrification and mineralization in soils (Pouyat and Turechek 2001; Aber et al. 2003; Boggs et al. 2005), higher forest floor (leaf litter) N concentrations and soil N content (Boggs et al. 2005; Fang et al. 2011), greater uptake of N by plants (Aber et al. 2003; McNeil et al. 2007) and greater rates of productivity and net C uptake by some tree species (Thomas et al. 2010). While urban areas do have a capacity to retain N within lawns, large stature plants, soils and groundwater (Groffman and Pouyat 2009; Raciti et al. 2011), impervious areas within urban areas do not allow for easy penetration of N into the soils. In an opportunistic New York City study measuring soil chemistry under pavement, Raciti et al. (2012a) found that soil N content is decreased by 95 percent, likely due to isolation from above ground biological and atmospheric inputs.

Conversely, if urban N pools become saturated and N is no longer a limiting nutrient, excess N can be lost to the atmosphere as ammonia or as nitrate in waterways (Groffman et al. 2002). The leaching of excess N and P from urban systems can result in eutrophication within water bodies (Howarth et al. 1996; Carpenter et al. 1998). The addition of nutrients into water bodies can initially stimulate biological growth and the accumulation of biomass, but there can also be massive biological die-offs due to the reduction in dissolved oxygen levels caused by the decomposition of the organic matter. Eutrophication associated with excess N is more common in marine and estuary systems, while P-driven eutrophication tends to occur in freshwater systems (Schindler 2006).

Given the important role that reactive N compounds have on both air quality and ecosystems (Table 13.1), the US Clean Air Act of 1990 regulates NO_2 explicitly as well as

Urban nutrient cycling

CO and O_3 which are involved in reactive chemistry with other N compounds. Through effective regulation, smog and acid rain have become less problematic in the US, but it is still an important issue globally. In countries like India and China, where industrial and automobile activity has been growing rapidly, emissions reduction technologies are currently not as extensively used.

This characterization of urban N inputs and understanding of the urban N cycle is limited by sparse observational networks. Monitoring networks, like the National Atmospheric Deposition Program (NADP), National Trends Network (NTN) and the Clean Air Status and Trends Network (CASTNET), provide continuous, long-term data on N deposition. However, similar to the C observing networks, the N networks are intentionally located away from urban areas and point sources of pollution in order to capture regional background trends. While background changes are clearly important for monitoring overall air quality, it is also important to improve our characterization of N hotspots. Further, existing networked observations are largely limited to North America and Western Europe. Nascent efforts are underway to create an urban N monitoring network, but integration into the broader networks is challenging due to contributions from point sources. Steep increases in N inputs with increasing urbanization intensity have been observed (e.g. Groffman et al. 2002; Rao et al. 2014), but the contributions of landscape management activities, land use and traffic have not been well quantified. Further, the fate of additional N within vegetated and aquatic ecosystems needs to be better quantified.

The phosphorus cycle

In contrast to the cycling of C and N, phosphorus (P) does not have a significant gaseous component. Phosphorus is made available to the biosphere through the weathering of rocks, particularly calcium phosphate minerals. Nitrogen and phosphorus are considered to be the most important limiting nutrients for vegetation in terrestrial ecosystems (Vitousek 1984). Analogous to the global N cycle, humans have dramatically enhanced the availability of P through mining of phosphate rocks for agricultural fertilizers. Humans have approximately tripled the global mobilization of P (Smil 2000). Residential and agricultural fertilizers contain N:P:K (K, being potassium) ratios in differing proportions reflecting the specific site or crop needs. The relative nutrient limitation of an ecosystem depends on the soil type, vegetation and age of the soils. Vitousek (1984) find that plants growing on young soils tend to be N limited, while vegetation on older, highly weathered soils is often P limited. Broadly extrapolating, temperate ecosystems are typically N limited, while tropical ecosystems tend to be more limited by P.

The major modifications to the global P cycle occur outside of urban areas, but urban areas can concentrate P within soils and waterways (Fissore et al. 2011). A detailed material flow analysis in Minneapolis, MN finds that most of the P imported into cities comes through chemical detergents and food for humans and pets (Baker 2011). The Minneapolis study find that 65 percent of the imported P remains in the city in landfills, septic systems and sludge. Excess P can dramatically impair water quality and biodiversity in lakes and estuaries. Urban wastewater treatment practices have improved in recent decades to include tertiary treatments to remove more P and N in an attempt to control eutrophication. Several states in the US have passed laws banning detergents containing P, leading to an eventual voluntary industry phase out of phosphorus-based detergents in the US (Litke 1999). However, in many parts of the world, there is still very limited wastewater treatment.

Urban waste

Cities currently generate about 1.3 billion tons of solid waste globally each year, what we typically think of as 'trash,' and that amount is expected to reach 2.2 billion tons by 2025 (Hoornweg and Bhada-Tata 2012). The volume of waste is growing faster than the rate of urban population growth, with higher incomes concomitant with urbanization associated with greater amounts of solid waste production. In higher-income regions, the management of waste is one of the most important services a city provides and often constitutes a significant portion of municipal financial budgets. When waste collection is unavailable or too expensive, trash is often burned in open piles or in makeshift combustors such as metal cans. While the air quality impacts of burning trash depend on the material composition, burning temperature and the availability of oxygen during combustion, the human health impacts can be very significant (Bond et al. 2004). Burning trash in barrels often occurs at <250°C, resulting in much higher dioxin (a hormone disruptor), particulate matter (associated with respiratory problems) and carbon monoxide (reduces blood oxygen) emissions. Locally, trash burning can have very significant human health impacts; globally, trash incineration impacts overall air quality and climate.

While we might think of 'throwing *away* waste and trash,' it all has a destination that will impact residence time, leakage rates and the unintended environmental consequences. For example, open landfills that are uncovered and unlined allow leachate, a liquid formed by the decomposition of organic matter and breakdown of inorganic materials, to soak into the soil and potentially contaminate drinking water supplies. Alternatively, waste incineration, which typically costs several times as much as landfill disposal, may not contaminate urban waters, but does release secondary waste (e.g. mercury, particulates, SO_2, dioxins, lead and NO_2) to the atmosphere. Further, incineration results in ash and by-pass waste that still needs to be dealt with. Waste disposal currently accounts for nearly 25 percent of anthropogenic CH_4 emissions (Kirschke et al. 2013); CH_4 is 21 times more efficient than CO_2 for warming the climate. Carbon capture of landfill CH_4 for reuse as energy is possible, but is not yet a widespread practice due to cost and retrofit requirements. There is no ideal waste disposal solution, except to reduce consumption and reuse/recycle materials where possible. The impacts of waste on nutrient cycling can be very large, but are also highly variable locally depending on the composition of the waste (organic-to-inorganic ratios) and the methods of disposal.

Conclusion

Urbanization, and all of the biogeochemical changes associated therein, is a fundamental driver of global environmental change. Urbanization is multifaceted and does not have a predefined biogeochemical trajectory or end point. Urban ecosystems are intimately coupled human–natural systems, and their dynamic interactions often include non-linearities, thresholds and path dependencies. To advance understanding of urban biogeochemistry, we must build from current scientific understanding, but also consider the dynamic aspects of urbanization and the potential for unique urban biogeochemistry. The lack of cross-sectional and longitudinal urban observational data is currently the biggest obstacle to advancing urban biogeochemical theory. This paucity of data is largely a byproduct of the perception that urban ecosystems have limited ecological value because they are heavily modified by humans and are relatively small in size (see Grimm et al., Chapter 14 in this volume). However, urban areas are rapidly evolving in their spatial configuration and are growing in spatial extent, often concentrating both pools and fluxes of nutrients, and the ecology of cities is becoming ever more germane to people's lives, the development of local to regional greenhouse gas and eutrophication management strategies.

While this chapter reviews the current understanding of urban influences on C, N and P cycling, most of this understanding is derived from a limited number of cities in temperate regions, largely in high-income countries. Given the heterogeneity in urban form and the large role that behavioral preferences and municipal capacity have on biogeochemical cycling, we need more observational studies in order to fundamentally advance understanding of urban biogeochemistry. Kaye et al. (2006) introduce the concept of a 'new urban biogeochemistry,' suggesting that existing conceptual and predictive models, developed primarily in unmanaged and agricultural ecosystems, work poorly in urban ecosystems because they do not include human biogeochemical controls on the system such as impervious surface proliferation and engineered aqueous flow paths. Urban influences and populations have grown dramatically since Kaye et al. made their case, but only incremental progress has been made advancing urban biogeochemistry models. More observational and process-level studies are needed from a wider breadth of global cities to advance the funda-mental understanding of how urban development influences biogeochemical cycles both locally and globally. Urban environmental sustainability is a broad community goal, but we must advance our understanding of the underlying urban biogeochemistry for effective and targeted policies.

Key summary points

- Urban areas have a disproportionately large impact on global biogeochemical cycles due to the concentration of people and activities.
- Urbanization influences the rates of nutrient inputs, transformations and outputs through changing environmental conditions and direct human/industrial actions.
- Improved understanding of urban biogeochemistry is critical for developing urban environmental sustainability policies.

Key research and data needs

- Expanded cross-sectional and longitudinal urban biogeochemistry observational data.
- Characterization of the spatial and temporal variation in biogeochemical fluxes across biomes and different urban development forms and intensities.
- Biogeochemical models need to better capture direct human modifications to nutrient cycling rates to improve their overall performance.

References

Aber, J.D., Goodale, C.L., Ollinger, S.V. et al. (2003) Is nitrogen deposition altering the nitrogen status of northeastern forests? *Bioscience* 53(4): 375–389.

Angel, S., Parent, J., Civco, D.L., Blei, A., and Potere, D. (2011) The dimensions of global urban expansion: Estimates and projections for all countries, 2000–2050. *Progress in Planning* 75(2): 53–107.

Baker, L.A. (2011) Can urban P conservation help to prevent the brown devolution? *Chemosphere* 84: 779–784.

Baker, L.A., Hope, D., Xu, Y., Edmonds, J., and Lauver, L. (2001) Nitrogen balance for the central Arizona-Phoenix (CAP) ecosystem. *Ecosystems* 4: 582–602.

Briber, B.M., Hutyra, L.R., Dunn, A.L., Raciti, S.M., and Munger, J.W. (2013) Variations in atmospheric CO2 and carbon fluxes across a Boston, MA urban gradient. *Land* 2(3): 304–327.

Boggs, J.L., et al. (2005) Tree growth, foliar chemistry, and nitrogen cycling across a nitrogen deposition gradient in southern Appalachian deciduous forests. *Can. J. For. Res.* 35: 1901–1913.

Bond, T.C., Streets, D.G., Yarber, K.F., Nelson, S.M., Woo, J.-H., and Klimont, Z. (2004) A technology-based global inventory of black and organic carbon emissions from combustion. *Journal of Geophysical Research* 109(D14203). doi:10.1029/2003JD003697.

Brown, D.G., Johnson, K.M., Loveland, T.R., and Theobald, D.M. (2005) Rural land-use trends in the conterminous United States. *Ecological Applications* 15: 1851–1963.

Cannell, M.G.R., Milne, R., Hargreaves, K.J., et al. (1999) National inventories of terrestrial carbon sources and sinks: The UK experience. *Climatic Change* 42: 505–530.

Carpenter, S.R., Caraco, N.F., Correll, D.L., Howarth, R.W., Sharpley, A.N., and Smith, V.H. (1998) Nonpoint pollution of surface waters with phosphorus and nitrogen. *Ecological Applications* 8(3): 559–568.

Churkina, G., Brown, D.G., and Keoleian, G. (2010) Carbon stored in human settlements: The conterminous United States. *Global Change Biology* 16: 135–143.

Cleveland, C.C. and Liptzin, D. (2007) C:N:P soichiometry in soil: is there a "Redfield ratio" for the microbial biomass? *Biogeochemistry* 85(3): 235–252.

Coutts, A.M., Beringer, J., and Tapper, N.J. (2007) Characteristics influencing the variability of urban CO2 fluxes in Melbourne, Australia. *Atmospheric Environment* 41: 51–62.

Davidson, E.A., et al. (2010) Nitrogen in runoff from residential roads in a coastal area. *Water Air Soil Pollution* 210: 3–13.

Day, T.A., Gober, P., Xiong, F.S., and Wentz, E.A. (2002) Temporal patterns in near-surface CO2 concentrations over contrasting vegetation types in the Phoenix metropolitan area. *Agricultural and Forest Meteorology* 110: 229–245.

DeFries, R.S., Rudel, T., Uriarte, M., and Hansen, M. (2010) Deforestation driven by urban population growth and agricultural trade in the twenty-first century. *Nature Geosciences* 3: 178–181.

Driscoll, C.T., Whitall, D., Aber, J.D. et al. (2003) Nitrogen pollution in the Northeastern United States: Sources, effects, and management options. *Bioscience* 53(4): 357–374.

Easterling, D.R., Meehl, G.A., Parmesan, C., Changnon, S.A., Karl, T.R., and Mearns, L.O. (2000) Climate extremes: Observation, modeling, and impacts. *Science* 289: 2068–2074.

Elser, J.J., et al. (2000) Biological stoichiometry from genes to ecosystems. *Ecology Letters* 3(6): 540–550.

Emrath, P. (2013) Characteristics of homes started in 2012: Size increase continues. National Association of Home Builders. www.nahb.org/generic.aspx?sectionID=734&genericContentID=213414&channelID=311. Accessed October 28, 2013.

Energy Information Administration (EIA) (2013) International energy outlook 2013, *DOE/EIA-0484*, U.S. Energy Inf. Admin., Off. of Energy Analysis, U.S. Dep. of Energy, Washington, D. C.

Fang, Y.T., Yoh, M., Koba, K., et al. (2011) Nitrogen deposition and forest nitrogen cycling along an urban-rural transect in Southern China. *Global Change Biology* 17(2): 872–885.

Fenn, M.E., Baron, J.S., Allen, E.B. et al. (2003) Ecological effects of nitrogen deposition in the Western United States. *Bioscience* 53(4): 404–420.

Fissore, C., et al. (2011) Carbon, nitrogen, and phosphorus fluxes in household ecosystems in the Minneapolis-St. Paul, Minnesota, urban region. *Ecological Applications* 21: 619–639.

Foley, J.A., et al. (2005) Global consequences of land use. *Science* 309: 570–574.

Galloway, J.N., et al. (2008) Transformation of the nitrogen cycle: Recent trends, questions, and potential solutions. *Science* 320: 889–892.

Gately, C.K., Hutyra, L.R., Sue Wing, I., and Brondfield, M.N. (2013) A bottom-up approach to on-road CO2 emissions estimate: Improved spatial accuracy and applications for regional planning. *Environmental Science and Technology* 47(5): 2423–2430.

George, K., Ziska, L.H., Bunce, J.A., and Quebedeaux, B. (2007) Elevated atmospheric CO2 concentration and temperature across an urban-rural transect. *Atmospheric Environment* 41: 7654–7665.

Gregg, J.S., Andres, R.J., and Marland, G. (2008) China: Emissions pattern of the world leader in CO2 emissions from fossil fuel consumption and cement production. *Geophys. Res. Lett.* 35: L08806.

Groffman, P.M., and Pouyat, R.V. (2009) Methane uptake in urban forests and lawns. *Environment Science and Technology* 43: 5229–5235.

Groffman, P.M., Boulware, N.J., Zipperer, W.C., Pouyat, R.V., and Band, L.E. (2002) Soil nitrogen processes in urban riparian zones. *Environmental Science and Technology* 36: 4547–4552.

Urban nutrient cycling

Grimmond, C.S.B. (2006) Progress in measuring and observing the urban atmosphere. *Theoretical and Applied Climatology* 84: 3–22.

Hoornweg, D., and Bhada-Tata, P. (2012) What a waste: A global review of solid waste management. Urban development series; knowledge papers no. 15. Washington, DC, World Bank.

Howarth, R.W., et al. (1996) Regional nitrogen budgets and riverine N&P fluxes for the drainages to the North Atlantic Ocean: Natural and human influences. *Biogeochemistry* 35: 75–139.

Hutyra, L.R., Yoon, B., and Alberti, A. (2011) Terrestrial carbon stocks across a gradient of urbanization: A study of the Seattle, WA region. *Global Change Biology* 17: 783–797.

Hutyra, L.R., Duren, R., Gurney, K.R., Grimm, N., Kort, E.A., Larson, E., and Shrestha, G. (2014) Urbanization and the carbon cycle: Current capabilities and research outlook from the natural sciences perspective. *Earth's Future*, n/a–n/a, doi:10.1002/2014EF000255.

Idso, C.D., Idso, S.B., and Balling, R.C. (2001) An intensive two-week study of an urban CO2 dome in Phoenix, Arizona, USA. *Atmospheric Environment* 35: 995–1000.

IPCC (2007) *Climate Change 2007: Synthesis Report. Contribution Contribution of Working Groups I, II and III to the Fourth Assessment Report of the Intergovernmental Panel on Climate Change*, edited by R. Pachauri and A. Reisinger, IPCC, Geneva, Switzerland.

IPCC (2013) *Climate Change 2013: The Physical Science Basis. Working Group I Contribution to the Fifth Assessment Report of the Intergovernmental Panel on Climate Change*, edited by T.F. Stocker, D. Qin, G.K. Plattner, M. Tignor, S.K. Allen, J. Boschung, A. Nauels, Y. Xia, V. Bex, and P.M. Midgley, Cambridge University Press, Cambridge, UK.

Jacob, D.J. (2000) Hetergoeneous chemistry and tropospheric ozone. *Atmospheric Environment* 34: 2131–2159.

Jenerette, G.D., Harlan, S.L., Brazel, A., Jones, N., Larsen, L., and Stefanov, W.L. (2007) Regional relationships between surface temperature, vegetation, and human settlement in a rapidly urbanizing ecosystem. *Landscape Ecology* 22: 353–365.

Kaye, J.P., et al. (2006) A distinct urban biogeochemistry? *Trends in Ecology and Evolution* 21: 192–199.

Keeling, C.D., Piper, S.C., Bacastow, R.B., Wahlen, W., Whorf, T.P., Heimann, M., and Meijer, H.A. (2001) Exchanges of atmospheric CO2 and 13CO2 with the terrestrial biosphere and oceans from 1978 to 2000. Global aspects, SIO Reference Series, No. 01-06, Scripps Institution of Oceanography, San Diego, 88 pages.

Kennedy, C. (2012) Comment on article "Is there a metabolism of an urban ecosystem?" by Golubiewski. *Ambio* 41: 765–766.

Kennedy, C., Cuddihy, J., and Engel-Yan, J. (2007) The changing metabolism of cities. *Journal of Industrial Ecology* 11(2): 43–59.

Kirschke, S., et al. (2013) Three decades of global methane sources and sinks. *Nature Geosci, advance online publication*, doi:10.1038/ngeo1955.

Le Quéré, et al. (2013) The global carbon budget: 1959–2011. *Earth System Science Data Discussion* 5: 165–185.

LeBauer, D.S., and Treseder, K.K. (2008). Nitrogen limitation of net primary productivity in terrestrial ecosystems is globally distributed. *Ecology* 89(2): 371–379.

Lewis, D., Kaye, J.P., Gries, C., Kinzig, A., and Redman, C. (2006) Agrarian legacy in soils of urbanizing aridlands. *Global Change Biology* 12: 703–709.

Litke, D.W. (1999) Review of phosphorus control in the United States and their effect on water quality. Water-Resources Investigations Report 99-4007, US Geological Survey, Denver.

McNeil, B.E., Read, J.M., and Driscoll, C.T. (2007) Foliar nitrogen responses to elevated atmospheric nitrogen deposition in nine temperate forest canopy species. *Environment Science and Technology* 41: 5191–5197.

Marty, B. (2012) The origins and concentrations of water, carbon, nitrogen, and noble gases on Earth. *Earth and Planetary Science Letters* 313/314: 56–66.

Molina, M.J., and Rowland, F.S. (1974) Statospheric sink for chloroflouromethanes – chlorine atomic-catalysed destruction of ozone. *Nature* 249: 810–812.

NRC Committee on Methods for Estimating Greenhouse Gas Emissions (2010) *Verifying Greenhouse Gas Emissions: Method to Support International Climate Agreements*. National Research Council. The National Academy Press, Washington, D.C.

Oke, T.R. (1982) The energetic basis of the urban heat island. *Q. J. R. Meteorol. Soc.* 108: 1–24.

Parmesan, C. (2006) Ecological and evolutionary responses to recent climate change. *Annual Review of Ecology Evolution and Systematics* 37: 637–669.

Peters, G.P., Davis, S.J., and Andrew, R. (2012) A synthesis of carbon in international trade. *Biogeosciences* 9(8): 3247–3276.

Peylin P., et al. (2009) Importance of fossil fuel emission uncertainties over Europe for CO2 modeling: model intercomparison. *Atmos. Chem. Phys. Discuss.* 9: 7457–7503.

Pickett, S.T.A., et al. (2011) Urban ecological systems: Scientific foundations and a decade of progress. *Journal of Environmental Management* 92: 331–362.

Pouyat, R.V., and Turechek, W.W. (2001) Short- and long-term effects of site factors on net N-mineralization and nitrification rates along an urban-to-rural gradient. *Urban Ecosystems* 5: 159–178.

Pouyat, R.V., Russell-Anelli, J., Yesilonis, I.D., and Groffman, P.M. (2003) Soil carbon in urban forest ecosystems, pp. 347–362. In J.M. Kimble et al. (ed.) *The potential of U.S. forest soils to sequester carbon and mitigate the greenhouse effect.* CRC Press, Boca Raton, FL.

Pouyat, R.V., Yesilonis, I., and Nowak, D. (2006) Carbon storage by urban soils in the United States. *Journal of Environmental Quality* 35: 1566–1575.

Raciti, S.M., Groffman, P.M., Jenkins, J.C., et al. (2011) Accumulation of carbon and nitrogen in residential soils with different land use histories. *Ecosystems* 14(2): 287–297.

Raciti, S.M., Hutyra, L.R., and Finzi, A.C. (2012a) Depleted soil carbon and nitrogen stocks under impervious surfaces. *Environment Pollution* 164: 248–251.

Raciti, S.M., Hutyra, L.R., Rao, P., and Finzi, A.C. (2012b) Soil and vegetation carbon in urban ecosystems: The importance of importance of urban definition and scale. *Ecological Applications* 22(3): 1015–1035.

Rao, P., Hutyra, L.R., Raciti, S.M., and Templer, P.H. (2014), Atmospheric nitrogen inputs and losses along an urbanization gradient in the Boston metropolitan region. *Biogeochemistry* 121: 229–245.

Schindler, D.W. (2006) Recent advances in the understanding and management of eutrophication. *Limnology and Oceanography* 51: 356–363.

Schlesinger, W.H., and Bernhardt, E.S. (2013) *Biogeochemistry: An analysis of global environmental change*, 3rd edition. Academic Press, Waltham, MA, USA and Kidlington, Oxford, UK.

Sinha, R.P., and Hader, D.P. (2002) UV-induced DNA damage and repair: A review. *Photochemical and Photobiological Sciences* 1(4): 225–236.

Smil, V. (2000) Phosphorus in the environment: Natural flows and human interferences. *Annual Review of Energy and the Environment* 25: 53–88.

Sterner, R.W., and Elser, J.J. (2002) *Ecological stoichiometry: The biology of elements from molecules to the biosphere.* Princeton University Press, Princeton, NJ.

Tebaldi, C., Hayhoe, K., Arblaster, J.M., and Meehl, G.A. (2006) Going to the extremes. *Climatic Change* 79: 185–211.

Thomas, R.Q., et al. (2010) Increased tree carbon storage in response to nitrogen deposition in the US. *Nature Geoscience* 3(1): 13–17.

Vitousek, P.M. (1984) Litterfall, nutrient cycling, and nutrient limitation in tropical forests. *Ecology* 65: 285–298.

Vitousek, P.M., Mooney, H.A., Lubchenco, J., and Melillo, J. (1997) Human domination of Earth's ecosystems. *Science* 277: 494–499.

Zhang, X.Y., Friedl, M.A., Schaaf, C.B., Strahler, A.H., and Schneider, A. (2004) The footprint of urban climates on vegetation phenology. *Geophysical Research Letters* 31: L12209.

PART II

Impacts of global environmental change on urban systems and urbanization processes

The chapters in Part I focus on the pathways through which urbanization drives processes of global environmental change by drawing upon our understanding of urbanization trends, linkages between urbanization and resource demand, urbanization influences on behaviors and effects on climate. Part II explores the bi-directional nature of these interactions—the pathways through which global environmental change impacts urban systems and processes of urbanization. A major thread throughout this section is how global climate change impacts urban societies and the environmental systems that support human livelihoods and well-being. Chapters in this section explore a) *how climate change and related global scale processes impact urban conditions* including vulnerable sites, sectors and populations.

The 21st century is increasingly uncertain and unpredictable with respect to weather and climate. More urban areas are exposed to extreme events at a higher frequency and intensity, which is often attributed to climate change. Particularly in low- and middle-income countries, where social and physical infrastructure is lacking or absent, the impacts of climate change are of great concern. Moreover, many of the world's megacities with high population densities characterized by a prevalence of informal settlements and slums are located in sensitive coastal ecosystems, which are at high risk from flooding, storm surges and sea level rise. Prolonged or extreme heat and cold, dry spells and drought are other such changes experienced in many parts of the world. Chapters addressing this first topic provide an understanding of where and how climate change impacts urban populations, which has important implications for planning and design of resilient cities (Young, Chapter 16). Such an understanding requires appropriate frameworks that take into account the local contexts which shape the factors that contribute to vulnerability and risk of urban populations and the environments in which they live (Romero-Lankao and Gnatz, Chapter 15). This also includes an analysis of the gendered impacts of climate impacts and further, recognizing that cities are well positioned to tackle climate change issues together with social concerns to promote equity in urban areas (Alber and Cahoon, Chapter 21).

Second, this section addresses b) *competition for resources*, i.e. how global environmental change creates the conditions under which this occurs, which often leads to situations of instability of supply between and within cities. Just as urbanization puts pressure on resources, environmental changes such as climate change can often add additional stress to already limited resources in urban areas. Flooding from prolonged and heavy rainfall, for example, can threaten water quality, availability and supply in modified urban environments (Murphy, Chapter 18; Pfister et al., Chapter 17). Food and agricultural systems are equally sensitive to changing climates and resulting price fluctuations in the market, which can often lead to situations of food insecurity, particularly in low- and middle-income nations where coping mechanisms are weak and supporting infrastructure inadequate (Cohen and Garrett, Chapter 20).

The third topic to be addressed in this section concerns c) *changes in ecosystem* services as a result of the ways in which global environmental change influences their supply and quality within cities and their extended metropolitan regions. How ecosystem services are defined and valued in urban areas is still not well understood, but advancements in understanding the benefits, costs and tradeoffs of such services offers a great opportunity for designing and re-building cities with ecosystems in mind that support greater resilience to climate impacts (Grimm et al., Chapter 14). This further extends to how use of such services known as 'urban greening' can support human health and well-being in cities. However, climate change, with changing temperatures and rainfall availability modifying the ecosystem services provided by green infrastructure, creates challenges for responsive decision-making and management strategies (Trundle and McEvoy, Chapter 19).

In Part II, the chapters identify and discuss the broad range of pathways through which global environmental change affects urban systems. Emphasis is placed on the context-specific nature of impacts, as they will depend on a number of factors, including socio-economic, political and institutional forces and processes that shape the adaptability and resilience to the coupled interactions of urbanization and environmental change.

14

A BROADER FRAMING OF ECOSYSTEM SERVICES IN CITIES

Benefits and challenges of built, natural or hybrid system function

Nancy B. Grimm, Elizabeth M. Cook,
Rebecca L. Hale and David M. Iwaniec

People derive benefits from ecosystems, their components and their processes. These benefits, called ecosystem services, are often taken for granted (i.e. the cool shade provided by a tree) or sometimes beyond everyday perception (e.g. the 'purifying' role of rivers in reducing pollutant inputs). The diverse suite of services proffered by natural ecosystems is categorized by the Millennium Ecosystem Assessment (MEA, 2005) and others (e.g. de Groot et al., 2002) into provisioning (production functions), regulating (regulation functions) and cultural (information functions) services. A fourth MEA category, supporting services, is not considered here because these are the ecosystem processes that underpin the above mentioned services. Likewise, biodiversity cannot be considered a service except insofar as it provides aesthetic enjoyment benefits to people, although it may or may not underpin the ecosystem processes that yield services. deGroot et al. (2002) also recognize habitat functions, i.e. the places for organisms to live and reproduce, as services that ecosystems provide.

However services are categorized, the provision of diverse benefits to society has represented a strong argument for preserving natural ecosystems and accounting explicitly for their value to people, in terms of money, jobs, lives, or damage or harm avoided. Indeed, these two arguments in favor of ecosystems have spawned much research and motivated international efforts like the International Platform on Biodiversity and Ecosystem Services (IPBES) and The Economics of Ecosystems and Biodiversity (TEEB). At the same time, the conservation community, at least in part, is coming to recognize that an exclusive focus on ever-dwindling (even mythological?) pristine ecosystems does not serve the ecosystem services argument, as many partly altered or even wholly designed ecosystems are also contributing benefits for people. What about the services that are more explicitly manipulated by engineers or designers? Human agency has not only inadvertently altered nature's services, but has also intentionally taken on the challenge of either amplifying or replacing those services with the use of designed or built infrastructure (Andersson et al., 2014).

At the extreme of this type of designed system is the city. Cities are now home to over half (and soon, two-thirds) of the human population (UN, 2012). There, intentionally designed systems are often conceived as replacements for ecosystems, with functions that replace the ecosystem functions that underpin services. The extent to which nature is 'pushed back' from urban places and processes has varied over time, and in recent years nature is seeing a resurgence in the form of 'green infrastructure' (Tzoulas et al., 2007; Schäffler and Swilling, 2013; Hansen and Pauleit, 2014; Andersson et al., 2014).

This chapter explores the concept of ecosystem services as it may be applied to cities. If cities are ecosystems, as urban ecologists have argued, then are not all services that they provide 'ecosystem services'? How are the dichotomies of intentional vs. unintentional, amenities vs. disamenities, designed vs. natural and gray vs. green to be treated? Cities are hybrid systems (Swyngedouw, 1996; Alberti, 2008, pp. 17–19) and hence are the services delivered by their component subsystems and processes best viewed as hybrid services? Replaceability and ecosystem-based adaptation to climate change in cities, concepts that draw from the idea of services, are also discussed herein. Finally, this new concept of ecosystem services in cities is used to explore how cities can plan for, cope with, adapt to or increase resilience to climate change.

The ecosystem conundrum and a solution

Cities are ecosystems. Although early research on cities in the US was focused on the use of ecology as a source of theory in understanding the organization of people in cities (Park et al., 1925; Hollingshead, 1940) and in Europe on the ecology of organisms in cities (Sukopp, 1990, 2008), modern urban ecology recognizes that like any ecosystem, cities are bounded pieces of Earth with interacting biotic and abiotic elements (Pickett et al., 2001; Grimm et al., 2000). They can be characterized by their structure (e.g. vegetation abundance, food webs, human demography, building density and others) and function (e.g. energy flows, material cycles and information exchange). Recognizing cities as social-ecological-technological systems (SETS) expands the domain and requires additional expertise, but remains fundamentally an *ecosystem approach*.

Because cities are ecosystems, any services that they provide—i.e. benefits that the urban population derives from cities and their component systems—must be ecosystem services. This generates a conundrum. How useful is a concept that applies generically to all facets of a complex SETS including such things as the transportation system and wastewater treatment plants as well as parks and playgrounds? Certainly, it is helpful to differentiate between those components that represent nature in the city and those that are completely constructed; i.e. the green compared to the gray. Exploring only the green parts of ecosystem services is unnecessarily restrictive and not representative of the entirety of the urban ecosystem. Moreover, examining only green spaces misses opportunities to consider how a continuum from gray to green, or the blending or hybridization of those extremes (see Figure 14.1), could enhance value to people.

A second challenge is the definition of ecosystem services: the benefits people derive from ecosystems. If the social system is not benefiting from the patterns and processes of the ecosystem or some component of the ecosystem, then it is not a service. Researchers understand very little of how people perceive nature's services, what drives change in these perceptions and how they are connected to people's actions. For instance, most people do not perceive regulating services such as water quality improvement that may result from ecosystem processes like denitrification. However, most people can enjoy the microclimate modulation afforded by a shady grove of trees on a hot summer's day. In both cases, there is a benefit to people. Thus, a service (a shady grove or a riparian wetland in an uninhabited

Ecosystem services in cities

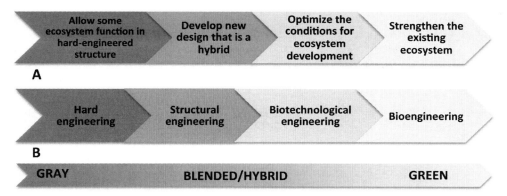

Figure 14.1 Gradients of infrastructure solutions appropriate for cities, including some common terms used to describe them (Hulsman et al., 2011; Naylor et al., 2011)

landscape), though performing the same ecosystem functions as in a city (production of leaves, transformations of nutrients), may not be providing an ecosystem service. A direct connection between perceptions about the benefits of shade trees and homeowners' or city managers' decisions to plant one or a 'million' trees (respectively) is easy to envision, whereas efforts to restore wetlands may fail because the connection between that action and a water quality benefit is unclear to most.

Human-built systems (technological systems or infrastructure) also perform functions, providing the services for which they are specifically designed (see Table 14.1). Such services are referred to here as 'intentional' services; often they are seen as replacements for or improvements upon nature's services. Along a continuum between ecological system services and technological system services are 'hybrid' services, i.e. services resulting from the blending of ecological and built components of hybrid ecosystems.

What is meant by 'hybrid' ecosystems? In conservation science, the term has been applied to ecosystems that, when exposed to biotic or abiotic drivers, exceed historic ranges of variability while retaining some characteristics of their historic state (Hobbs et al., 2009). These hybrid ecosystems occupy a state that is intermediate between 'historic' and 'novel'

Table 14.1 Examples of built (technological) ecosystem components, their functions and related services provided to urban inhabitants

Ecosystem component	Function	Service
Transportation network	Facilitating human movements	Provision of roadways, railways and transport systems
Water delivery infrastructure	Water fluxes	Provision of water to users
Stormwater infrastructure	Water fluxes	Protection from flooding
Wastewater infrastructure	Water and sewage fluxes; physical and biogeochemical transformations	Sanitation; removal of wastes; improvement of water quality
Energy supply infrastructure	Heating; cooling; other work	Regulation of microclimate; provision of power for manufacturing, etc.
Housing and buildings	Structure and architecture	Provision of habitat

ecosystems and are considered to be in a reversible state (Hobbs et al., 2009). In a similar vein, restoration ecologists view the goals of restoration efforts to represent hybrids between some often unknown pre-degradation state (Hobbs et al.'s historic state) and the current state (e.g. Lake et al., 2007; Palmer et al., 2005). In climate change adaptation, strategies that are intermediate between ecosystem-based adaptation (where the services of ecosystems are relied upon to reduce exposure or provide coping or adaptive capacity) (Jones et al., 2012) and engineering-based adaptation (involving human-built structures) are considered hybrid approaches (CBD, 2009; Munroe et al., 2012). Finally, the term has recently been invoked in a discussion of novel and hybrid ecosystems of cities (Hobbs et al., 2009) and the services they provide (Perring et al., 2013). Alberti (2008) maintains that urban ecosystems are themselves hybrid ecosystems because their properties are a consequence of coupled human and ecological dynamics. Urban ecologists, city planners and managers use 'green' (and more recently 'blue') and 'gray' to describe ecosystem-based and engineered infrastructure (respectively).

Here hybrid approaches and hybrid systems are considered to be intermediate between the two extremes of engineered/built/technological (hereafter, gray) and ecological/'natural' (hereafter, green) (see Figure 14.1), but without reference to any historic or pre-urban ecosystems or processes. The term is thus used in a way that is distinct from its most common usage (Hobbs et al., 2009). Ecosystem services, or more simply, services, can arise from functions of built, natural or hybrid systems. A simple conceptual framework for urban services is presented here that begins with this continuum and incorporates the role of people, both as recipients and moderators of services and builders and designers of infrastructure. People also indirectly affect ecosystems (the green end of the spectrum) through their activities.

A final concern about the application of ecosystem services concepts in urban ecosystems is the issue of scale. Cities are notoriously dependent on external ecosystems for natural resources; thus, urbanites benefit from external services to an extent that is opaque to most. Provisioning of food, fiber and building materials relies upon external primary production functions via supply chains that are the invisible bonds connecting cities to external systems (see Güneralp, Chapter 6 in this volume, for a more detailed discussion on resource chains and the construction/operation of the built environment) (Hellweg and Milà i Canals, 2014; Hoekstra and Wiedmann, 2014). Ecosystems far from the urban sites of production of wastes may absorb or remove those wastes. However, in the case of greenhouse gases, this removal is not sufficient to prevent their increasing concentrations in the atmosphere. Thus, services in urban ecosystems similarly have a long reach: the 50 largest cities in the world draw water from over 40 percent of the Earth's terrestrial surface (McDonald et al., 2014).

Diverse services in cities: dichotomies and shades of green

The contention raised previously in this chapter is that services in cities are not limited to 'natural' features. However, those services provided by remnant ecosystems (e.g. uncultivated lands, nature preserves, wetlands and streams) are important assets; if they can be protected, remnant ecosystems may provide higher quality and more diverse services than the ecosystems that replace them, often for a lower cost. However, remnant ecosystems in cities often are: (1) directly threatened by urbanization, such as land use changes and increased fragmentation and eutrophication (Saunders et al., 1991); (2) indirectly threatened by human activities, such as air quality impacts on areas within or adjacent to the city (Grimm et al., 2000; Cook, 2014); or, (3) affected by de-urbanization, i.e. when previously managed lands are no longer maintained (Kremer et al., 2013; McPhearson et al., 2013). Importantly, the consequences of

Ecosystem services in cities

degradation are that these remnant ecosystems no longer provide the services people currently value and may even be dangerous to human well-being (disservices).

Disservices are those outcomes of ecosystem processes that are directly or indirectly harmful to people. For example, an urban estuary that receives high nutrient and organic loading from combined sewer overflows during storms may experience harmful algal blooms. This disservice may be seen as a consequence of either the failure of the ecosystem to modulate those inputs or the uni-dimensionality of the services provided by the stormwater conveyance system (i.e. to carry water out of the city without providing opportunities to improve its quality). The same ecosystem may be producing services and disservices. An example is a flood management wetland installed to slow the flow of water during storms and improve its quality (providing more than one benefit). It may be providing habitat for waterfowl, birds and native plants (service), but also for undesired species such as mosquitos (disservice). This discussion illustrates the importance of evaluating the full suite of benefits and costs associated with any management option, whether it is preservation of an existing ecosystem, construction of gray infrastructure or installation of low-impact stormwater management features that are to some extent hybrids (i.e. mid-ranges of Figure 14.1).

Another challenge is to provide services not only now, but also into the future, in the face of continued environmental and social-ecological change (Chapin, 2009). Restoration initiatives have proven successful in slowing down degradation and ameliorating losses. However, with few exceptions, it is more difficult and costly to restore ecosystems than to protect against their loss in the first place (Royal Society, 2014). The dynamics of uncertainty and complexity, such as legacy effects and time lags, can be difficult to manage for and 100 percent recovery is not considered plausible (Ramalho and Hobbs, 2012; de Groot et al., 2013).

Instead of restoring the ecosystem structure, many cities opt for replacement. For example, a wetland may be replaced with a buried pipe, a retention basin or a green space to maintain rainwater drainage. Replacement involves a loss of one or more services in return for other services. Replacement is rarely able to provide all of the original services and an equivalent supply and quality of those services (Hansen and Pauleit, 2014). Furthermore, unintended services and disservices should be expected. It is likely that gray, green and hybrid replacements differ in their ability to provide services. Replacement will result in the loss of some benefits, yet others may be greatly enhanced (e.g. pipes can rapidly transport water to treatment facilities or out of the city). Where designed infrastructure excels is at providing the services for which it is specifically intended (i.e. a single or a few functions; Table 14.1). It is based upon risk reduction approaches that assume a known set of probabilities and are judged fail-safe (Park et al., 2013), at least to a level of acceptable risk. However, an uncertain future (where probabilities are unknown) calls for a new paradigm recognizing that redundancy and flexibility are hallmarks of resilient systems. Flexibility implies redundancy of function, i.e. a suite of alternatives that can fill in required functions in case of small failures in another component. Flexibility allows the incorporation of uncertainty—about future conditions as well as future human decisions—rather than assuming a known set of probabilities. Future infrastructure and hybrid systems need to be designed to excel at being multi-functional (Hansen and Pauleit, 2014) and with planned redundancy to minimize the risk of systemic failure (i.e. safe-to-fail).

A final consideration on urban services is whether they can contribute to human well-being in an equitable way. It is unclear how gray, hybrid and green replacements affect the distribution of services, but we do know that not all residents have equal access to services that sustain human health, quality of life and livelihood in a healthy environment. The challenge for this century is to create urban ecosystems where all residents have access to the diverse services delivered by urban gray to green systems. In order to get there, a growing body of knowledge on services is required: (1) What are the services being delivered and how are they delivered in different urban ecosystems? (2) How do urban stakeholders value services (both economic and non-economic valuation approaches)? (3) Who decides which services are recognized and how they are prioritized, and where do benefits and uses accrue (conversely, also for disservices)? (4) What actions, actor/institution roles and visions are necessary to improve access to services? (5) What are the governance and knowledge barriers and how can they be overcome?

Inclusive processes are crucial in developing this body of knowledge, since the consideration of services is socially dependent, that is, different stakeholders may identify and prioritize services differently (Chan et al., 2012). For example, whether fruit trees on public land are used for provisioning may depend on regulations, cultural norms or needs, regardless of the potential for provisioning. To foster urban sustainability and resilience, residents, urban planners, designers and other stakeholders need to co-identify the wide range of services that are necessary and desirable.

Urban social-ecological system resilience to global change

Ongoing and future global environmental changes are pervasive, yet diverse, and may affect ecosystem services and urban resilience in many ways. Global environmental change (GEC) includes sustained and gradual changes in mean environmental variables, such as sea level and air temperature as well as the increased frequency and severity of extreme events, such as heatwaves and flooding (Kunkel et al., 2003, 2010; IPCC, 2012). Human activities drive urbanization and land use/land cover change, which also are considered GECs: both in terms of their widespread distribution and their global impact (Foley et al., 2005; Grimm et al., 2008). Importantly, shifts in populations and demography may fundamentally change ecosystem services as a result of changing perceptions, experiences and culture. Ecosystems, and the services they provide, can be a source of resilience against GECs (Jones et al., 2012); yet, major sources of resilience in social-ecological systems are social learning and adaptive management (Adger et al., 2005; Folke et al., 2002; Tompkins and Adger, 2004). This highlights the diverse and multi-faceted aspects of urban services discussed previously: services are culturally dependent; can be provided by ecological, built and social aspects of the system; and are affected by global changes themselves. How global change will affect services will depend on: (1) the type of global change; (2) the identity of the service; (3) the people receiving the service; and, (4) how the service is currently provided (what structures/ processes provide the service and whether there is redundancy). Here, the example of coastal flooding in New York City (NYC) is used to illustrate these ideas and explore some lessons for designing urban ecosystems for resilience to GEC.

Coastal flooding has been a major concern in many coastal cities including NYC and it is caused by multiple global and local changes (e.g. Gornitz et al., 2001; Adger et al., 2005; Rosenzweig et al., 2011). In NYC, for example, global changes of concern are sea level rise and the increased frequency and size of storm surges (Gornitz et al., 2001; Rosenzweig et al., 2011). Meanwhile, local wetlands have been lost to urbanization and other local development (Gornitz et al., 2001), further decreasing resilience to storms. Cultural aspects such as preferences for waterfront housing and development, also create challenges for resilience planning, while economic incentives may not encourage the best development patterns for resilience. Finally, consider how flood protection is currently provided and how it could be provided in the future. Is there redundancy in ecosystem service provision? Are some structures more resilient than others? In NYC, flood protection is provided by beaches, wetlands and sea walls (Gornitz et al., 2001; Rosenzweig et al., 2011). Future adaptations to flood risk may include a mix of gray and green solutions (also called 'hard' and 'soft' solutions). Gray solutions may include stormwalls and storm surge barriers, while green solutions may include wetland and dune restoration and beach nourishment (adding sand) (Rosenzweig et al., 2011; Arkema et al., 2013). Furthermore, other proposed adaptations involve reducing the vulnerability of infrastructure and people to coastal flooding, for example, by flood proofing or elevating infrastructure, and building redundancy into water supply infrastructure, or by changing zoning and design policies so that fewer people live in the flood zone (Rosenzweig et al., 2011). These different strategies vary in their resilience to global changes. Planners are explicitly considering flexible and adaptable solutions to coastal flooding, highlighting that the level of ecosystem service provision may not be affected by global change, though the means of provision may.

This example highlights the need for broad, flexible and hybrid approaches to designing integrated SETS solutions with resilience to GEC in mind. Municipal planners increasingly recognize the need to incorporate a diversity of structural components such as the green, hybrid and gray infrastructure as well as engagement from a diversity of cultural and institutional stakeholders (Monsalves Gavilan et al., 2013). Intentionally designed and managed hybrid systems can integrate multi-functional or flexible infrastructure to provide services that build resilience of cities to global changes (Anderies, 2014; Royal Society, 2014). Hybrid systems may create redundancies and flexibility within the system, which reduces the risk of failure and increases resilience. For example, NYC, in addressing coastal flooding, is using a combination of green, gray and social tools to increase resilience.

Strategies to increase urban resilience to global changes also must incorporate adaptive management, which accounts for both the interconnections with other global changes and the tradeoffs between intended and unintended consequences of approaches. Furthermore, there is a large body of literature that suggests that adaptability and flexibility must be not only a characteristic of infrastructure designs, but also of the coupled social-ecological system as a whole (e.g. Adger et al., 2005; Folke et al., 2002; Tompkins and Adger, 2004). In addition to building capacity of the (hybrid–gray–green) physical infrastructure, these strategies must consider the diversity of values and needs and the equitable distribution of those services—especially with underrepresented and vulnerable populations.

Design with ecosystems and services in mind

This chapter has discussed how ecosystem services (or, more simply, services) are defined in cities, and how the diversity of services delivered can be depicted along a gradient from gray to green (see Figure 14.1). There are urgent needs in this urban century to provide new infrastructure in rapidly growing urban regions of the developing world, which often lack the financial resources to build the monolithic hard engineered infrastructure that characterizes urbanization in the developed world, and to seek alternative solutions to aging infrastructure in existing cities as it is replaced. Awareness of the potential benefits, costs and tradeoffs of urban services along the gray–green spectrum can lead to solutions that enhance resilience, by opting for the flexibility, redundancy and multi-functionality that hybrid systems can provide. Although there is a place for the preservation and restoration of open space in cities, ecologists must not research and promote these approaches to the exclusion of working with managers, engineers and the public to derive such hybrid systems that incorporate ecosystem properties and processes (i.e. ecosystem-based approaches).

Incorporating an ecosystem understanding into the building and planning of cities is essential for these complex SETS to develop along sustainable trajectories. Such pathways are concerned with economic benefit and feasibility, ecological values and social equity for current and future residents of these systems. While one or a few solutions will not be appropriate for all cases—indeed, solutions must be context dependent, where contexts include socio-political, economic, biophysical and geographical constraints and opportunities. However, the broad principles embodied in an ecosystem approach, the concept of services, resilience, flexibility, adaptive management, inclusive planning and new paradigms appropriate for an uncertain future will help cities follow more sustainable trajectories.

Conclusions and needs for future research and practice

- Focusing on 'natural' ecosystem services is not representative of the entirety of services provided in cities; rather, an expanded ecosystem services framework is needed to describe the diversity of services provided by green (natural), gray (built) and hybrid components of the urban ecosystem.
- Ecosystem services are modulated by peoples' values and needs (the translation of functions to services) and can arise from functions of built, natural or hybrid systems.
- Future hybrid systems need to be designed to excel at being multi-functional, safe-to-fail, resilient to GECs and their diverse benefits must be accessible to all residents.
- Future research on ecosystem services in cities must incorporate gray and hybrid systems to be relevant to urban design, planning, increased resilience to global changes and the wide range of stakeholder needs and values.
- In response to this broader framing of services, co-development of strategies by research-ers and practitioners should identify and prioritize the full suite of context-dependent services that are necessary and desirable in complex social-ecological-technological systems.
- Practitioners should adopt flexible approaches to account for changing values and needs (including the needs of future residents) under rapidly changing conditions (i.e. demo-graphic shifts, global environment changes and technological innovations).

References

Adger, W. N., Hughes, T. P., Folke, C., Carpenter, S. R., and Rockström, J. (2005). Social-ecological resilience to coastal disasters. *Science*, *309*(5737), 1036–1039.

Alberti, M. (2008). *Advances in Urban Ecology: Integrating Humans and Ecological Processes in Urban Ecosystems*. Springer, New York.

Anderies, J. M. (2014). Embedding built environments in social–ecological systems: resilience-based design principles. *Building Research & Information: The International Journal of Research, Development and Demonstration*, *42*(2) (March), 130–142.

Andersson, E., Barthel, S., Borgström, S., Colding, J., Elmqvist, T., Folke, C., and Gren, Å. (2014). Reconnecting cities to the biosphere: Stewardship of green infrastructure and urban ecosystem services. *Ambio*, *43*(4), 445–453.

Arkema, K. K., Guannel, G., Verutes, G., Wood, S. A., Guerry, A., Ruckelshaus, M., . . . Silver, J. M. (2013). Coastal habitats shield people and property from sea-level rise and storms. *Nature Climate Change*, *3*(10), 913–918.

CBD, Convention on Biological Diversity. (2009). Biodiversity and climate-change adaptation. In *Connecting Biodiversity and Climate Change Mitigation and Adaptation: Report of the Second Ad Hoc Technical Expert Group on Biodiversity and Climate Change*, Technical Series, Volume 41. Montreal: Secretariat of the Convention on Biological Diversity.

Ecosystem services in cities

Chan, K., Satterfield, T., and Goldstein, J. (2012). Rethinking ecosystem services to better address and navigate cultural values. *Ecological Economics*, *74*, 8–18.

Chapin III, F. S. (2009). Managing ecosystems sustainably: The key role of resilience. In *Principles of Ecosystem Stewardship* (pp. 29–53). Springer, New York.

Cook, E. M. (2014). Direct and indirect ecological consequences of human activities in urban and native ecosystems. *Arizona State University. ProQuest Dissertation*.

Foley, J. A., DeFries, R., Asner, G. P., Barford, C., Bonan, G., Carpenter, S. R., . . . and Snyder, P. K. (2005). Global consequences of land use. *Science*, *309*(5734), 570–574.

Folke, C., Carpenter, S., Elmqvist, T., Gunderson, L., Holling, C. S., and Walker, B. (2002). Resilience and sustainable development: Building adaptive capacity in a world of transformations. *Ambio*, *31*(5), 437–440.

Gornitz, V., Couch, S., and Hartig, E. K. (2001). Impacts of sea level rise in the New York City metropolitan area. *Global and Planetary Change*, *32*(1), 61–88.

Grimm, N. B., Grove, J. M., Pickett, S. T. A., and Redman, C. L. (2000). Integrated approaches to long term studies of urban ecological systems. *BioScience*, *50*(7), 571–584.

Grimm, N. B., Faeth, S. H., Golubiewski, N. E., Redman, C. L., Wu, J., Bai, X., and Briggs, J. M. (2008). Global change and the ecology of cities. *Science*, *319*(5864), 756–760.

de Groot, R. S., Wilson, M. A., and Boumans, R. M. (2002). A typology for the classification, description and valuation of ecosystem functions, goods and services. *Ecological Economics*, *41*(3), 393–408.

de Groot, R. S., Blignaut, J., Ploeg, S., Aronson, J., Elmqvist, T., and Farley, J. (2013). Benefits of investing in ecosystem restoration. *Conservation Biology*, *27*(6), 1286–1293.

Hansen, R., and Pauleit, S. (2014). From multifunctionality to multiple ecosystem services? A conceptual framework for multifunctionality in green infrastructure planning for urban areas. *Ambio*, *43*(4), 516–529.

Hellweg, S., and Milà i Canals, L. (2014). Emerging approaches, challenges and opportunities in life cycle assessment. *Science*, *344*(6188), 1109–1113. doi:10.1126/science.1248361

Hobbs, R. J., Higgs, E., and Harris, J. A. (2009). Novel ecosystems: Implications for conservation and restoration. *Trends in Ecology & Evolution*, *24*, 599–605. http://dx.doi.org/10.1016/j.tree.2009.05.012

Hoekstra, A. Y., and Wiedmann, T. O. (2014). Humanity's unsustainable environmental footprint. *Science*, *344*(6188), 1114–1117. doi:10.1126/science.1248365

Hollingshead, A. B. (1940). Human ecology and human society. *Ecological Monographs*, *10*(3), 354–366.

Hulsman, H., van der Meulen, M., and van Wesenbeeck, B. K. (2011). Green adaptation: Making use of ecosystem services for infrastructure solutions in developing countries. *Report to Conservation International*. Deltares, The Netherlands.

IPCC, Intergovernmental Panel on Climate Change. (2012). *Managing the Risks of Extreme Events and Disasters to Advance Climate Change Adaptation. A Special Report of Working Groups I and II of the Intergovernmental Panel on Climate Change*. Field, C. B., Barros, V., Stocker, T. F., Qin, d., Dokken, D. K., Ebi, K. L., . . . Midgley, P. M. (eds.). Cambridge University Press, New York, 582 pp.

Jones, H. P., Hole, D. G., and Zavaleta, E. S. (2012). Harnessing nature to help people adapt to climate change. *Nature Climate Change*, *2*(7), 504–509.

Kremer, P., Hamstead, Z. A., and McPhearson, T. (2013). A social–ecological assessment of vacant lots in New York City. *Landscape and Urban Planning*, *120*, 218–233.

Kunkel, K. E., Easterling, D. R., Redmond, K., and Hubbard, K. (2003). Temporal variations of extreme precipitation events in the United States: 1895–2000. *Geophysical Research Letters*, *30*(1900), doi: 10.1029/2003GL018052

Kunkel, K. E., Easterling, D. R., Kristovich, D. A., Gleason, B., Stoecker, L., and Smith, R. (2010). Recent increases in US heavy precipitation associated with tropical cyclones. *Geophysical Research Letters*, *37*(244706), doi: 10.1029/2010GL045164

Lake, P. S., Bond, N., and Reich, P. (2007). Linking ecological theory with stream restoration. *Freshwater Biology*, *52*(4), 597–615.

McDonald, R. I., Weber, K., Padowski, J., Florke, M., Schneider, C., Green, P. A., . . . Montgomery, M. (2014). Water on an urban planet: Urbanization and the reach of urban water infrastructure. *Global Environmental Change*, *27*, 96–105.

McPhearson, T., Kremer, P., and Hamstead, Z. A. (2013). Mapping ecosystem services in New York City: Applying a social–ecological approach in urban vacant land. *Ecosystem Services*, *5*, 11–26.

MEA, Millennium Ecosystem Assessment. (2005). *Ecosystems and Human Well-Being: Synthesis*. Island, Washington, DC.

Monsalves Gavilan, P., Pincheira-Ulbrich, J., and Rojo Mendoza, F. (2013). Climate change and its effects on urban spaces in Chile: A summary of research carried out in the period 2000–2012. *Atmosfera*, *26*(4), 5417–5566.

Munroe, R., Roe, D., Doswald, N., Spencer, T., Möller, I., Vira, B., . . . and Stephens, J. (2012). Review of the evidence base for ecosystem-based approaches for adaptation to climate change. *Environmental Evidence*, *1*(1), 13.

Naylor, L. A., Venn, O., Coombes, M. A., Jackson, J., and Thompson, R. C. (2011). Including ecological enhancements in the planning, design and construction of hard coastal structures: a process guide. *Report to the Environment Agency (PID 110461)*. University of Exeter, 66 pp.

Palmer, M. A., Bernhardt, E. S., Allan, J. D., Lake, P. S., Alexander, G., Brooks, S., . . . and Sudduth, E. (2005). Standards for ecologically successful river restoration. *Journal of Applied Ecology*, *42*(2), 208–217.

Park, J., Seager, T. P., Rao, P. S. C., Convertino, M., and Linkov, I. (2013). Integrating risk and resilience approaches to catastrophe management in engineering systems. *Risk Analysis*, *33*(3), 356–367. doi:10.1111/j.1539-6924.2012.01885.x

Park, R. E., Burgess, E. W., and McKenzie, R. D. (1925). *The City*. Chicago University Press, Chicago.

Perring, M. P., Manning, P., Hobbs, R. J., Lugo, A. E., Ramalho, C. E., and Standish, R. J. (2013). Novel urban ecosystems and ecosystem services. *Novel Ecosystems: Intervening in the New Ecological World Order* (pp. 310–325). Wiley-Blackwell, Oxford.

Pickett, S. T., Cadenasso, M. L., Grove, J. M., Nilon, C. H., Pouyat, R. V., Zipperer, W. C., and Costanza, R. (2001). Urban ecological systems: Linking terrestrial ecological, physical, and socioeconomic components of metropolitan areas. *Annual Review of Ecology and Systematics*, *32*, 127–157.

Ramalho, C. E., and Hobbs, R. J. (2012). Time for a change: Dynamic urban ecology. *Trends in Ecology & Evolution*, *27*(3), 179–188.

Rosenzweig, C., Solecki, W. D., Blake, R., Bowman, M., Faris, C., Gornitz, V., . . . and Zimmerman, R. (2011). Developing coastal adaptation to climate change in the New York City infrastructure-shed: Process, approach, tools, and strategies. *Climatic Change*, *106*(1), 93–127.

Royal Society. (2014). Resilience to extreme weather. The Royal Society Science Policy Centre report 02/14, issued: November 2014 DES3400.

Saunders D. A., Hobbs, R. J., and Margules, C. R. (1991). Biological consequences of ecosystem fragmentation: A review. *Conservation Biology*, *5*, 18–32.

Schäffler, A., and Swilling, M. (2013). Valuing green infrastructure in an urban environment under pressure—The Johannesburg case. *Ecological Economics*, *86*, 246–257.

Sukopp, H. (1990). Urban ecology and its application in Europe. In *Urban Ecology: Plants and Plant Communities in Urban Environments* (pp. 1–22). SPB Academic Publishers, The Hague (The Netherlands).

Sukopp, H. (2008). On the early history of urban ecology in Europe. In *Urban Ecology* (pp. 79–97). Springer US.

Swyngedouw, E. (1996). The city as a hybrid: On nature, society and cyborg urbanization. *Capitalism Nature Socialism*, *7*(2), 65–80.

Tompkins, E. L., and Adger, W. (2004). Does adaptive management of natural resources enhance resilience to climate change? *Ecology and Society*, *9*(2), 10.

Tzoulas, K., Korpela, K., Venn, S., Yli-Pelkonen, V., Kaźmierczak, A., Niemela, J., and James, P. (2007). Promoting ecosystem and human health in urban areas using Green Infrastructure: A literature review. *Landscape and Urban Planning*, *81*(3), 167–178.

UN, United Nations (2012). World urbanization prospects, the 2011 revision. United Nations, *ESA/P/WP/224*.

15
URBANIZATION, VULNERABILITY AND RISK

Patricia Romero-Lankao and Daniel Gnatz

Greenhouse gas (GHG) emissions are at unprecedented highs and continue to increase, rapidly causing changes in all components of the climate system. These changes are likely to push average global temperatures beyond the 2°C limit that has been set as a threshold in many models and discussions. Yet, slowing this change will require substantial and sustained reductions of GHG emissions through international cooperation and coordination that seems out of reach within current world governance frameworks. Even if such coordinated and sustained efforts are undertaken today, the world will probably not see relief from the impacts of climate change until 2050. Since an increase in average global temperature beyond the 2°C threshold is almost assured, efforts to adapt and to reduce risk and vulnerability are as important to human survival as the reduction of GHG emissions.

There must be a profound shift in humankind's traditional ways of understanding and responding to the climate challenge. This new way of understanding and responding, however, will require bringing together diverse knowledge and power domains—an undertaking that creates its own set of seemingly intractable complications. As documented by scholars studying diverse fields of human endeavor, from pure scientific inquiry to government and to the private sector, one of the most difficult problems in creating change lies in moving people beyond the paradigms, tools and analytical systems they learn during their academic training and professionalization (Aylett, 2013).

Scholarship on urban vulnerability and risk has, by no means, offered an exception to this trend. While research on risk has grown considerably during recent years, it has consisted primarily of case studies based on the assumption that urban vulnerability depends on context. Furthermore, risk depending on context and scholarship on risk has offered often limited models or conflicting theories and paradigms that tend to shed light on only certain aspects of the problem, while other areas remain in the dark (Romero-Lankao et al., 2012).

Rapid urbanization and urban development pathways are frequently perceived as central determinants of vulnerability and risk in urban areas. However, existing scholarship on urban vulnerability has yet to determine the mechanisms by which urbanization and urban areas as places influence vulnerability, as development occurs within particular socio-ecological and institutional contexts.

This chapter has three objectives. The first section outlines the diversity of approaches addressing the different dimensions and determinants of urban vulnerability and risk. This aims to foster an understanding that while many of the approaches have strengths, an integrated approach and a common set of tools that will enhance urban response capacity within multiple contexts remains lacking. The second section addresses the gap in the analysis of the interactions between urbanization and risk through a brief description of some of the key mechanisms by which vulnerability and risk are shaped by dynamics of urbanization, creating unique socio-ecological histories, opportunities and constraints. The final section moves away from this outline of current approaches and gaps to sketch out some of the necessary components of a more integrated and interdisciplinary understanding of how 'environmental' processes, such as climate variability and change, and 'societal' processes such as urbanization interact with hazard exposure and vulnerability to generate common patterns and differences in urban risk within and across urban areas.

Conceptualizing urban vulnerability and risk

Research on urban risk has grown considerably in recent years. The concept of urban risk, however, is still characterized by interdisciplinary differences in definition and scope. Thywissen (2006), for example, identifies 25 definitions of risk. Risk can be defined, for example, as the likelihood of occurrence of an urban hazard; the possibility of loss, injury and other impacts; or the probability of the occurrence of an adverse event and the probable magnitude of its consequences (Shrader-Frechette, 1991). Risk scholarship is still dominated by narrow approaches such as those focusing on how changes in an environmental hazard or combination of hazards (e.g. temperature extremes, air pollution and precipitation extremes) relate to such outcomes (risk proxies) as mortality, morbidity and economic damage, and on how demographic factors such as age and gender mediate the relationship between the urban hazard and risk (O'Brien et al., 2007; Romero-Lankao and Qin, 2011).

Studies on urban vulnerability tend to portray it as the degree to which a city, population, infrastructure or economic sector (i.e. a system of concern) is susceptible to and unable to cope with the adverse effects of hazards or stresses such as heatwaves, storms and political instability (Revi et al., 2014). Urban vulnerability is a relational concept that captures a complex and dynamic reality. Besides referring to the possibility that a system may be negatively affected by a hazard or stress, it is also a relative property defining both the sensitivity and the capacity to cope with that stressor. Therefore, vulnerability cannot be defined by the hazard alone, nor can it be represented strictly by internal properties of the system being stressed. Instead, it must be looked at as an interaction of these factors. For the most part, as is the case with scholarship on urban risk, scholarship on urban vulnerability consists primarily of case studies and analyses based on conflicting theories and paradigms. This section will briefly outline the concepts, research questions, dimensions and indicators of urban vulnerability and risk promulgated by three lineages on urban vulnerability: *vulnerability as impact* or top-down (the most commonly applied approach) (O'Brien et al., 2007; Romero-Lankao et al., 2012), *inherent or contextual vulnerability* and *urban resilience* (see Table 15.1).

Urban vulnerability as impact

The *urban vulnerability as impact* approach perceives vulnerability as an outcome (e.g. mortality, economic damage) from the exposure to hazards and the sensitivity of urban systems. This paradigm has its origins in the disaster risk and climate communities (Cardona et al., 2012). Studies within this paradigm are concerned with the relationship between a hazard and an impact while controlling for confounding factors. These studies often explore, for instance,

Urbanization, vulnerability and risk

Table 15.1 Examples of research questions on urban vulnerability

Research question	Vulnerability as impact	Inherent vulnerability	Urban resilience
1. What is the relationship between a hazard (temperature) and an outcome (mortality) excluding additional factors (see question 2)?	XX		
2. Which additional factors affect the relationship between temperature and human health?	XXX	X	X
a. Which factors make people more sensitive to a hazard?	XXX	X	X
b. Which factors influence people's ability to adapt to or cope with a hazard?	XX	X	X
c. What are the structural drivers (e.g. socio-economic inequality, political power) of vulnerability to temperature-related hazards?		XXX	X
3. How do biophysical parameters (e.g. air pollution) affect the hazard—outcome relationship?	XX		
4. How does/will climate change affect temperature–health relationships?	X		
5. Which factors influence hazards and their distribution (e.g. urban form, land cover, heat islands)?	X		X
6. How do people perceive vulnerability and risk?	X	X	X
7. What are existing and potential adaptation options?	X	X	X

Source: Based on Romero-Lankao et al. (2012). Symbols refer to levels at which the question is explored by each lineage with X=lowest and XXX=highest.

how changes in temperature, air pollution and precipitation relate to impacts such as mortality and economic damage. When dealing with climate change, this lineage applies the so-called top-down impact assessments, a scaled-down version of global climate change scenarios to urban centers in order to model how parameters such as temperature and precipitation will evolve in the future. Future climate hazards, along with their effects and adaptation options, are estimated under particular climate change scenarios.

The *urban vulnerability as impact* approach has shed light on the nature of the urban hazard–impact relationship. For instance, research on urban air pollution and temperature extremes find that the risks of adverse human health impacts depend on key considerations. The first is the nature of the hazards to which urban populations are exposed. For example, while the health effects of air pollution show a general linear form, temperature-mortality links follow a V- or J-shaped relationship curve, with deaths increasing as temperatures fall below or rise above certain threshold values. The second is given by multiple factors influencing exposure, sensitivity and capacity to respond such as age, pre-existing medical conditions, education, physiological acclimatization or access to air conditioning and other home amenities (Makri and Stilianakis, 2008; O'Neill, 2005). However, because most urban vulnerability as impact studies look at populations at the city-scale, they are often unable to shed light on intra-urban inequalities or on questions such as: how and why specific urban neighborhoods, populations or sectors within cities are differentially affected; whether decision-makers and populations are receptive to adaptation options and motivated to make necessary changes; whether they possess necessary skills, awareness and assets to be able to adapt; and how their potential response choices are constrained by the socio-economic, political and environmental circumstances in which they live and operate.

Inherent urban vulnerability

Some of the aforementioned questions have been explored by the research communities focusing on *inherent* or *contextual urban vulnerability* and on *urban resilience* (see the following section), which look at populations at the community or neighborhood scale. The different thrusts of research on *inherent urban vulnerability*, which have evolved from livelihoods, political ecology, disaster risk and environmental justice approaches, shed light on adaptive capacity and on the structural drivers creating differences in urban populations' vulnerability and risk (Pelling, 1999). These lineages tend to differ in their emphasis on entry points for analysis and intervention. A livelihoods approach focuses on the dynamic building of assets and options at the individual, family or community level (e.g. self-help/affordable housing) as a fundamental mechanism to cope with the hazards they encounter (Moser, 1998). However, it is difficult to scale up these actions to the city level and to determine how much livelihood strategies pursued by individuals help them reduce their vulnerability compared to efforts at city, government or governance levels such as land use planning and service provision. Other scholars point to the role these levels play in shaping adaptive capacity by promoting economic growth and poverty reduction, and by creating structural determinants of adaptive capacity such as urban infrastructures and services, land and housing and public emergency-response systems (Parnell et al., 2007; Romero-Lankao et al., 2014b).

Inherent vulnerability perspectives shed light on the influence of class and socio-spatial segregation on urban risks and vulnerabilities through various mechanisms. Economic elites are able to monopolize the best land and enjoy the rewards of environmental amenities such as clean air, safe drinking water, open space and tree shade (Harlan et al., 2007; Harlan and Ruddell, 2011; Collins and Bolin, 2009). A dynamic relationship exists between the generation of hazards such as floods and air pollution on the one hand, and the creation of vulnerability on the other, where institutions and social actors play a key role (Aragón-Durand, 2007; Manuel-Navarrete et al., 2011). Lastly, although wealthy sectors are moving into risk prone coastal and forested areas (Collins and Bolin, 2009), with certain hazards such as air pollution, affecting both rich and poor alike (Romero-Lankao et al., 2013), climate risks tend to disproportionally burden the poor or otherwise marginalized populations (Revi et al., 2014).

Despite offering a picture of how direct and underlying socio-economic and institutional factors of urban vulnerability change over time, *inherent urban vulnerability* studies stop short of offering an entire causal sequence that may show how inequities and hazards correspond with differential impacts, susceptibility and adaptation capacities. For instance, while some studies on human health find that poverty, income and deprivation relate with higher risks of mortality from exposure to air pollution and temperature (Johnson et al., 2009), other studies find that these factors either have no significant effect (Stafoggia et al., 2006) or have inconsistent effects, i.e. sometimes being positively and other times negatively related (D'Ippoliti et al., 2010). These inconsistencies may arise because it is problematic to merge hazards as diverse as air pollution, temperature dynamics, toxic waste and floods without a careful understanding of their nature and dynamics. For instance, due to their physical characteristics, toxic wastes can be dumped in poor neighborhoods with relative ease, but it is difficult to control the movement of air pollution and temperature. Although wealthy residents often live in the more vegetated suburban areas and farther away from heavy industries and freeways, air pollutants and extreme weather do not know boundaries and do not stop when they reach the limits of wealthy neighborhoods, cities and even countries. Extreme examples of this are plumes of airborne pollutants that originate in Mexico City and travel to the Gulf of Mexico or those that originate in Asia and journey to North America (Tie et al., 2009).

Urban resilience

The *urban resilience* approach underscores that urban populations and places are not only vulnerable, but also have the ability to bounce back (or bounce forward), i.e. to recover from and even take advantage of some stresses. Urban resilience points to:

> the capacity of [urban] social, economic and environmental systems to cope with a hazardous event or trend or disturbance, responding or reorganizing in ways that maintain their essential function, identity and structure, while also maintaining the capacity for adaptation, learning and transformation.
>
> *(Field et al., 2014, 40)*

From a socio-ecological systems tradition, urban centers exhibit non- and multi-equilibrium dynamics that can amplify or attenuate the impacts of hazards. Urban activities invariably alter their own environment and that of their hinterlands. Because of this, urban socio-economic and environmental patterns and histories may move in one of two directions: towards increasing environmental degradation and reduced resilience (e.g. irreversible overexploitation and degradation of water resources and inflexibility and ineffectiveness of management systems) or towards an increasing ability of urban populations and urban-relevant decision-makers to repair damage, sustain the environment and foster the ability of urban actors to build capacity for learning and adaptation as well as capitalize on the learning opportunities that might be opened by a disaster.

Cities are diverse; Dhaka (Bangladesh) and Boulder (Colorado, US) are such examples. Since 1991, when a hurricane slammed Bangladesh killing at least 138,000 people and leaving ten million people homeless, efforts were undertaken, promoted by local authorities, national governments and international organizations, to decrease the risk from tropical cyclones. These efforts including the development of an early warning system and the construction of public shelters to host evacuees, were tested by the impact of Cyclone Sidr, which hit Bangladesh in 2007. Although between eight and ten million Bangladeshis were exposed to Sidr, perhaps the strongest cyclone to hit the country since 1991, there was a 32-fold reduction in the death toll (i.e. 4,234 people compared to 138,000), proving Bangladesh's capacity for learning and adaptation (UN-Habitat, 2011).

The historic flood of September 11–18, 2013 in Boulder (Colorado, USA)—which killed ten people, resulted in 18,000 evacuees, the destruction of 688 homes and damages to an additional 9,900 homes—brought into view many of the interconnections and interdependencies between urban vulnerability and resilience (MacClune et al., 2014). Although Boulder was exposed to a flood estimated in the 1-in-25 to 1-in-100 year range, the city's Greenways Program allowed green areas to function as storm-water routes to mitigate flood damage. Impact damages were also conditioned upon historic development pathways and social, political and economic factors in the city. Apartments impacted by sewage upwelling, for instance, were below-grade and frequently occupied by lower-income families and university students. Although city utility staff was aware of the need to upgrade the sewage drainage system, the cost of such an improvement was prohibitively high. Together with a fear of potential litigation, this led to either inaction or minimal action, which increased city-wide vulnerability to the floods. Wastewater treatment plants in two neighboring cities, Lyons and Longmont, had to be shut down and it was only through the resourcefulness of operators that Boulder's plant remained in operation. Six of seven key roads that follow creeks up mountain canyons failed and left affected populations isolated and unable to leave flood-damaged areas. Yet, even amidst the near chaos and extensive damages wrought by the flood, strong pre-existing relationships and a culture of cooperation were key resources that sped up response and enabled effective recovery, for example, through learning from previous experiences, such as the Four Mile Fire of 2010.

There is no doubt that the urban resilience lineage has enriched the urban vulnerability discussion by exploring how the dynamics of cities as coupled human-environmental systems allow or constrain their ability to respond to hazards and stresses (Harlan et al., 2007; Uejio et al., 2011). Nevertheless, there are difficulties in translating and contextualizing the concept of resilience from an analysis of urban vulnerability to global climate and environmental change, as is reflected in the relatively few studies exploring these issues (Leichenko, 2011). In a socio-ecological systems context, urban resilience appears as a passive attribute of urban communities and regions in the larger systems of the bio-physical world and global environmental change. This scope-limited view misses the fact that rather than being merely passive receptors of stresses and hazards, authorities, NGOs, grassroots organizations, individuals and households are actively responding to and shaping the stresses and opportunities associated with global climate and environmental change. Furthermore, both the *urban resilience* and *inherent urban vulnerability* approaches are limited in their spatial and temporal scopes. As such, they may provide valuable information on a particular context, but they lack an ability to move beyond case studies and uncover broader trends in urban vulnerability and risk.

Taking stock

Existing knowledge on urban vulnerability and risk has depended fundamentally on the questions asked and how existing lineages go about answering those questions. Thus, knowledge within each lineage is a function of the components of vulnerability and risk on which it focuses and of the methods and data it uses or has available. For instance, although the *urban vulnerability as impact* paradigm has made important contributions to the hazard–impact relationships, it tends to produce a set of explanatory variables that are tightly constrained by the availability of data, particularly in low- and middle-income countries, and it omits any attempt to gain ethnographic knowledge of behavioral norms, social networks and risk perceptions that are equally relevant to understanding urban populations' vulnerability. In a similar way, despite offering a picture of how direct and underlying socio-economic and institutional factors of urban vulnerability and risk change over time (O'Brien et al., 2007), *inherent urban vulnerability* studies stop short of providing an entire causal sequence that may reveal the mechanisms by which hazards interact with inequities to create differential impacts and susceptibility.

This underscores how scale can influence a study's findings. Here, scale may be defined as the scope or extent of a spatial, temporal, quantitative or analytical dimension used to measure urban vulnerability and risk. The three approaches outlined above differ in their spatial or analytical scale, but the time scale of a study is equally important. *Inherent vulnerability* studies that include a longer-term perspective find that long-term global and national pressures (e.g. construction of infrastructure to provide flood control for coastal property, but not for urban populations) crystalize in local conditions defining vulnerability to urban flood hazard. In Guyana, for instance, reduced access to assets springing from "the gendered and ethnic nature of social systems, partisan politics and an underdeveloped civil society" (Pelling, 1999, p. 259) has led to inadequacy of sea level infrastructure protection for the urban poor majority.

Many studies on *urban vulnerability as impact* tend to examine single events or short time periods of less than ten years, thereby missing any examination of how long-term processes such as increasing average temperatures at the city level, interact with broader and more subtle socio-economic and urbanization trends. During the 20th century, for example, an epidemiologic transition took place in many high-income countries, which, together with the use of air conditioning and heating systems (Carson et al., 2006), helps explain why disease and vulnerability to cold and heat have decreased even while populations have aged, despite the increase in average temperatures in many urban centers. However, there may be limits to the mortality reductions associated with this transition. After reaching a presently unknown threshold, these effects might be offset by the increases in temperatures and extremes brought on by climate change. Regardless of its outcome under future conditions, however, the transition's importance in enhancing adaptive capacity in urban centers of some nations should not be underestimated. Without a doubt, a more integrated approach to the multiple dimensions and determinants of urban vulnerability and risk is needed to shed light on currently unknown interactions and threshold levels, and to develop more interdisciplinary and integrated approaches. The final section of this chapter will revisit this issue.

Current understandings of urbanization and urban areas and implications for urban vulnerability and risk

Before exploring the roles urbanization and urban centers play, with their unique place-based features and qualities, in shaping vulnerability and risk, a consideration of the definitions of 'urbanization' and 'urban' will be useful. Both the process of urbanization and the urban areas it produces belong to a set of worldwide mega-phenomena that are profoundly altering the relationship between human beings and the environment. Human–environment relations are consequently affecting vulnerability and risk in complex and accelerating ways. However, little agreement exists among national and international bodies about the definition of what urban means and what makes a population urban. For example, around half of the 228 countries for which the United Nations has data use administrative definitions of urban (e.g. such as living in the capital city); 22 percent use size and density; 17 percent use functional characteristics (e.g. economic activity); 9.7 percent have no definition of urban; and the rest define all (e.g. Singapore) or none (e.g. Polynesian countries) of their population as urban (Galea and Vlahov, 2005).

Urbanization is defined here as a series of interconnected changes (transitions) defining how humans interact with each other and the environment (Marcotullio et al., 2014; Romero-Lankao et al., 2014a). These changes, which have been studied by diverse disciplines, include: an increased number of people living in urban areas; changes in lifestyles and cultures; economic shifts from primary activities such as agriculture or mineral extraction to manufacturing and services; the land use expansion of urban areas and associated infrastructures; and the transformations of ecosystems, surface energy balance, habitats and hydrological systems. Within the milieu of the unique place-based features and qualities where it occurs (including historic and cultural features), urbanization gives rise to urban areas with unique physical, socio-economic and institutional characteristics with implications for vulnerability and risk. These characteristic differences can be considered with the following examples related to land use, biogeochemical cycles, climate, biodiversity, sheer concentration of people and assets, and socio-economic and institutional factors.

Urbanization is a key driver of land fragmentation (Seto et al., 2011) and loss of biodiversity (Grimm et al., 2008). Differences in land quality, land use and functional characteristics of land determine urban vulnerability and risk in different and not yet fully understood ways. For instance, changes in vegetation cover are one of the factors influencing the risk of floods, rainfall triggered landslides and wildfires near or in urban centers (Braimoh and Onishi, 2007; Smyth and Royle, 2000). Changes in land use may cause changes in land-surface physical characteristics (e.g. surface albedo), which have implications for water-related hazards such as droughts and floods, as precipitation can be enhanced or reduced depending on climate regime, geographic location and regional patterns of land, energy and water use.

Urban areas are key drivers of changes in carbon, water and other biogeochemical cycles (Pataki et al., 2006), but they are also vulnerable to extreme temperatures, air pollution, water degradation and other hazard-risks associated with these changes. Particularly when combined with adverse weather conditions (e.g. heatwaves caused by climate change), high levels of air pollution, for instance, are known to increase the risks of negative health impacts on human populations (Bell et al., 2008). However, the aggregate of health impacts from air quality and temperature changes become especially critical in rapidly growing low- and middle-income countries (Kan et al., 2008; Romero-Lankao et al., 2013; Revi et al., 2014). Hazards such as floods and droughts are another example. History reveals that vulnerability of urban water systems and their users can result from such long-term socio-environmental processes, as the construction and operation of water infrastructure and land use changes (e.g. increased impervious cover) are driven by urbanization, among other factors.

At the local level, climate changes affecting urban areas can be exacerbated by the effects of urbanization. These effects, such as the urban heat island (UHI), might eclipse the outcomes of global climate change (Ntelekos et al., 2010). The UHI, which varies across and within cities, often based upon affluence and urban planning (Miao et al., 2009), can increase human health risks differently across the urban–rural gradient and within cities, mostly due to both physical and socio-economic factors such as land-cover patterns, city size and the ratio of impervious surfaces to areas covered by vegetation and/or water (Grimm et al., 2008; Harlan and Ruddell, 2011). Not of lesser importance are intra-urban socio-spatial differences in access to assets and options including access to air conditioning and green and open space. Based on these differences, lower socio-economic and ethnic minority groups are more likely to live in warmer neighborhoods with greater exposure to heat stress and higher vulnerability (Harlan et al., 2007).

Urbanization and urban areas relate to vulnerability and risk in ways that have been explored primarily using case study and lessons-learned approaches (Satterthwaite et al., 2007; Revi et al., 2014). These studies find that urban settlements with a long history of investment in housing, urban infrastructure and services (such as in many high-income countries) and in public emergency response (such as Cuba) as well as those with economic/financial losses much reduced by insurance, are relatively more resilient to cope with the impacts of climate change. Yet, in high-income countries, where buildings and infrastructure are built to withstand extreme and uncommon weather events (such as a 100-year flood), urban areas can still be overwhelmed by the increased intensity of storms, heatwaves and other hazards (Romero-Lankao et al., 2014b). Urban centers facing adaptation deficits, with problems including the lack of adequate roads, piped water supplies and other infrastructures and services, compound these dangers, as they are depended upon in the event of severe weather. Without considering the future impacts of global warming, the populations and infrastructures

Urbanization, vulnerability and risk

of urban settlements already showing adaptive deficits are vulnerable within the current range of climate variability (UN-Habitat, 2011). Literature predicts that the impacts of global climate change will converge on these vulnerable areas with particular severity.

The reason for this heightening of risk and vulnerability is that the concentration of people, infrastructures, industries and wastes in urban places has two implications. On the one hand, urban populations can be vulnerable to extreme weather events or other hazards with the potential to become disasters. For instance, a high concentration of people and economic activity can generate risk when residential and industrial areas lack space for evacuation and emergency vehicle access, when high-income populations are lured by coastal properties in low-lying coastal zones or when lower-income groups, lacking the means to access safer land, settle on sites at risk from floods or landslides. However, because of the strong interconnection of services and activities in these areas, urban settlements can also increase the risk of 'concatenated hazards' (Satterthwaite et al., 2007). This means that a heavy storm can lead to a secondary hazard (e.g. floods creating water supply contamination) or that heatwaves will overlap with pollution events leading to compounded impacts that make urban disaster risk management highly complex. In another example, urban density has been found to play a role in exacerbating exposure to hazards such as the UHI, extreme heat and flooding. Hence, affluence may become a factor affecting urban vulnerability, as affluent areas are less densely settled, have relatively more open green space, decreased impermeable surfaces and better infrastructure and services. Affluent areas therefore often have lower mean temperatures (Harlan et al., 2007), a lower flood potential (Romero-Lankao et al., 2014b) and a reduced overall risk.

Counter-balancing the increased risk in urban areas, the concentration of people, infrastructures and economic activities creates economies of scale or proximity that may lead to a lower per capita cost of better watershed management, warning systems and other measures to prevent and lessen disaster risks. Furthermore, by developing policies focused on enhancing sustainability and resilience, and moving from disaster response to disaster preparedness, urban settlements can increase their effectiveness at coping with climate hazards.

Scholarship on urbanization and vulnerability has focused mostly on the global and national distributions of the current and future exposure of urban areas to climate hazards (Balk et al., 2009; McGranahan et al., 2007). However, other dimensions of urban vulnerability, such as sensitivity and capacity, have been insufficiently explored. To address the gap, Garschagen and Romero-Lankao (2013) applied national level data to ten country groups sharing similar patterns of urbanization and socio-economic development to explore the associations between these country groups and selected indicators of exposure, sensitivity, coping capacity and adaptive capacity. They find that while country groups are at similar risks from exposure to hazards, countries with rapid urbanization and economic growth face greater challenges with respect to sensitivity and lack of capacity (see Figure 15.1). For instance, country groups with rapid urban growth (groups six, eight and ten, which are mainly in Asia and Africa) have the highest levels of sensitivity and the lowest capacity levels. In fact, these countries show significantly higher sensitivity and lower capacity values than the groups with similar current income and urbanization levels, but less dynamic urban growth (notably groups seven and nine).

These correlative findings, however, should not lead one to conclude that rapid urbanization is a sufficient condition to cause high sensitivity and low capacity. Other authors find, for example, a series of capacity-enhancing effects of urbanization, particularly among emerging economies and transition countries. Yet, in cases where rapid urbanization is not matched by inclusive economic growth and governance structures, the challenge of building response

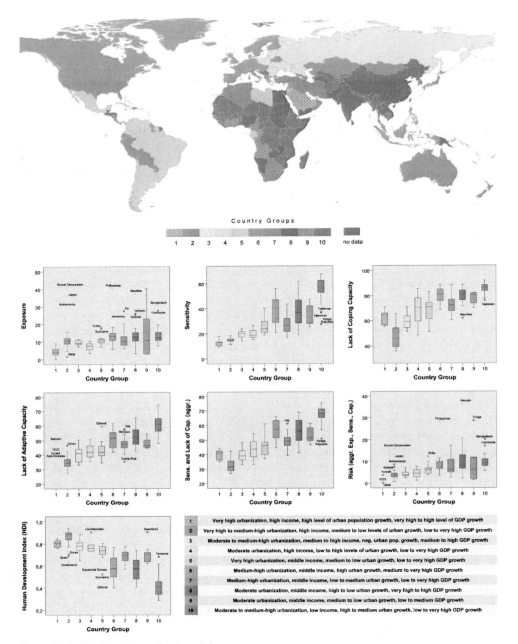

Figure 15.1 Urbanization and vulnerability

Source: Garschagen and Romero-Lankao, 2013.

Note: The map on the top illustrates the global distribution of the countries in the ten country groups. The hatching in groups one, six, eight and ten indicates that these groups are particularly dynamic in terms of urban and economic growth. The box plots illustrate the associations between the country groups (x-axes) and the different sub-indices from the World Risk Index and other data (y-axes), i.e. starting from the left, exposure, sensitivity and lack of adaptive capacity. The boxes are demarcated by the first and third quartile and provide the median. Outliers (i.e. countries with values greater than 1.5 interquartile ranges away from the first or the third quartile) are named in the box plots and can be best read by zooming in using the digital version. The whiskers indicate the lowest and highest values that are not outliers. The table on the bottom summarizes the main characteristics of the countries' groups. A detailed, colored version of this figure can be found on the book's website, www.routledge.com/9780415732260

capacity can be overwhelming (e.g. in many countries in group ten). As noted previously, urbanization may also offer opportunities for disaster risk reduction, and urban population density can be utilized for increasing the efficiency and effectiveness of emergency response programs, hydraulic infrastructure or risk-sensitive land use zoning (UN-Habitat, 2011). Conversely, slower urbanization processes can bring forth tremendous challenges for designing and financing efficient adaptation strategies, as observed in country group three (see Figure 15.1), which consists mostly of ex-members or successor states of the Soviet Union and features a high proportion of shrinking cities with decreasing densities.

Towards an integrative framework for climate change adaptation

By virtue of the existence of different lineages of research on the interactions between urbanization and risk, there are rich opportunities for synthesis and convergence in our understanding of the nature and the linkages between the key dimensions and determinants involved. Recently, efforts to integrate urban-relevant knowledge from the disaster risk management, climate change and development communities have moved urban hazard/vulnerability research towards more integrated frameworks (Cardona et al., 2012; IPCC, 2012; Harlan and Ruddell, 2011; Uejio et al., 2011). This section presents one such effort through an integrated framework that has already been applied to some cities. This framework attempts to account for the mechanisms by which an interaction of physical and social factors such as climate variability and change interacting with societal and urbanization processes, gives rise to the unique expression of urban vulnerability and risk within a local context.

Urban risk is defined in this framework as the possibility of loss, injury and other outcomes resulting from two mechanisms: the convergence of hazards with the vulnerabilities or capacities of exposed populations and the interaction of broader societal and environmental processes such as urbanization, climate change and governance structures (see Figure 15.2).

Urban risk is the result of exposure to hazards as mediated by particular capabilities to perceive and respond to these hazards (*adaptation capacity*). *Hazards* are probable perturbations and stresses to which urban populations, economic activities and infrastructures are exposed.

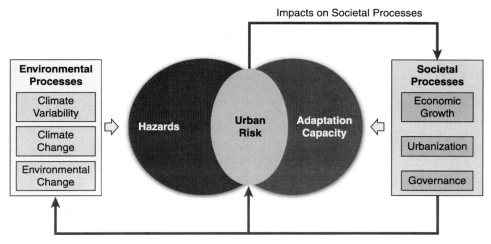

Figure 15.2 Urban risk, its dimensions and its drivers

Source: Romero-Lankao et al. (2014b) based on IPCC (2012).

These hazards can be one-off extreme events of short duration such as floods, heatwaves or storm surges. Some hazards may last days, hours or no more than a few minutes and often strike with little warning. They can also be slow-onset events, such as sea level rise or a range of subtle, 'everyday threats' that are the product of a variety of stress mechanisms (e.g. landslides resulting from land use changes induced by urbanization, fires resulting from fire suppression policies (Romero-Lankao et al., 2014b) or precarious electrical connections). The social construction of risk affects not only a population's adaptation capacity, but also the hazards to which it is exposed (see Figure 15.2).

Adaptation capacity is a population's or urban area's ability to perceive risk and to avoid or lessen the negative consequences of the multiple hazards to which it is exposed. This capacity can be affected by individual characteristics that can make household members sensitive such as age or social status, or on household, neighborhood and city level access to resources, assets and options such as education, income, house quality and social capital (e.g. individual levels of social trust, participation in networks and family support). It may also be based on governmental or institutional support approaches such as emergency response systems and recovery policies. The determinants and attributes of adaptation capacity vary across scale. For instance, while adaptation capacity at the household level is determined by the interaction of age, pre-existing medical conditions and other individual characteristics such as a household's access to safe housing or social networks, at the neighborhood level it is determined by the quality of infrastructural and built-environment characteristics of the area (e.g. density and percentage of green areas) (Hardoy and Barrero, 2014).

Furthermore, urban vulnerability and risk are driven by two broader processes: *environmental processes* such as changes in precipitation, temperature, water availability, sea level rise and other components of climate and environmental change, which are having (and will continue to have) an effect on the extent and location of physical hazards for urban areas, and by *societal processes* such as *urbanization* (already described earlier in the section on current understandings of urbanization) and *governance*. Here, urban governance is defined as the set of formal and informal rules, rule-making systems and actor-networks at all levels (from local to global) that are established to steer cities towards preventing, mitigating and adapting to climate change (Biermann et al., 2010).

These interacting processes have systemic, yet often indirect, effects on the extent to which urban communities are able to effectively perceive and respond to hazards. For instance, Romero-Lankao et al. (2014b) find that middle- and low-income neighborhoods in the Latin American cities of Bogota, Buenos Aires, Mexico City and Santiago are exposed to multiple hazards and that spatially unbounded hazards such as air pollution, can create risk across socio-economic lines. Risks remain largely determined by the formal status of affected populations. For example, much risk is determined by whether a neighborhood is zoned and formally sanctioned or illegal or unauthorized by the local authorities. This is because zoning and legality influence whether residents can legally connect to electricity networks or have access to clean drinking water and drainage systems thus creating increased risks of loss of life or livelihood and property damage from fires, floods and other hazards in many informal neighborhoods (see also Revi et al., 2014).

By using integrative approaches such as the one proposed and exemplified above, a better understanding can be gained of the interplay between the multiple dimensions and determinants of urban vulnerability and risk. While more narrowly focused studies have helped identify many of the numerous parts in play, a cohesive picture of the dynamic whole created by the interaction of these parts remains lacking. Through the application of holistic approaches and frameworks (e.g. as demonstrated in this chapter), scholars working across traditions and disciplines might be able to create an integrative knowledge that will aid in the design and implementation of more effective adaptation actions (see Redman, Chapter 34 in this volume, for a discussion on the multiple approaches to the sustainability challenge).

Adaptation actions need to consider not only the local hazards to which a population is exposed, but also an array of local supports and challenges arising from family and social structures and from the larger societal dynamics that may support or undermine the capacity to respond. Urbanization is one such large societal force; however, it is one that is fueled by local conditions and the imperatives of individual lives, livelihoods, hopes and challenges. In order to understand the whole, there is the need to look at the parts, but to understand the parts, there is the need to look back and see the whole. It is only through such integrative approaches that a greater understanding of urban vulnerability and risk can be achieved. In an increasingly urban world with greater hazards being ushered in by climate change, scholars need to bring together their knowledge systems in a search for integrative solutions.

Key messages

- Scholarship on urban vulnerability and risk has grown considerably in recent years.
- Existing and sometimes incompatible approaches have shed light on different dimensions of these multifaceted issues (e.g. the dynamics of heatwaves risk perception and adaptation capacity).
- This chapter describes one of the existing efforts to integrate urban-relevant knowledge from different communities (e.g. disaster risk, climate change, development) into an integrated framework on urbanization, vulnerability and risk.
- Research is needed, however, that teases out both cross-cutting and context-specific mechanisms by which urbanization and urban areas determine vulnerability and risk.
- Adaptation actions need to consider not only the local hazards to which a population is exposed, but also an array of local supports and challenges arising from multiple levels: household, neighborhood, city and even national and international.

References

Aragón-Durand, Fernando. 2007. "Urbanisation and Flood Vulnerability in the Peri-Urban Interface of Mexico City." *Disasters* 31 (4): 477–494.

Aylett, Alex. 2013. "The Socio-Institutional Dynamics of Urban Climate Governance: A Comparative Analysis of Innovation and Change in Durban (KZN, South Africa) and Portland (OR, USA)." *Urban Studies* 50 (7): 1386–1402.

Balk, Deborah, Mark R Montgomery, Gordon McGranahan, Donghwan Kim, Valentina Mara, Megan Todd, Thomas Buettner, and Audrey Dorélien. 2009. "Mapping Urban Settlements and the Risks of Climate Change in Africa, Asia and South America." In *Population Dynamics and Climate Change*, edited by Guzmán, J M, G Martine, G McGranahan, D Schensul, and C Tacoli, 80–103. London: IIED.

Bell, Michelle L, Marie S O'Neill, Nalini Ranjit, Victor H Borja-Aburto, Luis A Cifuentes, and Nelson C Gouveia. 2008. "Vulnerability to Heat-Related Mortality in Latin America: A Case-Crossover Study in Sao Paulo, Brazil, Santiago, Chile and Mexico City, Mexico." *International Journal of Epidemiology* 37 (4): 796–804.

Biermann, Frank, Michele M Betsill, Joyeeta Gupta, Norichika Kanie, Louis Lebel, Diana Liverman, Heike Schroeder, Bernd Siebenhüner, and Ruben Zondervan. 2010. "Earth System Governance: A Research Framework." *International Environmental Agreements: Politics, Law and Economics* 10 (4): 277–98.

Braimoh, Ademola K, and Takashi Onishi. 2007. "Spatial Determinants of Urban Land Use Change in Lagos, Nigeria." *Land Use Policy* 24 (2): 502–15.

Cardona, O D, M K van Aalst, J Birkmann, M Fordham, G McGregor, R Perez, R S Pulwarty, E L F Schipper, and B T Sinh. 2012. "Determinants of Risk: Exposure and Vulnerability." In *Managing the Risks of Extreme Events and Disasters to Advance Climate Change Adaptation. A Special Report of Working Groups I and II of the Intergovernmental Panel on Climate Change (IPCC)*, edited by Field, C B, V Barros, T F Stocker, D Qin, D J Dokken, K L Ebi, M D Mastrandrea, K J Mach, G- K Plattner, S K Allen, M Tignor, and P M Midgley, 65–108. Cambridge, UK, and New York, NY, USA: Cambridge University Press.

Carson, Claire, Shakoor Hajat, Ben Armstrong, and Paul Wilkinson. 2006. "Declining Vulnerability to Temperature-Related Mortality in London over the 20th Century." *American Journal of Epidemiology* 164 (1): 77–84.

Collins, Timothy W, and Bob Bolin. 2009. "Situating Hazard Vulnerability: People's Negotiations with Wildfire Environments in the US Southwest." *Environmental Management* 44 (3): 441–55.

D'Ippoliti, Daniela, Paola Michelozzi, Claudia Marino, Francesca de'Donato, Bettina Menne, Klea Katsouyanni . . . C A Perucci. 2010. "The Impact of Heat Waves on Mortality in 9 European Cities: Results from the EuroHEAT Project." *Environmental Health* 9, 37. doi:10.1186/1476-069X-9-37.

Field, Christopher B, Maarten van Aalst, Neil Adger, Douglas Arent, Jonathon Barnett, Richard, Betts, Eren Bilir, Joern Birkmann, JoAnn Carmin, Dave Chadee, Andrew Challinor, Monalisa Chatterjee, Wolfgang Cramer, Debra Davidson, Yuka Estrada, Jean-Pierre Gattuso, Yasuaki Hijioka, Ove Hoegh-Guldberg, He-Qing Huang, Gregory Insarov, Roger Jones, Sari Kovats, Patricia Romero-Lankao, Joan Nymand Larsen, Iñigo Losada, et al. 2014. "Technical Summary." In *Climate Change 2014: Impacts, Adaptation, and Vulnerability. Part A: Global and Sectoral Aspects. Contribution of Working Group II to the Fifth Assessment Report of the Intergovernmental Panel on Climate Change*, edited by Field, C B, V R Barros, D J Dokken, K J Mach, M D Mastrandrea, T E Bilir, M Chatterjee, K L Ebi, Y O Estrada, R C Genova, B Girma, E S Kissel, A N Levy, S MacCracken, P R Mastrandrea, and L L White, 35–94. Cambridge, UK, and New York, NY, USA: Cambridge University Press.

Galea, Sandro, and David Vlahov. 2005. "Urban Health: Evidence, Challenges, and Directions." *Annu. Rev. Public Health* 26: 341–65.

Garschagen, Matthias, and Patricia Romero-Lankao. 2013. "Exploring the Relationships between Urbanization Trends and Climate Change Vulnerability." *Climatic Change*, August, 1–16. doi:10.1007/s10584-013-0812-6.

Grimm, Nancy B, Stanley H Faeth, Nancy E Golubiewski, Charles L Redman, Jianguo Wu, Xuemei Bai, and John M Briggs. 2008. "Global Change and the Ecology of Cities." *Science* 319 (5864): 756–60.

Hardoy, Jorgelina, and Luz Stella Velásquez Barrero. 2014. "Re-Thinking 'Biomanizales': Addressing Climate Change Adaptation in Manizales, Colombia." *Environment and Urbanization* 26 (1): 53–68. doi:10.1177/0956247813518687.

Harlan, Sharon L, and Darren M Ruddell. 2011. "Climate Change and Health in Cities: Impacts of Heat and Air Pollution and Potential Co-Benefits from Mitigation and Adaptation." *Current Opinion in Environmental Sustainability* 3 (3): 126–34. doi:10.1016/j.cosust.2011.01.001.

Harlan, Sharon L, Anthony J Brazel, G Darrel Jenerette, Nancy S Jones, Larissa Larsen, Lela Prashad, and William L Stefanov. 2007. "In the Shade of Affluence: The Inequitable Distribution of the Urban Heat Island." *Research in Social Problems and Public Policy* 15: 173–202.

IPCC. 2012. *Managing the Risks of Extreme Events and Disasters to Advance Climate Change Adaptation: Special Report of the Intergovernmental Panel on Climate Change*, edited by Field, C B, V R Barros, T F Stocker, D Qin, D J Dokken, K L Ebi, M D Mastrandrea, K J Mach, G -K Plattner, S K Allen, M Tignor, and P M Midgley. Cambridge, UK, and New York, NY, USA: Cambridge University Press, 582pp.

Johnson, Daniel P, Jeffrey S Wilson, and George C Luber. 2009. "Socioeconomic Indicators of Heat-Related Health Risk Supplemented with Remotely Sensed Data." *International Journal of Health Geographics* 8 (1): 57. doi:10.1186/1476-072X-8-57.

Kan, Haidong, Stephanie J London, Guohai Chen, Yunhui Zhang, Guixiang Song, Naiqing Zhao, Lili Jiang, and Bingheng Chen. 2008. "Season, Sex, Age, and Education as Modifiers of the Effects of Outdoor Air Pollution on Daily Mortality in Shanghai, China: The Public Health and Air Pollution in Asia (PAPA) Study." *Environmental Health Perspectives* 116 (9): 1183.

Leichenko, Robin. 2011. "Climate Change and Urban Resilience." *Current Opinion in Environmental Sustainability* 3 (3): 164–68. doi:10.1016/j.cosust.2010.12.014.

MacClune, K, C Allan, K Venkateswaran, and L Sabbag. 2014. *Floods in Boulder: A Study of Resilience.* Boulder CO: ISET-International. http://i-s-e-t.org/resources/case-studies/floods-in-boulder.html.

McGranahan, Gordon, Deborah Balk, and Bridget Anderson. 2007. "The Rising Tide: Assessing the Risks of Climate Change and Human Settlements in Low Elevation Coastal Zones." *Environment and Urbanization* 19 (1): 17–37.

Makri, Anna, and Nikolaos I Stilianakis. 2008. "Vulnerability to Air Pollution Health Effects." *International Journal of Hygiene and Environmental Health* 211 (3–4): 326–36. doi:10.1016/j.ijheh.2007.06.005.

Manuel-Navarrete, David, Mark Pelling, and Michael Redclift. 2011. "Critical Adaptation to Hurricanes in the Mexican Caribbean: Development Visions, Governance Structures, and Coping Strategies." *Global Environmental Change* 21 (1): 249–58.

Marcotullio, Peter J, Sara Hughes, Andrea Sarzynski, Stephanie Pincetl, Landy Sanchez Peña, Patricia Romero-Lankao, Daniel Runfola, and Karen C Seto. 2014. "Urbanization and the Carbon Cycle: Contributions from Social Science." *Earth's Future*, August. doi:10.1002/2014EF000257.

Miao, Shiguang, Fei Chen, Margaret A LeMone, Mukul Tewari, Qingchun Li, and Yingchun Wang. 2009. "An Observational and Modeling Study of Characteristics of Urban Heat Island and Boundary Layer Structures in Beijing." *Journal of Applied Meteorology and Climatology* 48 (3): 484–501.

Moser, C O N. 1998. "The Asset Vulnerability Framework: Reassessing Urban Poverty Reduction Strategies." *World Development* 26 (1): 1–19.

Ntelekos, Alexandros A, Michael Oppenheimer, James A Smith, and Andrew J Miller. 2010. "Urbanization, Climate Change and Flood Policy in the United States." *Climatic Change* 103 (3–4): 597–616. doi:10.1007/s10584-009-9789-6.

O'Brien, K, S Eriksen, L P Nygaard, and A Schjolden. 2007. "Why Different Interpretations of Vulnerability Matter in Climate Change Discourses." *Climate Policy* 7 (1): 73–88.

O'Neill, M S. 2005. "Disparities by Race in Heat-Related Mortality in Four US Cities: The Role of Air Conditioning Prevalence." *Journal of Urban Health: Bulletin of the New York Academy of Medicine* 82 (2): 191–97. doi:10.1093/jurban/jti043.

Parnell, Susan, David Simon, and Coleen Vogel. 2007. "Global Environmental Change: Conceptualising the Growing Challenge for Cities in Poor Countries." *Area* 39 (3): 357–69.

Pataki, D E, R J Alig, A S Fung, N E Golubiewski, C A Kennedy, E G McPherson, D J Nowak, R V Pouyat, and P Romero-Lankao. 2006. "Urban Ecosystems and the North American Carbon Cycle." *Global Change Biology* 12 (11): 2092–2102.

Pelling, Mark. 1999. "The Political Ecology of Flood Hazard in Urban Guyana." *Geoforum* 30 (3): 249–61.

Revi, A, D E Satterthwaite, F Aragón-Durand, J Corfee-Morlot, R B R Kiunsi, M Pelling, D C Roberts, and W Solecki. 2014. "Urban Areas." In *Climate Change 2014: Impacts, Adaptation, and Vulnerability. Part A: Global and Sectoral Aspects. Contribution of Working Group II to the Fifth Assessment Report of the Intergovernmental Panel on Climate Change*, edited by Field, C B, V R Barros, D J Dokken, K J Mach, M D Mastrandrea, T E Bilir, M Chatterjee, K L Ebi, Y O Estrada, R C Genova, B Girma, E S Kissel, A N Levy, S MacCracken, P R Mastrandrea, and L L White, 535–612. Cambridge, UK and New York, NY, USA: Cambridge University Press.

Romero-Lankao, Patricia, and Hua Qin. 2011. "Conceptualizing Urban Vulnerability to Global Climate and Environmental Change." *Current Opinion in Environmental Sustainability* 3 (3): 142–49. doi:10.1016/j.cosust.2010.12.016.

Romero-Lankao, Patricia, Hua Qin, and Katie Dickinson. 2012. "Vulnerability to Temperature-Related Hazards: A Meta-Analysis and Meta-Knowledge Approach." *Global Environmental Change* 22 (3): 670–83.

Romero-Lankao, Patricia, Hua Qin, and Mercy Borbor-Cordova. 2013. "Exploration of Health Risks Related to Air Pollution and Temperature in Three Latin American Cities." *Social Science & Medicine* 83 (0): 110–18. doi:10.1016/j.socscimed.2013.01.009.

Romero-Lankao, Patricia, Kevin Gurney, Karen Seto, Mikhail Chester, Riley M Duren, Sara Hughes, Lucy R Hutyra, et al. 2014a. "A Critical Knowledge Pathway to Low-Carbon, Sustainable Futures: Integrated Understanding of Urbanization, Urban Areas and Carbon." *Earth's Future*, August. doi:10.1002/2014EF000258.

Romero-Lankao, Patricia, Sara Hughes, Hua Qin, Jorgelina Hardoy, Angélica Rosas-Huerta, Roxana Borquez, and Andrea Lampis. 2014b. "Scale, Urban Risk and Adaptation Capacity in Neighborhoods of Latin American Cities." *Habitat International* 42 (0): 224–35. doi:10.1016/j.habitatint.2013.12.008.

Satterthwaite, David, Saleemul Huq, H Reid, M Pelling, and P Romero-Lankao. 2007. *Adapting to Climate Change in Urban Areas*. London: IIED.

Seto, Karen C, Michail Fragkias, Burak Güneralp, and Michael K Reilly. 2011. "A Meta-Analysis of Global Urban Land Expansion." *PLoS ONE* 6 (8): e23777.

Shrader-Frechette, Kristin. 1991. "Reductionist Approaches to Risk." In *Acceptable Evidence: Science and Values in Risk Management*, edited by Mayo, D G, and R D Hollander, 218–248. New York, NY, USA: Oxford University Press.

Smyth, Conor G, and Stephen A Royle. 2000. "Urban Landslide Hazards: Incidence and Causative Factors in Niteroi, Rio de Janeiro State, Brazil." *Applied Geography* 20 (2): 95–118.

Stafoggia, Massimo, Francesco Forastiere, Daniele Agostini, Annibale Biggeri, Luigi Bisanti, Ennio Cadum, Nicola Caranci, et al. 2006. "Vulnerability to Heat-Related Mortality: A Multicity, Population-Based, Case-Crossover Analysis." *Epidemiology* 17 (3): 315–23. doi:10.1097/01. ede.0000208477.36665.34.

Thywissen, Katharina. 2006. "Components of Risk." *Source (Studies of the University: Research, Counsel, Education)* No 2.

Tie, Xue-xi, Sasha Madronich, G Li, Zhuming Ying, Andrew Weinheimer, Eric Apel, and Teresa Campos. 2009. "Simulation of Mexico City Plumes during the MIRAGE-Mex Field Campaign Using the WRF-Chem Model." *Atmospheric Chemistry and Physics* 9 (14): 4621–38.

Uejio, Christopher K, Olga V Wilhelmi, Jay S Golden, David M Mills, Sam P Gulino, and Jason P Samenow. 2011. "Intra-Urban Societal Vulnerability to Extreme Heat: The Role of Heat Exposure and the Built Environment, Socioeconomics, and Neighborhood Stability." *Health & Place* 17 (2): 498–507.

UN-Habitat. 2011. *Cities and Climate Change: Global Report on Human Settlements*.

16

EXTREME EVENTS AND THEIR IMPACTS ON URBAN AREAS

Andrea Ferraz Young

The climate system, which is influenced by land use/land cover, oceans and water bodies, the atmosphere, cryosphere and biosphere (Pielke Sr. et al., 2002), affects physical and biological systems on all continents and in most oceans. Urban areas, especially in developing countries, are particularly vulnerable to frequent climate variability and associated extreme events (Castán and Bulkeley, 2013; Hunt and Watkiss, 2011; Romero-Lankao and Dodman, 2011; Rosenzweig et al., 2011b). Recent high-impact weather events have lead researchers to question whether their frequency and intensity have changed and will continue to change over time (Zwiers et al., 2012). Projections reveal that it is very likely that effects of atmospheric pollution (i.e. disease) (Yuming et al., 2009) as well as extreme weather events such as heavy rainfall, warm spells and extreme heat, drought, intense storm surges and sea level rise will increase in frequency, intensity and duration as a result of climate change (Rosenzweig et al., 2011b; Hunt and Watkiss, 2011; Romero-Lankao and Dodman, 2011; IPCC, 2014).

It is clear that climate is also transforming the cities in which we live, with the geographical locations of observed changes showing consistency with spatial patterns of atmospheric and warming trends. Urban areas worldwide, e.g. New York City (US) (Rosenzweig et al., 2011a), São Paulo (Brazil) (Young, 2013), Rio de Janeiro (Brazil) (De Sherbinin et al., 2007), Shanghai (China) (Wang and Zheng, 2013), Beijing (China) (Yuming et al., 2009) and London (UK) (Armstrong et al., 2011) are all experiencing multiple climate changes and resulting impacts. Climate impacts interact with context-specific urban conditions (e.g. social, economic and environmental stressors) exacerbating and compounding risks to individual and household well-being (Leichenko and Thomas, 2012, Rosenzweig et al., 2011b; Hunt and Watkiss, 2011; Romero-Lankao and Dodman, 2011). These interactions not only have widespread negative impacts on people (e.g. health, well-being, livelihoods and assets), but also on local and national economies and ecosystems (IPCC, 2014). Together, climate variability and urbanization pressures create profound impacts across a broad spectrum of infrastructure (i.e. water and energy supply, sanitation and drainage, transport and telecommunications), services (i.e. health care and emergency services), socio-economic conditions (i.e. income inequality), the built environment and ecosystem services (Hunt and Watkiss, 2011; Romero-Lankao and Dodman, 2011; Rosenzweig et al., 2011b; IPCC, 2012, 2014).

As urban areas continue to experience seasonal shifts, rising temperatures (Armstrong et al., 2011), fluctuations in rainfall patterns, sea level rise and storm surges (Yuzva, 2012), risks including heat stress, water scarcity and worsening air pollution also increase. Disaster situations can occur when hazards (i.e. floods, landslide and drought) combine with conditions of vulnerability (i.e. poverty and socio-economic inequalities) and insufficient capacity or measures to reduce the negative consequences of risk (UNISDR, 2014). In rapidly urbanizing countries, the combination of structural poverty and unequal concentration of income, the absence of infrastructure (i.e. decaying and sub-standard), high population densities and the centralization of economic assets and commercial and industrial activities heighten urban vulnerabilities (Kreimer et al., 2003).

This chapter begins with an overview of urban climate risks in order to understand how cities are being impacted by specific extreme events. Thereafter, the chapter examines the actions that cities have undertaken in the face of extreme human, environmental and economic loss that could be opportunities for shared learning, in order to effectively respond to future risks. The chapter concludes by making the link between extreme events and the potential for a transition towards a more sustainable future while emphasizing that even in the presence of multiple stressors, policy relevant information from ongoing scientific research, experiential knowledge and observation provide opportunities for future adaptation.

Climate variability and extreme events

Some of the largest climate impacts on cities are associated with extreme events such as drought, heatwaves, intense precipitation, coastal storms and cyclones, and are critical components of climate impact assessments (IPCC, 2014). Over the last decade, a significant number of these events have caused large losses of human life, for example, Typhoon Bopha in the Philippines (1,901 deaths in 2012), flooding and landslides in Brazil (900 deaths in 2011), Storm Nargis in Myanmar (138,366 deaths in 2008) and Storm Stan in Guatemala (1,513 deaths in 2005) as well as tremendous economic losses (Easterling et al., 2012), e.g. Hurricane Sandy in 2012 (~50 billion USD), flooding in Thailand in 2011 (~41 billion USD) and Hurricane Katrina in 2005 (~147 billion USD) (United Nations, 2014).

Climate projections indicate a rise in temperature for the majority of cities around the world, e.g. New York City (US), London (UK) and Toronto (Canada) (Rosenzweig et al., 2011b). Global mean surface air temperatures over land and oceans have increased, and measurements show a continuing increase of heat content in the oceans. It can be expected that droughts will be more frequent and intense as a warmer atmosphere is expected to hold more moisture. Sea level rise and stronger tropical storms may further increase flood risk, particularly when high tides combine with storm surges and/or high river flows. Low-elevation coastal zones are particularly at risk of flooding and storm damage as a result of climate change (McGranahan et al., 2007, 2008). Heat and cold waves, intense rainfall and cold and dry spells, among other extreme events, have distinct effects on nations with varied impacts across sectors—water, agriculture, food security, forestry, health and tourism (Marengo et al., 2009).

The processes affecting climate can exhibit considerable natural variability, much of which can be represented by simple (e.g. unimodal or power law) distributions, but many components of the climate system also exhibit multiple states such as the glacial-interglacial cycles and certain modes of internal variability, such as El Niño-Southern Oscillation (ENSO) (IPCC, 2013). Delhi (India) is one city where annual mean temperature is projected to increase over the next century, but certain aspects of its humid subtropical climate are markedly different from many other humid subtropical cities, such as São Paulo, where there are dry winters and rainy summers (Marengo et al., 2013; IPCC, 2014). Tokyo (Japan) and Brisbane (Australia) experience dust storms as a result of monsoonal climate patterns and have relatively dry winters but often-prolonged spells of very hot weather (Shepherd et al., 2002). Furthermore, warming is expected to increase with distance from the Equator. Inland or continental regions are expected to warm more than coastal regions because they experience climate-moderating influences from the oceans; this explains why more warming is expected in Toronto than London despite comparable latitudes (Shepherd et al., 2002). Warming will also generally be greatest in winter of extra-tropical regions such as the United States, but there is a greater variability in these climate change projections and in the potential for disasters (Shepherd et al., 2002).

Moreover, some cities are expected to see increases in precipitation while others are projected to experience declines. When precipitation does occur, it will tend to be more intense, basically concentrated in extreme events (Rosenzweig et al., 2011b; IPCC, 2014). In mid-latitudes such as Tokyo, New York City (NYC) and Toronto most of the precipitation will be in the form of rainfall. However, London is expected to experience decreases in precipitation, which could lead to drier summers (Rosenzweig et al., 2011b). Similarly, other cities located between mid-latitudes and the subtropics are expected to experience greater aridity, e.g. Melbourne (Australia) (Shepherd et al., 2002). Others in tropical latitudes such as Manaus (Brazil) (tropical rainforest climate—Af) will experience more precipitation while others less such as São Paulo and Rio de Janeiro (tropical wet and dry—Aw). This depends on the wet and rainy season and the influence of Pacific Decadal Oscillation (PDO) or sea surface temperature anomalies (SSTA) over the Pacific that interact with ENSO combined with stochastic atmospheric variability (INPE, 2015). Extreme precipitation events can have important effects on urban areas, for example, in the São Paulo Metropolitan Area (SPMA), Brazil (Marengo et al., 2013) where flash floods associated with intense precipitation, even during periods of brief rainfall, may be destructive.

Sea level rise

As the oceans warm and expand and ice sheets continue to melt, the distribution of tropical storms, cyclones (Webster et al., 2005; Knutson et al., 2010) and hurricanes are expected to have significant impacts on coastal patterns and processes (Michener et al., 1997). Coastal cities are expected to suffer the effects of sea level rise most; however, the rate at which this occurs will vary due to height of the ocean, influence of the ocean currents, water temperature, salinity intrusion, influence of wind, air pressure (Shepherd et al., 2002), soil erosion, shifting dunes and the characteristics of estuaries, deltas and mangroves (Muehe, 2011). According to UN-Habitat (2009), up to 3,351 cities around the world are located in low-lying coastal zones that may be affected by rising sea levels. The impacts will depend on factors such as exposure and vulnerability and the extent to which society is adapted to particular extreme events (Goodess, 2013; Romero-Lankao and Gnatz, Chapter 15 in this volume).

Changes in precipitation and sea level rise will have important consequences for coastal ecosystems; increases in precipitation and runoff may respectively increase the risk of coastal flooding (Kennedy et al., 2002). Rosenzweig et al. (2011a) affirm that heavy precipitation events could potentially increase the risk of flash floods in NYC. Sea level rise will also increase risk of storm surge-related flooding, enhance vulnerability of energy infrastructure located in coastal areas and threaten transportation and telecommunications facilities. Flooding can produce higher pollutant levels in water supplies, inundate wastewater treatment plants, saturate coastal lands and wetland habitats and flood key rail lines, roadways and transportation hubs. Coastal elevation data have been widely used to identify the potential effects of sea level rise; however, the complex dynamic processes of coastal ecosystems along with rapid development along coastal regions can influence the assessments (Grech et al., 2012; Poulter and Halpin, 2008; Muehe and Klumb-Oliveira, 2014; Muehe, 2011). For example, some coastal cities across the world like Shanghai are sinking due to the effects of groundwater extraction and compaction of the soils from urban construction (Ye-Shuang et al., 2012), while others, such as Rio de Janeiro, are suffering from coastal erosion and/or receding shoreline (Muehe and Klumb-Oliveira, 2014; Muehe, 2011).

It is important to highlight that projections for cities are not always accurate in various regions, especially where there are strong seasonal precipitation cycles and where baseline measures do not exist, as is the case in Manaus, São Paulo and Rio de Janeiro. Furthermore, it is often difficult to obtain projections for cities worldwide even with downscaled, climate models and using ten-years smoothing. For some cases, there remains considerable variability and no statistically significant trends. Nevertheless, in some cases like NYC, efforts have been taken to project sea level rise by taking into account thermal expansion, meltwater from glaciers, local land subsidence and local water surface elevation components (Horton and Rosenzweig, 2010).

Weather-related hazards combined with human activities and environmental degradation lead to greater erosion, flooding and salinization of surface waters (Kreimer et al., 2003). Sea or saltwater intrusion is the encroachment of saline water into fresh groundwater in regions with coastal aquifer settings (Werner and Simmons, 2009). The rise in sea levels leads to increased saline water intrusion on estuaries, an important component of the complex and dynamic coastal watershed. The wetlands bordering estuaries perform valuable functions including water quality, flood protection and water storage (Werner and Simmons, 2009). Efforts to understand large-scale climatic changes, the multi-decadal PDO, the inter-annual ENSO and SSTA on ecological processes such as fisheries biomass yields in Large Marine Ecosystems (LMEs) indicate the presence of emergent trends induced by global warming (Sherman et al., 2009), which will affect coral reefs, fish and fisheries (productivity and trophic relationship of food chain), pollution and ecosystem health, economic productivity and urban consumption (Gherardi et al., 2010).

Research suggests that regional factors will influence relative sea level rise for specific coastlines around the world, which depend on land elevation changes that occur as a result of subsidence (sinking) or uplift (rising). Therefore, sea level rise depends on changes in currents, winds, salinity and water temperatures as well as proximity to thinning ice sheets (Karl et al., 2009). Evidence reveals that the persistence of coastal wetlands will be determined by the interactions of climatic and anthropogenic effects, especially human response to rising sea levels and the effects on resource exploitation, pollution and water use from further human encroachment on coastal wetlands (Michener et al., 1997). In many parts of the world, developers have drained wetlands, removing the buffer of protection against tidal floods. In this case, the water drains more rapidly from the urban built-up landscape, increasing peak flows (McGranahan et al., 2008). Future regional patterns of sea level variations reveal that vulnerable areas along the eastern coast of the United States will likely be affected by a more rapid sea level rise because of ocean circulation changes and static effects as well as Southeast Asia, given the high exposure of populations to risks of flooding and erosion (Cazenave and Le Cozannet, 2013).

Flood zones

According to the IPCC (2014), nearly 200 million people live in coastal flood zones; in South Asia alone the number exceeds 60 million people. The catastrophes involving flood events

Extreme events and urban areas

have affected different regions around the world and highlight the vulnerability of cities. Economic losses from damage associated with floods are quite high, e.g. 9.5 billion USD in Pakistan (floods in July–August, 2010) and 30 billion AUD in Australia (floods in 2010). Many are vulnerable in terms of infrastructure related to transport, access (routes) and energy supplies pre- and post-events. Japan experienced a total loss of over 300 billion USD (Tohoku earthquake and tsunami in 2011) and in response has developed modern warning systems, but many actions are still necessary (UNISDR, 2014).

In the urban environment, rainwater falls on rough and impervious surfaces causing increased volumes of rapid runoff, which enter watercourses soon after a rainfall event. This causes increased peak flows that often lead to flooding because of overfilled channels. Land use characteristics along with economic and demographic changes underpin increased human vulnerability to extreme hydro-meteorological conditions (Baldassarre et al., 2010). Urban dwellers, especially low-income populations living in settlements along flood plains, are at risk and especially vulnerable to flood hazards (McGranahan et al., 2008). This is quite prevalent in the developing world, as widespread and unplanned informal settlements often emerge in high-risk locations such as flood-prone areas. Increased risk creates significant challenges for local authorities to help communities achieve economic development, but it can also elicit opportunities to building urban resilience.

Research examining the effects of urbanization on flood peak distributions shows that there is a significant increase in flood magnitude during periods of rapid urbanization (Villarini et al., 2009), with changes affecting both the central tendency and the scattering of annual flood peak magnitudes. Cunha et al. (2011) demonstrate that land cover strongly affects the hydrological response of rivers and, consequently, the frequency and magnitude of floods, with the effects being scale dependent. The environmental consequences of flooding, however, can be extremely complex and difficult to assess due to their large spatial extent, multiple sources and potential effects on nearly all components of the environment. It is extremely difficult and complex to specify how well landscapes and streams accommodate natural events because this process is influenced by modifications of terrain and drainage systems as well as vegetative changes, alterations to stream channels such as channelization, floodplains occupation and confinement of river systems and a large list of other land use changes (EPA, 2013). Particularly, in the SPMA, flooding and landslide risks will increase with extreme events like storms and heavy rainfall as well as urbanization processes linked to deforestation and soil compaction. This is also heavily influenced by other socio-economic factors such as income inequality and the underlying structural causes of social injustice and urban inequality (Young, 2013).

Many large cities such as New York, Shanghai, Rio de Janeiro and Tokyo are affected by coastal flooding which is caused by extreme tidal conditions that naturally occur either individually or in combination (Cayan et al., 2008): (1) astronomical tides or high tide levels—variations in tidal levels due to gravitational effects of the sun and moon that can result in higher sea levels; (2) storm surge—an increase in sea level above tidal level caused by low atmospheric pressure which may be exacerbated by the wind acting on the sea; and (3) wave action due to sea level pressure (barometric level) and wind stress fluctuations—dependent on wind speed and direction, local topography and exposure. In Shanghai, storm surges are far less likely to cause flooding than they are in places such as Dhaka (Bangladesh) and Kolkata (India). There and in other parts of Asia, serious flooding is attributed more often to earthquakes and tropical cyclones. For example, in December 2004, an earthquake off the Sumatra coast led to a series of tsunamis in the Indian Ocean that claimed the lives of more than 227,000 people (Telford and Cosgrave, 2007). A tsunami is a series of traveling ocean waves of extremely long length generated by disturbances associated primarily with earthquakes.

Damaging tsunamis are rare but can become potentially catastrophic events, presenting a danger to local people and the regional economy. As a tsunami travels across the ocean, it has potential energy to cause damage by flooding on land thousands of miles away and many hours after the source event occurs. Local tsunamis can be caused by offshore faults or coastal and submarine landslides, which have the potential to cause high flood damage because of great wave heights (State of California Seismic Safety Commission, 2005). Seismic activity has ranged from barely detectable earthquakes and tsunamis that cause no damage, to large-scale land and sea movements capable of destroying cities through the flood force and causing widespread regional disruption. Takeda (2011) reports on the strong earthquake followed by tsunami in Fukushima, Japan, that flooded the nuclear power station causing explosions of the reactors, and destruction of local infrastructure. The damage was severe, particularly in areas where there is no engineered coastal protection (Bird et al., 2007). Currently, the international research on the impacts of earthquakes and tsunamis is limited; additional research that distinguishes between manmade and natural disasters is needed in order to increase knowledge of the occurrence and causes of each type of disaster and their interactions to ensure that disaster preparedness is heightened (Goff and McFadgen, 2003).

Likewise, there is much uncertainty as to how the frequency and strength of tropical cyclones and hurricanes will affect different urban areas. Coastal storms can be devastating for urban areas with particularly high flood risk (e.g. NYC), given their dense population, reliance on transportation and energy infrastructure (Lane et al., 2013). Generally, the costliest hurricanes provoke storm surges that can rupture levees on drainage and navigation canals leading to catastrophic floods and land devastation, e.g. in New Orleans (US) (2005) and NYC (2012), causing flooding of streets, tunnels and subway lines and interrupting power in and around the city (Shepherd et al., 2002). In New Orleans during Hurricane Katrina (2005), a lack of emergency preparedness and planning by corrections officials led to chaos during the storm (Myers, 2007).

> After Sandy made landfall, hundreds of thousands of NYC residents initially lost power. However, even after the electric grid had been largely restored, many residential buildings in storm-inundated areas still lacked electric power, heat or running water, often because of salt water flood damage to buildings' electrical and heating systems. Many people who did not evacuate in advance of the storm sheltered in places with housing conditions that lacked one or more of these essential services.
>
> *(Lane et al., 2013:4)*

Landslides

Landslide risk is likely to remain a highly challenging issue now and into the future as urban poverty, climate variability and environmental degradation expose vulnerable populations to an entirely new scale of devastation (UNISDR, 2014). Disasters resulting from tropical cyclones, windstorms and related landslides affect populations concentrated in urban sites the most, with 366,000 people affected every year by landslides (CRED, 2014). Large populations of the poor remain at risk, as they often settle on unstable slopes and steep terrain, on areas previously affected by and prone to future landslides. Asia is most affected by landslides; the Americas, on the other hand, have suffered more deaths; and Europe bears the most economic losses with an average damage of almost 23 million USD per landslide event (UNISDR, 2014). The highest risk levels occur in middle- and low-income countries that have not adequately planned or regulated urban growth.

Greater rainfall in some areas will trigger landslides, with consequent disruption to agriculture, urban settlements, commerce and transport (Reichenbach and Gunther, 2010). South America will be at greater landslide risk in the future as more people live in unsafe urban settlements (i.e. slums located on mountains and hills). Capital venture and public irresponsibility can all too often increase the probability and severity of disasters, for example, by destroying forests and re-purposing the land thereafter. Angra dos Reis (2010), Bumba Hill in Rio de Janeiro (2010) and Teresopolis (2011) represent recent disaster realities in Brazil (United Nations, 2014). The capital of Liguria (Genoa), on the Italian Riviera at the base of the Alps in the foothills, is another example of a city that is particularly vulnerable to floods and landslides. It is well-known since the end of the 19th century for the Lemeglio landslide and its large effects on the transport (railway) systems. However, due to a lack of analysis and guidance, the population was not prepared when disasters struck in 2011 that left seven dead (Faccini et al., 2015). More recently the region has been equipped with geotechnical and hydrogeological monitoring tools to collect data and model movement of the landslide (Faccini et al., 2015).

Landslide is often a recurrent phenomenon responsible for casualties, destruction of assets, infrastructures and economic loss (Guzzetti, 2000). There is much research addressing landslide risk analysis, e.g. Chau et al. (2004); Hervás and Bobrowsky (2009); Ardizzone et al. (2008); Guzzetti et al. (2009); Reichenback and Gunther (2010); Brunetti et al. (2010); Martelloni et al. (2011); and Guzzetti (2000). Landslides can manifest in many different forms including rock falls, rockslides, rock avalanches, debris flows, soils slips and mud flows (Chau et al., 2004). Slope movements, especially rock falls and rotational slides, are the most frequently observed phenomena. Damage and fatalities have been caused by single catastrophic failures and also by widespread landsliding (Cardinali et al., 2002). In Italy, the most diffuse landslides are rock falls, especially multiple rock falls. Central Italy is characterized by widespread seismicity felt widely throughout Rome (Guzzetti et al., 2009). Landslide events are generally associated with a trigger such as an earthquake, a large storm, a rapid snowmelt or a volcanic eruption (Malamud et al., 2004). In Pakistan, a recent quake triggered landslides, burying or wiping out roads in many areas within the North-West Frontier Province (NWFP) and Pakistan-Administered Kashmir. Many highways were blocked at several points, hindering access and relief efforts (UNISDR, 2014). The landslide caused by a devastating earthquake in Nepal (2015) killed more than 3,000 and left tens of thousands homeless. The U.S. Geological Survey (2015) affirms the damage costs to be approximately 10 billion USD.

A study by Martelloni et al. (2011) hypothesizes that anomalous or extreme volumes of rainfall are responsible for triggering landslides. In this case, "the rainfall series is analyzed, and multiples of the standard deviation (σ) are used as thresholds to discriminate between ordinary and extraordinary rainfall events." Recent heavy rains (2015) provoked a landslide at the ancient Italian site of Pompeii, affecting area falls within the 'Great Pompeii' joint EU-Italian restoration project. Such events largely depend on the typology and the environmental context (Ardizzone et al., 2002). Historical records, local geology, lithology, geomorphology, soil structure, hydrological conditions, form and type of vegetation and climate/weather must be considered in estimations of landslide hazards (Chau et al., 2004).

The intensification of rainfall associated to falls, flows and earth mass mechanisms as well as irregular land use in hilly areas at risk will intensify landslides with strong threats, resulting in a large number of disasters mainly in densely populated cities. In Italy, since rainfall represents the most common trigger factor, civil protection agencies are setting up warning systems based on the interaction between rainfall and landslides (Martelloni et al., 2011). Models rely upon the understanding of the physical laws controlling slope instability and attempt to extend spatially the simplified stability models widely adopted in geotechnical engineering (Brunetti et al., 2010). However, it is difficult to define the exact spatial and temporal variation of the many involved factors (rainfall variation in space and in time, effect of vegetation, mechanic and hydraulic properties of both bedrock and soil layer) (Martelloni et al., 2011). Furthermore, factors contributing to uncertainty cannot be readily evaluated and explicitly incorporated in the subsequent phases of assessing the hazards and risks affecting human activities (Ardizzone et al., 2002). For this reason, the identification and mapping of landslide deposits are an intrinsically difficult and subjective operation that requires a great effort to minimize inherent uncertainty (Ardizzone et al., 2002). Landslides cannot be perfectly predicted, but people living in landslide-prone areas can be warned in advance if there is a warning system in place for measuring rainfall levels.

Drought

Drought is an extreme event with serious consequences for millions of people in urban areas. The inter-annual relationship between ENSO and the global climate is not stationary and can be modified by the PDO, but how patterns of drought are changing and how the global land distribution of the dry–wet changes associated with the combination of ENSO and the PDO remain unclear (IPCC, 2014). Defining the frequency characteristics of the decadal variability and the sequence of events in the different ocean basins, especially with respect to the central tropical Pacific, represents a critical component of pursuing the implications of an 'ENSO-like' decadal variability (Cobb et al., 2001:2209). Drought risks are not only associated with deficient rainfall caused by climate variability, but also with urban poverty and agriculture vulnerability (e.g. land use and agricultural practices), poor water and soil management and ineffective governance (UNISDR, 2014). Increased drought in some regions, e.g. of Asia and the Americas will lead to land degradation, crop damage and reduced yields as well as livestock deaths and increased wildfire risks. Agricultural-dependent societies will face food and water shortages, malnutrition and increased disease, with many being forced to migrate (UNISDR, 2014).

Southeastern Brazil is currently experiencing a record drought since 2014, causing concerns about electricity rationing, drinking-water shortages or another season of damaged export crops, which accounts for 60 percent of the country's gross domestic product. Record-high temperatures and drought (the most severe in at least 80 years) are punishing this region, with climate challenges threating economic recovery (SOMAR, 2015). As a consequence, the metropolitan areas such as São Paulo and Rio de Janeiro are facing water scarcity. Despite several afternoons of violent summer (2014) rainstorms in São Paulo, the Cantareira reservoir has remained at only 6.7 percent capacity. The rapid increase in drought risk is a non-linear response and given the dramatic potential evaporation increase possible with expected increases in summer temperatures, the situation could become more serious. In the last year, hospital admissions in São Paulo have increased with patients suffering from respiratory conditions, cardiovascular and other chronic diseases. Children with pre-existing health conditions and older adults have all been affected, leading to increased emergency medical services calls and emergency department visits from patients who rely on medical equipment and other health care facilities (Stauffer, 2015).

Private weather forecasting (SOMAR, 2015) warned of irregular rainfall in the center-west soy belt and the southeast as well as an atmospheric blockage preventing a cold front from advancing over the key producing regions of the world's exports of coffee, sugar, soy and beef. Summer rainfalls are not likely to bring reservoirs back to comfortable levels in Southeastern Brazil and will likely provoke floods in relatively small intervals. In the best scenario, this region could recover lost volume from the muddy reserves at the bottom of São Paulo's main reservoir, responsible for supplying potable water and 70 percent of the country's hydroelectric generation. The situation has been dramatic over the last year (2014–2015) with drought exacerbating flood risk in the SPMA, affecting human life and property (e.g. electrocution and fallen trees on automobiles), underground systems (e.g. halted train systems) and power and utility outages of electrical infrastructure (Stauffer, 2015).

Even so, the government has ruled out electricity rationing, as power distributors turn to expensive alternatives such as thermal energy and steep fines for above-average water use. The rising energy prices have increased costs for electrical distributors, and the government is considering a one billion USD loan from state-run banks to help companies cover costs, which would be the third such loan in less than one year (Stauffer, 2015). In the United States, especially in parts of the San Joaquin Valley and Southern California, 2012–2014 were also dry water years statewide; precipitation in some areas of the state is tracking at the driest on record. Statewide reservoir storage going into the wet season was about 75 percent of the average for that time of year, with evident impacts of three consecutive dry years on statewide groundwater levels (National Drought Mitigation Center, 2015).

Increased losses and responses to extreme events

The majority of reports, in the past decade, from Joint Scientific Committee for the World Weather Research Program; World Meteorological Organization (WMO); Commission for Atmospheric Sciences; Intergovernmental Panel on Climate Change (IPCC); and International Polar Year Commission show that storms, floods and droughts are among the most recurrent weather-, climate- and water-related hazards around the world. In order to define the hazards associated with extreme weather events, it is important to consider three factors: frequency, intensity (or magnitude) and persistence at the regional scale (Alexander and Arblaster, 2009; Tebaldi et al., 2006). Furthermore, the distribution of mortality and economic losses from these hazards varies according to region. For example, the main contributors to loss of human life have been droughts in Africa; storms in Asia, Central America, North America, the Caribbean and the South Pacific; floods in South America; and heatwaves in Europe (United Nations, 2014). For centuries, floods have been one of the most destructive disasters (Cunha et al., 2011), such as those experienced in Pakistan in August 2010, which displaced over 15 million people (Goodess, 2013). Damages from flooding in the winter and fall seasons have been widespread in the UK and Western Europe (Lavers et al., 2011); also flood-related fatalities and associated economic losses (Baldassarre et al., 2010) have increased dramatically over the past half-century in Africa.

Through both direct pathways (e.g. physical damage caused by floods) and indirect pathways (e.g. water shortages and spread of disease), extreme weather events have the potential to influence human decisions (Goodess, 2013). Generally, the urbanization process, particularly in rapidly urbanizing regions, is not well accompanied by spatial planning, sustainable industrialization, economic growth and welfare considerations or investments in environmental services and infrastructure, leading to poverty (social exclusion, inequalities or service gaps) and extensive informal settlements (Letema, 2012). For example, in cities of

Africa (i.e. Nairobi and Kisumu, Kenya and Kampala, Uganda), wastewater is discharged into Lake Victoria. To ensure smooth operation, the sanitation system requires high water flows resulting in an appropriate gravity-based system, minimum pipe diameters, a high number of household connections, sewage passing by both sides of the streets, minimum speed and depth, adequate slope of sewers, pumping stations at different stages of the sewer network and monitoring of the system (Letema, 2012). A large portion of economic losses is attributed to floods and waterborne outbreaks, since flooded areas and ditches, latrines and septic tanks are key reservoirs that perpetuate cholera, malaria, dengue and yellow fever in urban areas. Infectious disease outbreaks are also precipitated by the high population density found within these areas, with overcrowding triggering epidemic-prone infections like pertussis and influenza (WHO, 2014; UN-Habitat, 2009). This is especially a concern for East Africa where about 50–70 percent of the urban population lives in informal settlements that are neither planned nor serviced (UN-Habitat, 2008).

De Sherbinin et al. (2007) examined three cities that are particularly vulnerable to flooding hazards: Mumbai (India), Rio de Janeiro (Brazil) and Shanghai (China). Each of these cities has urban agglomerations in excess of ten million people, a lack of dykes (on the coast) and poor sanitation and waste treatment, which create a particular bundle of stresses that 'collide' with socio-economic characteristics (infrastructure, economic sectors and services) and environmental conditions (geologic, hydrologic, atmospheric and ecological processes). In Brazil (e.g. São Paulo, Rio de Janeiro), the flood and landslide risks are also associated with housing conditions, safe water and sanitation facilities, income and health, which are the main structural causes of social inequalities. Income inequality and absence of health care is accompanied by many differences in conditions of well-being at the individual and community levels, which may adversely influence the risks (Young, 2014).

Floods and environmental degradation are exacerbated in different South American cities ranging from Buenos Aires (Argentina), Arequipa (Peru) and Caracas (Venezuela), where urbanization has already reached unprecedented levels. This generates dramatic, complex and dynamic human-induced changes in ecosystems with a multiplicity of degradation ranges from loss of life to ecosystem collapse (Wolanski, 2006). During the period of 1970–2012, South America experienced about 696 disasters that resulted in 54,995 deaths and 71.8 billion USD in economic damages. Most of the reported disasters were related to weather, climate and water extremes involving floods (57 percent) and landslides (16 percent) (United Nations, 2014). In the SPMA, during the last three decades, the freshwater springs have disappeared with the urbanization process and consequent use of asphalt, concrete and building construction. In promising development and infrastructure, the state and private enterprises ultimately invaded the areas of protection under law, ignoring the interconnectivity of Atlantic forests, rainfall and water sources with the river and drainage system. This ultimately resulted in the degradation of the springs, which now no longer exist and implicate water scarcity (Nobre et al., 2011). Normally, springs reduce to about 40 percent of their potential water capacity during droughts (Tucci and Mendes, 2006), but given the current situation, the metropolitan reservoir (Cantareira System) is working with just 6 percent of capacity (SOMAR, 2015).

An essential starting point for reducing disaster risks in Brazil has been the development of a National Center on Disaster (CEMADEN) in order to produce a quantitative assessment that combines information on hazards with information on exposures and vulnerabilities of populations or assets such as infrastructure and homes on illegal settlements, and agricultural production. The Monitoring Impacts System of Climate Change (SISMOI) aims to share information, knowledge and good practices as the basis for providing policy and management advice on urban development.

Extreme events and urban areas

In Europe, 1,352 disasters in Italy, France, Spain, Germany, Russia and other countries caused 149,959 deaths and 375.7 billion USD in economic damages during the 1970–2012 period. Although floods (38 percent) and storms (30 percent) were the most reported causes of disasters, extreme temperatures led to the highest proportion of deaths (94 percent), with 72,210 deaths during the 2003 European heatwave and 55,736 during the 2010 heatwave in the Russian Federation. In contrast, floods and storms accounted for most of the economic losses during the time period (United Nations, 2014). The mortality per day increased in percentage, according to heatwave effects in the study period (1990–2002 and 2004) linked to respiratory, cardiovascular and cerebrovascular mortality (D'Ippoliti et al., 2010). Reports show that more women than men died during the European heatwave in 2003 (UNISDR, 2014). In their exposition, D'Ippoliti et al. (2010:3) argue:

> The number of deaths per day in the age group 65 + ranged from 12 in Valencia to 117 in London. Cardiovascular causes accounted for about 40–50 percent of the total, except in Paris (34.1%), Budapest (60.9%) and Athens (51.6%). Cerebrovascular mortality was about 10% of total mortality, with the lowest proportion observed in Munich and Valencia (8.3%) and the highest in Athens (17.7%).

In France, the public warning system 'Vigilance' was developed as part of the country's revised emergency planning and response mechanisms to include heat/health warnings following the intense heatwave in 2003, which now also includes river flood risk warnings following the major flood in 2007.

In North America, Central America and the Caribbean, in the same period from 1970 to 2012, 1,631 registered disasters caused the loss of 71,246 lives and economic damages of 1,008.5 billion USD. The hydrometeorological and climate-related disasters are attributed to storms (55 percent) and floods (30 percent). Storms were reported to be the greatest cause of accidents (72 percent) and economic loss (79 percent) (United Nations, 2014). With over 30 years of regional cooperation in tropical cyclone forecasting and warnings, facilitated by the WMO, the Central American and Caribbean regions have demonstrated the benefits of regional cooperation to reduce the impacts of tropical cyclones and other related hazards (UN-Habitat, 2008; United Nations, 2014). For example, in the Caribbean region, extensive cooperation in disaster risk management has been developed under the Comprehensive Disaster Management project of the Caribbean Disaster and Emergency Management Agency, underpinned by the Hyogo Framework for Action 2005–2015. The hazard models use historical data and forward-looking modeling and forecasting of environmental conditions such as tropical cyclones, rainfall, soil moisture and river basin hydrology (United Nations, 2014).

The opposite situation is occurring, for example, in the state of California (US) with the worst water crisis in a century. In the third consecutive year of extremely dry conditions, the precipitation and snowpack are a small fraction of their normal averages, reservoirs are at very low levels and rivers have severely diminished flows (National Drought Mitigation Center, 2015). The California Department of Food and Agriculture (CDFA) has awarded 5.8 million USD for 70 different projects in the second phase of a program to implement on-farm water irrigation systems that reduce water and energy use, thereby reducing greenhouse gas emissions (GHGs) as part of emergency drought legislation (SB 103) signed in early 2014 by Governor Brown, who authorized to distribute as much as ten million USD for eligible projects in cooperation with the Department of Water Resources and the State Water Board (National Drought Mitigation Center, 2015). The Guide Committee to Assist Urban Water Suppliers prepared a 2015 Urban Water Management Plan, which has been developed by the California

Department of Water Resources to assist urban water suppliers to prepare and adopt plans that meet the requirements of the water code and provide useful information to the public about the supplier and its current and future water management programs.

The South-West Pacific also experienced a large number of disasters; 1,156 recorded during 1970–2012 that resulted in 54,684 deaths and 118.4 billion USD in economic losses. The majority of these disasters were caused by storms (46 percent) and floods (38 percent) (United Nations, 2014). The threats to water resources in the Asia-Pacific region reveal a complex reality and raise many concerns. The urban areas and/or ecosystems that have overlapping challenges such as poor access to water and sanitation, limited water availability and deteriorating water quality also have an increased risk and exposure to climate change and water-related disasters (UNICEF/WHO, 2012). The Asia-Pacific is home to 60 percent of the world's population, but it has only 36 percent of water resources. About 480 million people still lacked access to improved water resources in 2008, and per capita water availability is the lowest in the world. Furthermore, an average 20,451 people were killed between 2000–2009 by water-related disasters, excluding victims of tsunamis (UN-DESA, 2014).

Between 1990 and 2008, significant achievements were made in meeting the Millennium Development Goals (MDGs) through access to safe drinking water. However, progress has generally been slower in providing an appropriate sanitation service, with the exception of Northeast and Southeast Asia (UN-DESA, 2014). The population practicing open defecation in Southeast Asia decreased from 141 million in 1990, to 83 million in 2008, but the significant number of people using a shared sanitation facility increased from 16 million in 1990 to 25 million in 2008 (UN-DESA, 2014).

On the other hand, India has provided more than 166 million people with access to sanitation since 1995, which represents more people than the population of Japan and Canada combined. Strategic investments via more equitable distribution of public and private resources are likely to have the highest impact on reducing health inequalities and consequently on climate change risks (UN-DESA, 2014). It is necessary to design interventions based on factors that can cause damages in urban areas while capturing positive impacts and minimizing negative ones. As stated by Castán and Bulkeley (2013), it is important to realize that while local governments are leading the management of organizations in distinct cities such as São Paulo (collection of waste for recycling and reuse), San Joaquin Valley (energy and water conservation measures) and Bombay (photovoltaic research), many other agents (i.e. industrial entrepreneurs) are intervening by either leading experiments or as partnerships.

Discussion

This chapter has reviewed some of the complex interactions of climate variability marked by ocean changes and weather patterns in different continents, which has resulted in irregular and extreme events such as storms, floods and drought and significant impacts on different urban areas worldwide.

It is clear that more people will be at greater risk in the future as global populations increase and concentrate in urban settlements, especially on the coast, where exposure to cyclones, storms and floods will impact many. Moreover, as the urban poor continue to concentrate in illegal settlements and urban slums, the number of disaster victims from flooding and landslides will also increase since these are often located in areas with the highest risk.

While knowledge of how disasters unfold in urban areas remains incomplete, we have learned from events such as Hurricanes Katrina and Sandy in New Orleans and NYC,

Extreme events and urban areas

respectively, and from drought events and water scarcity in California (US), São Paulo (Brazil), Beijing and Shanghai (China), how the infrastructure, services and society's environmental interventions are affected depending on climate variability and on specific environmental characteristics. It is, therefore, necessary to develop a different vision about urban form and environmental interactions based on ecological services and social-technological interactions. In this context, some key messages are equally important:

- Investment in public infrastructure and a more equitable distribution of public and private resources will have the biggest impact on reducing disasters.
- Governments have the primary responsibility of protecting their citizens against disasters, preserving the human rights of disaster-affected populations and victims.
- Public–private partnerships have increasingly emerged as key features in adaptation to climate change and related governance; further investigation is needed into what drives these partnerships and experimentation, the factors hindering action, effectiveness on the ground and impacts.
- There are contradictions that need to be overcome immediately; for example, notwithstanding acknowledgement of human rights protection as being a critical element of humanitarian strategies in emergency response to disasters, the longer-term aspects of a human rights-based approach to prevention through urban planning and disaster mitigation is still limited.

Some needs for the future related to climate-related disasters in urban areas are:

- The mainstreaming of emergency planning and response mechanisms into urban strategic planning and policies (i.e. population growth; housing and public transport conditions; environmental pollution; socio-spatial inequality and poverty; migration and racial discrimination).
- Integration of water sanitation and water supply systems (i.e. waste management, sewage, water sources and supply, reuse and protection) that are flexible to a changing urban landscape (such as the cut-off and/or partial cut-off effect of underground aquifers, the decrease in the groundwater level due to leakage of underground infrastructure and the reduction in recharge of groundwater from surrounding areas).
- Ongoing risk and vulnerability assessments of urban and intra-urban populations to climate change impacts along with preparedness (observational/monitoring research; early warning systems) measures to develop and prioritize short- to longer-term preventative actions.

References

Alexander, L. V. and Arblaster, J. M. (2009) Assessing trends in observed and modeled climate extremes over Australia in relation to future projections. *Int. J. Climatol.*, vol. 29. pp. 417–435.

Ardizzone, F., Cardinali, M., Carrara, A. Guzzetti, F. and Reichenbach, P. (2002) Natural Hazards and Earth System Sciences Impact of mapping errors on the reliability of landslide hazard maps. *Natural Hazards and Earth System Sciences*, European Geophysical Society. vol. 2, pp. 3–14.

Ardizzone, F., Cardinali, M., Guzzetti, F. and Reichenbach, P. (2008) Landslide hazard assessment: Estimation and risk evaluation at the basin scale. Proceedings of The First World Landslide Forum, Tokyo. United Nations University, November, ISDR, pp.70–74.

Armstrong, B. G., Chalabi, Z., Fenn, B., Hajat, S., Kovats, S., Milojevic, A. and Wilkinson, P. (2011) Association of mortality with high temperatures in a temperate climate: England and Wales. *J. Epidemiol. Community Health*, vol. 65, no. 4, pp. 340–345.

Baldassarre, G., Di Montanari, A., Lins, H., Koutsoyiannis, D., Brandimarte, L. and Blöschl, G. (2010) Flood fatalities in Africa: From diagnosis to mitigation. *Geophysical Research Letters*, vol. 37, L22402.

Bird, M., Cowie, S., Hawkes, A., Horton, B., MacGregor, C., Ong, J. E., Hwai, A. T. S., Sa, T. T. and Yasin, Z. (2007) Indian Ocean tsunamis: Environmental and socio-economic impacts in Langkawi, Malaysia. *The Geographical Journal*, vol. 173, no. 2, pp. 103–117.

Brunetti, M. T., Peruccacci, S., Rossi, M., Luciani, S., Valigi, D. and Guzzetti, F. (2010) Rainfall thresholds for the possible occurrence of landslides in Italy. *Natural Hazards Earth System Science*, vol. 10, pp. 447–458.

Cardinali, M., Reichenbach, P., Guzzetti, F., Ardizzone, F., Antonini, G., Galli, M., Cacciano, M., Castellani, M. and Salvati, P. (2002) A geomorphological approach to the estimation of landslide hazards and risks in Umbria, Central Italy. European Geophysical Society. *Natural Hazards and Earth System Sciences*, vol. 2, pp. 57–72.

Castán, B. V. and Bulkeley, H. (2013) A survey of urban climate change experiments in 100 cities *Glob. Environ. Change*, vol. 23, no. 1, pp. 92–102.

Cayan, D. R., Bromirski P. D., Hayhoe, K., Tyree, M., Dettinger, M. D. and Flick, R. E. (2008) Climate change projections of sea level extremes along the California coast. *Climatic Change*, vol. 87 (suppl. 1), pp. S57–S73.

Cazenave, A. and Le Cozannet, G. (2013) Sea level rise and its coastal impacts. *Earth's Future*, vol. 2, pp. 15–34.

Chau, K. T., Sze, Y. L., Fung, M. K., Wong, W. Y., Fong, E. L. and Chan, L. C. P. (2004) Landslide hazard analysis for Hong Kong using landslide inventory and GIS. *Computers and Geosciences*, Elsevier Science, vol. 30, pp. 429–443.

Cobb, K. M., Charles, C. D. and Hunter, D. E. (2001) A central tropical Pacific coral demonstrates Pacific, Indian, and Atlantic decadal climate connections. *Geographical Research Letters*, vol. 28, no. 11, pp. 2209–2212.

CRED (2014) World disasters report 2014 – From risk to resilience. www.ifrc.org/en/publications-and-reports/world-disasters-report/world-disasters-report-2014/

Cunha, L. K., Krajewski, W. F., Mantilla, R. and Cunha, L. (2011) A framework for flood risk assessment under nonstationary conditions or in the absence of historical data. *J. Flood Risk Management*, vol. 4, pp. 3–22.

De Sherbinin, A., Schiller, A. and Pulsipher, A. (2007) The vulnerability of global cities to climate hazards. *Environment & Urbanization Copyright*, International Institute for Environment and Development (IIED), vol, 19, no. 1, pp. 39–64.

D'Ippoliti, D., Michelozzi, P., Marino, C., de'Donato, F., Menne, B., Katsouyanni, K., Kirchmayer, U., Analitis, A., Medina-Ramón, M., Paldy, A., Atkinson, R., Kovats, S., Bisanti, L., Schneider, A., Lefranc, A., Iñiguez, C. and Perucci, C. A. (2010) The impact of heat waves on mortality in 9 European cities: results from the Euro HEAT project. *Environmental Health*, vol. 9, no. 37, 9 pp.

Easterling, D. R., Meehl, G. A., Parmesan, C., Changnon, S. A., Karl, T. R. and Mearns, L. O. (2012) Climate extremes: Observations, modeling, and impacts. *Science*, vol. 289, p. 2068.

EPA (2013) Floods and stability: The influence of major flooding on river stability/sediment. http://water.epa.gov/scitech/datait/tools/warsss/examples.cfm, accessed in October, 2013.

Faccini, F., Crispini, L., Federico L., Robbiano, A. and Roccati, A. (2015) New interpretation of Lemeglio Coastal Landslide (Liguria, Italy) based on field survey and integrated monitoring activities. *Engineering Geology for Society and Territory*, vol. 2, pp. 227–231.

Gherardi, D. F. M., Paes, E. T., Soares, H. C., Pezzi, L. P. and Kayano, M. T. (2010) Differences between spatial patterns of climate variability and large marine ecosystems in the western South Atlantic. *Pan-American Journal of Aquatic Sciences*, vol. 5, no. 2, pp. 310–319.

Goff, J. R. and McFadgen, B. G. (2003) Large earthquakes and the abandonment of prehistoric coastal settlements in 15th century New Zealand. *Geoarchaeology: An International Journal*, vol. 18, no. 6, pp. 609–623.

Goodess, C. M. (2013) How is the frequency, location and severity of extreme events likely to change up to 2060? *Environmental Science and Policy*, vol. 27, pp. s4–s14.

Grech, A., Miller, C. K., Erftemeijer, P., Fonseca, M., McKenzie, L., Rasheed, M., Taylor, H. and Coles, R. (2012) A comparison of threats, vulnerabilities and management approaches in global seagrass bioregions. *Environ. Res.*, vol. 7, no. 2, 8pp.

Guzzetti, F. (2000) Landslide fatalities and evaluation of landslide risk in Italy. *Engineering Geology, Elsevier Science*, vol. 58, pp. 89–107.

Extreme events and urban areas

Guzzetti, F., Esposito, E., Balducci, V., Porfido, S., Cardinali, M., Crescenzo, V., Fiorucci, F., Sacchi, M., Ardizzone, F., Mordini, A., Reichenbach, P. and Rossi, M. (2009) Central Italy Seismic Sequences-Induced Landsliding: 1997–1998 Umbria-Marche and 2008–2009 L'Aquila Cases. The Next Generation of Research on Earthquake-induced Landslides: An International Conference in Commemoration of 10th Anniversary of the Chi-Chi Earthquake, Proceedings Conference, Published by the National Central University, September National Central University, Taiwan. Number 300.

Hervás, J. and Bobrowsky, P. (2009) Mapping: Inventories, susceptibility, hazard and risk. In: *Landslides – Disaster Risk Reduction*. Ed. Sassa, K. and Canuti, P. Springer-Verlag, Berlin Heidelberg, pp. 321–349

Horton, R. and Rosenzweig, C. (2010) Climate risk information. In: *New York City Panel on Climate Change. 2010. Climate Change Adaptation in New York City: Building a Risk Management Response*. Ed. Rosenzweig, C. and Solecki, W. Prepared for use by the New York City Climate Change Adaptation Task Force. Annals of the New York Academy of Science 2010, New York, NY, USA, pp. 145–226.

Hunt, A. and Watkiss, P. (2011) Climate change impacts and adaptation in cities: A review of the literature. *Climatic Change*, vol. 104, no. 1, pp. 13–49.

INPE Instituto Nacional de Pesquisas Espaciais (2015) Fenômeno El Niño com Fraca Intensidade Contínua em Desenvolvimento no Pacífico Equatorial – (In Portuguese) http://enos.cptec.inpe.br/

IPCC (2012) Managing the Risks of Extreme Events and Disasters to Advance Climate Change Adaptation. Special Report of Intergovernmental Panel on Climate Change. Ed. Field, C., Barros, V., Stoker, T. F. and Dahe, Q. Cambridge University Press, Cambridge, New York, Melbourne, Madrid, Cape Town, Singapore, São Paulo, Delhi, Tokyo, Mexico City.

IPCC (2013) Climate Change 2013: The Physical Science Basis. Working Group I Contribution to the IPCC 5th Assessment Report – Changes to the Underlying Scientific/Technical Assessment (IPCC-XXVI/Doc.4). [This document lists the changes necessary to ensure consistency between the full Report and the Summary for Policymakers, which was approved line-by-line by 12th Working Group I Session and accepted by the Panel at its 36th Session.]

IPCC (2014) AR5 The Fifth Report Climate Change: Impact, Adaptation, and Vulnerability. Part A: Global and Sectorial Aspects. Contribution of Working Group II to the Fifth Assessment Report of the Intergovernmental Panel on Climate Change. Ed. Field, C. B., Barros, V. R., Dokken, D. J., Mach, K. J., Mastrandrea, M. D., Bilir, T. E., Chatterjee, M., Ebi, K. L., Estrada, Y. O., Genova, R. C., Girma, B., Kissel, E. S., Levy, A. N., MacCracken, S., Mastrandrea, P. R. and White, L. L. Cambridge University Press, Cambridge and New York, pp. 1–32.

Karl, T. R., Melillo, J. M. and Peterson, T. C. (2009) *Global Climate Change Impacts in the United States*. Exit EPA Disclaimer. Thomas R. United States Global Change Research Program. Cambridge University Press, New York.

Kennedy, V. S., Twilley, R. R., Kleypas, J. A., Cowan Jr., J. H. and Hare, S. R. (2002) Coastal and Marine Ecosystems and Global Climate Change: Potential Effects on U.S. Resources. Report Prepared for the Pew Center on Global Climate Change. August. 52 pp.

Kreimer, A., Arnold, M. and Carlin, A. (2003) Building Safer Cities: The Future of Disaster Risk. The World Bank Disaster Management Facility. Disaster Risk Management Series N. 3/27211.

Knutson, T. R., McBride, J. L., Chan, J., Emanuel, K., Holland, G., Landsea, C., Held, I., Kossin, J. P., Srivastava, A. K. and Sugi, M. (2010) Tropical cyclones and climate change. *Nature Geoscience*, vol. 3, pp. 157–163.

Lane, K., Charles-Guzman, K., Wheeler, K., Graber, Z. A. N. and Matte, T. (2013) Health effects of coastal storms and flooding in urban areas: A review and vulnerability assessment. *Journal of Environmental and Public Health*. Review Article Volume 2013, Article ID 913064, 13 pp.

Lavers, D. A., Allan, R. P., Wood, E. F., Villarini, G., Brayshaw, D. J. and Wade, A. J. (2011) Winter floods in Britain are connected to atmospheric rivers. *Geophysical Research Letter*, vol. 38, L-23803.

Leichenko R. M. and Thomas A. (2012) Coastal cities and regions in a changing climate: Economic impacts, risks and vulnerabilities. *Geography Compass*, vol. 6, no. 6, pp. 327–339.

Letema, S. C. (2012) Assessing sanitary mixtures in East African Cities. *Environment Police Series*, vol. 6. Wageningen Academic, The Netherlands.

McGranahan, G., Balk, D. and Anderson, B. (2007) The rising tide: Assessing the risks of climate change and human settlements in low elevation coastal zones. *Environment & Urbanization*, International Institute for Environment and Development (IIED), vol. 19, no. 1, pp. 17–37.

McGranahan, G., Balk, D. and Anderson, B. (2008) Risk of climate change for urban settlements in low elevation coastal zones. In: *The New Global Frontier: Urbanization, Poverty and Environment in the 21st Century.* Ed. Martine, G., McGranahan, G., Montgomery, M. and Castilla, R. F. UNFPA and IIED. Earthscan in UK and USA.

Malamud, B. D., Turcotte, D. L., Guzzetti, F. and Reichnbach, P. (2004) Landslide, earthquakes, and erosion. *Earth and Planetary Science Letter,* Elsevier Science, vol. 229, pp. 45–59.

Marengo, J. A., Jones, R., Alves, L. M. and Valverde, M. C. (2009) Future change of temperature and precipitation extremes in South America as derived from the PRECIS regional climate modeling system. *Int. J. Climatol,* vol. 29, no. 15, pp. 2241–2255.

Marengo, J. A., Valverde, M. C. and Obregon G. O. (2013) Observed and projected changes in rainfall extremes in the Metropolitan Area of São Paulo, *Clim. Res.,* vol. 57, pp. 61–72.

Martelloni, G., Segoni, S., Fanti, R. and Catani, F. (2011) Rainfall thresholds for the forecasting of landslide occurrence at regional scale, Landslides – Original Paper. [This article is published with open access at Springerlink.com, April, 11pp. http://link.springer.com/article/10.1007/s10346-011-0308-2.]

Michener, W. K., Blood, E. R., Bildstein, K. L., Brinson, M. M. and Gardner L. R. (1997) Climate change hurricanes and tropical storms, and rising sea level in coastal wetlands. *Ecological Applications,* vol. 7, pp. 770–801.

Muehe, D. (2011) Erosão costeira – tendência ou eventos extremos? O litoral entre Rio de Janeiro e Cabo Frio, Brasil. *Revista de Gestão Costeira Integrada,* vol. 11, pp. 315–325. (In Portuguese.)

Muehe, D. and Klumb-Oliveira, L. (2014) Deslocamento da linha de costa versus mobilidade praial. *Quaternary and Environmental Geosciences,* vol. 5, pp. 121–124. (In Portuguese.)

Myers, C. A. (2007) Population change and social vulnerability in the wake of disaster: The case of hurricanes Katrina and Rita. Thesis – Faculty of the Louisiana State University and Agricultural and Mechanical College. Louisiana Tech University, USA.

National Drought Mitigation Center (2015) Guidebook for 2015 Urban Water Management Plans. California Department of Water Resources (DWR). http://drought.unl.edu/Planning/Planning InfoByState/StatePlanning.aspx?st=ca

Nobre, C. A., Young, A. F., Saldiva, P. H., Marengo, J. A., Nobre, A. D., Ogura, A., Thomaz, O., Parraga, G. O. O., Silva, G. C. M., da Valverde, M., Silva, G., Silveira, A. C. and Rodrigues, G. de O. (2011) Vulnerabilidade das Megacidades Brasileiras às Mudanças Climáticas: Região Metropolitana de São Paulo. Ed. Carlos A. Nobre e Andrea F. Young. Relatório Final (Final Report). INPE, UNICAMP. USP, IPT. São Paulo, Brasil, 178 pp.

Pielke Sr., R. A., Marland, G., Betts, R. A., Chase, T. N., Eastman, J. L., Niles, J. O., Niyogi, D. D. S. and Running, S. W. (2002) The influence of land-use change and landscape dynamics on the climate system: Relevance to climate-change policy beyond the radiative effect of greenhouse gases. *Philosophical Transactions of the Royal Society A,* vol. 360, no. 1797, pp. 1705–1719.

Poulter, B. and Halpin, P. N. (2008) Raster modeling of coastal flooding from sea-level rise. *International Journal of Geographical Information Science,* vol. 22, no. 2, pp. 167–182.

Reichenbach, P. and Gunther, A. (2010) Methods and strategies to evaluate landslide hazard and risk, *Natural Hazards and Earth System Sciences,* vol. 10, pp. 2197–2198.

Romero-Lankao, P. and Dodman, D. (2011) Cities in transition: Transforming urban centers from hotbeds of GHG emissions and vulnerability to seedbeds of sustainability and resilience. *Current Opinion in Environmental Sustainability,* vol. 3, pp. 113–120.

Rosenzweig, C., Solecki, W. and DeGaetano, A. (2011a) Responding to Climate Change in New York State: The ClimAID Integrated Assessment for Effective Climate Change Adaptation in New York State, Final Report. NYSERDA-Report 11-18.

Rosenzweig, C., Solecki, W., Hammer, S. A. and Mehrotra, S. (2011b) *Climate change and cities: First assessment report of the urban climate change research network.* Cambridge University Press, New York.

Shepherd, J. M., Pierce, H. and Negri, A. J. (2002) Rainfall modification by major urban areas: Observations from spaceborne rain radar on the TRMM satellite. *Journal of Applied Methodology,* vol. 41, pp. 689–771.

Sherman, K., Belkin, I. M., Friedland, K. D., O'Reilly, J. and Hyde, K. (2009) Accelerated warming and emergent trends in fisheries biomass yields of the world's large marine ecosystems. *Ambio,* vol. 4, pp. 215–224.

SOMAR (2015) Central de Notícias SOMAR Meteorologia. in Portuguese http://somarmeteorologia. com.br/site/setores/energia

Extreme events and urban areas

Stauffer, C. (2015) Brazilians hoard water, prepare for possible drastic rationing. http://uk.reuters.com/article/2015/02/11/us-brazil-drought

State of California Seismic Safety Commission. (2005) The Tsunami Threat to California. Report. Executive Summary, California, USA, 16pp.

Takeda, M. (2011) Mental health care and East Japan Great Earthquake. *Psychiatry and Clinical Neurosciences*, vol. 65, pp. 207–212.

Tebaldi, C., Hayhoe, K., Arblaster, J. M. and Meehl, G. A. (2006) Going to the extremes an intercomparison of model-simulated historical and future changes in extreme events. *Climatic Change*, vol. 79, pp. 185–211.

Telford, J. and Cosgrave, J. (2007) The international humanitarian system and the 2004 Indian Ocean earthquake and tsunamis. *Disasters*, Overseas Development Institute. Published by Blackwell Publishing, Oxford, UK and Malden, US, vol. 31, no. 1, pp. 1–28.

Tucci, C. E. M. and Mendes, C. A. (2006) Avaliação Ambiental Integrada de Bacia Hidrográfica. Esplanada dos Ministérios, Report. Secretaria de Qualidade Ambiental – Rhama Consultoria Ambiental. In Portuguese. 302p.

UN-DESA (2014) Revision of the world urbanization prospects. Report-Major 2014 Publications. www.un.org/en/development/desa/publications/2014.html

UN-Habitat (2008) State of the world's Cities 2008/2009 – Harmonious cities. The State of the World's Cities Report. United Nations.

UN-Habitat (2009) Report of the international tripartite conference on urban challenges and poverty reduction in African, Caribbean and Pacific Countries. Nairobi (Africa), pp. 8–10.

UNICEF United Nations Children's Fund and the WHO World Health Organization (2012) Progress on drinking water and sanitation: 2012 Update and permission to reproduce or translate their publications – whether for sale or for non-commercial distribution. Applications and enquiries should be addressed to UNICEF, Division of Communication, New York. www.who.int/about/licensing/copyright_form/en/index.html

UNISDR (2014) The International Strategy for Disaster Reduction (ISDR) – Strategic framework adopted by United Nations Member States in 2000. [The guidebook – written by Brigitte Leoni from UNISDR, in collaboration with Tim Radford, a former journalist with *The Guardian*, Mark Schulman, a UNISDR consultant, and with support from a number of international journalists from Thomson Reuter's AlertNet, the BBC, Vietnam TV and Tempo in Jakarta.]

United Nations (2014) Atlas of mortality and economic losses from weather climate and water extremes (1970–2012). Publications Board World Meteorological Organization (WMO). WMO publications, 48 pp.

U.S. Geological Survey (2015) Magnitude 7.8 earthquake in Nepal and aftershocks. The USGS is providing up-to-date information. Science Features. U.S. Department of the Interior U.S. Geological Survey. www.usgs.gov; www.usgs.gov/blogs/features/usgs_top_story/magnitude-7-8-earthquake-in-nepal/

Villarini, G., Serinaldi, F., Smith, A. J. and Krajewski, W. F. (2009) On the stationary of annual flood peaks in the continental United States during the 20th century. *Water Resources Research*, vol. 45, no. 8.

Wang, W. and Zheng, G. (2013) *China's climate change policies*. Deputy Editors, Pan, J., Luo, Y., Chen, Y. and Chen, H. Earthscan, Abingdon, Oxon, UK and New York, USA, 280pp.

Webster, P. J., Holland G. J., Curry J. A. and Chang H. R. (2005) Changes in tropical cyclone number, duration, and intensity in a warming environment. *Science*, vol. 309, no. 5742, pp. 1844–1846.

Werner, A. D. and Simmons, C. T. (2009) Impact of sea-level rise on sea water intrusion in coastal aquifers. *Ground Water*, vol. 47, no. 2, pp. 197–204.

WHO (2014) World Health Organization. Report: World Health Statistics 2014. www.who.int, 180 pp.

Wolanski, E. (2006) *The environment in Asia Pacific Harbors*. Springer, the Netherlands, 494 pp.

Ye-Shuang, X., Lei, M., Yan-Jun, D. and Shui-Long, S. (2012) Analysis of urbanisation-induced land subsidence in Shanghai. *Natural Hazards*, vol. 63, no. 2, pp. 1255–1267.

Young, A. F. (2013) Urban expansion and environmental risk in the São Paulo Metropolitan Area. *Climate Research*, vol. 57, pp. 73–80. Inter-Research, www.int-res.com

Young, A. F. (2014) Mapping of areas vulnerable to leptospirosis based on the analysis of floods in the São Paulo metropolitan area. Flooding: Risk factors, environmental impacts and management strategies. Natural disaster research, prediction and mitigation. Nova Science Publishers, New York, pp. 35–46.

Yuming, G., Yuping, J., Xiaochuan, P., Liqun, L. and Erich Wichmannb, H. (2009) The association between fine particulate air pollution and hospital emergency room visits for cardiovascular diseases in Beijing. *China Science of The Total Environment*, vol. 407, no. 17, pp. 4826–4830.

Yuzva, K. (2012) Introduction: urban risk and assessing vulnerability at the local level. In: *Resilient Cities 2. Cities and Adaptation to Climate Change. Proceedings of the Global Forum 2011.* Ed. Otto-Zimmermann, K. Springer, Bonn, Germany, pp. 11–14.

Zwiers, F. W., Hegerl, G. C., Min, S. -K. and Zhang, X. (2012) Historical Context. In: *Explaining Extreme Events of 2011 from a Climate Perspective.* Ed. Peterson, T. C., Stott, P. A. and Herring, S. *Bull, Amer. Meteorol. Soc.*, vol. 93, pp. 1043–1047.

17
WATER SUPPLY AND URBAN WATER AVAILABILITY

Stephan Pfister, Stefan Schultze and Stefanie Hellweg

This chapter unfolds the challenges of urbanization for water management and the resulting environmental pressures related to water use. The first section offers an analysis of options for urban water management with a special focus on emerging and developing economies, where most of future urbanization will take place. Second, the water supply and associated problems of major US cities are presented as examples of water supply schemes established in a developed economy and to illustrate the related environmental issues. These US city water supply examples present great opportunities for learning and avoiding mistakes in other urban areas. The third section provides an examination of population dynamics under the influence of climate change in the coming decades in order to predict the location of new urban areas and the related water consumption globally. Finally, the chapter combines this assessment with a water scarcity analysis to identify hotspots of urban water problems and to discuss consequences for agriculture and rural areas.

Urban water management

Water supply is an integral component of life and part of the socio-economic system in urban areas. With fast-paced urbanization occurring in many regions of emerging and developing countries, urban water management presents a major challenge for the future. There is still a large share of the human population suffering from a lack of access to safe water or sanitation (UN, 2013a), and many are urban dwellers. In the past, large cities relied on large-scale centralized infrastructures to ensure water supply, often at high economic costs. Currently, different management methods exist that focus on decentralized or centralized infrastructure to facilitate water supply and wastewater treatment; however, there is no single global best solution (Marlow et al., 2013; Rauch and Morgenroth, 2013). Traditional concepts need to evolve beyond pure cost estimates in order to account for future societal needs associated with water scarcity and with concerns about resource efficiency and other environmental problems (Hering et al., 2013). Decentralized systems have great potential to meet future needs, as they are more robust and adaptable to changing demand, if properly planned and managed (Sitzenfrei et al., 2013). For a more detailed account of urban water demand and associated socio-ecological issues in urban areas refer to Gober and Quay (Chapter 7 in this volume).

Beyond water availability issues, quality must also be considered (Murphy, Chapter 18 in this volume). In some places water is contaminated with naturally occurring arsenic or fluoride, while the most significant water quality-related issue is fecal contamination due to lack of proper sanitation (Johnston et al., 2011). Contamination of water resources is especially significant in highly populated regions and in developing countries with limited economic capacities for wastewater management. In low-income cities (e.g. Dakar, Senegal), sewer systems are often too expensive and alternative fecal sludge management such as installation of septic tanks and transport of sludge for treatment by trucks are required to provide affordable, yet effective, solutions (Dodane et al., 2012). The use of treated wastewater for the purpose of increasing water availability in rivers (Halaburka et al., 2013) or for irrigation can help improve the urban water cycle and reduce impacts on the socio-ecological environment. Such alternative options are ones with which engineers and politicians are often unfamiliar (Luthy, 2013).

Climate change and population dynamics

While the lack of proper management or insufficient economic means are often responsible for urban system failures, these problems become even more critical in regions with low water availability and high population density. Climate change is a key driver of water scarcity in some urban areas, particularly in the case of Melbourne (Grant et al., 2013), where adaptations to water management practices are required. However, in most cases, population change will be the main driver for water scarcity (McDonald et al., 2011; Pfister et al., 2011b). This is especially true in urban areas of developing countries, where the majority of future population growth is expected to occur (WHO, 2013). There is a clear tendency for cities in low-income countries with higher population growth rates to withdraw water resources from nearby areas (McDonald et al., 2014). Financial and infrastructural restrictions in such fast-growing urban areas can aggravate problems of water scarcity. In fact, in many locations, water supply is more of an infrastructural issue than a physical scarcity issue (Molden, 2007). Increased income and consequent changes in household structure can lead to infrastructural improvements, but they can, at the same time, exacerbate water scarcity because of the resulting higher demand.

McDonald et al. (2011) model future water demand and water scarcity by evaluating local and regional water availability per capita. Using a moderate urban water demand of 100 liters per capita per day, they analyze renewable water availability within a radius of 100 km of the urban area. Their results reveal that by 2050, globally, one billion urban dwellers will face perennial water shortage if water is sourced from within the urban extent, and 145 million if the water is sourced within a 100-km radius. Furthermore, approximately 3 and 1.3 billion urban dwellers will be affected by seasonal water shortages if water is sourced within the urban extent or a 100-km radius, respectively (McDonald et al., 2011). In regions with seasonal water scarcity, water storage facilities can aid in its management. In the case of perennial water scarcity, water may be transported over long distances from locations with more abundant water resources; it can be obtained from desalination or by depleting groundwater reservoirs. However, such solutions are accompanied by environmental impacts, which should be accounted for in management plans.

Water scarcity and water footprinting

Historically, water scarcity has been considered an engineering problem. When included in the context of sustainability analyses, however, additional integral aspects must be considered

Water supply and urban water availability

such as the protection of resources and ecosystems as well as the needs of other water users. Initial approaches to characterize global water scarcity have defined water scarcity or stress as an exceedance of a fixed ratio of water use, e.g. 40 percent of the annual renewable water resources (Alcamo et al., 2003), or by considering environmental water requirements (EWR) such as the necessary volume of water that must remain in the river to sustain freshwater ecosystems (Smakhtin et al., 2004). Both approaches establish that extreme water scarcity exists even before the total amount of available water is used or is unavailable for other uses, leading to situations which can lead to spatial and temporal water distribution problems, difficulties in ecological functions and increased human competition. Another approach involves analyzing the absolute water availability per capita (including rain water) with respect to needs for agricultural production (e.g. Falkenmark and Rockstrom, 2004) and establishes that water stress would exist in all cases where subsistence farming is not possible. In light of regional and global food markets, however, the indicator resulting from such an approach might be too narrow in scope and, hence, not the most appropriate for urban water needs.

Recently, specific factors that account for regional water scarcity have been developed in the domain of water footprinting, which requires proper reporting of a water footprint based on the new ISO 14046 standard (ISO, 2014) (see Box 17.1). Kounina et al. (2013) recently published a review of different water scarcity indicators. There is no single best approach to characterize water scarcity on a global level, but the one most widely applied is the water stress index (WSI), ranging from 0.01 to 1 (Pfister et al., 2009). The WSI is based on withdrawal-to-availability data at the watershed level (Alcamo et al., 2003) and accounts for storage capacity as well as monthly and annual variability of water availability. More recent works include water quality aspects (Boulay et al., 2011) and a monthly assessment of the WSI (Pfister and Bayer, 2013), both of which are intended for broad application in the future.

Water footprint analysis has also been promoted by a non-governmental organization called Water Footprint Network, with a slightly different definition (Hoekstra et al., 2011) compared to the ISO 14046, which has led to much scientific debate (Pfister and

---

Box 17.1 ISO 14046: The international water footprint standard

The new international standard on water footprinting (ISO, 2014) provides the principles for properly analyzing and reporting the overall impacts related to water use of products or services. It involves a life cycle perspective (i.e. including the whole value chain) and addresses impacts on water resources from water consumption and pollution (i.e. water degradation). A comprehensive water footprint assessment includes all aspects affecting water resources, while a water footprint focusing only on water consumption and related scarcity effects is referred to as a 'water scarcity footprint'. The ISO standard also acknowledges other types of non-comprehensive water footprints such as the 'water eutrophication footprint' (focusing on eutrophying emissions only), but ISO's main applications will more likely focus on either a comprehensive water footprint or water scarcity footprint assessment. A water footprint can either constitute a stand-alone report or be presented as part of a wider environmental assessment and consists of an inventory (where all water consumption and pollution flows of the product system are reported in physical units) and an impact assessment analysis (where physical flows are characterized in terms of their potential environmental impacts).

Hellweg, 2009; Boulay et al., 2013; Pfister and Ridoutt, 2013). The principal components of accounting for local water use and availability are, however, common to both approaches.

In general, a water footprint assesses the environmental impact of water consumption over the life cycle of a product or service (including the supply chain and disposal). Water consumption (consumptive use) includes freshwater that is evaporated, incorporated in products and waste, transferred into different watersheds or discharged into the sea after use (Falkenmark and Rockstrom, 2004). Water degradation (pollution) is also included in order to analyze the environmental impacts of wastewater treatment systems including those on downstream water supply (Ridoutt and Pfister, 2013).

Water footprinting methods have focused primarily on the agricultural sector, which is responsible for 85 percent of all human water consumption (Shiklomanov, 2003). Another area of focus has been energy production, which is an indispensable component of the production of consumer goods and is responsible for a significant portion of overall water consumption, especially when it comes to hydropower (Mekonnen and Hoekstra, 2011; Pfister et al., 2011a). Direct water consumption in urban areas is considerably low compared to agricultural uses. Global average agricultural water consumption amounts to ~500–1,000 liters per day/per capita, depending on how it is estimated (Rost et al., 2008; Liu et al., 2009; Pfister et al., 2011b). Direct water use in urban areas ranges from 100 liters per day/per capita (in emerging economies) to 600 liters per day/per capita in modern cities in industrialized countries (Shiklomanov, 1999). The consumptive share is between 5 percent and 60 percent and typically higher if less water is used (since excess water, e.g. used for showering or flushing toilets, is released back to the watershed after treatment with minor losses). This results in a per capita water consumption rate of 30–60 liters per day. At this rate of water consumption and in combination with high population density, local water scarcity or even depletion is expected in many major urban areas. Many watersheds of large urban areas are water stressed and must source from distant water supply systems, as is the case for a number of US cities. On a global scale there is no evidence of water scarcity (Rockström et al., 2009), yet water withdrawals must be reduced by an average of 40–50 percent globally, in order to avoid severe localized water stress (Ridoutt and Pfister, 2010). The distribution of water scarcity is a crucial problem. While some regions can accommodate additional water consumption, many regions are already beyond the level of moderate water scarcity. On the one hand, urbanization exacerbates these issues of water supply. On the other hand, water supply itself can be a limiting factor for urban growth in some areas (see Figure 17.1).

Examples of current urban water systems and water scarcity

As outlined above, there is no standard water supply system or a single best-practice solution to tackle water scarcity. However, the possibility exists that, without looking for a more appropriate option, expensive and often ill-chosen urban water systems installed in Europe or North America are also introduced in other regions. In order to show the extent and problems of existing water supply systems of major urban areas, Staeubli (2012) analyzes the water supply and availability of five US cities with different systems and associated problems (see Table 17.1). The cities are: New York (NY), Los Angeles (CA), San Francisco (CA), Dallas (TX) and Miami (FL). New York and Los Angeles are the largest US cities, selected for their high population densities and histories. Dallas and Miami are among the top ten metropolitan areas and represent smaller cities/urban agglomerations. Miami is interesting due to its complete dependence on groundwater sources, while Dallas is one of the largest metropolitan

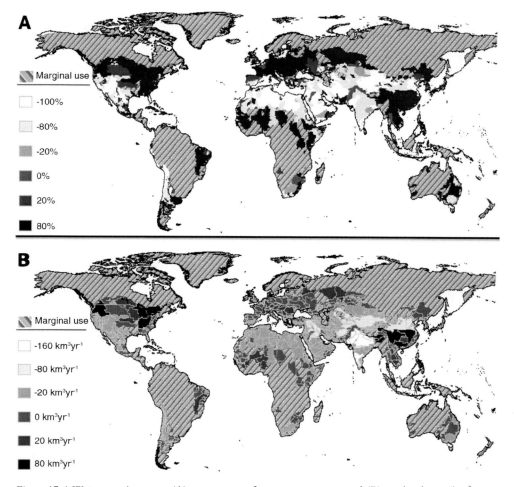

Figure 17.1 Water scarcity maps: (A) percentage of current water use and (B) total volume (km³ per year) that must be reduced in each watershed to meet a stabilization target of moderate water stress: WSI < 0.5

Source: Adapted from Ridoutt and Pfister (2010).
Note: Negative values indicate that the water use needs to be reduced. Watersheds with withdrawal-to-availability ratios <5 percent are hatched (water resources are currently developed only to a limited extent). All white and light grey shaded watersheds face severe water scarcity.

areas in the US without access to a major surface water body. San Francisco's geographical setting and current water supply system are similar to those of Los Angeles, but its future urban water management plans are different.

Water transfers

The lower water scarcity of water supply catchments does not mean that the Los Angeles supply is more sustainable (see Table 17.2). On the contrary, it has a much higher dependence on infrastructure. Maintenance and operation of a long-distance water supply has large

Stephan Pfister et al.

Table 17.1 Summary of the five metropolitan water supply systems

	New York	Los Angeles	San Francisco	Dallas	Miami
Population	9 Mio	4 Mio	2.5 Mio	2.4 Mio	2.2 Mio
Total water withdrawal (m^3/yr)	1.8E+09	8.2E+08	2.9E+08	6.8E+08	4.7E+08
Per capita water withdrawal (m^3/day/capita)	0.55	0.57	0.32	0.78	0.6
Total wastewater flow (m^3/yr)	1.9E+09	7.0E+08	3.7 E+08	5.5E+08	4.1E+08
Per capita wastewater flow (m^3/day/capita)	0.59	0.48	0.41	0.63	0.51
The water's origin	95% imported, 5% is local groundwater	88% imported, 11% groundwater, 1% recycled water	85% imported, 12% local surface water, 3% groundwater	80% surface water, 10% reuse, 7% groundwater	100% groundwater
Total storage capacity	$2.2{\star}10^9$ m^3 (equals 1.2 years of withdrawals)	Shared reservoirs with other cities	$1.56{\star}10^9$ m^3 (equals 5.4 years of withdrawal)	Shared reservoirs with other cities	$4.3{\star}10^5$ m^3 (equals <1 day of withdrawals)

Source: Adapted from Staeubli (2012).

Table 17.2 Resulting water stress indices based on two different approaches: the WSI of the watershed in which the city is located (no distinction of water origins) and the WSI of the watershed of external reservoirs

	Watershed	New York	Los Angeles	San Francisco	Dallas	Miami
Water supply share	City	5%	11%	15%	100%	100%
	External reservoir	95%	89%	85%	0%	0%
Water stress index (WSI)★	City	0.89	1.00	1.00	0.88	1.00
	External reservoir	0.31	0.77	1	–	–
	Combined (effective WSI)	**0.34**	**0.79**	**1.00**	**0.88**	**1.00**

Note: The effective WSI represents the withdrawal weighted mean of the WSI from the different water sources and is recommended as the best indicator.

★ WSI ranges from 0.01 (lowest water scarcity) to 1 (highest water scarcity level)

economic costs and, therefore, a higher risk of supply failure if the system is not properly maintained. Additionally, the water supplied over a large area has high energy costs and water losses. For instance, Southern California requires ~2.4 kWh per m^3 of water supplied and has a loss rate of ~5 percent (Cohen et al., 2004). This energy demand is close to the requirements of state-of-the-art seawater desalination, which is currently ~3kWh per m^3 (Fritzmann et al., 2007). Nevertheless, hydropower production at water supply storage dams has the potential

Box 17.2   Water stress of current water supply in five major US cities

New York and San Francisco source their water from watersheds some 200–300 km away, a distance exceeded in the case of Los Angeles. Dallas, on the other hand, sources most of its water from groundwater aquifers (~70 km away), while Miami relies on local groundwater sources. An overview of the systems is provided in Table 17.1. Based on a global model (Pfister et al., 2009), all five cities have a WSI >0.88, which means they are at the edge of extreme water scarcity. However, the underlying global hydrological models generally do not account for water transfers and therefore overestimate water scarcity in cities that source water from other watersheds. To tackle this issue, the WSI of the five cities' water supply systems was analyzed based on the origins of their tap water supply. The WSI of the cities' reservoir watersheds were quantified and then combined according to their contribution to the city's water supply. The results (Table 17.2) show a significant decrease of the effective water scarcity for some cities, especially for the case of New York. Interestingly, Miami's water supply system has a higher WSI than Los Angeles, although it has ~3 times higher precipitation. It is evident from these results that long-distance water transfers can alleviate water scarcity, but at the expense of high transport costs.

---

to reduce the net energy consumption allocated to transport to ~0.3 kWh per m^3 (Cohen et al., 2004).

Groundwater, wastewater, rainwater and desalination

Local groundwater overuse, as in the case of Miami, can lead to severe problems in the long-term, such as saltwater intrusion and sinkholes. In this case, water reuse or desalination of seawater might be a more sustainable solution. In Miami, there is potential for ~400 million m^3 per year of wastewater to be purified and used for water supply. An additional 0.8 kWh is needed to clean 1 m^3 of sewage water to quality suitable for tap water reuse (Seah et al., 2008), which is four times the energy use for standard wastewater treatment (ecoinvent Centre, 2008).

For Miami, this additional energy would lead to an energy demand of <1 percent of total electricity, an acceptable price to pay for a vital resource. However, the societal acceptance of wastewater reuse for drinking water purposes needs to be examined beforehand. Singapore began implementing direct reuse for tap water in 2003, successfully branding it as 'NEWater' (PUB, 2013). Another potential option is the use of rainwater harvesting techniques. In Miami, currently ~30 percent of the rainwater goes into the sewer system. Rainwater harvesting could supply up to 60 percent of precipitation water (UN-HABITAT, 2005), which would contribute almost 20 percent to Miami's water supply.

Desalination is much more expensive, but represents a solution for coastal cities with appropriate economic means, as shown in the case of Melbourne (Grant et al., 2013). However, desalination plants are planned for neither Los Angeles nor San Francisco, despite the fact that the water system infrastructure is already quite expensive and there is a persistent competition with agricultural and environmental uses (State of California and the Resources Agency, 2006). For the case of New York, the protection of the reservoir has reduced economic costs and is deemed a success story (Postel et al., 1996). The situation is much less tense in New York as compared to California, where water is naturally scarce and irrigated agriculture is a big economic sector.

Stephan Pfister et al.

Future supply options

Assuming stable political conditions, it is not expected that any of these cities will face water supply failures due to changes in allocation rights. Additionally, willingness to pay and prices for urban water are typically much higher than for agricultural water uses. Therefore, 'agricultural water' sourced from water rights of agricultural land might also be sold for urban use (Yoo et al., 2012). A focus on water savings practices has led to a large decrease of water use in San Francisco and Dallas and stands for a promising action on the demand side. While Dallas residents reduced their per capita water use by ~20 percent by minimizing water losses and optimizing sprinkler systems (Parks, 2011), Dallas still has the highest per capita water use of the five cities. On the other hand, San Francisco reduced water use by ~50 percent in the last four decades (Ritchie, 2011) and presents a better example for sustainable water use in metropolitan areas. The future plans of Dallas, Los Angeles and Miami focus mainly on conservation and reuse, while San Francisco and New York intend to strengthen their reservoir infrastructure. This demonstrates how slowly change takes place in developed urban water supply systems throughout the US. In developing and emerging economies the situation is different, as illustrated by the case of Chennai, India (Srinivasan et al., 2013). This fast-growing urban area is in a rural–urban transition and is highly dynamic in its use of water resources, therefore demanding implementation of more suitable urban water systems. New urban areas typically use decentralized groundwater wells, which make water supply management even more complex. In developing countries, where the majority of urban growth is expected, an integrated assessment of future population growth and the required water supply is necessary including centralized and decentralized systems (Srinivasan et al., 2013).

Urbanization and effects on water scarcity

As previously discussed, major cities need substantial investments in water supply savings schemes, even in regions of moderate water scarcity. Centralized systems are traditionally relied upon to support population growth and infrastructure development. Large water transfers over long distances have helped to secure urban water supply at a high price and should not be the model for developing and emerging urban economies, which require more sustainable and efficient solutions. Water supply systems need to consider local conditions of water supply and urban development. Engineering solutions therefore need to include urban planning and vice-versa, since the spatial distribution of water users co-determines water supply systems and related environmental impacts (Srinivasan et al., 2013). Another important aspect is drought resilience, which is often low in fast-growing cities and needs new planning strategies (Srinivasan et al., 2013). One way of avoiding water supply disasters is to recognize upcoming water scarcity issues and take counter-action through water savings promotion or through influential economic entities in such regions.

To identify hotspots of future water supply issues on a global level, population densities of the year 2000 were analyzed on a ~10 km spatial resolution (FAO, 2007), coupled with projected future population densities based on national population outlooks (UN, 2013b) and spatially explicit predictions of urbanization in 2030 (Seto et al., 2012). The population data was split into five classes of population density ranges, which were matched with national population numbers to determine the average population density of each class in each country. The highest population density class has an open range and accommodates any excess population that could not be allocated within the upper boundaries of the preceding population density classes. Urban areas in 2030 were determined based on population

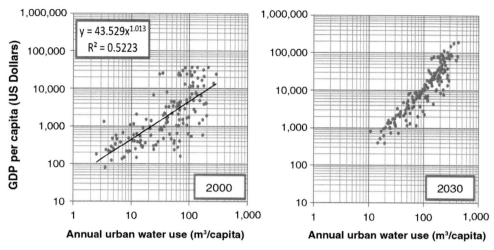

Figure 17.2 Regression analysis of water use and GDP for individual countries (left) and projection of water consumption for 2030 (right)

Source: Adapted from Schultze (2014).

predictions (UN, 2013b; WHO, 2013), assuming a constant population density in urban areas for each country. These additional urban areas in each country were allocated to the spatially explicit prediction of 'probable urban extent' modeled by Seto et al. (2012), starting with the highest probabilities. Further details are provided in Schultze (2014).

After estimating the increased urban population, the increased per capita water consumption rate was used to determine the additional urban water consumption in 2030. Regression relationships between current urban water consumption rates (FAO, 2008) and GDP on a country level were used to predict future water consumption based on future GDP estimates (IMF, 2013; PwC, 2013) as shown in Figure 17.2. The resulting average urban water use in 2030 is ~220 liters per person per day, which is about double what is considered the minimal amount (100 liters per capita per day) for a comfortable urban life (McDonald et al., 2011) and could be reduced with appropriate measures. Combining this data with the urban population densities calculated for each grid cell provides information on the urban water use in 2030 including its distribution (see Figure 17.3). Many new urban areas are predicted in watersheds which are already water scarce: 46 percent of these have a WSI >0.5, which means severe water stress, and the average WSI of the ~400 billion m^3 global urban water use in 2030 is 0.52. Forecasts for the Indian subcontinent, the Middle East and Northeastern China are grave, as their water supply issues will become even more severe. Many regions in Africa, North America and South America will experience a similar level of severe water scarcity.

While urban growth implies additional water consumption, urban growth demands might replace agricultural water use and consumption, resulting in a local net effect of water savings such as predicted in Egypt, Northern India, Pakistan and northeastern China (see Figure 17.3B). In these regions, annual agricultural water consumption can be in the order of 10^6 m^3 per km^2, which is comparable to water use in urban areas with >10,000 persons/km^2—a very high population density that is twice that of Hong Kong. Therefore, predicted urban water use is often lower than that of current irrigation. This effect is also observed for groundwater use in

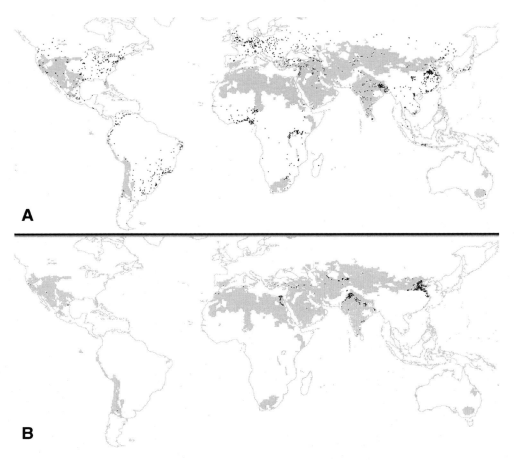

Figure 17.3 Effect of new urban areas in 2030 on water consumption compared to current water consumption by agriculture: (A) new urban areas leading to higher water consumption and (B) new urban areas in 2030 leading to reduced water consumption

Source: Adapted from Schultze (2014).
Note: The grey shaded areas indicate high water scarcity (WSI>0.5).

the case study of Chennai (Srinivasan et al., 2013). Since intense irrigation is concentrated in drier climates, in many cases urban areas in 2030 are expected to have higher water use than the agricultural water consumption in the year 2000. Nevertheless, agricultural activities might be moved to neighboring areas and even increase water pressure within the same region. If agricultural production is kept at current rates in the same watersheds, increased urban water use is expected to lead to additional water consumption and water stress in the respective watersheds and, consequently, to conflicts between urban and rural users.

Feeding the wealthier global population in 2050 is predicted to increase water consumption by an amount in the order of 50–100 percent, increasing water stress in many areas (Pfister et al., 2011c). This is especially relevant for urban areas with local food supply such as Delhi, where the average transport of irrigated crops from producer to consumer is 200 km and there is competition between the closely linked urban and rural water resources. This is in contrast to cities such as Berlin, where the average irrigated crop is produced 4,200 km away (Hoff et al., 2013).

Conclusions

Many new frameworks and methods for urban water management are now available. However, billions of people still lack safe water supply. Infrastructure is a key element for water supply, especially where fast-paced urban development is expected in the next decades. While infrastructure and management are mainly socio-economic issues, physical water scarcity must also be included in the analysis. Many urban dwellers may not immediately suffer from increased scarcity, but the poor tend to be disproportionately burdened.

Urban expansion is expected to put pressure on agricultural water use and on the environment. The areas projected to experience severe urban water management challenges are located in water scarce areas and areas with high agricultural water use. Farmers in these locations who might experience a lack of access to water due to increased prices will suffer from reduced yields and income, thus driving further rural–urban migration with low prospects for employment. Consequently, urban expansion is expected to put pressure on the regional food supply, but will mainly be a problem for low-income countries, since wealthier cities typically source food from far away (Hoff et al., 2013).

Cities in areas of high physical water scarcity have several options to tackle the problem. While resource depletion of groundwater or lakes is problematic from a sustainability point of view, wastewater reuse and efficiency increases in domestic water use represent large improvement potentials in many regions. This is shown in the cases of San Francisco or Melbourne, where education and policies reduced urban water use by half within a decade (Grant et al., 2013). Additionally, coastal areas can enhance water availability through seawater desalination at the relatively high cost of ~1 USD/m^3. Reuse of wastewater for urban water supply is another option, particularly suitable for inland cities with lower energy demand, but it is faced with barriers to public acceptance (Lahnsteiner and Lempert, 2007). Use of rainwater has low energy and infrastructural demand and should therefore be immediately considered as a suitable water supply option.

Traditional solutions such as long-distance water supply can be problematic as they shift water stress, such as in the case of the Colorado River and Los Angeles. However, if from a political and socio-economic perspective no other solutions are possible, it can still mitigate water stress, as shown for New York, Los Angeles and San Francisco. The largest project of this kind is the Chinese South–North water transfer. This project aims at improving water availability for urban and agricultural purposes in northern China and, as such, is purported to have an overall benefit for the environment. While there is an overall benefit regarding water scarcity, environmental impacts will be shifted from the North to the South (Lin et al., 2012). Moreover, these centralized systems of water transfer or desalination/wastewater reuse have relatively high energy demand, which should not only be considered in terms of cost, but also in terms of the associated environmental burden.

The energy use of a water supply system can be significantly large in terms of total energy consumption and can therefore contribute a large share to greenhouse gas emissions where power supply is based on coal, gas or oil. The tradeoff between energy use and water efficiency in industrial applications should also be considered, given that water reuse in industrial processes generally requires additional energy or even treatment processes. Water savings is therefore most efficient when it involves low-tech solutions such as low-flow toilets or showerheads, all of which can be achieved through education (Grant et al., 2013) and potentially through financial incentives or laws.

From a water footprint perspective, the highest water use of urban dwellers is associated with food consumption, which is briefly addressed in the previous sections. However, the increasing global share of urban population will lead to lower numbers of farmers and potentially higher industrialization of agriculture. Productivity increases have been shown to correlate with urbanization (though correlation does not imply causation), so increased production efficiency might be expected. From a global perspective, industrialization of the agricultural market might also lead to cultivation in the agricultural regions most suitable for efficient production. Land grabbing (large-scale acquisition of agricultural land in developing countries by foreign countries or international companies), a controversial issue as shown by reports in Braun and Meinzen-Dick (2013), can have major adverse effects, especially if additional land is used for increasing company affluence where communities are deprived of their local products. Irrespective of urbanization, water stress can be largely controlled, even without increasing deforestation, if food waste is reduced and agriculture is focused on areas suitable for rain-fed agriculture (Pfister et al., 2011c). The overall per capita water consumption and related environmental impacts could be reduced and the additional water consumption and stress could be limited to ~10 percent. It is evident from the above discussion that urban water supply should also embrace rural water supply and adopt a water footprint perspective, which includes managing water consumption in the supply chain of urban consumption of goods, especially food.

Global maps of water availability and water scarcity have high uncertainties (Pfister and Hellweg, 2011) and therefore are intended to be used for global analyses or for comparison purposes, but not for local assessments or planning. In particular, issues related to groundwater use are poorly represented in global datasets. Global analyses might help not only the identification of hotspots from a water footprint perspective regarding water use-related impacts of urban dwellers on the environment, but also the identification of potential water supply failure risks. In this context, a contribution analysis of direct urban water use to the total environmental impact including water consumption in food and power production, can be used to highlight relevant problems in the consumption pattern. While management of the supply chain is very difficult and typically not controlled by the consumer, environmentally conscious product choices and adjustment of diets in one place can have large effects on water stress and water supply security in other places.

Urban water supply challenges should then be addressed on an individual basis for each area, taking into account detailed local data and considering the geographical, political and socio-economic context. As illustrated here, a variety of measures exist to overcome scarcity issues, with different costs, environmental impacts and areas of potential application. Above all, decentralized solutions and wastewater reuse have a high and unlevered potential to contribute towards lowering the water supply risk for an increasing urban population and to reduce environmental impacts.

Key messages

- Urban water supply traditionally depends on heavy infrastructure at high costs.
- Urbanization largely impacts the water cycle including surrounding areas.
- Future challenges are expected in both developing and emerging economies.
- Alternative, region-specific water supply systems are required for cost-effective solutions in new urban areas.

Future research agenda

- Analyze major urban areas regarding current and future water supply risks, taking into account the current structure and future development models.

- Improve hydrological models to assess water availability on high spatial and temporal detail for different water sources present within the region.
- Create coupled models of urban expansion and water supply systems based on hydrological, engineering and spatial planning parameters.
- Analyze the effect of urban expansion on agriculture and related water scarcity problems.

References

Alcamo, J., P. Doll, T. Henrichs, F. Kaspar, B. Lehner, T. Rosch and S. Siebert (2003). "Global Estimates of Water Withdrawals and Availability Under Current and Future 'Business-as-Usual' Conditions." *Hydrological Sciences Journal-Journal Des Sciences Hydrologiques* 48(3): 339–348.

Boulay, A.-M., C. Bulle, J.-B. Bayart, L. Deschênes and M. Margni (2011). "Regional Characterization of Freshwater Use in LCA: Modeling Direct Impacts on Human Health." *Environmental Science & Technology* 45(20): 8948–8957.

Boulay, A.-M., A. Y. Hoekstra and S. Vionnet (2013). "Complementarities of Water-Focused Life Cycle Assessment and Water Footprint Assessment." *Environmental Science & Technology* 47(21): 11926–11927.

Braun, J. v. and R. Meinzen-Dick (2013). "Land Grabbing" by Foreign Investors in Developing Countries: Risks and Opportunities. Washington DC, IFPRI Policy Brief, IFRPI.

Cohen, R., G. Wolff and B. Nelson (2004). Energy Down the Drain: The Hidden Costs of California's Water Supply. Oakland, CA, NRDC.

Dodane, P.-H., M. Mbéguéré, O. Sow and L. Strande (2012). "Capital and Operating Costs of Full-Scale Fecal Sludge Management and Wastewater Treatment Systems in Dakar, Senegal." *Environmental Science & Technology* 46(7): 3705–3711.

ecoinvent Centre (2008, February 20). "ecoinvent data v2.01" from www.ecoinvent.org.

Falkenmark, M. and J. Rockstrom (2004). *Balancing Water for Humans and Nature: The New Approach in Ecohydrology*. London, Earthscan.

FAO (2007). FAO Geonetwork: POPULATION DENSITY (PERSONS/SQKM); Edition 3.6, FAO.

FAO (2008, 22 February). "Aquastat: Main Country Database" from www.fao.org/nr/water/aquastat/data/query/index.html.

Fritzmann, C., J. Lowenberg, T. Wintgens and T. Melin (2007). "State-of-the-Art of Reverse Osmosis Desalination." *Desalination* 216(1–3): 1–76.

Grant, S. B., T. D. Fletcher, D. Feldman, J.-D. Saphores, P. L. M. Cook, M. Stewardson, K. Low, K. Burry and A. J. Hamilton (2013). "Adapting Urban Water Systems to a Changing Climate: Lessons from the Millennium Drought in Southeast Australia." *Environmental Science & Technology* 47(19): 10727–10734.

Halaburka, B. J., J. E. Lawrence, H. N. Bischel, J. Hsiao, M. H. Plumlee, V. H. Resh and R. G. Luthy (2013). "Economic and Ecological Costs and Benefits of Streamflow Augmentation Using Recycled Water in a California Coastal Stream." *Environmental Science & Technology* 47(19): 10735–10743.

Hering, J. G., T. D. Waite, R. G. Luthy, J. E. Drewes and D. L. Sedlak (2013). "A Changing Framework for Urban Water Systems." *Environmental Science & Technology* 47(19): 10721–10726.

Hoekstra, A. Y., A. K. Chapagain, M. M. Aldaya and M. M. Mekonnen (2011). *The Water Footprint Assessment Manual: Setting the Global Standard*. London, Earthscan.

Hoff, H., P. Döll, M. Fader, D. Gerten, S. Hauser and S. Siebert (2013). "Water Footprints of Cities – Indicators for Sustainable Consumption and Production." *Hydrol. Earth Syst. Sci. Discuss.* 10(2): 2601–2639.

IMF (2013). World Economic and Financial Surveys. World Economic Outlook Database, International Monetary Fund.

ISO (2014). ISO 14046 Water Footprint – Principles, Requirements and Guidelines. International Organization for Standardization.

Johnston, R. B., M. Berg, C. A. Johnson, E. Tilley and J. G. Hering (2011). "Water and Sanitation in Developing Countries: Geochemical Aspects of Quality and Treatment." *Elements* 7(3): 163–168.

Kounina, A., M. Margni, J.-B. Bayart, A.-M. Boulay, M. Berger, C. Bulle, R. Frischknecht, A. Koehler, L. Milà i Canals, M. Motoshita, M. Núñez, G. Peters, S. Pfister, B. Ridoutt, R. Zelm, F. Verones and

S. Humbert (2013). "Review of Methods Addressing Freshwater Use in Life Cycle Inventory and Impact Assessment." *The International Journal of Life Cycle Assessment* 18(3): 707–721.

Lahnsteiner, J. and G. Lempert (2007) "Water Management in Windhoek, Namibia." *Water Science & Technology* 55(1–2). doi:10.2166/wst.2007.022

Lin, C., S. Suh and S. Pfister (2012). "Does South-to-North Water Transfer Reduce the Environmental Impact of Water Consumption in China?" *Journal of Industrial Ecology* 16(4): 647–654.

Liu, J. G., A. J. B. Zehnder and H. Yang (2009). "Global Consumptive Water Use for Crop Production: The Importance of Green Water and Virtual Water." *Water Resources Research* 45.

Luthy, R. G. (2013). "Design Options for a More Sustainable Urban Water Environment." *Environmental Science & Technology* 47(19): 10719–10720.

McDonald, R. I., P. Green, D. Balk, B. M. Fekete, C. Revenga, M. Todd and M. Montgomery (2011). "Urban Growth, Climate Change, and Freshwater Availability." *Proceedings of the National Academy of Sciences* 108(15): 6312–6317.

McDonald, R. I., K. Weber, J. Padowski, M. Flörke, C. Schneider, P. A. Green, T. Gleeson, S. Eckman, B. Lehner, D. Balk, T. Boucher, G. Grill and M. Montgomery (2014). "Water on an Urban Planet: Urbanization and the Reach of Urban Water Infrastructure." Global *Environmental Change* 27: 96–105.

Marlow, D. R., M. Moglia, S. Cook and D. J. Beale (2013). "Towards Sustainable Urban Water Management: A Critical Reassessment." *Water Research* 47(20): 7150–7161.

Mekonnen, M. M. and A. Y. Hoekstra (2011). The Water Footprint of Electricity from Hydropower. Value of Water Research Report Series UNESCO-IHE.

Molden, D. (2007). *Water for Food, Water for Life: A Comprehensive Assessment of Water Management in Agriculture*. London, Earthscan and IWMI (International Water Management Institute).

Parks, J. (2011). *Region C Water Planning Group*. Wylie, Texas Water Development Board.

Pfister, S. and S. Hellweg (2009). "The Water 'Shoesize' vs. Footprint of Bioenergy." *Proceedings of the National Academy of Sciences of the United States of America* 106(35): E93–E94.

Pfister, S. and S. Hellweg (2011). Surface Water Use – Human Health Impacts. Report of the LC-IMPACT project (EC: FP7). www.ifu.ethz.ch/ESD/downloads/Uncertainty_water_LCIA.pdf.

Pfister, S. and P. Bayer (2013). "Monthly Water Stress: Spatially and Temporally Explicit Consumptive Water Footprint of Global Crop Production." *Journal of Cleaner Production* 73: 52–62.

Pfister, S. and B. G. Ridoutt (2013). "Water Footprint: Pitfalls on Common Ground." *Environmental Science & Technology* 48(1): 4–4.

Pfister, S., A. Koehler and S. Hellweg (2009). "Assessing the Environmental Impacts of Freshwater Consumption in LCA." *Environmental Science & Technology* 43(11): 4098–4104.

Pfister, S., D. Saner and A. Koehler (2011a). "The Environmental Relevance of Freshwater Consumption in Global Power Production." *The International Journal of Life Cycle Assessment* 16(6): 580–591.

Pfister, S., P. Bayer, A. Koehler and S. Hellweg (2011b). "Environmental Impacts of Water Use in Global Crop Production: Hotspots and Trade-Offs with Land Use." *Environmental Science & Technology* 45(13): 5761–5768.

Pfister, S., P. Bayer, A. Koehler and S. Hellweg (2011c). "Projected Water Consumption in Future Global Agriculture: Scenarios and Related Impacts." *Sci. Total Environ.* 409(20): 4206–4216.

Postel, S. L., G. C. Daily and P. R. Ehrlich (1996). "Human Appropriation of Renewable Fresh Water." *Science* 271(5250): 785–788.

PUB (2013). Our Water, Our Future. PUB.

PwC (2013). World in 2050 – The BRICs and Beyond: Prospects, Challenges and Opportunities. PricewaterhouseCoopers LLP, UK.

Rauch, W. and E. Morgenroth (2013). "Urban Water Management to Increase Sustainability of Cities." *Water Research* 47(20): 7149.

Ridoutt, B. G. and S. Pfister (2010). "Reducing Humanity's Water Footprint." *Environmental Science & Technology* 44(16): 6019–6021.

Ridoutt, B. and S. Pfister (2013). "A New Water Footprint Calculation Method Integrating Consumptive and Degradative Water Use Into a Single Stand-Alone Weighted Indicator." *The International Journal of Life Cycle Assessment* 18(1): 204–207.

Ritchie, S. (2011). 2010 Urban Water Management Plan. San Francisco, San Francisco Public Utilities Commission.

Rockström, J., W. Steffen, K. Noone, A. Persson, F. S. Chapin, E. F. Lambin, T. M. Lenton, M. Scheffer, C. Folke, H. J. Schellnhuber, B. Nykvist, C. A. de Wit, T. Hughes, S. van der Leeuw, H. Rodhe, S.

Sorlin, P. K. Snyder, R. Costanza, U. Svedin, M. Falkenmark, L. Karlberg, R. W. Corell, V. J. Fabry, J. Hansen, B. Walker, D. Liverman, K. Richardson, P. Crutzen and J. A. Foley (2009). "A Safe Operating Space for Humanity." *Nature* 461(7263): 472–475.

Rost, S., D. Gerten, A. Bondeau, W. Lucht, J. Rohwer and S. Schaphoff (2008). "Agricultural Green and Blue Water Consumption and its Influence on the Global Water System." *Water Resources Research* 44(9), 17pp.

Schultze, S. (2014). Future Urban Water Use and Consequences on Water Scarcity (in German). BSc Thesis, February 2014, ETH Zurich, Switzerland.

Seah, H., T. P. Tan, M. L. Chong and J. Leong (2008). "NEWater—Multi Safety Barrier Approach for Indirect Potable Use." *Water Science & Technology: Water Supply* 8(5): 573–588.

Seto, K. C., B. Güneralp and L. R. Hutyra (2012). "Global Forecasts of Urban Expansion to 2030 and Direct Impacts on Biodiversity and Carbon Pools." *Proceedings of the National Academy of Sciences* 109(40): 16083–16088; published ahead of print September 17, 2012, doi:10.1073/pnas.1211658109.

Shiklomanov, A. I. (1999). World Water Resources at the Beginning of the 21st Century International Hydrological Programme. St. Petersburg, State Hydrological Institute (SHI)/UNESCO.

Shiklomanov, I. A. (2003). *World Water Resources at the Beginning of the 21st Century.* Cambridge, Cambridge University Press.

Sitzenfrei, R., M. Möderl and W. Rauch (2013). "Assessing the Impact of Transitions From Centralised to Decentralised Water Solutions on Existing Infrastructures – Integrated City-Scale Analysis with VIBe." *Water Research* 47(20): 7251–7263.

Smakhtin, V., C. Revenga and P. Doll (2004). "A Pilot Global Assessment of Environmental Water Requirements and Scarcity." *Water International* 29(3): 307–317.

Srinivasan, V., K. C. Seto, R. Emerson and S. M. Gorelick (2013). "The Impact of Urbanization on Water Vulnerability: A Coupled Human–Environment System Approach for Chennai, India." *Global Environmental Change* 23(1): 229–239.

Staeubli, A. (2012). Improving Water Footprint Approaches for City Level Water Stress Assessment. MSc, ETH Zurich.

State of California and the Resources Agency (2006). Hetch Hetchy Restoration Study.

UN (2013a). The Millennium Development Goals Report 2013. New York, United Nations.

UN (2013b). World Population Prospects: The 2012 Revision; Medium Fertility. P. D. United Nations, Department of Economic and Social Affairs.

UN-HABITAT (2005). Blue Drop Series on Rainwater Harvesting and Utilisation – Book 2: Beneficiairies and Capacity Builders. Nairobi, Kenya UN-HABITAT, Water, Sanitation and Infrastructure Branch.

WHO (2013). Global Health Observatory (GHO) – Urban Population Growth. WHO.

Yoo, J., C. Perrings, A. Kinzig, J. Abbot, S. Simonit, J. P. Connors and P. J. Maliszewski (2012). The Value of Water Rights in Agricultural Properties in the Phoenix Active Management Area. Agricultural and Applied Economics Association's 2012 AAEA Annual Meeting. Seattle, Washington.

18

URBAN WATER QUALITY

Conor Murphy

Urbanization has substantial impacts on catchment hydrology, fundamentally altering both the hydraulic regime and quality of water resources. Indeed, many authors have identified the 'urban stream syndrome' in recognition of the degrading influence of urbanization on receiving waters (Booth and Jackson, 1997; Walsh et al., 2005; Meyer et al., 2005). Walsh et al. (2005) associate the urban stream syndrome with increased hydrological flashiness (faster storm response), consistently observed losses in sensitive aquatic species and degradation of water quality through increases in nutrients and contaminants and a loss of organic matter. Such changes in the physical, chemical and biotic conditions of urban streams impact the ecosystem functions upon which societies depend (Palmer et al., 2004; Meyer et al., 2005).

Coupled with the pressures associated with the growth of urban areas, climate change and population growth are likely to interact in complex ways with changes in land use to exacerbate pressures on water quality. Trends in water quality are often not a simple function of population growth (Duh et al., 2008), with variable trends in different water quality indicators experienced in specific cities. Most problematic in terms of population growth is rapid urbanization, where large increases in population may outstrip the capacity of wastewater infrastructure to cope with increased demands, thereby increasing nitrification of surrounding surface waters. Climate change is likely to impact urban water quality in a myriad of ways through, for example, increases in timing and intensity of rainfall, increases in water temperatures, changes in the flow regime and the experience of flooding extremes and low flows.

This chapter provides an overview of and interactions between these challenges beginning with an outline of the links between urbanization and water quality, concentrating on how urbanization alters both point and non-point sources of pollution and discusses how changes in the hydrological regime of urban areas can lead to degradation in water quality. The second section reviews the understanding of how climate change is likely to impact water quality, drawing on studies that assess changes in water quality in relation to observed climate variability and change and on those that aim to project future impacts of climate change on water quality. The third section discusses the outcomes of studies that aim to assess, through integrated modeling approaches, the interactions and feedbacks between urbanization and climate change and their combined impacts on water quality. The chapter concludes with some thoughts on emerging issues relating to uncertainty in future projections of change in urban water quality.

Urbanization and water quality

Sources of water pollution can be characterized as either point or non-point sources. Point sources include industrial or municipal effluent. Non-point sources derive from diffuse sources including wash off of contaminants from urban surfaces such as transport routes and other impervious zones. Duh et al. (2008) highlight that while major point sources have been largely controlled in many developed country cities, they remain major sources of water pollution in developing regions such as Southeast Asia and Africa, due to inadequate infrastructure, uncontrolled population growth and development of informal settlements generally lacking in basic sanitary infrastructure.

Nyenje et al. (2010) highlight the alarming deterioration of surface waters in rapidly developing cities of sub-Saharan Africa due to the widespread problem of eutrophication and enhanced nutrient content—primarily nitrogen and phosphorus. This is driven by emissions of wastewaters from industries and urban areas, which are often discharged untreated into surface or groundwaters. Nyenje et al. (2010) find that around 63 percent of the total urban population relies on either septic tanks or pit latrines for sanitation. On average, over 80 percent of the wastewaters produced in large cities in sub-Saharan Africa are untreated and are either discharged into the soil via on-site sanitation systems or directly into rivers and lakes. In particular, the nutrient load produced from informal settlements or slums is extremely high, alarming even, given that on average, 50 percent of sub-Saharan Africa's population lives in such settlements (Nyenje et al., 2010).

Hydrology—the master controlling variable

Hydrology is often described as the master controlling variable of urbanization's impact on water (Fletcher et al., 2013). The main hydrological process altered by urbanization is a reduction in infiltration due to the impervious nature of the urban fabric, resulting in increased surface runoff (see Figure 18.1). A high concentration of impermeable surfaces and artificial drainage systems drives alterations in hydrological responses through shorter lag times between the onset of precipitation and resultant runoff peaks and increases in the total

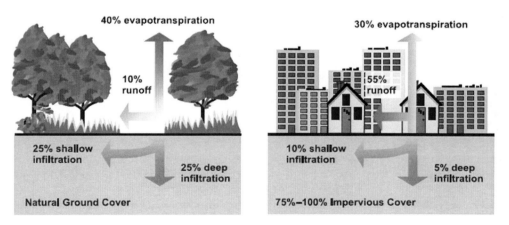

Figure 18.1 Key changes in runoff processes between idealized natural and urbanized conditions
Source: US EPA (2003).

volume of runoff received by urban streams (Shuster et al., 2005). This, in turn, increases the 'flashiness' of response of a catchment to a given input of rainfall, while also increasing flood peaks, the volume and variability of runoff (Dunne and Leopold, 1978). Increases in the total impervious area have been strongly associated with declines in ecosystem functions that are critical to aquatic biological communities (Klein, 1979; Arnold and Gibbons, 1996). Indeed, impervious cover is widely used as the leading indicator of urbanization's effect on hydrology in both empirical and modeling studies.

However, the relationship between impermeable surface and runoff is variable and depends on the extent, type and interconnectedness of impervious cover. Due to low vegetation cover and the abundance of impervious surfaces, urbanized areas have unique microclimates (Grimmond et al., 2010; Grimmond et al., Chapter 12 in this volume). Therefore, understanding the nature and spatial arrangement of impervious cover plays a dominant role in determining the influence of urbanization on catchment hydrology. Many studies in urban hydrology show that total impervious area does not affect runoff as much as effective impervious area or the proportion of impervious cover that is directly connected to the stream network (Brabec et al., 2002). While historically, many urban surfaces such as pedestrian paths and roads were thought to be impermeable, recent site specific studies have highlighted that significant infiltration can occur in parking lots and streets (Ragab et al., 2003a, 2003b; Ramier et al., 2011). Additionally, the rainfall-runoff responses from pervious areas in urban areas can be variable due to compaction, the volume of traffic, the loss of vegetation and the presence of artificial pipes and trenches (Fletcher et al., 2013).

Studies have shown, even in suburban areas, that increases in peak flows and reductions in the recession times of storm events are associated with increased urban cover (Burns et al., 2005). However, the catchment specific effects of urbanization on floods can vary largely on a catchment-by-catchment basis, while the effect is consistent across floods of varying magnitude. Rose and Peters (2001) estimate that flood peaks are between 30 percent and more than 100 percent greater in urbanized catchments compared to less- and non-urbanized catchments. Hawley and Bledsoe (2011) show that with greater area of impervious surface, all geomorphologically important flows for test catchments in Southern California increase in magnitude, frequency and duration.

The urbanization effect on subsurface hydrology is also variable. Local characteristics such as geology, vegetation and topography are important in determining whether baseflows in urbanizing catchments increase or decrease. There is often a complex spatial interaction of processes where reductions in infiltration and lower groundwater levels in part of a catchment can be offset by decreases in vegetation, reduced losses through evapotranspiration and greater infiltration in other locations.

Water quality and urbanization

Two primary mechanisms are present through which urban water quality is degraded: i) an increase in pollutants through urban land uses and associated human activities and ii) increased mobilization and transport of pollutants due to increased surface runoff and the heightened hydraulic efficiency of the urban environment (Fletcher et al., 2013). In urban areas, processing of natural pollutants by soils is prevented or mitigated by impervious surfaces through limiting infiltration and percolation and facilitates the direct conveyance of pollutants to waterways, especially during storm events (Arnold and Gibbons, 1996).

A large and growing number of studies have found statistically significant spatial and temporal relationships between a wide range of water quality indicators and urbanization

Urban water quality

(Brett et al., 2005). Increases in impervious cover and runoff are shown to directly impact the transport of non-point source pollutants including pathogens, nutrients, toxic contaminants and sediment to urban water courses (Hurd and Civco, 2004). Increases in impervious cover alters the chemical and biochemical characteristics of runoff from urban areas, with increased quantities of stormwater leading to higher pollutant wash-off, changes in sedimentation and increased pollutant loads carried by stormwater (Owens and Walling, 2002). Sonzogni et al. (1980) find that suspended solids and nutrients from urban areas range from ten to 100 times greater than loads from equivalent undisturbed land. Primary pollutant pathways include wet and dry atmospheric deposition and wash-off of contaminants deposited on the ground and other surfaces. In the absence of point source discharges, Klein (1979) emphasizes the importance of heightened runoff, increases in the magnitude and frequency of flood events, accelerated channel erosion and alteration of the composition of stream beds as the dominant factors which affect urban stream quality.

In recent years, there has also been a proliferation of studies examining the spatio-temporal relationships between urbanization and water quality. For example, Yang et al. (2008) study the regional impacts of urbanization in Hanyang, China, and conclude that rapid urbanization correlates with a rapid increase in runoff and non-point source pollutants. Ren et al. (2003) investigated the relationship between water quality and urbanization together with changing land use patterns across the entire city of Shanghai over the period 1947–1996. Results show that rapid urbanization corresponds with a rapid decline in water quality, particularly in industrial areas. Chang (2008) finds that in the Han River basin in South Korea, spatial and temporal variations in water quality are associated with urban development and locally important natural factors. Changes in urban land are found to be the most important variable in explaining changes in biological oxygen demand, total phosphorus and nitrogen concentrations. Tong and Chen (2002) examine the spatial relationships between land cover, the flow regime and water quality in the state of Ohio, US, with results indicating that higher levels of nitrogen and phosphorus are associated with impervious urban areas compared to other land surfaces considered. Groffman et al. (2004) find that urban and suburban watersheds in Baltimore, Maryland, have much higher nitrogen losses than completely forested watersheds. While yields from urban and suburban watersheds are lower than those from agricultural watersheds, the retention of nitrogen in suburban watersheds is high, with inputs driven by lawn fertilizer and atmospheric deposition. Tu et al. (2007) examined the impacts of urban sprawl in eastern Massachusetts on water quality by assessing the relationships between water quality, land cover and changes in population over a 30-year period. Water quality indicators (including specific conductance, dissolved ions and dissolved solids) are found to be highly spatially correlated with indicators of urban sprawl.

Finally, a number of studies examine the threshold of urbanization, represented by increases in impervious surface, at which degradation of water quality becomes apparent. In a study of 27 watersheds in Maryland, Klein (1979) finds that degraded water quality is first evidenced with an imperviousness of as little as 12 percent, but does not become severe until a threshold of approximately 30 percent is reached. Schoonover et al. (2005), in a study of a rapidly urbanizing catchment in western Georgia, show that within 16 sub-catchments, nutrient and fecal coliform concentrations within catchments with impervious surfaces greater than 5 percent often exceed those in other catchments during both baseflow and stormflow. However, Shuster et al. (2005) conclude that it is difficult to identify a single threshold of impervious cover that is robustly associated with hydrological and environmental impacts critical to ecosystem level responses outside of site specific contexts.

Climate change and water quality

While much research has been conducted on the impacts of climate change on the hydrological cycle (e.g. changes in runoff, flow regimes and groundwater storage) at global, regional, national and local scales, comparatively little research has explored the implications of climate change on water quality. Water quality can be impacted by many climatic factors such as air and water temperature, precipitation amounts and intensities and the occurrence of extreme events (Jiménez Cisneros et al., 2014; Kundzewicz and Krysanova, 2010). Additionally, changes in water quality during storms, snowmelt and periods of elevated air temperature and drought can cause conditions that exceed thresholds for ecosystem tolerance and lead to degradation in water quality (Murdoch et al., 2000). However, the assessment and attribution of changes in water quality is highly complex, given the many confounding factors that interact at various temporal and spatial scales and the lack of long records of observations through which longer-term variability can be contextualized (Kundzewicz and Krysanova, 2010). Furthermore, the time-steps at which measurements of water quality are taken (often bi-weekly or longer) increase the likelihood that extreme changes in water quality can be missed in the assessment of trends (Krysanova et al., 1989). With the exception of water temperatures, there is little evidence for consistent climate-related trends in water quality parameters in rivers, lakes and groundwater due to strong human impacts (Bates et al., 2008).

Given the close links between water quality and hydro–climatology, future impacts of climate change on water quality should be considered in the context of changes in temperature and precipitation, along with changes in the flow regime (Whitehead et al., 2009). Climate change has the potential to impact average flow conditions, flow velocities and hydraulic characteristics of rivers and to bring about changes in the frequency, magnitude and duration of extreme events. Changes in rainfall amounts, timings and intensities have implications for nutrient washout, changes in erosion, suspended sediment and sediment yields for individual catchments. Climate change impacts on water quality in urban areas are also dependent on the design standards of critical water supply infrastructure. Such infrastructure may be designed and built to cope with ranges of climatic variability; however, if variability increases or extreme events overwhelm critical infrastructure, negative impacts on water quality can be expected.

Changes in air temperature have a deteriorating effect on river water quality with associated increases in biological oxygen demand and suspended solids and decreases in dissolved oxygen (Ozaki et al., 2003), while warming is also known to accelerate nutrient cycling (Kundzewicz and Krysanova, 2010). Climate change has the potential to impact water quality through physical, chemical and biological changes brought about by changes in water temperature, turbidity, changes in pH and the concentrations of pollutants, and through changes in biodiversity and species abundance (Ducharne, 2008; Mohseni et al., 2003). In a review of the literature, Delpla et al. (2009) highlight that dissolved organic matter, micro–pollutants and pathogens are susceptible to a rise in concentration as a consequence of temperature increase (water, air and soil) and heavy rainfalls in temperate countries. Changes in the magnitude and duration of drought conditions, changes in rainfall and temperature are known to affect stream pH through higher maximum values and increases in average conditions. Lower dissolved oxygen solubility and concentrations have been associated with drought conditions and increases in temperatures (Komatsu et al., 2007; Prathumratana et al., 2008).

A number of studies assess the impacts of extreme hydrological events (floods and droughts) on water quality. Such assessments provide important analogues due to associated increases in the frequency and magnitude of such extremes in a warmer world. However, such studies

Urban water quality

tend to be catchment or region specific, and it is difficult to draw larger-scale conclusions. The response of individual catchments is highly variable and is determined to a large extent by physical catchment properties, land uses and management strategies. By analyzing observed water quality data, van Vliet and Zwolsman (2008) assess the effects of drought on water quality in the Meuse River in Western Europe and compare water quality as depicted by 24 representative parameters during extreme events to normal reference conditions.

Extreme low-flow conditions are found to be associated with deterioration in water quality, with an increase in pH, a decrease in dissolved oxygen concentrations and the development of conditions favorable to algal blooms through increases in water temperature, long residence times and high nutrient concentrations. Hrdinka et al. (2012) consider the role of extremes of flooding and drought on water quality in the Czech Republic and show that while droughts are considered to be more detrimental to water quality as a result of their duration and spatial scale, flood events can have a more marked impact on water quality, even if for a short period of time. During low-flow extremes, increases in water temperature, decreases in dissolved oxygen and increases in dissolved solids and phosphates can be found. Low concentrations of suspended solids and metals are also associated with low-flow volumes. During the flood event, large increases in the concentration of metals, specific organic compounds, fecal coliform and nitrates are evident, due to washout from surrounding lands. In Finland, Saarinen et al. (2010) examined long-term changes in alkalinity and acidity of nine rivers draining into the Gulf of Bothnia over the period 1913–2007. The authors find that autumn–winter discharges are key factors in explaining acidity of river waters, particularly where catchments were dominated by acidic soils. With climate change likely to increase riverflow in the study area, particularly in winter, acidity problems are likely to increase.

The assessment of future impacts of climate change on water quality is a challenging task also because it often involves the integration of complex, process-based models, which in many cases are insufficiently calibrated due to a lack of sufficient data. Given such complex modeling chains, future simulations of water quality are subject to large uncertainties. In a simulation of future impacts of climate change in the UK, Cox and Whitehead (2009) find that by the end of the 21st century (2080s), in the river Thames, the saturation concentration for dissolved oxygen will be reduced through enhanced biological oxygen demand and increased temperatures. Additionally, Conlan et al. (2007) highlight the potential for increased nitrate concentrations under reduced flow conditions with conditions favorable for the development of algal blooms. This could further affect dissolved oxygen concentrations.

In an important addition to the literature, van Vliet et al. (2013) assessed the future impact of climate change on global riverflows and river water temperatures, identifying regions of the world where water quality may become critical for freshwater ecosystems and water use. The study used a physically-based hydrological-water temperature modeling framework forced with an ensemble of bias-corrected output from three Global Climate Models (GCMs) forced using two greenhouse gas (GHG) emissions scenarios (SRES A2—medium high emissions; and SRES B1—a more conservative emissions scenario) to explore future changes in daily riverflow and temperature, together with their interactions under extreme conditions.

The results of van Vilet et al. (2013) show that global mean and high (95th percentile) river water temperatures are projected to increase by 1.0–2.2°C under the SRES A2 scenario by the end of the century (2071–2100), relevant to 1971–2000. Largest increases are projected for the US, Europe, eastern China and regions of southern Africa and Australia. Critically, the greatest exposures to water quality deterioration are found for regions where largest increases in water temperature coincide with the time of year when greatest reductions in low flows are projected, with climate change expected to significantly impact water quality in the southeastern US, Europe, eastern China, southern Africa and southern Australia. This work highlights the importance of considering both the direct impacts of warming and the indirect impacts of changes in river discharge in assessing future climate change impacts on water temperature.

A small number of studies assess climate change impacts on water quality for specific catchments or regions. The first of two illustrative examples includes a study by Ducharne (2008), which assesses the sensitivity of water quality to climate change for the Seine River in France. Over the period 1993–1999, variance in stream temperature was found to be closely associated to the lagged moving average of air temperature. Using both a biogeochemical model and a physically-based land surface model, changes in water temperature and riverflow were used to assess changes in biogeochemical water quality in Paris. Using a future climate change scenario, results indicate increased phytoplankton growth and decreases in dissolved oxygen. Importantly, Ducharne (2008) highlights the complex interactions that occur between changes in riverflow, water temperature and effluents from human activities. For the Seine, climate change impacts on water quality are found to be highly sensitive to simulated changes in point source inputs and even change the processes through which future impacts on water quality are likely to manifest themselves.

Second, Ahmadi et al. (2014) assessed catchment scale impacts of climate change on non-point source pollutants including phosphorous, nitrogen and pesticide fluxes for an agricultural catchment in the midwestern US using the Soil Water and Assessment Tool (SWAT) for the period 2015–2099. Following calibration under observed conditions, SWAT was forced with 112 statistically downscaled climate change projections representing low to high GHG emissions scenarios. While little change is found for precipitation, streamflow, sediment and total nutrient loads over the century, the proportion of dissolved to total nutrients increases significantly with important implications for management of water supplies and ecosystem processes.

Combined impacts of climate change and urbanization on hydrology

It is widely acknowledged that climate change will interact with other aspects of global change, including the continued growth of urban areas. Given the importance of understanding these interactions, a number of recent studies, predominantly in the US, use modeling to better understand how future changes in climate are likely to interact with changes in urbanization across different spatial and temporal scales and with a view to informing adaptation strategies. In many of these studies, detailed below, future changes in climate are identified as having a larger impact on water quality than the assessed scenarios of land use change/urbanization. This larger contribution of climate change is due to the effect of changes in the hydrological regime acting as the controlling variable in determining water quality. At smaller spatial scales, the mix of urban development types tends to have an influence on water quality with important opportunities for mitigation strategies at the sub-catchment level. However, the combined effects of urbanization and climate change leads to greater uncertainties in attribution of drivers of change in water quality, creating challenges for identifying potentially successful adaptation strategies.

In one of the earliest studies to explore the importance of combined effects of climate change and urbanization on water quality at the catchment scale, Chang (2004) investigates the potential changes in nitrogen and phosphorus loads under scenarios of climate change, urban growth and combined influences for the Conestoga River Basin in Pennsylvania. A geographic information systems (GIS)-based hydro-chemical model was used to assess the

Urban water quality

sensitivity of the catchment to projected changes by 2030. Under climate change, which is assessed using only one scenario, mean annual nitrogen and phosphorus loads are found to increase in spring months which are wetter, but decrease in the fall due to reductions in precipitation. However, when climate change and scenarios of urbanization are combined, the projected mean annual nitrogen loads increase by a further 50 percent in urbanizing sub-catchments.

Similar work also assesses the individual and combined impacts of climate and land use change on the discharge and water quality for watersheds in metropolitan Boston and surrounds in eastern Massachusetts (Tu, 2009). Climate change scenarios, based on one GCM forced using three SRES emissions scenarios, were employed to project changes in climatic variables for the period 2005–2024. Land use scenarios were developed to represent the conversion of forested to developed land, primarily residential use under constant, current and doubled rates. Tu (2009) finds that the impact of urbanization on riverflows is enhanced by climate change with increasing monthly flows in winter and late fall and decreases in flows for summer and early fall. The combined impacts of climate and land development are found to have important implications for nitrogen loads in winter and late fall, resulting in the amplification of loads beyond the level associated with just climate change or land development alone. Uncertain responses of nitrogen loads are found for other months where land use and climate change impacts are shown to have different directions of change on nitrogen loads. When assessed individually, climate change is found to have a greater influence on the streamflow than land cover changes, while both climate and land use change affect nitrogen loads (Tu, 2009).

Wilson and Weng (2011) examine the impacts of urban land cover and climate change on water quality for the Des Plaines River basin, Illinois, for the period 2010–2030 using three future land use scenarios representing low-density residential growth, normal urban growth and commercial growth. Climate models comprising the Coupled Model Intercomparison Project Phase 3 (CMIP3) project, forced with B1 and A1B SRES scenario groups, were used to inform future climate change impacts. The SWAT was employed to estimate total suspended solids and phosphorus concentrates under assumed scenarios. Their study highlights that future land use and climate changes have the potential to dramatically change the concentration levels of total suspended sediments and phosphorus at both the catchment and sub-catchment scales.

Climate change is found to exert a larger impact on the concentration of pollutants than the potential impact of land use change, with the authors highlighting the sensitivity of results to the emissions scenarios used to represent future climate change. For their study catchment, increases in late winter and early spring discharge under the A1B scenario are associated with reduced phosphorus loading through activation of dilution, but also highlight that the response is heavily reliant on the emissions scenario used. The type of development is also found to influence changes in water quality—but to a lesser extent than climate change— with the expansion of middle- and high-density residential land use associated with reductions in late winter and early spring loadings and the concentration of total suspended solids. The mix of types of urban development is also found to alter projected phosphorus levels with increased open spaces and vegetation, together with low-density residential areas serving as a mitigation strategy at the sub-catchment level.

Chung et al. (2011) examine the impacts of climate change and urbanization on the quantity and quality of riverflow for an urban catchment in South Korea using statistically downscaled output from a single GCM forced with the A1B and A2 emissions scenario together with scenarios representing changes in impervious cover. Using the Hydrological

Simulation Program—Fortran (HSPF) hydrological model, the impacts of climate change and urbanization on both water quality and quantity were evaluated for low flows, high flows and biochemical oxygen demand (BOD). Climate change impacts on low and high flows are found to be greater than those on BOD. The impacts of urbanization are found to be much greater for water quality than changes in the flow regime, having a larger impact on BOD, and climate change having a greater impact on flow rates. Additionally, the authors highlight that coupling the effects of urbanization and climate change leads to greater uncertainty than when each driver is considered in isolation, creating challenges for identifying the potential success of adaptation strategies (Chung et al., 2011).

Praskievicz and Chang (2009) also examine the relative contributions of climate and land use change in determining the quantity and quality of water resources in the Tualatin Basin in northwest Oregon. Uncertainty in future climate scenarios was approached using output from seven GCMs; three of which were forced with the A1B SRES emission scenario, three with the B1 scenario and one GCM was forced with both scenarios. Two land use scenarios represent conservation (i.e. constraining urban growth in existing urban areas; a high weighting of the provision of ecological services; and conservation and restoration of natural vegetation and wetlands) and development (i.e. dominated by market-led solutions to land management; relaxation of zoning regulations; and expanded urban development) storylines.

Using the Better Assessment Science Integrating point and Non-point Sources (BASINS) model to simulate hydrological response, combined scenarios of enhanced climate change and urban development reveal winter flows increasing by up to 71 percent and summer flows decreasing by as much as 48 percent. In agreement with previous studies, climate change is found to be more significant than urbanization in determining hydrological response. Additionally, Praskievicz and Chang (2009) show that water quality parameters (in this case suspended sediment and orthophosphate loads) are highly dependent on changes in the flow regime. Changes in loadings due to climate and urbanization closely track the signature of changes in hydrological regime, with the dominant pattern of change being increased seasonality. This case study highlights that increases in precipitation intensity coupled with increased impervious surface area will facilitate the increased flushing of pollutants into river courses during winter storm events, while lower summer flows are likely to constrain the capacity for dilution.

The coupling of climate change and urbanization impacts is also assessed for urban water systems by a small number of studies. As an illustrative example, Astaraie-Imani et al. (2012), studying an urban water system in the UK, highlight that changes in the characteristics of precipitation in a warmer world are likely to lead to a deterioration of water quality in urban rivers. In particular, Astaraie-Imani et al. (2012) emphasize the role of increases in rainfall depth over and above increased intensity, with more frequent and significant breaches of environmental thresholds for both dissolved oxygen (DO) and free ammonia (AMM) associated with the former. Interestingly, this work also highlights that per capita water consumption is the most significant parameter in driving urbanization effects on water quality due to the relationship between increased consumption, more extreme dry weather flows and combined sewer overflows, which increased AMM concentrations and depletion of DO in this case study. Semadeni-Davies et al. (2008a; 2008b) explore the impacts of different urbanization storylines under climate change for a combined sewer system in Helsingborg, Sweden. Using climate change signals from a regional climate model, and urbanization scenarios representing a continuation of current population changes and water management practices, it is found that city growth and increases in precipitation are set to worsen drainage problems for the city through increased stormwater flow and a reduction in system capacity. For further discussion on water quantity and accessibility in light of climate change and urbanization see Pfister et al., Chapter 17 in this volume.

Conclusion—the challenge of uncertainty

The last decade has witnessed a rapid advance in the understanding of how pressures of global environmental change, particularly climate change and urbanization, affect water quality in urban settings. Central to this progress has been the increased capacity for modeling such complex systems due to increased data availability, better process understanding and increased computing power. Indeed, such developments have also seen a move toward increasingly integrating various components of the urban water cycle into formal model structures, with a view to more informed management of urban environments. It is also clear that such models will continue to play an increasingly important role in understanding the feedbacks and interactions between pressures such as climate change and rapid urbanization over the coming decades. With models becoming increasingly important for decision-making and adapting to change at local levels, it is critical that the issue of uncertainty in model projections is more fully addressed, particularly in the context of understanding the coupled nature of pressures from urbanization and climate change.

Uncertainty is an inherent component of any modeling study, but is particularly prevalent in the context of understanding climate change impacts at the catchment scale. Fletcher et al. (2013) highlight that despite the very clear rationale for the application of integrated models for urban water management, uncertainty analysis remains a challenge for both research and practice. In the case of urban water models, important sources of uncertainty are derived from model structure, the non-uniqueness of model parameters, the instability of model parameters and uncertainty in input and calibration data. In assessing future climate change impacts, there is also a well recognized cascade of uncertainty that propagates through impact assessments with major uncertainties associated with the use of different GHG emissions scenarios, the use of different global climate models and downscaling or regionalization techniques (Bastola et al., 2011a; 2011b). It is also apparent that the contribution of various sources of uncertainty—impact model, climate model, emission scenario, etc. is also context specific (Bastola et al., 2011a). Therefore, while it remains a challenge to capture observed rainfall at temporal and spatial scales necessary for modeling the response of urban water courses, the projection of rainfall changes, particularly extremes, under changing climate conditions remains a very challenging task (Pielke and Wilby, 2012).

In approaching the challenge of uncertainty in future impacts, recent developments offer new pathways to explore the sensitivity of urban watercourses to climate change, particularly in identifying thresholds of change that may prove problematic in context-specific cases. In particular, Prudhomme et al. (2010) identify a scenario-neutral approach for exploring the sensitivity of catchments to ranges of plausible future changes, rather than time varying outcomes of individual scenarios. The scenario-neutral framework could be used to identify climate change risk for specific water quality metrics, to assess the utility and robustness of different adaptation options and for combining climatic with non-climatic pressures such as urbanization. A similar vulnerability-led approach to assessing climate change impacts is offered by Brown and Wilby (2012). This also has high potential for application in the context of understanding vulnerabilities of water quality in urban streams to climate change.

In addition to frameworks for dealing with uncertainty in modeling, increased emphasis must also be placed on monitoring the effects of urbanization and climate change on water quality. As highlighted by Duh et al. (2008), the impact of urban form on water quality remains an emerging research theme. As well as providing increased information for calibrating complex models, longitudinal monitoring studies that attempt to link changes in urban form to changes in water quality need to be implemented, particularly in developing countries, and maintained where already in existence.

Summary

Several summary conclusions can be derived from this assessment. They include the following:

- Urbanization alters runoff-generating mechanisms through reducing infiltration and evapotranspiration, and increasing runoff and flood peaks to produce a 'flashier' hydrological response. Such hydrological changes play an important role in determining water quality in urban areas. Understanding the nature and spatial arrangement of effective impervious cover plays a key role in determining the influence of urbanization on catchment hydrology.
- Urban areas have greater available pollutants through specific urban land uses and associated human activities. Natural pollutant-processing by soils is prevented by impervious surfaces, while increases in impervious cover alter the chemical and biochemical characteristics of runoff from urban areas. Increased quantities of stormwater lead to higher pollutant wash-off, changes in sedimentation and increased pollutant loads carried by stormwater.
- Climate change has the potential to reduce water quality in urban areas through changes in flow conditions, velocities and changes in the frequency, magnitude and duration of extreme events. More intense rainfall has implications for nutrient washout, changes in erosion, suspended sediment and sediment yield. Changes in air temperature have a deteriorating effect on river water quality, with associated increases in biological oxygen demand, suspended solids and decreases in dissolved oxygen. Warming is also known to accelerate nutrient cycling. Climate change impacts on water quality in urban areas are also dependent on the design standards of critical water supply infrastructure.
- The interactions of climate change and urbanization will impact water quality in complex ways. Climate change will alter the hydrological regime acting as the controlling variable in determining water quality. The rate and type of urbanization will have important influences on water quality at local scales.

Key research questions

- The impact of urban form on water quality remains an emerging research theme. Little work has assessed the impacts of urban form on water quality in developing countries where the pressures of urbanization and deteriorating water quality are often greatest.
- The application of integrated urban hydrology models will continue to play an important role in understanding the feedbacks and interactions between pressures such as climate change and rapid urbanization over the coming decades.
- With models becoming increasingly important for decision-making and adapting to change at local levels, it is critical that the issue of uncertainty in model projections is more fully addressed.
- Novel frameworks exist for decision-making under uncertainty. Such frameworks are well-suited to the problem of managing urbanization and global environmental change

Urban water quality

impacts on water quality, where critical thresholds can be identified. With increasingly complex interactions of drivers of change in urban water quality, such tools will be important for assessing the sensitivity of policy and management decisions to uncertainties.

References

Ahmadi, M., Records, R. and Arabi, M. (2014) Impact of climate change on diffuse pollutant fluxes at the watershed scale. *Hydrological Processes*, 28(4), 1962–1972. DOI: 10.1002/hyp.9723

Arnold Jr., C.L. and Gibbons, C.J. (1996) Impervious surface coverage: The emergence of a key environmental indicator. *Journal of the American Planning Organisation*, 62, 243–258.

Astaraie-Imani, M., Kapelan, Z., Fu, G. and Butler, D. (2012) Assessing the combined effects of urbanisation and climate change on the river water quality in an integrated urban wastewater system in the UK. *Journal of Environmental Management*, 112, 1–9.

Bastola, S., Murphy, C. and Sweeney, J. (2011a) The role of hydrological modeling uncertainties in climate change impact assessments of Irish river catchments. *Advances in Water Resources*, 34(5), 562–576.

Bastola, S., Murphy, C. and Sweeney, J. (2011b) Evaluation of the transferability of hydrological model parameters for simulations under changed climatic conditions. *Hydrology and Earth System Sciences Discussions*, 8(3), 5891–5915.

Bates, B.C., Kundzewicz, Z.W., Wu, S. and Palutikof, J.P. (Eds.) (2008) *Climate change and water.* Technical Paper of the Intergovernmental Panel on Climate Change. IPCC Secretariat, Geneva, p. 210.

Booth D.B. and Jackson C.R. (1997) Urbanization of aquatic systems – degradation thresholds, stormwater detention, and the limits of mitigation. *Journal of the American Water Resource Association*, 33, 1077–1090.

Brabec, E., Schulte, S. and Richards, P.L. (2002) Impervious surface and water quality: A review of current literature and its implications for watershed planning. *Journal of Planning Literature*, 16, 499–514.

Brett, M.T., Arhonditsis, G.B. and Mueller, S.E. (2005) Non-point-source impacts on stream nutrient concentrations along a forest to urban gradient. *Environmental Management*, 35, 330–342.

Brown, C. and Wilby, R.L. (2012) An alternate approach to assessing climate risks. *Eos, Transactions of the American Geophysical Union*, 93(41), 401–402.

Burns, D., Vitvar, T., McDonnell, J., Hassett, J., Duncan, J. and Kendall, C. (2005) Effects of suburban development on runoff generation in the Croton River basin, New York, USA. *Journal of Hydrology*, 311, 266–281.

Chang, H. (2004) Water quality impacts of climate and land use changes in southeastern Pennsylvania. *The Professional Geographer,* 56, 240–257.

Chang, H. (2008) Spatial analysis of water quality trends in the Han River basin, South Korea. *Water Research*, 42(13), 3285–3304.

Chung, E., Park, K. and Lee, K.S. (2011) The relative impacts of climate change and urbanization on the hydrologic response of a Korean urban watershed. *Hydrological Processes*, 25, 544–560.

Conlan, K., Wade, T., Ormerod, S., Lane, S., Durance, I. and Yu, D. (2007) *Preparing for climate change impacts on freshwater ecosystems, PRINCE: results.* Environment Agency Science Report SC030300/SR, Bristol, UK.

Cox, B.A. and Whitehead, P.G. (2009) Potential impacts of climate change on dissolved oxygen in the River Thames. *Hydrology Research*, 40(2–3), 138–152.

Delpla, I., Jung, A.-V., Baures, E., Clement, M. and Thomas, O. (2009) Impacts of climate change on surface water quality in relation to drinking water production. *Environment International*, 35, 1225–1233.

Ducharne, A. (2008) Importance of stream temperature to climate change impact on water quality. *Hydrology and Earth System Sciences*, 12, 797–810.

Duh, J.-D., Shandas, V., Chang, H. and George, L.A. (2008) Rates of urbanization and the resilience of air and water quality. *Science of the Total Environment*, 400, 238–256.

Dunne, T. and Leopold, A. (1978) *Water in environmental planning.* W.H. Freeman, San Francisco, CA.

Fletcher, T.D., Andrieu, H. and Hamel, P. (2013) Understanding, management and modelling of urban hydrology and its consequences for receiving waters: A state of the art. *Advances in Water Resources*, 51, 261–279.

Grimmond, C.S.B., Blackett, M., Best, M.J., Barlow, J., Baik, J.J., Belcher, S.E., et al. (2010) The international urban energy balance models comparison project: First results from phase 1. *Journal of Applied Meteorology and Climatology*, 49, 1268–1292.

Groffman, P.M., Law, N.L., Belt, K.T., Band, L.E. and Fisher, G.T. (2004) Nitrogen fluxes and retention in urban watershed ecosystems. *Ecosystems*, 7, 393–403.

Hawley, R.J. and Bledsoe, B.P. (2011) How do flow peaks and durations change in suburbanizing semi-arid watersheds? A southern California case study. *Journal of Hydrology*, 405, 69–82.

Hrdinka, T., Oldrich, N., Hanslik, E. and Rieder, M. (2012) Possible impacts of floods and droughts on water quality. *Journal of Hydro-environmental Research*, 6(2), 145–150.

Hurd, J.D. and Civco, D.L. (2004). *Temporal characterization of impervious surfaces for the State of Connecticut.* ASPRS Annual Conference Proceedings, Denver, Colorado, May 2004.

Jiménez Cisneros, B.E., Oki, T., Arnell, N.W., Benito, G., Cogley, J.G., Döll, P., Jiang, T. and Mwakalila, S.S. (2014) Freshwater resources. In: *Climate change 2014: Impacts, adaptation, and vulnerability. Part A: Global and sectoral aspects. Contribution of working group II to the fifth assessment report of the Intergovernmental Panel on Climate Change* [Field, C.B., V.R. Barros, D.J. Dokken, K.J. Mach, M.D. Mastrandrea, T.E. Bilir, M. Chatterjee, K.L. Ebi, Y.O. Estrada, R.C. Genova, B. Girma, E.S. Kissel, A.N. Levy, S. MacCracken, P.R. Mastrandrea and L.L. White (Eds.)]. Cambridge University Press, Cambridge, UK and New York, NY, US, pp. 229–269.

Klein, R.D. (1979) Urbanization and stream quality impairment. *Journal of the American Water Resources Association*, 15(4), 948–963.

Komatsu, E., Fukushima, T. and Harasawa, H. (2007) A modeling approach to forecast the effect of long-term climate change on lake water quality. *Ecological Modelling*, 209, 351–366.

Krysanova, V., Meiner, A., Roosaare, J. and Vasilyev, A. (1989) Simulation modeling of the coastal waters pollution from agricultural watershed. *Ecological Modelling*, 49, 7–29.

Kundzewicz, Z.W. and Krysanova, V. (2010) Climate change and stream water quality in the multi-factor context. *Climatic Change*, 103, 353–362.

Meyer, J.L., Paul, M.J. and Taulbee, W.K. (2005) Stream ecosystem function in urban landscapes. *J. N. Am. Benthol. Soc.*, 24, 601–612.

Mohseni, O., Stefan, H.G. and Eaton, J.G. (2003) Global warming and potential changes in fish habitat in US streams. *Climatic Change*, 59, 389–409.

Murdoch, P.S., Baron, J.S. and Miller, T.L. (2000) Potential effects of climate chance on surface-water quality in North America. *Journal of the American Water Resources Association*, 36, 347–366.

Nyenje, P.M., Foppen, J.W., Uhlenbrook, S., Kulabako, R. and Muwanga, A. (2010) Eutrophication and nutrient release in urban areas of sub-Sarahan Africa – A review. *Science of the Total Environment*, 408, 447–455.

Owens, P.N. and Walling, D.E. (2002) The phosphorus content of fluvial sediment in rural and industrialized river basins. *Water Research*, 36(3), 685–701.

Ozaki, N., Fukushima, T., Harasawa, H., Kojiri, T., Kawashima, K. and Ono, M. (2003) Statistical analyses on the effects of air temperature fluctuations on river water qualities. *Hydrological Processes*, 17, 2837–2853.

Palmer, M.E., Bernhardt, E., Chornesky, S., Collins, A., Dobson, C., Duke, B., Gold, R., Jacobson, S., Kingland, R., Kranz, M., Mappin, K.L., Martines, E., Micheli, J., Morse, M., Pace, M., Pascaul, S., Palumbi, O., Reichmand, A., Simons, A., Townsend, A. and Turner, M. (2004) Ecology for a crowded planet. *Science*, 304, 1251–1252.

Pielke, R.A. and Wilby, R.L. (2012) Regional climate downscaling: What's the point? *Eos, Transactions of the American Geophysical Union*, 93(5), 52–53.

Praskievicz, S. and Chang, H. (2009) A review of hydrologic modelling of basin-scale climate change and urban development impacts, *Progress in Physical Geography*, 33(5), 650–671.

Prathumratana, L., Sthiannopkao, S. and Kim, K.W. (2008) The relationship of climatic and hydrological parameters to surface water quality in the lower Mekong River. *Environment International*, 34, 860–866.

Prudhomme, C., Wilby, R.L., Crooks, S., Kay, A.L. and Reynard, N.S. (2010) Scenario-neutral approach to climate change impact studies: Application to flood risk. *Journal of Hydrology*, 390, 198–209.

Ragab, R., Bromley, J., Rosier, P., Cooper, J.D. and Gash, J.H.C. (2003a) Experimental study of water fluxes in a residential area: 1. Rainfall, roof runoff and evaporation: The effect of slope and aspect. *Hydrological Processes*, 17, 2409–2422.

Urban water quality

Ragab, R., Rosier, P., Dixon, A., Bromley, J. and Cooper, J.D. (2003b) Experimental study of water fluxes in a residential area: 2. Road infiltration, runoff and evaporation. *Hydrological Processes*, 17, 2423–2437.

Ramier, D., Berthier, E. and Andrieu, H. (2011) The hydrological behaviour of urban streets: Long-term observations and modelling of runoff losses and rainfall, runoff transformation. *Hydrological Processes*, 25, 2161–2178.

Ren, W.W., Zhong, Y., Meligrana, J., Anderson, B., Watt, W.E, Chen, J.K. and Leung, H.L. (2003) Urbanization, land use, and water quality in Shanghai 1947–1996. *Environment International*, 29(5), 649–659.

Rose, S. and Peters, N.E. (2001) Effects of urbanization on streamflow in the Atlanta area (Georgia, USA): A comparative hydrological approach. *Hydrological Processes*, 15, 1441–1457.

Saarinen, T., Vuori, K.-M., Alasaarela, E. and Klove, B. (2010) Long-term trends and variation of acidity, CODMn and colour in coastal rivers of Western Finland in relation to climate and hydrology. *Science of the Total Environment*, 408, 5019–5027.

Schoonover, J.E., Lockaby, B.G. and Pan, S. (2005) Changes in chemical and physical properties of stream water across and urban-rural gradient in western Georgia. *Urban Ecosystems*, 8, 107–124.

Semadeni-Davies A., Hernebring, C., Svensson, G., and Gustafsson, L.G. (2008a) The impacts of climate change and urbanisation on drainage in Helsingborg, Sweden: Suburban stormwater. *Journal of Hydrology*, 350, 114–125.

Semadeni-Davies, A., Hernebring, C., Svensson, G. and Gustafsson, L.G. (2008b) The impacts of climate change and urbanisation on drainage in Helsenborg, Sweden: Combined sewer system. *Journal of Hydrology*, 350, 100–113.

Shuster, W.D., Bonta, J., Thurston, H., Warnemuende, E. and Smith, D.R. (2005) Impacts of impervious surface on watershed hydrology: A review. *Urban Water Journal*, 2(4), 263–275.

Sonzogni, W.C., Chesters, G., Coote, D.R., Jeffs, D.N., Konrad, J.C., Ostry, R.C. and Robinson, J.B. (1980) Pollution from land runoff. *Environmental Science & Technology*, 14(2), 148–153.

Tong, S.T.Y and Chen, W. (2002) Modelling the relationship between land use and surface water quality. *Journal of Environmental Management*, 66, 377–393.

Tu, J. (2009) Combined impact of climate and land use changes on streamflow and water quality in eastern Massachusetts, USA. *Journal of Hydrology*, 379, 268–283.

Tu, J., Xia, Z., Clarke, K.C. and Frei, A. (2007) Impact of urban sprawl on water quality in eastern Massachusetts, USA. *Environmental Management*, 40, 183–200.

US EPA (2003) *Protecting water quality from urban runoff*. United States Environmental Protection Agency, EPA 841-F-03-003.

van Vliet, M.T.H. and Zwolsman, J.J.G. (2008) Impact of summer droughts on the water quality of the Meuse river. *Journal of Hydrology*, 353, 1–17.

van Vliet, M.T.H., Franssen, W.H.P., Yearsley, J.R., Ludwig, F., Haddeland, I., Lettenmaier, D.P. and Kabat, P. (2013) Global river discharge and water temperature under climate change. *Global Environmental Change*, 23, 450–464.

Walsh, C.J., Roy, A.H., Feminella, J.W., Cottingham, P.D., Groffman, P.M. and Morgan R.P. (2005) The urban stream syndrome: Current knowledge and the search for a cure. *J. N. Am. Benthol. Soc.*, 24, 706–723.

Whitehead, P.G., Wilby, R.L., Battarbee, R.W., Kernan, M. and Wade, A.J. (2009) A review of the potential impacts of climate change on surface water quality. *Hydrological Sciences Journal*, 54, 101–123.

Wilson, C.O. and Weng, Q. (2011) Simulating the impacts of future land use and climate changes on surface water quality in the Des Plaines River watershed, Chicago Metropolitan Statistical Area, Illinois. *Science of the Total Environment*, 409(20), 4387–4405.

Yang, L., Ma, K., Guo, Q. and Bai, X. (2008). Evaluating long-term hydrological impacts of regional urbanisation in Hanyang, China, using a GIS model and remote sensing. *International Journal of Sustainable Development & World Ecology*, 15(4), 350–356.

19

URBAN GREENING, HUMAN HEALTH AND WELL-BEING

Alexei Trundle and Darryn McEvoy

Human health and well-being are inextricably linked to the quality of their surrounding environmental conditions. In cities, the predominantly artificial settings that encompass urban inhabitants are critiqued through global rankings of 'livability' and quality of life. Cities are given low marks for detrimental attributes—such as air pollution and water contamination—which correlate to adverse health clusters in their respective populations (van Kamp and Leidelmeijer, 2003). Beyond urban boundaries the capacity for ecosystem services to sustain and enhance the global population's well-being and health is now in question. Of nine fundamental Earth system processes, human activity has already contributed to three 'safe' operating boundaries being crossed—thresholds of biodiversity loss, nitrogen cycle extraction and atmospheric carbon dioxide concentration (Rockström et al., 2009). Urban areas, as consumption nodes of our socio-economic networks, are intensifying these global patterns, acting as biodiversity dead zones, augmenting climate change through urban 'heat islands' and increasing nutrient runoff into waterways and catchments (Wu, 2014). The consequences of these changes for human health at a global scale are largely unknown; however, the need to balance human consumption with the ecosystem services upon which human health and well-being depends is self-evident.

Urban greening—the enhancement, integration and spread of different forms of vegetation within cities—is not a new concept. Although the footprints of ancient parks, gardens and trees lack the durability and consequently the archeological evidence of stone and metal structures, records of ornamental, functional and agricultural urban greenery date back millennia (Stark and Ossa, 2007). As with other urban features, the extent and form of these 'green infrastructure' elements have fluctuated over time, varied between cities and co-evolved technologically with changing and varied urban typologies. Equally, understandings of their social, economic and environmental value have been constructed through culturally and temporally specific lenses (compare, for example, the influence of post-war egalitarianism on the Garden City movement with the aesthetic form and meditative purpose of traditional Japanese Gardens [Batchelor, 1969; Ignatieva et al., 2010]). As a result, the diverse potential benefits of urban greening, the multifaceted relationship between the environment and human health and the scalable synergies of urban resilience and global sustainability are constantly evolving and being re-framed.

This chapter reviews evidence from a number of cities that are attempting to apply urban greening across a diverse range of socio-cultural, climatic, economic and morphological settings. In particular, findings from a recent collaborative and cross-disciplinary study into green infrastructure implementation in Melbourne, Australia—focused on examining its potential role in countering the urban heat island effect—are drawn upon, eliciting practical evidence of the challenges of targeting urban greening at different scales within the urban form (Bosomworth et al., 2013). Finally, the interconnected roles of allotment-, precinct- and municipal-scale urban greening will be set within the global resilience agenda and in the context of the systemic pressures discussed above.

Building the health case

Ulrich's 1986 review of human responses to vegetation and landscapes represents a seminal assessment of the potential benefits that can be derived from urban greening and concludes by noting that if a "tangible valuation of the aesthetic and psychological benefits" of green infrastructure is to eventually be formulated, any theoretical framework or model will first need to substantiate the relationships between "visual landscape characteristics, aesthetic and emotional responses and various behaviors" (Ulrich, 1986, p. 42). While components of this equation have been researched extensively over the last three decades, the difficulty in systematically quantifying and causally linking diverse urban vegetation typologies to health metrics, filtering confounding characteristics and extrapolating case study derived observations into wider models or theories has limited development of strong conceptual frameworks (Tzoulas et al., 2007, p. 170). As a result, measures of urban greenery's health benefits are incomplete, with quantified findings often not transferable between vegetation typologies, climates and cities or not replicable within spatially divergent divisions within a single metropolitan extent. This partiality can exclude not only positive, but also some negative health consequences such as biogenic hydrocarbon generation, increased wildfire risk in peri-urban zones and heightened anxiety levels in users of overgrown public areas (Taha, 1996; Bixler and Floyd, 1997; Buxton et al., 2011).

The empirical, theoretical and epidemiological evidence that is emerging from a wide array of methodological approaches and disciplines nonetheless provides a compelling—albeit incomplete—case for implementing urban greening programs on health grounds alone (Tzoulas et al., 2007, p. 171). This is not to disregard the significance of other co-benefits associated with urban greening, or the importance of developing a multi-sectoral case for green infrastructure. Rather, if complex structural modifications such as urban greening are to be effectively mainstreamed into urban policies, sector-specific rationales must be developed to provide a foundation for collaboration across departmental, legislative and policy divisions (Roe and Mell, 2013). Further, given the un-quantifiable and risk-based nature of some benefits, the complexity of distributing economic values and the interconnected nature of these drivers (see Romero-Lankao and Gnatz, Chapter 15 in this volume, for a broader discussion of urbanization, vulnerability and risk), a focus on health, well-being and urban livability is more likely to provide a strong basis for initiating a broader urban greening policy agenda (Choumert and Salanié, 2008).

A number of conceptual models have been proposed for explaining the relationship between ecosystem and human health (Bedimo-Rung et al., 2005; Lovell and Taylor, 2013). However, for the purpose of this chapter, a simplified representation of green infrastructure's contribution to human health and well-being has been developed by the authors based upon those themes identified in peer-reviewed literature (Figure 19.1). Two axes act to functionally

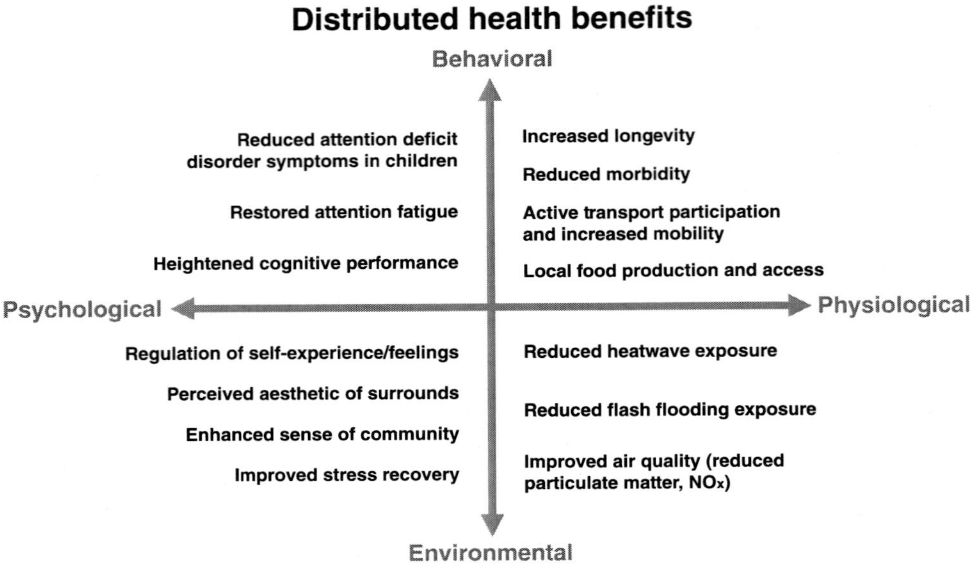

Distributed health benefits

Behavioral

Reduced attention deficit disorder symptoms in children

Increased longevity

Reduced morbidity

Restored attention fatigue

Active transport participation and increased mobility

Heightened cognitive performance

Local food production and access

Psychological ← → **Physiological**

Regulation of self-experience/feelings

Reduced heatwave exposure

Perceived aesthetic of surrounds

Reduced flash flooding exposure

Enhanced sense of community

Improved stress recovery

Improved air quality (reduced particulate matter, NOx)

Environmental

Figure 19.1 Categorization of health benefits

separate these benefits. The first represents divisions based upon the form of the health impact itself (psychological or physiological). The second represents the nature of the impact vector (behavioral or environmental). These axes are not intended to be definitive binaries (or exclusive categories for the health benefits depicted) but rather provide a practical framework for policy development.

Countering health risk: environmental exposure

Over the last decade, a series of heatwaves across Europe, North America and Australia have highlighted the health risks associated with extreme and prolonged heat events, with one Pan-European event in 2003 estimated to have resulted in over 30,000 excess mortalities across 19 countries (Robine et al., 2008). The disproportionate concentration of these deaths in urban areas is in large part a product of a series of processes known collectively as the urban heat island (UHI) effect. Although various aspects of urbanization contribute to UHI, the replacement of vegetation and permeable surfaces with hard or 'gray' infrastructure is consistently the foremost factor. This reduces the capacity for natural cooling through evapotranspiration and shading whilst simultaneously promoting an urban form that absorbs, traps and even generates heat (Kleerekoper et al., 2012).

Australia's highly variable climate is widely recognized for its severe wildfires, droughts and floods; however, heatwaves also result in prolonged spikes in mortality and morbidity. Indeed since European colonization, heatwaves are estimated to have killed more people in Australia than any other natural hazard, excluding disease (Coates, 1996). One contemporary example is the 2009 'Black Saturday' bushfires: while 173 people were killed in Melbourne's peri-urban fringe and rural surrounds as a result of fires, 374 citizens died due to the associated extreme heat conditions (Cordner et al., 2011; QUT, 2010). Notably, the decade-long 'Millennium Drought' preceding these events further exacerbated Melbourne's UHI, with restrictive water use policy responses resulting in urban forest decline and replacement of large areas of the city's vegetation with hard surfaces (van Dijk et al., 2013).

Estimates of the cooling capacity of urban greenery are species, scale and typology specific (being highly sensitive to factors such as irrigation regimes, infrastructure proximity and surrounding vegetation extent). In Melbourne, for example, a doubling of streetscape vegetation, coupled with 50 percent coverage of green roofs on Central Business District (CBD) buildings has been modeled to yield a relative reduction in daily summer maximum air temperature of 0.7°C (Chen, 2012). However, using a different metric, a study assessing the wider metropolitan area of Greater Manchester in the UK finds that a 10 percent increase in urban vegetation could result in a reduction of maximum surface temperatures by 2.2°C in high-density residential areas (Gill et al., 2007). Focusing instead on internal conditions, a third building-specific study in La Rochelle, France, identifies that the installation of green roofs could result in both summer indoor maximum and mean air temperatures being approximately 2°C cooler due to the combination of solar shading and evapotranspiration (Jaffal et al., 2012). Variations in the temporal distribution, location and period of temperature measurement thus demonstrate the difficulty in establishing a singular case for urban greening's health benefits during extreme heat events, wherein the behavioral responses of individuals and the subsequent consequences for community health are not yet well understood.

Thresholds for excess mortality and morbidity also vary between cities (due to variables such as local climatically adapted built environs and learnt human behavior), with morbidity threshold daily mean temperatures differing by 3°C between the sub-tropical Australian city of Brisbane and the dry continental climate zoned capital Canberra (Loughnan et al., 2013). While the extent of cooling and the corresponding relationship with human health impacts are demonstrably variable, it is nonetheless consistently evident that cooling through urban greening can yield quantitative benefits to human health by reducing exposure to heat extremes, effectively counteracting localized heat retention.

Additional environmentally driven, physiological benefits include the filtering of airborne particulates and reduced exposure to flash flooding. A study in Portland, Oregon, calculates that the city's street trees remove 92g of PM_{10} annually, reducing the occurrence of respiratory illness by an estimated 18 percent (City of Portland, 2010). Other research conducted in Brussels, Belgium, finds that the city's intensive green roofs retain up to 75 percent of rainfall that reaches them (Mentens et al., 2006), reducing the speed of storm water runoff and the potential for flash flooding (Figure 19.2). As with heat reduction, however, the health benefits of these urban ecosystem services are heavily dependent on the existing urban fabric, each city's climatic context and the behavioral patterns of its residents.

Direct health benefits and livability

A number of studies find that in addition to reducing risk, higher environmental exposure to urban greenery can improve psychological condition and overall well-being. A recent systematic review by Bowler et al. (2010) identifies 25 published research projects that demonstrate this relationship, whereby health conditions were assessed as a function of exposure to synthetic and/or natural urban areas. Across a range of methodologies and hypotheses contained within this body of research, more than half of the studies demonstrate improvements in participants' emotional state following activity in a vegetated space when tested against control measures prior to any such activity. Benefits included heightened feelings of energy as well as reduced levels of anxiety, anger, fatigue and sadness (Bowler et al., 2010). A number of studies also identify symptoms of improved psychological condition.

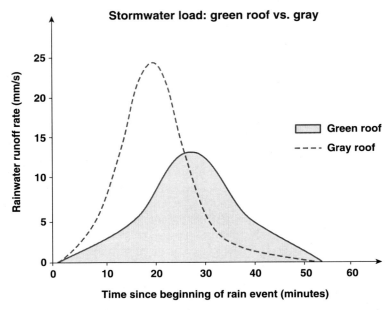

Figure 19.2 Green roof runoff benefits

Source: Illustration based on Czemiel Berndtsson (2010).

For example, one study demonstrates reduced levels of salivary cortisol, corresponding with lower cerebral activity in the prefrontal area, a proxy for relaxation and reduced stress (Park et al. 2007, p. 123).

The contribution of urban greenery to the overall livability of cities has also been linked to a community's resilience and collective identity. Although confounding factors make such observations extremely difficult to quantify, there is strong epidemiological evidence that communities with more green space have higher perceived levels of individual health (Maas et al., 2006). Similarly, assessments of community cohesiveness and well-being are highly subjective and have limited capacity to causally demonstrate benefits or be extrapolated or assessed across a significant sample size (Lee and Maheswaran, 2011). Interview-based research conducted in Kentlands and Orchard Village, US, identifies that natural features such as street trees and open space can play a central role in developing community identity (Kim and Kaplan 2004, p. 335). Detroit, US, also provides a contemporary case study of the role that 'Urban Green Commons' can play in strengthening community resilience, providing informal employment in the face of economic and demographic recession, while simultaneously supplementing food supplies and fresh food intake (Colding and Barthel, 2013, pp. 161–162).

Secondary health benefits: behavior change, cognitive enhancement and well-being

While the environmental risk reductions and associated whole-population benefits discussed above present a compelling generalized case for re-foresting and greening our cities, attempts to assess behavioral benefits highlight the difficulty in sufficiently scaling such evidence down

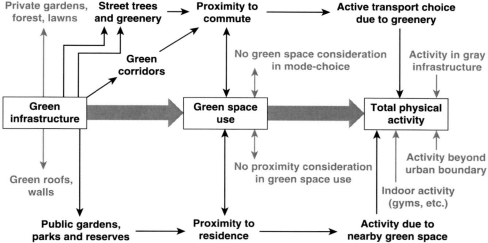

Figure 19.3 Linking greening and physical activity levels

to justify the practical implementation of specific urban greening projects. Figure 19.3 illustrates the complexity involved in isolating and causally linking green infrastructure to one such hypothesized benefit: heightened physical activity. Although this and other similar links have been widely cited (Bedimo-Rung et al., 2005; Bowler et al., 2010), confounding factors make quantifying such impacts extremely challenging. When quantifying increased active transport, for example, only certain forms of green infrastructure are located along commuter and recreation corridors, with private greenery, isolated parks and reserves contributing little to choice of transport mode (Fraser and Lock, 2011). Equally, use of green space for purposive exercise activities is limited by its size, form and level of public access (Coombes et al., 2010).

There is a strong link between individual physical activity levels and green space use. However, the relationship between proximity to green space and use of the space is both heavily dependent on subjective definitions of accessibility as well as the scale and fractions of distance applied in individual methodologies (Kaczynski and Henderson 2007, p. 345). As shown in Figure 19.3, the partial substitution of exercise in 'natural' urban environments for activity in synthetic ones such as gymnasiums is also evident in areas of high urban density (Bedimo-Rung et al., 2005). Subsequently, a number of literature reviews find the assertion of links between urban greening and improved total exercise behavior patterns to be either lacking individual quantification and methodological rigor (Hillsdon et al., 2006; Lee and Maheswaran, 2011) or collectively inconclusive (Kaczynski and Henderson, 2007; Bowler et al., 2010).

Empirical analysis has nonetheless been extensively cited by local governments (in some instances transferred from other urban settings) as part of the rationale for urban forest policies, parks and recreational plans and associated strategies (Scottish Government, 2007; City of Melbourne, 2012). If more defensible and wide-reaching policies are to be developed and irrefutable precinct- or development-level business cases constructed, the broader system within which typologies of green infrastructure are implemented must first be better understood. The links between the extent of urban greening in its various forms, usage by the surrounding population and the proximate boundaries of any such use must also be understood as being determined primarily by individual behavior rather than overarching environmental conditions or risks.

Shifting urban landscapes: the climate factor

Existing levels of environmental risk, city-wide health and community well-being—coupled with increasing evidence for enhanced behavioral health—present a strong empirical case for green infrastructure enhancement in many cities. However, when faced with the dynamic global pressures presented in the introduction to this chapter, the health case for urban greening becomes both more urgent and exponentially more challenging to implement. One such external variable is human-induced climate change, with global greenhouse gas emissions estimated to be tracking above both the 'high emissions' (8.5 Representative Concentration Pathway—a measure of projected watts per square meter of additional radiative forcing by 2100) and its precursory SRES scenario A1FI (a 'Fossil Intensive', globally connected and economically growing world) (Peters et al., 2012, p. 5). If these trends continue unabated, this trajectory is likely to result in average annual global temperatures increasing 4.0–6.1°C by 2100 (Rogelj et al., 2012). Within cities, such changes will compound UHI-based warming and amplify extreme heat risk (Wu, 2014).

Cities are already facing climatic changes: at a global scale, average temperatures have already increased 0.85°C since 1880 (IPCC, 2013, p. 3), while analysis of decadal trends over the second half of the 20th century reveal that an increase in the frequency and extremity of hotter temperatures is already underway, as illustrated in Figure 19.4. This shift is accelerating

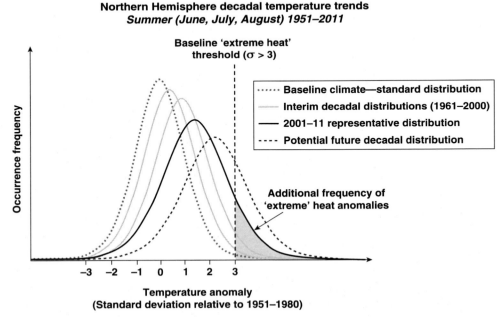

Figure 19.4 Smoothed recurrence of land-based extreme heat in the Northern Hemisphere 1951–2011

Source: Illustration based on analysis by Hansen et al. (2012).

the frequency of heat-related excess morbidity and mortality, a dynamic probability that is projected to increase substantially.

Modeling of mortality rates in Australian extreme heat events suggests that by 2050 Melbourne may experience an annual average number of heat-related deaths five times greater than would otherwise be expected without climate change. This model is based upon the A1B 'medium emissions' scenario, driven by a 300 percent increase in the number of extreme heat days expected annually (PwC, 2011). Another study based on modeling by the Commonwealth Scientific and Industrial Research Organisation (CSIRO) suggests that under a 'high' A1FI emissions scenario, the number of consecutive three-day periods experienced by the city with maximum temperatures over 37°C is likely to increase from the historic average of one per year to eight by 2100, while the number of single days over 40°C is likely to increase from four to 20 (AECOM, 2011). Although specific probabilities vary between cities, these projections are consistent with the observed trends in the Northern Hemisphere, further bolstering the need for preventative health risk mitigation actions such as cooling through urban re-forestation.

Climate change is also complicating the capacity to address such risks through urban greening, with changing temperatures and rainfall frequency modifying the ecosystem services provided by green infrastructure—challenges compounded by responsive management strategies and species selection by urban inhabitants and decision-makers. The 'Millennium Drought experienced across much of southeastern Australia provides a precursory example of how a sustained step change in precipitation can severely disrupt existing water management techniques, with an impending depletion of the region's largest city's water supply in the early 2000s resulting in increasingly restrictive regulation of potable water. Although the city's average storm water runoff over this period exceeded its total water use, the inability to rapidly or preemptively implement more sophisticated water re-use, storage or reclamation strategies resulted in extensive urban forest decline, an uptake in drought-tolerant species as well as artificial turf and hard-surface replacement of vegetative areas (van Dijk et al., 2013). The need to integrate and reconceptualize water availability is thus central to the ability to enhance urban greening, with a balance needing to be struck between urban greenery's drought and temperature tolerance and its ability to deliver evapotranspirative cooling, leaf coverage and biodiversity. In the case of heat-related morbidity, achieving such a balance will be crucial if both an overall health benefit and maintenance or improvement of well-being is to be achieved.

It is important to note that when considered within a city-specific context, the interactions between green infrastructure, health and a changing climate are far more complex than drawing a simple correlation between extreme heat recurrence and population mortality and morbidity. For example, it is possible that in cold climate cities the reduction in extreme 'cold'-related mortality and morbidity may exceed the risk presented by anomalously high temperature limits relative to future temperature distributions, although quantitative evidence of the relationship between cold-related mortality is currently weak (Haines et al., 2006). Further, even in temperate and tropical cities, where the bulk of the world's urban population currently resides, human populations and built environments will, to a limited extent, autonomously adapt over the longer-term, with human heat risk thresholds increasing as the standard distribution of temperatures shifts. New localized definitions of 'extreme' heat are therefore created as the baseline climate shifts (Mora et al., 2013). The difference in heat-related morbidity thresholds between Australian cities reflects this, whereby individual physiology, community behavior and built infrastructure differ between localities (Loughnan et al., 2013).

Consideration of such climate-related health risks must therefore be placed within existing local climatic contexts as well as the ongoing changes to these conditions over time.

Shifting urban landscapes: increasing populations and consumption rates

In many cities, climate change (both in terms of observed trends as well as projected future scenarios) is only one of a multiplicity of drivers of urban dynamics. Others, such as infrastructure development, socio-economic transformation and population growth, often take planning precedence. Products of these changes—such as infill, peri-urban growth and consumption-driven lifestyle demands—interact inextricably with urban greenery, limiting new green infrastructure opportunities, whilst simultaneously increasing the demand and need for its services. As with the recent climatic step changes observed in the Australian context, New York's rapid growth throughout the 19th century provides a prescient example of the challenges in addressing such human change retrospectively, while the recent 'retro-greening' of the city's High Line demonstrates one of the 21st century's more innovative and opportunistic urban greening case studies.

Central Park was not featured in the original 1811 Commissioners' Plan for New York, which instead depended upon a number of smaller remnant areas and parks distributed throughout Manhattan's gridded streetscape (Rosenzweig and Blackmar, 1998). While more than half of these parks were shrunk or removed in the subsequent decade due to government views that the plan was over-committed to open space (Heckscher, 2008, p. 9), a further multiplication of the city's population, perceived potential for land value increases by the urban elite and widespread support for a new form of social gathering space led to re-invigorated interest in urban greenery and, in turn, Central Park's conception (Rosenzweig and Blackmar, 1998, p. 36). Re-establishment of such a large area of green space came at a cost, however, with more than 260 African American and Irish migrant residents being forcibly evicted under Eminent Decree (Wall et al., 2008, p. 98). The park's establishment came at significant monetary cost also, with the $10 million investment three times the size of the city's then annual budget (Rosenzweig and Blackmar, 1998, p. 18).

Central Park's value—both more broadly and in terms of the health and well-being of Manhattan's workers, visitors and residents—is unquestionable today; however, two lessons remain applicable in the analysis of its re-establishment. First, the costs and the benefits of urban greening are distributed unequally—and in some cases exclusively—between a city's citizens; however, as a city grows, green space gains a broader social value, which exceeds such allocation challenges and inequities. Second, the ability to retrofit green infrastructure—whether to improve health, social well-being or even property prices—becomes increasingly technologically and economically challenging as a city increases in density. The New York High Line—an elevated inner-city railway that has been converted to a linear park—is one such example; however, opportunities to retrofit disused infrastructure are rare (in the case of the High Line, occurring due to a decline in lower Manhattan's warehousing, food processing and storage sector, coupled with a shift to road freight) (Foster, 2010).

Rapidly growing and heavily infilled cities around the globe are instead increasingly dependent on complex relationships with the private sector, whereby vertical and rooftop surfaces of existing private allotments and buildings are retrofitted through green infrastructure incentives, while new buildings and developments are 'greened' through either regulation, incentives or a combination of both (Carter and Fowler, 2008). For example, in Tokyo, an ordinance requires 'flat-roofed' buildings with a footprint greater than 1,000m^2 to include at least 20 percent green roof coverage, while Minneapolis provides a 100 percent storm water utility fee credit where green roofs replace impervious surfaces (Carter and Fowler, 2008, p. 154). As discussed earlier in this chapter, the ability for such private, micro-scale and often publicly inaccessible infrastructure to collectively equate to tangible health benefits beyond their immediate occupants is limited. Further, potentially scalable public health outcomes, such as urban heat reduction and improved air quality, generate minimal direct benefit for the private owners (Landry and Chakraborty, 2009).

Countering the de-privatization of green infrastructure and open space

While publicly owned green space is demonstrably pressured by growing populations and their appetite for housing, increased population density does not necessarily preclude privately owned green infrastructure. Well-designed urban growth can maintain, and even increase, the total ground-level green space potential within both 'greenfields' peri-urban precinct plans and middle- and inner-city re-development areas. In Figure 19.5, the three different layouts of buildings have the same population densities but different available space for vegetation and green infrastructure. The single-family, detached homes (Figure 19.5, left) provide the least amount of space for green infrastructure, whereas the multi-story high-rise (Figure 19.5, right) has the most. In practice, however, private investment rationales for procurement of expensive inner-city allotments, coupled with outer-city public demand for low-density detached housing, often result in declining private green space coupled with a heightened dependency on public parks and streetscapes (Brunner and Cozens, 2013). Analysis of Melbourne, Australia, shows that in some municipalities concerted local government policies to increase total urban greenery measured by 'canopy-cover' are effectively being offset by private increases in hardscape in response to infill and potable-water restrictions. In one municipality, a 38 percent increase in public trees over a ten-year period was almost entirely offset by a 4 percent decline in the private realm, resulting in a net increase of only 2 percent (VLSAC, 2011). In greenfields development a similar reduction in green space was identified: an 'average' Melbournian house built in 2007 covered 34.5 percent of its allotment, compared with 21 percent in 1990, reflecting market-driven demand to maximize measureable floor area (Goodman et al., 2010). Greenfields and infill design with integrated public urban greenspace is thus demonstrably preferable to retrofitting heavily privatized suburbia, wherein green infrastructure effectively operates as a market externality within significant physical constraints and economic pressures (Choumert and Salanié, 2008).

Figure 19.5 Equivalent floor-area densities in three different layouts: low-rise single-story homes (left); multi-story medium-rise (middle); high-rise towers (right)

Source: Adapted from Chen et al. (2008) in Seto et al. (2014).

Greening: an 'alternative' policy approach with cross-cutting community benefits

Urban greening is demonstrably capable of delivering a range of tangible health outcomes as well as further benefits that are either not yet effectively quantified, causally linked or in some cases fundamentally qualitative in nature. These benefits not only occur in tandem with positive outcomes in other sectors (such as improved energy efficiency and property value increases), but they can also counter increasing environmental and social stressors such as heatwaves, obesity and malnutrition (Chen, 2012; Coombes et al., 2010; Maxwell et al., 1999). However, there is limited longitudinal evidence of net increases in total urban green infrastructure on a city-wide basis during periods of population growth, with the possible exception of some arid-zone cities where suburbia has replaced areas of sparse vegetation (even Phoenix, Arizona, has increased its urban heat island effect over time, with residential areas having replaced surrounding agricultural allotments (Stabler et al., 2005)). The question therefore remains: if urban greenery has such demonstrable benefits, is its enhancement being impeded by political, social or institutional barriers or are these benefits being (or perceived to be) more effectively or efficiently delivered through other urban features or typologies?

From a policy perspective, establishment of precinct- and allotment-level evidence may be the central enabling factor in increasing uptake of green infrastructure by stakeholders within the private sector as well as those within the public realm who lack jurisdiction at a metropolitan scale. Dissecting the relatively well-evidenced and simplistic example of health benefits resulting from reduced vulnerability to extreme heat, these challenges become manifest. First, the ability to correlate allotment-level green infrastructure investment (both in terms of implementation costs and ongoing maintenance) to a tangible reduction in a city's macro-scale heat footprint, such that substantive policy action can be justified, is contestable. Similarly, equivalent private cooling benefits at a micro-scale can be generated through conventional approaches such as air-conditioning units, investments in insulation and even emergent technologies such as high-albedo 'Cool Roof' surfacing (Zinzi and Agnoli, 2012). At a municipal level, UHI characteristics vary sufficiently across metropolitan areas, so as to create distinct gradients in heat vulnerability (when measured by both morbidity and mortality rates). In Melbourne's 2009 heatwave, diurnal surface temperatures in the city's western suburbs were 5°C higher than the CBD, while nocturnal UHI surface temperature patterns conversely resulted in inner-city temperatures exceeding those in the outer-west by almost 10°C (QUT, 2010). Municipal support for urban greening was, therefore, found to be mixed, and in the case of rapid-growth areas, support for greening was outweighed by demands for housing affordability and the short-term demand for additional allotment sales (Zhu and Zhang, 2008). Therefore, the ability to distribute benefits and costs across actors and scales, even when looking at a single benefit, remains limited and can result in urban greenery either being replaced by infill or reduced to enable more profitable greenfields development.

Compiling the co-benefits that can be derived from green infrastructure across multiple policy areas presents as much of a challenge as the issues of scalability and cost-benefit distribution discussed above. Community gardens provide a tangible example of a greening approach that crosses all four of the distributed health and well-being categories in Figure 19.1: allowing local food production, generating a unique space for enhancing community cohesion, reducing urban heat and providing tangible benefits for individual allotment owners. However, each of these benefit categories operates through distinct structural and

Urban greening, human health and well-being

theoretical frameworks, as reflected by the disciplinary division of psychology and medicine, or the methodological divergence between vulnerability assessments of environmental heat exposure and targeted behavior-change programs for improving individual nutrition (Smith and Joyce, 2012). Beyond the health sector, these metrics and disciplinary languages need to be integrated with policy frameworks for energy efficiency, environmental biodiversity as well as qualitative measures of livability and quality of place. As a consequence, efforts to equitably distribute city-wide greening costs, while maximizing cross-sectoral engagement and support, require a highly sophisticated, multifaceted policy mechanism.

If such a cross-cutting approach can be developed, greening could provide an alternative policy approach for generating health, well-being or other benefits. Thus far, however, it has largely been applied in cities with either stable or declining populations (such as Detroit, US, and Stuttgart, Germany), with the notable exception of a few metropolises with considerable political capital or planning control. Singapore, which identifies itself as a 'City within a Garden,' is one such example (Colding and Barthel, 2013; Mees et al., 2013; Henderson, 2013). Outside of these settings, the rates of population growth experienced by many global cities are resulting in the addition of more 'gray' infrastructure than 'green.' Consequently many urban planners depend heavily on the greening of publicly owned areas such as parks, reserves and streetscapes at the expense of the more challenging alternative—generating evidence-based regulations and incentives for the much more expansive areas of privately held land (Bengston et al., 2004).

Conclusion

Urban greenery contributes to the health and well-being of the inhabitants of cities; however, the extent of this contribution, who benefits most from it and the relationship between these benefits and different green infrastructure typologies, is yet to be sufficiently conveyed to urban planners, practitioners and developers. As such, green infrastructure's measurable value is generally outweighed by the significant opportunity costs associated with alterative land uses, particularly in rapid-growth cities. Further, a substantial disconnect is evident between the private, micro-scale investments required in fragmented private ownership environments and the wider social health and well-being benefits generated by urban greenery. Mechanisms for distributing green infrastructure costs and benefits across sectors, public–private spheres and land-management scales will need to be created if the reductions in urban vegetation being currently observed in rapid-growth cities are to be reversed in an equitable and proactive manner. Alternatives, such as the retrospective re-forestation of New York City's Central Park, are likely to significantly increase urban greening's short-term social and economic costs. Proactive measures are urgently needed if health and well-being outcomes are to be achieved as the global population continues to urbanize, causing cities to grow and increase in density. The experience of southeastern Australia in the first decade of the 21st century provides a precursory warning of the limits to responsive—rather than proactive—action. In the case of green infrastructure, investments in street trees take decades to mature and provide their full potential benefits, while the concurrence of a decade-long drought demonstrates the importance of understanding the capacity to green cities within the context of wider environmental constraints and resource availability.

Climate change is already increasing the frequency of extreme heat events, which are becoming more intense and longer lasting. This global driver is interacting with each city's unique climatology (a product of both UHI and regional climatic contexts) as well as its broader socio-economic and demographic drivers, particularly population growth. In the face of such complex and interconnected pressures, the enhancement and application of equally multifaceted and cross-cutting urban features—such as green infrastructure—may be the most effective approach to address 'wicked' urban problems, which are reducing urban citizens' health and well-being. If the barriers discussed throughout this chapter are not able to be addressed through existing sectoral and governmental structures, more transformative urban change may be required, whereby new governance arrangements, urban forms and technologies are applied, a process potentially catalyzed by the cascading impacts of similarly transformative localized shifts in climate (Mora et al., 2013).

The terminology used in this chapter—such as 'green infrastructure,' 'urban greenery' and 'green space'—has been applied interchangeably, reflecting the lack of discursive consensus in the academic domain as well as the weakness of broader public understandings of the concept (Bosomworth et al., 2013). Given the different health, well-being and other benefits associated with different vegetative typologies (in terms of biological make up, location and extent), interlinked characterizations of urban greenery—defined by purposive outputs rather than composition—will assist in correlating health and other benefits with different vegetative forms, a key step in establishing any causal relationship. Until this evidence base can be effectively standardized and quantified, the economic rationality that drives much of urban development will continue to externalize urban greenery's value, thereby generating one local 'Tragedy of the Commons' (urban green space) triggered by another (population growth) (Hardin, 1968). From a behavioral perspective, however, green space is effectively being de-privatized, with high-density urban inhabitants depending heavily upon public green space for recreational use, ecosystem services and less tangible benefits such as quality of place and productivity.

Human health and well-being have provided the primary policy levers for investment in green infrastructure across many cities to date. As environmental health risks increase and behavioral health problems such as obesity and stress become more prevalent, the capacity to 'buy-back' privatized assets for urban greening is simultaneously being reduced. As demonstrated in Figure 19.5, different urban designs and layouts—whether high-rise or detached housing—have the potential to maintain and potentially enhance urban green space. However, as evidenced in the rapidly growing city of Melbourne, Australia, demands for floor space, affordable housing and development profitability often take short-term precedence over the longer-term contributions of green infrastructure to urban resilience and livability. Smart urban planning—proactively adapted to the future demographic, climatic and infrastructural pressures that cities will inevitably face—will be crucial if the resilience of these urban habitats and the health and well-being of their inhabitants are to be sustained into the future.

Key messages

- Urban greenery's health benefits vary depending upon land tenure (public or private), access and the typologies of both the urban environment and the green infrastructure itself.
- Generalizing these benefits across case study evidence can be challenging, as they can be highly context-specific, in some cases difficult to causally link or fundamentally qualitative in nature.
- Complex mechanisms operating across sectors, beneficiary groups and those who maintain or install green infrastructure within cities will need to be developed in order to rationalize and expand widespread urban greenery investments.

Urban greening, human health and well-being

Forward-looking research questions

- How do urban green infrastructure incentives, regulations and other mechanisms differ in effectiveness relative to the land tenure of the investments that they produce?
- At a precinct-scale, what are the systems of scale that operate between the individual and community health benefits derived from urban greenery?
- Under a variable and changing climate, shifting water availability within urban environments is likely to drastically modify the nature and extent of urban greenery and its associated ecosystem services including health-related benefits. What thresholds and mutual interdependencies exist that might restrict or change green infrastructure-based health outcomes into the future?

References

AECOM (AECOM Australia) (2011). *Adaptation of Melbourne's Metropolitan Rail Network in Response to Climate Change*. Melbourne: Department of Climate Change and Energy Efficiency.

Batchelor, P. (1969). The origin of the garden city concept of urban form. *Journal of the Society of Architectural Historians*, 28(3), 184–200.

Bedimo-Rung, A.L., Mowen, A.J. and Cohen, D.A. (2005). The significance of parks to physical activity and public health: a conceptual model. *American Journal of Preventative Medicine*, 28(2S2), 159–168.

Bengston, D.N., Fletcher, J.O. and Nelson, K.C. (2004). Public policies for managing urban growth and protecting open space: policy instruments and lessons learned in the United States. *Landscape and Urban Planning*, 69, 271–286.

Bixler, R.D. and Floyd, M.F. (1997). Nature is scary, disgusting, and uncomfortable. *Environment and Behavior*, 29(4), 443–467.

Bosomworth, K., Trundle, A. and McEvoy, D. (2013). *Responding to the Urban Heat Island: A Policy and Institutional Analysis, Final Report*. Melbourne: Victorian Centre for Climate Change Adaptation Research.

Bowler, D.E., Buyung-Ali, L.M., Knight, T.M. and Pullin, A.S. (2010). A systematic review of evidence for the added benefits to health of exposure to natural environments. *BMC Public Health*, 10(456), 1–10.

Brunner, J. and Cozens, P. (2013). "Where have all the trees gone?" Urban consolidation and the demise of urban vegetation: a case study from Western Australia. *Planning Practice and Research*, 28(2), 231–255.

Buxton, M., Haynes, R., Mercer, D. and Butt, A. (2011). Vulnerability to bushfire risk at Melbourne's urban fringe: the failure of regulatory land use planning. *Geographical Research*, 49(1), 1–12.

Carter, T. and Fowler, L. (2008). Establishing green roof infrastructure through environmental policy instruments. *Environmental Management*, 42(1), 151–164.

Chen, D. (2012). Mitigating extreme summer temperatures with vegetation. *Nursery Papers Technical*, (5), 1-4.

Chen, H., Jia, B. and Lau, S.S.Y. (2008). Sustainable urban form for Chinese compact cities: challenges of a rapid urbanized economy. *Habitat Int.*, 32, 28–40.

Choumert, J. and Salanié, J. (2008). Provision of urban green spaces: some insights from economics. *Landscape Research*, 33(3), 331–345.

City of Portland (2010). *Portland's Green Infrastructure: Quantifying the Health, Energy, and Community Livability Benefits*. Portland: City of Portland.

City of Melbourne (2012). *Open Space Strategy: Planning for Future Growth*. Melbourne CBD: City of Melbourne.

Coates, L. (1996). An overview of fatalities from some natural hazards in Australia. *Paper presented at Conference on Natural Disaster Reduction*. Barton, Australia.

Colding, J. and Barthel, S. (2013). The potential of "Urban Green Commons" in the resilience building of cities. *Ecological Economics*, 86, 156–166.

Coombes, E., Jones, A.P. and Hillsdon, M. (2010). The relationship of physical activity and overweight to objectively measured green space accessibility and use. *Social Science & Medicine*, 70(6), 816–822.

Cordner, S.M., Woodford, N. and Bassed, R. (2011). Forensic aspects of the 2009 Victorian Bushfires Disaster. *Forensic Science International*, 205(1–3), 2–7.

Czemiel Berndtsson, J. (2010). Green roof performance towards management of runoff water quantity and quality: a review. *Ecological Engineering*, 36(4), 351–360.

Foster, J. (2010). Off track, in nature: constructing ecology on old rail lines in Paris and New York. *Nature and Culture*, 5(3), 316–337.

Fraser, S.D.S. and Lock, K. (2011). Cycling for transport and public health: a systematic review of the effect of the environment on cycling. *European Journal of Public Health*, 21(6), 738–743.

Gill, S.E., Handley, J.F., Ennos, A.R. and Pauleit, S. (2007). Adapting cities for climate change: the role of the green infrastructure. *Built Environment*, 33(1), 115–133.

Goodman, R., Buxton, M., Chhetri, P., Taylor, E. and Wood, G. (2010). *Planning and the Characteristics of Housing Supply in Melbourne – Final Report*. Melbourne: Australian Housing and Urban Research Institute.

Haines, A., Kovats, R.S., Campbell-Lendrum, D. and Corvalan, C. (2006). Climate change and human health: impacts, vulnerability and public health. *Public Health*, 120(7), 585–596.

Hansen, J., Sato, M. and Ruedy, R. (2012). Perception of climate change. *Proceedings of the National Academy of Sciences of the United States of America*, 109(37), 2415–2423.

Hardin, G. (1968). The tragedy of the commons. *Science*, 162(3859), 1243–1248.

Heckscher, M.H. (2008). *Creating Central Park*. New Haven, Connecticut: Yale University Press.

Henderson, J.C. (2013). Urban parks and green spaces in Singapore. *Managing Leisure*, 18(3), 213–225.

Hillsdon, M., Panter, J., Foster, C. and Jones, A. (2006). The relationship between access and quality of urban green space with population physical activity. *Public Health*, 120(12), 1127–1132.

Ignatieva, M., Stewart, G.H. and Meurk, C. (2010). Planning and design of ecological networks in urban areas. *Landscape and Ecological Engineering*, 7(1), 17–25.

IPCC (Intergovernmental Panel on Climate Change) (2013). Summary for policymakers. In T.F. Stocker et al., eds. *Climate Change 2013: The Physical Science Basis. Contribution of Working Group 1 to the Fifth Assessment Report of the Intergovernmental Panel on Climate Change*. Cambridge, UK and New York, USA: Cambridge University Press.

Jaffal, I., Ouldboukhitine, S.E. and Belarbi, R. (2012). A comprehensive study of the impact of green roofs on building energy performance. *Renewable Energy*, 43, 157–164.

Kaczynski, A.T. and Henderson, K. (2007). Environmental correlates of physical activity: a review of evidence about parks and recreation. *Leisure Sciences*, 29(4), 315–354.

Kim, J. and Kaplan, R. (2004). Physical and psychological factors in sense of community: new urbanist Kentlands and nearby orchard village. *Environment & Behavior*, 36(3), 313–340.

Kleerekoper, L., van Esch, M. and Salcedo, T.B. (2012). How to make a city climate-proof, addressing the urban heat island effect. *Resources, Conservation and Recycling*, 64, 30–38.

Landry, S.M. and Chakraborty, J. (2009). Street trees and equity: evaluating the spatial distribution of an urban amenity. *Environment and Planning A*, 41(11), 2651–2670.

Lee, A.C.K. and Maheswaran, R. (2011). The health benefits of urban green spaces: a review of the evidence. *Journal of Public Health*, 33(2), 212–222.

Loughnan, M.E., Tapper, N.J., Phan, T., Lynch, K. and McInnes, J.A. (2013). A spatial vulnerability analysis of urban populations during extreme heat events in Australian capital cities, National Climate Change Adaptation Research Facility, Gold Coast, 128 pp. www.nccarf.edu.au/publications/spatial-vulnerability-urban-extreme-heat-events

Lovell, S.T. and Taylor, J.R. (2013). Supplying urban ecosystem services through multifunctional green infrastructure in the United States. *Landscape Ecology*, 28(8), 1447–1463.

Maas, J., Verheij, R.A., Groenewegen, P.P., de Vries, S. and Spreeuwengerg, P. (2006). Green space, urbanity, and health: how strong is the relation? *Journal of Epidemiology and Community Health*, 60(7), 587–592.

Maxwell, D., Levin, C. and Csete, J. (1999). Does urban agriculture help prevent malnutrition? Evidence from Kampala. *Food Policy*, 23(5), 411–424.

Mees, H.L.P., Driessen, P., Runhaar, H. and Stamatelos, J. (2013). Who governs climate adaptation? Getting green roofs for stormwater retention off the ground. *Journal of Environmental Planning and Management*, 56(6), 802–825.

Mentens, J., Raes, D. and Hermy, M. (2006). Green roofs as a tool for solving the rainwater runoff problem in the urbanized 21st century? *Landscape and Urban Planning*, 77(3), 217–226.

Mora, C., Frazier, A.G., Longman, R.J., Dacks, R.S., Walton, M.M., Tong, E.J., Sanchez, J.J., Kaiser, L.R., Stender, Y.O., Anderson, J.M., Ambrosino, C.M., Fernandez-Silva, I., Giuseffi, L.M. and Giambelluca,

T.W. (2013). The projected timing of climate departure from recent variability. *Nature*, 502(7470), 183–187.

QUT (Queensland University of Technology) (2010). *Impacts and Adaptation Responses of Infrastructure and Communities to Heatwaves: The Southern Australian Experience of 2009* 1st ed., Gold Coast, Australia: National Climate Change Adaptation Research Facility.

Park, B.-J., Tsunetsugu, Y., Kasetani, T., Hirano, H., Kagawa, T., Sato, M. and Miyazaki, Y. (2007). Physiological effects of Shinrin-yoku (taking in the atmosphere of the forest) using salivary cortisol and cerebral activity as indicators. *Journal of Physiological Anthropology*, 26(2), 123–128.

Peters, G.P., Andrew, R.M., Boden, T., Canadell, J.G., Ciais, P., Le Quere, C., Marland, G., Rauach, M.R. and Wilson, C. (2012). The challenge to keep global warming below 2 °C. *Nature Climate Change*, 3(1), 4–6.

PwC (Pricewaterhouse Coopers Australia) (2011). *Protecting Human Health and Safety during Severe and Extreme Weather Events: A National Framework*. Canberra: Commonwealth Government of Australia.

Robine, J.-M., Cheung, S.L.K., Le Roy, S., van Oyen, H., Griffiths, C., Michel, J.-P. and Herrmann, F.R. (2008). Death toll exceeded 70,000 in Europe during the summer of 2003. *Comptes Rendus Biologies*, 331(2), 171–8.

Rockström, J., Steffen, W., Noone, K., Persson, A., Stuart Chapin, F., Lambin, E.F., Lenton, T.M., Scheffer, M., Folke, C., Joachim Schellnhuber, H., Nykvist, B., de Wit, C.A., Hughes, T., van der leeuw, S., Rodhe, H., Sorlin, S., Snyder, P.K., Costanza, R., Svedin, U., Falkenmark, M., Karlberg, L., Corell, R.W., Fabry, V.J., Hansen, J., Walker, B., Liverman, D., Richardson, K., Crutzen, P. and Foley, J.A. (2009). A safe operating space for humanity. *Nature*, 461, 472–475.

Roe, M. and Mell, I. (2013). Negotiating value and priorities: evaluating the demands of green infrastructure development. *Journal of Environmental Planning and Management*, 56(5), 650–673.

Rogelj, J., Meinshausen, M. and Knutti, R. (2012). Global warming under old and new scenarios using IPCC climate sensitivity range estimates. *Nature Climate Change*, 2(4), 248–253.

Rosenzweig, R. and Blackmar, E. (1998). *The Park and the People: A History of Central Park*. Ithaca, NY: Cornell University Press.

Scottish Government (2007). *Scottish Planning Policy: Open Space and Physical Activity*. Edinburgh: Scottish Government.

Seto K.C., Dhakal, S., Bigio, A., Blanco, H., Delgado, G.C., Dewar, D., Huang, L., Inaba, A., Kansal, A., Lwasa, S., McMahon, J.E., Müller, D.B., Murakami, J., Nagendra, H. and Ramaswami, A. (2014). Human settlements, infrastructure and spatial planning. In O. Edenhofer, R. Pichs-Madruga, Y. Sokona, E. Farahani, S. Kadner, K. Seyboth, A. Adler, I. Baum, S. Brunner, P. Eickemeier, B. Kriemann, J. Savolainen, S. Schlömer, C. von Stechow, T. Zwickel and J.C. Minx, eds. *Climate Change 2014: Mitigation of Climate Change. Contribution of Working Group III to the Fifth Assessment Report of the Intergovernmental Panel on Climate Change*. Cambridge, UK and New York, USA: Cambridge University Press.

Smith, K.E. and Joyce, K.E. (2012). Capturing complex realities: understanding efforts to achieve evidence-based policy and practice in public health. *Evidence & Policy: A Journal of Research, Debate and Practice*, 8(1), 57–78.

Stabler, L.B., Martin, C.A. and Brazel, A.J. (2005). Microclimates in a desert city were related to land use and vegetation index. *Urban Forestry & Urban Greening*, 3, 137–147.

Stark, B.L. and Ossa, A. (2007). Ancient settlement, urban gardening and environment in the gulf lowlands of Mexico. *Latin American Antiquity*, 18(4), 385–406.

Taha, H. (1996). Modeling impacts of increased urban vegetation on ozone air quality in the South Coast Air Basin. *Atmospheric Environment*, 30(20), 3423–3430.

Tzoulas, K., Korpela, K., Venn, S., Yli-Pelkonen, V., Kazmierczak, A., Niemela, J. and James, P. (2007). Promoting ecosystem and human health in urban areas using Green Infrastructure: a literature review. *Landscape and Urban Planning*, 81(3), 167–178.

Ulrich, R.S. (1986). Human responses to vegetation and landscapes. *Landscape and Urban Planning*, 13, 29–44.

van Dijk, A., Beck, H.E., Crosbie, R.S., de Jeu, R.A.N., Liu, Y.Y., Podger, G.M., Timbal, B. and Viney, N.R. (2013). The Millennium Drought in southeast Australia (2001–2009): natural and human causes and implications for water resources, ecosystems, economy, and society. *Water Resources Research*, 49(2), 1040–1057.

van Kamp, I. and Leidelmeijer, K. (2003). Urban environmental quality and human well-being: towards a conceptual framework and demarcation of concepts; a literature study. *Landscape and Urban Planning*, 65, 5–18.

VLSAC (Victorian Local Sustainability Advisory Committee) (2011). *Victorian Local Sustainability Accord – Urban Forestry Background Issues Paper.* Melbourne: Victorian Government.

Wall, D., Rothschild, N. and Copeland, C. (2008). Seneca Village and Little Africa: two African American communities in antebellum New York City. *Historical Archaeology*, 42(1), 97–107.

Wu, J. (2014). Urban ecology and sustainability: the state-of-the-science and future directions. *Landscape and Urban Planning*, 125, 209–221.

Zhu, P. and Zhang, Y. (2008). Demand for urban forests in United States cities. *Landscape and Urban Planning*, 84(3–4), 293–300.

Zinzi, M. and Agnoli, S. (2012). Cool and green roofs. An energy and comfort comparison between passive cooling and mitigation urban heat island techniques for residential buildings in the Mediterranean region. *Energy and Buildings*, 55, 66–76.

20

FOOD PRICE VOLATILITY AND URBAN FOOD INSECURITY

Marc J. Cohen and James L. Garrett

Since 2007, global food prices have exhibited sharp increases and declines in both nominal and real terms. Despite these fluctuations, real prices have consistently remained higher than those during the relatively stable period of 1990–2006 (Figure 20.1). A recent study contends that "food price volatility (FPV) has become the new norm: people have come to expect food prices to rapidly rise and fall, though nobody knows by how much or when" (Hossain et al., 2013, p. 5).

What are the implications of higher and more volatile global food prices given increasing urban poverty and food insecurity in many low- and middle-income cities? Urban food insecurity differs in several respects from that in rural areas. In particular, urban dwellers at all income levels have fewer opportunities to produce their own food and thus rely primarily on food purchases, while low-income city dwellers often devote the lion's share of their household expenditures to procuring food. As a result, FPV has unique consequences for urban populations in low-income countries; for example, their diets often differ notably from those in rural areas (Garrett, 2000), resulting in the adoption of different coping strategies compared to their rural counterparts. Moreover, policy-makers need different types of responses to meet these unique challenges.

To address the global food price spikes of 2007–2008 and growing FPV, policy-makers have agreed on a number of action plans, including the UN Comprehensive Framework for Action in 2008 (updated in 2010) (HLTF, 2008; HTLF, 2010) and the L'Aquila Food Security Initiative agreed upon at the 2009 Group of Eight Summit (G8 Summit, 2009). Although these plans duly note poor urban dwellers' vulnerability to high food prices, their policy prescriptions focus primarily on rural food production constraints, food stocks and macroeconomic measures. Action in these areas potentially contributes indirectly to urban food security over the long-term, but policy-makers and analysts have paid much less attention to measures that would directly improve urban food security.

This chapter argues that although poverty in low-income countries is often deeper and more widespread in rural rather than urban areas, the disproportionate attention given to rural populations in policy responses to FPV since 2007 is misplaced. Accordingly, this chapter first outlines the pathways through which increases in food prices impact city residents, followed by evidence of impacts and how they have played out in recent years. The last section discusses current policy responses and makes suggestions for enhancing urban food security.

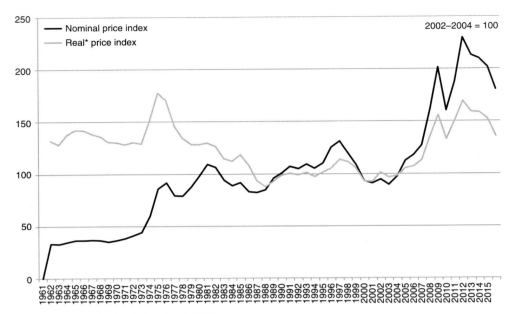

Figure 20.1 Food price index, 1961–2015

Source: FAO (2015).
Notes: *The real price index is the nominal price index deflated by the World Bank Manufactures Unit Value Index. 2015 data are through February.

Contextualizing urban poverty and food insecurity

Although the majority of the world's poor live in rural areas (IFAD, 2010), rapid urbanization is also increasingly pulling poverty and food insecurity into cities (Chen and Ravallion, 2007). The numbers of poor urban residents, from market towns to megacities, are substantial. As of 2010, more than one of every five of the world's extremely poor people (with incomes equivalent to less than 1.25 USD per day) live in urban areas—264 million people in all (Olinto et al., 2013). A study of extreme poverty finds that in 12 of 18 low- and middle-income countries, all of which had nationally representative household surveys taken between 1996 and 2003, urban food insecurity equals or exceeds rural levels (Figure 20.2) (Ahmed et al., 2007). Therefore, focusing exclusively or primarily on rural food insecurity and agriculture when food price spikes occur misses a large part of the problem.

According to the Food and Agriculture Organization of the United Nations (FAO), a state of 'food security' exists when "all people, at all times, have physical and economic access to sufficient, safe and nutritious food that meets their dietary needs and food preferences for an active and healthy life" (FAO, 2006, p. 1). Food security has four dimensions: food availability (provided via domestic production, commercial imports or food aid); food access (via production, purchase, government transfer programs or private charity); food utilization (through adequate diet, clean water, safe sanitation and adequate healthcare to reach a state of nutritional well-being); and food stability (the absence of sudden shocks or cyclical events that cause a loss of access to food) (FAO, 2006). Food insecurity exists when one of any of the aforementioned four dimensions is not met.

While food is likely to be available in urban markets, it may not be affordable to all. Food is the largest item in low-income urban household budgets. Other necessities such as housing,

Food price volatility and urban food insecurity

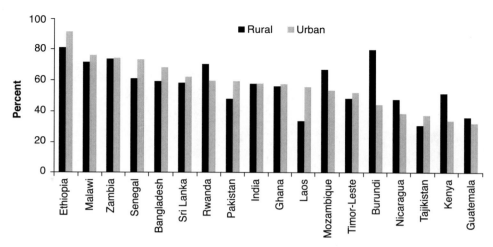

Figure 20.2 Rural and urban incidences of hunger (calorie deficiency)
Source: Ahmed et al. (2007).

transportation, healthcare and education also factor into urban household expenditures. Urban food security reflects household circumstances because household members generally pool their resources and consume food together. While urban incomes may be relatively higher than those in rural areas, these higher incomes may not compensate for higher food prices in urban markets and other urban household budget needs. Ahmed et al. (2007) find that in 20 low- and middle-income countries, the food share of extremely poor urban households' expenditures ranges from 48 percent in Guatemala to 74 percent in Tajikistan. In all but two of the countries, the proportion of the urban household budget spent on food exceeds half. In contrast, low-income urban households in the United States allocate 12 percent of their budgets to food.

In addition to lack of access to adequate food for healthy diets due to high prices, food utilization is an important dimension that contributes to food insecurity. Nutritional well-being results from the interaction of food availability and accessibility with health and care (Figure 20.3). With regard to the latter, urban women are more likely to work outside the home than their rural counterparts, may have less time to care for children and may have difficulty obtaining childcare assistance. As a result, they tend to end breastfeeding two to three months earlier, often depriving their children of nutrients and immunity to disease

---

Box 20.1   Food purchases of urban dwellers in developing countries

Urban dwellers purchase the majority of the food they consume. In Ghana, for example, urban residents purchase 92 percent of their food supplies. In Egypt, the figure is 95 percent. Residents of Lima, Peru, purchase 91 percent of their food. Worldwide, over 97 percent of poor urban households are net food purchasers. In Guatemala, this rises to 98 percent; 99 percent in Malawi; and 100 percent in Vietnam (Garrett, 2002; FAO, 2008a). Because urban dwellers are net food purchasers, urban households have a lower capacity to buffer shocks by producing their own food than their rural counterparts.

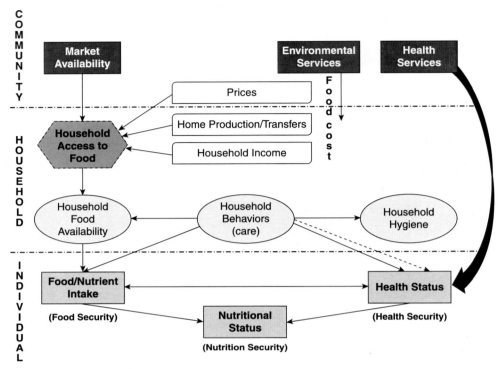

Figure 20.3 Determinants of food, nutrition and health security

Source: Adapted by the authors from UNICEF (1998).

(Ruel, 2000). Furthermore, low-income urban dwellers frequently live in unhealthy conditions. Globally, between 860 million and 1.2 billion people live in slum conditions (HLTF, 2010; UN-Habitat, 2013). As Satterthwaite (2013, p. 1) points out:

> [O]ne in four people in urban areas lives in poor quality and usually overcrowded housing without access to safe, sufficient water, good-quality sanitation and drainage. Their [neighborhoods] lack health care, schools, emergency services and the rule of law. Most of these people live in informal settlements and are at risk from eviction. Moreover, most thriving cities concentrate unsustainable production systems and consumption patterns, taking us closer to dangerous climate change.

In addition to health, care and sanitation, access to clean water may also contribute to nutritional well-being and the capacity to properly grow and prepare food with implications for urban populations. Across low-income countries, only 5 percent of the urban population lacks access to an 'improved' source of drinking water. 'Unimproved' sources would include surface water, unprotected wells and springs, tanker trucks or carts and sole reliance on bottled water. However, the figure rises to 17 percent in Sub-Saharan Africa, the poorest region globally (UNICEF and WHO, 2012). Across low-income countries, 10 percent of the urban population also lacks 'improved' sanitation facilities such as sewers or septic tanks. In many of these places, pit latrines without slabs are the norm. This figure rises to 18 percent in South

Asia and 26 percent in sub-Saharan Africa, the regions that form the center of gravity of global hunger (UNICEF and WHO, 2012).

Sources of urban vulnerability to food price volatility

To better understand poor urban dwellers' vulnerability to rising food prices, it is also important to understand the nature of urban employment, urban food consumption patterns and urban food markets (Garrett, 2000). All of these influence the state of food security in cities and are discussed here in greater detail.

Urban employment

Urban employment data defy stereotypes. A survey of five rapidly urbanizing countries in Africa, Asia and Latin America found urban unemployment rates of 10 percent or less—high, but not spectacularly so (Garrett, 2005). Labor participation rates of the urban poor in formal-sector jobs mirror those of other income levels. The urban poor do not hold many jobs simultaneously, nor does informal employment necessarily predominate in low-income cities, as is sometimes assumed. However, workers do not tend to be day laborers and jobs are often casual, insecure, uncertain and low-paying (Garrett, 2005; Ersado, 2002; Maxwell et al., 2000). Examples include street vending, rickshaw driving, construction and factory work. Employment insecurity and low wages contribute to the vulnerability of the urban poor to global food price fluctuations.

Urban–rural linkages are important sources of urban employment. Even in megacities, the food and agriculture system provides employment opportunities in transport, street vending, retailing, wholesaling and manufacturing. In smaller cities, links to rural agriculture are even more fundamental. Merchants and mechanics provide inputs and tools for farmers. Traders dynamically connect town and countryside. In addition, some urban dwellers travel to the urban periphery to farm. In Egypt and Malawi, 10 percent of non-metropolitan urban residents work in agriculture (Garrett, 2005). City folk also sometimes seek seasonal rural jobs, e.g. in Colombia, they provide coffee harvest labor. Moreover, long-term migrants from rural areas may keep close ties to former homes to hedge against bad times. For example, in Haiti following the 2010 earthquake, 600,000 people migrated from the urban quake zone to shelter with rural family and friends. In Botswana, half the low-income city dwellers keep land or cattle in rural areas (Cohen, 2010; Garrett, 2000, 2005; Tacoli, 2000).

Box 20.2   Employment and incomes for urban dwellers of rapidly urbanizing countries

On average, 50–80 percent of working-age men are in the labor force in rapidly urbanizing countries, with much lower rates for women. In Egypt, Ghana and Peru, 5–10 percent of urban children report having or seeking jobs, with boys much more likely than girls to hold a job. In Bangladeshi slums, 15–20 percent of children work (Garrett, 2005). Few studies have been done to date to capture the significance of illegal incomes and informal employment, which may be significant in some cities and have implications for food security. This varies by country; in Accra, Ghana, 53 percent of the workforce earns its living from informal or self-employment (Maxwell et al., 2000). In Egypt and Malawi, however, 70 percent or more of city jobs pay wages or salaries (Garrett, 2005).

Urban food consumption patterns

Worldwide, a nutrition transition has accompanied urbanization and economic development, as higher relative incomes in urban areas permit purchase of higher-cost, value-added processed foods. Urban dwellers consume more fats, sugars and salt as well as high-calorie foods. This, combined with more sedentary lifestyles, heightens the risk of chronic diseases such as diabetes and heart disease amongst urban populations (Popkin, 2000). In addition, population concentration in urban areas has allowed for marketing efficiencies. At the same time, increased employment outside the home, which raises the opportunity cost of time, has increased demand for ready-to-eat and easily prepared foods. Thus, urban food consumers have shifted from staples such as sorghum, millet, maize and root crops (cassava, yams, sweet potatoes and potatoes) to rice and wheat (Popkin, 2000; Kearney, 2010; Beddington, 2010; Godfray et al., 2010; Delgado et al., 2003). Unlike cassava and other root crops, rice, wheat and maize are widely traded on international markets, which further ties the prices of these crops to fluctuations in global markets and heightens low-income urban dwellers' vulnerability to global price fluctuations. Urban populations also consume more meat and dairy products and a greater variety of fruits and vegetables than poor rural dwellers (Garrett, 2002). The increased dietary diversity enhances micronutrient intakes, but also contains more saturated and trans-fats, sugar and salt and lower amounts of fiber (Garrett, 2002). For further discussion on the interactions of urbanization with food systems and the global environment, see Murray et al., Chapter 2 in this volume.

Urban food markets

Low-income urban residents tend to purchase their food at local markets or neighborhood kiosks, but these are frequently small in scale and distributed geographically (Aragrande and Argenti, 2001; Argenti, 2000). Wholesale urban markets, when they exist in low- and mid-income cities, often are old and located in city centers. Storage facilities are frequently inadequate or badly managed, and refrigeration is scarce, thus costs of food in these markets may be inflated (Aragrande and Argenti, 2001; Argenti, 2000). The more recent trend of domestic and transnational supermarket chains emerging in many lower-income countries is displacing traditional retailers, especially those in local markets. These stores may help overcome infrastructure inefficiencies by utilizing large purchasing and distribution networks, but evidence is mixed on whether large chains lower prices for food, especially staples. While supermarket chains are continuing to increase their retail share, the traditional retail sector and local markets still dominate urban food marketing (Reardon and Timmer, 2008). In general, multinational supermarket chains have little presence in low-income countries; for example, they have yet to reach poor urban neighborhoods in Sub-Saharan Africa. In any case, poor urban dwellers may not have the transportation or cash needed for bulk purchases at supermarkets (Reardon and Timmer, 2008), perhaps resulting in the emergence of a purchasing pattern similar to that of some wealthy countries, i.e. upper- and middle-income consumers shop at chains and get better prices and quality; meanwhile, low-income inner-city dwellers with limited options purchase daily at street vendors and small stores that may offer credit (Reardon and Timmer, 2008).

Street vendors are often an important part of poor urban communities' daily routine, and vending is also an important source of low-income urban employment opportunities, especially for women. Poor consumers can purchase nutritious food from street vendors in small quantities and save on preparation time. However, these foods may be relatively expensive. Consumption of street foods varies by country and city. Urban Nigerians devote as much as half their food purchases to street foods. Residents of all socio-economic status in Bamako, Mali, obtain an average of 275 calories per day from these foods and allocate over 20 percent of their food budgets to procuring them (Ag Bendech et al., 2000). In Accra, street foods account for 40 percent of low-income families' food purchases and even 25 percent of food purchases for families in high-income brackets. Over half of spending is by or for children (Maxwell et al., 2000). Increased reliance on street food consumption is sometimes tied to increases in food and cooking fuel costs. The price of the latter tends to rise more slowly because of production economies of scale. But poor infrastructure (i.e. water and sanitation), inadequate vendor hygiene training and weak or arbitrary enforcement of food safety regulations (if these exist) can create risks (FAO, 1997; Tinker, 1997).

Recent effects of soaring food prices on urban hunger

Between early 2003 and the second quarter of 2008, global nominal prices of wheat, maize and milk tripled, beef and poultry prices doubled and rice prices ballooned five-fold. In real terms, overall food prices jumped 64 percent between 2002 and mid-2008. Causes of the increases included (among other factors): rising energy prices; diversion of food crops to production of biofuel; commodity price speculation; long-term neglect of agriculture as a public investment priority; natural resource degradation; changing weather patterns; rising incomes; urbanization and dysfunctional trade policies such as export embargoes.[1] After peaking in mid-2008, global cereal prices declined 30–40 percent in the third quarter of the year, due to the worldwide recession, good weather and farmers' production responses to higher prices. Most analysts do not believe that prices will return to the low levels of the early 2000s between now and 2020. This is due to continued strong demand for energy and for cereals for food, feed and fuel as well as to structural land and water constraints and likely food production impacts of climate change (FAO, 2008a, b; von Braun, 2008a, b).

Transmission of global food prices into developing country markets varies greatly and depends on degree of import dependence, transport costs, market structures and domestic price policies. In Tanzania, local prices reflected 81 percent of the international maize price increase between 2003 and early 2008, but markets in Surabaya, Indonesia, reflected only 32 percent. Rice prices in Ghana and the Philippines increased by about 50 percent of the global rise (von Braun, 2008a, b). Even within the same sub-region, price transmission can vary dramatically. In July 2008, prices of rice, the main staple, jumped more than 100 percent over the average July price of 2002–2007 in Dakar, Senegal. Senegal is a leading global rice importer. However, in nearby Bamako, Mali, and Ougadougou, Burkina Faso, which rely much less on imports, the price of maize, the local staple, rose only 24 percent and 5 percent, respectively (Egal et al., 2010).

More recently, as food prices have remained above pre-2007 levels, low-income city residents have reported facing trade-offs between food and other necessities, such as education (in Guatemala) and housing (in Kenya and elsewhere) (Hossain et al., 2013). A study of 23 communities in ten developing countries found that while urban wages in both the formal and informal sectors had generally increased between 2011 and 2012, self-employed low-income urbanites did not see their incomes increase (Hossain et al., 2013). Poor households tend to suffer more than others from price increases because they spend a bigger share of their incomes on food. Female-headed households experience larger proportional declines in well-being because their food expenditures tend to account for an even greater share of their resources (FAO, 2008a).

Box 20.3 The effects of soaring food prices on food security and nutrition

Measurable negative effects on urban food security and nutrition have resulted from soaring food prices:

- Globally, the World Bank estimates that 95 percent of the urban income losses, due to the food price spike, came from those who were already poor before the shock (Baker, 2008).
- In November 2008, a balanced diet supplying daily micronutrient needs cost nearly twice as much in Guatemala City as consuming just tortillas and cooking oil, making healthy eating difficult for poor residents (von Braun, 2008b).
- In Cambodia, rice prices increased 100 percent between May 2007 and May 2008. The urban poor were among those most adversely affected as the cost of petrol, water and cooking fuel also skyrocketed. A survey found that 12 percent of Cambodian households (1.7 million people) were food-insecure in mid-2008; the figure was 18 percent among mostly urban female-headed households (CDRI, 2008).
- In Ethiopia, the food price shocks of 2006–2008 reduced purchasing power and increased the number of people living in poverty in urban areas (by as much as 29 percent), particularly in smaller cities (Ticci, 2012). In 2008, Bellatu Bakane, a mother of three in Addis Ababa, remarked, "I get angry because every time I go [to the market] food prices are higher" (FAO, 2008b).

Protests and violence

Higher food prices can also threaten political stability. Scores of countries saw food price protests during 2007–2008. Almost all the demonstrations took place in cities, and several turned violent (FAO, 2008b). According to a study by Bellemare (2012), increased food prices are strongly associated with social unrest, but FPV in and of itself is not. Evidence on the class basis of the 2007–2008 protests is unclear. Some analysts argue that the protests mostly involved middle-class consumers (von Braun, 2008a). Others point to desperation among both poor and middle-class people (FAO, 2008a). It is not unusual for low-income urban dwellers to support opposition to the national or municipal government (FANTA-II, 2008).

While food price protests may involve the poor, these protests are also likely entangled with broader political agendas. Too often, organized crime groups manipulate the political grievances of low-income people to advance their own goals. In Kenya and Haiti, where protests turned violent, criminal elements have a strong urban presence (UN-Habitat, 2007; World Bank, 2006). Peaceful protests in smaller Haitian towns and villages in April 2008 turned into looting and deadly clashes with police and UN troops in the capital, Port-au-Prince. The violence resulted when organized crime groups opposed to the government and UN mission paid slum dwellers to riot. This led to a parliamentary vote of no-confidence in the cabinet and a period of political deadlock (Rights and Democracy and GRAMIR, 2008; Cohen, 2013).[2]

Obviously, policy-makers favor programs that protect city dwellers from the full brunt of food price increases largely because these pacify urban discontent. The Egyptian government boosted spending on its already expensive food subsidy system in 2007–2008 partly to avoid the bloody riots that followed abortive efforts to cut the subsidies in 1977 (Trego, 2012).

Food price volatility and urban food insecurity

However, such poorly targeted subsidies are fiscally unsustainable and often offer the poor limited benefits (HLTF, 2008).

Coping with food price shocks

In light of the vulnerabilities outlined earlier, poor urban dwellers have employed a number of strategies to cope with food price shocks. These include reducing their spending on other necessities and altering food consumption—eating less, shifting to cheaper foods that may be less nutritious or adjusting intra-household distribution. Mothers often forego food when it is scarce, and boys frequently get larger rations than girls. Poor urban households buy food on credit, seek it from neighbors and rely on food distribution programs (Baker, 2008). During past food price spikes, poor households reduced consumption of relatively higher-priced animal source foods, fruits, vegetables and pulses in favor of cheaper, non-processed staples. When poor Indonesian consumers reduced non-staple purchases as prices rose in the late 1990s, iron deficiency increased among young children and their mothers. Higher rice prices likewise spurred malnutrition in Bangladesh (FAO, 2008b). Another strategy is to increase income. Even poor families may be able to send more household members to work, especially women and children, albeit at a potential cost to childcare, health, education and, sometimes, safety. Many urban poor have little room to maneuver, so coping mechanisms may decrease their food security. However, households in smaller cities and market towns often have stronger connections to agriculture. They may grow their own food to mitigate price increase effects, and may even benefit from higher prices through sales.[3]

In response to the 2007–2008 food price spike, poor urban households in developing countries employed all of the above strategies. In cities around the world, higher food prices reduced migrants' remittances back to their rural families (Redwood, 2009; Baker, 2008). In Cambodia, half the households responding to a survey reported that they had cut back on food consumption (CDRI, 2008). In Dhaka, Bangladesh, media accounts in April 2008 indicated that the poor had cut out one daily meal; stopped eating meat, fish and eggs; and could no longer save any money (WHO, 2008). Many poor urban Ethiopians skipped meals and eliminated eggs and vegetables (FAO, 2008a). In Burkina Faso, survey respondents reported increasing food expenditures from 50–60 percent to 75 percent of household budgets. Debt rose, and 40 percent of households surveyed in Niger said they took food loans (FAO, 2008b). In Manila, Philippines, with inflation at 10 percent in May 2008, security guard Leonardo Zafra said he borrowed money at high interest to feed his family, as his daily wage could not cover food, education and utilities (FAO, 2008a). In South Africa, some low-income people engaged in prostitution to afford food (Redwood, 2009).

Urban agriculture

Some urban households grow crops and raise livestock in order to produce a portion of their own food and supplement their incomes. Many urban farmers are women. Urban and peri-urban agriculture has advantages, including low costs, sales near the point of production, producers who respond to market demand and potential contributions to environmental protection. Urban farming systems recycle liquid and solid wastes, but this can be a mixed blessing, as it may lead to soil and water pollution and threaten food safety (Redwood, 2009). Urban and peri-urban agriculture can also facilitate climate change adaptation (see Griffith

Box 20.4 Urban agriculture participation and practice

As much as 40 percent of the population of some cities in Africa engages in urban or peri-urban farming; in Latin America the figure is as high as 50 percent (IFPRI, 2002; Bryld, 2003). In Hanoi, Vietnam, 18 percent of the land is farmed, as is 35 percent in Quito, Ecuador. In Rosario, Argentina's third largest city, 80 percent of the land is vacant and 10,000 city residents farm (Redwood, 2009; IDRC, 2008). By one estimate, 200 million city dwellers produce food for urban markets, accounting for 15–20 percent of global food production (van Veenhuizen, 2006). In West Africa, 20 million households (20 percent of the urban population) farm, providing 60–100 percent of their cities' fresh vegetables (Baker, 2008). However, less than 15 percent of Accra's households engage in agriculture, and those that do cover an average of just 7 percent of household food needs. While urban farmers do produce much of Accra's fresh vegetable supply, wealthier residents consume the bulk of this (Ruel, 2003).

et al., Chapter 28 in this volume) and boost local production of food. Local food production reduces imports, which in turn can lower energy requirements and greenhouse gas emissions (De Zeeuw et al., 2011).

Urban agriculture tends to be part of the unregulated and unmonitored informal economy. Hence, there are no systematic studies of its economic value. Some cities have enacted urban agriculture policies. For example, in Kampala, Uganda, urban agriculture is the most significant land use, and nearly half of city households produce some of their own food. The city council established a regulatory framework following broad stakeholder consultation (IDRC, undated; Cole et al., 2008). Unfortunately, participatory and transparent policy-making is rare, as too many municipal governments regard urban agriculture with hostility. Even when rules are in place, as in Harare, Zimbabwe, they are often poorly enforced and not well known (IDRC, undated; Redwood, 2009; Cole et al., 2008).

International policy prescriptions: necessary but not sufficient for urban food security

The food price spike of 2007–2008 and ongoing FPV have drawn serious attention from global policy-makers. They have backed their action plans with pledges of billions of dollars in assistance to developing countries (G8 Summit, 2009). The plans (HLTF, 2008, 2010; G8 Summit, 2009) have focused primarily on boosting the productivity of smallholder agriculture, enhancing emergency assistance and social protection systems and supporting appropriate macroeconomic and trade policies. Emergency aid and stronger social protection programs can clearly help moderate the effects of higher food prices for the urban poor and macroeconomic and trade policy adjustments may indirectly help if they have food price stabilizing effects. Productivity gains in agriculture may help lower average prices and improve food system efficiency over time to the benefit of urban consumers, but the effects in cities are neither direct nor immediate.

The global policy consensus on how to address hunger and high food prices generally endorses more direct urban-oriented measures, such as greater support for urban agriculture, market development incentives, sustainability and urban development policies and support for secure livelihoods. But the action plans consign this discussion to sidebars, footnotes and annexes (HLTF, 2008, 2010). In order to effectively address urban hunger, policies and programs should better reflect the urban context.

Shaping an urban public policy response

Urban-oriented responses to high and volatile food prices in developing countries need to cushion shocks and provide short-term income supplements. National government policies are particularly important, but municipal governments have a role to play both in ensuring effective policy implementation and, where resources permit, in implementing their own policies and programs. The quality of urban governance therefore matters greatly. Based on the authors' research and a review of the urban food security literature, we conclude that efforts to mitigate food insecurity should center on three public policy areas: preserving income; moderating price increases; and, strengthening interventions to maintain dietary diversity and avoiding harmful coping strategies.

Income preservation

Government price interventions may distort market price signals; they tend to be regressive, expensive and politically difficult to remove. Instead, policies should aim at short-term income improvements (Baker, 2008). Programs should expand easily when need increases and also have mechanisms to scale back as the crisis recedes. It is desirable to make use of existing social programs (such as cash transfer or employment guarantee schemes) to serve these purposes, as citizens will be familiar with the eligibility and exit criteria. It is also especially important to provide unemployed youth with job training and connections to employers (Frankenberger et al., 2000). The presence of a large pool of unemployed young people—especially young men—can contribute to political unrest (USAID, 2005). Since the urban environment facilitates rapid program expansion, the government can use mass media to communicate availability and requirements. It should be possible to use the banking system or vending kiosks and supermarkets to assist in implementation. Targeting assistance via means-testing is often easier in cities, where income is usually wage or salary-based. Geographical targeting is more difficult, as a neighborhood may house both affluent and low-income households. All this, of course, assumes the existence of an effective urban safety net. In some low-income countries, such as Haiti, such programs are in their infancy (World Bank, 2006), but in the past decade, many low-income country governments have strengthened social protection (ILO, 2014).

Subsidizing production costs including energy, or removing existing taxes are alternate approaches to income preservation. These would require government selection of beneficiary products or industries, and efforts to remove the subsidy or re-apply the taxes will likely face fierce opposition. Ultimately, part of the solution is to lift people out of poverty and provide them with access to financial services. This requires investment in infrastructure, institutions and human capital. It also means ensuring land and housing security, so that crises do not cause poor people to lose assets or social capital.

Moderating price increases

Policies needed to moderate price increases require careful design to avoid negative repercussions. Moreover, time may be needed to implement them. In the longer-term, interventions that prevent producers from seeing true market prices will dampen production response, keeping prices higher. Wedges separating consumer and producer prices are often fiscally unsustainable. Tax and trade policies such as temporary import tariff reductions, can moderate price increases, but may also distort price signals. Governments can work through international bodies to discourage food export embargoes. These can increase price volatility, reduce production incentives and promote black markets.

Food distribution policies could also contribute to price stability. The logistics of food distribution are easier in urban areas, and food distribution can have additional benefits, i.e. providing foods fortified with micronutrients. In Peru, for example, community kitchens in the 1990s used food aid commodities and complementary items purchased in bulk to provide low-cost meals. The kitchens also increased independence and empowerment for the women who organized and managed them (IFPRI, 2002). Food and cash distribution programs require nimble international organizations that can respond quickly and with sufficient quantities of food and funds where and when needed. In June 2008, donors pledged 12.3 billion USD to address the food crisis, but by the following October, they had only delivered one billion USD (Cavero and Galián, 2008). Without improvements in international crisis response such as a globally coordinated system of national food reserves or a cash pool, international organizations will have to continue to issue ad hoc appeals for each crisis.

National, sub-national and municipal authorities should seek to enhance urban markets' efficiency. Reduced costs will help lower food prices. Authorities must ensure that urban wholesale and retail markets are well planned and managed, with adequate access and infrastructure. The availability of microcredit can help traders improve hygiene and storage. Dialogues involving consumer, trader and transport organizations can facilitate market efficiencies and dispute resolution (FAO, undated; FAO, 1999). In addition, municipal governments can promote farmers' markets with clear, non-burdensome regulations and logistical support. This can enhance urban–rural links with shorter and more cost-efficient supply chains (FAO, 1999).

Over the longer-term, investment in infrastructure and production technologies will also help lower costs. Investment in and incentives for increased use of information technologies (also easier in cities) will broaden knowledge of costs and prices. This, along with transparent regulations, can encourage fair and open competition and reduce how much price increases are passed on to consumers. Investment in energy efficiency can also help production, processing and marketing systems respond to increased food demand with less vulnerability to oil price spikes.

Strengthening safety nets and coping mechanisms

Other actions could strengthen urban safety nets and coping mechanisms while protecting food and nutrition security (particularly dietary diversity) and human capital. Food banks may be appropriate in urban settings, but are not widespread in developing countries. Municipal governments often do not know how to incorporate urban agriculture into planning, or worry about negative environmental consequences. As a result, potential urban agriculture contributions to food supplies, dietary diversity and incomes go unrealized. City authorities should develop an enabling regulatory framework that incorporates urban agriculture into the formal economy; ensures land tenure security; and addresses food safety, health and environmental concerns. Cities should formulate policy through broad stakeholder consultations. It is valuable to have a municipal department of food and agriculture, as in Kampala, Addis Ababa and other cities (IDRC, undated; Redwood, 2009; Cole et al., 2008). Given the importance of street foods, municipal governments should also train vendors in hygiene, adequately

Food price volatility and urban food insecurity

and consistently enforce regulations and improve infrastructure. Collaborating with vendor associations can help with training and regulatory compliance (Maxwell et al., 2000).

In addition, it is important to improve urban social cohesion. City folk can develop strong social capital but often outside geographical boundaries. Some neighborhoods have strong civic associations, but bonds also form along ethnic, religious or political lines. Municipal government and civil society organizations should help low-income people develop the capacity to organize and articulate demands. They should also strengthen the municipality's capability to respond to citizens. Efforts to improve water and sanitation often help establish trust and mechanisms for further cooperation, especially when community members plan and manage projects. Stronger social networks may facilitate access to credit or food sharing (IFPRI, 2002; Frankenberger et al., 2000; Garrett, 2000).

Conclusion

The urban-focused policies examined in this chapter are an essential supplement to the global action plans for addressing high and volatile food prices. Of particular importance are preparedness measures, particularly establishing temporary social protection programs that can quickly enroll shock-affected people or increase benefit levels and target transfers based on need. Preparedness will also reduce the likelihood that when a crisis strikes, governments will turn to quick fixes that may have unintended negative consequences over the long-term such as export embargoes or poorly targeted subsidies. Additionally, monitoring and evaluation facilitates timely action. Engaging beneficiaries in monitoring and program design and management helps ensure effectiveness. Furthermore, international cooperation and coordination is required. This will ensure coherent action; help to avoid measures in one country that harm others (such as incentives to use food crops to create biofuels, which have contributed to higher global food prices); and avoid reactive global pledging efforts and delays in action.

Key messages

- Although poverty and food insecurity in low-income countries remain concentrated in rural areas, rapid urbanization is transforming these into urban issues.
- Much of the global policy response to the 2007–2008 food price crisis had an undue rural bias. Greater global policy attention needs to focus on urban food insecurity.
- City dwellers in low-income countries depend on purchases to secure their food and have fewer opportunities to produce food for themselves, as compared to rural people.
- Consequently, policies aimed at addressing the effects of FPV on low-income and vulnerable people in urban areas need to focus not only on boosting food production, but also on income, prices and social safety nets.

Future research

Further research is needed to support policies that promote sustainable urban food security and build resilience against future shocks including the likely effects of climate change. In particular, more knowledge is needed about:

- urban labor markets, in order for policies to foster secure livelihoods;
- the economic value of urban agriculture, to help scale up successes and facilitate climate change adaptation;

Notes

1 There is a sizeable literature on this topic, and the discussion is somewhat contentious. See, for example, FAO (2008a); Timmer (2008); von Braun (2008a, b); Clapp and Cohen (2009); Headey and Fan (2010); and Cohen and Smale (2012).
2 Additional information provided by co-authors of Rights and Democracy and GRAMIR (2008) via personal communications.
3 This is our observation based on 28 combined years of research in developing country cities and towns.

References

Ag Bendech, M., Chauliac, M., Gerbouin Rérolle, P., Kante, N., and Malvy, D.J.M. (2000). Les enjeux de la consommation alimentaire en milieu urbain à Bamako. *Santé Publique*, *12*, 45–53. Retrieved from http://fulltext.bdsp.ehesp.fr/Sfsp/SantePublique/2000/1/IMP_BENDECH_ps.pdf.
Ahmed, A.U., Hill, R.V., Smith, L.C., Wiesmann, D.M., and Frankenberger, T. (2007). The world's most deprived: Characteristics and causes of extreme poverty and hunger. *2020 Vision for Food, Agriculture, and the Environment Discussion Paper 43*. Retrieved from www.ifpri.org/sites/default/files/publications/vp43.pdf.
Aragrande, M. and Argenti, O. (2001). Studying food supply and distribution systems to cities in developing countries and countries in transition. *Methodological and Operational Guide*. Rome: FAO.
Argenti, O. (2000). Feeding the cities: Food supply and distribution. Overview. In J.L. Garrett and M.T. Ruel (Eds.), Achieving urban food and nutrition security in the developing world (brief 5 of 10). *2020 Vision for Food, Agriculture, and the Environment Focus 3*. Retrieved from www.ifpri.org/sites/default/files/publications/focus03.pdf.
Baker, J.L. (2008). Impacts of financial, food, and fuel crisis on the urban poor. *Directions in Urban Development*. Retrieved from http://siteresources.worldbank.org/INTURBANDEVELOPMENT/Resources/336387–1226422021646/directions_2.pdf?resourceurlname=directions_2.pdf.
Beddington, J. (2010). Global food and farming futures, *Phil. Trans. R. Soc. B.*, *365*(1554), 2767 doi:10.1098/rstb.2010.0181.
Bellemare, M.F. (2012). Rising food prices, food price volatility, and social unrest. Paper presented at the 2012 annual meeting of the American Political Science Association. Retrieved from http://papers.ssrn.com/sol3/papers.cfm?abstract_id=2107132.
Bryld, E. (2003). Potentials, problems, and policy implications for urban agriculture in developing countries. *Agriculture and Human Values*, 79–86.
Cavero, T. and Galián, C. (2008). Double-edged prices: Lessons from the food price crisis, 10 actions developing countries should take. *Oxfam Briefing Paper 121*.
CDRI (Cambodia Development Resource Institute). (2008). *Impact of High Food Prices in Cambodia*. Phnom Penh: CRDI.
Chen, S. and Ravallion, M. (2007). Absolute poverty measures for the developing world, 1981–2004. *World Bank Policy Research Working Paper 4211*. Retrieved from www-wds.worldbank.org/external/default/WDSContentServer/IW3P/IB/2007/04/16/000016406_20070416104010/Rendered/PDF/wps4211.pdf.
Clapp, J. and Cohen, M.J. (Eds.). (2009). *The Global Food Crisis: Governance Challenges and Opportunities*. Waterloo, Canada: Wilfrid Laurier University Press.
Cohen, M. (2010). Planting now: Agricultural challenges and opportunities for Haiti's reconstruction. *Oxfam Briefing Paper 140*. Retrieved from www.oxfamamerica.org/files/planting-now.pdf.
Cohen, M.J. (2013). *Diri nasyonal ou diri Miami?* Food, agriculture, and US-Haiti relations. *Food Security*, *5*, 597–606.

Food price volatility and urban food insecurity

Cohen, M.J. and Smale, M. (Eds.). (2012). *Global Food-Price Shocks and Poor People: Themes and Case Studies*. Oxford: Routledge.

Cole, D., Lee-Smith, D., and Nasinyama, G. (2008). *Healthy City Harvests: Generating Evidence to Guide Policy on Urban Agriculture*. Lima and Kampala: Urban Harvest and Makerere University Press.

De Zeeuw, H. et al. (2011). The role of urban agriculture in building resilient cities in developing countries. *The Journal of Agricultural Science, 149*, 153–163.

Delgado, C.L. et al. (2003). *Fish to 2020. Supply and Demand in Changing Global Markets*. Washington, DC: International Food Policy Research Institute.

Egal, F., Thiam, I., and Cohen, M.J. (2010). Soaring food prices, climate change and bioenergy: New challenges for food security and nutrition. *SCN News, 38s*, 2–6. Retrieved from www.unscn.org/files/Publications/SCN_News/Supplement_ECOWAS_scnnews38_final.pdf.

Ersado, L. (2002). *Livelihood Strategies in Urban and Rural Areas: Activity and Income Diversification in Zimbabwe. Mimeo*. Washington, DC: International Food Policy Research Institute (IFPRI).

FANTA-II (Food and Nutrition Technical Assistance II Project). (2008). *Emergencies in Urban Settings: A Technical Review of Food-Based Program Options*. Washington, DC: Academy for Educational Development.

FAO (Food and Agriculture Organization of the United Nations). (Undated). *Feeding the Cities*. Rome: FAO.

FAO. (1997). Street foods. *FAO Food and Nutrition Paper 63*. Rome: FAO.

FAO. (1999). *Spotlight: Urban food marketing*. Rome: FAO.

FAO. (2006). Food security. *Policy Brief, 2*. Retrieved from ftp://ftp.fao.org/es/ESA/policybriefs/pb_02.pdf.

FAO. (2008a). *The State of Food Insecurity in the World 2008*. Rome: FAO.

FAO. (2008b). Soaring food prices: Facts, perspectives, impacts, and actions required. Information Paper prepared for the High-Level Conference on World Food Security: The Challenges of Climate Change and Bioenergy, Rome, 3–5 June. Rome: FAO. Retrieved from www.fao.org/fileadmin/user_upload/foodclimate/HLCdocs/HLC08-inf-1-E.pdf.

FAO. (2015). *FAO Food Price Index*. Retrieved from www.fao.org/worldfoodsituation/foodpricesindex/en/.

Frankenberger, T.R., Garrett, J.L., and Downen, J. (2000). Programming for urban food and nutrition security. In J.L. Garrett and M.T. Ruel (Eds.) Achieving urban food and nutrition security in the developing world (brief 10 of 10). *2020 Vision for Food, Agriculture, and the Environment Focus 3*. Retrieved from www.ifpri.org/sites/default/files/publications/focus03.pdf.

G8 Summit. (2009). *"L'Aquila" Joint Statement on Global Food Security: L'Aquila Food Security Initiative (AFSI)*. Retrieved from www.g8.utoronto.ca/summit/2009laquila/2009-food.pdf.

Garrett, J. (2002). Livelihoods in the city: Challenges and options for the urban poor. In *Toward Eliminating Urban Poverty*, Seminar Series, USAID, Washington, DC, pp. 29–38.

Garrett, J. (2005). Living life: Overlooked aspects of urban employment. In U. Kracht and M. Schulz (Eds.), *Food and Nutrition Security in the Process of Globalization and Urbanization* (pp. 512–529). Münster, Germany: Lit.

Garrett, J.L. (2000). Overview. In J.L. Garrett and M.T. Ruel (Eds.), Achieving urban food and nutrition security in the developing world (brief 1 of 10). *2020 Vision for Food, Agriculture, and the Environment Focus 3*. Retrieved from www.ifpri.org/sites/default/files/publications/focus03.pdf.

Godfray, H.C.J., Crute, I.R., Haddad, L., Lawrence, D., Muir, J.F., Nisbett, N., Pretty, J., Robinson, S., Toulmin, C., and Whiteley, R. (2010). The future of the global food system, *Phil. Trans. R. Soc. Lond. B. Biol. Sci., 365*(1554), 2769–2777, doi: 10.1098/rstb.2010.0180.

Headey, D. and Fan, S. (2010). Reflections on the global food crisis: How did it happen? How has it hurt? And how can we prevent the next one? *Research Monograph 165*, Washington, DC: IFPRI..

HLTF (UN High-level Taskforce on the Global Food Security Crisis). (2008). *Comprehensive Framework for Action*. New York: United Nations. Retrieved from www.un.org/en/issues/food/taskforce/Documentation/CFA%20Web.pdf.

HLTF. (2010). *Updated Comprehensive Framework for Action*. New York: United Nations. Retrieved from http://un-foodsecurity.org/sites/default/files/UCFA_English.pdf.

Hossain, N., King, R., and Kelbert, A. (2013). Squeezed: Life in a time of food price volatility, year 1 results. *Joint Agency Research Report*. Retrieved from www.oxfam.org/sites/www.oxfam.org/files/rr-squeezed-food-price-volatility-year-one-230513-en.pdf.

IDRC (International Development Research Centre). (Undated). *Achieving the Millennium Development Goals One Neighbourhood at a Time*. Ottawa: IDRC.

IDRC. (2008). *Shaping Liveable Cities: Stories of Progress Around the World*. Ottawa: IDRC.

IFAD (International Fund for Agricultural Development). (2010). *Rural Poverty Report 2011*. Rome: IFAD.

IFPRI (International Food Policy Research Institute). (2002). Living in the city: Challenges and options for the urban poor. *IFPRI Issue Brief 9*.

ILO (International Labor Organization). (2014). *World Social Protection Report 2014–15: Building Economic Recovery, Inclusive Development and Social Justice*. Geneva: ILO.

Kearney, J. (2010). Food consumption trends and drivers. *Phil. Trans. R. Soc. B.*, *365*, 2793–2807.

Maxwell, D., Levin, C., Amar-Klemesu, M., Ruel, M., Morris, S., and Ahiadeke, C. (2000). Urban livelihoods and food and nutrition security in greater Accra, Ghana. Research Report No. 112. Washington, DC: IFPRI.

Olinto, P., Beegle, K., Sobrado, C., and Uematsuedro, H.I. (2013). The state of the poor: Where are the poor, where is extreme poverty harder to end, and what is the current profile of the world's poor? *Economic Premise, 125*. Retrieved from http://siteresources.worldbank.org/EXTPREMNET/Resources/EP125.pdf.

Popkin, B.M. (2000). Urbanization and the nutrition transition. In J.L. Garrett and M.T. Ruel (Eds.), Achieving urban food and nutrition security in the developing world (brief 7 of 10). *2020 Vision for Food, Agriculture, and the Environment Focus 3*. Retrieved from www.ifpri.org/sites/default/files/publications/focus03.pdf.

Reardon, T., and Timmer, C.P. (2008). The rise of supermarkets in the global food system. In J. von Braun and E. Diaz-Bonilla (Eds.), *Globalization of Food and Agriculture and the Poor* (pp. 189–213). New Delhi: Oxford University Press.

Redwood, M. (2009). Urban agriculture and changing food markets. In J. Clapp and M.J. Cohen (Eds.), *The Global Food Crisis: Governance Challenges and Opportunities* (pp. 205–215). Waterloo, Canada: Wilfrid Laurier University Press.

Rights and Democracy and GRAMIR (Groupe de Recherche et d'Appui au Milieu Rural). (2008). *The Human Right to Food in Haiti*. Montréal and Port-au-Prince: Rights and Democracy and GRAMIR, Montréal and Port-au-Prince.

Ruel, M.T. (2000). Urbanization in Latin America: Constraints and opportunities for child feeding and care. *Food and Nutrition Bulletin*, *21*, 12–24.

Ruel, M.T. (2003). Women and children getting by in urban Accra. *Food Consumption and Nutrition Division City Profiles*. Washington, DC: IFPRI.

Satterthwaite, D. (2013). A future urban poor groups want: Addressing inequalities and governance post-2015. *Briefing*. Retrieved from http://pubs.iied.org/pdfs/17155IIED.pdf.

Tacoli, C. (2000). Rural-urban interdependence. In J.L. Garrett and M.T. Ruel (Eds.), Achieving urban food and nutrition security in the developing world (brief 3 of 10). *2020 Vision for Food, Agriculture, and the Environment Focus 3*. Retrieved from www.ifpri.org/sites/default/files/publications/focus03.pdf.

Ticci, E. (2012). Can inflation be a good thing for the poor? Evidence from Ethiopia. In M.J. Cohen and M. Smale (Eds.), *Global Food-Price Shocks and the Poor: Themes and Case Studies* (pp. 175–187). Oxford: Routledge.

Timmer, C.P. (2008). Causes of high food prices. *ADB Economics Working Paper 128*.

Tinker, I. (1997). *Street Foods: Urban Food and Employment in Developing Countries*. Oxford: Oxford University Press.

Trego, R. (2012). The functioning of the Egyptian food-subsidy system during food-price shocks. In M.J. Cohen and M. Smale (Eds.), *Global Food-Price Shocks and the Poor: Themes and Case Studies* (pp. 212–224). Oxford: Routledge.

UN-Habitat (United Nations Human Settlements Programme). (2007). *Enhancing Urban Safety and Security: Global Report on Human Settlements 2007*. London: Earthscan.

UN-Habitat. (2013). *Streets as Public Spaces and Drivers of Urban Prosperity*. Nairobi: UN-Habitat.

UNICEF (UN Children's Fund). (1998). *The State of the World's Children 1998*. Oxford: Oxford University Press for UNICEF.

UNICEF and WHO (World Health Organization). (2012). *Progress on Drinking Water and Sanitation, 2012 Update*. New York: UNICEF. Retrieved from www.unicef.org/media/files/JMPreport2012.pdf.

USAID (US Agency for International Development). (2005). *Youth and Conflict: A Toolkit for Intervention*. Washington, DC: USAID. Retrieved from http://pdf.usaid.gov/pdf_docs/pnadb336.pdf.

van Veenhuizen, R. (Ed.). (2006). *Cities Farming for the Future: Urban Agriculture for Green and Productive Cities*. Leusden, the Netherlands: Resource Centre on Urban Agriculture and Food Security (RUAF) Foundation, IDRC and International Institute for Rural Reconstruction.

von Braun, J. (2008a). High food prices: The what, who, and how of proposed policy actions. *IFPRI Policy Brief.*

von Braun, J. (2008b). Food and financial crises: Implications for agriculture and the poor. *Food Policy Report 20*. Washington, DC: IFPRI.

WHO. (2008). Rising food insecurity: Health and nutrition implications for the South-East Asia Region. Draft Discussion Paper for the Southeast Asia Nutrition Research-cum-Action Network Meeting. Mimeo.

World Bank. (2006). *Social Resilience and State Fragility in Haiti: A Country Social Analysis*. Report No. 36069-HT. Washington, DC: The World Bank.

21

URBANIZATION AND GLOBAL ENVIRONMENTAL CHANGE

From a gender and equity perspective

Gotelind Alber and Kate Cahoon

Understanding of the complex challenges of urban existence in light of global environmental change is constantly expanding. Particularly in the context of a rapidly changing climate, the specific challenges that cities face are increasingly well documented (see other chapters in this volume), just as the linkages between climate change and gender are receiving a growing amount of attention (Terry, 2009, Dankelman, 2010, Skinner, 2011, Alston and Whittenbury, 2013). For some time already, a considerable amount of material has dealt with the interaction between urbanization processes and gender, addressing poverty, health, migration and population (see, for example, Tacoli, 2012, Tacoli and Satterthwaite, 2013, UN-HABITAT, 2010).[1] Yet, what about the interlinkages between these various thematic areas: climate change, urbanization and gender? While each has received attention in its own right, the nexus between climate change and gender as it relates to urban settlements remains underexplored. As one of only a handful of publications dealing explicitly with this topic points out, cities, gender and climate change are hardly ever found mentioned together on one page (Alber, 2010; Khosla and Masaud, 2010). Although hundreds of cities are already engaged in climate policy—often working together in international and national networks— the vast majority do not yet have a comprehensive climate policy in place, let alone a gender-sensitive response to climate change (Alber, 2010).

This chapter aims to highlight how cities are uniquely positioned to address both climate challenges and social concerns including gender equity. In addition to pointing out *why* a gender analysis of climate policy is necessary and *what* this entails, it will provide a number of suggestions as to *how* this can be done. This includes how gaps in knowledge concerning gender-sensitive approaches to urban climate policy can be addressed, with the goal of working towards a growing number of low carbon, climate resilient, equitable, gender-just and inclusive cities in the near future.

Why is a gender analysis necessary?

The motivation for adopting a gender lens is twofold: first, addressing gender issues in climate policy at all levels, including the local level is a matter of equity. While much of the material dealing with equity issues in the context of climate change has largely focused on differences *between* countries, in terms of contribution to emissions, historical

310

Urbanization and GEC: gender and equity perspectives

responsibility and future development opportunities, it is evident that significant differences exist within countries, and in particular, within urban areas (Mattoo and Subramanian, 2010). Indeed, a range of inequalities and injustices are particularly apparent at the local level (see Boone and Klinsky, Chapter 22 in this volume). Research has shown that in most cities around the world, the divide between the privileged and underprivileged is as large as the global divide between industrialized and low-income countries (UN-HABITAT, 2008). This often means that a small percentage of the population has access to more space—for generous housing, mobility and recreation—while the majority is crowded together in densely populated areas. The poorest groups such as slum dwellers usually have much smaller carbon footprints than the wealthy, and often live in areas most exposed to climate hazards, such as landslide or flood prone areas (UNDP, 2008). Furthermore, even within slums or low-income populations, considerable socio-economic, ethnic, race, gender, religious and age stratification exists. The poorest of the poor are often women, especially single parents and women-headed households from minority communities (Khosla and Masaud, 2010). It seems obvious that these variations result in differing experiences of climate change for different segments of the population, although this is an aspect that is often neglected in the wider climate debate and policy processes. Indeed, climate change has been shown to impact differently on men and women, just as planned legislation, policies and programs have different implications for men and women (Skinner, 2011, EIGE, 2012). Climate change can create or worsen disparities, as is the case when girls cannot attend schools because they are expected to help their mothers with water and fuel collection. This becomes even more time-consuming during floods or droughts, resulting in an increasing gender gap in education. By the same token, political responses to climate change can also have negative impacts on women. A much-cited example is increased biofuel production, which has been shown to contribute to increasing food prices, thereby hampering women's efforts to feed their families (Alvarez, 2013, Rossi and Lambrou, 2008). Moreover, policies to foster biofuel production might lead to a gender bias in the benefits of job creation. For example, a case study in Mozambique indicates that such shifts to commercial agriculture might have resulted in increased capacity to raise income within communities, but at the same time, they have often excluded women, who are often already over-burdened by their time commitments related to family care, subsistence farming and other income-generating activities (Arndt et al., 2011).

While these are merely a small number of examples, they serve to highlight the need for gender dimensions to be considered, in order to avoid worsening the situation for women. Also important from an equity standpoint is that women and men need to be equally involved in planning and decision-making. Existing standards and norms need to be reviewed and addressed as a means of working towards gender equality and to help eliminate discrimination against women. A gender perspective emphasizes that working towards gender equality is an issue in its own right, which needs to be taken into consideration in all areas including climate policy.

Second, addressing gender dimensions can improve the effectiveness and efficiency of climate policy. Policies which respond to the needs of both women and men, and consider their capacities, preferences and ability to access resources, services and infrastructure in the various sectors (energy, transport, water management, disaster risk reduction, etc.) will be more acceptable and viable as a general rule. By the same token, if gender issues are not taken into account, climate change policies and measures run the risk of failing to reach crucial target groups. Thus, in order to be successful, climate responses must speak to the social context in which they are implemented. This means addressing the different needs and opportunities of men and women—related to their different social standing, and differing access to resources.

What does a gender analysis reveal?

Awareness of this necessity to address social dimensions when dealing with climate change has increased within recent years (UN, 2011). This is reflected in the burgeoning literature addressing gender dimensions of climate change, which builds on the substantial amount of work done in the past on the relationship between women, gender and environment as well as in the field of women and development (UN, 2011, EIGE, 2012). The focus of the overwhelming majority gender-based analysis, however, has been on the 'vulnerability' of women to climate impacts (see, for instance, Alston and Whittenbury, 2013, Dankelman, 2010, Patt et al., 2009) or increasingly, the need to include women in decision-making (UN Women and MRFCJ, 2013). Similarly, the few case studies available on gender-sensitive responses to climate change are predominantly set in rural contexts, highlighting primarily the gender dimensions of adaptation, such as women's leading role in community-based adaptation, or evaluating the REDD (Reducing Emissions from Deforestation and Degradation) mechanism from a gender perspective (USAID, 2011).

This narrow approach has been criticized for its tendency to portray women either as 'victims' of climate change or as 'agents of change', with an unjustified focus on low-income countries (Arora-Jonsson, 2011, see also GenderCC, 2009). More recent publications have tried to redress this by looking not only at the gendered impacts of climate change, but also at the gender dimensions of policies to mitigate climate change such as the gender-specific consumption patterns in high-income countries, which have been shown to affect contributions to greenhouse gas (GHG) emissions (Dankelman, 2010, Skinner, 2011). The European Institute for Gender Equality (EIGE) report 'Gender Equality and Climate Change' (2012) focusing on the 27 European Union member states is an important contribution to this end, highlighting that gender dimensions remain of crucial importance in high-income countries, even in contexts where considerable progress has already been made on gender equality. Yet, publications on the gender dimensions of urban climate policy, in particular, in high-income countries are rare. Notable exceptions, which address urban contexts in these countries, include two recent Swedish studies that look at gendering local adaption (Edvardsson Björnberg and Hansson, 2013) and the gendered dimensions of climate change response in Swedish municipalities (Dymén et al., 2013).

These developments suggest that a gender-based analysis of climate change needs to be nuanced and multi-dimensional. For one thing, the notion of 'gender' is in itself quite complex. The term is generally used to describe the socially and culturally constructed roles, expectations, identities and behaviors that a given society considers appropriate for men and women (UN Women, n.d.). Thus, while 'sex' is usually used to describe the physical differences between women and men, based on their sexual and reproductive functions, 'gender' is not fixed – it varies across different cultures and societies and changes over time (UN Women, n.d.). Moreover, gender differentials are also interwoven with other factors including ethnicity income, class, age, education and so forth. Social categories are not independent from one another, they overlap and are mutually reinforcing. Thus, in order to understand the multiple layers of discrimination, which might exist, the concept of intersectionality is useful. This also means that considerable differences exist within given categories, making it difficult to generalize about either 'women' as a group, or 'men' as a group, because lived realities of both men and women are highly diverse.

Urbanization and GEC: gender and equity perspectives

Nevertheless, it is still possible to draw a number of conclusions by adopting a gender lens, as socially constructed roles and identities, and underlying power relations are affecting the way women and men experience and respond to climate change (Skinner, 2011). While the purpose of this chapter is not to provide a detailed analysis of gender inequalities, it is worth noting that the disparities as well as the roles and responsibilities mentioned here in the context of climate change stem from ongoing structural injustices and unequal power relations between men and women. This is reflected in the prevalence of androcentric systems, which favor the male perspective and neglect women's identities, attitudes and behavior or regard these as deviations from the 'norm' (Jarvis et al., 2009, Röhr et al., 2008). This also relates to the gendered division of labor which persists in all countries in the world, meaning that women are more likely to be given the primary responsibility for family care including the provision of food, caring for children, elderly and sick family members (Miranda, 2011). In the event of climate change, and in post-disaster situations particularly, this workload may increase, resulting in an even greater 'care' burden for women. Policies to mitigate climate change also have the potential to increase the work burden of the family care-givers, as is the case with some energy saving measures, utilizing biomass-based heating systems or separating waste (Carlsson-Kanyama and Lindén, 2007, Organo et al., 2012). Furthermore, in most countries, educational gaps and occupational segregation continue to entrench gender-based discrimination and differentials in income (the gender 'pay gap') and assets. Other barriers to gender equality include a widespread bias in decision-making, reflected in the under-representation of women in both public entities and the private sector. Legal and socio-cultural discrimination also mean that in many places, women are not guaranteed basic rights or are unable to own or inherit land, or that they face restrictions on their mobility or their activities outside the home. A gender-aware approach to climate change needs to address these inequalities, for "it is vital that women's rights and gender equality are promoted, rather than undermined, through all climate change policies and interventions" (Skinner, 2011, p. 9).

According to the EIGE report mentioned above, there are two main aspects of climate policy that need to be examined from a gender perspective. The first is the question of power and participation: who is planning and deciding on climate policy, and how is this conducted? Factors to consider include the share of women and men in decision-making positions, in climate change research and technology developments, and in international negotiations as well as in participatory processes in general. It is worth considering whether a more equal 'gender balance' might shape debates or result in different priorities and outcomes, or perhaps even a more representative or holistic approach to climate policy (see, for example, Kronsell, 2013, Chattopadhyay and Duflo, 2004). The second component relates to the various aspects of climate change: who is affected the most by its impacts; who emits what amounts of GHGs and for what purposes; how is climate change perceived; which mitigation and adaptation solutions are preferred; and, how are various social groups impacted by policies and measures? While these different aspects are no doubt interlinked, this can be broadly summarized for the purpose of a gender analysis into the following dimensions to be addressed: vulnerability to the impacts of climate change, the causes of climate change and responses to it in terms of policy.

Gendered impacts

While there is a tendency in climate change literature to think of vulnerability in terms of physical exposure and sensitivity, climate vulnerability has strong social dimensions because it is linked to access to resources and is rooted in the construction of everyday social space or social existence (Ahmed and Fajber, 2009). Therefore, the impacts of climate change affect people differently based on their sensitivity and their capacity to respond (explored in detail in UNFCCC, 2007). The Hyogo framework thus defines vulnerability as a "set of conditions determined by physical, social, or economic and environmental factors or processes which increase the susceptibility of a community to the impacts of hazards" (UNISDR, 2005). These conditions can include climate variation, which is strongly related to location. For example, coastal cities have a higher exposure to sea level rise and storm surges (van Staden, 2013, Hallegatte et al., 2013), and within these cities, low-lying settlements and those located in direct proximity to the sea are most prone to flooding (IPCC, 2014). Vulnerability is also related to sensitivity, which is the degree to which a community is affected by these impacts, or is responsive to climate stimuli (van Staden, 2013, Hallegatte et al., 2013). For example, fragile houses in informal settlements are more sensitive to storms and flooding than robust, well-built houses.

Adaptive capacity also plays a key role. This is defined by the Intergovernmental Panel on Climate Change (IPCC) as "the ability of a system to adjust to climate change to moderate potential damages, to take advantage of opportunities, or to cope with the consequences," with systems being either natural ecosystems or human systems such as individuals, households and communities (IPCC, 2001). This depends largely on access to, and control over, various kinds of resources including reliable income and financial resources, natural resources, infrastructure and services such as energy, water and sanitation. Community networks, social support and access to information on climate risks, knowledge and skills are further major factors. Evidently, poverty substantially contributes to vulnerability. Poor citizens' adaptive capacity is much lower, largely because they lack the assets to protect themselves against crisis. Their settlements are also often more exposed to climate hazards such as landslide or flooding; in particular, informal and high-density settlements are highly vulnerable. Furthermore, as urban populations tend to be more dependent on basic infrastructure and services to secure their livelihoods, they have fewer options to cope with disruptions.

Having understood that vulnerability and adaptive capacity is connected to a range of social factors, it becomes important to further this analysis with a consideration of gender issues. Although it is difficult to come across reliable statistics, it is generally accepted that on a whole, more women are living in poverty than men (see, for example, UN DESA, 2010). In particular at the household level, female-headed households are more likely to be poor than male households of the same kind, particularly in the case of households of lone mothers with young children (UN DESA, 2010). In urban areas, the proportion of female-headed households is greater than in rural areas, and often they make up a disproportionately high share of slum dwellers, lacking resources and services, and security. It is likely that considerable intra-household inequality exists between men and women, given that in many regions women are less likely to have a cash income, and a significant proportion of married women have no say in how their cash earnings are spent (UN DESA, 2010). Given that a lack of income and assets increases vulnerability and limits people's ability to provide adequate self-protection, female-headed households, for example, often experience particular hardship. As one source focusing on Bangladesh highlights, this can mean that the ability of women to create safe conditions in the face of impending floods or cyclones is reduced (Cannon, 2009, p. 11).

Thus, even within low-income populations, women often have particular vulnerabilities as a result of gender-related inequalities. Poverty is just part of the picture: gender-differentiated roles and responsibilities in families and households, and gender-segregated labor markets also contribute to the differentiated vulnerabilities of women and men to the effects of climate change (EIGE, 2012). A higher proportion of women work in the informal sector and are therefore subject to greater income insecurity, in particular, during disruptions due to disasters. For example, women are more likely to earn their income from activities at home and face considerable losses if their house or equipment is destroyed in a flood. In certain contexts, climate-related shortages of energy, water and food can also mean an increased workload for women, as a result of the role that they are required to adopt within the family (Skinner, 2011). Similarly, socially defined roles often require women to care for children as well as the elderly and the sick. This can become more demanding when climate change impacts negatively on the health and psychological well-being of individuals and communities. In some parts of the world, such as in Bangladesh, social norms impede women's access to early warning systems during extreme weather events as well as to emergency post-disaster services. Shelters might also be less accessible for women due to mobility constraints and childcare responsibilities, or less suited to accommodate them in terms of their hygienic and safety requirements (Ahmed et al., 2007, Alber, 2010, Enarson, 2000). Similarly, reproductive and maternal health issues are also linked to vulnerability; the need for sanitation, for instance, can be more difficult to address in the context of climate impacts.

Gendered contributions to climate change

Climate change doesn't just affect men and women differently; gender is also a relevant consideration when seeking to identify (and address) the causes of climate change. While data on a global level is still lacking, research in Europe has shown that women and men's contributions to GHGs do in fact differ, both in terms of the quantity, and the quality or purpose of consumption (Räty and Carlsson-Kanyama, 2009). These differences stem from prevailing gender roles and identities, which affect behavior and consumption patterns (Räty and Carlsson-Kanyama, 2009).

In general, carbon footprints are directly correlated with income and spending—those who are well-off live in larger dwellings, own more energy consuming devices, drive larger motorized vehicles, and consume more goods and services. How does this relate to gender? A number of attempts have been made to show that the carbon footprints of women and men differ. Statistics from the transport sector, for example, in high-, middle- and low-income countries show that men are more likely to travel to work and for leisure, while women tend to work closer to or at home and make more shopping trips (EIGE, 2012, Tran and Schlyter, 2010, Oldrup and Romer Christensen, 2007). In general, women travel shorter distances and have more limited access to motorized means of transport, which in some contexts means that they walk more and over longer distances than men. A case study on Hanoi, for example, reveals that in this city, women walk short distances and use bicycles for longer distances, whereas men use motorbikes (Tran and Schlyter, 2010). Moreover, men use cars more often including for short distances, and even for very short trips, while women are more strongly dependent on public transport services (PAKLIM, 2013, Tran and Schlyter, 2010, Oldrup and Romer Christensen, 2007). These differences provide some explanation for men's higher energy consumption, which combined with their higher income, spending and food preferences, results on average in a larger carbon footprint compared to that of women. It is worth noting, however, that this does not necessarily mean that women are more 'climate-friendly'—in many cases they merely lack the means to generate a large carbon footprint, as a result of more limited access to energy and transport services, or financial means for consumption. In order to successfully mitigate the causes of climate change, it is important to understand how GHG emissions occur including at the consumer level. For this reason, policy-makers need to be aware of such gender-based differences, so that they can tailor their climate policy interventions and develop effective and equitable responses.

Gendered responses to climate change

Gender is an important consideration at the response level for two main reasons. As highlighted earlier, policy responses are likely to be more effective if they consider gender dimensions. First, this can relate to capabilities to mitigate climate change at the individual level, as "the portfolio of options women and men have to convert to a low carbon lifestyle and to invest in energy efficiency or renewable energy installations are shaped by education, gender roles, division of labor in the household, and income" (EIGE, 2012, p. 25). In many parts of the world, women are responsible for the provision of water and food for their families and are therefore more dependent on essential resources and services such as cooking fuels, electricity, mobility and water. This has been analyzed in a recent case study in Delhi, India, for example, where some 20 percent of the city's inhabitants live in informal settlements. Women living in these settlements already spend huge amounts of time collecting water. With the impacts of climate change in these areas, the situation of these women is likely to worsen, unless action is taken to address the challenges of urbanization and climate change (Kher et al., 2015).

Energy needs also vary according to how much time is spent in the home and which tasks need to be carried out. For example, women tend to spend more time doing unpaid work in the home (such as household chores, or caring for a child or other adults), and are thus more reliant on energy for household devices as well as the thermal comfort of their homes and indoor air quality, which has consequences for heating systems and energy consumption (EIGE, 2012). This also means that women are more strongly affected by scarcity of energy, or 'fuel poverty', even in high-income countries. It has also been noted that women due to their lower socio-economic status are less likely to own their own house or be in the position to make decisions relating to major investments such as heating systems, renovations or household appliances (EIGE, 2012).

Yet, while women often lack the opportunity to carry out energy efficiency-related improvements or investments, it has been observed that they are often more prepared and willing to make changes. Research in EU countries suggests that women tend to be more concerned about climate change than men and are more likely to take personal action to mitigate climate change (European Commission, 2009, European Commission and European Parliament, 2009, BMU, 2006). Again, this is less an indication of any inherent 'greenness' of women, but it can be seen as a reflection of socially constructed roles and expectations such as that women 'care' for the fate of future generations (Hemmati and Röhr, 2007, 2009; MacGregor, 2010).

Second, as already mentioned, the implementation of response measures can also impact differently on men and women. Social dimensions, such as gender, need to be considered in order to maximize positive effects and importantly, to avoid adverse socio-economic impacts. Even though many climate policy options can produce important co-benefits—for instance, savings on energy (and energy bills!)—they also have the potential to disadvantage certain groups, particularly if policy relies mainly on financial instruments such as taxation or market-based mechanisms. This can mean, for example, that low-income households are faced with higher prices for energy, or that women are excluded from the benefits of job creation if these jobs are in sectors where women are underrepresented, e.g. the renewable energy industry. Gender experts have suggested that this is likely to occur unless specific action is taken to encourage women to enter this business sector and improve their opportunities to flourish (Röhr and Ruggieri, 2008, BIDS, 2014).

While these two aspects—differing capacities to adapt and mitigate, and differing impacts of political responses to climate change—need to be considered from a gender perspective in order to successfully carry out these climate responses, an additional factor relates to the policy-making process itself. The ongoing lack of consultation and underrepresentation of women in decision-making has been shown to be more entrenched in climate policy than in other policy areas (UN Women and MRFCJ, 2013). In the energy and transport sectors, for instance, the proportion of women tends to be critically low. An analysis by the Climate Alliance of European Cities (2005) indicates that in positions relevant to climate change policy such as urban energy and transport planning, the share of women is considerably lower than men, particularly in executive positions.

This is also the case at the international level, with women's representation in some bodies of the United Nations Framework Convention on Climate Change (UNFCCC) remaining even recently as low as 10–11 percent (UN Women and MRFCJ, 2013, p. 8). While a recent decision at the 18th Conference of the Parties aims to address ways to promote gender balance and improve the participation of women in UNFCCC negotiations (UNFCCC, 2012), those working in the field of gender and climate change have pointed out that this is merely one step—albeit an important one—in the direction of gender equality in climate policy-making (GenderCC, 2013). At the national and local level, in essentially all parts of the world, much remains to be done in order to address gender imbalances and inequalities, and to critically analyze the diverse situations and needs of both women and men.

Closing the gap: gender-sensitive approaches to urban climate policy

Given the complexity of the various issues which are revealed by adopting a gender lens on climate change, it would be surprising if the following questions did not emerge at some stage: how can urban policy be expected to take all of these factors into account, when cities are already facing the challenge of dealing with climate change? Furthermore, if they are prepared to do this, how can policy-makers address gaps in knowledge when it comes to gender-sensitive approaches? A number of answers can be provided to these questions, in addition to the argument made at the beginning of this chapter that climate policy, which addresses gender dimensions, is likely to be more effective and more equitable. Cities are increasingly recognized as essential actors of climate change and are well placed to contribute to emissions cuts, and to develop strategic responses to adapt and build resilience.

Urban policies have the capacity to integrate social issues such as poverty alleviation and gender equality into climate responses. As awareness of the social dimensions of climate change increases, it becomes clearer that involving communities is necessary to create change; indeed, the more representative urban climate policies are—and the more attentive to social and gender relations—the more acceptable, viable and useful they are likely to be to the people they are targeting (for more detail, see Alber, 2015). Local governments are in a particularly good position to tackle both the causes of climate change and the structures which perpetuate gender inequality, which, when combined, can represent an important step in overcoming the 'fossil patriarchy', as it has been described by gender activists (GenderCC, 2012).

Furthermore, if cities address both climate change and social inequalities, the implementation of gender-sensitive policies can bring important co-benefits including (but not limited to) improved air quality and health, livability of cities, job creation and enhanced resilience. In order to tap into these benefits, and in light of the inequalities regarding access to energy, transport and other services at the city level, mitigation policies should avoid focusing merely on carbon or short-term GHG emissions reductions. Instead, cities would be advised to pursue a broad multi-dimensional approach with the aim of becoming low-carbon as well as more resilient, equitable, gender-just and inclusive. This involves responding to the different needs of men and women, addressing both poverty and affluence in terms of housing, energy, mobility and consumption, and also taking into consideration the care economy and informal economy, in which women are disproportionately involved. While these dimensions are quite complex, a number of concrete examples of what can be done at an urban level are provided below.

One of the priorities for gender-sensitive urban climate policies is building resilience. In particular for cities in many middle- and low-income countries, adaptation to the impacts of climate change is imperative. Technical measures such as improving physical infrastructure are necessary, but they should not be prioritized over building resilience among communities and neighborhoods. In order for this to be addressed, a change of perspective is required. Rather than focusing solely on climate phenomena or technical responses, policies must address social relations and provide more attention to non-technical, 'soft' approaches. These would include efforts to improve the livelihoods of citizens as well as food security, housing and basic infrastructure for energy, water and sanitation, mobility, knowledge and skills, and cohesion in the community. A differentiated approach is crucial in order to understand gender-related dimensions. One example of many would be providing for an adequate resolution of vulnerability maps during risk and vulnerability assessments (Alber, 2010, Ahmed, 2010), which looks not only at neighborhoods, but also at households in order to capture intra-household disparities.

Specific sectors requiring attention in this context are water and sanitation, and disaster management. Here, cities can rely on a wealth of often underemphasized practical experience, and the resources which already exist on gender-sensitive approaches provided by international organizations and networks such as the Gender and Water Alliance (GWA) and the Gender and Disaster Network (GDN).[2] In low- and middle-income countries, upgrading informal settlements in collaboration with slum dwellers' groups is crucial, rather than merely relocating slums to remote areas where the exclusion of marginalized groups—particularly women—might be exacerbated by a lack of opportunities for livelihood and access to jobs.

Another priority is urban planning, as it is a core area and very specific field of action for local governments. It is also highly relevant both in terms of climate change and gender equality. Land use patterns, spatial structures and urban form are decisive in defining the resilience on one hand, and the future carbon intensity of societies on the other hand. Urban development and planning can, for instance, avoid the creation of flood prone locations and contribute to reducing the heat island effect. In terms of mitigation, working towards compact cities and mixed-use quarters allows for short distances and climate-friendly mobility, thereby supporting the mobility of citizens who lack access to motorized transport, such as women. Urban transport is also a core area for climate change mitigation, given the strong gender dimension and extensive evidence on gender-differentiated trip patterns and mobility modes. Working towards accessible, affordable and safe public transport systems in cities therefore has the potential to contribute to both GHG emissions reductions and gender equality.

A further priority for urban mitigation is certainly energy consumption. Gender-sensitive urban energy policy should especially address the residential sector by developing strategies to reduce energy consumption through energy efficiency and sufficiency, and at the same time, improving access and affordability for underprivileged groups. Moreover, gendered preferences need to be considered in the policy-making process as well as differences between men and women when it comes to access to information and affinity to technologies (see, for instance, Alber, 2013).

Interestingly, some more recent projects have attempted to combine both mitigation efforts and improve social conditions for urban inhabitants, as highlighted by the following case study from Bangladesh, where new energy efficient Hybrid Hoffman Kilns were installed in several brick manufacturing operations. The aim was to reduce air pollution and GHG

Box 21.1 Key questions used to conduct a Gender Impact Assessment (GIA) of climate policies and measures (see Alber, 2015)

In a first step, some simple questions help to determine whether a policy or measure will have an impact on gender equality. In order to gain representative responses, a balanced participation of women and men should be ensured and gender experts should also be involved.

- Does the policy or measure concern one or more target groups and will it affect the daily life of the population or specific groups?
- Does the policy or measure affect gender differences in regard to rights, resources, participation, values and norms?

A detailed analysis helps to improve the policy or measure in terms of its effects on gender equality. Some general questions to ask are:

- What impact does the policy or measure have on gender equality?
- Does the policy or measure affect equality policy objectives? In particular, does a program or policy initiative affect women and men differently, and might it lead to positive or negative impacts on gender equality?
- What data and knowledge are available to assess the impacts of the measure on gender equality, e.g. sex-disaggregated data?
- Who are the actors involved in the development of the initiative, and which additional experts and groups should be involved? Is there a gender balance in the group of actors?
- Is there a need for further (sex-disaggregated) data, information and research?

Some questions more specific to climate and energy policy could be:

- Intra-household dynamics: If policies and measures are addressing households, who is actually targeted? Who is consuming energy for what purposes?
- Behavior: If measures are addressing consumption habits, whose behavior is to be changed?
- Care work: Who is responsible for family care, and how can care work be taken into consideration without reinforcing gender stereotypes?
- Time scarcity: Do policies and measures lead to an additional work burden, and who is affected?
- Promotional programs: If incentives are provided, who might benefit, e.g. only homeowners, or also tenants?
- Job creation: If additional jobs are created, who might benefit? Would accompanying measures be necessary to ensure that women and men benefit equally?

emissions. As the brickmaking industry in Bangladesh is characterized by informal labor or low employment standards, a transformation from the informal to formal sector took place that included financial investment as well as measures to improve occupational health and safety standards. A large number of women working in this sector were able to gain access to formal jobs with a reasonable wage all year round, and enjoy benefits such as health facilities, availability of drinking water, the provision of washrooms and appropriate clothing. This kind of holistic approach is a good example of how local governments and cities can conduct mitigation measures, while also improving the resilience and social conditions of inhabitants. It should be noted that these social co-benefits do not come automatically, but rather need additional efforts to negotiate with the sector and funding institutions to ensure that labor rights and social dimensions are addressed, particularly when it comes to gender dimensions (for more information, see UNDP, n.d.).

In all these fields, a range of existing methods and tools can be applied by cities to assess and improve their climate policy in terms of gender. These include GIAs, gender budgeting and gender mainstreaming. The following set of criteria, initially developed for urban transport projects, provides an illustration of how projects and programs can be assessed for their gender impacts in a number of crucial areas (Spitzner et al., 2007):

- Care economy (unpaid care-work for the family and community): Does the policy take into account the requirements of care-work adequately, which is mostly done by women (needs of time, transport, energy, etc.)?
- Resources: Do the financial resources and measures of a project benefit women to the same extent as men? Does the project lead to a more balanced distribution of public resources?
- Women in decision-making: To what extent does the policy contribute to increasing women's influence in policy design, planning and decision-making processes?
- Androcentrism: Does the policy reinforce the model of male lifestyles and ways of thinking while those of women are seen as 'different'? Or does it help to challenge the widespread generalization of the male experience and perspective?
- Symbolic order: Does the policy or project contribute to changing gender-biased power relations and allocation of duties?
- Harassment: Does the policy contribute to reducing the harassment of women? Does it contribute to relieving women of threats, restrictions and sanctions?

As the EIGE (2012) report suggests, such gender-sensitive criteria are applicable to a variety of contexts, in all countries across the development spectrum.

Conclusions and recommendations

This chapter has highlighted the importance of considering the nexus between gender, climate change and cities. In conclusion:

- There is a growing amount of evidence of the gender dimensions of both mitigation and adaptation. Research suggests that addressing these dimensions can make urban climate policy more effective and efficient as well as improve its acceptance and viability.
- Therefore, cities should pursue an integrated approach and enter into commitments to work towards becoming low carbon, climate resilient, equitable, gender-just and inclusive.

- For this to be implemented, improvements in gender balance are required, along with the development of gender-sensitive policies and measures, which take into account the different needs of women and men due to power relations, societal roles, the division of labor, etc.

To support such endeavors:

- City networks can play a pivotal role in their work with local governments (Gugler, 2004; Kern and Alber, 2009). They are able to address the linkages between gender and climate change and sensitize cities on the need to integrate both aspects into their policy. Methods and tools providing practical guidance for local governments to address climate change should also be revised, in order to reflect this need.
- National governments can assist by providing guidance on urban strategies, policies and measures as well as offering incentives to stimulate action. They can recommend methodologies that integrate gender considerations, and apply criteria for incentives that involve gender-balanced participation and gender-sensitive priorities and approaches.
- Although much of the research on gender and climate change to date has focused on rural areas, it is clear that the urban context needs to be urgently addressed. More efforts are required to more fully understand the linkages between gender and climate change in urban areas. This includes addressing the lack of sex-disaggregated data available on vulnerability to the impacts of climate change as well as, crucially, on the drivers of climate change (such as energy consumption and mobility), and on the socio-economic impacts of policies. Yet, while data help to highlight gender dimensions and differences, this is not sufficient alone. An analysis of the underlying causes is required in order to effectively address inequality and discrimination, at the same time as addressing climate change.

Notes

1 For a critical introduction to gendered urban analysis, see also Jarvis et al. (2009).
2 For more information, see Gender Water Alliance and UNDP (2006) and Gender and Disaster Network (2009) as well as ISDR, UNDP and IUCN (2009).

References

Ahmed, A. U. (Ed.). (2010). *Reducing Vulnerability to Climate Change: The Pioneering Example of Community Based Adaptation in Bangladesh*. Centre for Global Change (CGC) House in association with CARE Bangladesh: Dhaka.

Ahmed, A. U., Neelormi, S., and Adri, N. (2007). *Climate Change in Bangladesh: Concerns Regarding Women and Special Vulnerable Groups*. Jointly published by UNDP, Climate Change Cell, DFID, BASTOB and Centre for Global Change (CGC): Dhaka.

Ahmed, S., and Fajber, E. (2009). Engendering adaptation to climate variability in Gujarat, India. In G. Terry (Ed.), *Climate Change and Gender Justice* (pp. 39–56). Practical Action Publishing Ltd; Oxfam GB.

Alber, G. (2010). *Gender, Cities and Climate Change*. UN-HABITAT: Nairobi, Kenya.

Alber, G. (2013). Gendered access to green power: motivations and barriers for changing the energy provider. In M. Alston and K. Whittenbury (Eds.), *Research, Action and Policy: Addressing the Gendered Impacts of Climate Change* (pp. 135–148). Dordrecht: Springer Netherlands.

Alber, G. 2015. *Gender and Urban Climate Policy – Gender-Sensitive Policies Make a Difference: Guidebook*. Eschborn, Bonn: Deutsche Gesellschaft für Internationale Zusammenarbeit (GIZ) GmbH in collaboration with United Nations Human Settlements Programme (UN-Habitat) and GenderCC– Women for Climate Justice.

Alston, M., and Whittenbury, K. (Eds.). (2013). *Research, Action and Policy: Addressing the Gendered Impacts of Climate Change*. Dordrecht: Springer Netherlands.

Alvarez, I. (2013). *Increasing the Gender Gap: The Impacts of the Bioeconomy and Markets in Environmental Services on Women*. Global Forest Coalition. Retrieved from http://globalforestcoalition.org/wp-content/uploads/2013/06/INCREASING-THE-GENDER-GAP-FINAL.pdf.

Arndt, C., Benfica, R., and Thurlow, J. (2011). Gender implications of biofuels expansion in Africa: the case of Mozambique. *World Development, 39*(9), 1649–1662.

Arora-Jonsson, S. (2011). Virtue and vulnerability: discourses on women, gender and climate change. *Global Environmental Change, 21*(2), 744–751.

BIDS (Bangladesh Institute of Development Studies). (2014). *Integration of Women into Grameen Shakti's Clean Energy Program in Bangladesh: Research – completed*. Website. Retrieved from www.bids.org.bd/Research.php.

BMU (Bundesministerium für Umwelt, Naturschutz und Reaktorsicherheit). (2006). *Umweltbewusstsein in Deutschland 2008. Ergebnisse einer repräsentativen Bevölkerungsumfrage*. Berlin.

Cannon, T. (2009). Gender and climate hazards in Bangladesh. In G. Terry (Ed.), *Climate Change and Gender Justice* (pp. 11–18). Practical Action Publishing Ltd; Oxfam GB.

Carlsson-Kanyama, A., and Lindén, A. L. (2007). Energy efficiency in residences – challenges for women and men in the North. *Energy Policy, 35*, 2163–2172.

Chattopadhyay, R., and Duflo, E. (2004). Women as policy makers: evidence from a randomized policy experiment in India. *Econometrica: Journal of the Econometric Society, 72*(5), 1409.

Climate Alliance of European Cities. (2005). *Climate for Change. Data – Facts – Arguments*. Frankfurt am Main.

Dankelman, I. (Ed.). (2010). *Gender and Climate Change: An Introduction*. Washington, DC: Earthscan.

Dymén, C., Andersson, M., and Langlais, R. (2013). Gendered dimensions of climate change response in Swedish municipalities. *Local Environment, 18*(9), 1066–1078.

Edvardsson Björnberg, K., and Hansson, S. O. (2013). Gendering local climate adaptation. *Local Environment, 18*(2), 217–232.

EIGE (European Institute for Gender Equality). (2012). *Gender Equality and Climate Change: Report: Review of the Implementation in the EU of area K of the Beijing Platform for Action: Women and the Environment*. Retrieved from www.eige.europa.eu/content/document/gender-equality-and-climate-change-report.

Enarson, E. P. (2000). *Gender and Natural Disasters. Working Paper: Vol. 1*. Geneva: ILO.

European Commission. (2009). *Europeans' Attitudes Towards Climate Change. Special Eurobarometer 322. Eurobarometer*. Brussels.

European Commission and European Parliament. (2009). *Europeans' Attitudes Towards Climate Change. Special Eurobarometer 313. Eurobarometer: Vol. 313*. Brussels.

Gender and Disaster Network. (2009). *Gender and Disaster Sourcebook*. Newcastle upon Tyne: Gender and Disaster Network. Retrieved from www.gdnonline.org/sourcebook/.

Gender Water Alliance and UNDP (United Nations Development Programme). (2006). *Resource Guide: Mainstreaming Gender in Water Management*. Nairobi.

GenderCC. (2009). *Submission Paper on Gender and Climate Change: Parliament Public Hearings on Political, Economic, Legal, Gender and Social Impacts of Climate Change, 17–18 November 2009*.

GenderCC. (2012). *Ironically Gender in the Fossil Patriarchy*. Retrieved from www.gendercc.net/fileadmin/inhalte/Dokumente/UNFCCC_conferences/COP18/Press_statement_Doha_GenderCC.pdf.

GenderCC. (2013). *Submission in Response to Decision 23/CP.18*. Retrieved from http://unfccc.int/resource/docs/2013/smsn/ngo/376.pdf.

Gugler, J. (2004). *World Cities Beyond the West: Globalization, Development, and Inequality*. Cambridge, New York: Cambridge University Press.

Hallegatte, S., Green, C., Nicholls, R. J., and Corfee-Morlot, J. (2013). Future flood losses in major coastal cities. *Nature Clim. Change, 3*(9), 802–806.

Hemmati, M., and Röhr, U. (2007). A huge challenge and a narrow discourse. In *Women and Environments* (pp. 5–9). Women and Gender Studies Institute, New College, University of Toronto.

Hemmati, M., and Röhr, U. (2009). Engendering the climate change negotiations: experiences, challenges and steps forward. In G. Terry (Ed.), *Climate Change and Gender Justice* (pp. 155–168). Practical Action Publishing Ltd; Oxfam GB.

IPCC (Intergovernmental Panel on Climate Change) (2001). *IPCC Third Assessment Report: Summary for Policy Makers: Working Group II: Impacts, Adaptation, Vulnerability*. Geneva, Switzerland.

Urbanization and GEC: gender and equity perspectives

IPCC (Intergovernmental Panel on Climate Change). (2014). *IPCC Fifth Assessment Report, WG II: Climate Change 2014. Impacts, Adaptation and Vulnerability.* Geneva, Switzerland. Retrieved from http://ipcc-wg2.gov/AR5/.

ISDR, UNDP and IUCN. (2009). *Making Disaster Risk Reduction Gender-Sensitive. Policy and Practical Guidelines.* Geneva: UNISDR, UNDP and IUCN.

Jarvis, H., Kantor, P., and Cloke, J. (2009). *Cities and Gender. Routledge Critical Introductions to Urbanism and the City.* London, New York: Routledge.

Kern, K., and Alber, G. (2009). Governing Climate Change in Cities: Modes of Urban Climate Governance in Multi-level Systems. In *Competitive Cities and Climate Change.* OECD Conference Proceedings Milan, Italy, 9-10 OCTOBER 2008 (pp. 171–196). Paris: OECD. Retrieved from http://www.oecd.org/gov/urbandevelopment/milanproceedings

Kher, J., Aggarwal, S., and Punhani, G. (2015). Vulnerability of poor urban women to climate-linked water insecurities at the household level: a case study of slums in Delhi. *Indian Journal of Gender Studies, 22*(1), 15–40.

Khosla, P., and Masaud, A. (2010). Cities, climate change and gender: a brief overview. In I. Dankelman (Ed.), *Gender and Climate Change. An Introduction* (pp. 78–96). Washington, DC: Earthscan.

Kronsell, A. (2013). Gender and transition in climate governance. *Environmental Innovation and Societal Transitions, 7*(1), 1–15.

MacGregor, S. (2010). 'Gender and climate change': from impacts to discourses. *Journal of the Indian Ocean Region, 6*(2), 223–238.

Mattoo, A., and Subramanian, A. (2010). *Equity in Climate Change: An Analytical Review. Policy Research Working Paper: Vol. 5383.* Washington, D.C: World Bank, Development Research Group, Trade and Integration Team.

Miranda, V. (2011). Cooking, caring and volunteering: unpaid work around the world. In *OECD Social, Employment and Migration Working Papers*, No. 116. OECD Publishing: Paris. DOI: http://dx.doi.org/10.1787/5kghrjm8s142-en

Oldrup, H., and Romer Christensen, H. (2007). *TRANSGEN: Gender Mainstreaming European Transport Research and Policies. Building the Knowledge Base and Mapping Good Practices.* Copenhagen: Co-ordination for Gender Studies. University of Copenhagen. Retrieved from http://koensforskning.soc.ku.dk/projekter/transgen/eu-rapport-transgen.pdf/.

Organo, V., Head, L., and Waitt, G. (2012). Who does the work in sustainable households? A time and gender analysis in New South Wales, Australia. *Gender, Place & Culture, 20*(5), 559–577.

Patt, A. G., Dazé, A., and Suarez, P. (2009). Gender and climate change vulnerability: what's the problem, what's the solution? In *Distributional Impacts of Climate Change and Disasters. Concepts and Cases.* Cheltenham, UK: Edward Elger.

PAKLIM (Policy Advice for Environment and Climate Change) (2013). Gender Assessment in Urban Transportation. Case Study: Semarang City. Retrieved from www.paklim.org.

Räty, R., and Carlsson-Kanyama, A. (2009). Energy consumption by gender in some European countries. *Energy Policy, 38*, 646–649.

Röhr, U., and Ruggieri, D. (2008). *Erneuerbare Energien – ein Arbeitsmarkt für Frauen!* LIFE e.V. (Hg): Berlin, Germany.

Röhr, U., Spitzner, M., Stiefel, E., and Winterfeld, U. von. (2008). *Gender Justice as the Basis for Sustainable Climate Policies. A Feminist Background Paper.* Bonn, Berlin. Retrieved from http://genanet.de/fileadmin/user_upload/dokumente/Themen/Klima/Background_Paper_Gender_Justice_CC_EN_2008.pdf.

Rossi, A., and Lambrou, Y. (2008). *Gender and Equity Issues in Liquid Biofuels Production: Minimizing the Risks to Maximize the Opportunities.* Food and Agriculture Organization of the United Nations: Rome.

Skinner, E. (2011). *Gender and Climate Change: Overview Report.* Brighton, UK: Bridge Cutting Edge Pack.

Spitzner, M., Weiler, F., Rahmah, A., and Turner, J. (2007). *Städtische Mobilität und Gender Förderung des öffentlichen Regionalverkehrs im Großraum Jakarta.* KfW-Entwicklungsbank: Frankfurt am Main.

Tacoli, C. (2012). *Urbanization, Gender and Urban Poverty: Paid Work and Unpaid Carework in the City. Urbanization and Emerging Population Issues Working Paper: Vol. 7.* London, UK, New York, NY, USA: Human Settlements Group, International Institute for Environment and Development; Population and Development Branch, United Nations Population Fund.

Tacoli, C., and Satterthwaite, D. (2013). Gender and urban change. *Environment and Urbanization*, *25*(1), 3–8.

Terry, G. (Ed.). (2009). *Climate Change and Gender Justice*: Practical Action Publishing Ltd; Oxfam GB.

Tran, H. A., and Schlyter, A. (2010). Gender and class in urban transport: the cases of Xian and Hanoi. *Environment and Urbanization*, *22*(1), 139–155.

UN (United Nations). (2011). *The Social Dimensions of Climate Change*. Retrieved from www.ilo.org/wcmsp5/groups/public/---ed_emp/---emp_ent/documents/publication/wcms_169567.pdf.

UN DESA (United Nations Department of Economic and Social Affairs). (2010). *The World's Women: Trends and Statistics* (No. UN Doc. ST/ESA/STAT/SER.K/19). New York. Retrieved from http://unstats.un.org/unsd/demographic/products/Worldswomen/WW_full%20report_color.pdf.

UN Women. (n.d.). *OSAGI Gender Mainstreaming – Concepts and Definitions*. Retrieved from www.un.org/womenwatch/osagi/conceptsandefinitions.htm.

UN Women and MRFCJ (Mary Robinson Foundation). (2013). *The Full View: Advancing the Goal of Gender Balance in Multilateral and Intergovernmental Processes*.

UNDP (United Nations Development Programme). (n.d.). *GREEN Brick (Improving Kiln Efficiency in Brick Making Industry)*. Retrieved from www.bd.undp.org/content/bangladesh/en/home/operations/projects/environment_and_energy/improving-kiln-efficiency-in-brick-making-industry-/.

UNDP (United Nations Development Programme). (2008). *Human Development Report 2007/2008: Fighting Climate Change: Human Solidarity in a Divided World*: Palgrave Macmillan. Retrieved from http://hdr.undp.org/sites/default/files/reports/268/hdr_20072008_en_complete.pdf.

UNFCCC (United Nations Framework Convention on Climate Change). (2007). *Climate Change: Impacts, Vulnerabilities, and Adaptation in Developing Countries*. Retrieved from http://unfccc.int/resource/docs/publications/impacts.pdf

UNFCCC (United Nations Framework Convention on Climate Change). (2012). *Promoting Gender Balance and Improving the Participation of Women in UNFCCC Negotiations and in the Representation of Parties in Bodies Established Pursuant to the Convention or the Kyoto Protocol* (No. Decision 23/CP.18).

UN-HABITAT. (2008). *State of the World's Cities: Harmonious Cities*. London, Sterling, VA, Nairobi, Kenya.

UN-HABITAT. (2010). *Gender Equality for Smarter Cities, Challenges and Progress*. Nairobi, Kenya.

UNISDR (United Nations International Strategy for Disaster Reduction). (2005). *Hyogo Framework for Action (HFA) 2005–2015: Building the Resilience of Nations and Communities to Disasters*. Retrieved from www.unisdr.org/we/coordinate/hfa.

USAID. (2011). *Getting REDD+ Right for Women: An Analysis of the Barriers and Opportunities for Women's Participation in the REDD+ Sector in Asia*. Retrieved from www.leafasia.org/sites/default/files/resources/GenderLit_GettingREDDRightWomen.pdf.

van Staden, R. (2013). *Climate Change: Implications for Cities: Key Findings from the Intergovernmental Panel on Climate Change Fifth Assessment Report*. Retrieved from http://unfccc.int/cc_inet/cc_inet/six_elements/public_awareness/items/3529.php?displayPool=1743.

PART III

Urban responses to global environmental change

Part III addresses the wide-ranging responses to global environmental changes, which are manifest in cities across multiple regions and in various stages of their urbanization trajectories. Whereas Part I and Part II of this volume largely center on the bi-directional processes and interactions of urbanization and global change, Part III turns to the interactions and responses within the urban system specifically, addressing how the interactions between human and physical systems shape the impacts and responses of global environmental change as well as how the livelihoods of urban communities are affected by such impacts.

A major topic explored within this third part is a) *urban climate action*, which includes both climate change mitigation and adaptation in urban areas of varying size and levels of economic development. This part offers a comparison of mitigation actions and associated challenges and opportunities in developed cities with high levels of wealth (Mohareb et al., Chapter 26) with those in low-income, small and medium-sized cities (Lwasa, Chapter 27). It also includes how adaptation responses in such rapidly urbanizing cities are being approached and explores the opportunities for building more robust urban systems that integrate adaptation and mitigation along with urban development priorities (Sánchez-Rodríguez, Chapter 24). Part of this challenge concerns the ability and preparedness of urban institutional and governance structures within urban areas to effectively tackle the confounding challenges of a changing and unpredictable climate (Shi et al., Chapter 23).

Part III also addresses the ways in which societies and associated decision-making structures are b) *creating resilient cities*. This includes the actions taken and good practice examples of urban responses, for example, spatial planning for climate change (Huang and Wang, Chapter 25), how urban agriculture can support urban climate resilience (Griffith et al., Chapter 28) and the incorporation of 'softer approaches' into urban planning such as the integration of ecosystem services and biodiversity into comprehensive strategies (McDonald, Chapter 29). A common thread throughout these chapters concerns the need for more flexible, integrative and cross-boundary/cross-sectoral approaches to policymaking and decision-making that incorporate global environmental change concerns with other developmental goals, which are a high priority, particularly in low- and middle-income cities.

The remaining chapters within Part III explore c) *sustainable urban and infrastructure engineering*, which captures the advantage of innovation in cities where a number of sustainability solutions already exist or where new ones can be practiced and tested. Urban

Urban responses to GEC

areas are where both social and technical infrastructures have the potential to be reimagined to support a transition to a post-carbon city (Pincetl, Chapter 33) and where the industrial process can work symbiotically to share and reduce resource use and environmental externalities (Chertow et al., Chapter 31). Such infrastructure will also need to be increasingly resilient for adapting to extreme environmental disturbances, which can be expected to continue in our future (Zimmerman, Chapter 32).

Cities have the potential to be the catalysts for a transformation towards a lower carbon future and there is evidence of this occurring, e.g. through green infrastructure promotion and other green economic measures that can vary along the spectrum from more incremental changes to fundamental system shifts (Simon, Chapter 30). It is clear that urban areas and their socio-political and technical infrastructure systems will have a great influence on the global environment and arguably that the urban transition already underway should be guided by principles of justice towards a more equitable and livable sustainable global urban future (Boone and Klinsky, Chapter 22).

The chapters in Part III demonstrate that there is not a one-size-fits-all approach, rather many different responses towards building more robust and resilient cities that can be shared and adapted across multiple scales, landscapes and regions.

22
ENVIRONMENTAL JUSTICE AND TRANSITIONS TO A SUSTAINABLE URBAN FUTURE

Christopher G. Boone and Sonja Klinsky

Cities are distinguished from rural areas in part by their high-density and diversity of people, built environment and land uses. 'Heterogeneous' is one term used to describe the unevenness of urban landscape, where even over short distances, the material components of a city (e.g. buildings, green space, roads), the activities that occur or the social characteristics of people who live there can change quite dramatically. These uneven landscapes result in clusters of potential hazards, such as toxic emissions from industry, or unwanted activities near residential communities such as meat packing plants or idling diesel truck stops. Since cities are socially and spatially segregated by neighborhoods (Sampson 2012; Boone 2013), this means that clusters of some groups of people may potentially bear a greater burden of urban hazards than other groups.

Environmental justice as an academic field and activist movement evolved to address the uneven and unjust distribution of environmental benefits and burdens. This chapter examines how environmental justice can infuse sustainability thought and practice in ways that lead to a just transition to an urban planet. Transitions can bring opportunity, but may also result in painful consequences. Sustainability is about intervening in systems to accelerate transitions to a plausible, but normative future. The future we want includes the principle of achieving intra- and inter-generational equity, but sustainability should be more than about outcomes. The central argument is that lessons from environmental justice about the need for fairness of processes—not only outcomes—are useful for ensuring that sustainability transitions do not undermine or threaten justice principles on the pathway to a better future.

This chapter is organized into four sections beginning with a discussion of the urbanization transition which draws on examples of other transitions to demonstrate how radical change is possible. The second section provides an account of the foundations of environmental justice scholarship and an elaboration on how it has evolved, revealing how it can amplify the normative dimensions of sustainability in the third section. The concluding section offers a brief discussion on the importance of justice for guiding a desirable transition towards the transformation of a sustainable urban future.

The global urban transition

History shows that significant transitions are possible and these radical changes can have far reaching impacts on human beings and the environment. In a span of just three human lifespans, roughly 200 years, the world has experienced a demographic transition, energy transition and economic transition that have altered the human condition and our relationship with the planet. In the United States in 1800, birth rates were high, but life could be brutally short; people depended on animals, impounded water and wood for energy; and the economy was based on agriculture and resource extraction. Today, in the United States, families are not large enough to replace the current generation but people can expect to enjoy long lives; we are utterly dependent on fossil fuels for energy; and the economy is based mainly on services. The implications of these transitions are multi-faceted and complex, but they have contributed to, among other concerns, rising energy and material demands, global climate change, biodiversity loss and increasing disparities of human well-being (Rockstrom et al. 2011).

The shift to an urban world is an equally profound transition. Although cities have existed for at least 10,000 years, only recently could a majority of people live in urban centers. England became the first urban country, meaning more than half of its population lived in cities, in 1851. The United States did not reach the urban threshold until 1920. Now that half of humanity lives in cities and nearly all of the projected three billion in population growth by 2050 is expected to occur in urban environments, it is critically important—as the transition is underway—to think about sustainable pathways forward. This is no easy goal, especially since many of the current sustainability challenges are the result of living in highly urbanized societies. Cities consume 65 percent of the world's energy and generate 70 percent of global greenhouse gas (GHG) emissions (Solecki et al. 2013). In China, people who move from the countryside to its burgeoning cities double their energy consumption and carbon emissions (Fan et al. 2013). Higher incomes in cities compared to rural areas mean greater demand for resources and higher production of wastes, which threaten the health of the world's ecosystems (Grimm et al. 2008). Inadequate housing and infrastructure in burgeoning cities will likely lead to a plethora of slums and informal housing, putting vulnerable populations at risk and further exacerbating environmental injustices (Davis, 2006; Boone and Fragkias 2013).

It will be very difficult if not perilous to maintain the 20th century model of urbanization. A sustainable transition depends on promoting and guiding the best assets of urban life— innovation, opportunities for collaboration and exchange, an educated and healthy citizenry, diversity of people and opportunities, and concentration of financial, human and social capital—to help people build a desirable, sustainable future. Global urbanization is happening, and will happen on a grand scale. It would be unwise to simply stand on the sidelines and watch it unfold—sustainability depends on the ability and willingness to intervene, rather than hope or wait for the system to correct itself (Miller et al. 2014).

A fundamental principle of sustainability is that action and intervention is necessary in order to avoid potentially catastrophic change. Scarcity of fossil fuels, for instance, may eventually force a transition to a renewable energy portfolio, but the danger in waiting for price signals is that environmental damage and human suffering will occur as a result of increased and persistent carbon dioxide in the atmosphere. Sea level rise is already underway (Jacob et al. 2012) and many of the world's cities, located in low coastal elevation zones, are especially vulnerable to damage from rising oceans, storm surges (likely exacerbated by climate change) and an inability or unwillingness to plan for climate change hazards (McGranahan et al. 2007). If municipalities pay heed to early warning signals, careful planning can save human lives, property and resources. Rather than waiting for crises (such as Hurricane Sandy in 2012 or the devastating European heatwave of 2003), cities can accelerate a transition to a new, more desirable and resilient state (Barron et al. 2012).

In most wealthy, industrialized countries, urban populations have reached what appears to be an upper plateau of approximately 80 percent of total population. Many of the challenges of sustainable urbanization in these regions will focus on how to retrofit what is already in place. Most new urban growth over the next 50 years will be in Asia and Africa, not in the megacities that attract most attention, but in cities of less than 500,000 in population (UNFPA 2007). Before these cities dot the landscape, there is a huge opportunity to rethink what cities should be, how they should function and how they can support, rather than hinder, global sustainability.

Urban centers created in this century do not have to—and indeed should not—follow the models of cities created in the industrial era of the last century. New York, London and Tokyo invested billions of dollars in concrete, asphalt, steel and cables to make the industrial city function. The sunk costs of hard or gray infrastructure make it difficult to try innovative ways to service the city. New cities built around the idea of green infrastructure using ecosystem services to make cities livable and healthy is a way to 'leap frog' the traditional pathway. For instance, foresting watersheds can be a more cost-effective way to maintain water quality than an energy intensive water treatment plant. A forested watershed has other co-benefits such as recreation space, wildlife habitat and flood control, that make a green infrastructure strategy an attractive proposition (Copeland 2014). In Bangkok, urban designers are exploring how 'waterscape urbanism' and 'liquid perception' can improve urban resilience in a city threatened with flooding and a loss of culturally relevant landscapes (Thaitakoo et al. 2013). As canals are buried and agriculture gets pushed to the periphery, Bangkok has become increasingly solid, less porous, more vulnerable to storm damage and less manageable using indigenous knowledge and local wisdom. Maintaining and rebuilding the liquid landscape of Bangkok is one strategy for creating resilience and ensuring a sustainable future that draws on centuries of living with and in an 'aqua-city'.

Ecological design principles are useful for new design, but also as a way of repairing degraded urban environments (Steiner 2014). Many cities built in the 19th and 20th centuries to service an industrial economy are now struggling to make use of contaminated properties or brownfields, rather than expanding into greenfield sites on the periphery. Ecological design can remediate contaminants from these properties and find new creative ways to make them livable and attractive. One example is the Don Valley Brick Works in Toronto, Canada. A quarry and brick-making factory for more than a century, the Toronto and Region Conservation Authority expropriated the land in 1987 and began restoration in 1994 with a Canadian non-profit organization called Evergreen. The abandoned buildings and the site were transformed into a community environment center that includes waking paths, classrooms, art exhibits and office space for environmental organizations (www.evergreen.ca).

The Don Valley Brickworks is something to celebrate, but it took a great deal of time and money to turn a former eyesore and environmental hazard into a winning design. New cities can leap frog painful and expensive processes of retro-fitting by investing in urban design that elevates human well-being and ecological integrity, and includes explicit attention to reducing and preventing environmental injustices.

Environmental justice: local to global

Environmental justice started as a grassroots movement in the United States in the late 1970s and early 1980s to confront unfair environmental burdens, especially on ethnic and racial minority groups. A number of studies that examine the spatial distribution of toxic emitters or hazardous waste facilities find that race or ethnicity—rather than income—of nearby residents is the best social predictor for the location and concentration of unwanted land uses. After nearly three decades as a national issue, the patterns of environmental injustice along racial and ethnic lines found in early studies have largely persisted (Mohai and Saha 2007).

From an initial focus on the unfair distribution of 'environmental bads,' environmental justice has also examined distribution of 'environmental goods'. Health and well-being depend on protection from hazards but also the ability to enjoy the beneficial services from the environment such as clean air and water, green space, or access to healthy, nutritious and affordable food (Boone et al. 2009). Literature from the public health community, for example, shows that living within a short five to ten minute walking distance of a park (generally considered an environmental good) is a strong predictor of daily physical activity (Bedimo-Rung et al. 2005). Nearby access to community garden space, farmers' markets or other forms of urban agriculture can also reduce some of the symptoms of food deserts including increased correlation with obesity (Brewis 2010). Environmental justice is a powerful way to frame the issue of how environmental goods are distributed among groups, especially in heterogeneous urban areas (Larsen and Gilliland 2008; Ibes 2015).

Analyzing the distribution of environmental bads and goods, also known as 'distributive equity,' has remained a fundamental approach in environmental justice scholarship and can serve as a galvanizing force of activism. When racial or ethnic minority communities in the United States discover they are living with a disproportionate number or concentration of toxic facilities, this can be a powerful motivator for action and change. However, environmental justice is also about fairness of process. Regardless of the distributive justice of environmental good and bads, if residents or stakeholders do not have a voice in decision-making processes that have environmental implications, that is an equally, if not greater, form of injustice. The same is true if environmental laws are unevenly applied or if stakeholders are not recognized in institutional and governance decisions (Schlosberg 2007). These forms of environmental justice are called 'process' or 'procedural equity.'

In the last decade, the environmental justice movement has become increasingly international, with legislation and activism bringing to light information that allows scholars and others to examine the distributive and procedural equity of environmental goods and bads as well as fair application of environmental laws (Agyeman 2014; Boone and Fragkias 2013). Transboundary flows of pollutants, hazardous waste and contaminants have led to the development of global scale environmental justice movements (Pellow 2007). Several infamous cases of hazardous waste dumping in poor countries such as the transfer of 18,000 barrels of toxic waste from Italy to Nigeria in 1987–88 for a fee of $100 per month (Liu 1992), have promoted a global level environmental justice framing of such issues. In response to international dumping of hazardous wastes, sometimes called 'toxic colonialism,' the United Nations Environment Programme drafted the Basel Convention on the Control of Transboundary Movements of Hazardous Wastes in 1989 (Kummer 1992), which has been adopted broadly around the world. Health and environmental hazards created by electronic waste (e-waste) recycling has also been addressed as a global level environmental justice issue (Smith et al. 2006; Xing et al. 2009). Since the recycling of e-waste can be a health hazard, the European Commission (2014) requires sellers of electronics to take back any used products free of charge to be recycled properly. In the United States, several states have passed legislation to mandate electronics recycling, but the federal government has yet to succeed in passing national level legislation (RERA 2011).

Similarly, climate policy has a strong justice component stemming from the basic pattern that those who have caused the least GHG emissions often face the greatest climate impacts while simultaneously experiencing significant human development challenges (see Alber and

Cahoon Chapter 21 in this volume, for a related discussion on global environmental change, gender and equity). Recognition of the justice dimensions of climate change has spurred a wealth of scholarship (Shue 1992; Gardiner et al. 2010), influenced the initial UNFCCC text (UNFCCC 1992) and has continued to shape international negotiations (Morgan and Waskow 2014).

However, the multi-faceted nature of climate change and potential tension between human development achievement and climate goals complicates justice claims. Although the majority of cumulative emissions were produced in industrialized countries, avoiding a greater than 2°C temperature change will require emissions reductions from countries still facing domestic development challenges. The tight association between inexpensive fossil energy and conventional human development raises difficult trade-offs—across non-fungible costs and benefits and often for the most vulnerable populations—within these countries (Winkler et al. 2011). A forced choice between immediate well-being improvements and long-term avoidance of climate impacts can itself be seen as an injustice, and makes it difficult to identify which policy options would be most fair in non-ideal circumstances. Decisions that appear just at one level of aggregation using one set of metrics may appear unjust if evaluated using different metrics, aggregation levels or time scales. These ambiguities do not negate the core concerns or insights of environmental justice, but do demand nuanced analyses.

Climate policy decisions include where and how emissions should be reduced as well as how to deal with spatially specific, but globally interconnected, climate impacts. The majority of early scholarship on climate justice focused on mitigation, but it was soon realized that adaptation also raised diverse justice issues (Adger et al. 2006). Large international scale issues such as funding for adaptation (Dellink et al. 2009), along with locally specific elements of adaptation planning require thought from a justice perspective. Akin to environmental justice work focused on environmental goods, examining adaptation through a justice lens requires questioning who has access to which kinds of resources, and how this is connected to broader patterns of privilege and power.

Determining fair process in the climate context poses a more profound challenge to environmental justice. Environmental justice scholarship has turned to fair process as a central means for avoiding and remedying injustices (Schlosberg 2007), but it is not clear what 'fair process' should look like across the multiple spatial and temporal scales relevant to climate decision-making. Climate-related decisions that are relatively bounded in scope, such as designing a stormwater management system or developing a reforestation program, resonate with existing environmental justice and community engagement practices sensitive to local power dynamics and differentiated vulnerabilities. Fair process gets more complicated as the decision scope becomes broader or when the implications will be very diffuse (Klinsky 2015). For instance, what does fair process look like when designing policies explicitly intended to create diffuse, economically transformative effects such as shifts in industrial tax rules or subsidies to particular forms of new technology? These challenges are amplified by the strongly inter-generational nature of climate decisions. Is a fair process even possible when some of the most central stakeholders cannot, by definition, participate? Again, these considerations do not mean that fair process should not be considered, but they do suggest that careful thought needs to be invested in defining what this looks like in any particular decision context.

Sustainability grounded in justice

Sustainability is improving human well-being and ensuring social justice for present and future generations while safeguarding the planet's life-supporting ecosystems. It is more than just surviving—sustainability is founded on moral arguments of justice, fairness, and right and wrong (Boone 2014). The normative motivations may be grounded in looking after future generations, especially one's own children and grandchildren, or for ameliorating present day inequities such as the growing income gaps between rich and poor, simply because it is the *right* thing to do. Paying attention to *inter-* and *intra-*generational equity can also be motivated by practical considerations. Gross inequities in the present generation can undermine social and political stability, or hamper the capacity of people to develop to their fullest potential (Sen 2009). While the causes of social protests are complex, the Arab Spring—revolutionary movements that spread across the Middle East starting in 2010—was motivated and fueled, in part, by persistently high youth unemployment and high costs of living (Cammett and Diwan 2014). Inequitable access to scarce water resources, exacerbated by climate change, may be a contributing factor to ongoing conflicts in Syria (Beck 2014). Widening disparities between rich and poor have alarmed decision-makers worldwide including the World Economic Forum (2014), because it may threaten social and economic security. Similarly, ongoing disputes about fairness have consistently hampered the progress of international efforts to develop strong agreements on climate change. In a political negotiation among sovereign states, no country is likely to agree to something it fundamentally feels is unfair.

For individuals, groups, organizations or governments, the sustainability motivations may differ, but blend inter- and intra-generational equity along with a moral sense of fairness with practical considerations for what is necessary to ensure stability and survival. Over the last decade, cities around the world have developed sustainability plans that address intra-generational equity concerns such as affordable housing and poverty alleviation as well as inter-generational issues of climate change and resource conservation. New York's sustainability plan includes goals to reduce carbon emissions by 30 percent by 2030 (addressing global scale and future generations) and to clean up contaminated vacant land found in primarily poor and minority neighborhoods (focused on local scale and current generation) (PlaNYC 2014). Cape Town, South Africa, has developed a draft set of sustainability indicators that include local and intra-generational equity metrics of per capita green space and number of households with access to adequate sanitation. It also has set global and inter-generational equity metrics such as square kilometers of vegetation types required to achieve biodiversity targets (City of Cape Town 2014). Equity is increasingly a goal and objective of urban planning that can be tracked using specified metrics.

Sustainability plans that incorporate equity metrics demonstrate that the principles of environmental justice are beginning to inform sustainable urbanization strategies. However, many of the sustainability metrics focus on local concerns that, though justifiable, may conflict with the sustainability objectives of other cities and the goal of a sustainable urban transition on a planetary scale. Well-intentioned recycling programs for electronics, for example, can translate into hazardous living conditions for workers in low-income cities exposed to toxins while breaking up circuit boards for tiny amounts of precious metals (Xing et al. 2009). Cities around the world are 'teleconnected' to one another, meaning that an action at one place can have a rapid impact on other cities even at great distances (Liu et al. 2015). For a sustainable urban transition, scholars and decision-makers need to take into account the teleconnected systems of cities that function on a global scale (Seto et al. 2012). Sustainability at the gross expense of others is inequitable and unjust, and could ultimately undermine the ability of the world to function as an urban planet.

Sustainability transformation of an urbanizing planet

The world is reaching another critical transition point and it is difficult to predict how it will emerge. Barring nuclear catastrophe, the promise of sustainability is that the future can be one where human well-being is supported by vibrant, healthy ecosystems. But a difficulty with transitions is that it accentuates vulnerabilities and is most painful for those who lack the resources or adaptive capacity to survive and thrive while they occur. It is critical to pay special attention to the most vulnerable populations to ensure that such groups are not left behind and can adapt to changing conditions. In other words, the transition process should be just and the outcome, the transformation that occurs on the other side, must be just as well. A low-carbon urban planet that safeguards ecosystems should be just not only for moral reasons, but also if it is to endure as a fundamental transformation in the way human beings inhabit, use and depend on the planet's life support systems. Environmental justice should not be treated as an outcome of an urbanizing planet, but as a principle to ensure a sustainable transition to a desired and enduring future.

Main points of this chapter

- History shows that significant transitions that alter human life and relationships with the planet are possible including the present transition to an urban world.
- Sustainability is about intervening in systems to accelerate transitions to a desirable future.
- Justice is a foundational principle for sustainability.
- Environmental justice has important lessons for sustainability, especially the need to make sure that processes including the transition process are just and fair, not solely the expressed outcome of a desirable future.

Recommendations for future research

- Improve understanding of how local environmental justice action can aggregate to achieve sustainable, just outcomes and processes on a global scale.
- Analyze the efficacy of climate justice and energy justice to achieve local to global just outcomes and processes.
- Explore how teleconnections can accelerate rather than undermine transitions to a sustainable urban future.

References

Adger, W.N., Paavola, J., and Huq S. 2006. Toward Justice in Adaptation to Climate Change. In *Fairness in Adaptation to Climate Change*, eds. W.N. Adger, J. Paavola, S. Huq, and M.J. Mace, 1–19. Cambridge, MA: MIT Press.

Agyeman, J. 2014. Global Environmental Justice or Le Droit Au Monde? *Geoforum* 54 (July): 236–238. doi:10.1016/j.geoforum.2012.12.021.

Barron, S. et al. 2012. A Climate Change Adaptation Planning Process for Low-Lying, Communities Vulnerable to Sea Level Rise. *Sustainability*, 4(9): 2176–2208.

Beck, A. 2014. Drought, Dams, and Survival: Linking Water to Conflict and Cooperation in Syria's Civil War. *International Affairs Forum*, 5(1) (January 2): 11–22.

Bedimo-Rung, A.L., Mowen, A.J., and Cohen, D.A. 2005. The Significance of Parks to Physical Activity and Public Health: A Conceptual Model. *American Journal of Preventive Medicine*, 28(2):159–168.

Boone, C.G. 2013. Social Contexts of Contemporary Urban Systems. In *Resilience In Ecology and Urban Design: Synergies For Theory And Practice In The Urban Century*, eds. S. Pickett, M. Cadenasso and B. McGrath, 47–69. Netherlands: Springer Press.

Boone, C.G. 2014. Equity and Justice in the City. In *North American Odyssey: Historical Geographies for the Twenty-First Century*, eds. C.E. Colten and G.L. Buckley, 413–428. Lanham, MD, USA and Plymouth, UK: Rowman & Littlefield.

Boone, C.G. and Fragkias, M. 2013. Connecting Environmental Justice, Sustainability, and Vulnerability. In *Urbanization and Sustainability – Linking Ecology, Environmental Justice, and Global Environmental Change*, eds. C.G. Boone and M. Fragkias, 49–59. Netherlands: Springer Press.

Boone, C.G., Buckley, G.L., Grove, J.M., and Sister, C. 2009. Parks and People: An Environmental Justice Inquiry in Baltimore, Maryland. *Annals of the Association of American Geographers, 99*(4) (October): 767–787.

Brewis, A.A. 2010. *Obesity: Cultural and Biocultural Perspectives*. Piscataway, NJ: Rutgers University Press.

Cammett, M. and Diwan, I. 2014. *The Political Economy of the Arab Uprisings*. Boulder, CO: Westview Press.

City of Cape Town. 2014. City of Cape Town Sustainability Report: Phase 1 Draft Set of Indicators. Available online (last accessed November 12, 2014): www.capetown.gov.za/en/EnvironmentalResource Management/publications/Documents/Sustainability_Indicators_Report_2004_1642007153240_465.pdf

Copeland, C. 2014. Green Infrastructure and Issues in Managing Urban Stormwater. Congressional Research Service, R43131. Available online (last accessed November 13, 2014): http://digital.library.unt.edu/ark:/67531/metadc284504/m1/1/high_res_d/R43131_2014Mar21.pdf

Davis, M. 2006. *Planet of the Slums*. New York: Verso.

Dellink, R., den Elzen, M., Aiking, H., Bergsma, E., Berkhout, F., Dekker, T. and Gupta, J. 2009. Sharing the Burden of Financing Adaptation to Climate Change. *Global Environmental Change, 19*(4): 411–421.

European Commission. 2014. Waste Electrical and Electronic Equipment (WEEE). Available online: http://ec.europa.eu/environment/waste/weee/index_en.htm

Fan, J.-L., Liao, H., Liang, Q.-M., Tatano, H., Liu, C.-F., and Wei, Y.-M. 2013. Residential Carbon Emission Evolutions in Urban–Rural Divided China: An End-Use and Behavior Analysis. *Applied Energy, 101* (January): 323–332. doi:10.1016/j.apenergy.2012.01.020.

Gardiner, S., Caney, S., Jamieson, D. and Shue H. (eds.) 2010. *Climate Ethics: Essential Readings*. New York: Oxford University Press.

Grimm, N.B., Faeth, S.H., Golubiewski, N.E., and Redman, C.L. 2008. Global Change and the Ecology of Cities. *Science, 319*(5864): 756–760.

Ibes, D.C. 2015. A Multi-Dimensional Classification and Equity Analysis of an Urban Park System: A Novel Methodology and Case Study Application. *Landscape and Urban Planning, 137*(0) (May): 122–137. doi:http://dx.doi.org/10.1016/j.landurbplan.2014.12.014.

Jacob, T. et al., 2012. Recent Contributions of Glaciers and Ice Caps to Sea Level Rise. *Nature, 482*(7386): 514–518.

Klinsky, S. 2015. Justice and Boundary Setting in Greenhouse Gas Cap and Trade Policy: A Case Study of the Western Climate Initiative. *Annals of the Association of American Geographers, 105*(1):105–122.

Kummer, K. 1992. The International Regulation of Transboundary Traffic in Hazardous Wastes: The 1989 Basel Convention. *International & Comparative Law Quarterly, 41*(03): 530–562.

Larsen, K. and Gilliland, J., 2008. Mapping the Evolution of "Food Deserts" in a Canadian City: Supermarket Accessibility in London, Ontario, 1961–2005. *International Journal of Health Geographics, 7*: 16.

Liu, J., Mooney, H., Hull, V., Davis, S.J., Gaskell, J., Hertel, T., Lubchenco, J. et al. 2015. Systems Integration for Global Sustainability. *Science, 347*(6225) (February 27). doi:10.1126/science.1258832.

Liu, S.F. 1992. The Koko Incident: Developing International Norms for the Transboundary Movement of Hazardous Waste. *Journal Natural Resources & Environmental Law. 8*: 121–154.

McGranahan, G., Balk, D., and Anderson, B. 2007. The Rising Tide: Assessing the Risks of Climate Change and Human Settlements in Low Elevation Coastal Zones. *Environment and Urbanization, 19*(1): 17–37.

Miller, T.R., Wiek, A., Sarewitz, D., Robinson, J., Olsson, L., Kriebel, D., and Loorbach, D. 2014. The Future of Sustainability Science: A Solutions-Oriented Research Agenda. *Sustainability Science, 9*(2): 239–246. doi:10.1007/s11625-013-0224-6.

Mohai, P. and Saha, R. 2007. Racial Inequality in the Distribution of Hazardous Waste: A National-Level Reassessment. *Social Problems, 54*(3): 343–370.

Justice and a sustainable urban future

Morgan, J. and Waskow, D. 2014. A New Look at Climate Equity in the UNFCCC. *Climate Policy*, *14*(1):17–22.

Pellow, D.N. 2007. *Resisting Global Toxics: Transnational Movements for Environmental Justice*. Cambridge, MA: MIT Press.

PlaNYC (2014). PROGRESS REPORT: Sustainability & Resiliency 2014. The City of New York. Available online: www.nyc.gov/html/planyc2030/downloads/pdf/140422_PlaNYCP-Report_FINAL_Web.pdf

RERA (Responsible Electronics Recycling Act). 2011. H.R. 2284, 112 Cong. Available online: www.govtrack.us/congress/bills/112/hr2284/text

Rockstrom, J., Steffen, W., Noone, K., Persson, A., Chapin, F.S., Lambin, E. F., Lenton, T. M. et al. 2011. A Safe Operating Space for Humanity. *Nature*, *461*(7263) (September 24): 472–475.

Sampson, R.J. 2012. *Great American City: Chicago and the Enduring Neighborhood Effect*. Chicago: University of Chicago Press.

Schlosberg, D. 2007. *Defining Environmental Justice: Theories, Movements, and Nature*. Oxford: Oxford University Press.

Sen, A.K., 2009. *The Idea of Justice*. Cambridge, MA: The Belknap Press.

Seto, K.C. et al., 2012. Urban Land Teleconnections and Sustainability. *Proceedings of the National Academy of Sciences*, *109*(20): 7687–7692.

Shue, H. 1992. The Unavoidability of Justice. In *The International Politics of the Environment: Actors, Interests and Institutions*, eds. A. Hurrell and B. Kingsbury, 373–397. Oxford: Clarendon Press.

Smith, T., Sonnenfeld, D.A. and Pellow, D.N. 2006. *Challenging the Chip: Labor Rights and Environmental Justice in the Global Electronics Industry*. Philadelphia: Temple University Press.

Solecki, W., Seto, K.C., and Marcotullio, P.J. 2013. It's Time for an Urbanization Science. *Environment: Science and Policy for Sustainable Development*, *55*(1) (January 1): 12–17.

Steiner, F. 2014. Frontiers in Urban Ecological Design and Planning Research. *Landscape and Urban Planning*, *125*(0) (May): 304–311. doi:http://dx.doi.org/10.1016/j.landurbplan.2014.01.023.

Thaitakoo, D., McGrath, B., Srinthanyarat, S., and Palopakon, Y. 2013. Bangkok: The Ecology and Design of an Aqua-City. In *Resilience in Ecology and Design: Linking Theory and Practice for Sustainable Cities*, eds. S.T.A. Pickett, M.L. Cadenasso and B. McGrath, 427–442. Dordrecht: Springer.

UNFCCC. 1992. United Nations Framework Convention on Climate Change. Available online: http://unfccc.int/resource/docs/convkp/conveng.pdf

UNFPA (United Nations Population Fund). 2007. *State of the World Population: Unleashing the Potential of Urban Growth*.

Winkler, H., Jayaraman, T., Pan, J., Santihago de Oliveira, A., Zhang, Y., Sant, G., Gonzalez Miguez, J.D., Letete, T., Marquand, A., and Raubenheimer, S. 2011. *Equitable Access to Sustainable Development: A Paper by Experts from BASIC Countries*. BASIC Expert Group: Beijing, Brasilia, Cape Town, Mumbai.

World Economic Forum. 2014. Global Risks 2014, Ninth Edition. Available online: www3.weforum.org/docs/WEF_GlobalRisks_Report_2014.pdf

Xing, G.H., Chan, J.K.Y., and Leung, A.O.W. 2009. Environmental Impact and Human Exposure to PCBs in Guiyu, an Electronic Waste Recycling Site in China. *Environment International*, *35*(1): 76–82.

23

GLOBAL PATTERNS OF ADAPTATION PLANNING

Results from a global survey

Linda Shi, Eric Chu and JoAnn Carmin

Cities worldwide are increasingly aware of how climate change is affecting urban systems, services and quality of life (Field et al., 2014). Since urban areas are the home of the majority of the world's population and economic assets (Struyk and Giddings, 2009; McKinsey Global Institute, 2011), many cities are beginning to take leadership roles in recognizing climate vulnerabilities and adapting to impacts (Carmin et al., 2012a; Rosenzweig et al., 2010). The agglomeration economies generated by concentrated urban assets and populations, however, coincide with the historic location of many cities near river valleys, deltas, coastlines and other regions that have rising hazard risk probabilities under climate change (Lall and Deichmann, 2009). This is especially the case in the Global South, where fast growing cities are home to large numbers of people without sufficient access to infrastructure, services and housing security (Satterthwaite et al., 2007).

International organizations, national and regional governments, and research institutes are increasingly publishing adaptation guidelines, toolkits and handbooks to assist cities in planning and implementing adaptation projects and programs (Andonova et al., 2009; Lee, 2013). While many studies have examined city initiatives on a case study scale, less is known about patterns of local experiences of climate change, different adaptation challenges and the causes of variation in adaptive capacity worldwide. The diverse pathways to adaptation and lack of an empirical grounding for adaptation needs and practices pose a challenge to policy-makers and funders seeking to evaluate how best to target technical and financial assistance. As adaptation efforts begin to spread beyond pilot cities and 'early adapters,' the question of how to scale-up and catalyze adaptation planning becomes even more pressing (Field et al., 2014).

This chapter draws on the insights of a global survey of 468 cities in 2011 on their experiences with climate impacts, motivations for and challenges to adaptation, and strategies and status of adaptation action. The analysis addresses the following questions: how do variations in city size (as measured by population) relate to local governance capacity and, in turn, affect cities' abilities to plan and implement adaptation projects and programs? The results indicate that differences in city size are associated with variations in motivations to adapt, progress in conducting assessments, approaches to adaptation planning and implementation, methods of external engagement and access to adaptation finance. This suggests that

Global patterns of adaptation planning

external funders and policy-makers can better leverage their assistance by targeting technical and funding support that takes into account the legal, fiscal and human capacity constraints and opportunities available to different classes of cities.

Theory and practice of climate adaptation planning

In addition to rising concentrations of people and economic assets, cities also face high socio-economic vulnerability and exposure to natural hazards (Brecht et al., 2012). Given these trends, the prevailing view is that adaptation planning must begin with initial assessments of risks and socio-economic vulnerability. Hazard probability risk assessments investigate the likelihood and impact of natural hazards across a city's geography (Füssel, 2007), while socio-economic and institutional vulnerability assessments recognize the social factors affecting exposure and resilience to natural events (Smit and Wandel, 2006). Some frameworks such as the World Bank's 'Urban Risk Assessment' (Dickson et al., 2012) integrate the two approaches to obtain an overall urban risk profile. The lack of downscaled and accurate scientific data is often seen as inhibiting assessments and, therefore, constraining local adaptation.

Translating assessments into action requires cities to implement structural investments (such as engineered sea walls, infrastructure upgrades or restored wetlands), institutional reforms (such as changes to policies, laws and economic incentives) and societal measures (such as public education, livelihood replacement and behavior change) (Carmin et al., 2015). Assessing, planning and implementing adaptation strategies place a significant burden on city staff and finances at a time when resources are often dedicated to other pressing social service and infrastructure mandates (Measham et al., 2011). In response, scholars recommend mainstreaming adaptation planning into local land use planning and sector work streams, with emphasis on integrating adaptation with disaster risk reduction and economic development (Birkmann and Teichman, 2010; Halsnæs and Trærup, 2009; Huq and Reid, 2009; Mercer, 2010). Mainstreaming prevents adaptation from becoming a solely environmental issue, reduces the risks of agenda sidelining and lowers administrative and implementation costs (Agrawala and van Aalst, 2008; Nelson et al., 2007). By leveraging new climate finance and support, cities can use climate adaptation to address underlying socio-economic vulnerabilities and development needs (Pelling, 2010).

To date, adaptation has been incremental, with an emphasis on climate-proofing infrastructure, rather than more fundamental institutional and societal change reforms aimed at reducing overall urban and social vulnerability (Ayers and Dodman, 2010; Bierbaum et al., 2013). Significant barriers to adaptation include the lack of information, assessment and planning staff know-how, financial resources, political leadership and national or regional support (Carmin, et al., 2013; Amundsen et al., 2010; Moser, 2009). In addition, the stress on mainstreaming and identifying win–win solutions has led, in some instances, to nearer-term priorities co-opting the adaptation agenda such as in the case of promoting real estate development in at-risk areas (Haughton and Counsell, 2004). Mal-adaptation practices that increase social or systemic vulnerability can also take place such as when adaptation increases greenhouses gas emissions, disproportionately burdens the poor and most vulnerable, or reduces future flexibility to adapt (Barnett and O'Neill, 2010).

In recent years, the number of climate adaptation guidance documents has grown rapidly. These guidance documents vary in their focus and level of specificity, particularly depending on whether they target the Global North versus Global South (see Table 23.1). This phenomenon can be attributed to the continuous discovery of different local adaptation

Linda Shi et al.

Table 23.1 Examples of adaptation planning guidance documents in the Global North and South

Author	Title	Date
Global North		
Green and Blue Space Adaptation for Urban Areas and Eco Towns	*Adaptation Action Planning Toolkit: Planning for a Changing Climate across Europe*	2011
Davoudi, Crawford and Mehmood	*Planning for Climate Change: Strategies for Mitigation and Adaptation for Spatial Planners*	2009
Environment Canada	*Canadian Communities' Guidebook for Adaptation to Climate Change*	2008
U.S. Environmental Protection Agency	*Adaptation Strategies Guide for Water Utilities*	2012
U.S. EPA-Region 9	*Climate Change Handbook for Regional Water Planning*	2011
American Public Health Association	*Climate Change: Mastering the Public Health Role, a Practical Guidebook*	2011
National Wildlife Federation	*Restoring the Great Lakes' Coastal Future: Technical Guidance for the Design and Implementation of Climate-Smart Restoration Projects*	2011
Wildlife Conservation Society	*Climate Change Planning for the Great Plains*	2010
ICLEI-Local Governments for Sustainability	*Preparing for Climate Change: A Guidebook for Local, Regional, and State Governments*	2007
Global South		
UN-HABITAT	*Planning for Climate Change: A Strategic, Values-Based Approach for Urban Planners*	2014
World Bank	*Urban Risk Assessment*	2012
World Bank	*Guide to Climate Change Adaptation in Cities*	2011
United Nations Environmental Programme (UNEP)	*Handbook on Methods of Climate Change Impact Assessment and Adaptation Strategies*	1998
Asia Disaster Preparedness Center	*Integrating Disaster Risk Management into Climate Change Adaptation*	2013
Asian Development Bank	*Increasing Climate Change Resilience of Urban Water Infrastructure*	2013
World Bank	*A Workbook on Planning for Urban Resilience in the Face of Disasters: Adapting Experiences from Vietnam's Cities to Other Cities*	2012
United Nations Development Programme (UNDP)/UNEP	*Mainstreaming Climate Change Adaptation into Development Planning: A Guide for Practitioners*	2011
Rockefeller Foundation	*Planning for Urban Climate Resilience: Framework and Examples from the Asian Cities Climate Change Resilience Network*	2010
UNDP	*Gender, Climate Change and Community-Based Adaptation*	2010
World Health Organization/Pan American Health Organization	*Protecting Health from Climate Change: Vulnerability and Adaptation Assessment*	2010

needs, the charting of diverse pathways to adaptation planning and implementation or, less optimistically, the lack of an empirical understanding of when and which cities rely on which types of strategies (Anguelovski and Carmin, 2011).

Global patterns of adaptation planning

In the Global North, adaptation guidelines and manuals are complemented by those that target specific ecosystems, sectors and scales. These documents tend to be produced by different public and private actors including national government agencies (or in Europe, by transnational entities like the European Union), subnational governments as well as by nongovernmental organizations (NGOs). In contrast, guidance documents for the Global South feature very different substantive emphases. Some countries of the Global South are beginning to produce national climate impact assessments and action plans, but there are few examples of nation-specific toolkits guiding cities on how to develop adaptation plans according to local contexts. Where these exist, they are usually developed with international support. Reflecting on the academic discourse on the importance of adaptation integration under resource-constrained conditions, guidance for the Global South tends to emphasize mainstreaming adaptation into traditional development spheres (Anguelovski et al., 2014).

The differences between planning guidelines available to countries in the Global North and Global South point to several assumptions about the way cities in these different regions adapt. First, the gap in adaptive capacity between these countries is manifested in the degree to which governments and nongovernmental stakeholders can develop technical guidelines for their constituents. The lack of resources dedicated to technical guideline development in the global South forces cities to draw on guidelines for 'developing countries,' which assumes that cities, regardless of region, level of development, geography or size, can benefit from the same processes and strategies. Second, they assume that cities in wealthier nations benefit from guidance on a more sectoral approach, while those in less developed nations prefer to (or should) take a mainstreaming, integrated approach. But do these assumptions fit actual local government needs? How can the experiences of early adapter cities inform our understanding of how to better target technical and financial assistance to catalyze local adaptation?

Survey methodology

To answer these questions, this chapter draws on a 2011 global survey of cities that were then members of ICLEI-Local Governments for Sustainability. ICLEI is an association of over 1,200 cities in 86 countries that have committed to undertake climate and sustainability action. Survey response rates by region are presented in Table 23.2 (for detailed survey methodology, see Carmin et al., 2012b). The sample may be slightly skewed due to the self-selection effect of ICLEI member cities and of those that chose to respond to the survey as well as the disproportionate number of responses from US cities relative to other regions. Still, while recognizing that the data does not speak for all cities, the descriptive analysis provided in this chapter is nevertheless useful in shedding light on the drivers, motivations, strategies and challenges among cities that have embarked on climate adaptation. Unless otherwise noted, the statistics cited refer to those local governments responding to the survey and not to local governments or cities in general.

Summary of survey results

An earlier report, 'Progress and Challenges in Urban Climate Adaptation Planning: Results of a Global Survey' (Carmin et al., 2012b), explores in detail the variations in city experiences of climate change impacts and adaptation progress. That report showed that while 80 percent of cities have perceived changes in their local climate, the types of changes and impacts

Table 23.2 Types of cities surveyed by population size and topography

	Survey Respondents						
Region \ Population	Over 1 million (large)	500,000 to 1 million (medium)	100,000 to 500,000 (medium)	Less than 100,000 (small)	Total response (% of Total)	Cities contacted	Response rate
Africa	3	1	3	3	10 (2%)	29	34%
Asia	9	6	8	1	24 (5%)	108	22%
Australia and New Zealand	0	1	10	32	43 (9%)	122	35%
Canada	1	4	6	15	26 (6%)	41	63%
Europe	1	3	22	21	47 (10%)	154	31%
Latin America	10	2	7	1	20 (4%)	43	47%
United States	13	18	73	194	298 (64%)	578	52%
Total	37	35	129	267	468	1,075	44%
(% of Total)	(8%)	(7%)	(28%)	(57%)			

vary by cities' coastal, inland, lowland and highland geographies. For 21 percent of cities, the frequency of natural disasters has increased. Disaster-related fatalities have risen in 10 percent of cities worldwide, with a much higher rate in Asia (33 percent), Latin America (27 percent) and Africa (25 percent). Heat is a particular challenge for Asian and African cities, while drought is an important issue in Australia.

Nearly 70 percent of cities say that they have initiated (or plan to undertake) a risk or vulnerability assessment, with an equal proportion having already begun to plan for concrete adaptation actions. Local initiatives have taken place under resource-constrained contexts, and, in most cases, without national support. Despite common perceptions, the lack of data and technical know-how is primarily a concern among cities of the Global South rather than North. Instead, the vast majority of cities report difficulty obtaining financial resources, allocating staff time, communicating the nature of adaptation, and generating interest among local stakeholders (Carmin et al., 2012b).

The 2012 report also pointed to regional variations in adaptation strategies. Cities in Europe and Asia are more likely to have met with national agencies in comparison to the US, while local public support is strongest in Canada. Canadian, Latin American and Asian cities favor forming taskforces or advisory groups and partnerships to support adaptation planning. Cities in Canada, Europe, Africa and the US often partner with other cities, while those in Asia and Latin America more often partner with NGOs and community groups. Only 11 percent of respondents have formed partnerships with the private sector, with Africa having the highest level of engagement (29 percent) (see Roberts, Chapter 35 in this volume, for further discussion of urban governance and institutions in the case of Durban, South Africa).

This chapter advances the earlier analysis by examining whether local adaptation governance varies with city size. The patterns that emerge help shed light on whether technical and fiscal assistance can be better targeted by taking these urban characteristics into account. The analysis raises examples of smaller cities that have managed to overcome capacity barriers to pursue adaptation planning and implementation, and the approaches they pursued to overcome such constraints.

Global patterns of adaptation planning

Variations by city size: motivations, progress and challenges in adaptation planning

Motivations for local adaptation planning

Existing analyses of early adapters have identified two categories of motivators, namely endogenous considerations (such as local leadership or civil society pressure) and exogenous forces (such as opportunities to showcase the city as a global leader in sustainability) (Carmin et al., 2012a; Anguelovski et al., 2014). Survey respondents overwhelmingly emphasize endogenous factors related to long-term quality of life, development and resilience. Of the 280 cities that answered this question, 66 percent cite being prepared for the future and 45 percent cite reducing hazard impacts as key priorities. While these are the top motivations across all city sizes, medium and small cities are more likely to cite additional motivators. For instance, 48 percent of cities with 500,000 to one million residents and 37 percent of smaller cities (fewer than 500,000 residents) cite community livability as a key motivator. Thirty-two percent of cities with 100,000 to 500,000 residents and 29 percent of those with fewer than 100,000 residents also emphasize helping achieve local government goals as a factor, highlighting the greater importance of integrative planning with decreasing city size (Carmin et al., 2012a). Securing municipal services for the longer-term is a motivator for 21 percent of cities overall, with greater importance (29 percent) among cities with 100,000 to 500,000 people.

Economic considerations such as gaining a competitive advantage, advancing economic development, building a reputation, obtaining external financing or keeping up with peers are rarely cited as top motivating factors. This may suggest that adaptation continues to be seen more as part of disaster preparedness or environmental protection agendas, especially since adaptation initiatives are led by environmental departments and sustainability offices in 43 percent to 49 percent of the cities surveyed. Demonstrating regional leadership is a motivating factor for 29 percent of small cities. Contrary to the emphasis placed in the literature on reducing the vulnerability of the poorest to climate change (Adger, 2006; Adger et al., 2009; Dodman and Satterthwaite, 2009; Tol et al., 2004), poverty and inequality reduction is cited as a top three motivating factor for adaptation by only ten cities including only three from the Global South.

Progress in conducting assessments

Progress in conducting assessments varies more by city population than region or level of development. Cities with over one million people are much more likely (95 percent) to have an assessment in the works, and 75 percent have completed or are presently in the process of assessing risks and vulnerability. City progress with assessments declines with city size. Among cities with fewer than 100,000 residents, 35 percent have completed or are conducting an assessment, and 33 percent plan to conduct an assessment (see Figure 23.1). Among the 87 cities that have completed assessments, by far the most prevalent concerns are storm water runoff (65 percent) and change in demand for storm water management (61 percent), followed by changes to electricity demand (41 percent), and potential disruptions to transportation and waste management (30 percent).

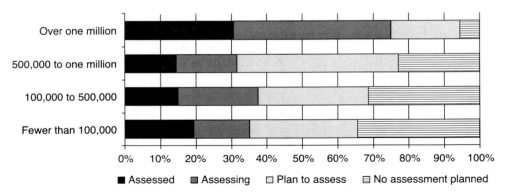

Figure 23.1 Percentages of cities conducting risk or vulnerability assessments by city size

Mainstream or standalone: approaches to adaptation planning

Around 68 percent of cities have initiated some sort of adaptation planning and, of the 273 cities that answered this question, 58 percent have undertaken some combination of developing strategic, adaptation, or sector plans, and/or integrating adaptation into existing community or sector plans. Overall, 50 percent of cities are developing or have drafted strategic plans, as compared to 19 percent using detailed adaptation plans and 18 percent using sector plans. Whatever plan they develop, 46 percent of cities choose to integrate it into existing community plans such as comprehensive land use plans, and 20 percent of cities integrate adaptation plans into existing sector plans.

With these global trends, a city uses different approaches depending on its size. Large cities are more likely (61 percent) to develop standalone or strategic plans without integrating them into existing sector or community plans, while 42 percent to 45 percent of cities under one million have chosen this approach (see Table 23.3). Conversely, 11 percent of large cities say they are integrating climate issues directly into existing community or sector plans without first developing a climate plan, while 27 to 32 percent of cities under one million are taking a purely integrative approach. Around 26 to 29 percent of all cities are undertaking a hybrid approach by developing both specialized adaptation plans and integrating these newly developed plans into existing plans. In addition, 68 percent of large cities formed special groups such as taskforces and committees to manage adaptation initiatives, compared to 49 percent of small cities. While 40 percent of local governments in large cities formed partnerships during adaptation planning (for instance with other cities, NGOs or the private sector), 67 percent of local governments with 100,000 to 500,000 residents reached out to these sectors. Despite the emphasis on mainstreaming adaptation approaches into development planning (Anguelovski et al., 2014; Carmin et al., 2013), it appears that many cities, particularly large ones, are developing standalone adaptation plans without further integration, at least in the planning stages.

Mobilizing resources for adaptation planning

The capacity to mobilize human resources to conduct risk and vulnerability assessments poses a major challenge particularly for smaller cities. Medium-sized cities rely most on resources from within local governments to conduct assessments (64 percent of cities 500,000 to one million, and 60 percent of cities 100,000 to 500,000 residents), with researchers and consultants

Global patterns of adaptation planning

Table 23.3 Plan development, integration or both by city size

	Plan (strategic, adaptation or sector)	Integrate into community or sector plan	Both plan and integrate
Over one million	17 (61%)	3 (11%)	8 (29%)
500,000 to one million	10 (45%)	6 (27%)	6 (27%)
100,000 to 500,000	34 (42%)	26 (32%)	21 (26%)
Fewer than 100,000	60 (42%)	43 (30%)	39 (27%)

serving a minor role. In contrast, 36 percent of large cities rely on local government and 56 percent draw on state or regional agencies, academics and international NGOs or transnational bodies (such as the European Union). The engagement of higher-tier agencies may be due to the greater importance of these cities to national and international economies as well as these cities being the location of major universities and research institutions. For small cities, a diverse set of actors, including international NGOs (28 percent), state or regional agencies (27 percent), community-based organizations (15 percent) and academic institutions (11 percent) work together to assist and conduct assessments. The near non-existent role of local government among small cities may reflect their lack of technical capacities or financial resources as well as the potential for economies of scale in regional risk assessments.

Among large cities, 86 percent have initiated adaptation planning, compared to 73 percent of medium cities and 63 percent of small cities. Different public actors have taken the lead in adaptation planning worldwide, including the mayor's office, town or city managers' and councilors' offices, departments of environment, planning, public works, emergency management, economic development and engineering as well as utilities, nonprofits and citizen groups. Cities with over one million or between 500,000 to one million residents are much more likely (54 percent and 50 percent, respectively) to have a dedicated climate office, as compared to cities with 100,000 to 500,000 (28 percent) or less than 100,000 (20 percent) residents. Large cities are nearly three times as likely to have a dedicated climate office compared to small cities. The offices of the mayor, town manager and city councilors lead adaptation in about 20 percent to 30 percent of cities. These offices are in a strong position to provide coordination and integration across local agencies, and are more likely to emphasize plan integration than environmental offices in small cities.

Obtaining funding and staff resources are cited as challenges by 85 percent of respondents worldwide, ranging from 70 percent in Asia, Europe and Canada to 90 percent in other regions. Most local governments (60 percent) are not receiving any financial support for their adaptation planning. National governments are the most common source of funding, although this supports only 24 percent of surveyed cities. Subnational resources including from within local governments, support 9 percent of cities. International funds from the multilateral banks, different United Nations agencies and development organizations are particularly important for Africa (21 percent), Asia (17 percent) and Latin America (17 percent) as well as Europe (18 percent) in the form of the European Union. In the US, rates of support from private philanthropic foundations (17 percent) and nonprofits (16 percent) are high compared to other regions.

Funding sources also vary by city size, as shown in Figure 23.2, which clearly indicates that cities pursue funding portfolios across public, private, local and international sources. Local governments, regardless of size, do not rely on a single, dominant source of funds for

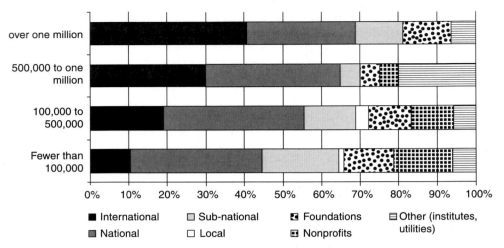

Figure 23.2 Proportional sources of funding for climate adaptation for local governments by size

Note: International funders include development and UN agencies and multilateral development banks. Philanthropic foundations and nonprofits include both domestic and international entities. This figure represents portfolios of funding sources across an entire group of cities by size, not the amount of funding any particular city or group of cities receives from each source. The fact that 73 percent of the small cities sample is located in the US contributes to the lack of international funding in this group.

adaptation planning and implementation, although the diversity of sources increases as city size shrinks. International support is a source of funding for 41 percent of cities over one million. This source declines in prevalence as city size declines, serving as a source for only 10 percent of cities under 100,000. National funding is a source for 28 percent of cities over one million and increases in prevalence as city size decreases, serving as a source for 34 percent of cities under 100,000. Surprisingly, funds generated through local taxes, permit fees and public service user charges are not significant for most cities of any size. Although larger cities are usually considered to be more fiscally autonomous given their larger property tax base, few cities above 500,000 draw on such local funds. Among cities with 500,000 to one million residents, funds from research universities and public utilities are especially important, and serve as a source for 20 percent of cities in this size category.

In many countries, smaller cities are prohibited from directly receiving funds from international development agencies, making them more likely to seek alternative sources from domestic and foreign philanthropic foundations, and nonprofits. Foundations and nonprofits fund similar proportions of cities across the different size groups, with about a 15 percent difference between large and small cities. This may be explained by the fact that some international funds are being channeled through these entities both to get around intergovernmental fiscal transfer restrictions and because of international preferences for engaging civil society. The increasing reliance on subnational funds in small cities may also reflect the prevalence of US cities in this group and their reliance on state and federal grants.

Interpreting trends in city size, governance capacity and adaptation progress

The key takeaway from the survey is that many elements of local government adaptive capacity (as measured by adaptation planning progress and perceived challenges) are associated

Global patterns of adaptation planning

with city size. As a result, urban adaptation planners and policy-makers must pay greater attention to city size and target technical and funding support according to local realities.

Of course, city size is not the only determinant of a city's willingness to adapt to climate change, but among motivated cities, size can be a useful proxy for gauging administrative capacity and the appropriateness of different adaptation strategies. Though not consistent across all cities, local governments whose jurisdictions include larger population sizes often enjoy higher levels of fiscal and political autonomy, more diverse urban development mandates, larger concentrations of supportive citizens and businesses, more diverse approaches to local revenue generation and better access to national and international support because of their reputational prominence. As planning and decision-making authority is devolved from national to local governments, as in the case of many developmental and environmental sustainability-related issues (Betsill and Bulkeley, 2006; Bulkeley, 2010), larger cities tend to have a higher capacity of agglomerating staffing and financial resources. By contrast, smaller cities tend to be more capacity constrained and rely to a much greater extent on intergovernmental transfers (Bird and Slack, 2004). While a few small cities have managed to overcome capacity deficits to become early adapters, local governance capacity, in general, is often directly related to a local government's ability to conceive, plan and implement climate adaptation projects and programs.

According to the survey results, city size has implications for local government planning, finance and governance. First, in terms of urban planning processes, large cities have higher rates of beginning assessments and adaptation planning (95 percent and 86 percent respectively) compared to small cities (68 percent and 63 percent respectively). Size is also associated with capacity to mobilize human and fiscal resources for and approaches to mainstreaming or integrating adaptation planning. Contrary to significant academic and policy attention placed on mainstreaming adaptation, most cities are developing standalone climate plans during the early stages and only some cities are taking either hybrid or purely integrative approaches. Many cities find it difficult to mainstream adaptation given that they often also find it challenging to communicate the importance of adaptation to others. In other words, the larger the city, the more likely they are to begin with a planning-only approach, while small and medium cities tend to emphasize integration and partnership formation.

Second, while cities have made significant progress in initiating local assessments and adaptation planning, the survey responses indicate that they will face major difficulties in advancing towards implementation under current resource constraints. Every city indicated that lack of financing for adaptation is the most significant barrier, with many also highlighting the lack of national or even local political support and understanding. Almost no respondent received local government funding for adaptation, making access to external funding critical to adaptation planning and implementation. Given the reliance on external funding, large cities are 1.5 times more likely to receive national and subnational funding support than small cities, and four times more likely to benefit from international funding. While large cities draw on a few major sources of external technical and financial support, in general, small cities have to seek out support from a diverse set of organizations, with a greater emphasis on subnational funding, and private and philanthropic sources.

Third, in terms of multilevel governance, the progress and constraints in local adaptation should be seen not only as dependent on local contexts, but also on the level of adaptation taking place at various scales of government and within different sectors (Adger et al., 2009; Mukheibir et al., 2013). The importance of city size in local government capacity and the variation in financial and human resources and approaches between large and small cities are likely due to national policies for intergovernmental relations, which delineate the authority and responsibility over fiscal resources at each level of government. Local governments are creatures of the state that, sooner or later, run into limits to their adaptation capacity due to statutory constraints on their authority over infrastructure planning, and their lack of autonomy over lifecycle planning and resource deployment (Measham et al., 2011; Mukheibir et al., 2013).

The implication for practitioners is that the emphasis on local leadership in adaptation planning will need to give way to a greater focus on developing national frameworks, coordinating climate data and adaptation policies, identifying obligations, incentives, and funding timeframes, and facilitating regional cooperation to address cross-jurisdictional adaptation needs and economies of scale (Rosenzweig et al., 2010; Mukheibir et al., 2013). Given that national support, policies and attitudes affect local inclination towards, capacity for, and strategies of adaptation, the variations in political support across different levels of government is an important consideration in external efforts to promote local adaptation (Carmin et al., 2012a). These political contexts shape cities' adaptation approaches and preferences, suggesting a greater need for national-level development of urban technical guidelines that take into account the political receptivity for particular forms of integration or implementation (Measham et al., 2011).

Finally, despite constraints noted in this chapter, some municipalities have overcome these obstacles by establishing partnerships with other sectors and levels of government. For instance, the towns of Marshfield (population of 25,000), Duxbury (15,000) and Scituate (18,000) near Boston, US, came together and applied for regional grants to conduct assessments of vulnerability to sea level rise. In other cases, international actors developed pilot sites for adaptation projects in second- and third-tier cities with manageable scales of challenges and a minimal bar of governance capacity. International organizations bring resources to cities that may not otherwise have benefited and, since many of these are private or philanthropic in nature, can directly intervene and engage at the local level. Examples of these external interventions include the Rockefeller Foundation's Asian Cities Climate Change Resilience Network and their climate resilience building projects in cities such as Hat Yai in Thailand (population of 160,000), Shimla in India (170,000) and Quy Nhon in Vietnam (280,000). Other examples include UN-HABITAT's Cities and Climate Change Initiative's interventions in Esmeraldas, Ecuador (165,000) and Walvis Bay, Namibia (85,000). These dedicated capacity support and funding mechanisms have helped cities overcome institutional and capacity barriers. However, because of the extensive financial and staffing costs associated with intensive individualized city engagement, the unique experiences of these pilot cities are unlikely to be scaled up to a large number of towns and municipalities.

Next steps for advancing local climate adaptation

As adaptation initiatives move beyond early adapters, a large number of medium-sized and, to a lesser extent, small-sized cities are planning to undertake adaptation assessments. Existing external adaptation toolkits and guidance documents emphasize how to conduct assessments and develop adaptation plans (in the Global North) or mainstream adaptation (in the Global South). As urban adaptive capacity often varies according to population size and bureaucratic capacity, efforts to catalyze and support local action can and should take into account variations in local contexts, and levels of fiscal and administrative autonomy. In particular, there is a need for technical guidance to:

- *Match the level of assessment and planning with the target audience.* The fact that almost no local government of a small city in this survey undertook an assessment in-house suggests that assessment requirements are too complex or are too unrealistic at this scale.

Global patterns of adaptation planning

Efforts may be better spent building the capacity of nongovernmental actors, regional agencies and research or academic institutions that can support local governments.

- *Recognize the diversity of local adaptation strategies, and support both sectoral and mainstream approaches.* There is a strong need, particularly in the Global South, to develop country-specific urban adaptation guidelines, to establish distinct climate offices and to generate sector-specific guidance, given that many cities, especially larger ones, are more likely to develop standalone and sector plans.
- *Develop the technical support to meet the most cited local adaptation challenges.* While most guidance documents focus on obtaining data and conducting assessments, most cities find these to be less important challenges. Instead, the vast majority of cities world-wide cite financing as a challenge, even where they take a mainstreaming approach.
- *Identify distinct financing strategies for large, medium and small cities.* Catalyzing adaptation implementation will require not only technical assistance, but also stronger fiscal support for adaptation, particularly for small and medium cities that have had difficulty accessing international funding to complement national sources of support.

Practitioners and scholars have an opportunity to strengthen the multilevel governance of adaptation planning and implementation. There has been limited emphasis on subnational support for local adaptation, which may offer economies of scale for assessments and planning activities as well as the proper implementation and investment authority over important adaptation responses. In many cases, cities have compensated for the lack of available information, technical assistance and funding by connecting with peer city networks, universities and private foundations, with some even reaching out to the business community. Given the number of cities that have initiated adaptation, soon these cities will require capital financing and political support to implement hard investments, non-structural actions and institutional reforms identified in the assessments and plans. Only some of these costs will be able to be mainstreamed into existing budgets before they require additional funding to avoid cutting into the resources of other local priorities.

These emerging challenges to understanding local urban climate impacts, socio-economic vulnerability and adaptive capacity suggest that future research must take into account the following questions:

- What are viable alternatives to the single-city adaptation assessment model that can facilitate the scaling-up of adaptation planning, implementation and governance?
- How does a country's intergovernmental fiscal and political context shape local adaptation strategies, options and outcomes? How can policy-makers transform these institutional constraints into opportunities for sustained and effective engagement with cities?
- How can external actors more effectively frame and direct their capacity and financial support mechanisms, while acknowledging the spatial, economic, political and social variations that exist in cities around the world?

To support adaptation planning, one must look beyond the city scale and towards the roles and responsibilities of subnational and national governments in mobilizing fiscal and political resources. Far in advance of reaching social limits to adaptation (Adger et al., 2009; Dow et al., 2013), cities, particularly those under 100,000 people, will run up against fiscal limits of adaptation. In the absence of major available international funds, understanding how existing intergovernmental relations and fiscal decentralization can accommodate looming adaptation needs remains a critical research gap.

Acknowledgments

The authors acknowledge the ICLEI member communities and representatives from the cities who generously gave their time to participate in this research, and to ICLEI-Local Governments for Sustainability for collaborating on this project. This research was funded by a grant from the United States National Science Foundation (#0926349), with additional support from the Department of Urban Studies and Planning at the Massachusetts Institute of Technology.

References

Adger, W. N. (2006). Vulnerability. *Global Environmental Change, 16*(3), 268–281.

Adger, W. N., Dessai, S., Goulden, M., Hulme, M., Lorenzoni, I., Nelson, D. R., . . . Wreford, A. (2009). Are There Social Limits to Adaptation to Climate Change? *Climatic Change, 93*(3–4), 335–354.

Agrawala, S., and van Aalst, M. (2008). Adapting Development Cooperation to Adapt to Climate Change. *Climate Policy, 8*(2), 183–193.

Amundsen, H., Berglund, F., and Westskog, H. (2010). Overcoming Barriers to Climate Change Adaptation—A Question of Multilevel Governance? *Environment and Planning C: Government and Policy, 28*(2), 276–289.

Andonova, L. B., Betsill, M. M., and Bulkeley, H. (2009). Transnational Climate Governance. *Global Environmental Politics, 9*(2), 52–73.

Anguelovski, I., and Carmin, J. (2011). Something Borrowed, Everything New: Innovation and Institutionalization in Urban Climate Governance. *Current Opinion in Environmental Sustainability, 3*(3), 169–175.

Anguelovski, I., Chu, E., and Carmin, J. (2014). Variations in Approaches to Urban Climate Adaptation: Experiences and Experimentation from the Global South. *Global Environmental Change, 27,* 156–167.

Ayers, J., and Dodman, D. (2010). Climate Change Adaptation and Development I: The State of the Debate. *Progress in Development Studies, 10*(2), 161–168.

Barnett, J., and O'Neill, S. (2010). Maladaptation. *Global Environmental Change, 20*(2), 211–213.

Betsill, M. M., and Bulkeley, H. (2006). Cities and the Multilevel Governance of Global Climate Change. *Global Governance: A Review of Multilateralism and International Organizations, 12*(2), 141–159.

Bierbaum, R., Smith, J. B., Lee, A., Blair, M., Carter, L., Chapin, F. S., . . . Verduzco, L. (2013). A Comprehensive Review of Climate Adaptation in the United States: More than Before, but Less than Needed. *Mitigation and Adaptation Strategies for Global Change, 18*(3), 361–406.

Bird, R. M., and Slack, E. (2004). *Fiscal Aspects of Metropolitan Governance* (ITP Paper No. 401). Toronto: University of Toronto.

Birkmann, J., and Teichman, K. (2010). Integrating Disaster Risk Reduction and Climate Change Adaptation: Key Challenges—Scales, Knowledge, and Norms. *Sustainability Science, 5*(2), 171–184.

Brecht, H., Dasgupta, S., Laplante, B., Murray, S., and Wheeler, D. (2012). Sea-Level Rise and Storm Surges: High Stakes for a Small Number of Developing Countries. *The Journal of Environment and Development, 21*(1), 102–138.

Bulkeley, H. (2010). Cities and the Governing of Climate Change. *Annual Review of Environment and Resources, 35*(1), 229–253.

Carmin, J., Anguelovski, I., and Roberts, D. (2012a). Urban Climate Adaptation in the Global South Planning in an Emerging Policy Domain. *Journal of Planning Education and Research, 32*(1), 18–32.

Carmin, J., Nadkarni, N., and Rhie, C. (2012b). *Progress and Challenges in Urban Climate Adaptation Planning: Results of a Global Survey.* Cambridge, MA: MIT.

Carmin, J., Dodman, D., and Chu, E. (2013). Urban Climate Adaptation and Leadership: From Conceptual Understanding to Practical Action. OECD Regional Development Working Paper 2013/26. OECD Publishing.

Carmin, J., Tierney, K., Chu, E., Hunter, L., Roberts, J. T., and Shi, L. (2015). Adaptation to Climate Change. In R. E. Dunlap and R. J. Brule (Eds.), *Sociological Perspectives on Climate Change.* Oxford and New York: Oxford University Press.

Dickson, E., Baker, J. L., Hoornweg, D., and Tiwari, A. (2012). *Urban Risk Assessments: Understanding Disaster and Climate Risk in Cities.* Washington, D.C.: World Bank.

Global patterns of adaptation planning

Dodman, D., and Satterthwaite, D. (2009). Institutional Capacity, Climate Change Adaptation and the Urban Poor. *IDS Bulletin, 39*(4), 67–74.

Dow, K., Berkhout, F., Preston, B. L., Klein, R. J., Midgley, G., and Shaw, M. R. (2013). Limits to Adaptation. *Nature Climate Change, 3*(4), 305–307.

Field, C. B., Barros, V., Dokken, D. J., Mach, K. J., Mastrandrea, M. D., Bilir, T. E., . . . White, L. L. (Eds.). (2014). Summary for Policymakers. In *Climate Change 2014: Impacts, Adaptation, and Vulnerability. Part A: Global and Sectoral Aspects. Contribution of Working Group II to the Fifth Assessment Report of the Intergovernmental Panel on Climate Change* (pp. 1–32). Cambridge, UK, and New York, USA: Cambridge University Press.

Füssel, H.-M. (2007). Adaptation Planning for Climate Change: Concepts, Assessment Approaches, and Key Lessons. *Sustainability Science, 2*(2), 265–275.

Halsnæs, K., and Trærup, S. (2009). Development and Climate Change: A Mainstreaming Approach for Assessing Economic, Social, and Environmental Impacts of Adaptation Measures. *Environmental Management, 43*(5), 765–778.

Haughton, G., and Counsell, D. (2004). Regions and Sustainable Development: Regional Planning Matters. *The Geographical Journal, 170*(2), 135–145.

Huq, S., and Reid, H. (2009). Mainstreaming Adaptation in Development. *IDS Bulletin, 35*(3), 15–21.

Lall, S. V., and Deichmann, U. (2009). *Density and Disasters: Economics of Urban Hazard Risk* (Policy Research Working Paper No. 5161). Washington, D.C.: World Bank.

Lee, T. (2013). Global Cities and Transnational Climate Change Networks. *Global Environmental Politics, 13*(1), 108–127.

McKinsey Global Institute. (2011). *Urban World: Mapping the Economic Power of Cities*. New York: McKinsey and Co.

Measham, T. G., Preston, B. L., Smith, T. F., Brooke, C., Gorddard, R., Withycombe, G., and Morrison, C. (2011). Adapting to Climate Change through Local Municipal Planning: Barriers and Challenges. *Mitigation and Adaptation Strategies for Global Change, 16*(8), 889–909.

Mercer, J. (2010). Disaster Risk Reduction or Climate Change Adaptation: Are We Reinventing the Wheel? *Journal of International Development, 22*(2), 247–264.

Moser, S. C. (2009). Governance and the Art of Overcoming Barriers to Adaptation. *Magazine of the International Human Dimensions Programme on Global Environmental Change, 3*, 31–36.

Mukheibir, P., Kuruppu, N., Gero, A., and Herriman, J. (2013). Overcoming Cross-Scale Challenges to Climate Change Adaptation for Local Government: A Focus on Australia. *Climatic Change, 121*(2), 271–283.

Nelson, D. R., Adger, W. N., and Brown, K. (2007). Adaptation to Environmental Change: Contributions of a Resilience Framework. *Annual Review of Environment and Resources, 32*, 395–419.

Pelling, M. (2010). *Adaptation to Climate Change: From Resilience to Transformation*. London: Routledge.

Rosenzweig, C., Solecki, W., Hammer, S. A., and Mehrotra, S. (2010). Cities Lead the Way in Climate-Change Action. *Nature, 467*(7318), 909–911.

Satterthwaite, D., Huq, S., Pelling, M., Reid, H., and Romero-Lankao, P. (2007). Building Climate Change Resilience in Urban Areas and Among Urban Populations in Low- and Middle-Income Nations. Commissioned report for the Rockefeller Foundation. London: International Institute for Environment and Development (IIED).

Smit, B., and Wandel, J. (2006). Adaptation, Adaptive Capacity and Vulnerability. *Global Environmental Change, 16*(3), 282–292.

Struyk, R. J., and Giddings, S. (2009). *The Challenge of an Urban World: An Opportunity for U.S. Foreign Assistance* (White Paper). Washington, D.C.: International Housing Coalition.

Tol, R. S. J., Downing, T. E., Kuik, O. J., and Smith, J. B. (2004). Distributional Aspects of Climate Change Impacts. *Global Environmental Change, 14*(3), 259–272.

24

ADAPTATION TO CLIMATE CHANGE IN RAPIDLY GROWING CITIES

Roberto Sánchez-Rodríguez

This chapter reflects on the challenges and opportunities for climate change adaptation (CCA) in rapidly growing cities. These cities are particularly important in the discussion of adaptation to climate change, since this is where a significant part of future urban growth will take place (Martine et al., 2007; Seto et al., 2010; UN-HABITAT, 2013)—primarily in Asia (mostly in China and India), Africa and to a lesser extent, in Latin America (Seto et al., 2010). Rapidly growing cities face significant challenges and development pressures that create obstacles for adaptation, but at the same time, offer unique opportunities to introduce adaptation strategies and actions with significant benefits for their inhabitants in the short-, medium- and long-term. Timing is an essential factor, as the dynamic growth of these cities creates a limited timeframe to develop adaptation strategies and actions at an affordable cost. Delays or inaction can result in high social, economic and environmental costs to urban inhabitants.

This chapter highlights the importance of operational approaches for adaptation planning and implementation that are inclusive of both formal and informal urban growth and the adaptation needs for each of these processes. Focusing on formal urban or informal growth independently of one another can lead to incomplete perspectives that will likely create conflicts and contradictions among adaptation responses. Furthermore, current institutional structures and cultures create obstacles to adaptation planning and implementation. Although financial resources, capacity building and risk communications are essential elements of CCA, successful strategies and actions depend deeply on the institutional dimension. This chapter seeks to raise awareness of these critical issues in order to transcend rhetoric to operational approaches.

The case of rapidly growing cities in the context of climate change

Rapidly growing cities are often associated with fast population growth, physical expansion and economic growth. They are confronted with development pressures with few human, technical and financial resources and information to respond. These cities are burdened with problems of unemployment and underemployment, environmental degradation, deficiencies in urban services and housing, deterioration of existing infrastructure, problems in securing access to natural resources needed for urban functions (water, energy, food, construction materials), deficient governance, incomplete planning and implementation, and violence (Moser and McIlwaine, 2006; Martine et al., 2007; UN-HABITAT, 2013). Climate change can further exacerbate these challenges, as it is closely interacts with national and local key social processes in societies and the distribution of power within institutions that manage growth (Adger et al., 2005; Thomas and Twyman, 2005).

Despite these challenges, cities have multiple benefits such as offering a better chance for people to live fuller lives, poverty reduction, minimized environmental damage and, in the case of climate change, reduced vulnerabilities and the potential for adaptation with the support of proper policies and actions (Martine et al., 2007; World Bank, 2010; Sánchez-Rodríguez, 2013; UN-HABITAT, 2013). Even cities with severe development pressures have encouraging examples of bottom-up projects that improve livelihoods in poor communities (Carolini, 2007; Satterthwaite et al., 2007), enhance environmental conditions (Simon et al., 2013; UN-HABITAT, 2013) and increase resilience to climate variability and change (ISDR et al., 2010; UNDP, 2010).

A number of international organizations consider CCA an imperative for developing countries and their cities (UNDP, 2010; World Bank, 2010, 2012; UN-HABITAT, 2011). However, only a very limited number of rapidly growing cities have created formal adaptation strategies, plans and actions to climate change. Even in these cases, it is uncertain how efficient these strategies and actions are and will be during the next decades. So far, little to no attention has been placed on monitoring and evaluating climate change responses—an integral component of adaptation planning and implementation. The key issue for the construction of operational approaches incorporating inclusive and balanced growth is to take advantage of the opportunities offered by cities. Moreover, focusing on structural conditions of urban growth and how these interact with climate change is instrumental in constructing development paths that meet adaptation needs. An important component in this process is to conceptualize CCA not only as an environmental problem, but fundamentally as a development challenge.

Indeed, the size, form, structure and function of urban areas and their future growth trajectories are critical elements to be considered in the discussion of climate change. The construction of formal urban space results from top-down actions managed by a framework of state control (urban planning and economic incentives) and the real estate market. However, in rapidly growing cities the creation of urban space is also characterized by the dynamic expansion of informal settlements outside this framework, where their incorporation into the real estate market and formal sector occurs over time. Informal growth results from complex social processes operating at different geographical scales that drive local growth; the limited capacity of local and national governments to meet the demand of housing and urban services of their inhabitants; and social agency. The UN-Habitat estimates that 900 million people lived in informal urban settlements in 2010, but there are significant regional differences. For example, data reveals that informal settlements create 62 percent of urban growth in sub-Saharan Africa, 32 percent in Asia and 24 percent in Latin America (UN-HABITAT, 2013). These informal settlements are often located in hazardous areas, vulnerable to the impacts of hydro-meteorological and climatic events. Hence, addressing CCA in rapidly growing cities requires strategies capable of addressing both formal and informal urban growth processes (Moser and Satterthwaite, 2008; ISDR et al., 2010; World Bank, 2011, 2012; UN-HABITAT, 2011).

The construction of operational approaches to adaptation

Current urban adaptation strategies are often based on the construction of defensive infrastructure (i.e. resilient infrastructure) to flooding and other hazards associated with climate change (Lowe et al., 2009; Rotterdam, 2012; New York, 2011; Chicago, 2012; Matthews, 2012). Although defensive infrastructure is an important component of building resilient cities, adaptation strategies to climate change must be inclusive of formal and informal growth as well as address the underlying vulnerabilities of urban communities within the context of local development. Mainstreaming adaptation strategies in urban policies will require attention to mechanisms often found in urban planning (building codes, land use permits, zoning, etc.) and through economic incentives common in the regulation of the urban built environment. These are top-down actions that are managed by the formal framework of state control (urban planning). However, the vulnerabilities to multiple and cumulative development challenges are often neglected by formal planning (Hull, 1998; Watson, 2009) and adaptation strategies are prone to aggravate this problem if this is not addressed. Despite the fact that informal settlements are among the communities with the highest vulnerabilities and most urgently in need of development actions including adaptation to climate change (Satterthwaite et al., 2007; Moser and Satterthwaite, 2008; World Bank, 2010), these communities are often invisible to policies orienting urban growth and neglected in the distribution of resources for development (Watson, 2009).

Central to the construction of operational approaches is enabling policy-makers and stakeholders to recognize the different interests and claims on assets and the roles and responsibilities of different social and institutional actors. Focusing on livelihoods to connect vulnerability, adaptation and development through social agency and bottom-up initiatives is particularly relevant in informal communities (Moser and Satterthwaite, 2008; World Bank, 2010; UN-HABITAT, 2011). Hence, cities must consider the importance of multidimensional approaches, inclusive governance and placing adaptation to climate change within development challenges and opportunities. Recognizing the social agency of individuals and communities, along with combining bottom-up and top-down initiatives, are useful approaches to addressing vulnerabilities; reducing the impact of limited local financial, technical and human resources; and opening opportunities for adaptation.

A review of the literature reveals that a significant number and diversity of studies support the construction of adaptation strategies and actions in directing the growth of cities over the last decade. Diverse international organizations have shown a particular interest in this direction, publishing methodological guidelines to assist cities in the design of adaptation. It is worth noting that each organization orients its guidelines according to its area of interest. For example, the World Bank orients its methodology to connect adaptation and poverty reduction (World Bank, 2010). Guidelines and tools to evaluate community risks, livelihoods and adaptation have been developed by the International Red Cross, CARE, the International Institute for Sustainable Development, OXFAM and other organizations (IFRC, 2007; CARE, 2012; IISD, 2012). UNDP, UN-HABITAT and USAID address adaptation actions within the context of urban planning (USAID, 2007; UNDP, 2010; UN-HABITAT, 2011; World Bank, 2012) and the UN International Strategy for Risk Reduction seeks to connect adaptation to climate change with disaster risk reduction (DRR) (ISDR et al., 2010). All of these organizations have supported local initiatives in developing countries, particularly in small- to medium-sized cities (i.e. ≤ two million inhabitants).

Although there are common elements to some of these methodologies, there are also differences that can be confusing for local decision-makers, planners and stakeholders. Attention to community adaptation strategies and bottom-up initiatives are important components in creating inclusive adaptation actions in informal settlements. Unfortunately, there is little effort to integrate the strengths of these methodologies; however, a publication oriented to cities in Latin America is a contribution in this direction (Sánchez-Rodríguez,

Adaptation in rapidly growing cities

2013). It distinguishes between operational approaches for adaptation planning and those for adaptation implementation, while considering monitoring and evaluation as an integral part of both processes.

It is important to highlight that the construction of operational approaches can be obstructed or facilitated by the institutional dimension of CCA, but some useful elements to consider for rapidly growing cities include (Sánchez-Rodríguez, 2013):

- distinguishing between adaptation planning and implementation as two different processes that might include the participation of a different set of actors, different goals, methods, time frameworks and resources;
- securing the participation of stakeholders and creating work teams;
- identifying leadership and creating an integral and multidimensional vision of the adaptation process;
- adopting an approach with incremental steps and realistic goals;
- identifying 'low-regret' adaptation actions in the early stages of the process and framing adaptation within the context of local development;
- identifying funding sources and integrating a monitoring and evaluation strategy in the early stages of adaptation planning and implementation; and,
- creating a dynamic communication strategy for the adaptation process.

Multi-scalar and multi-sectoral coordination to overcome institutional challenges

While there is growing recognition that adaptation planning is essential (Wilbanks and Kates, 2010; Ford et al., 2011), the magnitude of the institutional challenges of linking adaptation planning and implementation are only now becoming recognized. Fast-paced urban and population growth and the slow development of their institutions often create conflict in cities. There are also gaps in understanding and evaluating how institutional networks operate and how they are connected to climate change. For example, research on CCA processes in South Africa and Latin America show that few institutions fully understand the implications of adaptation to climate change and the nature of their respective roles and responsibilities in this context (Koch et al., 2007; Roberts, 2010; Carmin et al., 2012). Studies suggest that adaptation to climate change can challenge the hierarchical manner in which local governments have operated up to now and stress the importance of strengthening local institutions in order to implement adaptation in urban areas. The institutional dimension of CCA is a key theme that requires broader attention beyond the scope of this chapter; however, this chapter will address two major problems highlighted in the literature: the need to create multi-scale and multi-sector coordination in building adaptation processes. These issues are important for cities seeking to adapt to climate change in both developed and developing countries alike, but critical for rapidly growing cities.

In Europe, the United States, Australia and Latin America, municipal-level studies document the challenges to create adaptation plans with multidimensional and integrated approaches (Roberts, 2010; Tompkins et al., 2010; Glaas and Juhola, 2013; Vammen Larsen et al., 2012; Carmin et al., 2012). These studies document the structure of municipal governments as being comprised of professional silos without much interdepartmental coordination. Each silo has a defined work area (economic development, water services, parks and recreation, urban planning, etc.) as well as its own internal norms, cultures and procedures that may hinder horizontal coordination with other departments. A study of Danish municipalities expands this point, suggesting that climate change does not possess clear institutional characteristics as a professional municipal area. Rather, it is viewed as a void with no clear rules and norms, according to which politics are to be conducted and policy measures agreed upon. The study highlights the interdisciplinary nature of climate change affecting most municipal sectors and the little experience municipalities have in assuming the coordinating role that is required. This dilemma complicates the integration of the interdisciplinary element of climate change into the bureaucratic organization (Vammen Larsen et al., 2012).

This lack of a formalized institutional character and the cross-scale and cross-sector nature of CCA are highlighted further in a number of cases worldwide. For example, Anguelovski and Carmin (2011) focus on institutions of urban climate governance, highlighting the ways in which public, private and civil society actors and institutions articulate climate goals, exercise influence and authority and manage urban climate planning and implementation processes. They document urban areas that tend to formalize and institutionalize their work through the establishment of dedicated climate units, either within a relevant department or as a separate and cross-cutting office. However, few local governments have the resources and know-how to institutionalize adaptation to climate change. Despite successful efforts in the case of Durban, South Africa, Roberts (2010) notes cross-sectoral difficulties, as not all departments have the same capacity or ability to mainstream adaptation activities into their ongoing work. Meanwhile, the cross-scalar difficulties are highlighted by Tompkins et al. (2010) who argue there is little real evidence of CCA initiatives trickling down to the local government level in the United Kingdom.

One factor for the lack of cross-scalar and cross-sector integration may be due to a lack of information and no clear formal structure to initiate CCA plans through already established institutions. For example, a study on the role of experts and decision-makers building adaptation to climate change in the Florida Keys describes how these individuals are currently operating with limited information and lack a formal institutional framework necessary to shape and execute adaptation measures on an urgent basis (Mozumder et al., 2011). Despite the recognition of the importance of climate change impacts, very few experts and decision-makers report that their respective agencies have developed formal adaptation plans. Other studies suggest that few changes have been made in forecasts, plans, design criteria, investment decisions, budgets or staffing patterns in response to climate risks (Repetto, 2008; Berrang-Ford et al., 2011).

It is clear that multi-scalar coordination is essential to create adaptation strategies; in particular, for rapidly growing cities, given the development pressures they face (including climate change, and their limited resources to address it). However, deficiencies in the coordination among national, subnational and local governments are a common problem obstructing adaptation of cities in both developed and developing countries (Wilson, 2006; Storbjörk, 2007; Roberts, 2010; Measham et al., 2011; Mukheibir et al., 2013). Studies document complaints by local governments about a lack of guidance from national governments to help identify their role and responsibilities in developing adaptation strategies and the lack of adequate financial and technical support to implement them. Nonetheless, some cities have created their own climate change legislation (e.g. Sao Paulo, Brazil and Mexico City, Mexico) in an effort to strengthen their activities, but with little impact on adaptation planning. This is because the multi-scale coordination required depends on the political will in national and local governments and has been a major bottleneck in the construction of inclusive development and adaptation to climate change.

The role of planning

Mainstreaming CCA in cities is increasingly recognized in the literature as a critical step for sustainable development, but the challenge remains in its operationalization within the current structures and operational cultures of local institutions, as discussed above. The growing literature on CCA frequently assumes planning will be capable of guiding, coordinating and delivering adaptation. Planning has been highlighted as a key tool for urban development, DRR and CCA, but it requires systemic changes in its operational structure to fulfill this role (Wilson, 2006; Blanco and Alberti, 2009; Roberts, 2010; Measham et al., 2011; Preston et al., 2011; Carmin et al., 2012; Sánchez-Rodríguez, 2012). Particularly important are changes strengthening planning capacity to address complexity, uncertainty, multi-dimensional and multilevel coordination (Tompkins et al., 2010; Huntjens et al., 2011).

Planning is considered a societal tool to create order among activities and interests driving growth in cities, to reduce conflicts among them and to promote the well-being of inhabitants (Blair, 1973). This model has prevailed since the early schemes of urban planning more than a century ago. However, a number of scholars have pointed out the limitations of that model including its difficulty in ordering complex urban systems. Blair (1973) highlights the lack of capacity of urban planning to solve conflicts among economic, political and socio-cultural priorities in the urban space. He stresses its mainly physical focus (the built environment) while criticizing the techno-bureaucratic approach of planning that addresses problems ad hoc, with little scientific knowledge and without self-critique or a synoptic view of the problems. Moreover, Blair argues that the planning profession operates without updating its methodological and conceptual frameworks in order to meet the continuous changes in urban society, and proposes an alternative approach to urban planning based on multidimensional and interdisciplinary perspectives. By the same token, Simmie (2003) criticizes the techno-administrative function of urban planning, incapable of controlling and orienting urban growth as well as the undesired consequences of concentrating the benefits of planning in some social groups and its negative costs on others. Furthermore, Bridge (2007) questions the assumption that urban planning would impose order to the inherent chaos of the city. He criticizes the dominant assumption within urban planning that the city can be directed by an instrumental rationality. These deficiencies in planning are common in rapidly growing cities.

The discussion hitherto suggests that the urgency of adaptation to climate change aggravates these problems. Blair's multidimensional proposal for urban planning becomes more complicated when the biophysical and environmental dimensions of adaptation to climate change are incorporated as part of the challenges of planning in the 21st century. Although this is an issue requiring a broader discussion than what is provided in this chapter, the remainder of this section highlights four issues that are particularly relevant in rapidly growing cities.

Adaptation planning as a learning process

Part of the literature on adaptation suggests the importance of considering adaptation planning as a learning process likely to require regular revisiting of development policies, plans and projects, as climate and socio-economic conditions change (Hinkel et al., 2009, Hofmann et al., 2011). This approach is particularly relevant for rapidly growing cities. Considering adaptation planning as a learning process involves regarding adaptation as just the start of a policy process rather than its culmination (Hulme et al., 2009). Hence, framing adaptation within the discussion of local development facilitates connecting climate change with changes in local socio-economic and physical conditions and national and regional processes. This is an issue a number of international organizations have emphasized (UNDP, 2010; UN-HABITAT, 2011; World Bank, 2010, 2011, 2012). But, unfortunately, not enough attention has been provided to the common elements between development and adaptation and how they can be combined in adaptation strategies, plans and actions. This issue is discussed in more detail below.

Although there is little attention in the literature to the potential benefits of planning as a social learning process for adaptation, some research on planning has addressed the issue (Pelling et al., 2008). For example, Holden (2008) suggests that social learning is a relevant but under-investigated feature of planning and a critical part in the adaptation of innovations. The understanding of why and how learning takes place is needed in the theory and practice of planning to improve the impact and efficiency of the plan, improve the transferability of best practices, increase public support and translate the learning into new plans. However, Holden notes the few analytical tools to assess how and when learning is taking place among different professional and public communities. Considering adaptation planning as a social learning process would allow for periodical adjustments in order to reduce the uncertainty of the impacts of climate change, identify societal needs to adapt to them (Frommer, 2009) and reduce the risk of maladaptation (Adger et al., 2009).

Linking adaptation to development

Several authors consider it a critical task to incorporate local knowledge and experiences into multidimensional and multi-scale approaches that can better guide the construction of adaptation responses to climate change and integrate them with development strategies (Ewing et al., 2008; Hodson and Marvin, 2009). Dovers (2009) stresses the need for connecting climate adaptation more closely to existing policy and agendas, knowledge, risks and issues that communities already face. He emphasizes the important role of planning in connecting adaptation and development needs and challenges. Other authors support the contention that adaptation takes place as a response to multiple stimuli, not just climate (Adger et al., 2009; Tompkins et al., 2010).

It is worth noting that despite the fact that social change is a central element of development, there is perhaps not enough attention focused on livelihoods in development studies to connect adaptation, vulnerability and development (Paavola, 2008; Sánchez-Rodríguez, 2009). A critical task of integrating that knowledge into multidimensional and multi-scale approaches guiding the construction of adaptation responses is to frame CCA within development strategies (Ewing et al., 2008; Hodson and Marvin, 2009; Moser and Satterthwaite, 2008). Several contributions that are useful in this direction suggest a framework of pro-poor asset adaptation for climate change as a conceptual and operational framework. Moser (2008), for example, proposes a second-generation, asset-based policy as a means to sustain current poverty reduction policies focusing on the provision of housing, urban services and infrastructure, health, education and microfinance as tools in adaptation strategies. Adger and Barnett (2009) argue that the social context in which adaptation takes place is a key element to measure the success of adaptation and the trade-offs that may be involved. Along these lines, Barnett and Campbell (2009) believe that community values must be taken into account if adaptation planning is to be effective, efficient, legitimate and equitable. Moreover, Sánchez-Rodríguez (2012) highlights the need to build operational approaches of adaptation planning, recognizing structural socio-economic conditions in low- and middle-income countries. Focusing on 'low-regrets' actions can be a useful tool framing adaptation planning and implementation within the context of local development (Hallegatte, 2009; UNDP, 2010).

Adaptation and disaster risk reduction

The UN Hyogo Convention (2005–2015) has developed extended support to DRR by national and local governments in developed and developing countries. The experiences and lessons learned thus far from DRR efforts can help rapidly growing cities develop CCA (ISDR et al., 2010; ISDR, 2011; Solecki et al., 2011). CCA and DRR share similar objectives, but they are managed by separate agencies and different legal frameworks in many countries. The lack of coordination between these two agendas duplicates efforts in most cities in both developed and developing countries.

The lessons learned from DRR reported in the literature suggest benefits for CCA, particularly considering the difficulties cities face to begin an adaptation process. Local support for adaptation to climate change can be created linking the two approaches together (Few et al., 2007; Rosenzweig et al., 2011; Etkin et al., 2012), as highlighted by the IPCC (2012) Special Report on Managing the Risk of Extreme Events and Disasters to Advance Climate Change Adaptation (SREX), which notes the complementary aspects between DDR and CCA. In fact, the connection of climate adaptation and development discussed above can be influenced by how the issue is framed. For example, to the extent that adaptation is viewed as a public safety issue, it may have greater resonance within local governments (Measham et al., 2011) as well as support and acceptance by stakeholders and decision-makers (Dovers, 2009; Sovacool et al., 2012). Addressing adaptation to climate variability and change can help cities reduce the uncertainty of the frequency and location of impacts (Lowe et al., 2009; Matthews, 2012), and it can help cities address the stressors driving social vulnerability to climate variability and climate change (Turnbull and Turvill, 2012).

However, caution should be taken against overemphasizing a hazards approach in adaptation planning. A number of authors have expressed concern that a disproportionate focus on the impacts of climate change can neglect addressing the drivers of vulnerability, limiting its contribution to create adaptation (Orlove, 2009; Lemos et al., 2007; Romero-Lanko and Gnatz, Chapter 15 in this volume). Boyd and Juhola (2009) express concern about how the debate of climate change is dominated by impacts-led approaches that focus on climate risks rather than on human vulnerability. Ribot (2011) notes that adaptation can fail to address causality by focusing on responses and reducing attention to the underlying social causes of risk. He highlights the importance of understanding multi-scale causes of vulnerability to better identify the adequate dimensions of adaptation actions and planning. Other authors propose that adaptive capacity building can be delivered through a two-tiered approach that focuses on developing effective disaster reduction to climate-related hazards and implementing policy reforms that address deeper structural inequalities that are often at the heart of entrenched vulnerabilities to climate change (Lemos et al., 2007; Dovers, 2009; Boyd and Juhola, 2009).

Monitoring and evaluation

Two important tools of learning in adaptation planning and implementation are monitoring and evaluation, particularly critical in the case of rapidly growing cities given their dynamic pace of growth and the need to periodically adjust and correct adaptation strategies and actions (Preston et al., 2011; Sánchez-Rodríguez, 2013). Although some literature recognizes the importance of evaluation in adaptation, this is an area that is under-researched and requires significant work to go beyond the simple evaluation criteria that have been developed to date (Doria et al., 2009; Arnell, 2010). Saavedra and Budd (2009) suggest that the institutional arrangements for the evaluation of adaptation processes, policies and measures are still in their developmental infancy. Evaluation and monitoring are often advocated within adaptation decision-making frameworks, but methods for undertaking such work are rarely articulated. Moreover, adaptation plans frequently fail to acknowledge the importance of core design principles for adaptation policies and measures such as efficacy, efficiency and equity. Adger and Barnett (2009) argue that the metrics that may be used to determine the goals of adaptation, the measures of success and trade-offs that may be involved can be understood only in terms of the social context in which adaptation takes place. Communities value things differently, and this must be taken into account if adaptation is to be effective, efficient, legitimate and equitable (Barnett and Campbell, 2009). Similarly, Arnell (2010) highlights that local circumstances significantly affect what adaptation options are considered feasible, what information is likely to be used, what assessment techniques are adopted and, crucially, how adaptation decisions are actually made. This implies that it will be difficult to make generalized assessments of the potential contributions of adaptation to managing the risks posed by climate change and to construct generalized models of the adaptation process.

Other research extends this discussion of evaluation by calling attention to the interpretation of key concepts in adaptation such as adaptive capacity (Engle, 2011) and vulnerability (Hinkel, 2011). Engel calls attention to the limited effort to evaluate adaptive capacity across vulnerability and resilience frameworks and to improve the understanding of adaptive capacity dynamics. It is important to identify what builds adaptive capacity and what functions as limits and barriers to adaptation. Hinkel questions the use of vulnerability as a concept to identify mitigation targets of vulnerability, raising awareness about the importance of adaptation to guide the allocation of adaptation funds, monitoring of adaptation policy and conducting scientific research. Vulnerability measurement can be misleading, as it raises false expectations. These and other contributions in the literature (Adger et al., 2009; Tompkins et al., 2010; Preston et al., 2011) move the discussion of adaptation planning and implementation towards a better understanding of the elements needed to operationalize adaptation to present and future climate impacts.

Final considerations

This chapter has highlighted the limited understanding of how CCA planning and implementation is actually taking place. The majority of CCA studies report on assessments of potential vulnerability of social and natural systems to the negative impacts of climate change and describe an intention to act rather than concrete adaptation actions (Ford et al., 2011; Tompkins et al., 2010). For example, there are very few published case studies revealing how adaptation to climate change is actually being delivered or the barriers that will influence how adaptation takes place (Arnell, 2010). Tompkins et al. (2010) question whether the observed adjustments and changes to perceived climate risks represent evidence of a societal shift towards a well-adapting society or if they are merely unconnected actions of individuals motivated by different stimuli.

However, there is a growing body of knowledge on adaptation to climate change that can help rapidly growing cities develop their strategies, plans and actions. This chapter intended to make a contribution in this direction by addressing elements that local authorities and stakeholders can consider in creating adaptation processes in their cities: a) questioning the assumption that current structures and operational cultures of planning will be able to meet the needs of adaptation; b) the importance of developing operational approaches to adaptation planning and implementation, according to the conditions of each city; and, c) the need to

Adaptation in rapidly growing cities

provide attention to the institutional changes required to build efficient, equitable and effective adaptation responses to climate change.

The chapter has also highlighted elements that support the creation of useful approaches for adaptation planning and implementation suggested in the literature including: a) adaptation as a learning process; b) framing adaptation within the context of development; c) integrating CCA and DRR and design; and d) implementation of monitoring and evaluation as an integral component of the adaptation process. Planning is expected to create multidimensional and integrated perspectives of these complex challenges, but as discussed, it is often a technical tool focusing on the physical dimension of urban growth. Rapidly growing cities would benefit from considering the institutional arrangement needed to create multidimensional perspectives in the planning process of adaptation and its implementation. It is worth stressing that these institutional arrangements might be different for planning and implementation processes.

As remarked previously, timing is an essential factor for CCA in rapidly growing cities. Their dynamic growth creates a limited timeframe to develop adaptation strategies and actions. Current urban policy decisions have the potential to either constrain or facilitate future urban adaptation strategies. Further delays in developing and implementing adaptation actions can have severe consequences for millions of urban inhabitants and national economies. Attention to how operational approaches for adaptation planning and implementation can be constructed represents a contribution in this direction.

Key messages

- Despite the increasing attention to CCA by international organizations, national and local governments, there is still limited knowledge of how types of adaptation responses are actually taking place.
- Adaptation to climate change must transcend from conceptual approaches to operational approaches designed to meet the needs of rapidly growing cities.
- Planning is expected to create multidimensional and integrated perspectives of these complex challenges, but it is often a technical tool focusing on the physical dimension of urban growth.
- Rapidly growing cities would benefit from considering the institutional arrangement needed to create multidimensional perspectives in the planning process of adaptation and its implementation.

Future agenda items

- Place attention on the particular conditions of rapidly growing cities with particular emphasis on the development of operational approaches.
- Develop a better understanding of the institutional dimensions of CCA within the context of local development.
- Create multidimensional approaches of planning able to meet the challenges of rapidly growing cities and CCA in the 21st century.

References

Adger, N., and Barnett, J. (2009). Four reasons for concern about adaptation to climate change. *Environment and Planning A*, 41(12), 2800–2805.

Adger, W. N., Arnell, N. W., and Tompkins, E. L. (2005). Successful adaptation to climate change across scales. *Global Environmental Change*, 15(2), 77–86.

Adger, W. N., Dessai, S., Goulden, M., Hulme, M., Lorenzoni, I., Nelson, D. R., Naess, L. O., Wolf, J., and Wreford, A. (2009). Are there social limits to adaptation to climate change? *Climatic Change*, 93(3–4), 335–354.

Anguelovski, I., and Carmin, J. (2011). Something borrowed, everything new: Innovation and institutionalization in urban climate governance. *Current Opinion in Environmental Sustainability*, 3(3), 169–175.

Arnell, N. (2010). Adapting to climate change: An evolving research agenda. *Climatic Change*, 100, 107–111.

Barnett, J., and Campbell, J. (2009). *Climate Change and Small Island States: Power, Knowledge and the South Pacific.* London, Earthscan.

Berrang-Ford, L., Ford, J., and Patterson, J. (2011). Are we adapting to climate change? *Global Environmental Change*, 21(1), 25–33.

Blair, T. (1973). *The Poverty of Planning.* London, Macdonald Publishers.

Blanco, H., and Alberti, M. (2009). Chapter 2. Building capacity to adapt to climate change through planning. *Progress in Planning*, 71(3), 153–205.

Boyd, E., and Juhola, S. (2009). Stepping up to the climate change: Opportunities in re-conceptualising development futures. *Journal of International Development*, 21(6), 792–804.

Bridge, G. (2007). City senses: On the radical possibilities of pragmatism in geography. *Geoforum*, doi: 10.1016/j.geoforum. 2007.02.004.

CARE (2012). *Tackling the Limits to Adaptation: An International Framework to Address 'Loss and Damage' from Climate Change Impacts.* Geneva, Switzerland, CARE International, ActionAid International, World Wildlife Fund.

Carmin, J., Anguelovski, I., and Roberts, D. (2012). Urban climate adaptation in the global south: Planning in an emerging policy domain. *Journal of Planning Education and Research*, 32(1), 18–32.

Carolini, G. (2007). Organizations of the urban poor and equitable urban development: Process and product. In *The New Global Frontier. Urbanization, Poverty and Environment in the 21st Century*, ed. Martine, G., McGranahan, G., Montgomery, M., and Fernandez-Castilla, R., 133–150. London, Earthscan.

Chicago (2012). *Chicago Climate Action Plan.* Available at www.chicagoclimateaction.org/.

Doria, M., Boyd, E., Tompkins, E. L., and Adger, W. N. (2009). Using expert elicitation to define successful adaptation to climate change. *Environmental Science and Policy*, 12, 810–819.

Dovers, S. (2009). Normalizing adaptation. Editorial. *Global Environmental Change*, 19, 4–6.

Engle, N. L. (2011). Adaptive capacity and its assessment. *Global Environmental Change*, 21(2): 647–656.

Etkin, D., Medalye, J., and Higuchi, K. (2012). Climate warming and natural disaster management: An exploration of the issues. *Climatic Change*, 112 (3–4), 585–599.

Ewing, R., Bartholomew, K., Winkelman, S., Walters, J., and Anderson, G. (2008). Urban development and climate change. *Journal of Urbanism: International Research on Placemaking and Urban Sustainability*, 1(3), 201–216.

Few, R., Brown, K., and Tompkins, E. L. (2007). Public participation and climate change adaptation: Avoiding the illusion of inclusion. *Climate Policy*, 7(1), 46–59.

Ford, J. D., Berrang-Ford, L., and Paterson, J. (2011). A systematic review of observed climate change adaptation in developed nations. *Climatic Change*, 106, 327–336.

Frommer, B. (2009). Climate change and the resilient society: Utopia or realistic option for German regions? *Natural Hazards*, 58(1), 85–101.

Glaas, E., and Juhola, S. (2013). New levels of climate adaptation policy: Analyzing the institutional interplay in the Baltic Sea region. *Sustainability*, 5, 256–275.

Hallegatte, S. (2009). Strategies to adapt to an uncertain climate change. *Global Environment Change*, 19(2), 240–247.

Hinkel, J. (2011). Indicators of vulnerability and adaptive capacity: Towards a clarification of the science–policy interface. *Global Environmental Change*, 21(1), 198–208.

Hinkel, J., Bisaro, S., Downing, T., Hofmann, M. E., Lonsdale, K., Mcevoy, D., and Tabara, D. (2009). Learning to adapt. Narratives of decision makers adapting to climate change. In *Making Climate Change Work for US: European Perspectives on Adaptation and Mitigation Strategies*, ed. Hulme, M., and Neufeldt, H., 113–134. Cambridge, UK, Cambridge University Press.

Adaptation in rapidly growing cities

Hodson, M., and Marvin, S. (2009). Urban ecological security: A new urban paradigm? *International Journal of Urban and Regional Research*, 33(1), 193–215.

Hofmann, M., Hinkel, J., and Wrobel, M. (2011). Classifying knowledge on climate change impacts, adaptation and vulnerability in Europe for informing adaptation research and decision-making: A conceptual meta-analysis. *Global Environmental Change*, 21(3), 1106–1116.

Holden, M. (2008). Social learning in planning: Seattle's sustainable development codebooks. *Progress in Planning*, 69, 1–40.

Hull, A. (1998). The development plan as a vehicle to unlock development potential. *Cities*, 15(5), 327–335.

Hulme, M., Neufeldt, H., Colyer, H., and Ritchie, A. (Eds.) (2009). *Adaptation and Mitigation Strategies: Supporting European Climate Policy.* The final report from the ADAM Project. Revised June 2009. Tyndall Centre for Climate Change Research, University of East Anglia, Norwich, UK.

Huntjens, P., Pahl-Wostl, C., Flachner, Z., Neto, S., Koskova, R., Schlueter, M., NabideKiti, I., and Dickens, C. (2011). Adaptive water management and policy learning in a changing climate. A formal comparative analysis of eight water management regimes in Europe, Asia, and Africa. *Environmental Policy Governance*, 21(3), 145–163.

IFRC (2007). Red Cross/Red Crescent Climate Guide. Available at www.ifrc.org/Global/Publications/disasters/climate%20change/climate-guide.pdf.

IISD (2012). *CRiSTAL User's Manual Version 5.* Community-based risk screening tool – adaptation and livelihoods. Manitoba, Canada, The International Institute for Sustainable Development, IISD, IUCN, Helvetias, SEI.

IPCC (2012). *Managing the Risks of Extreme Events and Disasters to Advance Climate Change Adaptation.* A special report of working groups I and II of the Intergovernmental Panel on Climate Change. Ed. C. B. Field, V. Barros, T. F. Stocker, D. Qin, D. J. Dokken, K. L. Ebi, M. D. Mastrandrea, K. J. Mach, G.-K. Plattner, S. K. Allen, M. Tignor, and P. M. Midgley. Cambridge, UK, and New York, NY, USA, Cambridge University Press, 582 pp.

ISDR (2011). *Hyogo Framework for Action 2005–2015: Building the Resilience of Nations and Communities to Disasters.* Mid-term review 2010–2011. Geneva, Switzerland, The United Nations Secretariat of the International Strategy for Disaster Reduction (ISDR).

ISDR, ITC, and UNDP (2010). *Local Governments and Disaster Risk Reduction: Good Practices and Lessons Learned.* A contribution to the "Making Cities Resilient" Campaign. Geneva, Switzerland, The United Nations Secretariat of the International Strategy for Disaster Reduction (ISDR), 86 pp.

Koch, I. C., Vogel, C., and Patel, Z. (2007). Institutional dynamics and climate change adaptation in South Africa. *Mitigation and Adaptation Strategies for Global Change*, 12(8), 1323–1339.

Lemos, M. C., Boyd, E., Tompkins, E., Osbahr, H., and Liverman, D. (2007). Developing adaptation and adapting development. *Ecology and Society*, 12(2), 26 [online]. Available at www.ecologyandsociety.org/vol12/iss2/art26/.

Lowe, A., Foster, J., and Winkelman, S. (2009). *Ask the Climate Question: Adapting to Climate Change Impacts in Urban Regions.* Washington, D.C., Center for Clean Air Policy, 44 pp.

Martine, G., McGranahan, G., Montgomery, M., and Fernandez-Castilla, R. (2007). Introduction. In *The New Global Frontier. Urbanization, Poverty and Environment in the 21st Century*, ed. Martine, G., McGranahan, G., Montgomery, M., and Fernandez-Castilla, R., 1–16. London, Earthscan.

Matthews, T. (2012). Responding to climate change as a transformative stressor through metro-regional planning. *Local Environment*, 17(10), 1089–1103.

Measham, T. G., Preston, B. L., Smith, T. F., Brooke, C., Gorddard, R., Withycombe, G., and Morrison, C. (2011). Adapting to climate change through local municipal planning: Barriers and challenges. *Mitigation and Adaptation Strategies for Global Change*, 16(8), 889–909.

Moser, C. (2008). Assets and livelihoods: A framework for asset-based social policy. In *Assets, Livelihoods, and Social Policy*, ed. Moser, C. and Dani, A., 44–84. Washington, D.C., The World Bank.

Moser, C., and McIlwaine, C. (2006). Latin American urban violence as a development concern: Towards a framework for violence reduction. *World Development*, 34(1), 89–112.

Moser, C., and Satterthwaite, D. (2008). Towards pro-poor adaptation to climate change in the urban centers of low and middle-income countries. *Climate Change and Cities Discussion*, Paper 3, IIED.

Mozumder, P., Flugman, E., and Randhir, T. (2011). Adaptation behavior in the face of climate change: Survey responses from experts and decision makers serving the Florida Keys. *Ocean and Coastal Management*, 54, 37–44.

Mukheibir, P., Kuruppu, N., Gero, A., and Herriman, J. (2013). *Cross-Scale Barriers to Climate Change Adaptation in Local Government, Australia*. Gold Coast, Australia, National Climate Change Adaptation Research Facility, 95 pp.

New York (2011). PlaNYC. Available at www.nyc.gov/html/planyc2030/html/theplan/the-plan.shtml.

Orlove, B. (2009). The past, the present and some possible futures of adaptation. In *Adapting to Climate Change: Thresholds, Values, Governance*, ed. Adger, W. N., Lorenzoni, I., and O'Brien, K., 131–162. Cambridge, UK, Cambridge University Press.

Paavola, J. (2008). Livelihoods, vulnerability and adaptation to climate change: Lessons from Morogoro, Tanzania. *Environmental Science & Policy*, 11, 642–654.

Pelling, M., High, C., Dearing, J., and Smith, D. (2008). Shadow spaces for social learning: A relational understanding of adaptive capacity to climate change within organizations. *Environment and Planning A*, 40, 867–884.

Preston, B. L., Westaway, R. M., and Yuen, E. J. (2011). Climate adaptation planning in practice: An evaluation of adaptation plans from three developed nations. *Mitigation and Adaptation Strategies for Global Change*, 16, 407–438.

Repetto, R. (2008). *The Climate Crisis and the Adaptation Myth*. Working paper No. 13, New Haven, USA, Yale School of Forestry and Environmental Studies.

Ribot, J. (2011). Vulnerability before adaptation: Towards transformative climate action. *Global Environmental Change*, 21, 1160–1162.

Roberts, D. (2010). Prioritizing climate change adaptation and local level resilience in Durban, South Africa. *Environment and Urbanization*, 22, 397–413.

Rotterdam (2012) *Rotterdam Climate Initiative*. Available at www.rotterdamclimateinitiative.nl/nl/top/home. Accessed on May 15, 2012.

Rosenzweig, C., Soleki, W., Hammer, S., and Mehrota, S. (Eds.) (2011). *Climate Change and Cities. First Assessment of the Urban Climate Change Research Network*. Cambridge, UK, Cambridge University Press.

Saavedra, C., and Budd, W. (2009). Climate change and environmental planning: Working to build community resilience and adaptive capacity in Washington State, USA. *Habitat International*, 246–252.

Sánchez-Rodríguez, R. (2009). Learning to adapt to climate change in urban areas: A review of recent contributions. *Current Opinion in Environmental Sustainability*, 1, 201–206.

Sánchez-Rodríguez, R. (2012). Understanding and improving urban responses to climate change. Reflections for an operational approach to adaptation in low and middle-income countries. In *Cities and Climate Change. Responding to an Urgent Agenda*, ed. Hoornweg, D., Freire, M., Lee, M. J., Bhada-Tata, P., and Yuen, B., 452–469. Washington, D.C., The World Bank.

Sánchez-Rodríguez, R. (2013). Vulnerabilidad y adaptación al cambio climático. In *Respuestas Urbanas al Cambio Climático en América Latina* [Sánchez-Rodríguez, R., Ed.]. Instituto Inter Americano de Cambio Global (IAI) y Comisión Económica de Naciones Unidas para América Latina (CEPAL), 71–118. Santiago de Chile, CEPAL.

Satterthwaite, D., Saleemul, H., Pelling, M., Reid, H., and Lankau-Romero, P. (2007). *Adapting to Climate Change in Urban Areas: Possibilities and Constraints in Low-And-Middle-Income Nations*. Human Settlements Discussion Paper: Climate Change and Cities 1, London, IIED.

Seto, K., Sánchez-Rodríguez, R., and Fragkias, M. (2010). The new geography of contemporary urbanization and the environment. *Annual Review of Environment and Resources*, 35, 167–194.

Simmie, J. (2003). *Planning at the Crossroads*. London, University College London.

Simon, D., Fragkias, M., Leichenko, R., Sánchez-Rodríguez, R., Seto, K., and Solecki, W. (2013). *The Green Economy and the Prosperity of Cities*. UN-HABITAT Background Paper for the State of World Cities Report 2012/3, p. 48.

Solecki, W., Leichenko, R., and O'Brien, K. (2011). Climate change adaptation strategies and disaster risk reduction in cities: Connections, contentions, and synergies. *Current Opinion in Environmental Sustainability*, 3(3), 135–141.

Sovacool, B. K., D'Agostino, A. L., Meenawat, H., and Rawlani, A. (2012). Expert views of climate change adaptation in least developed Asia. *Journal of Environmental Management*, 97, 78–88.

Storbjörk, S. (2007). Governing climate adaptation in the local arena: Challenges of risk management and planning in Sweden. *Local Environment*, 12 (5), 457–469.

Thomas, D., and Twyman, C. (2005). Equity and justice in climate change adaptation amongst natural-resources-dependent societies. *Global Environmental Change*, 15, 115–124.

Adaptation in rapidly growing cities

Tompkins, E. L., Adger, W. N., Boyd, E., Nicholson-Cole, S., Weatherhead, K., and Arnell, A. (2010). Observed adaptation to climate change: UK evidence of transition to a well-adapting society. *Global Environmental Change*, 20(4), 627–635.

Turnbull, M., and Turvill, E. (2012). *Análisis de la capacidad de participación y de la vulnerabilidad. Guía para profesionales*. Un recurso de OXFAM para la reducción de riesgos y la adaptación al cambio climático. Oxford, Reino Unido. OXFAM.

UNDP (2010). *Designing Climate Change Adaptation Initiatives*. A UNDP toolkit for practitioners. UNDP Bureau of Development Policies.

UN-HABITAT (2011). *Planning for Climate Change. A Strategic Values Based Approach for Urban Planners*. For field testing and piloting in training. Version 1. United Nations Human Settlements Programme. UN-HABITAT.

UN-HABITAT (2013). *State of the World Cities 2012–2013. Prosperity of Cities*. New York, Routledge, United Nations Human Settlements Programme, Earthscan.

USAID (2007). *Adapting to Climate Variability and Change*. A Guidance Manual for Development Planning. Washington, D.C., United States International Development Agency (USAID).

Vammen Larsen, S., Kørnøv, L., and Wejs, A. (2012). Mind the gap in SEA: An institutional perspective on why assessment of synergies amongst climate change mitigation, adaptation and other policy areas are missing. *Environmental Impact Assessment Review*, 33, 32–40.

Watson, V. (2009). The planned city sweeps the poor away: Urban planning and 21st century urbanization. *Progress in Planning*, 72, 151–193.

Wilbanks, T. J., and Kates, R. W. (2010). Beyond adapting to climate change: Embedding adaptation in responses to multiple threats and stresses. *Annals of the Association of American Geographers*, 100(4), 719–728.

Wilson, E. (2006). Adapting to climate change at the local level: The spatial response, *Local Environment*, 11(6), 609–625.

World Bank (2010). *Participatory Scenario Development Approaches for Identifying Pro-Poor Adaptation Options*. Discussion Paper, 18. Washington, D.C., The World Bank.

World Bank (2011). *Guide to Climate Change Adaptation in Cities*. Washington, D.C., The World Bank.

World Bank (2012). *Building Urban Resilience: Principles, Tools and Practice*. Washington, D.C., The World Bank.

25

SPATIAL PLANNING

An integrative approach to climate change response

Shu-Li Huang and Szu-Hua Wang

Despite worldwide efforts to mitigate climate change, greenhouse gas (GHG) emissions between 2000 and 2010 have been higher than in the previous three decades. Overcoming substantial technological, economic and institutional challenges will be necessary in order to limit global warming to 2°C (Hartmann et al., 2013). Most strategies for coping with climate change and other global environmental changes are formulated and embedded in different development sectors such as energy, transit, agriculture, etc. Since these different sectors address land use and urbanization at specific geographical scales, both climate change adaptation and mitigation strategies can vary in spatial dimension. This, in turn, means that implementation occurs at either regional or local levels.

Spatial planning is an approach for responding to climate change that focuses on the interactions of land use and climate at local or regional scales (Biesbroek et al., 2009; Crawford and French, 2008). This approach is concerned with the coordination or integration of policies of different sectors through land-based strategies and goes beyond traditional regulatory-focused land use planning (Cullingworth and Nadin, 2006; Office of the Deputy Prime Minister, 2005). Relevant disciplines include land use planning, urban planning, regional planning and environmental planning. To be effective, spatial planning must have the capability to coordinate different sectors that manage, for example, transportation systems (Dewar, 2011; Grazi et al., 2008), local economic development (Gibbs, 2002; Todes, 2008; UNECE, 2008), housing (Bareman, 2002; Jones et al., 2012; Gkartzios and Scott, 2013), agriculture (Blom and Paulissen, 2007; Beukes et al., 2008; UNECE, 2008) and water systems (Wiering and Immink, 2006; Carter, 2007; Woltjer and Al, 2013).

This chapter reflects on the increasing attention given to climate change and associated discussions on the role spatial planning might have in minimizing both its causes and the consequences (Campbell, 2006; Bulkeley, 2006). The chapter begins by elaborating on how this might occur, followed by a discussion of the challenges and opportunities for the integration of mitigation and adaptation strategies into spatial planning. The final section draws conclusions and suggests future research needs to enhance the capacity for spatial planning to address urban and climate change concerns.

Spatial planning as an important feedback to climate change

The Office of the Deputy Prime Minister (UK) (2004) suggests that the built environment, infrastructure, local decision-making, rural environments and land use should be included in comprehensive planning efforts, in order to respond more effectively to the impacts of climate change. Spatial planning demonstrates potential for contributing to climate change targets by functioning as a switchboard for mitigation, adaptation and sustainable development efforts (Biesbroek et al., 2009; Jackson, 2006), and by offering an opportunity to shift the policy foundation of climate change to one with a strong spatial dimension (Bulkeley, 2006). It can largely influence the spatial allocation of land use activities, investment in infrastructure and the preservation of open space when utilized by the public sector (local, regional or national). Moreover, it offers opportunities to mitigate GHG emissions with respect to their sources and sinks, increase adaptive capacity through the rational allocation of land uses and to manage natural resources to enhance the protection of green infrastructure (i.e. the use of vegetation, soils and water for more natural management of urban environments) and, thereby, human well-being (Veraart et al., 2007; Campbell, 2006; Bulkeley, 2006). It can also incorporate land use control techniques such as zoning, transit-oriented development (TOD) or open space conservation for mitigating GHG emissions as well as provide insight to climate change adaptation through spatially explicit vulnerability assessments. The strength of spatial planning, when taken in the context of climate change, is that it can provide a lens through which the trade-offs and synergies between mitigation and adaptation strategies are assessed. Figure 25.1 shows the relationships of climate change, urban land use and spatial planning. Activities related to urban land use ranging from transportation, industrial development, energy use, etc., are the main sources of GHG emissions and contributors to climate change.

Early efforts to address climate change erroneously viewed mitigation and adaptation strategies as two fundamentally different approaches to the same problem. It is now widely recognized that this mindset ignores possible synergies and trade-offs between both types of

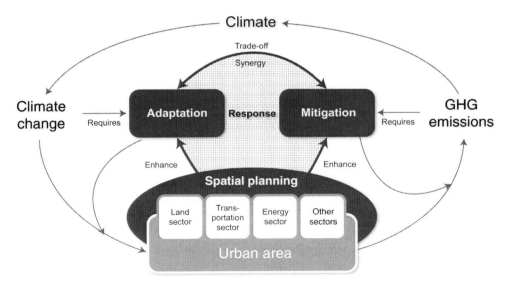

Figure 25.1 Spatial planning framework for integrative urban climate change strategies

strategies (Biesbroek et al., 2009). Although differences exist in their spatial and temporal characteristics as well as approaches to implementation, complicating their integration with spatial planning efforts, the links between climate change and sustainable development have raised awareness to the fact that coping with climate change is not a single political or scientific problem (Biesbroek et al., 2009). Despite the challenges, different sectoral objectives can be combined through spatial planning efforts, offering a more suitable, integrated approach to addressing climate change.

Toward a low carbon future through urban spatial organization

Although energy efficiency in the construction and transportation sectors has been a major focus in most, if not all, mitigation plans, it is important to realize that the spatial organization of urban land use activities, transport systems and GHG emissions are closely interrelated (Bulkeley, 2006). Spatial planning has begun to play an increasingly important role in mitigation by providing energy efficient settlements and promoting the utilization of renewable energy resources (Wende et al., 2010). The goal is to promote arrangements of urban land use activities that encourage and promote the movement of a region or nation toward becoming low carbon through its reduction in GHG emissions. For the first time since the first assessment report of the Intergovernmental Panel on Climate Change (IPCC) in 1990, Working Group III of the IPCC Fifth Assessment Report (AR5) produced Chapter 12, 'Human Settlements, Infrastructure, and Spatial Planning,' focusing on issues related to reducing urban sprawl, energy consumption and automobile dependency. The spatial planning section of that chapter reviews and discusses strategies from different geographic scales at the: a) macro-scale including regional allocation of development, urban containment through encouraging inward growth of cities and a regional jobs–housing balance to minimize travel distance; b) meso-scale strategies including corridor-level growth management and networked TODs; and c) micro-scale strategies such as urban regeneration to revitalize urban centers, neighborhood design and a 'new urbanism' to conserve energy as well as TOD to encourage use of public transit (Seto et al., 2014).

Urban climate change mitigation can be promoted through climate-sensitive urban planning to efficiently and effectively implement strategies for reducing GHG emissions. For example, the European Environmental Agency (EEA) indicates that spatial planning could help make destinations more accessible and journeys shorter, thereby reducing GHG (including carbon dioxide [CO_2], methane [CH_4], nitrous oxide [N_2O] and fluorinated gases) emissions (EEA, 2009; Wilson and Piper, 2010). As a result, the European Union has adopted strategic policies to promote a modal shift in and between cities promoting rail transport, short-sea shipping and the use of inland waterways (Wilson and Piper, 2010). Singapore's convenient mass-transit, mixed-use development (shops near residences) and disincentives for driving result in 53 percent of the population using mass transit, 31 percent using private vehicles or taxis and 15 percent not commuting or using 'other' modes of transportation (e.g. bicycles) (Singapore Department of Statistics, 2006).

Additionally, urban form is associated with having potentially significant impacts on energy use, resource consumption and the ability of a region to mitigate climate change (Blanco et al., 2011). Key urban form characteristics that influence GHG emissions are density, land use mix, connectivity and accessibility (Seto et al., 2014). For example, the separation of urban land use activities through zoning has influenced transport demand and the consequent GHG emissions in urban areas. Similarly, economic activities and densities of urban land use activities as they are ascribed in urban land use plans influence GHG

Spatial planning for urban areas

emissions through changes in both direct and indirect energy consumption (Dodman, 2011). Technological advancements in spatial planning can aid in the creation of new urban forms that contribute to a lower carbon future (Crawford and French, 2008). Imperatively, the pattern of urban development must promote sustainable transportation (e.g. provide multiple public transportation options such as cycling and walking) while reducing the use of motor vehicles.

Higher urban densities have a tendency to decrease per capita gasoline consumption with a corresponding decrease in GHG emissions (Marcotullio et al., 2012). Evidence suggests that per capita daily vehicle-kilometers traveled are related to population density and other urban form attributes, hence sprawl reduction could play an important role in mitigation (Marshall, 2008). The structures, orientations and conditions of buildings and streetscapes also increase energy needs for cooling and heating buildings (Bulkeley, 2006; Blanco et al., 2011). Spatial organization (e.g. site selection, housing block arrangements, tree locations, etc.) and density of the urban settlement play an important role in a community's energy consumption (Wende et al., 2010; Dodman, 2011). Compact and higher urban densities (lower surface-to-volume ratios) are much more energy efficient than low-density buildings, suggesting that the urban building structure should be stipulated in legally binding land use plans. Wende et al. (2010) reveal that the heat demand of two eight-story row-houses is only 35.6 percent of that of 64 single-family homes.

Synergies of promoting public transit usage, reducing energy expenditure and enabling waste reuse can be achieved by combining TOD and green urbanism (Cevero and Sullican, 2011). Furthermore, the mixture of urban land uses is an important element of a compact city strategy, which can reduce the number and/or length of vehicle trips as well as energy use and associated GHG emissions (Blanco et al., 2011). Successful implementation of spatial planning strategies to reduce emissions at the city level will require effective and feasible policy instruments (Seto et al., 2014). Therefore, city governments should review and revise urban land regulations including zoning, subdivision regulations and building standards to ensure that land use plays an important role in reducing energy needs, enhancing energy efficiencies and climate change mitigation.

In addition to providing an implementation platform for integrating various sectoral strategies, land itself has benefits of carbon sequestration that should not be ignored (Zomer et al., 2008). The change in land cover from natural vegetation to impervious surfaces due to urbanization has not only destroyed carbon sinks but also intensified heat island effects (Blanco et al., 2011). As a result, open space delineation and protection should be a priority for both regional and local plans. According to Rose et al. (2011), land-based mitigation—agriculture, forestry and bio-energy—will contribute approximately 100 to 340 GtC equivalents over the next century (approximately 15–40 percent of total abatement). The traditional use of greenbelts to avoid urban expansion and protect green spaces, farmlands, forests, wetlands and watersheds can thus be an important means for climate change mitigation. The Ontario Greenbelt (covering approximately 750,000 hectares), for example, keeps 172 million tonnes of GHGs out of the atmosphere, locked away in the rich soil and vegetation of its wetlands and forests (Tomalty, 2012). Urban green space can also sequester carbon and reduce energy consumption for cooling purposes, contributing to the reduction of GHG emissions (Schwarz et al, 2011; Seto et al., 2014). Furthermore, increasing urban green space provides co-benefits of increasing property values and recreational space, stormwater regulation, reduction of air pollutants as well as supplying shade, cooling, rainwater interception, infiltration and biodiversity support (Seto et al., 2014).

A strategic guide to urban development and climate change adaptation

Climate change impacts have resulted in wide-ranging consequences for human societies and ecosystems and have put pressure on energy, water and other resources (IPCC, 2013). Adaptation to climate change usually includes actions related to the built environment, infrastructure, rural environments and land use. They can range from the orientation for passive solar gains to sustainable urban drainage to reduce flooding—efforts that not only reduce risk but aid in the conservation of valuable resources (Office of the Deputy Prime Minister, 2004). In the case of coastal megacities, climate change impacts can be significant due to increased vulnerability to flooding from sea level rise or storm surges. This, in turn, necessitates the development of risk management strategies, which may include improved infrastructure, the development of early warning systems and evacuation plans, increased disaster response and relief aid (Fuchs, 2010). Spatial planning can offer a strategic guide to development, providing communities with the tools to adapt to climate risks and offer greater protection from possible impacts such as flooding, erosion, water shortages and subsidence (SEERA, 2005; Bulkeley, 2006).

First, many urban areas cope with the direct impacts of climate change by making substantial adaptations. These include infrastructure investments (e.g. levees, sea barriers, dikes, drainage systems, hardening or relocation of infrastructure and utilities, etc.) and land use policies, such as zoning regulations, subdivision regulations and building codes (Sanchez-Rodriguez et al., 2008; Blanco et al., 2011; Ifeanyi and Ayadiulo, 2012). These short-term actions of investments in infrastructure and land use regulations can become long-term efforts for adapting to climate change within a spatial context (Biesbroek et al., 2009; Sanchez-Rodriguez et al., 2008). Non-substantial adaptations, which emphasize inter-sectoral coordination and integration of planning at different governmental levels, can include changes in formal institutions ranging from structural changes in governance to micro-adjustments in policy tools in urban areas (Bulkeley, 2006; Biesbroek et al., 2009).

Second, to address the impacts of climate change, spatial planners must modify traditional administrative structures to which they are accustomed and concentrate on development activities within the context of the existing urban areas (Biesbroek et al., 2009). In doing so, local planning bodies would have greater capacity to adopt the precautionary principle in formulating policies that are more resource efficient given the potential change and uncertainty in the supply of resources, such as land and water (Office of the Deputy Prime Minister, 2004). This puts a higher priority on cooperation among different sectors, making such efforts an integral part of the planning process (Bajec, 2011). Government institutional structures and different sectoral policies must take climate change responses into account in spatial planning practices locally and comprehensively for adapting to the future impacts of climate changes (Biesbroek et al., 2009; Bulkeley, 2006; Betsill and Bulkeley, 2006). In addition, effective mainstreaming strategies for climate change adaptation should be coordinated among the multi-level spatial planning authorities (local, regional and national level) as well as between agencies and other sectors (Betsill and Bulkeley, 2006; Bulkeley, 2006; Bajec, 2011).

Lastly, integrated vulnerability assessments and integrated management of protection, rescue and relief systems should be adopted in the spatial planning process (Birkmann, 2007; Downes et al., 2010). Risk and vulnerability assessments at the urban scale should be illustrated by incorporating spatially explicit information on past extreme events and disasters and assessing vulnerability by mapping the hazard of possible future events (UNHABITAT, 2011; Schmidt-Thome and Klein, 2011) and infrastructures at risk (Fussel, 2007; Carmin et al., 2009; Schmidt-Thome and Klein, 2011). Urban planners need to collaborate with interdisciplinary

Spatial planning for urban areas

teams of natural scientists to produce model–based impact analyses of climate change prior to the preparation of adaptation plans (Blanco et al., 2011). Roy and Blaschke (2011), for example, use a grid–based method to transfer relevant data to geographic information systems (GIS) maps to spatially assess vulnerability informing zoning and the delineation of conservation areas. Land use planning can help avoid future investment in risky urban development schemes by developing resilience to the predicted impacts of climate change and by identifying and working to reduce urban social vulnerability to extreme events related to climate variability and climate change (Sanchez-Rodriguez et al., 2008; Schmidt-Thome and Klein, 2011). Adapting to climate change is an important national policy in Taiwan (CEPD, 2012) and since 2011, 17 municipalities have completed local adaptation plans using a strategic spatial planning approach and by building consensus and committing themselves to addressing the impacts and key issues of climate change (see Box 25.1).

Box 25.1   Consensus building for local adaptation strategies in Taipei through strategic spatial planning

In addition to climate change mitigation strategies, climate change adaptation has become an important national policy in Taiwan. While formulating the national climate change adaptation strategy, 'Adaptation Strategy to Climate Change in Taiwan' (CEPD, 2012), adopted by the Executive Yuan of Taiwan in 2012, initiatives for local adaptation strategies were also proposed, with Taipei selected as one of two pilot city projects tasked to formulate 'Local Climate Change Adaptation Planning Guidelines'.

Partnership for planning Taipei's adaptation strategy to climate change

A partnership amongst the national government, an expert committee, Taipei City government and local planning team was created to promote and administer the planning effort of Taipei's adaptation plan. Taiwan's National Development Council provided funding support for the Taipei City government to contract out a planning team for carrying out the process. An adaptation committee was established and chaired by the deputy mayor of Taipei to coordinate different sectors within the city government, while an expert committee was formed under the National Development Council to supervise the Taipei City government on formulation procedures and to assist in the operations of the adaptation committee. Meetings between the National Development Council, expert committee, adaptation committee of Taipei City and local planning team were held regularly to discuss the issues of concern and reach consensus on the adaptation strategies.

Priority setting of key issues within Taipei's adaptation plan

In order to identify the most important or immediate concerns associated with climate change adaptation for Taipei, the local planning team conducted interviews and distributed questionnaires with results presented to the adaptation committee in order to reach consensus amongst the members. The three most urgent sectorial issues identified were:

- safety and reliability of infrastructure;
- dependable supply of water resources;
- minimum loss and damage from natural hazards.

Four additional issues were identified as requiring continuous and long-term strategic planning for adapting to climate change impacts:

- industrial transformation and energy supply;
- food security and biodiversity;
- land use;
- urban health.

Strategic spatial planning

Existing planning regulations in Taipei are not only ineffective for responding to the urgent needs of adapting to climate change, but also create conflicts between sectors. However, using a strategic planning approach, the adaptation strategy was formulated according to a shared vision: 'Creating a capital city with low vulnerability by 2030'. As a result, 185 actions were proposed by different departments throughout the city government and the priorities were set based on the opportunities for co-benefits of each action across sectors. In total, 30 actions were prioritized, with each having more than five co-benefits linked to other sectors. The strategic planning approach in this case played an important role in fostering consensus building between different departments within the Taipei City government and further necessitates a strong regional coalition between Taipei City, New Taipei City and Keelung City for the strategy's implementation.

Source: Chen and Lee (2013)

Challenges and opportunities of spatial planning for climate change

Climate change strategies from different sectors involve a variety of stakeholders, objectives and capacities. With an increasing awareness that humans continue to influence the changing climate, disciplinary coping mechanisms have been shifting toward more transdisciplinary strategies (Biesbroek et al., 2009). Among the wide variety of options available, spatial planning not only contributes to the development of integrated strategies, but it also functions as a mechanism for implementation at local and regional levels (Biesbroek et al., 2009). Consequently, what differentiates these approaches from individual sectoral approaches is that the former integrates multiple sectors while viewing urban space as a whole, thereby enabling win–win solutions that are not possible under single-sector policies alone.

Although a well-established approach to the spatial allocation of land use and investments in many industrialized societies, there remain a number of issues and challenges that must be addressed for spatial planning to be an effective tool for responding to climate change. The approach has two main statutory levers: (1) forward-thinking planning through spatial development plans at regional and local levels and (2) controls on development projects and proposals through permit systems (Crawford and French, 2008). Integrating climate protection with spatial planning involves addressing a number of competing and conflicting planning issues that go well beyond the conventional scope of local and regional planning (Bulkeley, 2006). But, as it is conceived, spatial planning can assist in the implementation of financial, institutional and technical means to mitigate climate change and related impacts (Biesbroek et al., 2009). Instead of the blueprint, the inclusion of climate protection will raise the issues of synergies and trade-offs between environmental, economic and social dimensions. Ideally, it should direct the arrangement and allocation of land use activities as well as investment in

Spatial planning for urban areas

different sectors. In addition, integrated spatial planning should be able to balance critical infrastructures with diversified local economic opportunities and will be central in providing options for cities experiencing fast economic and spatial growth. However, most spatial planning initiatives at the city level are still treated separately from other sectors within the government. Coordination remains challenging, especially when its location in the organizational hierarchy is not high enough, thereby limiting cross-sectoral collaboration. Furthermore, if government policies focus primarily on economic growth, and the development-oriented ministries (e.g. transportation) have more financial power than others, this can further complicate inter-sectoral coordination. While climate change concerns and commitments to reduce GHG emissions have begun to take shape in spatial planning objectives (Crawford and French, 2008), yet there is much pressure to facilitate large-scale housing developments over the next 20 years. For example, it is estimated that approximately 15–20 million square meters of housing will be built in the urban zones of China between 2005 and 2020 to accommodate new immigrants to cities (Li and Colombier, 2009).

In addition to urban governance and formal urban land use and transport planning invested in municipal authorities, private sector developers and non-governmental organizations (NGOs) are progressively incorporating sustainable urban development and spatial planning into their operations (Seto et al., 2014). Private sector developers are increasingly incorporating high energy efficiency ratings in housing developments, providing guidance to reduce GHG emissions, and exceeding codes and established standards for governing the nature of urban development. NGOs, such as the US Green Building Council, Korea Green Building Certification Criteria, the UK's Building Research Establishment Environmental Assessment Method and other industry groups, have also become important in shaping urban development, particularly in terms of regeneration and the refurbishment or retrofitting of existing buildings. Spatial planning can be used as a tool to engage private sectors and NGOs in more comprehensive climate change mitigation strategies.

Although climate protection is increasingly a policy issue, it is still unclear as to how spatial planning is perceived by governments in the climate change debate. It closely relates to climate protection, but still lacks pragmatic solutions to reduce climate change impacts (Bulkeley, 2006). The most important challenges involve developing and promoting spatial arrangements of land use activities that will move a region or nation toward becoming a low carbon and low risk society. Although protection of the global climate is not the primary objective of urban development, local land use plans should not only take local interests into consideration but also link urban planning practices with broader concerns of climate change (Wende et al., 2010). As local mitigation strategies are usually influenced by regional or national policies, which are directed by international agreements (e.g. United Nations Framework Convention on Climate Change and the Kyoto Protocol) for reducing GHG emissions, spatial planning can fill an important role in the implementation of measures at both regional and local levels.

Zoning ordinances are key tools for implementing local land use plans by regulating the types and intensities of land use activities that can be accommodated on a given piece of property. Conventional structural planning approaches often fail to integrate strategies from different sectors and lack the tools for trans-boundary implementation. Thus, the spatial planning concept is rooted in moving away from a conventional regulatory framework toward more proactive coordination between public and private investments with integration across different sectors (Crawford and French, 2008). Although there is evidence that local authorities are beginning to develop climate change strategies and seeking to integrate climate protection into spatial plans, progress has been slow. Due to the local autonomous nature of the land use planning process, the decision-making of integrating climate change considerations are limited to certain administrative areas. Furthermore, planning itself does not provide a mechanism to transfer policies and strategies into actions thus the rhetoric of climate change and spatial planning integration still must be translated into practice (Bulkeley, 2006).

An effective process must have the capacity to coordinate different sectors such as transportation systems, local economic development, housing, agriculture and water management (Halsnaes, 2006; Dewar, 2011; Grazi et al., 2008). The problems of urban sprawl and energy consumption cannot be resolved separately. Urban mitigation strategies for reducing GHGs should be approached from a regional perspective extending beyond urban boundaries. By preventing development in peri-urban areas and encouraging mixed-use zoning, travel distances can be reduced and lead to reductions in GHG emissions. In order to incorporate climate change policies over larger regional areas and across broader political boundaries, regional planning can promote higher efficiency of resource use in urban areas, reduce travel demands and preserve open space. This can, in turn, improve carbon sequestration and encourage the development and use of renewable energy and waste recycling (Halsnaes, 2006; Crawford and French, 2008). Planning for urban form by protecting valuable ecosystem services (e.g. regulating runoff) also has co-benefits and links well with climate change adaptation by increasing resilience to extreme climatic events (Chapin, 2009; Wang et al., 2012; McDonald, Chapter 29 in this volume). The delineation and protection of open space through the practice of regional land use planning can help regulate the local climate by preventing continued energy inefficiencies in urban built-up areas (Wende et al., 2010) and by providing carbon sinks.

Concluding remarks and ways forward for integrative climate change strategies

There is a growing recognition that mitigation strategies alone are insufficient and that both mitigation and adaptation measures are needed to mediate impacts from climate change at the local level (Biesbroek et al., 2009; Laukkonen et al., 2009). Adaptation and mitigation strategies do not always complement each other and can be counterproductive. One of the important roles for spatial planners is to coordinate local preferences and stakeholder initiatives with sectoral policies and strategies. Biesbroek et al. (2009) argue that the multi-level governance approach of spatial planning, which incorporates different spatial and temporal scales and stakeholders, can contribute to reducing the mitigation-adaptation dichotomy. Spatial planning functions to manage land use, develop well-functioning urban areas, distribute and protect resources and, at the same time, promote climate change mitigation and adaptation. However, the gap between spatial planning and the adoption of mitigation and adaptation strategies must be addressed. The development of methodologies and tools for comparisons are needed at the local level to analyze synergies and trade-offs between both strategies and to help assess and prioritize the cost-effectiveness of strategies (Laukkonen et al., 2009).

To effectuate implementation, instruments are needed for both mitigation and adaptation strategies. Adequate planning of the spatial organization of human settlements—including urban form—and rational allocation of land use activities are necessary for formulating efficient and effective strategies for reducing GHG emissions. It may be more effective for top-down approaches to mitigate climate change such as meeting the Kyoto targets and other emissions reduction objectives (Sovacool and Brown, 2010), but bottom-up approaches are likely to be more appropriate for adaptation, given the multitude of variables, context dependencies and cultural settings (Biesbroek et al., 2009). Clearly, however, both approaches are needed. To be truly effective, land use change under various climate change scenarios will need to be

Spatial planning for urban areas

simulated as a point of reference; assessment methods and tools will have to be adopted appropriately to assess and evaluate possible outputs and outcomes (qualities and quantities) of short-term and mid-term spatial plans and strategies for both mitigation and adaptation; and spatial planning efforts will need to be monitored and verified in order to realize a more sustainable future. For spatial planning to become effective in dealing with climate change it needs to become more strategic, integrative and action-oriented. The strategic spatial planning process needs to reflect a broad-based consensus among stakeholders and the adoption of development strategies targeted to cross-sectorial co-benefits. Consequently, the output of strategic spatial planning will not just be a spatial plan, but will include a set of interrelated strategies for an urban response to climate change. Special attention to the spatial aspects of planning can provide a better understanding of the relationships between mitigation and adaptation actions to avoid conflicts between these efforts (Sanchez-Rodriguez et al., 2008). The greatest opportunity to mitigate and adapt to climate change lies in the combined geographic scale that spatial planning can provide (Sovacool and Brown, 2010).

Key messages of this chapter

- Spatial planning can contribute both to mitigating GHG emissions and adapting to climate change at different geographic scales through coordination or integration of policies from different sectors.
- Trade-offs and synergies between mitigation and adaptation strategies can be framed through the integration of climate change issues with spatial planning.
- Progress on integrating climate protection into spatial planning strategies is still limited; proposed integration of climate change with spatial planning still must be translated into practice.
- The adequacy and effectiveness of spatial planning strategies and policy instruments to cope with climate change vary by city and institutional capacity.

Key research questions of spatial planning and climate change

- How can the paradigm of spatial planning shift toward a more transdisciplinary approach for coping with climate change?
- How can spatial planning be used as a tool to engage the private sector and NGOs to contribute to climate change mitigation?
- How has spatial planning impacted GHG emissions and subsequently the vulnerability of cities to climate change?
- What are the key policy instruments for spatial planning that can help communities mitigate GHG emissions and adapt to climate change?

References

Bajec, N.L. (2011). Integrating climate change adaptation policies in spatial development planning in Serbia: A challenging task ahead. *Spatium International Review*, 24, 1–8.

Bareman, P. (2002). The inspectorate of housing, spatial planning and the environment enforces legislation on the return of materials and packaging. In the 6th International Conference on Environmental Compliance and Enforcement, San Jose, US.

Betsill, M.M., and Bulkeley, H. (2006). Cities and the multilevel governance of global climate change. *Global Governance*, 12, 141–159.

Beukes, O., Carstens, J., de Vos, A., Joubert, M., Mansvelt, L., Reinten, E., and Wooldrige, J. (2008). Development of a spatial planning database and analysis of agriculture and tourist potential in the Strandveld region of the Overstrand Local. Stellenbosch: Municipality.

Biesbroek, G.R., Swart, R.J., and van der Knaap, W.G.M. (2009). The mitigation–adaptation dichotomy and the role of spatial planning. *Habitat International*, 33, 230–237.

Birkmann, J. (2007). Climate change and vulnerability: Challenges for spatial planning and civil protection. In The Forum DKKV/CEDIM: Disaster Reduction in Climate Change, Karlsruhe, Germany.

Blanco, H., McCarney, P., Oarnell, S., Schmidt, M., and Seto, K. (2011). The role of urban land in climate change. In C. Rozensweig, W. Solecki and S. Hammer (Eds.), *First Urban Climate Change Research Network (UCCRN) Assessment Report on Climate Change in Cities* (pp. 218–248). Cambridge: Cambridge University Press.

Blom, G., and Paulissen, M. (2007). How will climate change affect spatial planning in agriculture and nature? In The Conference of Climate Change Spatial Planning 2007, Amsterdam, the Netherlands.

Bulkeley, H. (2006) A changing climate for spatial planning. *Planning Theory and Practice*, 7(2), 203–214.

Campbell, H. (2006). Is the issue of climate change too big for spatial planning? *Planning Theory and Practice*, 7(2), 201–230.

Carmin, J., Rovers, D., and Anguelovski, I. (2009). Planning climate resilient cities: Early lessons from early adapters. Paper prepared for the World Bank, in The World Bank 5th Urban Research Symposium, Cities and Climate Change, Marseille, France.

Carter, J.G. (2007). Spatial planning, water and the water framework directive: Insights from theory and practice. *The Geographical Journal*, 173(4), 330–342.

CEPD (2012). Adaptation strategy to climate change in Taiwan. Taipei, Taiwan: National Development Council.

Cevero, R., and Sullican, C. (2011). Green TODs: Marrying transit-oriented development and green urbanism. *International Journal of Sustainable Development and World Ecology*, 18(3), 210–218.

Chapin III, F.S. (2009). Managing ecosystems sustainably: The key role of resilience. In *Principles of Ecosystem Stewardship* (pp. 29–53). New York: Springer.

Chen, W.-B., and Lee, C.-L. (2013). Global issues, local actions climate change adaptation of Taipei City: A process of communication and consensus. *ACT Newsletter*, 3(4), 20–21.

Crawford, J., and French, W. (2008). A low-carbon future: Spatial planning's role in enhancing technological innovation in the build environment. *Energy Policy*, 36(12), 4575–4579.

Cullingworth, B., and Nadin, V. (2006). *Town and Country Planning in the UK* (Fourteenth edition). London: Routledge.

Dewar, B. (2011). Caring about caring: An appreciative inquiry into compassionate relationship centered care (Unpublished PhD's dissertation). Edinburgh Napier University, UK.

Dodman, D. (2011). Forces driving urban greenhouse gas emissions. *Current Opinion in Environmental Sustainability*, 3, 121–125.

Downes, N., Storch, H., Moon, K., and Rujner, H. (2010). Urban sustainability in times of changing climate: The case of Ho Chi Minh City, Vietnam. In The 46th ISOCARP Congress 2010, Nairobi.

EEA (2009). EEA signals 2009: Key environmental issues facing Europe, European Environment Agency. Copenhagen: EEA.

Fuchs, R.J. (2010). Cities at risk: Asia's coastal cities in an age of climate change. *Asia-Pacific Issues*, 96, 1–2.

Fussel, H.M. (2007). Adaptation planning for climate change: Concepts, assessment approaches, and key lessons. *Sustainability Science*, 2, 265–275.

Gibbs, D. (2002). *Local Economic Development and the Environment*. London: Routledge.

Gkartzios, M., and Scott, M. (2013). Planning for rural housing in the Republic of Ireland: From national spatial strategies to development plans. *European Planning Studies*, 17(12), 1751–1780.

Grazi, F., van den Bergh, J.C., and van Ommeren, J.M. (2008). An empirical analysis of urban form, transport and global warming. *The Energy Journal*, 29(4), 97–122.

Halsnaes, K. (2006). Climate change and planning. *Planning Theory and Practice*, 7(2), 227–230.

Hartmann, D.L., Klein Tank, A.M.G., Rusticucci, M., Alexander, L.V., Brönnimann, S., Charabi, Y., Dentener, F.J., Dlugokencky, E.J., Easterling, D.R., Kaplan, A., Soden, B.J., Thorne, P.W., Wild, M., and Zhai, P.M. (2013). Observations: Atmosphere and surface. In T.F. Stocker, D. Qin, G.-K. Plattner, M. Tignor, S.K. Allen, J. Boschung, A. Nauels, Y. Xia, V. Bex and P.M. Midgley (Eds.), *Climate Change 2013: The Physical Science Basis. Contribution of Working Group I to the Fifth Assessment Report of the*

Spatial planning for urban areas

Intergovernmental Panel on Climate Change. Cambridge, UK, and New York, NY, USA: Cambridge University Press.

Ifeanyi, E., and Ayadiulo, R.U. (2012). Urbanization and the challenge of climate change in Nigeria cities: A review. *Journal of Environmental Science*, 1(6), 13–18.

IPCC. (2013). Summary for policymakers. In T.F. Stocker, D. Qin, G.-K. Plattner, M. Tignor, S.K. Allen, J. Boschung, A. Nauels, Y. Xia, V. Bex and P.M. Midgley (Eds.), *Climate Change 2013: The Physical Science Basis. Contribution of Working Group I to the Fifth Assessment Report of the Intergovernmental Panel on Climate Change*. Cambridge, UK, and New York, NY, USA: Cambridge University Press.

Jackson, R. (2006). The role of spatial planning in combating climate change. *Planning Research Network*, 1–18.

Jones, C., Coombes, M., and Wong, C. (2012). A system of national tiered housing-market areas and spatial planning. *Environment and Planning B: Planning and Design*, 39(3), 518–532.

Laukkonen, J., Blanco, P.K., Lenhart, J., Keiner, M., Cavric, B., and Kinuthia-Njienga, C. (2009). Combining climate change adaptation and mitigation measures at the local level. *Habitat International*, 33, 287–292.

Li, J., and Colombier, M. (2009). Managing carbon emissions in China through building energy efficiency. *The Journal of Environmental Management*, 90(8), 2436–2447.

Marcotullio P.J., Sarzynski, A., Albrecht, J., and Schulz, N. (2012). The geography of urban greenhouse gas emissions in Asia: A regional analysis. *Global Environmental Change*, 22, 944–958. doi: 10.1016 / j.gloenvcha.2012.07.002, ISSN: 0959-3780

Marshall, J.D. (2008, May 1). Reducing urban sprawl could play an important role in addressing climate change. *Environmental Science & Technology*, 42(9), 3133–3137.

Office of the Deputy Prime Minister (2004). Sustainability appraisal of regional spatial strategies and local development frameworks, Consultation Paper. London: Office of the Deputy Prime Minister.

Office of the Deputy Prime Minister (2005). Planning policy statement 1: Delivering sustainable development. London: Office of the Deputy Prime Minister.

Rose, S., Ahammad, H., Eickhout, B., Fisher, B., Kurosawa, A., Rao, S., Riahi, K., and Van Vuuren, D.P. (2011). Land-based mitigation in climate stabilization. *Energy Economics*, 34, 365–380.

Roy, D.C., and Blaschke, T. (2011). A grid-based approach for spatial vulnerability assessment to floods: A case study on the coastal area of Bangladesh. In GI4DM Conference, Antalya, Turkey.

Sanchez-Rodriguez, R., Fragkias, M., and Solecki, W. (2008). Urban response to climate change: A focus on the Americas, A Workshop Paper, Los Angeles: UGEC.

Schmidt-Thome, P., and Klein, J. (2011). Applying climate change adaptation in spatial planning processes. *Global Change*, 1, 177–192.

Schwarz, N., Bauer, A., and Haase, D. (2011). Assessing climate impacts of planning policies—An estimation for the urban region of Leipzig (Germany). *Environmental Impact Assessment Review*, 31(2), 97–111.

SEERA (2005). Clear vision of the South East. Draft South East Plan, Guildford: South East Regional Assembly.

Seto, K., Dhakal, S., Bigio, A., Blanco, H., Delgado, G.C., Dewar, D., Huang, L., Inaba, A., Kansal, A., Lwasa, S., McMahon, J., Müller, D., Murakami, J., Nagendra, H., and Ramaswami, A. (2014). Human Settlements, Infrastructure and Spatial Planning. In O. Edenhofer, R. Pichs-Madruga, Y. Sokona, E. Farahani, S. Kadner, K. Seyboth, A. Adler, I. Baum, S. Brunner, P. Eickemeier, B. Kriemann, J. Savolainen, S. Schlömer, C. von Stechow, T. Zwickel and J.C. Minx (Eds.), *Climate Change 2014: Mitigation of Climate Change Contribution of Working Group III to the Fifth Assessment Report of the Intergovernmental Panel on Climate Change*. Cambridge, UK and New York, NY, USA: Cambridge University Press.

Singapore Department of Statistics (2006). General household survey. Singapore: Singapore Government.

Sovacool, B.K., and Brown, M.A. (2010). Competing dimensions of energy security: An international perspective. *Annual Review of Environment and Resources*, 35, 77–108.

Todes, A. (2008). Rethinking spatial planning. *Town and Regional Planning*, 13, 9–13.

Tomalty, R. (2012). *Carbon in the Bank*. Toronto: David Suzuki Foundation.

UNECE (2008). Spatial planning: Key instrument for development and effective governance with special reference to countries in transition. Geneva, US: United Nations.

UNHABITAT (2011). Global reports on human settlements 2011: Cities and climate change: Policy directions. Washington D.C.: UNHABITAT.

Veraart, J.A., Brinkman, S., Klostermann, J.E.M., and Kabat, P. (2007). Climate changes Spatial Planning: Introduction to the Dutch national research programme, Wageningen UR, Wageningen.

Wang, S.-H., Huang, S.-L., and Budd, W.W. (2012). Resilience analysis of the interaction of between typhoons and land use change. *Landscape and Urban Planning*, 106, 303–315.

Wende, A., Johansson, P., Vollrath, R., Dyall-Smith, M., Oesterhelt, D., and Grininger, M. (2010). Structural and biochemical characterization of a halophilic archaeal alkaline phosphatase. *Journal of Molecular Biology*, 400, 52–62.

Wiering, M., and Immink, I. (2006). When water management meets spatial planning: A policy-arrangements perspective. *Environment and Planning C: Government and Policy*, 24, 423–438.

Wilson, E., and Piper, J. (2010). *Spatial Planning and Climate Change*. Abingdon: Routledge.

Woltjer, J., and Al, N. (2013). Integrating water management and spatial planning. *Journal of the American Planning Association*, 73(2), 211–222.

Zomer, R.J., Trabucco, A., Verchot, L.V., and Muys, B. (2008). Land area eligible for afforestation and reforestation within the Clean Development Mechanism: A global analysis of the impact of forest definition. *Mitigation and Adaptation Strategies for Global Change*, 13, 219–239.

26

CLIMATE CHANGE MITIGATION IN HIGH-INCOME CITIES

Eugene Mohareb, David Bristow and Sybil Derrible

High-income cities face significant challenges in mitigating anthropogenic climate change; constraints exist in their evolution towards low-carbon urban systems due to their mature infrastructure, established energy sources and recent uncertainty in economic growth. The extent of these challenges depends on individual economic, social and environmental contexts. Seto et al. (2014) present four principal drivers of urban emissions in the Inter-governmental Panel on Climate Change's Fifth Assessment Report: economic geography and income, socio-demographic factors, technology and infrastructure and urban form. Given that these drivers vary substantially across urban areas in high-income cities, emissions per capita also differ (see Figure 26.1). However, regardless of context, deep emissions reductions in these cities are necessary.

High-income cities are principal drivers of energy consumption within their national boundaries. Grubler et al. (2012) estimate that over 80 percent of energy use in OECD90[1] countries took place in their urban areas in 2005. Moreover, Elzen et al. (2013) calculate that industrialized nations have contributed 52 percent of all emissions between 1850–2010, while having only hosted 26.7 percent of the global population during that time. These industrialized countries have undergone rapid urbanization in the past 60 years, with the United Nations (UN) (2014) suggesting that populations living in cities increased from 54.6 percent to 77.1 percent between 1950 and 2010. Cities in industrialized nations also enable much of the world's economic activity, with 380 high-income cities contributing to 50 percent of GDP in 2007 (Mckinsey Global Institute, 2011); GDP per capita has been shown to correlate with greenhouse gas (GHG) emissions (Kennedy et al., 2014). When considering energy demand, history of GHG emissions, high levels of urbanization and large economies, a case can be made that high-income cities are substantial contributors to climate change, supporting the call for their leadership towards a low carbon future.

Leadership in climate change mitigation at the municipal level has often matched or exceeded that at higher levels of government, especially in North America (Kennedy et al., 2012). With organizations such as C40 Cities and the US Conference of Mayors as well as programs such as ICLEI's Cities for Climate Protection and Carbon, municipal decision-makers have been able to share knowledge on best practices and develop initiatives that reduce GHG emissions at the local level. In many cases, this approach has required supportive frameworks or programs from higher levels of government, though context-specific municipal

377

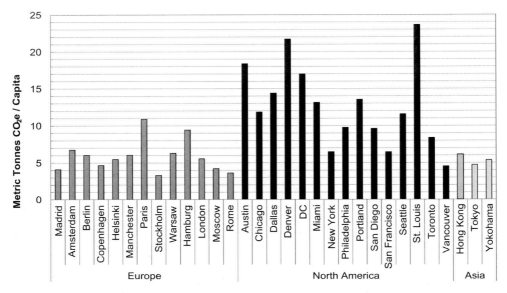

Figure 26.1 Per capita emissions for a variety of cities in industrialized countries
Source: CDP (2012).[2]

policy has been an important component for action. This chapter provides a discussion of the challenges associated with GHG mitigation in high-income cities and current efforts to reduce their contribution to climate change.

Identifying the sources of emissions

The identification of major emitting sectors in cities is a critical first step towards mitigation, as this information can then be used to estimate the importance of the aforementioned four drivers of GHG emissions. The building (residential and commercial), transportation (passenger and freight) and industrial sectors differ in the types of secondary energy resources (and associated emissions) upon which they rely; cities have recognized the usefulness of their quantification and have identified strategies that will produce significant GHG emissions reductions. Figure 26.2 presents summaries of GHG emissions of eight cities within industrialized countries, while Figure 26.3 presents the annual emissions reductions they have achieved over a given timeframe (Kennedy et al., 2012). The absolute and per capita values of emissions in these cities are shown in Table 26.1. Annual reductions in emissions have been observed in each sector, with consistency found in the waste sector (albeit generally a small component of total emissions, as demonstrated in Figure 26.2). The greatest challenge lies with the transportation sector.

Although many urban areas in industrialized countries have been able to achieve some success with GHG mitigation, a challenge remains with existing, and often aging infrastructure (Kennedy et al., 2014; Kennedy et al., 2012; Hodson et al., 2012). Building, transportation and energy infrastructure have been established based on the availability of low-cost sources of secondary energy that possess a relatively high energy density. Energy efficiency gains can lower energy demand substantially, but limitations on demand reduction are imposed by existing urban form and infrastructure systems, not to mention the rebound effect or Jevons

Climate change mitigation in high-income cities

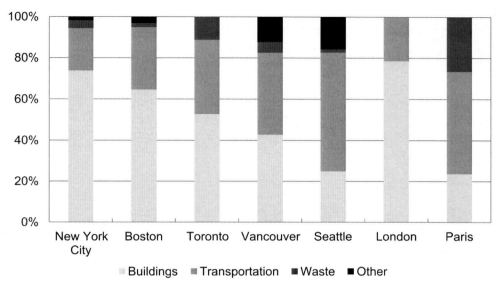

Figure 26.2 GHG emissions from seven high-income cities

Sources: PlaNYC (2013); City of Boston (2013a); City of Toronto (2013a); Government of British Columbia (2014); Stockholm Environment Institute (2014); City of London (2014); City of Paris (2011). Note that London is missing emissions for 'waste' and 'other', and Toronto is missing emissions for 'other'.

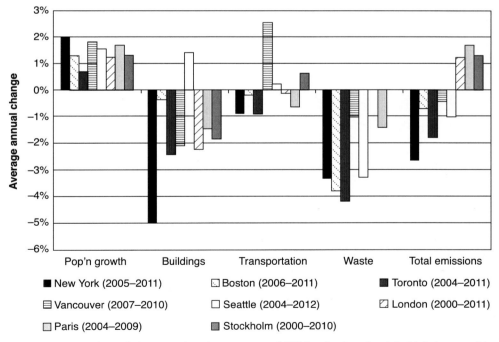

Figure 26.3 Annual population growth and average annual GHG reductions for eight high-income cities

Sources: PlaNYC (2013); City of Boston (2013a); City of Toronto (2013a); Government of British Columbia (2014); Stockholm Environment Institute (2014); City of London (2014); City of Paris (2011). Note: Averages are over the time periods displayed.

Eugene Mohareb et al.

Table 26.1 GHG emissions and reductions achieved from eight industrialized country cities

City (inventory year)	Population	GHG emissions (tonnes CO_2e)	Average annual emissions reductions from baseline* (tonnes CO_2e)			Baseline year
			Buildings	Transportation	Waste	
Toronto (2011)	2,615,060	23,258,000	360,000	−40,500	155,000	2004
New York City (2011)	8,175,133	53,400,000	1,200,000	105,000	86,000	2005
Vancouver (2010)	642,843	2,646,000	13,000	−12,000	1,000	2007
Seattle (2012)	634,535	6,132,000	−23,000	−9,000	5,000	2005
Boston (2011)	625,087	6,767,000	25,000	4,000	7,000	2006
London (2011)	8,204,100	39,905,000	934,000	12,000	N/A	2000
Paris (2009)	2,274,880	31,851,395	87,000	172,000	7,000	2004
Stockholm (2010)	847,073	3,118,000	46,000	-7,000	N/A	2000

Sources: PlaNYC (2013); City of Boston (2013a); City of Toronto (2013a); Government of British Columbia (2014); Stockholm Environment Institute (2014); City of London (2014); City of Paris (2011).
*negative reductions indicate an emissions increase in that sector.

paradox (i.e. some of the gain in efficiency is lost in demand increase) (Sorrell, 2007). If pressure increases for mature economies to provide a greater share of the global carbon budget to emerging economies, the legacy of this infrastructure imposes significant energy demands that must be met by low carbon energy sources. The next section of this chapter describes the nature of the mature infrastructure challenge to high-income cities in more detail.

The challenges: aging infrastructure and lock-in

Many cities in industrialized countries are currently experiencing major challenges with their aging infrastructure stock, which is not only based on the availability of fossil fuels, but is also frequently in need of replacement (FCM et al., 2012; ASCE, 2013; Rockefeller Foundation, 2014). At some point in their history, these cities have either experienced prolonged periods of economic growth or urban renewal that led to the undertaking of substantial infrastructure development; many of these projects are now either reaching or passed the end of their design lives, in addition to being ill-suited to a low carbon economy. This is notably the case for cities that rebuilt themselves after important events such as the great fires of London in 1666 and Chicago in 1871 or the modernization of Paris in the second half of the 19th century. This is also the case for a number of cities that have witnessed tremendous growth or redevelopment after World War II.

The infrastructure systems and urban form of high-income cities have often arisen during eras of much different energy realities than those of today (Mohareb et al., 2015). Conversely, historic or older cities, which may exhibit urban forms that are more conducive to public transportation, for example, may be constrained (spatially, legally, culturally, etc.) in retrofitting the built environment for active transportation or high-performance buildings (Seto et al., 2014; Ma et al., 2012). In established cities that experienced much of their development in the 20th century, the urban form and infrastructural legacy of an automobile-centric era has resulted in low population densities (which are difficult to superimpose with effective active or public transportation infrastructure) and single family homes. Presently, as these periods of

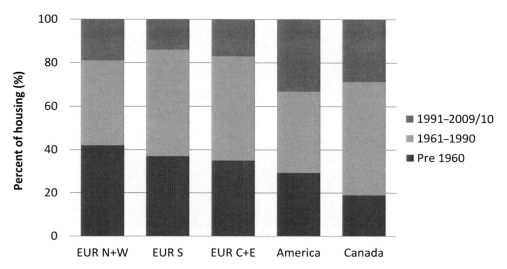

Figure 26.4 Distribution of residential housing stock by construction age for Europe and North America

Sources: OEE (2013) (Canadian data); EIA (2013) (American data); and Building Performance Institute Europe (2011) (European data). The Canadian estimates are from dwelling number counts assuming American floor space distributions.

Note: EUR N+W: Finland, Ireland, Austria, the Netherlands, Germany, France, Sweden, Denmark, United Kingdom; EUR S: Greece, Malta, Spain, Italy; EUR C+E: Estonia, Lithuania, Latvia, Hungary, Romania, Slovakia, Slovenia, Poland, Bulgaria, Czech Republic.

economic and population growth have waned in many cases, high-income cities are in a position where they lack the financial resources to maintain or renew this infrastructure; derelict and decaying urban expressways and neighborhoods in many North American cities serve as evidence.

Building stocks represent another example of challenges associated with aging infrastructure that varies according to location. Given that buildings are long-lived, and older buildings tend to have higher space conditioning energy demand per unit of area (Building Performance Institute Europe, 2014; OEE, 2013), a burden is placed on urban residents of older cities. Figure 26.4 demonstrates the building stock age observed in a number of industrialized countries and world regions. In all cases, much of the housing stock has been constructed prior to the 1960s; in the case of Northern and Western Europe, it is over 40 percent. In England, 56 percent of the housing stock was built before 1965, with over one third constructed prior to 1945 (Department of Communities and Local Government, 2015).

In addition, as integrated entities, aging infrastructure systems may unnecessarily burden one another. For instance, a water system in poor condition will require more electricity and may also structurally damage roads and buildings. This is the case in London (UK), where 30 percent of all drinking water supply distributed is reportedly lost in the ground because of leaky pipes (Kennedy et al., 2007).

Finally, this challenge extends beyond the simple replacement of infrastructure. Indeed, coupled with consistent urban growth, increasing environmental concerns and a society with constantly evolving needs, current infrastructure must be redesigned and adapted to contemporary challenges. At the moment, much hope falls within the general concept of the 'smart city' that has become popular (Allwinkle and Cruickshank, 2011; Grob, 2010), although it remains to be formally defined. With the advent of sensors, a 'smart city' is often illustrated with features such as smart energy and water meters or with the use of new technologies for travel demand management purposes. The extent to which these smart

Box 26.1 Transitions of urban centers' GHG emissions—the unique case of Singapore

It is revealing to consider how the sometimes generic and often radically unique context of Singapore's development affected its emissions over the last four decades (see below; World Bank, 2013). Singapore's development, similar in economic and physical form to many other cities, is characterized by an increasing use of energy as affluence and population rose. The effect on GHG emissions of this increasing energy usage was mitigated by an increasingly cleaner energy supply mix. Since the 1990s, Singapore's emissions from electricity and heat production have been in decline. Many developed cities of a similar population do not have as much GHG-intensive industries, but are also not as tightly constrained geographically, which limits energy-intensity of the urban form (see discussion in the section 'Land use and spatial planning' below). In Singapore, the carbon benefits from densification are at least partially offset by the land reclamation efforts. The industry mix and density are quite the opposite of many of the larger North American cities, for example, those that tend to eschew heavy industry as they develop and have the 'luxury' of space resulting in lower-density development. The case of Singapore is remarkable for the pace of its development and its recent decoupling of emissions from population and economic growth. However, there are many possible paths for urban development. Will there be just as many paths for urban GHG mitigation?

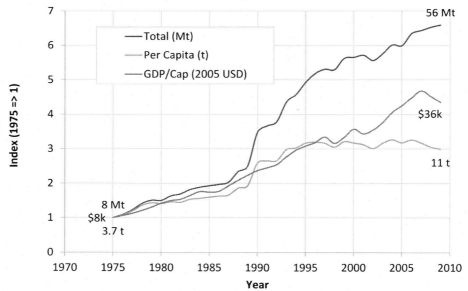

Figure 26.5 The relationship between Singapore's development (GDP per capita) and its GHG emissions over the last four decades (per capita energy use)

Source: World Bank, 2013

initiatives can assist in broadly reducing GHG emissions remains unclear, however, especially since these initiatives tend to provide the data upon which decisions can be made, as opposed to directly lowering energy usage.

Urban approaches to GHG mitigation

Urban governments have long understood that cities, being the foci of future population and economic growth, must leverage their power to begin the market transformation towards decarbonization. While it has been noted that the ability for municipalities to influence the reduction of GHGs in some sectors requires the assistance of higher levels of government, there is a great deal that can be accomplished at the urban scale (Arup and C40 Cities, 2011; Seto et al., 2014, p. 34). While higher levels of government may be politically constrained in their mitigation policy (or simply unwilling to take action), municipal governments are able to take appropriate measures at a scale that reflects the will of their electorate (Lutsey and Sperling, 2008). As pointed out by Seto et al. (2014), a city's greatest leverage in reducing GHG emissions is at the 'meso' scale such as policies to improve the efficiency of the municipally owned building stock or improving the level of service of public transportation. Municipalities also operate at a more manageable scale than national or state levels for the systemic solutions inherent in most mitigation measures such as the integration of land use and transportation planning or resource recovery (Grubler et al., 2012). Additionally, cities have direct influence over a number of measures that will be key to the long-term emissions reductions necessary to meet the target of 2°C of warming (such as land use planning, public transportation systems or waste management [Seto et al., 2014]).

Urban economic systems also have the potential to be much more resource efficient in their functioning when compared with rural areas. Cities are often the focal points of a nation's economic growth, providing incubators for new ideas and drawing together actors that can leverage their specializations to greater effect through collaboration (Jacobs, 1970). Given the proximity that arises through the increasing population density (stimulated by the increasing returns to scale that is typically observed), cities can find greater economic, infrastructural and energy efficiencies in serving their inhabitants (Carlino et al., 2007; Bettencourt et al., 2007). In addition, politically attractive co-benefits of GHG mitigation exist at the urban scale such as improved urban environments through increased greenery (aesthetics and comfort [Pikora et al., 2003; Rosenzweig et al., 2009]) as well as reduced health impacts of fossil fuel consumption (Harlan and Ruddell, 2011). For a more detailed discussion on urban greening and associated health benefits see Trundle and McEvoy, Chapter 19 in this volume.

The energy, building, transportation and municipal sectors have all been targeted through financial incentives, regulatory measures and direct investment. Prominent measures that have been taken in high-income cities are described below. It is important to point out that a multitude of mitigation options are available to cities within the industrialized world, with certain measures proving more successful than others. Kennedy et al. (2014) demonstrate how urban characteristics, such as population density and electricity grid emissions intensity, can be used in the development of infrastructure strategies that provide the most effective mitigation. For example, in a low-density city such as Los Angeles, focusing on the decarbonization of the electricity grid coupled with the electrification of transportation (including electric vehicles [EVs]) will likely result in lower transportation sector emissions than focusing on the expansion of public transportation options. Other characteristics such as land use mix, connectivity, climate (heating or cooling degree days), housing stock composition, industrial

Eugene Mohareb et al.

sector characteristics and infrastructure age, must also be examined in order for a city to identify the drivers of its emissions.

Energy

Given that most cities in industrialized nations have a long-standing reliance on fossil-based sources of primary energy, municipal governments have identified opportunities to reduce GHG emissions through encouraging the growth of renewable energy. This approach is particularly effective in cities in the industrialized world, since energy prices are often a relatively small component of disposable income, making them good candidates for early adoption of emerging renewable energy technologies (which have yet to reach grid parity with other electricity generation). A list of sample programs in North American, European and Australian cities are presented in Table 26.2.

A number of mechanisms can be applied to accelerate the transition to low carbon energy sources, including financing, direct municipal investments or legal mandates. Considering financial approaches, Property Assessed Clean Energy (PACE) programs provide a means to tie expenditures on renewable energy to the property itself (rather than the investor). Additionally, the upfront costs can also be avoided by providing the option of payment installments, which can, in turn, be based on actual cost savings realized (i.e. a 'pay-as-you-save' plan). The benefits of such schemes can reach beyond mitigation; one study estimates that four million USD in spending on PACE programs in the US has the potential to create ten million USD in economic output, one million USD in tax revenue and 60 jobs (ECONorthwest, 2011).

Direct municipal investments or purchase can include those made by the city or by its associated utilities. Cities such as Houston, TX, purchase a fraction of their total energy demand of the municipal operations in equivalent renewable energy certificates (RECs). Conversely, the utility provider in the City of Brussels awards green certificates (similar to RECs) to local small-scale producers of solar power, so that these certificates can then be sold on the open market (Energuide, 2013). An alternative to green certificates is 'white certificates', which can be credited to a utility customer based on reductions in energy demand (Transue and Felder, 2010). Additionally, investments in infrastructure such as combined heat and power (CHP) plants improve the efficiency of energy provision and reduce associated GHG emissions. In the case of Vaxjo, Sweden, the firing of such a plant with biomass, along with biomass-fired district heating plants, has resulted in an estimated reduction of 76 percent in GHG emissions from residential heating between 1993–2005 (City of Vaxjo, 2007). Switching from fossil sources to biomass can provide additional mitigation, though considerations of forest carbon recovery and feedstock must be made (McKechnie et al., 2011).

Regulations also provide valuable tools to reduce GHG emissions in the energy sector. Requirements for renewable energy installations on new developments (e.g. Lancaster, California) improve local expertise in installations and create a broader market for low carbon energy. These can also provide incentives to reduce energy demand; the Merton rule promotes both renewable energy technologies and lower total energy demand for a building, as lower demand facilitates fulfilling the requirement of meeting 10 percent of energy needs with onsite renewable energy generation (Merton Council, 2013). Feed-in tariffs provide an incentive for low-carbon energy development based on price certainty for producers that wish to sell to the grid. Some care/flexibility should be applied to ensure that a net carbon emissions reduction is actually achieved in a given jurisdiction, as renewable alternatives can increase emissions upstream (e.g. manufacturing) or onsite (e.g. additional concrete for

Climate change mitigation in high-income cities

Table 26.2 Municipal actions in established cities to mitigate GHG emissions in the energy generation/supply sector

Initiative	Description	Further reading
Renewable energy financing	Payments are made through property taxes, eliminating upfront costs that can frequently act as a barrier to investment, while alleviating concerns around reclaiming retrofit costs during the sale of a property. Sacramento's PACE represents one of the largest programs to allow financing of renewable energy and efficiency retrofits to both commercial and residential properties.	PACE Now (2013)
Municipal REC purchasing requirements	Numerous cities set purchasing requirements for RECs, supporting regional initiatives for renewable power generation. Houston, TX, will purchase 50 percent of its electricity between 2013–2015 through RECs.	City of Houston (2013)
District energy and CHP generation from biomass	Centralized energy systems have the potential for energy efficiency improvements relative to distributed systems, due to improved equipment sizing and load balancing. District heating systems provide a more efficient fuel means to provide space heating energy services, with estimated energy-use savings of 10–20 percent (Harvey, 2006; section 15.3.2). Vaxjo, Sweden, produced 53 percent of its energy from renewable sources (predominantly biomass) in 2010, using district energy plants for both heating and electricity—over 80 percent of heating demands were met through renewable energy sources.	City of Vaxjo (2011)
Renewable energy requirement for new construction	Popularized by the London borough of Merton, the council requires that 10 percent of new commercial buildings over 1,000m² generate 10 percent of their energy needs onsite. Barcelona requires that 60 percent of hot water in new, renovated or repurposed buildings be supplied by solar hot water. The City of Lancaster, California, became the first city in the US to require all new construction projects to incorporate solar PV as of 2014. The requirement specifies that 0.5–1.5kW be produced per unit constructed, depending on lot size and location.	City of Lancaster (2013)
Smart meters	To encourage electricity peak shaving, many municipal and regional electric utilities (such as those in Melbourne, Australia) require the installation of smart meters. These meters transmit energy demand for a given consumer at regular intervals throughout the day, allowing for appropriate price signals to be sent during peak demand (which is often supplied by carbon-intensive energy).	Government of Victoria (2013)
Feed-in tariffs	In order to incentivize renewable energy development, many municipalities offer feed-in tariffs through their municipal electric utilities. Los Angeles' Department of Water and Power, as one example, intends to allocate 100 MW of generation to local small- and large-scale solar power producers.	City of Los Angeles (2013)

structural integrity) or have a long-term carbon payback (e.g. biomass); these may not completely offset costs associated with conventional alternatives, especially as low carbon options diffuse into the energy supply.

Buildings

Buildings generally represent a significant opportunity to mitigate GHG emissions in established cities; much of the existing stock has been built to standards that are not only far below the level of energy performance required to achieve significant reductions in global GHG emissions, but also fall well short of current building codes due to their age. Municipal approaches have aimed to improve the GHG performance of new buildings (through building code revisions) and existing buildings (through regulations or subsidizing/financing of retrofit options) as well as to increase public awareness of the measures that can be taken to reduce demand. To make any substantial impact, however, current buildings will have to go through 'deep' retrofits, which can lower energy by more than 50 percent, in contrast with 'shallow' retrofits that amount to a 10 to 30 percent decrease (Lucon et al., 2014).

Improving energy literacy of businesses and residents is an important step being taken by many cities of the industrialized world. For example, San Francisco, Boston and New York City require regular reporting of energy consumption from large commercial buildings; this creates awareness of energy consumption, engagement with energy benchmarks within communities and encourages action to reduce energy costs. Moreover, the implementation of energy feedback systems in the residential sector improves the appreciation of behavior and technologies that can reduce energy costs as well as GHG emissions.

All levels of government have provided subsidies for energy efficiency upgrades within the building sector through rebate programs as an economic instrument towards market transformation (Lucon et al., 2014). However, uncertainty exists in the effectiveness of the provision of direct subsidies through rebate programs; free-ridership, rebound effects and financial burdens can impact the magnitude of energy savings from rebates (Galarraga et al., 2013). A summary of actions taken in the building sector is provided in Table 26.3.

Transportation

High-income cities, especially in North America and Australasia, exhibit urban forms that are generally dependent on private automobiles to provide transportation services (Newman and Kenworthy, 1989). However, facilitating the transition to low carbon transportation technologies has the potential to substantially reduce this, with an estimate for a low-density North American city suggesting an 80 percent reduction in GHG emissions is possible by 2050 (Mohareb and Kennedy, 2014). As observed in Figure 26.3, GHG emissions reductions in transportation are slower than in other sectors, given the slow turnover of the vehicle stock (the median vehicle service life in the US for cars and light trucks manufactured in 1990 was 16.9 and 15.5 years, respectively [Davis et al., 2013]), the length of time to complete public transportation projects (from inception to operation) and the relative inelasticity of transportation demand (Havranek et al., 2012). Additionally, considering the current low mode share of public transportation, a substantial shift from private passenger vehicles to public transportation would require substantial investments in new infrastructure in order to make a significant reduction in vehicle trips (Engel-Yan and Hollingworth, 2008). Finally, approaches to encourage shifts to active modes of transportation (i.e. walking,

Climate change mitigation in high-income cities

Table 26.3 Municipal actions in established cities to mitigate GHG emissions in the building sector

Initiative	Description	Further reading
High-energy performance standards for municipal building codes and ordinances	In an effort to improve energy performance of buildings some cities have developed more rigorous building codes than those enacted by higher levels of government. For example, Toronto's 'Green Standard' requires a significant energy efficiency improvement to low-rise residential buildings two years prior to the adoption of the same standard by the provincial government. Builders attaining an even higher level of building performance were provided with a 20 percent reduction in their development charges. Additionally, ordinances are being implemented that require more rigid energy efficiency standards for new construction (numerous municipalities in Illinois; Dallas, TX), properties undergoing significant retrofits (Berkeley, CA) and properties being sold (San Francisco, CA). In Germany, cities including Frankfurt, Leipzig and Hamburg require that new public buildings adhere to the Passive House standard (since 2007, 2008 and 2012, respectively); meanwhile, Freiburg has made the same requirement for all residential buildings since 2011. The City of Brussels has responded to the European Union's Energy Performance of Buildings Directive by mandating that construction and retrofits of residential, office and school buildings align with the Passive House standard as of January 2015.	City of Toronto (2013b); City of Boston (2013b); City of Berkeley (2013); City of San Francisco (2013a); NREL (2009); Passive House International (2014)
Addressing split incentives between tenants and property owners	Benefits to improve energy efficiency of rental units are often either realized by the renter (in the case where the tenant pays utility bills), reducing the incentive for property owners to improve the energy performance of these units. Cambridge, MA, is one jurisdiction attempting to remedy this issue through the provision of draft 'green leases', which allow increases in rental charges that amount to less than the expected savings in energy costs. Additionally, requiring energy bill information to be available for potential tenants is another approach to addressing the split incentive, as is required in the European Union and various US jurisdictions such as Seattle and New York City.	Williams (2008); Hsu (2014)
Mandatory certifications	Energy efficiency is an important component of internationally recognized green building certification systems. Cities including Chelmsford, UK, and Boston, MA, have recognized this (as well as the numerous other benefits related to productivity and property value) and now require that certain new developments meet Building Research Establishment Environmental Assessment Methodology (BREEAM) and Leadership in Energy and Environmental Design (LEED) standards, respectively. Additionally, smaller towns have also taken this approach for the residential sector, such as the Town of East Gwillimbury, Canada, which requires a national energy efficiency standard for all new residential construction.	Chelmsford City Council (2013); City of Boston (2013b); FCM (2013)

(continued)

Eugene Mohareb et al.

Table 26.3 (continued)

Initiative	Description	Further reading
Indirect and direct energy feedback systems	This approach informs the energy end-user on how their energy consumption compares with their historic usage and with the usage of their neighbors. Jurisdictions in Illinois and Minnesota have realized significant reductions in energy demand (6 percent and 2 percent, respectively), primarily by equipping their residents with more information on their energy use.	Allcott (2011); Harding and McNamara (2013)
Incentives for high-performance construction	Municipalities have also begun to provide incentives for developers that aim to exceed existing building codes and municipal buildings standards. These incentives include expedited approvals for developers or reduced development charges (numerous cities in California as well as Toronto).	City of Toronto (2013b); State of California (2010); Passive House International (2014);
Property-tied renewable energy or energy efficiency financing	In addressing the barrier imposed by high upfront costs of investments in GHG mitigation, municipalities and utilities offer financing for technologies that is linked to the property. This financing can take the form of pay-as-you-save billing (Toronto) or a long-term arrangement where a small charge is regularly applied to property tax or utility bills (such as Berlin's Energy Saving Partnership or Property Assessed Clean Energy in various US jurisdictions).	City of Toronto, (2013c); C40 Cities (2007); PACE Now (2013)
Energy audit subsidies or requirements	Cities have been able to provide basic energy audits for commercial and industrial buildings either at a subsidized rate or free of charge (Seattle). Additionally, certain cities have mandated audits upon sale (Austin) or at a particular frequency (San Francisco) as well as benchmarking (Boston and New York City) of annual energy consumption, with the potential to improve the monitoring and analysis of energy consumption towards reducing demand. The City of Stockton, CA, has set retrofit targets; if targets are not met by 2013, energy auditing will be mandatory for all home sales or when building permits exceeding 1,200 USD are issued.	City of Seattle (2013); BuildingRating (2014); ACEEE (2013); City of San Francisco (2013b); City of Stockton (2013)
Municipal carbon trading	Large cities may be able to direct resources to reduce building energy demand through a regional cap-and-trade scheme. Tokyo's cap-and-trade program (including facilities in its industrial and commercial sector) was able to realize a 13 percent reduction in GHG emissions (roughly 1.4 Mt) in its first year of operation.	Tokyo Metropolitan Government (2012)

cycling) are generally in the form of large-scale urban (re)development projects, with long-term results.

Attempts to either develop public transportation systems that can operate effectively within these constructs or reduce the carbon intensity of existing modes of transport have been made in many high-income cities. The US has experienced a surge in public transportation investment over the period between 1995–2012, with increases in capital funding from federal, state and municipal governments of 131 percent, 108 percent and 179

Climate change mitigation in high-income cities

percent, respectively (APTA, 2014); this amounted to annual capital funding of 17 million USD in 2012. These transportation modes have a significant energy and carbon benefit; for example, Poudenx and Merida (2007) suggest trolley buses and light rail transportation are eight times more energy efficient than personal vehicles, per passenger kilometer traveled. Additionally, enabling the evolution of markets for alternatively fueled vehicles through greening the municipal fleet or encouraging the adoption of EV infrastructure paves the way for long-term reductions in GHG emissions. Efforts to address travel behavior are also important, with a notable approach having been taken in the London borough of Sutton. Rewards (lotteries and gifts) are used to encourage active transportation; transportation planning advice and encouragement of travel planning is also provided to schools, businesses and residences (Borough of Sutton, 2010).

Ultimately, a concerted effort from transportation and urban planners is required to reduce the energy and GHG emissions associated with urban transportation. This is particularly important in cities expecting continued population growth. Infill development with mixed uses and transit-oriented design are also important components of long-term transportation emissions reduction strategies. A summary of prominent transportation-related actions taken in high-income cities is provided in Table 26.4.

Table 26.4 Municipal actions in established cities to mitigate GHG emissions in the transportation sector

Initiative	Description	Further reading
Investments in public transportation	Significant investments in public transit have the potential to reduce emissions through shifting transportation-mode shares from automobiles to buses, light rail and subways. Los Angeles, London and Madrid are all examples of cities investing in meeting transportation demand through public transit. Madrid estimates that its subway system emits 50.73 g CO_2e/passenger-km (compared to a Spanish 2015 model year passenger vehicle average of approximately 84 g CO_2e/passenger-km [European Commission, 2014; European Environmental Agency, 2010—uses 2010 estimate for occupancy]).	TRCA (2010); Metro Madrid (2013)
Intelligent transportation systems	Improving the efficiency and information systems of a transit network can boost ridership by providing a more reliable and attractive alternative to private automobiles. Innovations such as bus tracker systems and transit signal priority are examples of how these behavior changes can be promoted.	Zhang et al. (2011)
Pedestrianization and improved cycling infrastructure	Pedestrianization of many streets in Europe or in Times Square in New York City shift the convenience of using motorized transport to active transport. Focusing on cycling infrastructure has a similar impact (e.g. Copenhagen), with the ability to address longer trip lengths (such as commuters). Additionally, bicycle sharing programs make active modes of transportation available and convenient for use for point-to-point trips.	New York City (2013); City of Copenhagen (2011)

(continued)

Eugene Mohareb et al.

Table 26.4 (continued)

Initiative	Description	Further reading
Vehicle share programs	Zurich was the first city to host a car-sharing program in 1948 (Selbstfahrergemeinschaft), with many European cities experimenting since then and burgeoning into other global cities with greater success. Car shares have the potential to reduce upstream GHG emissions embodied in vehicles by reducing the total number of vehicles purchased as well as providing an indirect road-pricing approach to vehicle usage. Cities have assisted in these programs by providing parking spaces for these vehicles (such as Dailmer's Car2Go model in many North American and European cities).	Shaheen and Cohen (2007)
Electric vehicle (EV) charging infrastructure	In jurisdictions with low carbon electricity grids, the installation of EV charging infrastructure has the potential to mitigate community transportation GHG emissions. The presence of charging stations encourages the adoption of EVs, increasing visibility and reducing range anxiety associated with these technologies. Some jurisdictions that lead the adoption of EVs include Amsterdam and Copenhagen.	Hydro Quebec (2013); *New York Times* (2013); State of Green (n.d.)
Low-emission municipal fleets and taxis	Many municipalities demonstrate leadership in converting their own vehicle fleets to hybrid EVs and other lower-emissions vehicles. Additionally, numerous cities (e.g. Hong Kong, Barcelona and New York City) encourage EVs in private taxi fleets. Taxis are ideal candidates for EVs since they mostly operate in central business districts (CBDs), where congestion allows the exploitation of recursive breaking and sufficient density of charging stations is easier to attain. Some cities (Reykjavik, Oslo and Stockholm) also opt to use biofuels (some of which are sourced from municipal solid waste feedstocks) for their municipal fleets.	CFM (2010); Nordic Council of Ministers (2012)
Parking charges and road pricing	Incentivizing lower-carbon modes of transportation or carpooling by increasing the price of automobile usage and increasing its convenience is a common approach to reducing congestion. Parking prices are frequently set in cities to disincentivize automobile usage, presumably reducing vehicle emissions as well. The imposition of road pricing mechanisms in various cities (e.g. Singapore, Stockholm and London) has been shown to be effective in reducing congestion and, by extension, its related GHG emissions. Pay-as-you-drive insurance provides another means to price vehicle kilometers traveled, reducing travel behavior (increasingly available through insurance providers).	TRCA (2010)
High-occupancy vehicle (HOV) lanes and bus-on-shoulder (BoS)	HOV lanes and BoS transportation represent relatively low-cost approaches to improving public transit services. These address congestion and any associated GHG emissions from the net decrease in fuel consumption.	TRB (2013)

Municipal services

Emissions reductions from municipal activities provide leadership for the broader community, in addition to necessary participation in the early stages of a local market transformation to low-carbon technologies, services and behaviors. By adopting more efficient or alternatively powered vehicles, buildings, street lighting, or water and wastewater treatment and distribution operations, municipalities reduce their emissions and raise awareness of the importance of mitigation. Given the permanence of most municipal operations, investments in measures with longer-term payback periods also become easier to justify (e.g. deeper building retrofits) (Arup and C40 Cities, 2011). Municipalities use GHG inventories of their operations to identify the most significant opportunities to reduce both costs and emissions. In addition to approaches taken in municipally operated transportation and buildings (described in previous sections), options for mitigation are discussed here.

Table 26.5 Municipal actions in established cities to mitigate GHG emissions in the government services sector

Initiative	Description	Further reading
Source-separated organics	In order to prevent the release of methane from landfill operations and improve the diversion of waste, source-separated organics programs have been implemented.	City of Toronto (2009)
Landfill gas capture	Capturing methane produced by organic waste deposited in landfill sites is pursued to both reduce emissions and provide for a relatively clean source of renewable energy. As an example, Hong Kong displaces naphtha consumption by utilizing landfill gas in the production of electricity and heat.	Government of Hong Kong (2013)
Wastewater biogas capture and sludge management	Anaerobic digestion of wastewater sludge provides an opportunity for energy generation (biogas) and GHG mitigation. This biogas can be used within a facility or externally to meet heating or electricity needs.	Baltic Biogas Bus (2012)
Waste prevention	Programs to address residential and non-residential waste are using outreach and disincentive programs to reduce the amount of waste that is being generated. For example, studies on pay-as-you-throw programs in the Czech Republic and Sweden find that these initiatives result in an 8 percent and 20 percent reduction in waste sent to the landfill, respectively, compared to jurisdictions without these programs.	Arup and C40 Cities (2011); Dahlén and Lagerkvist (2010); European Commission (2008)
High-efficiency traffic lights and street lighting	Replacement of high-pressure sodium lighting with LED technology provides significant savings in GHG emissions, costs and maintenance.	City of Halifax (2011)
Innovative waste-to-energy	Though traditional waste-to-energy through incineration is common globally, innovative approaches that can convert solid waste to advanced biofuels are supported in many high-income cities (British Airways' GreenSky project in London; Enerkem in Edmonton, Canada; transportation fuels in multiple Nordic cities).	British Airways (2012); Enerkem (2014); Nordic Council of Ministers (2012)

Cities have relatively greater influence in controlling GHG emissions from waste. Dependence on sanitary landfills for waste disposal in many North American municipalities has led to a legacy of methane emissions (a GHG that is 34 times more potent in its radiative forcing than CO_2 over a 100-year time horizon) (Myhre et al., 2013), which would not have been released otherwise had these materials degraded in the presence of oxygen. Furthermore, substantial GHG emissions occurring outside municipal boundaries can be mitigated through recycling and waste prevention (USEPA, 2006). A number of approaches have been taken to reduce these, ranging from simple source reduction to end-of-pipe treatment methods. A common example of the latter has been to capture landfill gas and either flare it or utilize it for electricity or heat generation. Cities such as Toronto, Canada, have realized significant emissions reductions from capture and utilization. Additionally, by diverting biogenic carbon from landfills, some cities have been able to use waste as a resource to a greater degree by digesting this waste in bioreactors, producing a valuable soil amendment and more efficiently producing biogas (City of Toronto, 2009).

Waste-to-energy facilities are more commonly used in Europe, and cities in the region have recently employed more sophisticated approaches to improve energy yield and offset emissions elsewhere. For example, Copenhagen's waste-to-energy operations provide enough energy through its district energy system to meet the demands of 70,000 homes (DAC, 2014). The city also plans to divert more plastics (i.e. fossil carbon) from its incineration waste stream, which is estimated to reduce emissions by an additional 100,000 t CO_2e towards the goal of carbon neutrality by 2025 (City of Copenhagen, 2009). Additionally, biogenic carbon treated in wastewater treatment plants has also been exploited for its energy potential; in the case of the City of Stockholm, biogas from its wastewater treatment process is refined and utilized in its public transit buses (Baltic Biogas Bus, 2012).

Other municipal responsibilities (from administrative buildings to street lights) also present opportunities to reduce expenses and address climate change. Looking at street lights, cities in various industrialized nations have recognized the value in replacing high-pressure sodium lighting with light-emitting diode (LED) technology. The City of Halifax, Canada, has planned the replacement of its street lights with LEDs, with projected annual savings of 120,000 CAD and 1,000 t CO_2e (City of Halifax, 2011). A summary of measures undertaken are provided in Table 26.5.

Land use and spatial planning

Many developed cities face significant barriers to GHG mitigation imposed by their urban form. As touched upon in the transportation section, urban expansion since the early 20th century has often relied upon low-density growth with segregated land uses. This land use and spatial planning approach is characterized by high transportation energy demand, low floor area ratios and greater floor area per housing unit (Seto et al., 2014; Peiravian et al., 2014b). Improvements in land use and spatial planning can go a long way towards breaking free of the energy dependency observed in the other sectors discussed above. However, in this sense, developed cities are disadvantaged relative to those in the early stages of infrastructure growth; altering emissions intensities associated with the transportation and building sectors of established low-density, single-use urban areas requires a long-term vision, where infrastructure and development plans are well aligned with improvements in the urban form that are inherently lower carbon.

Seto et al. (2014) describe four principal elements of land use and spatial planning that require consideration towards the long-term goals of transportation demand reduction and

Climate change mitigation in high-income cities

intensification. These are density, land use, connectivity and accessibility. Achieving sufficient emissions reductions while reducing congestion requires consideration of these in relation to one another. For example, redeveloping a neighborhood in a low carbon city must aim to improve residential/commercial density, increase the diversity of work/recreational opportunities, incorporate finer grain blocks (to improve form for connectivity/walkability) and consider connectivity (for example, low carbon transportation modes allowing access to the CBD). One study concluded that efforts to double residential density, increase employment density, improve the mixture of land uses and encourage public transportation alternatives could reduce vehicle miles traveled per household by as much as 25 percent (National Research Council, 2009). Moreover, land use features including density and diversity have a direct and positive impact on providing an environment that is more friendly for walking (Peiravian et al. 2014a). A summary of prominent municipal actions in land use and spatial planning are detailed in Table 26.6.

Table 26.6 Municipal actions in established cities to mitigate GHG emissions in land use and spatial planning

Initiative	Description	Further reading
Smart growth or Compact City policies	Developed world cities have attempted to reverse the low-density development patterns of the 20th century through development changes that better reflect the infrastructure costs of different development types, and through the creation of strict zoning constraints to curb sprawl.	Lu et al. (2013); Council of Capital City Lord Mayors (2014)
Mixed-use development	Zoning strategies have been implemented in many cities to prevent further segregation of land uses. Some prominent examples of encouraging mixed land use include Singapore and Vancouver.	Walsh (2013); Urban Redevelopment Authority (2014)
Major regeneration or infill projects	Infill and regeneration exploit the increased property values of derelict industrial sites in close proximity to CBDs in developed cities, while encouraging lower-carbon building and planning approaches. Some prominent examples include Hammarby-Sjostad in Stockholm, Sweden, and Newstead and Teneriffe in Brisbane, Australia.	Hammarby-Sjostad (2014); Council of Capital City Lord Mayors (2014)
New urbanism/ transit-oriented design	New urbanist building and neighborhood designs are increasing in their uptake, with nearly 300 developments listed in the Congress for New Urbanism's (CNU) project database as of April 2015. These types of developments incorporate high-density, mixed-use and public/active transportation components, with potential for mitigation of building and transportation sector emissions. The CNU provides information on projects located around the world including Brisbane, Baltimore and Berlin.	Congress for New Urbanism (2014)
'Complete street' design and promotion	Streets that make room for the safe participation of all modes of transportation have been increasingly encouraged in recent decades, in addition to the advancement of segregated infrastructure. Prominent examples of complete street designs in developed cities include Copenhagen and Seattle.	Smart Growth America (2010); McCann and Rynne (2010); Woodcock et al. (2014)

Bottom-up initiatives

Many bottom-up initiatives on urban GHG emissions mitigation are taking place in European, North American and Oceanic cities. These are often promoted by community renewable energy co-operatives, not-for-profit organizations and neighborhood groups. Due to the authors' experiences, this section explores examples of such bottom-up initiatives from the Toronto, Canada area. Toronto's waterfront is the site of the first co-operatively owned and operated urban wind turbine in the province. This 600 kW turbine is estimated to offset the needs of around 800 homes annually, and has been operating since 2002. This initiative has since expanded to develop other turbines and a number of solar energy co-operatives, which allow members to purchase bonds in solar projects that are backed by government feed-in tariff programs.

Another form of bottom-up initiative is led by non-profit community action groups. One such group in Toronto, called Project Neutral, has been conducting carbon footprint surveys of homes in Toronto in 2012–2013. The project initially targeted two neighborhoods through community-led efforts, with the specific goal of eventually retrofitting these neighborhoods for carbon neutrality. The hope is that a repeatable model of urban neighborhood retrofit for carbon neutrality can be developed. Thus far, there are a number of achieved successes, including the completion of hundreds of household surveys to establish baseline emissions levels, the creation of several online tools for reporting household and neighborhood improvements and creation of a tool to help achieve carbon emissions improvements (Project Neutral, 2014; Naismith, 2014).

Year-on-year, the two pilot-selected neighborhoods show decreases, excluding single detached homes in one of the neighborhoods. This increase, however, may be because additional households joined in 2013, but are not differentiated in the reporting. Households outside of the pilot neighborhoods are allowed to fill out surveys, but do not receive the extra direct guidance and support of the initial neighborhoods such as the facilitation of neighborhood members going door-to-door to raise awareness. The survey respondents outside the pilot neighborhoods, however, show reductions. The findings are encouraging, and in time it will be possible to assess trends with more depth and hopefully correct for factors such as changing weather.

Together the co-operative and the community action groups share at least three things in common: (1) they were developed by dedicated citizens concerned with climate change; (2) they have benefited from a number of strategic partnerships with governmental and non-governmental organizations in the region; and (3) early on they have identified financial sustainability as an organizational need for success in achieving their mission.

Top-down initiatives

There are also examples of cities and regions in the industrialized world that have taken a 'master planning' approach to low carbon urban design. These new developments often include highly efficient housing, considerations for either public transportation or low-carbon transport and onsite renewable energy generation. Some prominent examples are provided in Table 26.7.

While these approaches can demonstrate the best practices for new construction and serve as valuable case studies for low-carbon development, their relevance to reconstruction of existing buildings in established cities is limited. Reducing emissions in developed countries by 2050 on the order of 80 percent from 1990 levels, as suggested by Gupta et al. (2007, Box 13.7), necessitates deep retrofitting of existing building stock.

Climate change mitigation in high-income cities

Table 26.7 Sample urban (re)development projects that include measures to achieve substantial GHG emissions reductions

	Key characteristics and challenges to success	Further reading
Hammarby-Sjostad, Stockholm, Sweden	Hammarby-Sjostad, an urban redevelopment project on a former industrial site, has received considerable attention for its well-integrated district- and building-scale renewable energy systems. Some key challenges to the further adoption of additional alternative energy systems are the concerns they pose to the economics of existing energy infrastructure systems and the interference through the monopoly of the energy retailer.	Pandis Iveroth et al. (2013); Hammarby-Sjostad (2014)
Beddington Zero Energy Development (BedZED), and One Brighton, UK	Completed in 2002, BedZED is a long-standing example of a low-carbon greenfield development. Providing 100 homes and workspace for 100 jobs, the original plan was for the neighborhood's low-energy demand to be met through onsite biomass CHP system. However, given the small scale of this operation, maintenance requirements rendered this approach uneconomical. Following BedZED, One Brighton was constructed to apply the lessons learned from BedZED, and suggested that houses in this community achieve a 60–80 percent reduction in GHG emissions relative to a typical UK home. Other lessons learned from the project include the importance of making low carbon lifestyles more convenient through design; the co-benefits of improved community cohesion; the value of energy service companies; and discussion of 'green facilities management' at the design stage.	BioRegional (2002); BioRegional (2009); BioRegional (2014)
Masdar City, UAE	Featuring its own photovoltaic (PV) array complemented by roof top panels, a concentrating solar plant, short streets that are cooled by a large wind tower, driverless EVs and sensor-controlled buildings, Masdar is an example of an attempt to create a new, low carbon city in an arid climate. Major challenges have included the seclusion of the site, the high proportion of commuters and inability to attract sufficient foreign investment to boost employment.	Wired (2013); Deutsche Welle (2013)
PlanIT Valley, Portugal	Endorsed by the Portuguese national government and the municipality of Paredes, this eco-city with a project population of 225,000 is intended to be an incubator for low carbon innovation. Enhanced building monitoring, grid-connected vehicles and other intelligent infrastructure systems will rely on millions of sensors to optimize their performance. Observers state concerns about how this type of top-down control system for an urban environment will adapt to the complexities of a working city.	Salon (2012); Living PlanIT (2014)

(continued)

Table 26.7 (continued)

	Key characteristics and challenges to success	Further reading
Songdo, South Korea	Hosting branches of four overseas universities, Songdo ultimately hopes to house 75,000 residents and be the destination for 300,000 commuters. Active transportation will be a key focus, with networks of cycle paths and pedestrian routes as well as an integrated underground waste collection system, which will supply a share of the site's energy needs. The district, which is planned to obtain LEED certification for neighborhood development, is facing early challenges in attracting businesses and investors to populate its office space as well as receiving criticism for the destruction of the unique wetland ecosystem and high eco-premiums that exclude low-income/low-revenue residents and businesses.	Songdo (2014); World Finance (2014); Shwayri (2013)

Addressing both adaptation and mitigation

Many high-income cities have recently recognized the need to adapt to climate change, given the consensus on the increased frequency of severe weather events on vulnerable urban populations (Zimmerman and Faris, 2011). When it comes to mobilizing political capital to address climate change, there is some concern that adaptation measures may take precedence over mitigation measures (Gore, 2013). The concern arises since adaptation measures—where benefits are local, clearly targeted and can be directly observed (with sufficient co-benefits that circumvent the need to emphasize the threat of climate change)—are more saleable than mitigation measures, where impacts are globally distributed and may lack popular support in some jurisdictions. Put another way, local measures that address adaptation-related concerns can readily be seen functioning in their purpose of insulating against moderate and extreme weather events. Meanwhile, mitigation strategies that address a city's GHG emissions appear as a drop in the atmospheric ocean where increasing radiative forcing is altering the climate; as a result, successful mitigation requires the elusive (up until now) cooperation of an unwieldy group of international actors in order to observe any effect (which will be gradual and over the long-term, escaping the notice of most urban inhabitants). This can make budgeting for adaptation strategies a politically-safe approach for any municipality, especially where hard infrastructure is highly visible and can be appreciated regardless of one's perspective on humanity's role in the changing climate. As the impacts of climate change worsen, this potential dichotomy could be exacerbated in the urgency to mitigate local impacts of extreme weather events.

The disproportionate role of industrialized nations in changing the composition of the atmosphere is illustrated by den Elzen et al. (2005), with nearly 80 percent of fossil energy CO_2 emissions between 1890–2000 attributable to industrialized (and former Soviet) nations. Additionally, many of the cities that are most vulnerable to extreme weather events are in earlier stages of economic growth, where per capita GHG emissions are relatively low compared with those in industrialized nations (Hallegatte et al., 2013; Dhakal, 2009; Sugar et al., 2012). Hence, mitigation activities may not be perceived as urgent in locations where the adaptation needs are most acutely felt. Stern (2006) demonstrates that while adaptation is necessary, the costs of mitigation are much lower than the recovery from severe weather events, and adaptation expenditures may not be equitably borne by those contributing to climate change. Therefore, efforts for adaptation and resiliency should be viewed as additional to and separate from mitigation efforts.

Some studies show that mitigation and adaptation issues can be addressed simultaneously through certain strategies, with prominent examples including improved building energy performance, expansion of urban forests and renewable energy systems (Sugar et al., 2013; Harlan and Ruddell, 2011; Laukkonen et al., 2009; Kennedy and Corfee-Morlot, 2012; Hamin and Gurran, 2009). While there may be more urgency for low- and middle-income cities to focus on the measures that address both sides of the climate change conundrum (see Lwasa, Chapter 27 in this volume, for this discussion), high-income cities with (typically) greater financial resources should lead in the early experimentation and deployment of mitigation measures.

Conclusion

In order to meet long-term global targets for GHG emissions reductions, cities in the industrialized world require the replacement of energy service technologies as well as the enabling infrastructure and urban form. Many of these cities are providing exemplary pathways for initiating the transition to low carbon systems, while others have created entire developments that are able to operate with minimal fossil energy inputs. There remains much to be done within municipal boundaries, especially with respect to existing building stocks that require deep retrofits to facilitate the use of low-density forms of energy that are intermittent. Moreover, while a radical change in transportation infrastructure is difficult, future land use planning policies may greatly contribute, notably by encouraging mixed-use plans that favor active modes of transportation. With the combined efforts of top-down approaches from all levels of government, bottom-up approaches to reductions can support mitigation projects and help high-income cities provide the leadership required of them in moving towards a low carbon global economy.

Ultimately, detailed regional long-term planning and monitoring are required for high-income cities to achieve deep reductions in their GHG emissions. A price on carbon is an essential component to effectively drive market transformation to low carbon systems. New low- and zero-carbon energy systems must replace existing carbon-intensive systems. Sprawling developments must be reimagined in order to support low carbon lifestyles. Continued strengthening of leadership at the municipal level is central to this effort in order to properly understand and implement the most effective options for mitigation in specific urban contexts. More research and experimentation at the city scale is required to achieve deep, permanent emissions reductions in order to surpass what can be achieved by targeting low-hanging fruit.

While many jurisdictions have exhibited leadership, numerous others have been hindered by the absence of political will and, even more threatening, the funding gap observed between infrastructure needs and public funds. Broad, durable acceptance of the scope and onus of mitigation requirements that rests with the industrialized world is necessary for significant action. Even more fundamental is a consistent approach to GHG quantification across all cities, to identify successful, scalable mitigation strategies though comparable, longitudinal data (employing methodologies such as the WRI/ICLEI/C40 Cities Global Protocol for Community Scale Emissions). Once stable global leadership is in place, cities can play their prominent role in guiding their residents, and indeed the world, towards a low carbon future.

Chapter summary

- Progress is being made in many high-income cities in mitigating GHG emissions including improvements in all major emissions sectors.
- High-income cities face the additional challenges of having established infrastructure, uncertain future economic growth and entrenched values that inhibit political will to rethink urban systems.
- Emissions reduction strategies within cities of the industrialized world span across sectors for energy provision, buildings, transportation, municipal services, land use and spatial planning.
- Concerns remain as to whether these reductions can be maintained for the long-term or if they are merely addressing low-hanging fruit with politically attractive co-benefits, in addition to the added demands for adaptation.
- Comprehensive frameworks for emissions reductions need to be incorporated at the city level.

Research needs

- Broad undertaking of longitudinal urban GHG quantification in cities of all scales and in all regions is needed, using a consistent methodology and with an effort to identify which emissions are locked-in and how this can be addressed.
- Deeper understanding of future GHG implications of current infrastructure decisions and redevelopment options for the city are needed at neighborhood- and street-scales along with an understanding of how to make these adaptable to mitigation solutions on the horizon.
- Further exploration as to the extent to which urban mitigation and adaptation activities have complementary goals as well as a broader quantification of the co-benefits of mitigation measures.

Notes

1 OECD90 includes countries in OECD Asia, Western Europe and North America.
2 It should be noted that the methodologies for GHG emissions inventories presented here use different allocation approaches as well as varying spatial and temporal boundaries. This explains some of the variation (see Ibrahim et al., 2012), though the trends are generally informative of development approaches within the regions above.

References

ACEEE (American Council for an Energy Efficient Economy), 2013. Austin Energy Conservation Audit and Disclosure Ordinance. [Online] Available: http://aceee.org/files/Case-Study--Austin-Energy-ECAD.pdf, accessed October 24, 2013.
Allcott, H., 2011. Social Norms and Energy Conservation. *Journal of Public Economics*, 95(9–10), 1082–1095.
Allwinkle, S., and Cruickshank, P., 2011. Creating Smart-er Cities – An Overview. *Journal of Urban Technology*, 18(2), 1–16.
APTA, 2014. Public Transportation Investment Background Data. [Online] Available: www.apta.com/resources/reportsandpublications/Documents/Public-Transportation-Investment-Background-Data.pdf, accessed October 19, 2013.
Arup and C40 Cities, 2011. Climate Action in Megacities – C40 Cities Baseline and Opportunities. [Online] Available: www.arup.com/news/2011_06_june/01_jun_11_c40_climate_action_megacities_sao_paulo.aspx

Climate change mitigation in high-income cities

ASCE (American Society of Civil Engineering), 2013. Infrastructure Report Card, 2013. [Online] Available: www.infrastructurereportcard.org/a/documents/2013-Report-Card.pdf, accessed August 23, 2014.

Baltic Biogas Bus, 2012. Production and Supply of Biogas in Stockholm. [Online] Available: www.balticbiogasbus.eu/web/Upload/Supply_of_biogas/Act_4_6/Production%20and%20supply%20of%20biogas%20in%20the%20Stockholm%20region.pdf, accessed April 23, 2015.

Bettencourt, L. M. A., Lobo, J., Helbing, D., Kühnert, C., and West, G. B. (2007). Growth, Innovation, Scaling, and the Pace of Life in Cities. *Proceedings of the National Academy of Sciences of the United States of America, 104*(17), 7301–7306. doi:10.1073/pnas.0610172104

BioRegional, 2002. BedZED Case Study Report. [Online] Available: www.bioregional.com/files/publications/BedZEDCaseStudyReport_Dec02.pdf, accessed October 19, 2013.

BioRegional, 2009. BedZED Seven Years On: The Impact of the UK's Best Known Eco-Village and its Residents. [Online] Available: www.bioregional.com/wp-content/uploads/2014/10/BedZED_seven_years_on.pdf, accessed September 11, 2014.

BioRegional, 2014. One Brighton Achieves Deep Carbon Cuts. [Online] Available: www.bioregional.com/news-views/news/onebrightonlcanewsrelease/, accessed September 12, 2014.

Borough of Sutton, 2010. Smart Travel Sutton – Third Annual Report. [Online] Available: http://epomm.eu/maxeva/uploads/STSthirdANNUALREPORT2010_V08.pdf, accessed April 23, 2015.

British Airways, 2012. GreenSky London Biofuel Plant Preparing for Lift-Off. [Online] Available: www.britishairways.com/en-gb/bamediacentre/newsarticles?articleID=20131227133451&articleType=LatestNews#.U_9Nf_ldVQ4, accessed August 28, 2014.

Building Performance Institute Europe, 2011. Europe's Buildings Under the Microscope. [Online] Available: www.europeanclimate.org/documents/LR_%20CbC_study.pdf, accessed August 23, 2014.

Building Performance Institute Europe, 2014. Data Hub for Energy Performance of Buildings. [Online] Available: www.buildingsdata.eu/, accessed August 27, 2014.

BuildingRating, 2014. US Benchmarking Policy Landscape. [Online] Available: www.buildingrating.org/graphic/us-benchmarking-policy-landscape, accessed August 28, 2014.

C40 Cities, 2007. Case Study – Energy Saving Partnership. [Online] Available: www.c40cities.org/system/resources/BAhbBlsHOgZmIi4yMDExLzEwLzI1LzIwXzAyXzEwXzQyMF9iZXJsa W5fZW5lcmd5LnBkZg/berlin_energy.pdf, accessed October 22, 2013.

Carlino, G. A., Chatterjee, S., and Hunt, R. M. (2007). Urban Density and the Rate of Invention. *Journal of Urban Economics, 61*(3), 389–419. doi:10.1016/j.jue.2006.08.003

CDP (Carbon Disclosure Project), 2012. Measure for Management – CDP Cities 2012 Global Report. [Online] Available: www.cdp.net/cdpresults/cdp-cities-2012-global-report.pdf, accessed August 21, 2014.

CFM (Canadian Federation of Municipalities), 2010. Enviro-Fleets: Reducing Emissions from Municipal Heavy-Duty Vehicles. [Online] Available: www.fcm.ca/Documents/reports/Reducing_emissions_from_municipal_heavy_duty_vehicle_EN.pdf, accessed October 24, 2013.

Chelmsford City Council, 2013. Assessing the Environmental Performance of Your Development. [Online] Available: http://consult.chelmsford.gov.uk/portal/building_for_tomorrow?pointId=s1361895302233, accessed October 20, 2013.

City of Berkeley, 2013. Residential Energy Conservation Ordinance. [Online] Available: www.ci.berkeley.ca.us/reco/, accessed October 20, 2013.

City of Boston, 2013a. Boston Community Greenhouse Gas Inventories (2013 Update). [Online] Available: www.cityofboston.gov/images_documents/updatedversionhg1_tcm3-38142.pdf, accessed September 19, 2014.

City of Boston, 2013b. Green Buildings – Article 37. [Online] Available: www.cityofboston.gov/images_documents/Article%2037%20Green%20Buildings%20LEED_tcm3-2760.pdf, accessed October 20, 2013.

City of Copenhagen, 2009. CPH 2025 Climate Plan. [Online] Available: http://bit.ly/1qO3fJa, accessed September 14, 2014.

City of Copenhagen, 2011. Good, Better, Best – The City of Copenhagen's Bicycle Strategy, 2011–2025. [Online] Available: http://kk.sites.itera.dk/apps/kk_pub2/pdf/823_Bg65v7UH2t.pdf, accessed October 24, 2013.

City of Halifax, 2011. Halifax's LED Streetlighting Project Sees 2,100 New Installations. [Online] Available: www.halifax.ca/mediaroom/pressrelease/pr2011/110214LEDStreetlightInstallations.html, accessed October 27, 2013.

City of Houston, 2013. City of Houston Increases Renewable Energy Purchase. [Online] Available: www.greenhoustontx.gov/pressrelease20130620.html, accessed October 14, 2013.

City of Lancaster, 2013. Outlook, June 2013. City Moves Forward with New Solar Initiative. [Online] Available: www.cityoflancasterca.org, accessed October 14, 2013.

City of London, 2014. London Greenhouse Gas and Energy Inventory. [Online] Available: http://data.london.gov.uk/datastore/package/leggi-2011, accessed August 31, 2014.

City of Los Angeles, 2013. Feed-in Tariff Program. [Online] Available: http://goo.gl/8eC8fQ, accessed October 19, 2013.

City of Paris, 2011. La Bilan Carbone de Paris [Online] Available: www.paris.fr/publications/publications-de-la-ville/guides-et-brochures/publication-des-resultats-du-bilan-carbone-de-paris/rub_6403_stand_31447_port_14438, accessed September 19, 2014.

City of San Francisco, 2013a. Residential Energy Conservation Ordinance. [Online] Available: www.sfenvironment.org/article/existing-buildings-other-than-major-renovations/residential-energy-conservation-ordinance, accessed October 20, 2013.

City of San Francisco, 2013b. Existing Commercial Building Energy Performance Ordinance. [Online] Available: www.sfenvironment.org/sites/default/files/fliers/files/sfe_gb_ecb_ordinance_overview.pdf, accessed October 24, 2013.

City of Seattle, 2013. Facility Assessment. [Online] Available: www.seattle.gov/light/conserve/business/cv5_ora.htm, accessed October 24, 2013.

City of Stockton, 2013. Residential Energy Efficiency Assessment and Retrofit Ordinance. [Online] Available: http://stockton.granicus.com/MetaViewer.php?meta_id=308373&view=&showpdf=1, accessed October 24, 2013.

City of Toronto, 2009. Staff Report on Utilization of Biogas. [Online] Available: www.toronto.ca/legdocs/mmis/2009/pw/bgrd/backgroundfile-24372.pdf, accessed October 27, 2013.

City of Toronto, 2013a. Summary of Toronto's 2011 GHG and Air Quality Pollutant Emissions Inventory. [Online] Available: www.toronto.ca/legdocs/mmis/2013/pe/bgrd/backgroundfile-57187.pdf, accessed September 19, 2014.

City of Toronto, 2013b. Toronto Green Standard. [Online] Available: www.toronto.ca/planning/environment/index.htm, accessed October 20, 2013.

City of Toronto, 2013c. Residential Energy Retrofit Program. [Online] Available: www.toronto.ca/teo/residential-energy-retrofit.htm, accessed October 27, 2013.

City of Vaxjo, 2007. Climate Strategy for Vaxjo. [Online] Available: www.energy-cities.eu/IMG/pdf/Vaxjo-climate_strategy.pdf, accessed October 13, 2013.

City of Vaxjo, 2011. Fossil Fuel Free Vaxjo. [Online] Available: www.vaxjo.se/Other-languages/Other-languages/Engelska—English1/Sustainable-development/Fossil-Fuel-Free-Vaxjo/, accessed October 14, 2013.

Congress for New Urbanism, 2014. Project Database. [Online] Available: www.cnu.org/resources/projects, accessed September 12, 2014.

Council of Capital City Lord Mayors, 2014. Unlocking Smart Growth in Australia's Capital Cities. [Online] Available: http://lordmayors.org/, accessed September 12, 2014.

DAC, 2014. Copenhagen: Waste-to-Energy Plants. [Online] Available: www.dac.dk/en/dac-cities/sustainable-cities/all-cases/waste/copenhagen-waste-to-energy-plants/, accessed September 14, 2014.

Dahlén, L., and Lagerkvist, A. (2010). Pay as you Throw: Strengths and Weaknesses of Weight-Based Billing in Household Waste Collection Systems in Sweden. *Waste Management (New York, N.Y.)*, *30*(1), 23–31. doi:10.1016/j.wasman.2009.09.022

Davis, S. C., Diegel, S. W., and Boundy, R. G., 2013. Transportation Energy Data Book – Edition 32. Oak Ridge National Laboratory, ORNL-6989.

den Elzen, M., Fuglestvedt, J., Hohne, N., Trudinger, C., Lowe, J., Matthews, B., Romstad, B., Pires de Campos, C., and Andronova, N., 2005. Analysing Countries' Contribution to Climate Change: Scientific and Policy-Related Choices. *Environmental Science & Policy*, 8, 614–636.

Department of Communities and Local Government, 2015. English Housing Survey, Headline Report, 2013–14. [Online] Available: www.gov.uk/government/uploads/system/uploads/attachment_data/file/406740/English_Housing_Survey_Headline_Report_2013-14.pdf, accessed April 23, 2015.

Deutsche Welle, 2013. Masdar Eco-City Rebounds After Setbacks. [Online] Available: www.dw.de/masdar-eco-city-rebounds-after-setbacks/a-16664316, accessed September 12, 2014.

Climate change mitigation in high-income cities

Dhakal, S. (2009). Urban Energy Use and Carbon Emissions from Cities in China and Policy Implications. *Energy Policy, 37*(11), 4208–4219. doi:10.1016/j.enpol.2009.05.020

ECONorthwest, 2011. Economic Impact Analysis of Property Assessed Clean Energy Programs (PACE). [Online] Available: http://pacenow.org/wp-content/uploads/2012/08/Economic-Impact-Analysis-of-Property-Assessed-Clean-Energy-Programs-PACE1.pdf, accessed October 19, 2013.

EIA (Energy Information Administration), 2013. 2009 RECS Survey Data [Online] Available: www.eia.gov/consumption/residential/data/2009/#undefined, accessed August 23, 2014.

Elzen, M. G. J., Olivier, J. G. J., Höhne, N., and Janssens-Maenhout, G. (2013). Countries' Contributions to Climate Change: Effect of Accounting for all Greenhouse Gases, Recent Trends, Basic Needs and Technological Progress. *Climatic Change, 121*(2), 397–412. doi:10.1007/s10584-013-0865-6

Energuide, 2013. Energy and Your Money. [Online] Available: www.energuide.be/en/questions-answers/category/energy-and-your-money/4/, accessed October 19, 2013.

Enerkem, 2014. Press Release – Launch of World's First Full-Scale Waste-to-Biofuels and Chemicals Facility. [Online] Available: www.enerkem.com/assets/files/News%20releases/EAB%20inauguration%20news%20release_FINAL_EN.pdf, accessed August 28, 2014.

Engel-Yan, J., and Hollingworth, B. J., 2008. Putting Transportation Emission Reduction Strategies in Perspective: Why Incremental Improvements Will Not Do. 2008 Annual Conference of Transportation Association of Canada, Toronto.

European Commission, 2008. Pay as you Throw Schemes Encourage Household Recycling. [Online] Available: http://ec.europa.eu/environment/integration/research/newsalert/pdf/133na1_en.pdf, accessed September 14, 2014.

European Commission, 2014. Reducing Climate Impact of Passenger Cars. [Online] Available: http://ec.europa.eu/clima/policies/transport/vehicles/cars/index_en.htm, accessed September 13, 2014.

European Environmental Agency, 2010. Occupancy Rates of Passenger Vehicles. [Online] Available: www.eea.europa.eu/data-and-maps/indicators/occupancy-rates-of-passenger-vehicles/occupancy-rates-of-passenger-vehicles-1, accessed September 13, 2014.

FCM (Federation of Canadian Municipalities), 2013. Town-Wide Energy Star Standards for New Homes. [Online] Available: www.fcm.ca/home/awards/fcm-sustainable-communities-awards/2007-winners/2007-residential-development.htm, accessed October 20, 2013.

FCM (Federation of Canadian Municipalities), Canadian Construction Association, Canadian Public Works Association, and Canadian Society of Civil Engineers, (2012). *Canadian Infrastructure Report Card, 2012* (Vol. 1, p. 75).

Galarraga, I., Abadie, L. M., and Ansuategi, A., 2013. Efficiency, Effectiveness and Implementation Feasibility of Energy Efficiency Rebates: The 'Renove' Plan in Spain. *Energy Economics, 40*(Supplement 1), S98–S107.

Gore, A., 2013. *The Future – Six Drivers of Global Change*. Random House, New York. Chapter 6 'The Edge'; Subsection 'Mitigation versus Adaptation'.

Government of British Columbia, 2014. Community Energy Emissions Inventory. [Online] Available: www.env.gov.bc.ca/cas/mitigation/ceei/, accessed September 19, 2014.

Government of Hong Kong, 2013. Landfill Gas Utilization. [Online] Available: www.epd.gov.hk/epd/english/environmentinhk/waste/prob_solutions/msw_lgu.html, accessed October 27, 2013.

Government of Victoria, 2013. Smart Meters. [Online] Available: www.smartmeters.vic.gov.au/, accessed October 28, 2013.

Grob, G. R., 2010. Future Transportation with Smart Grids and Sustainable Energy. *International Journal of Systems, Signals and Devices, 13*(4), 2–7.

Grubler A., Bai, X., Buettner, T., Dhakal, S., Fisk, D., Ichinose, T., Keirstead, J., Sammer, G., Satterthwaite, D., Schulz, N., Shah, N., Steinberger, J., and Weisz, H., (2012). Urban Energy Systems. In: GEA. (2012). *Global Energy Assessment – Toward a Sustainable Future*. Cambridge University Press, Cambridge, UK and New York, NY, USA and the International Institute for Applied Systems Analysis, Laxenburg, Austria. Retrieved from www.globalenergyassessment.org.

Gupta, S., Tirpak, D. A., Burger, N., Gupta, J., Höhne, N., Boncheva, A. I., Kanoan, G. M., Kolstad, C., Kruger, J. A., Michaelowa, A., Murase, S., Pershing, J., Saijo, T., and Sari, A., 2007. Policies, Instruments and Co-operative Arrangements. In: *Climate Change 2007: Mitigation. Contribution of Working Group III to the Fourth Assessment Report of the Intergovernmental Panel on Climate Change* [B. Metz, O. R. Davidson, P. R. Bosch, R. Dave, and L. A. Meyer (eds)], Cambridge, UK and New York, NY, USA, Cambridge University Press.

Hallegatte, S., Green, C., Nicholls, R. J., and Corfee-Morlot, J., (2013). Future Flood Losses in Major Coastal Cities. *Nature Climate Change*, *3*(9), 802–806. doi:10.1038/nclimate1979

Hamin, E. M., and Gurran, N., (2009). Urban Form and Climate Change: Balancing Adaptation and Mitigation in the U.S. and Australia. *Habitat International*, *33*(3), 238–245. doi:10.1016/j.habitatint.2008.10.005

Hammarby-Sjostad, 2014. The Hammarby Model. [Online] Available: www.hammarbysjostad.se/, accessed September 12, 2014.

Harding, M., and McNamara, P., 2013. Rewarding Energy Engagement. [Online] Available: http://smartgridcc.org/wp-content/uploads/2012/01/Stanford_CES-Evaluation_Draft.pdf, accessed October 20, 2013.

Harlan, S. L., and Ruddell, D. M., (2011). Climate Change and Health in Cities: Impacts of Heat and Air Pollution and Potential Co-Benefits from Mitigation and Adaptation. *Current Opinion in Environmental Sustainability*, *3*(3), 126–134. doi:10.1016/j.cosust.2011.01.001

Harvey, L. D. D., 2006. *A Handbook on Low-Energy Buildings and District-Energy Systems: Fundamentals, Techniques and Examples*. London, Earthscan.

Havranek, T., Irsova, Z., and Janda, K., 2012. Demand for Gasoline is More Price Inelastic than Commonly Thought. *Energy Economics*, *34*(1), 201–207.

Hodson, M., Marvin, S., Robinson, B., and Swilling, M., (2012). Reshaping Urban Infrastructure. *Journal of Industrial Ecology*, *16*(6), 789–800. doi:10.1111/j.1530-9290.2012.00559.x

Hsu, D., (2014). How Much Information Disclosure of Building Energy Performance is Necessary? *Energy Policy*, *64*, 263–272.

Hydro Quebec, 2013. City of Montreal Joins the Electric Circuit and will Install 80 Electric Vehicle Charging Stations. [Online] Available: http://news.hydroquebec.com/en/press-releases/414/the-city-of-montreal-joins-the-electric-circuit-and-will-install-80-electric-vehicle-charging-stations/#.UmmmlvkQaqg, accessed October 24, 2013.

Ibrahim, N., Sugar, L., and Hoornweg, D., (2012). Greenhouse Gas Emissions from Cities: Comparison of International Inventory Frameworks. *Local Environment: The International Journal of Justice and Sustainability*, *17*(2), 223–241.

Jacobs, J., (1970). *The Economy of Cities* (p. 288). New York, Vintage.

Kennedy, C. A., Cuddihy, J., and Engel-Yan, J., 2007. The Changing Metabolism of Cities. *Journal of Industrial Ecology*, *11*(2), 43–59.

Kennedy, C., and Corfee-Morlot, J. (2012). *Mobilising Investment in Low Carbon, Climate Resilient Infrastructure* (No. 46). Retrieved from http://dx.doi.org/10.1787/5k8zm3gxxmnq-en

Kennedy, C., Demoullin, S., and Mohareb, E. (2012). Cities Reducing their Greenhouse Gas Emissions. *Energy Policy*, *49*, 774–777.

Kennedy, C. A., Ibrahim, N., and Hoornweg, D. (2014). Low-Carbon Infrastructure Strategies for Cities. *Nature Climate Change*, *4*, 343–346.

Laukkonen, J., Blanco, P. K., Lenhart, J., Keiner, M., Cavric, B., and Kinuthia-Njenga, C. (2009). Combining Climate Change Adaptation and Mitigation Measures at the Local Level. *Habitat International*, *33*(3), 287–292. doi:10.1016/j.habitatint.2008.10.003

Living PlanIT, 2014. PlanIT Valley – The Living Laboratory and Benchmark for Future Communities. [Online] Available: http://living-planit.com/design_wins.htm, accessed September 12, 2014.

Lu, Z., Noonan, D., Crittenden, J., Jeong, H., and Wang, D., 2013. Use of Impact Fees to Incentivize Low-Impact Development and Promote Compact Growth. *Environmental Science and Technology*, *47*, 10744–10752.

Lucon, O., Ürge-Vorsatz, D., Ahmed, A. Z., Akbari, H., et al. (2014) Chapter 9: Buildings. In: *Climate Change 2014: Mitigation of Climate Change. Contribution of Working Group III to the Fifth Assessment Report of the Intergovernmental Panel on Climate Change*. [O. Edenhofer, R. Pichs-Madruga, Y. Sokona, E. Farahani, et al. (eds)], Cambridge, UK, Cambridge University Press. [Online]. Available from: http://report.mitigation2014.org/.

Lutsey, N., and Sperling, D., (2008). America's Bottom-Up Climate Change Mitigation Policy. *Energy Policy*, *36*(2), 673–685. doi:10.1016/j.enpol.2007.10.018

Ma, Z., Cooper, P., Daly, D., and Ledo, L. (2012). Existing Building Retrofits: Methodology and State-of-the-Art. *Energy and Buildings*, *55*, 889–902. doi:10.1016/j.enbuild.2012.08.018

McCann, B., and Rynne, S., 2010. Complete Streets – Best Practices and Policy for Implementation. [Online] Available: www.smartgrowthamerica.org/documents/cs/resources/cs-bestpractices-chapter5.pdf, accessed September 12, 2014.

McKechnie, J., Colombo, S., Chen, J., Mabee, W., and MacLean, H., 2011. Forest Bioenergy or Forest Carbon? Assessing Trade-offs in GHG Mitigation with Wood-Based Fuels. *Environmental Science and Technology, 45*, 789–795.

Mckinsey Global Institute. (2011). *Urban World: Mapping the Economic Power of Cities.* [Online] Available: www.mckinsey.com/insights/urbanization/urban_world.

Merton Council, 2013. The Merton Rule. [Online] Available: www.merton.gov.uk/environment/planning/planningpolicy/mertonrule.htm, accessed October 18, 2013.

Metro Madrid, 2013. Sustainability and Corporate Policiers. [Online] Available: www.metromadrid.es/en/conocenos/responsabilidad_corporativa/Huella_carbono/index.html, accessed October 24, 2013.

Mohareb, E. A., and Kennedy, C. A., 2014. Scenarios of Technology Adoption Towards Low-Carbon Cities. *Energy Policy, 66*, 685–693.

Mohareb, E., Derrible, S., and Peiravian, F. (2015). Intersections of Jane Jacobs' Conditions for Diversity and Low-Carbon Urban Systems: A Look at Four Global Cities. *Journal of Urban Planning and Development,* 10.1061/(ASCE)UP.1943-5444.0000287, 05015004.

Myhre, G., Shindell, D., Bréon, F.-M., Collins, W., Fuglestvedt, J., Huang, J., . . . Zhang, H., (2013). Anthropogenic and Natural Radiative Forcing. In: *Climate Change 2013: The Physical Science Basis. Contribution of Working Group I to the Fifth Assessment Report of the Intergovernmental Panel on Climate Change* (pp. 659–740). Cambridge, UK, Cambridge University Press.

Naismith, K., 2014. Personal Communication with Karen Naismith, Director, Project Neutral. September 9, 2014.

National Research Council, 2009. Driving and the Built Environment: The Effects of Compact Development on Motorized Travel, Energy Use and CO_2 Emissions. The National Academies Press, Washington D.C., USA, 257 pp. Available at: www.nap.edu/catalog.php?record_id=12747.

New York City, 2013. Pedestrians – Public Plazas. [Online] Available: www.nyc.gov/html/dot/html/pedestrians/public-plazas.shtml, accessed October 24, 2013.

New York Times, 2013. Plugging in, Dutch Put Electric Cars to the Test. February 9, 2013. [Online] Available: www.nytimes.com/2013/02/10/world/europe/dutch-put-electric-cars-to-the-test.html?pagewanted=all&_r=0, accessed September 13, 2014.

Newman, P. W. G., and Kenworthy, J. R., 1989. Gasoline Consumption and Cities. *Journal of the American Planning Association, 55*(1), 24–37.

Nordic Council of Ministers, 2012. Nordic Solutions for Sustainable Cities. [Online] Available: http://bit.ly/1s1TxW9, accessed August 28, 2014.

NREL (National Renewable Energy Laboratory), 2009. Energy Efficiency Policy in the US. [Online] Available: www.nrel.gov/docs/fy10osti/46532.pdf, accessed October 20, 2013.

OEE (Office of Energy Efficiency), 2013. Comprehensive Energy Use Table – Residential. [Online] Available: http://oee.nrcan.gc.ca/corporate/statistics/neud/dpa/trends_res_ca.cfm, accessed October 28, 2013.

PACE Now, 2013. PACE – Financing Energy Efficiency. [Online] Available: http://pacenow.org/resources/all-programs/, Accessed Oct 14, 2013.

Pandis Iveroth, S., Vernay, A.-L., Mulder, K. F., and Brandt, N. (2013). Implications of Systems Integration at the Urban Level: The Case of Hammarby Sjöstad, Stockholm. *Journal of Cleaner Production, 48*, 220–231. doi:10.1016/j.jclepro.2012.09.012

Passive House International, 2014. Passive House – Legislation [Online] Available: www.passivehouse-international.org/index.php?page_id=176, accessed August 28, 2014.

Peiravian, F., Kermanshah, A., and Derrible, S. (2014b). Spatial Data Analysis of Complex Urban Systems. *2014 IEEE International Conference on Big Data,* October 27–30, Washington D.C., USA.

Peiravian, F., Derrible, S., and Ijaz, F., 2014a, Development and application of the Pedestrian Environment Index (PEI), *Journal of Transport Geography,* 39:73–84. See: http://www.sciencedirect.com/science/article/pii/S0966692314001343

Pikora, T., Giles-Corti, B., Bull, F., Jamrozik, K., and Donovan, R., (2003). Developing a Framework for Assessment of the Environmental Determinants of Walking and Cycling. *Social Science & Medicine (1982), 56*(8), 1693–703. Available: www.ncbi.nlm.nih.gov/pubmed/12639586.

PlaNYC, 2013. Inventory of NYC GHG Emissions, December 2013. [Online] Available: www.nyc.gov/html/planyc/downloads/pdf/publications/NYC_GHG_Inventory_2013.pdf, accessed September 19, 2014.

Poudenx, P., and Merida, W., 2007. Energy Demand and Greenhouse Gas Emissions from Urban Passenger Transportation Versus Availability of Renewable Energy: The Example of the Canadian Lower Fraser Valley. *Energy, 32*(1), 1–9.

Project Neutral, 2014. Project Neutral. [Online] Available: http://projectneutral.org/, accessed September 9, 2014.

Rockefeller Foundation, 2014. Five Lessons from the First Round of 100 Resilient Cities Applicants. [Online] Available: www.100resilientcities.org/blog/entry/five-lessons-from-the-first-round-of-100-resilient-cities-applicants, accessed September 13, 2014.

Rosenzweig, C., Solecki, W. D., Cox, J., Hodges, S., Parshall, L., Lynn, B., . . . Dunstan, F. (2009). Mitigating New York City's Heat Island: Integrating Stakeholder Perspectives and Scientific Evaluation. *Bulletin of the American Meteorological Society*, *90*(9), 1297–1312. doi:10.1175/2009BAMS 2308.1

Salon, 2012. Science Fiction No More: The Perfect City is Under Construction. [Online] Available: www.salon.com/2012/04/28/science_fiction_no_more_the_perfect_city_is_under_construction/, accessed September 12, 2014.

Seto, K. C., Dhakal, S., Bigio, A., Blanco, H., Delgado, G. C., Dewar, D., . . . Ramaswami, A. (2014). Chapter 12: Human Settlements, Infrastructure and Spatial Planning. In: *Climate Change 2014: Mitigation of Climate Change. Contribution of Working Group III to the Fifth Assessment Report of the Intergovernmental Panel on Climate Change.* [O. Edenhofer, R. Pichs-Madruga, Y. Sokona, E. Farahani, et al. (eds)], Cambridge, UK, Cambridge University Press. [Online]. Available from: http://report. mitigation2014.org/.

Shaheen, S. A., and Cohen, A. P., (2007). Growth in Worldwide Carsharing: An International Comparison. *Transportation Research Record: Journal of the Transportation Research Board*, 1992, 81–89.

Shwayri, S. T., (2013). A Model Korean Ubiquitous Eco-City? The Politics of Making Songdo. *Journal of Urban Technology*, *20*(1), 39–55. doi:10.1080/10630732.2012.735409

Smart Growth America, 2010. Complete Streets – Lessons from Copenhagen [Online] Available: www.smartgrowthamerica.org/2010/07/01/complete-streets-lessons-from-copenhagen/, accessed September 12, 2014.

Songdo, 2014. The City. [Online] Available: www.songdo.com/, accessed September 14, 2014.

Sorrell, S., (2007). *The Rebound Effect: An Assessment of the Evidence for Economy-Wide Energy Savings from Improved Energy Efficiency.* [Online]. Available from: http://aida.econ.yale.edu/~nordhaus/homepage/documents/UK_ReboundEffectReport.pdf.

State of California, 2010. Department of Justice – State and Local Government Green Building Ordinances in California. [Online] Available: http://ag.ca.gov/globalwarming/pdf/green_building.pdf, accessed October 20, 2013.

State of Green, n.d. More Electric Vehicles in Copenhagen. [Online] Available: https://stateofgreen.com/en/profiles/city-of-copenhagen/solutions/more-electric-vehicles-in-copenhagen, accessed April 23, 2015.

Stern, N., 2006. Stern Review on the Economics of Climate Change. [Online] Available: http://webarchive.nationalarchives.gov.uk/+/http:/www.hm-treasury.gov.uk/sternreview_index.htm, accessed October 30, 2013.

Stockholm Environment Institute, 2014. 2012 Seattle Community GHG Emissions Inventory. [Online] Available: http://sei-international.org/mediamanager/documents/Publications/Climate/Seattle-2012-GHG-inventory-report.pdf, accessed September 19, 2014.

Sugar, L., Kennedy, C., and Leman, E., 2012. Greenhouse Gas Emissions from Chinese Cities. *Journal of Industrial Ecology*, *16*(4), 552–563. doi:10.1111/j.1530-9290.2012.00481.x

Sugar, L., Kennedy, C., and Hoornweg, D., 2013. Synergies Between Climate Change Adaptation and Mitigation in Development: Case Studies of Amman, Jakarta, and Dar es Salaam. *International Journal of Climate Change Strategies and Management*, *5*(1), 95–111.

Tokyo Metropolitan Government, 2012. Tokyo Cap-and-Trade Program – Results of the First Fiscal Year of Operation. [Online] Available: www.kankyo.metro.tokyo.jp/en/climate/attachement/Result%20of%20the%20First%20FY%20of%20the%20Tokyo%20CT%20Program_final.pdf, accessed October 24, 2013.

Transue, M., and Felder, F.A., 2010. Comparison of Energy Efficiency Incentive Programs: Rebates and White Certificates. *Utilities Policy*, *18*(2), 103–111.

TRB (Transportation Research Board), 2013. A Guide for Implementing Bus on Shoulder Systems. [Online] Available: www.trb.org/TCRP/Blurbs/166878.aspx, accessed October 24, 2013.

TRCA (Toronto Region Conservation Authority), 2010. Getting to Carbon Neutral. [Online] Available: www.utoronto.ca/sig/CarbonNeutralReport_May52010_FINAL.pdf, accessed October 29, 2013.

United Nations, 2014. World Urbanization Prospects, the 2014 revision. [Online] Available: http://esa.un.org/unpd/wup/, accessed August 21, 2014.

Urban Redevelopment Authority (Singapore), 2014. The Planning Act – Master Plan Written Statement. [Online] Available: www.ura.gov.sg/uol/~/media/User%20Defined/URA%20Online/master-plan/master-plan-2014/Written-Statement-2014.ashx, accessed September 9, 2014.

USEPA. (2006). *Solid Waste Management and Greenhouse Gases – A Life Cycle Assessment of Emissions and Sinks* (p. 170).

Walsh, R. M., 2013. The Origins of Vancouverism: A Historical Inquiry into the Architecture and Urban Form of Vancouver, British Columbia. Doctoral Thesis, University of Michigan. Available: http://deepblue.lib.umich.edu/handle/2027.42/97802, accessed September 12, 2014.

Williams, B., 2008. Overcoming Barriers to Energy Efficiency for Rental Housing. [Online] Available: http://dspace.mit.edu/bitstream/handle/1721.1/44348/276307447.pdf?sequence=1, accessed October 14, 2013.

Wired, 2013. Masdar – The Shifting Goalposts of Abu Dhabi's Ambitious Eco-City. [Online] Available: www.wired.co.uk/magazine/archive/2013/12/features/reality-hits-masdar, accessed September 12, 2014.

Woodcock, J., Tainio, M., Cheshire, J., O'Brien, O., and Goodman, A. (2014). Health Effects of the London Bicycle Sharing System: Health Impact Modelling Study. *BMJ (Clinical Research Ed.), 348*(February), g425. doi:10.1136/bmj.g425

World Bank, 2013. Singapore – Data. [Online] Available: http://data.worldbank.org/country/singapore, accessed October 30, 2013.

World Finance, 2014. Could Songdo be the World's Smartest City? [Online] Available: www.worldfinance.com/inward-investment/could-songdo-be-the-worlds-smartest-city, accessed September 12, 2014.

Zhang, J., Wang, F.-Y., Wang, K., Lin, W.-H., Xu, X., and Chen, C., 2011. Data-Driven Intelligent Transportation Systems: A Survey. *IEEE Transactions on Intelligent Transportation Systems, 12*(4), 1624–1639.

Zimmerman, R., and Faris, C. (2011). Climate Change Mitigation and Adaptation in North American Cities. *Current Opinion in Environmental Sustainability, 3*(3), 181–187. doi:10.1016/j.cosust.2010.12.004

27

CLIMATE CHANGE MITIGATION IN MEDIUM-SIZED, LOW-INCOME CITIES

Shuaib Lwasa

Due to the increasing concentration of greenhouse gases (GHGs) in the atmosphere, worldwide mitigation efforts are imperative for stabilizing carbon levels and slowing further warming of the planet (IPCC, 2014). Cities in low-income countries are projected to grow faster and therefore likely to contribute to GHG emissions in the next decades. This raises concern about future emissions and the need to transform urban development pathways following low carbon development strategies. The second wave of urbanization is occurring mostly in Southeast Asia and Africa (Fischer-Kowalski and Swilling, 2011; Swilling, 2010), where it is projected that the urban population will grow to 750 million in Africa and 2.6 billion in Asia by 2030 (UN-Habitat, 2009). Although by 2011 the urban majority was living in small cities of less than 0.5 million inhabitants, medium-sized cities are experiencing the fastest growth (UN-Habitat, 2009). UN-Habitat defines medium-sized cities as those with populations ranging from 0.5 million to two million inhabitants, although there are debates around the typologies of cities based on population size (UN-Habitat, 2009; Lacour and Puissant, 2008).

Despite their disproportional and low contribution to global GHGs, low- and middle-income countries are considering mitigation on a voluntary basis. Human settlements in these countries are now viewed as having potential to leapfrog the development paths characteristic of industrialized countries in favor of pathways that are of low-carbon intensity, which would mitigate GHG emissions and climate change impacts. Such cities have begun to profile GHG emissions and benchmark targets through Nationally Appropriate Mitigation Actions (NAMAs), a voluntary response to the Kyoto Protocol (Hänsel et al., 2014; Olsen, 2013). This chapter synthesizes the climate change mitigation potential, opportunities and limitations in low-income, medium-sized cities and analyzes existing urban climate change mitigation actions, plans, emerging lessons and constraints in rapidly urbanizing and industrializing cities.

The case of medium-sized, low-income cities

The growth of medium-sized cities, for example, Lusaka, Kampala, Dar es Salam and Addis Ababa, in Africa is driven and characterized by increasing wealth due to sustained economic growth, though the gross domestic product (GDP) is disproportionate.

However, increasing wealth is also characterized by social differentiation and spatial fragmentation in urban development and contributing to a surge in consumerism and intense resource utilization (Swilling, 2010). Medium-sized cities are also characterized by peri-urban growth by the annexing of previously small towns into the surrounding metropolitan regions (Planning, 2009; Qadeer, 2004). In some countries these metropolitan regions have extended beyond administrative boundaries to create urban corridors (UN-Habitat, 2013).

Material flows within the expanded regions are coupled with intense energy use in medium-sized cities driven by increasing population (UN-Habitat, 2008). The city regions continue to dominate industrial production, as they are characterized by expanding transportation systems, which are fossil fuel dependent, and associated provision of service infrastructure (Hodson et al., 2012). Sources of energy among these cities differ widely, but transportation and industry heavily rely on fossil fuel sources, which significantly contribute to GHG emissions (Ramaswami et al., 2012). For example, an estimated 35 percent of energy in Accra is derived from thermal power generation, which relies on petroleum (Okafor, 2008). Moreover, history reveals that as cities grow and expand, they often draw on resources from the hinterlands as well as distal places, a process that produces significant emissions (Ramaswami et al., 2012). Although literature on distal connections and relationships in the context of resource utilization is still limited, emerging research indicates that cities, through consumption, drive emissions in rural hinterlands, some of which may be far away from where the consumption occurs (Busck et al., 2006; Seto et al., 2011).

In recent years, cities have begun conducting GHG inventories of the key sectors contributing to emissions including waste, industrial processes and production (Dodman, 2009). Transportation in small- and medium-sized cities contributes less emissions when compared to the other sectors (Romero-Lankao and Dodman, 2011), but as medium-sized cities in low-income countries grow, emissions are likely to increase with the expansion of transportation infrastructure (AfDB, 2013). The key sectors in energy use and emissions generation are likely to include transportation, energy generation, waste management and treatment and industrialization (Asomani-Boateng and Haight, 1999; Kareem and Lwasa, 2011). Given this likelihood of future growth, low-income cities have begun to formulate climate action plans that include mitigation actions. The climate action plans are part of city-wide strategies or projects that link into the market-based carbon credit mechanisms (Olsen, 2013; Revi et al., 2014). For example, Kumasi, Ghana, values its urban forestry ecosystem for conservation, while Tubigon, Philippines, uses the ecoBudget tool to valuate the consumption of natural resources as part of mitigation efforts (ICLEI and GIZ, 2012). This ecoBudget tool assesses the biomass intake by the city and is the basis for controlled extraction of biomass in the peri-urban zone of the city.

Planning for climate mitigation: approaches and tools

In order to assess the potential for mitigation of climate change in cities, emissions must be calculated and benchmarked for targeted reduction. Thus, low-income and medium-sized cities have embarked on processes that enable authorities to understand emissions sources, which are necessary for identifying emissions reduction strategies and actions. There are different frameworks and approaches that low-income cities of medium size have utilized to formulate climate plans and mitigation actions. These can be categorized into four general approaches with embedded tools: a. *climate risk or vulnerability framework*; b. *GHG emissions inventories*; c. *climate-specific, action-targeted planning* such as NAMAs; and, d. *spatial planning and scenario approaches* for low carbon development (Boswell et al., 2010; C40 et al., 2012; de Boer, 2007; Ian et al., 2002).

Climate risk and vulnerability analysis

Following the Intergovernmental Panel on Climate Change (IPCC) vulnerability framework of 2001, climate risk and vulnerability analysis has been adapted to different climate risks and local contexts. For example, Dhaka (Bangladesh), the Niger Delta and Sososgon (Philippines) have all used this framework to identify urban climate risks and vulnerability of different social groups and livelihoods (Bulkeley and Betsill, 2013; Castán Broto and Bulkeley, 2013). The literature shows a proliferation of urban vulnerability assessments with city-specific risks, some of which include heatwaves, cold waves, flooding, extreme rainfall, drought, sea level rise and salt water intrusion, which can damage infrastructure and stress livelihoods (Bhattacharya, 2010; Haines et al., 2006; Ingrid et al., 2007; Kates et al., 2012). Vulnerability assessments recommend that cities conduct GHG inventories and establish GHG baselines to prepare mitigation action plans. The UN-Habitat Cities and Climate Change Initiative (CCCI) has conducted and reported vulnerability assessments in over 40 pilot cities, all of which are low-income and medium-sized (Cities and Climate Change – Initial lessons from UN-Habitat, 2011).

GHG emissions inventories

Projected climate impacts on cities reinforce the necessity of mitigation, but this requires an understanding of the sources and estimation of GHG emissions in cities. The GHG inventory is a critical pre-requisite for policy support and response toward mitigation, but must be supported by strategy development and implementation. This has prompted ICLEI, the World Resource Institute, UN-Habitat and others to develop tools for estimating emissions at the community-scale (C40 et al., 2012), many of which have been piloted and implemented. These include the IPCC inventory methodology, which analyzes emissions from combustion, land use conversion and industrial processes calculated by sector (IPCC – Task Force on National Greenhouse Gas Inventories, 2006). This inventory framework has been developed such that it can be adapted in different countries and is generally appropriate for national level emissions estimation. But this protocol does not give appropriate estimates at the city-regional level because it largely utilizes the consumption-based estimation.

The Global Protocol for Community-Scale Greenhouse Gas Emissions (GPC), a co-developed tool by the World Resources Institute, Global Carbon Project, ICLEI and UN-Habitat, is a city-focused tool for GHG inventory and calculation (ICLEI, 2009) adapted from various frameworks based on internationally accepted tools for quantifying the GHG emissions attributable to cities and local regions. This tool provides a framework for calculation and attribution of GHG emissions to regions, cities and local areas, which could be rural settlements. The GPC framework calculates emissions from three levels. These are: *Scope 1*, GHG emissions that occur within the territorial boundary of the city or local region; *Scope 2*, indirect emissions that occur outside of the city boundary as a result of activities that occur within the city (limited to electricity generation); and, *Scope 3*, any other indirect emissions and embodied emissions that occur outside of the city boundary, as a result of activities of the city. This city-appropriate framework has been undertaken across regions and emerging literature provides evidence of adoption by some low-income cities (Lwasa, 2013a). The GPC inventory in a number of cities has yielded emissions reports which are starting to emerge in literature (Timmermans et al., 2013).

Climate-specific, action-targeted planning

Nationally Appropriate Mitigation Actions (NAMAs) have evolved for non–Annex I[1] countries to develop voluntary actions to reduce GHG emissions and recent negotiations have evolved plans for UNFCCC Annex II countries to do the same (Olsen, 2013; Den Elzen and Höhne, 2008). Of all prioritized sectors, transportation and energy sectors have a direct impact on urban systems. NAMAs have been deposited by several countries to the UNFCCC repository for bankable projects, which seek funding for proposal development or project support.

The main objective of NAMAs is to enable industrializing countries willing to cut the Business As Usual (BAU) emissions levels to do so without a legal obligation under the UNFCCC. The origin of NAMAs traces back to the 2008 Bali Action Plan that initiated enhanced mitigation action under the UNFCCC framework. Negotiations have yielded to the NAMA concept, which underpins diverse approaches different countries can utilize to identify, prepare for and implement development strategies targeting lower GHG emissions (Boswell et al., 2010). NAMAs have an inherent principle of supporting transformational development in countries that implement emissions reduction strategies to change long-term emissions trends, while pursuing sustainable development. They have the potential to leverage large-scale emissions reductions, which are designed for transparent measurement, reporting and verification.

Two of three broad NAMA types have been designed in some low-income, medium-sized cities; these are Policy NAMAs, which are actions at the policy/regulatory level aimed at sectoral transitions that mainstream emissions reductions with long-term (over 20 years) cumulative effects or savings of emissions; and Project NAMAs, which are specific investments in technologies to achieve energy efficiency for emissions savings or reduction targets. Such projects can include landfill gas flare, compost plants, energy switching technology and mass-transit systems. A Project NAMA also has a short-time horizon for its implementation, with clear verifiable indicators, termed Monitoring Reporting and Verifiable indicators (MRV). A Program NAMA is a longer-term undertaking, which would include an up-scaled project to cover city-wide or sector-wide emissions reductions or savings; these have not yet been addressed in low-income cities.

In all types of NAMAs, the key elements are the baseline emissions, which are calculated as CO_2e, and emissions targets over the planned period. In addition, each NAMA must have a monitoring plan with verifiable targets. For developing countries, which are historically low emitting and where current emissions are still very low by comparison with developed countries, the issue of co-benefits is critical in formulation of the NAMAs, since mitigation as per the Kyoto Protocol was initially for Annex I countries (Peters and Hertwich, 2008). Thus, NAMA concept notes and proposals that are re-deposited with the UNFCCC must indicate clear co-benefits for each of the NAMAs as well as sector linkages. NAMAs are still within the development stage, but are designed for implementation through economic incentives (or disincentives) and by changing standards (ICLEI, 2009). Although not yet implemented, some cities have designed actions, such as feed-in tariffs (FiTs) for grid-connected renewable energy, emissions trading schemes and building codes that set standards for energy efficiency.

Spatial planning and scenario approaches

Planning for urban climate action has also benefited from a range of approaches and tools for city development planning and modeling (Jenks et al., 2000; Lwasa and Kinuthia-Njenga, 2012;

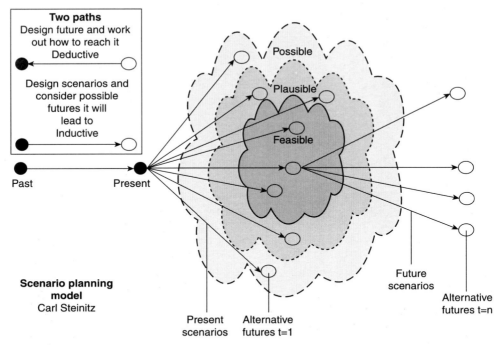

Figure 27.1 A generalized model of scenario planning

Source: Based upon a lecture given by Carl Steinitz, University of Amsterdam 2008 in Sliuzas et al. (2013).

Storch and Downes, 2011; Wegener, 2001). Spatial planning is a tool that can analyze demand and supply in current and future terms when coupled with scenarios in energy modeling (see Huang and Wang, Chapter 25 in this volume, for a discussion on urban spatial planning for climate change) and is an important strategy for reducing emissions (Crawford and French, 2008). Spatial planning provides a framework linking project-based mitigation actions, which are often policy-oriented responses at the city-region scale (Simon, 2012). Developing models and planning scenarios enable city managers to consider a range of options and uncertainties that influence future emissions from a city or city-region. This approach is essentially an inductive process that is based upon an empirical analysis of past changes and considers a range of possible, plausible or feasible futures as illustrated in Figure 27.1 (Sliuzas et al., 2013). This is in contrast to a typical design approach, which uses consumption projections to assess the future state of a city's emissions on the basis of which strategies are developed to reduce targeted emissions. This may or may not relate or be linked to spatial development plans (Wu et al., 2010). Such urban development plans have been designed for energy efficiency and utilization of renewable energy resources (Wende et al., 2010), decentralization of utilities and services as well as planning for mass-transit transportation systems, adjustment of building codes for energy efficiency and mixed use urban development patterns with mixed densities and ecological service areas.

Urban actions and strategies for climate change mitigation

City-wide strategies for energy efficiency are starting to emerge in medium-sized, low-income cities like Dar es Salaam (Holgate, 2007; Pierre and Gina, 2009). The responses, which are largely motivated by the growing carbon market, relate to the acceptance of bundled carbon reduction and/or savings projects. In some cases, the response has been influenced by experienced or anticipated climate change risks that have led to conducting emissions inventories and targeted emissions reduction (Wheeler, 2008). While some large cities (of more than two million people) have developed mitigation action plans, medium-sized cities have focused largely on adaptation to climate risks. Urban adaptation, as reported by Carmin et al. (2012) and Shi et al. (Chapter 23 in this volume), has multiple successes, but also constraints. In the case of low-income cities, the infrastructure deficit makes it difficult to adapt at the city-wide scale such that adaptations are mainly micro- to meso-scale interventions. The interactions and experiences at these scales have led to an evolution of strategies that couple adaptation with mitigation to harness the co-benefits and trade offs of mitigation that have a more adaptation-focused orientation (Bogner et al., 2008; Creutzig and He, 2009; Puppim de Oliveira et al., 2013; Seto et al., 2014). Thus, mitigation actions range from small-scale interventions focusing on energy efficiency, energy alternatives, Certified Emissions Reductions (CER) from verified projects on emissions reduction potential like municipal compost plants, landfill gas flare projects and emissions reduction strategies such as mass-transit systems.

Bus rapid transits (BRTs) are a popular urban development strategy that couples mitigation actions with enhancing urban transport services. BRT systems have been designed largely to enhance urban public transportation, but emissions reduction is an added element to the designs (Wilkinson et al., 2011). For example, the direct emissions reduction notwithstanding, the BRT in Dar es Salaam was designed to maximize co-benefits and is transforming urban form and coverage of public services to include the urban poor. BRTs, landfill gas flares, compost plants, education and awareness campaigns are strategies that have been designed as NAMAs or Clean Development Mechanism (CDM) projects to tap the Climate Fund through market-based schemes (Olsen, 2013; Silver, 2015).

Globally established market mechanisms such as emissions trading are becoming more accessible to low-income countries as a result of NAMAs and access to bi-lateral and multi-lateral funding, thereby supporting climate change mitigation actions. For example, low-income cities have responded to market-based carbon credit programs to design and implement several types of bundled carbon credit projects. Compost plants have been established to recycle organic waste with targets of reducing methane emissions that would otherwise come from decomposing wastes in landfills (Silver, 2015). The compost plants have been designed to capture organic wastes for compost at or before reaching landfills, which are sold on the market to farmers, as is the case in Mbale, Uganda. Although compost plants were preceded by feasibility studies, in some countries their economic success is in question, as market for the compost seems very limited. Urban and peri-urban agriculture—one of the targeted activities for compost utilization—still relies heavily on non-market valued compost or natural soil fertility (Lwasa et al., 2013). It is also reported that in some cities, no carbon credits have been generated to enable authorities to have access to the carbon credits transferable on the market. Besides, the value of the carbon credit per ton of CER has also plummeted in recent years from an average 12 Euros in 2011 per metric ton of CO_2e to an average 6 Euros in 2013 (although contradicting values have been reported); the falling price of carbon credits may discourage small-scale emissions reduction projects due to economic unfeasibility (Yoshizaki et al., 2013).

Many strategies have also taken a spatial dimension due to the nature of urban development and land use patterns. For example, cities of low-income countries are targeting a low-carbon future through spatial planning and coupling it with promotion of new technologies to create new urban forms (Crawford and French, 2008). This includes reconfiguring urban form to increase efficiency in mitigating GHG emissions and formulating development frameworks through spatial planning that integrate climate change considerations at local and regional scales (Baynes and Wiedmann, 2012; Calthorpe and Fulton, 2001; Wilson, 2009). Given that urban transportation is a major emitter of GHGs and appropriate urban form can lead to GHG emissions reductions by reducing travel distance (Grazi and van den Bergh, 2008), the spatial pattern of urban development must promote sustainable transportation that includes and is not limited to non-motorized transportation such as cycling and walking. It must also reduce the use of private vehicles, promoting carpooling and mass-transit systems. Compact and higher urban densities (lower surface-to-volume ratio) have a higher potential for energy efficiency than low-density buildings. These planning and policy responses must align with urban building codes in legally binding land use plans (Wende et al., 2010).

Lessons from emerging urban climate actions

Aligning mitigation and adaptation measures

There is a growing recognition that mitigation strategies alone are insufficient and that both mitigation and adaptation measures are needed to alleviate impacts from climate change (Biesbroek et al., 2009). Many urban climate action plans have elements of mitigation included (Wheeler, 2008), and technical and resource support from other cities and organizations have aided low-income cities in this process (C40 et al., 2012). However, instruments are needed to implement planning strategies for both mitigation and adaptation of climate change. Support for innovation has come from new research findings, as is the case with the exploding literature on green and white roofs and other new methodologies for tackling climate change risks and uncertainties in city plans (Walsh et al., 2011). Cities of low-income countries are also implementing both regulatory and market-oriented strategies that are integrated to reduce carbon emissions in several sectors. For example, urban and peri-urban agriculture is now being promoted to mitigate as well as adapt to climate change while building resilience (Lwasa et al., 2013). A key factor of the design and implementation of urban climate action plans in low-income countries is that planning and urban management institutions are missing or deficient in capacity and capability. Urban climate change mitigation efforts require institutional transformation. City-regional planning bodies and administrative institutions have been created for metropolitan regulations that manage urban growth or sprawl such as urban greenbelts, mixed land use and decentralization of urban infrastructure. But the success and lessons from these are yet to be realized.

Linking longer-term emissions reductions with development strategies

To accommodate future urbanization, cities must also conduct large-scale housing development for a growing population, and this warrants targets for reduction of GHG emissions over a longer time scale. Emerging and industrializing cities are attracting real estate development and infrastructure development to fill the deficit and thus will require additional materials as urbanization progresses. Similarly, buildings and transportation, a large proportion of which are in cities, consume the largest proportion of energy (Seto et al., 2014). The gradual reduction in use of non-renewable resources globally and increasing dependence on finite mineral resources has significant implications on future urban greening strategies in low-income cities (Newman et al., 2009). Medium-sized as well as small cities will continue to exert demand on the non-renewable resources especially in rapidly urbanizing regions (Calkins, 2009; Pulselli et al., 2008; Steinberger et al., 2010), however, several countries and cities have invested more recently in climate mitigation strategies in the transition to low carbon and green economies (Hodson and Marvin, 2009a; Suzuki et al., 2009). Outcomes of these actions and strategies for greening urban economies are yet to emerge since the process is in the early stages and on a relatively small-scale. However, a concern of industrializing countries is whether greening economies will lead to possible loss of jobs and reduced economic growth (European Commission, 2007; Fields, 2009; Getter et al., 2009; Hoornweg et al., 2011). For example, the materials sector, with a large proportion of activity in cities, has been rising between 1900 and 2005 by a factor of eight; the most significant growth comes from construction materials, ores/industrial minerals and biomass, hence the concern is that reductions in this sector or a shift to renewables will slow economic growth and reduce job opportunities (Krausmann et al., 2009; Steinberger et al., 2010).

Spatial planning: an integrated approach to mitigation

As mentioned above, spatial planning is utilized as a vehicle for configuration of land use activities for low carbon cities (De Zeeuw et al., 2011; Huang et al., 2011; Lwasa et al., 2013; Lwasa, 2013b). An integrated spatial planning approach that includes energy efficient transportation, mixed densities, reduced transport demand through shortened trips, decentralized urban sewerage infrastructure and utilization of ecosystem services are some of the policy mitigation strategies developed by cities in low-income countries (UN-Habitat, 2011). Some cities have formulated spatial planning frameworks that integrate waste-energy, CO_2 sequestration and transportation (UN-Habitat, 2011). Some of the strategies include specific programs to reduce energy consumption with incentives around alternative energy, waste-to-energy actions and institutional mitigation as role models to encourage city-wide adoption of mitigation actions. The other feature of climate action plans is the integration of urban sectors through instruments such as comprehensive spatial plans (Francis, 2011; Lwasa, 2013b; Onyenechere, 2010; Storch and Downes, 2011). The fruits of these plans are yet to be seen through monitoring and evaluation of the implementation. However, many still remain unimplemented due to financial, technical and institutional constraints. For most urban climate action plans, the investment requirements far exceed the capacity of municipal authorities' budgets and financial sources (Carraro and Massetti, 2012). Thus, urban climate actions in medium-sized and low-income cities extend from micro interventions at the household level, to meso-scale and macro-scale planning. There is increasing realization of the city's boundary as being fluid and extending into the hinterland, but without being further defined (Lwasa, 2014). This realization is spurring a city-regional level of planning actions that target emissions reduction from cities via regional transportation systems, waste-to-energy actions and decentralized utility infrastructure for energy efficiency.

Linking urban mitigation with economic growth: resource and impact decoupling

One of the lessons from the few urban mitigation actions is that urban jobs can be lost if mitigation strategies focus on drastic reduction in utilization of oil and its products, implying a significant reduction in GDP, or if industrial production processes and infrastructure are transformed to a greener system (Hodson and Marvin, 2010; Kamal-Chaoui and Roberts, 2009). However, in regions that are affected most such as Africa the loss is economic and predicted to be equivalent to 1.7 percent of Africa's GDP with a 1.5° C average increase in temperature (PACJA, 2009). Literature contends that with 0.5° C increase in temperature, costs are likely to double (Antrobus, 2011). Synergies between climate change mitigation and economic growth are important to counter possible loss of jobs and the slowing down of economic growth. But some examples of cities like Durban have combined ecosystem restoration with poverty reduction through the tree-preneur program that provides some income to urban poor who collect seedlings of indigenous trees that are planted by the municipality (Aylett, 2010).

The potential to link urban mitigation with economic growth and improved well-being, however, has not been harnessed adequately. Through the Resource Efficient Cities initiative of the United Nations Environment Programme (UNEP), programs have been developed first for climate change and economic growth as well as to decouple economic growth from environmental degradation (Arando Usón et al., 2011; Dhakal, 2010; Holden and Norland, 2005; Polimeni et al., 2009). The decoupling of resource utilization and economic growth at multiple scales is underway (Kamal-Chaoui and Roberts, 2009; OECD, 2010; Suzuki et al., 2009). Cities are targeting resource decoupling, which is less use of primary resources for the same economic output, and impact decoupling, which is reducing the relationship between economic output and negative environmental impacts (Seto et al., 2011). Some of the emerging lessons on resource decoupling are from systems transformation in production processes, reducing wastes and capturing resource value from residues and wastes of any nature. These can be organic but also inorganic wastes.

Impact decoupling in cities is mainly through payment for ecosystem services, as an engine of sustainable urban development (Hodson and Marvin, 2009b; Suzuki et al., 2009). Ecosystem services differ between and within cities due to diversity of ecologies. With a strong focus on rural environments for payments of ecosystem services, Reducing Emissions from Deforestation and Forest Degradation (REDD) and REDD+ programs have evolved in cities through programs for green rooftops, greening catchments and permaculture (Davies et al., 2011; Escobedo et al., 2006; Stoffberg et al., 2010). Evidence exists on green city carbon storage, carbon stocks in built materials of cities and potential along urban–rural gradients (Carter and Fowler, 2008; Fields, 2009). These programs and strategies have a potential to mitigate and increase resource efficiency, which can translate into positive economic growth while reducing future urban emissions (De Zeeuw et al., 2011). Investments in ecosystem services are required and these will include, but are not limited to, urban forestry, valued ecological systems management, catchment restoration and urban agriculture along the urban and peri-urban gradient across city regions. A number of cities have adapted a spatial development pattern with natural green patches interwoven with agricultural fields that provide ecosystem services of provisioning and regulation (Wilson, 2009). In these cities, programs have been developed with local communities to enhance effectiveness in mitigating climate change and increasing resilience. Ecosystem services that offer investment opportunities will be crucial to mitigate climate change and potentially increase economic growth.

Summary and pathways forward

Low-income cities vary in size from small- to medium-sized cities and mega cities. The bulk of future urbanization is likely to occur in small- and medium-sized cities. Some of these cities have conducted vulnerability assessments to determine climate risks while others have taken this further to conduct GHG inventories for targeting emissions reduction. The motivation for mitigation of climate change in cities is largely externally driven, though some internal city-driven initiatives also exist. Institutions, through the funding architecture of climate mitigation, have shaped the response, design and implementation of the urban mitigation actions. Two levels of mitigation actions include project-based, largely through the CDM, and policy-based, which leverage voluntary emissions reduction through NAMAs.

The frameworks for planning and implementing urban mitigation actions in low-income cities range from vulnerability assessments, GHG inventories, climate action plans and planning and scenario development for future emissions. This chapter raises three key issues, among others, as follows:

- Short-term and long-term mitigation strategies have been developed, some of which are undergoing implementation, while many are still at the project development and planning stage.
- Lessons from ongoing urban mitigation actions show that emissions reduction in low-income cities will largely be long-term as cities fill the infrastructure deficit, construct more buildings and provide better services for sustainable and inclusive cities.
- There is a dearth of literature on specific examples of actions for decoupling economic growth and environmental degradation and what the likely role will be of cities.

Future research therefore is needed on the follow key topics, among others:

- Scalable synergistic mitigation actions with co-benefits; there is need to quantify the co-benefits of mitigation at various scales.
- Sustainable cities are ideally conceptualized but there is limited knowledge on how sustainable cities can be tracked or evaluated towards progress, especially in relation to the Sustainable Development Goal on cities and human settlements.
- Ecology of cities with all possible urban green characteristics, as there is limited knowledge due to the context specific nature of ecology.

Note

1 Non-Annex I countries are Parties to the United Nations Framework Convention on Climate Change (UNFCCC) and mostly low-income developing countries classified according to the Kyoto Protocol as those without the obligation to reduce emissions (http://unfccc.int/parties_and_observers/items/2704.php).

References

AfDB (African Development Bank), 2013. An Integrated Approach to Infrastructure Provision in Africa (Economic Brief). African Development Bank.

Antrobus, D., 2011. Smart green cities: from modernization to resilience? *Urban Research & Practice* 4, 207–214. doi:10.1080/17535069.2011.579777

Arando Usón, A., Valero Capilla, A., Zabalza Bribián, I., and Scarpellini, Sabina, 2011. Energy efficiency in transport and mobility from an eco-efficiency viewpoint. *Energy* 36, 1916–1923. doi:10.1016/j.energy.2010.05.002

Asomani-Boateng, R., and Haight, M., 1999. Reusing organic solid waste in urban farming in African cities: a challenge for urban planners. *Third World Planning Review* 21, 411–428.

Aylett, A., 2010. Participatory planning, justice, and climate change in Durban, South Africa. *Environment and Planning A* 42, 99–115. doi:10.1068/a4274

Baynes, T.M., and Wiedmann, T., 2012. General approaches for assessing urban environmental sustainability. *Current Opinion in Environmental Sustainability* 4, 458–464. doi:10.1016/j.cosust.2012.09.003

Bhattacharya, S., 2010. Key vulnerabilities of human society in South Asia to climate change and adaptation issues and strategies, in: Mitra, A.P., and Sharma, C. (Eds.), *Global Environmental Changes in South Asia*. Springer Netherlands, Dordrecht, pp. 327–349.

Biesbroek, G.R., Swart, R.J., and van der Knaap, W.G.M., 2009. The mitigation–adaptation dichotomy and the role of spatial planning. *Habitat International* 33, 230–237. doi:10.1016/j.habitatint.2008.10.001

Bogner, J., Pipatti, R., Hashimoto, S., Diaz, C., Mareckova, K., Diaz, L., Kjeldsen, P., Monni, S., Faaij, A., Qingxian Gao, Tianzhu Zhang, Mohammed Abdelrafie Ahmed, Sutamihardja, R.T.M., and Gregory, R., 2008. Mitigation of global greenhouse gas emissions from waste: conclusions and strategies from the Intergovernmental Panel on Climate Change (IPCC) Fourth Assessment Report. Working Group III (Mitigation). Waste Management and Research 26, 11–32. doi:10.1177/0734242X07088433

Boswell, M.R., Greve, A.I., and Seale, T.L., 2010. An assessment of the link between greenhouse gas emissions inventories and climate action plans. *Journal of the American Planning Association* 76, 451–462. doi:10.1080/01944363.2010.503313

Bulkeley, H., and Betsill, M.M., 2013. Revisiting the urban politics of climate change. *Environmental Politics* 22, 136–154.

Busck, A.G., Kristensen, S.P., Præstholm, S., Reenberg, A., and Primdahl, J., 2006. Land system changes in the context of urbanisation: examples from the peri-urban area of Greater Copenhagen. *Geografisk Tidsskrift* 106, 21–34.

C40, ICLEI, and WRI, 2012. Global Protocol for Community-Scale Greenhouse gas Emissions (GPC) (No. Version 9).

Calkins, M., 2009. *Materials for Sustainable Sites*. John Wiley and Sons, New Jersey, US.

Calthorpe, P., and Fulton, W., 2001. *The Regional City: Planning for the End of Sprawl*. Island Press, Washington, D.C.

Carmin, J., Anguelovski, I., and Roberts, D., 2012. Urban climate adaptation in the global south: planning in an emerging policy domain. *Journal of Planning Education and Research* 32, 18–32.

Carraro, C., and Massetti, E., 2012. Beyond Copenhagen: a realistic climate policy in a fragmented world. *Climatic Change* 110, 523–542.

Carter, T., and Fowler, L., 2008. Establishing green roof infrastructure through environmental policy instruments. *Environmental Management* 42, 151–164. doi:10.1007/s00267-008-9095-5

Castán Broto, V., and Bulkeley, H., 2013. A survey of urban climate change experiments in 100 cities. *Global Environmental Change* 23, 92–102.

Cities and Climate Change – Initial lessons from UN-Habitat, 2011. UN Habitat, Nairobi.

Crawford, J., and French, W., 2008. A low-carbon future: spatial planning's role in enhancing technological innovation in the built environment. *Energy Policy* 36, 4575–4579. doi:10.1016/j.enpol.2008.09.008

Creutzig, F., and He, D., 2009. Climate change mitigation and co-benefits of feasible transport demand policies in Beijing. *Transportation Research Part D: Transport and Environment* 14, 120–131. doi:10.1016/j.trd.2008.11.007

Davies, Z.G., Edmondson, J.L., Heinemeyer, A., Leake, J.R., and Gaston, K.J., 2011. Mapping an urban ecosystem service: quantifying above-ground carbon storage at a city-wide scale. *Journal of Applied Ecology* 48, 1125–1134. doi:10.1111/j.1365-2664.2011.02021.x

de Boer, J., 2007. Framing climate change and spatial planning: how risk communication can be improved. *Water Science & Technology* 56, 71.

De Zeeuw, H., Van Veenhuizen, R., and Dubbeling, M., 2011. The role of urban agriculture in building resilient cities in developing countries. *The Journal of Agricultural Science* 149, 153–163.

den Elzen, M., and Höhne, N., 2008. Reductions of greenhouse gas emissions in Annex I and non-Annex I countries for meeting concentration stabilisation targets. *Climatic Change* 91, 249–274.

Dhakal, S., 2010. GHG emissions from urbanization and opportunities for urban carbon mitigation. *Current Opinion in Environmental Sustainability* 2, 277–283.

Dodman, D., 2009. Blaming cities for climate change? An analysis of urban greenhouse gas emissions inventories. *Environment and Urbanization* 21, 185–201. doi:10.1177/0956247809103016

Escobedo, F.J., Nowak, D.J., Wagner, J.E., De la Maza, C.L., Rodríguez, M., Crane, D.E., and Hernández, J., 2006. The socioeconomics and management of Santiago de Chile's public urban forests. *Urban Forestry & Urban Greening* 4, 105–114. doi:10.1016/j.ufug.2005.12.002

European Commission, 2007. Green Paper Towards a New Culture for Urban Mobility.

Fields, B., 2009. From green dots to greenways: planning in the age of climate change in post-Katrina New Orleans. *Journal of Urban Design* 14, 325–344. doi:10.1080/13574800903056515

Climate change mitigation in low-income cities

Fischer-Kowalski, M., and Swilling, M., 2011. Decoupling: natural resource use and environmental impacts from economic growth. United Nations Environment Programme.

Francis, O., 2011. Urban Planning: The Way Forward for Africa's Cities? Think Africa Press [WWW Document]. URL http://thinkafricapress.com/population-matters/impact-demographic-change-resources-and-urban-planning-africa

Getter, K.L., Rowe, D.B., Robertson, G.P., Cregg, B.M., and Andresen, J.A., 2009. Carbon sequestration potential of extensive green roofs. *Environ. Sci. Technol.* 43, 7564–7570. doi:10.1021/es901539x

Grazi, F., and van den Bergh, J.C.J.M., 2008. Spatial organization, transport, and climate change: comparing instruments of spatial planning and policy. *Ecological Economics* 67, 630–639. doi:10.1016/j.ecolecon.2008.01.014

Haines, A., Kovats, R., Campbell-Lendrum, D., and Corvalan, C., 2006. Climate change and human health: impacts, vulnerability, and mitigation. *The Lancet* 367, 2101–2109.

Hänsel, G., Röser, F., Hoehne, N., and Tilburg, X. van, Cameron, L., 2014. Annual status report on nationally appropriate mitigation actions (NAMAs). *Policy Studies* 2013, 2012.

Hodson, M., and Marvin, S., 2010. Can cities shape socio-technical transitions and how would we know if they were? *Research Policy* 39, 477–485. doi:10.1016/j.respol.2010.01.020

Hodson, M., and Marvin, S., 2009a. Cities mediating technological transitions: understanding visions, intermediation and consequences. *Technology Analysis & Strategic Management* 21, 515–534. doi:10.1080/09537320902819213

Hodson, M., and Marvin, S., 2009b. "Urban ecological security": a new urban paradigm? *International Journal of Urban and Regional Research* 33, 193–215. doi:10.1111/j.1468-2427.2009.00832.x

Hodson, M., Marvin, S., Robinson, B., and Swilling, M., 2012. Reshaping urban infrastructure. *Journal of Industrial Ecology* 16, 789–800.

Holden, E., and Norland, I., 2005. Three challenges for the compact city as a sustainable urban form: household consumption of energy and transport in eight residential areas in the greater Oslo Region. *Urban Studies* 42, 2145–2166. doi:10.1080/00420980500332064

Holgate, C., 2007. Factors and actors in climate change mitigation: a tale of two South African cities. *Local Environment* 12, 471–484. doi:10.1080/13549830701656994

Hoornweg, D., Sugar, L., and Gómez, C.L.T., 2011. Cities and greenhouse gas emissions: moving forward. *Environment and Urbanization* 23, 207–227. doi:10.1177/0956247810392270

Huang, S.-L., Chen, Y.-H., Kuo, F.-Y., and Wang, S.-H., 2011. Emergy-based evaluation of peri-urban ecosystem services. *Ecological Complexity* 8, 38–50. doi:10.1016/j.ecocom.2010.12.002

Ian, B., Saleemul, H., Bo, L., Olga, P., and Emma, L., 2002. From impacts assessment to adaptation priorities: the shaping of adaptation policy. *Climate Policy* 2, 145–159.

ICLEI, 2009. International Local Government GHG Emission Analysis Protocol (IEAP): Version 1.0. ICLEI – Local Governments for Sustainability, Bonn, Germany.

ICLEI, and GIZ, 2012. Green Urban Economy Concpetual Basis and Courses of Action, Discussion Paper. ed.

Ingrid, C., Coleen, V., and Zarina, P., 2007. Institutional dynamics and climate change adaptation in South Africa. *Mitigation and Adaptation Strategies for Global Change* 12, 1323–1339.

IPCC (Intergovernmental Panel on Climate Change) (Ed.), 2014. Climate Change 2013 – The Physical Science Basis: Working Group I Contribution to the Fifth Assessment Report of the Intergovernmental Panel on Climate Change. Cambridge University Press, Cambridge.

IPCC – Task Force on National Greenhouse Gas Inventories [WWW Document], 2006. URL www.ipcc-nggip.iges.or.jp/public/2006gl/vol1.html (accessed April 15, 2013).

Jenks, M., Jenks, M., and Burgess, R., 2000. *Compact Cities: Sustainable Urban Forms for Developing Countries*. Taylor and Francis, London and New York.

Kamal-Chaoui, L., and Roberts, A., 2009. Competitive Cities and Climate Change (OECD Regional Development Working Papers No. 2). OECD Publishing, Paris.

Kareem, B., and Lwasa, S., 2011. From dependency to interdependencies: the emergence of a socially rooted but commercial waste sector in Kampala City, Uganda. *African Journal of Environmental Science and Technology* 5, 136–142.

Kates, R.W., Travis, W.R., and Wilbanks, T.J., 2012. Transformational adaptation when incremental adaptations to climate change are insufficient. *Proceedings of the National Academy of Sciences* 109, 7156–7161.

Krausmann, F., Gingrich, S., Eisenmenger, N., Erb, K.-H., Haberl, H., and Fischer-Kowalski, M., 2009. Growth in global materials use, GDP and population during the 20th century. *Ecological Economics* 68, 2696–2705. doi:10.1016/j.ecolecon.2009.05.007

Lacour, C., and Puissant, S., 2008. Groupement de Recherches Economiques et Sociales. Cahier n 07.

Lwasa, S., 2013a. Greenhouse Gas Emissions Inventory for Kampala City and Metropolitan Region. Makerere University. URL http://mirror.unhabitat.org/downloads/docs/12220_1_595178.pdf.

Lwasa, S., 2013b. Planning innovation for better urban communities in sub-Saharan Africa: the education challenge and potential responses. *Town and Regional Planning* 60, 38–48.

Lwasa, S., 2014. Managing African urbanization in the context of environmental change. *INTERdisciplina* 2, 263–280.

Lwasa, S., and Kinuthia-Njenga, C., 2012. Reappraising Urban Planning and Urban Sustainability in East Africa, *Urban Development*, Dr. Serafeim Polyzos (Ed.), (pp. 3–22). InTech, URL www.intechopen.com/books/urban-development/reappraising-urban-planning-and-urban-sustainability-in-east-africa.

Lwasa, S., Mugagga, F., Wahab, B., Simon, D., Connors, J., and Griffith, C., 2013. Urban and peri-urban agriculture and forestry: transcending poverty alleviation to climate change mitigation and adaptation. *Urban Climate*. doi:10.1016/j.uclim.2013.10.007

Newman, P., Beatley, T., and Boyer, H., 2009. *Resilient Cities: Respondig to Peak Oil and Climate Change*. Island Press, Washington, D.C.

OECD, 2010. Contribution of cities to a green growth model, in: *Cities and Climate Change. Organisation for Economic Co-operation and Development Publishing*, pp. 145–168.

Okafor, E.E., 2008. Development crisis of power supply and implications for industrial sector in Nigeria. *Journal of Tribes and Tribals* 6, 83–92.

Olsen, K., 2013. NAMAs for sustainable development. *Mitigation Talks* 3, 4.

Onyenechere, E.C., 2010. Climate change and spatial planning concerns in Nigeria: remedial measures for more effective response. *Journal of Human Ecology* 32, 137–148.

PACJA. (2009). The Economic Cost of Climate Change in Africa. Pan African Climate Justice Alliance, Nairobi, Kenya.

Peters, G., and Hertwich, E., 2008. CO2 embodied in international trade with implications for global climate policy. *Environmental Science & Technology* 42, 1401–1407. doi:10.1021/es072023k

Pierre, M., and Gina, Z., 2009. Developing a Municipal Adaptation Plan (MAP) for climate change: the city of Cape Town, in: Bicknell, J., Dodman, D., and Satterthwaite, D. (Eds.), *Adapting Cities to Climate Change*. Earthscan, London, p. 397.

Planning, B., 2009. Urban revival: will Africa soon become an urban continent? *Urban Revival*.

Polimeni, J., Mayumi, K., Giampietro, M., and Alcott, B., 2009. *The Myth of Resource Efficiency: The Jevons Paradox*. Earthscan, London; Sterling, VA.

Pulselli, R., Simoncini, E., Ridolfi, R., and Bastianoni, S., 2008. Specific emergy of cement and concrete: an energy-based appraisal of building materials and their transport. *Ecological Indicators* 8, 647–656. doi:10.1016/j.ecolind.2007.10.001

Puppim de Oliveira, J.A., Doll, C.N.H., Kurniawan, T.A., Geng, Y., Kapshe, M., and Huisingh, D., 2013. Promoting win–win situations in climate change mitigation, local environmental quality and development in Asian cities through co-benefits. *Journal of Cleaner Production* 58, 1–6. doi:10.1016/j.jclepro.2013.08.011

Qadeer, M.A., 2004. Urbanization by implosion. *Habitat International* 28, 1–12. doi:10.1016/S0197-3975(02)00069-3

Ramaswami, A., Chavez, A., and Chertow, M., 2012. Carbon footprinting of cities and implications for analysis of urban material and energy flows. *Journal of Industrial Ecology* 16, 783–785. doi:10.1111/j.1530-9290.2012.00569.x

Revi, A., Satterthwaite, D.E., Aragón-Durand, F., Corfee-Morlot, J., Kiunsi, R.B.R., Pelling, M., Roberts, D.C., and Solecki, W., 2014. Urban areas, in: Field, C.B., Barros, V.R., Dokken, D.J., Mach, K.J., Mastrandrea, M.D., Bilir, T.E., Chatterjee, M., Ebi, K.L., Estrada, Y.O., Genova, R.C., Girma, B., Kissel, E.S., Levy, A.N., MacCracken, S., Mastrandrea, P.R., and White, L.L. (Eds.), *Climate Change 2014: Impacts, Adaptation, and Vulnerability. Part A: Global and Sectoral Aspects. Contribution of Working Group II to the Fifth Assessment Report of the Intergovernmental Panel of Climate Change*. Cambridge University Press, Cambridge, UK and New York, NY, USA, pp. 535–612.

Romero-Lankao, P., and Dodman, D., 2011. Cities in transition: transforming urban centers from hotbeds of GHG emissions and vulnerability to seedbeds of sustainability and resilience: Introduction and Editorial overview. *Current Opinion in Environmental Sustainability* 3, 113–120.

Climate change mitigation in low-income cities

Seto, K.C., Fragkias, M., Güneralp, B., and Reilly, M.K., 2011. A meta-analysis of global urban land expansion. *PLoS ONE* 6, e23777.

Seto K.C., Dhakal, S., Bigio, A., Blanco, H., Delgado, G.C., Dewar, D., Huang, L., Inaba, A., Kansal, A., Lwasa, S., McMahon, J.E., Müller, D.B., Murakami, J., Nagendra, H., and Ramaswami, A., 2014. Human settlements, infrastructure and spatial planning, in: Edenhofer, O., Pichs-Madruga, R., Sokona, Y., Farahani, E., Kadner, S., Seyboth, K., Adler, A., Baum, I., Brunner, S., Eickemeier, P., Kriemann, B., Savolainen, J., Schlömer, S., von Stechow, C., Zwickel, T., and Minx, J.C. (Eds.), *Climate Change 2014: Mitigation of Climate Change. Contribution of Working Group III to the Fifth Assessment Report of the Intergovernmental Panel on Climate Change*. Cambridge University Press, Cambridge, UK and New York, NY, USA.

Silver, J., 2015. The potentials of carbon markets for infrastructure investment in sub-Saharan urban Africa. *Current Opinion in Environmental Sustainability* 13, 25–31.

Simon, D., 2012. Climate and environmental change and the potential for greening African cities. *Local Economy* 28, 203–217. doi:10.1177/0269094212463674

Sliuzas, R., Flacke, J., and Jetten, V., (2013). Modelling urbanization and flooding in Kampala, Uganda. University of Twente. Retreived from: http://n-aerus.net/web/sat/workshops/2013/PDF/N-AERUS14_sliuzas%20et%20al%20Final_FINAL.pdf

Steinberger, J.K., Krausmann, F., and Eisenmenger, N., 2010. Global patterns of materials use: a socioeconomic and geophysical analysis. *Ecological Economics* 69, 1148–1158. doi:10.1016/j.ecolecon.2009.12.009

Stoffberg, G.H., van Rooyen, M.W., van der Linde, M.J., and Groeneveld, H.T., 2010. Carbon sequestration estimates of indigenous street trees in the City of Tshwane, South Africa. *Urban Forestry & Urban Greening* 9, 9–14. doi:10.1016/j.ufug.2009.09.004

Storch, H., and Downes, N.K., 2011. A scenario-based approach to assess Ho Chi Minh City's urban development strategies against the impact of climate change. *Cities* 28, 517–526.

Suzuki, H., Moffatt, S., Yabuki, N., and Dastur, A., 2009. *Eco2 Cities: Ecological Cities as Economic Cities*. World Bank, Washington, D.C.

Swilling, M., 2010. Africa 2050–Growth, Resource Productivity and Decoupling. Presented at the Policy brief for the 7th meeting of the International Panel for Sustainable Resource Management of the United Nations Environment Programme (UNEP).

Timmermans, R., van der Gon, H.D., Kuenen, J., Segers, A., Honoré, C., Perrussel, O., Builtjes, P., and Schaap, M., 2013. Quantification of the urban air pollution increment and its dependency on the use of down-scaled and bottom-up city emission inventories. *Urban Climate* 6, 44–62.

UN-Habitat, 2008. The State of African Cities. United Nations Human Settlements Programme, Nairobi.

UN-Habitat, 2009. Planning Sustainable Cities—Global Report on Human Settlements 2009. Earthscan, London.

UN-Habitat, 2011. Cities and Climate Chage: Global Report on Human Settlements, 2011. Earthscan/UN-Habitat.

UN-Habitat, 2013. State of the World's Cities 2012/2013: Prosperity of Cities. New York City, NY: Routledge for and on behalf of the United Nations Human Settlements Programme (UN-Habitat).

Walsh, C.L., Dawson, R.J., Hall, J.W., Barr, S.L., Batty, M., Bristow, A.L., Carney, S., Dagoumas, A.S., Ford, A.C., Harpham, C., Tight, M.R., Watters, H., and Zanni, A.M., 2011. Assessment of climate change mitigation and adaptation in cities. *Proceedings of the ICE – Urban Design and Planning* 164, 75–84. doi:10.1680/udap.2011.164.2.75

Wegener, M., 2001. New spatial planning models. *JAG* 3, 14.

Wende, W., Huelsmann, W., Marty, M., Penn-Bressel, G., and Bobylev, N., 2010. Climate protection and compact urban structures in spatial planning and local construction plans in Germany. *Land Use Policy* 27, 864–868. doi:10.1016/j.landusepol.2009.11.005

Wheeler, S.M., 2008. State and municipal climate change plans: the first generation. *Journal of the American Planning Association* 74, 481–496. doi:10.1080/01944360802377973

Wilkinson, P., Golub, A., Behrens, R., Ferro, P.S., and Schalekamp, H., 2011. Transformation of urban public transport systems in the global south. *International Handbook of Urban Policy: Issues in the Developing World* 3, 146.

Wilson, E., 2009. Multiple scales for environmental intervention: spatial planning and the environment under New Labour. *Planning Practice and Research* 24, 119–138. doi:10.1080/02697450902742205

Shuaib Lwasa

Wu, W., Simpson, A., and Maier, H., 2010. Accounting for greenhouse gas emissions in multiobjective genetic algorithm optimization of water distribution systems. *Journal of Water Resources Planning and Management* 136, 146–155. doi:10.1061/(ASCE)WR.1943-5452.0000020

Yoshizaki, T., Shirai, Y., Hassan, M.A., Baharuddin, A.S., Raja Abdullah, N.M., Sulaiman, A., and Busu, Z., 2013. Improved economic viability of integrated biogas energy and compost production for sustainable palm oil mill management. *Journal of Cleaner Production* 44, 1–7.

28

URBAN AND PERI-URBAN AGRICULTURE

Cultivating urban climate resilience

John P. Connors, Corrie A. Griffith, Camille L. Nolasco,
Bolanle Wahab and Frank Mugagga

One of the primary concerns with climate change is the impact it has on global food security. Changes in the global climate will have far reaching impacts on food systems, and changes in food systems invariably impact climate change (Liverman and Kapadia, 2010). While crop and climate models provide detailed forecasts of possible changes in food production, less is known about the interplay of urbanization with climate change and food security. As the global population continues to urbanize—70 percent of the global population is expected to live in urban areas by 2050 (OECD, 2012)—issues of poverty and food insecurity will also increasingly become urban phenomena.

Many cities worldwide are facing high rates of population growth that are outpacing their economic growth and ability to adequately provide access to clean water, employment, social services and reliable food supplies for millions of urban residents (UNDESA, 2014). Climate change adds yet another layer to these challenges; it exacerbates conditions that already threaten food security, e.g. changes in environmental conditions that could undermine food production or disrupt distribution infrastructure (UNCTAD, 2011). Food security and climate change vulnerability are interrelated and display feedbacks (IPCC, 2014), which necessitates that response strategies address both of these issues in tandem. Although urbanization is often associated with an economic shift away from agricultural production, promotion of urban and peri-urban agricultural (UPA) practices including horticulture, livestock, aquaculture and forestry promise to provide benefits by mitigating risk and enhancing adaptive capacity. UPA has the potential to support livelihoods, create jobs and provide coping mechanisms for shocks to food systems (Lwasa et al., 2009; Swalheim and Dodman, 2008), thus offering a mechanism to address the dual challenges of enhancing food security and reducing vulnerability to climate change. In addition to these immediate benefits, ecosystem services enhancement may both support livelihoods and mitigate climate change impacts.

In light of the convergent impacts of urbanization and climate change on food security, this chapter considers the potential of UPA to build the resilience of urban food systems and to reduce the vulnerability of urban populations to climate change. The following section begins with the ways in which the changing climate, urbanization processes and food systems interact and implicate global food security. This is followed by a discussion of numerous potential pathways through which UPA could contribute to urban climate resilience including some of the challenges to its application. This discussion is particularly concerned with cities in industrializing countries, where UPA is being increasingly recognized as an important strategy to respond to a number of these key challenges. The chapter concludes with some recommendations for future research and practice including planning and policy needs for more climate resilient cities.

Agricultural systems, urbanization and a changing climate: implications for food security

The challenges facing food systems will accelerate in the coming decades, as the demand for food is expected to increase twofold within the next 25–50 years. This will take place primarily in industrializing countries (Von Braun, 2009) where urbanization and growing incomes are also associated with shifts in dietary preferences. Projected rapid and unplanned urbanization will also continue to drive land use change in peri-urban areas and rural hinterlands including many areas dedicated to agricultural production (Seto et al., 2011). The dual processes will interact to create both challenges and opportunities for addressing global food security and climate change.

Despite definitional variations and multiple scales to which it can be applied (national, community or household/individual), 'food security' can be understood as a situation "when all people, at all times, have physical and economic access to sufficient, safe, and nutritious food to meet their dietary needs and food preferences for an active and healthy life" (FAO et al., 2013). Food security is a function of the access, availability and utilization of food, and thus shaped by various factors such as economic resources, stability of agricultural systems, patterns of consumption and cultural norms (Liverman and Kapadia, 2010). Globally, progress has been made towards reducing food insecurity, but this progress has varied regionally, and there are still millions of people who suffer from hunger and malnutrition (Porter et al., 2014; Ingram et al., 2010). In particular, progress has been slow in parts of Western Asia, Northern Africa and sub-Saharan Africa, with approximately one in eight people worldwide still suffering from chronic hunger (FAO et al., 2013).

Urban land use change and food production

The relationship between food systems and urbanization is complex (see Murray et al., Chapter 2 in this volume), with many interrelated factors that have implications for food security. Increasing urban populations have obvious impacts on centers of demand for agricultural products, but they also commonly have greater consumption per capita as urban populations transition to more meat-heavy diets. Changes in diet are, in part, associated with economic growth, but this does not imply that all urban residents fare better in terms of hunger and malnutrition. In fact, studies in low-income countries show that food-energy deficiencies in urban areas are the same or higher than in rural areas in 66 percent of cases; in large proportions of urban areas in sub-Saharan Africa, deficiencies are above 40 percent and in a few cases are above 60 percent (Ruel and Garrett, 2003; Ahmed et al., 2007). Understanding the interactions between urbanization, food production and adaptation of both rural and urban residents to food system changes requires a deeper look into complex rural–urban linkages, the actors involved as well as consideration of a changing and increasingly uncertain climate.

At the same time, the movement of people to urban areas, particularly in areas of lower density construction, often results in the conversion of agricultural lands to other uses, putting pressure on food production systems (Seto et al., 2011). In some cases, however, there has been little effect or even counter effects to agricultural land loss as a result of urbanization (see Deng, Chapter 3 in this volume for this discussion). Urban encroachment is most apparent in peri-urban areas, which are neither solely urban nor rural, but a combination of both in terms of market activities, land use and other socio-economic processes (Simon et al., 2006). In industrializing countries, these resource-rich zones play an important role in supporting the economies and lives of urban residents but are often under great pressure that contributes to ecosystem and biodiversity degradation, decreased water availability and quality, and environmental pollution (McGranahan et al., 2004).

Global climate change and food system impacts on urban populations

The pressures of urbanization on productive lands are compounded by the impacts of changing environmental conditions under climate change. Globally, mean temperatures are expected to increase by 1°C to 3°C, contributing to increased yields in temperate regions and yield losses in many tropical regions (IPCC, 2014). By the end of the 21st century, the tropics and the subtropics, which include the majority of industrializing countries and the future location of fast-paced urbanization, are highly likely to experience major changes in mean growing season temperatures. Models predict these will exceed the most extreme temperatures of the past century (Battisti and Naylor, 2009). In these regions, crop losses are estimated to be between 2.5 percent and 16 percent for every 1°C change in seasonal temperature (Lobell et al., 2008), with some of the most severe impacts in the already stressed areas of Africa and Latin America (Jones and Thornton, 2003). Estimating economic losses under climate change is difficult, with often highly variable results, but historical evidence suggests that climate has already contributed to five billion USD in losses to wheat, barley and maize between 1981–2002 (Lobell and Field, 2007). Although there was a global upward trend in crop production during this period, climate change significantly offset these trends in many countries (Lobell et al., 2011). These shifting patterns will affect yields on productive lands, but also alter the range of suitable lands for various crops, potentially bringing some new areas into production, while others become infeasible (Fischer et al., 2006). In general, tropical regions, particularly in sub-Saharan Africa, are likely to experience the greatest losses in production, while Northern latitudes will experience longer growing seasons that may increase production (Porter et al., 2014).

The effects of global climate change including irregularities in rainfall patterns as well as extreme weather events such as thunderstorms, heatwaves, heavy winds and floods devastate farmlands and can lead to food shortages due to crop failure. Although the impacts of climate change will vary around the world, changes in key production centers can send reverberations through the global markets, increasing the prices of many staple foods and affecting food security of distant but nested and teleconnected communities (Adger et al., 2009) including urban populations. The increasingly interconnected vulnerabilities of global food systems were revealed during the food price crisis of 2007–2008. During the period from January 2007 to June 2008, the FAO food price index increased by 63 percent and prices of rice increased by 166 percent (FAO, 2008). One of the driving factors behind the surge in food prices was environmental shocks, specifically floods in Canada and prolonged drought in Australia, two countries that together account for 35–40 percent of global cereal exports (FAO, 2008). These climatic factors, combined with increasing oil prices, changing diets, protectionist trade policies and broader financial crises, contributed to the surge (Ingram et al., 2010). These rapid and dramatic changes in food prices illustrate the evolving food security risks that result from globally connected food systems, linking food outcomes to economic and environmental changes in distant regions (see Cohen and Garrett, Chapter 20 in this volume). Urban populations that do not produce much of their food supplies are highly susceptible to these changes in prices. Peri-urban populations will experience similar shocks, but some producers may also benefit from greater prices for agricultural products.

Changes in productivity pose one set of challenges, but the distribution, processing and storage of food are also susceptible to global environmental change. Increasing extreme events and phenological changes may interfere with current harvest schedules, contributing to losses and increasing prices for urban populations. Once food is produced in rural agricultural areas it travels to urban areas through transport infrastructure networks to be stored and distributed through formal and informal systems before it finally reaches consumers. Transport infrastructure (roads and railways), which is often lacking or under-maintained in low-income cities, is highly vulnerable to extreme or increased temperatures, often resulting in short-term disruptions (Chen et al., 2010). For instance, floods damaged available link roads to some peri-urban and rural communities in some parts of Nigeria during the 2011 and 2012 floods making transportation of food products from rural to urban communities difficult. Coastal seaports, key nodes in the supply chain, are particularly vulnerable to sea level rise and extreme events, and coastal inundation could result from storm surges and river floods (IPCC SREX, 2012). In Vietnam, for example, past severe weather conditions have contributed to the loss of 80 percent of rice crops (Hodges et al., 2011). Warm and wet spells also affect food safety, reducing storage life and increasing the risk of food-borne illness (James and James, 2010). Incidence of food poisoning in some regions is already closely tied to seasonal variation in temperatures (Bentham and Langford, 2001) such as throughout Europe where above-average temperatures are associated with increased incidence of *salmonellosis* (Kovats et al., 2004). Impoverished and crowded areas with sanitation challenges are particularly vulnerable to food contamination, as increased incidence of flooding brings wastewater in contact with crops and irrigation waters. Throughout the production chain, from harvest to storage to transport, foodstuffs will be increasingly susceptible to these risks.

Faced with increasing climate risks and changes in human vulnerability to these risks due to urbanization, strategies are needed to enhance the resilience of urban populations. As urbanization transforms the environment and food systems, there is a need to assess these new risks and to develop strategies to mitigate and adapt to these changes. UPA has emerged as one possible strategy that can help to reduce the vulnerability of urban populations to climate change through the combined benefits for livelihoods, food security and ecosystem services.

Opportunities for climate resilience through urban and peri-urban agriculture

Climate resilience refers to a city's ability to respond to, resist and recover from changing climate conditions (Nelson et al., 2007). The greatest threats to urban populations may come from vulnerability and exposure to extreme events such as flooding, but the aforementioned impacts on global food supplies will also have major impacts on urban and peri-urban areas. The urban poor spend a major portion of their income on food and are particularly sensitive to changes in global food supplies and prices. Climate change threatens to increase the frequency and extremity of fluctuations in global food supplies, but a number of pathways exist for UPA to improve the adaptive capacity of city dwellers and the ecological environment in which they live (de Zeeuw et al., 2011). The co-benefits of UPA practices are increasingly recognized in urban resilience and poverty literature, revealing that it not only contributes to poverty alleviation and livelihood enhancement, but can also enhance economic and ecological

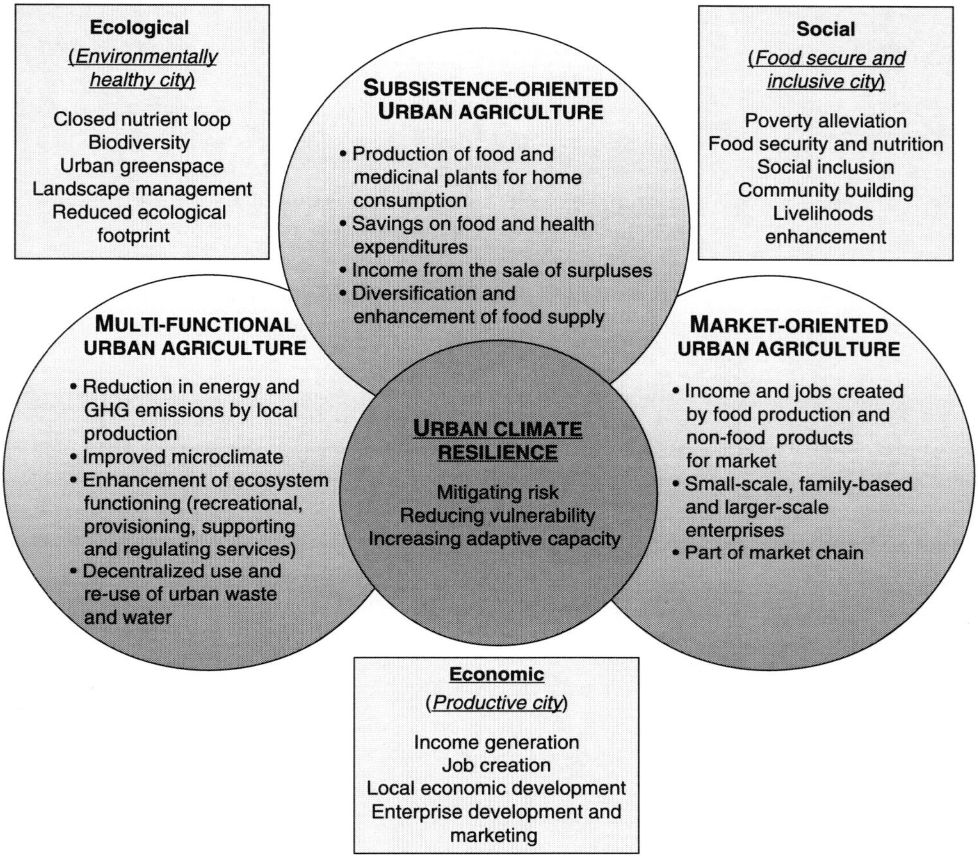

Urban and peri-urban agriculture

Ecological
(*Environmentally healthy city*)

Closed nutrient loop
Biodiversity
Urban greenspace
Landscape management
Reduced ecological footprint

SUBSISTENCE-ORIENTED URBAN AGRICULTURE
- Production of food and medicinal plants for home consumption
- Savings on food and health expenditures
- Income from the sale of surpluses
- Diversification and enhancement of food supply

Social
(*Food secure and inclusive city*)

Poverty alleviation
Food security and nutrition
Social inclusion
Community building
Livelihoods enhancement

MULTI-FUNCTIONAL URBAN AGRICULTURE
- Reduction in energy and GHG emissions by local production
- Improved microclimate
- Enhancement of ecosystem functioning (recreational, provisioning, supporting and regulating services)
- Decentralized use and re-use of urban waste and water

URBAN CLIMATE RESILIENCE

Mitigating risk
Reducing vulnerability
Increasing adaptive capacity

MARKET-ORIENTED URBAN AGRICULTURE
- Income and jobs created by food production and non-food products for market
- Small-scale, family-based and larger-scale enterprises
- Part of market chain

Economic
(*Productive city*)

Income generation
Job creation
Local economic development
Enterprise development and marketing

Figure 28.1 Main contributions from UPA across social, economic and ecological dimensions to support climate resilient cities

Source: Adapted from de Zeeuw et al. (2011).

resilience in light of climate change risks (Lee-Smith, 2010). Figure 28.1 depicts the multiple pathways through which UPA practices can enhance sustainability across the economic, social and environmental spheres, and how their convergence can help build climate resilience.

Livelihood enhancement

Urban and peri-urban agriculture can benefit climate resilience by supporting the livelihoods of urban and peri-urban residents, increasing the income of the most vulnerable populations and providing a reliable food supply. A 'livelihood' can be defined as "comprising the capabilities, assets and activities required for a means of living" (Chambers and Conway, 1988). The vulnerability of urban households is largely shaped by their particular livelihood strategies and the impacts of climate change on the different activities that support household incomes and food consumption. In the context of UPA, food production within the city supplements and diversifies the supply when shortages occur in neighboring rural areas. UPA not only enhances income, but also directly provides food, fuel and fiber. Thus, it has the potential to act as a buffer against periodic shocks and stresses affecting supply and price within the global food system. This is particularly important for the livelihoods of urban poor who are disproportionately affected by climate change.

UPA already plays a key role in the livelihoods of urban dwellers (Mkwambisi et al., 2011; Prain et al., 2010), and there is evidence that it provides a major portion of the income of vulnerable populations, i.e. the urban poor (Danso et al., 2002). High-value crops such as vegetables, tomatoes, poultry products, fruits and herbal crops supplement urban household incomes and represent a growing food supply enterprise (FAO, 1997). For example, as a result of a project led by the UN-Habitat Sustainable Cities Program in Cagaya de Oro, the Philippines, which has enabled land cultivation for urban vegetable production, 25 percent of vegetables produced by urban residents are consumed, while 68 percent are sold and the remaining 7 percent given away to neighbors or friends. Ultimately, this has not only enhanced food security, but is also a source of income generation for the urban poor (Holmer and Drescher, 2006; Hamilton et al., 2013). In Nepal, 52 percent of urban households generate income from urban farming, and in 13 percent of households income from urban agriculture exceeds 30 percent of the total (Zezza and Tasciotti, 2010). Estimates from Dar es Salaam, Tanzania, indicate some forms of livestock and dairy keeping along with vegetable production produce earnings 30 percent greater than the average salary (Mlozi et al, 2014). In Nairobi, during the 1990s, agriculture provided the highest self-employment earnings among small-scale enterprises and the third highest earnings in all of urban Kenya (House et al., 1993). The World Bank (2007) contends that the trend will continue, i.e. that the number of sub-Saharan Africans engaged in urban farming as a primary or secondary income source is not expected to decline (De Bon et al., 2010). Throughout sub-Saharan Africa, approximately 29 million households engage in urban crop production—other regional estimates include Asia (182 million), Latin America (39 million) and Eastern Europe (15 million) (Hamilton et al., 2013; Zezza and Tasciotti, 2010).

Social capital

In addition to noted positive psychological effects for farmers who have direct contact with plants, soil and the 'natural' environment (Kaplan, 1973; Brogan and James, 1980; Shoemaker, 2002), there are also associated social benefits of UPA practice to communities. Activities that enable urban farmers to interact with other farmers and stakeholders can facilitate social connections and strengthen ties within the community and associated social capacities (Brown and Jameton, 2000). Such social capital is believed to support adaptive capacity, particularly through helping social collectives and individuals respond to climate change (Pelling and High, 2005). Cooperative and collaborative endeavors to support production systems can enhance social capital, which plays a vital role in recovering from environmental shocks (Adger, 2010). The multiple benefits of urban agriculture (i.e. income generation, food security and health, environmental enhancement, etc.) provide an opportunity for vulnerable and disadvantaged groups (e.g. elderly, youth, women, migrants, immigrants, refugees, HIV-infected and the poor) to make strong positive social contributions and engage in commercial activities (de Zeeuw et al., 2011). In Lusaka, Zambia, for example, urban agriculture ranks highest over other types of informal activities in providing jobs for those who do not have skills (e.g. women, teenagers and retirees) required of the formal sector (Bishwapriya, 1985 in Smit et al., 2001). UPA programs have been initiated in multiple cities, such as in Addis Ababa, Ethiopia and Nakuru, Kenya, targeting the poor and HIV-affected to increase societal nutrition and health and income generation. In these cases, many thousands of women, orphans and vulnerable children have benefited through access to healthy food, new economic opportunities, stronger support networks and social linkages, and secure access to land and water (Karanja et al., 2010; DAI, 2006).

Risk reduction and mitigation potential

Transformations of urban food systems can reduce vulnerability to climate change by directly enhancing the livelihoods of urban residents through food supply and income generation, but they may also reduce exposure to risks by mitigating the effects of climate change. Through the maintenance of green open spaces, reduced impervious surfaces and greater vegetation cover, UPA and urban forestry can mitigate an array of environmental hazards resulting from climate change and compounding pressures of urbanization processes (Lwasa et al., 2014). These benefits are discussed in greater detail below.

Urban greening

Many of the most vulnerable to climate change live in slum and squatter settlements on steep hillsides, in poorly drained areas, flood plains or in low-lying coastal zones (UNFPA, 2007). Evidence suggests that UPA and agroforestry can aid in keeping those areas (e.g. coastal areas, flood plains or slopes) vulnerable to flooding or landslides free from construction, thus maintaining their natural functions (enhancing water storage and infiltration; reducing run off) and resulting in fewer floods and reduced impacts of prolonged and heavy rainfall (Dubbeling et al., 2009). In Dar es Salaam, Tanzania, areas that are temporarily flooded are utilized for UPA and can also commonly be found in valleys, plains and hillsides of cities within Kenya and Nigeria (Foeken and Owuor, 2008; Egbuna, 2009). Bogotá (Colombia) and São Paulo (Brazil) benefit immensely from the improved soil structure and porosity through efforts expanding agricultural practices in the cities, thereby avoiding landslides and flooding (EUKN, 2008; Barbizan and Giseke, 2011). As part of urban greening strategies, UPA can positively enhance micro-climate effects by reducing heat-related vulnerability through cooling. Carbon sequestration is also a dual benefit. Government-supported efforts are already underway in Australia to understand the extent and benefits of UPA practices in the context of resilience to climate-related heat stress through comparative case study/city research (Burton et al., 2013).

Micro-climate and weather extremes

Increasing vegetation in urban areas also helps to reduce temperatures locally, minimizing the effects of the urban heat island (UHI) and heatwave impacts, potentially reducing energy use for cooling (Tidball and Krasny, 2007). Increased evapotranspiration from crops and trees provides a cooling effect, while the uptake of water can help to regulate hydrological flows. Tree crops provide the additional benefit of shade, reducing land surface temperatures and facilitating rapid cooling at night time. These can also reduce heat- and wind-related vulnerabilities in urban settings. Increased tree coverage further regulates local climate by providing buffers to high wind velocities associated with intense storms. It can offer an economical approach to risk reduction by reducing the frequency of extreme events and also mitigating wind damage to infrastructure and the natural environment. A higher frequency of windstorms in Ibadan, Nigeria, for example, has been linked to the dramatic reduction in vegetative cover (approximately a 50 percent reduction in 2003), due to urbanization transforming the landscape (Adelekan, 2010). Agroforestry—agricultural systems that are primarily tree-based crops or that integrate trees into the crop and animal production—is recognized as a strategy for climate change mitigation (Nair et al., 2009), but the aforementioned examples show that it may also benefit climate adaptation in urban areas.

Ecological footprint

The greatest benefits of UPA are likely to be at local and regional scales, but the mainstreaming of such practices could also have implications for the mitigation of global climate change. According to Intergovernmental Panel on Climate Change (IPCC) 5th Assessment Report calculations (Smith et al., 2014), the global economic mitigation potential for agriculture (supply-side) is estimated at 7.2 to 11 Gt of CO_2-equivalent emissions per year by 2030. Demand-side measures, such as changes in diet and reductions of losses in the food supply chain, have a significant, but uncertain mitigation potential; estimates vary from 0.76–8.6 Gt of CO_2-equivalent emissions per year by 2050 (Smith et al., 2014). For certain crops, substituting imports with local production may reduce emissions associated with energy use for food processing and storage facilities by shortening the distance of food transport from the rural to the urban setting (de Zeeuw et al., 2011; Brown et al., 2012). In the case of Kesbewa, Sri Lanka, reducing the transport of vegetables by increasing local production, coupled with improvements in organic waste reuse, reduced greenhouse gas (GHG) emissions by an estimated 4,133 tons/year (University of Colombo, 2013). In the city of Almere, Netherlands, findings reveal that energy savings associated with replacing 20 percent of imported food with local production (within a 20km radius of the city) as well as promoting fossil fuel reduction in production, processing and cooling by renewable energy sources, GHG emissions reductions (27.1 Kt CO_2-equivalent) would equal the carbon sequestration of about 1,360 ha of forest or the emissions of 2,000 Dutch households (Jansma et al., 2012).

Water and nutrient recycling

UPA can help to close nutrient cycles through various practices including reduced transport and local composting practices, increased storage and infiltration of water and recycling of water and wastewater. The ability for both rural and urban farmers alike to adapt to changing climatic conditions through improving system inputs will be critical to ensure necessary yields and, thus save limited and/or threatened resources while contributing to sustained food security (Drechsel and Kunze, 2001; Schertenleib et al., 2004). For example, the composting of organic materials not only eliminates wastes in urban landfills that emit methane gas, but provides nutrient-rich material, which reduces reliance on chemical fertilizers, thereby aiding in protection of water quality, conservation and enhancement of soils and other ecosystem services, and offers alternatives for livestock feed (Deelstra and Girardet, 2000).

An example of a city's potential to 'close the nutrient loop' is the Sustainable Ibadan Project (SIP) in Nigeria. The SIP was established in 1994 and trains members of the community on the proper practice of waste usage. The project not only supplies recycled waste for urban compost, but also provides economic benefits for the community via income generation (Sridhar and Adeoye, 2003; Wahab et al., 2010). Studies on UPA's potential reveal that in Kumasi and Accra, Ghana, 230,000 to 250,000 and 255,000 to 366,000 tonnes of organic waste, respectively, are collected and made available for composting. The nutrient content of this waste in Accra alone is estimated at 3,500 to 5,300 tonnes per year of nitrogen, 1,700 to 2,600 tonnes per year of phosphorus and 760 to 1,100 tonnes per year of potassium;

Urban and peri-urban agriculture

these amounts could easily cover the entire nutrient demand of urban farming (Drechsel et al., 2010).

The recycling and reuse of wastewater in urban farming reduces the demand for fresh water supplies in competition with other urban uses and helps urban dwellers adapt to uncertain rainfall patterns as a result of climate change. It also reduces pollution caused by the discharge of wastewater into rivers and streams and leakage into aquifers (Buechler et al., 2006). Wastewater can also be recycled for irrigation of horticultural crops as well as forested plantations that offer the aforementioned co-benefits. A study by Thebo et al. (2014) offers a first estimate of the global extent of urban and peri-urban irrigated and rainfed croplands. They estimate up to 67 Mha total area of urban croplands (24 Mha irrigated and 44 Mha rainfed) and up to 456 Mha of total croplands when peri-urban areas within 20 km of urban extents are included (of which 130 Mha are irrigated and 327 Mha are rainfed croplands). Their findings reveal that rainfed agriculture is common in sub-Saharan Africa and more temperate, water-abundant regions such as Canada and Europe, whereas in more densely populated and/or water scarce regions, such as North Africa and South and East Asia, irrigated urban croplands are more prevalent (Thebo et al., 2014). These findings have important implications for water resource management, but necessitate a deeper look at the underlying dynamics and interactions of UPA with urbanization and global environmental change processes across multiple scales.

UPA in practice

City strategies incorporating UPA for coping with or buffering climate change impacts can be found worldwide and are increasingly found in both peer-reviewed and grey literature (Lwasa et al., 2014). Box 28.1 provides a few examples of how such practices support urban food security and livelihoods with benefits for populations to deal with transformations in global food supplies and prices, while providing additional income—all of which can reduce the sensitivity of urban populations to the effects of climate change.

Box 28.1   City strategies promoting urban food production and climate change risk reduction

- Durban (South Africa)—Productive green rooftops are encouraged as part of a strategy for storm water management, biodiversity and food production. In order to adapt to a changing climate, the municipality is also testing less water-intensive crops to replace maize while promoting community reforestation and management.
- Kesbewa (Sri Lanka)—The integration of urban agriculture into the Kesbewa Urban Development Plan (preserving low-lying lands for urban agriculture and designing such areas based on the results of pilot projects) aims to increase city livability and livelihoods through reduced vulnerability to climate change including flood risk and enhanced local food production.
- Nairobi (Kenya)—Regional climate models suggest that urban areas will become drier, placing even greater stress on available water. In order to combat environmental degradation and to promote sustainability, tree planting has been adopted to reduce erosion, enhance green cover and replenish the decreasing water table. The trees provide urban residents (both human and livestock) with food and medicine as well as wood fuel and construction material.

- Bobo Dioulasso (Burkina Faso)—In order to promote a more sustainable urban development model, the municipality agreed to preserve and protect the border zones between the city and its forests, and to preserve the greenways as areas with multifunctional and productive land uses. The main objective of the project is to demonstrate the contribution of greenways in reducing climate change impacts, while improving the living conditions of the neighboring population. Such increased diversification of food and income sources helps to increase the resilience of poor households.
- Bangkok (Thailand)—Urban agriculture is used as an alternative strategy for dealing with the urban food agenda during floods. Although inner-city farming is of small-scale and only contributes to a minor extent to an alternative food system, in many respects it is able to play an important role, specifically for the urban poor and marginalized groups. City farming has gained wide institutional support since 2010.
- Rosario (Argentina)—The Rosario Municipality has designated 400+ ha in and at the outskirts of the city for expansion of UPA in the near future. The main goals of the program are to contribute to food security and income generation, and are integrated into a detailed study on climate change impacts and adaptation.

Source: RUAF (2014).

Limitations and challenges of UPA

Health risks

Some of the approaches to UPA that promise to improve nutrient and water cycling, however, also raise another set of environmental and public health concerns, which cannot be overlooked (Dubbeling and Merzthal, 2006). As in large farms systems, inappropriate and excessive applications of agrochemicals like fertilizers and pesticides in UPA and leaching of animal excreta may pollute or contaminate ground water and fresh water bodies with residues such as high concentration of nitrates, resulting in contaminated water for human consumption (Smit et al, 2001; FAO and ETC, 2010). Farmers in urban and peri-urban areas often make use of low-cost and easily accessible resources such as organic municipal waste, sewage and market refuse as production inputs, which often contributes to the contamination of produce (Keraita and Drechsel, 2004; Amoah et al., 2005; Stasinos and Zabetakis, 2013). Similarly, livestock and poultry manures may transport pathogens (Drechsel et al., 2006). Although wastewater, excreta and urban organic waste constitute an accessible source of nutrients necessary for crops (such as nitrogen, phosphorus and potassium), the presence of bacteria, viruses or parasites present a great risk for the health of farmers and consumers (Lee-Smith and Prain, 2006).

Mostly in lower-income countries, even when water sources are readily available, many cities lack proper wastewater treatment, and irrigation waters may come from contaminated rivers or canals (Buechler et al., 2006; van Rooijen et al., 2010). The use of watering cans, usually filled from polluted ponds and streams, are common irrigation methods in West Africa but contribute to crop contamination (Drechsel et al., 2008). In Juiz de Fora, Brazil, there are cases of crop irrigation with ground water that is contaminated by a vicinal polluted river stream (Nolasco, 2009). Contaminated water may also infiltrate agricultural lands during floods, as is the case in Kampala, Uganda, where vegetables grown in urban wetland soils often are exposed to polluted floodwaters from Lake Victoria (Mbabazi et al., 2010).

Urban and peri-urban agriculture

Table 28.1 Common sources of urban and peri-urban agriculture contamination

Source	Contaminant(s)
Burning wastes	PAHs[1], dioxins
Coal ash	Molybdenum, sulfur
Commercial/industrial site use	PAHs[1], petroleum products, solvents, lead, other trace elements
Paint (prior to 1978)	Lead, barium, mercury
Pesticides	Arsenic, lead, mercury (historical use), chlordane and other chlorinated pesticides
High traffic areas	Lead, zinc, PAHs[1]
Treated lumber (prior to 2002)	Arsenic, chromium, copper
Petroleum spills/emissions	PAHs[1], benzene, toluene, xylene, lead
Manure	Coppers, zinc
Sewage sludge	Cadmium, copper, zinc, lead, POPs[2]
Plumbing fixtures	Lead

Source: Adapted from Heinegg et al. (2000) and Crozier et al. (2012).

[1]PAHs=polycyclic aromatic hydrocarbons
[2]POPs=persistent organic pollutants, a generalized grouping of organic compounds resistant to environmental degradation

In addition to the threats from polluted water sources, UPA must also contend with soil contamination (see Table 28.1). Sites often include marginal lands located near busy roadways and railroads or in vacant lots or brownfields, which are often contaminated with heavy metals such as cadmium (Cd) and lead (Pb) (Makokha et al., 2008). Heavy metals are most common in post-industrial and landfill sites, but can also accumulate from vehicle emissions, particularly in leafy vegetables grown in close proximity to main roads (Sharma et al., 2007). Agriculture in these contaminated lands poses risks for both the producers, who are exposed directly to the contaminants, and for the consumer, who may ingest these chemicals (Alam et al., 2003). In this manner, green leafy vegetables, which are more economically viable and provide more nutrients, become a potential source of heavy metal exposure in urban areas (Nabulo et al., 2006; 2010).

The spread of diseases is another large concern with UPA. Animal raising (poultry, chickens, goats, pigs, etc.) in higher densities associated with urban areas, and slaughtering and consumption of meat in inadequate sanitary conditions may increase the transmission of diseases from domestic animals to people (zoonosis) among urban dwellers (Lock and de Zeeuw, 2001). Vectors of diseases (such as mosquitos, rodents and other animals) may find ideal locations for their propagation in areas of UPA that are not clean and well groomed. Fallow cropland and water puddles, for example, may contribute to the spread of diseases (e.g. pathogenic organisms), especially in warmer climates (van Veenhuizen, 2012). UPA has also been attributed to increasing the incidence of malaria due to standing water from irrigation and concrete surface fish ponds, creating a habitat for mosquito larvae (Asenso-Okyere et al., 2009).

Policy and institutional challenges

Cities are fertile territory for intervention and planning of innovative strategies that aim to eradicate urban hunger and reduce poverty. As discussed, urban and peri-urban agriculture has numerous benefits, but to achieve them, UPA development must address its biggest barriers at the policy and institutional levels. Treatment of UPA by governments, other administrative officials and decision-makers is quite varied across countries and at multiple governance scales. In some countries, UPA activities are considered illegal; in others, it might be ignored, regulated or incentivized. In Brazil, for example, UPA has been incorporated as an instrument for promoting social development and food security under the Hunger Zero Programme at the federal level. At the municipal level, however, since it is largely an informal activity, it receives little support or investment from local governments (Nolasco, 2009). When it finds support from one department (like social development), it does not find the same support from health, agriculture or urban planning departments. This creates conflicts for the long-term success of UPA activities.

Another large concern surrounds appropriate space to engage in UPA activities. In densely urbanized areas space for food production is challenging; this is particularly the case in slum or squatter settlements, where vulnerable populations that could highly benefit from such activities reside. Due to other competing land uses associated with urbanization, production often takes place in open public spaces, flood plains, roadway medians and other accessible patches of land. This issue is further complicated by its lack of incorporation in city planning and conflicts of land tenure (Cofie et al., 2003; Schmidt, 2012). Without the recognition as a formal land use, both urban environmental and human health continue to be harmed (de Zeeuw et al., 2011).

If managed properly, the benefits of UPA have the potential to outweigh the risks. Urban farmers' education and government investment and control, for example, could help reduce the aforementioned health concerns (Cissé et al., 2005). There are several types of instruments available for supporting UPA development: legal (laws and regulations); economic (taxes, incentives and subsidies); communicative/educative (information that can foster change or policy effectiveness); and, urban planning and design (zoning for UPA protection areas, and incorporating gardening areas into urban and housing design). Cities that have moved towards integrating UPA into urban planning and zoning, laws and public policies, and sanitation regulations while promoting good farm practices have benefited from what UPA can best bring to cities. Rosario (Argentina), Havana (Cuba), Mumbai (India), Belo Horizonte (Brazil), Mexico City (Mexico), Quito (Ecuador), Lima (Peru), Lubumbashi (Democratic Republic of Congo) and Vancouver (Canada) are good examples of where UPA has been integrated into various city-to-national strategies and plans, and thus contribute to urban resilience building (Bakker et al., 2000; de Zeeuw and Dubbeling, 2008).

Recommendations for a future with urban farming

Data and research

The magnitude of UPA participation and contribution within the greater global food system remains uncertain due to a lack of data. Local governments and departments usually do not have data and information on UPA production. For example, in Brazil, the general data on food production do not include UPA practices (with rare exceptions), as the activity is often ignored by the agencies responsible (Nolasco, 2009). The informality of this activity often results in a lack of rigorous and consistent data on the current and real state of urban and peri-urban agriculture in most urban areas worldwide (Zezza and Tasciotti, 2010), thus creating difficulty for the much needed comprehensive surveys and inventories for regional and global assessments. Furthermore, many quantitative studies that exist in the literature,

Urban and peri-urban agriculture

which offer insight into contributions of urban agriculture (e.g. participation, employment and income, etc.), are outdated. Hence, from a research standpoint, further investigation over greater temporal scales, within and across cities, particularly given the diversity of social and environmental characteristics among urban areas, is necessary to better ascertain co-benefits.

In addition to these general research needs, specific research is needed to better understand the ecosystem services derived from UPA across the urban–rural gradients within and across cities. In particular, there is a need for further research to understand aggregate carbon sequestration benefits of UPA for climate change mitigation as well as quantitative assessments of organic waste recycling benefits for nutrient cycling and impacts on agricultural and environmental sustainability in the rural–urban continuum (Allison et al., 1998; GTZ/GFA, 1999; Drechsel and Kunze, 2001). The lack of knowledge surrounding these activities including consistent indicators and monitoring, prevents the incorporation of UPA activities in plans for climate change adaptation and mitigation. In order to support the construction of strong public policies that integrate UPA into urban development plans, food systems and urban ecology research agendas must incorporate UPA and produce information that can support decision-making towards more resilient cities.

Policy and institutional uptake

The multidimensionality or functionality of UPA (i.e. the diversity of actors involved, socio-ecological and political dynamics, objectives, geographies and scales of practice) allows it to quickly adapt to changing conditions and demands (van Veenhuizen and Danso, 2007). In this sense, the establishment of a multi-stakeholder platform for investigating the current state of UPA activities as well as the gaps and strengths of these activities within individual municipalities is crucial to the construction of policies and institutional arrangements that can enhance the benefits of UPA while reducing risks. The following must be part of integrative and multi-sectoral policies at both local, state and national levels: the incorporation of sewage and wastewater use guidelines; health risks reduction and monitoring measures; efficient models of collecting and composting solid organic waste; access to water suitable for irrigation and cleaning/washing produce and crops; land tenure for cultivation; and zoning for UPA areas.

A step in the right direction is the promotion of good agricultural practices (GAPs), which seek to enhance food safety and to reduce environmental externalities and to enhance the benefits of UPA for supporting ecosystem services (de Zeeuw et al., 2011). GAPs provide context specific identification of human and environmental health risks on a case by case basis, thereby aiding in their reduction through the development and implementation of evidence-based policies with multiple stakeholders. In Kampala, Uganda, for example, regulations have been developed for urban livestock and aquaculture sectors through cooperation amongst researchers, policy-makers, urban planners, health officials and direct stakeholders (Cole et al., 2008). To support these endeavors, extension programs will increase the capacity of UPA to enhance ecosystem services (Fischer et al., 2006) by providing education and training on reducing environmental risks (e.g. the WHO/FAO's Safe Guidelines for the Safe Use of Wastewater for Agricultural Production) (Hoornweg and Munro-Faure, 2008) and on farming practices that are more ecologically sound. For example, reduction/reuse of solid waste is a strongly promoted urban waste management strategy in the region by the local authority. In order to further encourage these activities, others have proposed the creation of depots for livestock waste for use in co-composting (i.e. fecal sludge and organic solid waste) and for food and other organic waste for livestock feed and co-composting (Lee-Smith, 2010).

As highlighted above, UPA faces challenges from existing policies, but it also provides promise to support multiple objectives in municipalities and could have a wider contribution if pathways are found through which it can be scaled up and out (e.g. intra-city, city to city and city to regional) (Lwasa et al., 2014). At local scales, greater attention to UPA can contribute to climate change adaptation, which is an inescapable component of sustainable urban development objectives (Stolze, 2010). Wider acceptance, land use formalization and integration into, for example, urban planning, ecosystem and landscape management and waste and wastewater management policies can help create more comprehensive approaches. At national and international scales, UPA must be further integrated into national development strategies in order to bring about the institutional transformation required for effective climate change adaptation and mitigation (UNCTAD, 2011). In combination, these research and institutional changes can support the specific needs discussed above, including technological advancements and expansion of extension services and educational programs.

Lastly, numerous regional and international/inter-governmental organizations (e.g. UN-HABITAT, IDRC, Resource Centers on Urban Agriculture and Food Security [RUAF], Urban Harvest, among others) already support food security and climate change research and policy development in many African countries and across the world (Lee-Smith, 2010). These diverse organizations have concomitantly provided various procedural suggestions, many of which have immediate relevance to the UPA agenda. In order to realize the potential of urban agriculture, particularly in the context of climate change resilience, efforts are required to link these disparate actors and to develop common, empirically informed agendas that support coalitions among these many programs already operating in Africa and other vulnerable countries that promote food security and climate change research and policy development.

Final thoughts

In the coming decades, the issues of food security, poverty and global environmental changes including climate change and urbanization will increasingly converge, necessitating a systems approach to understand these interlinked concerns. Urban sustainability must increasingly consider human and environmental processes at multiple scales and the implications of changes in agriculture in geographically distant areas. The World Meteorological Organization suggests that urban and indoor farming should be incorporated into strategies for building more resilient cities (de Zeeuw et al., 2011). However, urban and peri-urban agriculture is a complex area for policy, potentially involving conflicting goals and thus necessitating strategies and plans to evaluate and assess the inherent tradeoffs involved in this arena. The potential for win-win solutions are there but require a comprehensive framework for assessing the possible solutions and tradeoffs during the design phase (Lee-Smith, 2010). With future research and continued inclusion in policy dialogues at multiple scales, UPA promises to support the development of climate resilient cities.

Main messages

- Urban and peri-urban agriculture is a widespread practice around the world, particularly in low- and middle-income countries, that provides a number of benefits to individuals and communities through, foremost, the enhancement of livelihoods and food security.
- UPA is multi-faceted by nature and nested within socio-economic and ecological systems; strategies promoting UPA can provide multiple benefits for urban populations by increasing adaptive capacity and reducing risks associated with both the effects of urbanization processes and impacts of global climate change.

Urban and peri-urban agriculture

- UPA faces challenges to policy to greater or lesser extents, depending on the respective cultural, political and geographical contexts, but it provides promise to support multiple objectives in municipalities and would benefit from integration across multiple sectors to achieve common goals including resilience.
- Many community-based organizations, inter- and non-governmental organizations are working to support UPA through research and training, information sharing, capacity building and financing; thus, efforts are required to link these disparate actors and to develop common, empirically informed agendas that support coalitions among these many programs already operating in multiple regions.

Future research directions

- The biggest challenge for widespread acceptance and practice of UPA includes concern over human and environmental health risks; further research (i.e. on farmers' behavior, on the impact of polluted soil, air and water on the food grown in urban areas and on the benefits and constrains that UPA can bring) are necessary to devise innovative strategies that will institutionalize these practices.
- Consistent and high-quality data, both qualitative and quantitative, within and across cities are needed for both regional and global assessments of UPA practice and multiple benefits to populations and ecosystems.
- Further research is necessary to understand aggregate carbon sequestration benefits of UPA for climate change mitigation as well as quantitative assessments of organic waste recycling benefits for nutrient cycling and impacts on agricultural and environmental sustainability in the rural–urban continuum.
- The development and implementation of consistent indicators and monitoring is crucial for decision-making support and successful practice of UPA at multiple scales.

References

Adelekan, I.O. (2010). *Urbanization and Extreme Weather: Vulnerability of Indigenous Populations to Windstorms in Ibadan, Nigeria* (Paper presented at the International Conference on Urbanization and Global Environmental Change, Arizona State University, Tempe, Arizona). Retrieved from http://ugec2010.ugecproject.org/images/8/84/UGEC_2010_Session_A3_Adelekan_0151.pdf

Adger, W.N. (2010). Addressing barriers and social challenges of climate change adaptation. In: Kasperson, R.E., and Stern, P.C. (eds.) *Facilitating Climate Change Responses: Report on Knowledge from the Social and Behavioural Sciences.* Washington, DC: National Academy Press, 79–84.

Adger, W.N., Eakin, H., and Winkels, A. (2009). Nested and teleconnected vulnerabilities to environmental change. *Frontiers in Ecology and the Environment*, 7(3), 150–157.

Ahmed, A.U., Hill, R.V., Smith, L.C., Wiesmann, D. M., Frankenberger, T., Gulati, K., Quabili, W., and Yohannes, Y. (2007). *The World's Most Deprived: Characteristics and Causes of Extreme Poverty and Hunger.* Washington, DC: International Food Policy Research Institute (IFPRI).

Alam, M.G.M., Snow, E.T., and Tanaka, A. (2003). Arsenic and heavy metal contamination of vegetables grown in Samta village, Bangladesh. *Science of the Total Environment*, 308(1–3), 83–96.

Allison, M., Harris, P.J.C., Hofny-Collins, A.H., and Stevens. W. (1998). *A Review of the Use of Urban Waste in Peri-Urban Interface Production Systems.* Coventry, UK: The Henry Doubleday Research Association.

Amoah, P., Drechsel, P., and Abaidoo, R.C. (2005). Irrigated urban vegetable production in Ghana: sources of pathogen contamination and health risk elimination. *Irrigation and Drainage*, 54(S1), S49–S61.

Asenso-Okyere, K., Asante, F.A., Tarekegn, J., and Andam, K.S. (2009). *The Linkages Between Agriculture and Malária – Issues for Policy, Research and Capacity Strengthening* (IFPRI Discussion Paper 00861, May 2009). Washington, DC: International Food Policy Research Institute. Retrieved from www.ifpri.org/sites/default/files/publications/ifpridp00861.pdf

John P. Connors et al.

Bakker, N., Dubbeling, M., Guendel, S., SabelKoschella, U., and de Zeeuw, H. (eds.) (2000). *Growing Cities, Growing Food.* Feldafing, Germany: Urban Agriculture on the Policy Agenda, DSE.

Barbizan, T.S., and Giseke, U. (2011). *Integrating Urban and Peri-Urban Agriculture into Public Policies to Improve Urban Growth: São Paulo as a Case Study.* Technische Universität Berlin.

Battisti, D.S., and Naylor, R.L. (2009). Historical warnings of future food insecurity with unprecedented seasonal heat. *Science, 323*(5911), 240–244.

Bentham, G., and Langford, I.H. (2001). Environmental temperatures and the incidence of food poisoning in England and Wales. *International Journal of Biometeorology, 45*(1), 22–26.

Bishwapriya, S. (1985). Urban agriculture: who cultivates and why. A case study of Lusaka, Zambia. *Food and Nutrition Bulletin, 7*(3), 15–24.

Brogan D.R., and James, L.D. (1980). Physical environment correlates of psychosocial health among urban residents. *American Journal of Community Psychology, 8*(5), 507–522.

Brown, K.H., and Jameton, A.L. (2000). Public health implications of urban agriculture. *Journal of Public Health Policy, 21*(1), 20–39.

Brown, S., Grais, A., Ambagis, S., and Pearson, T. (2012). *Baseline GHG Emissions from the Agricultural Sector and Mitigation Potential in Countries of East and West Africa* (CCAFS Working Paper no. 13). Copenhagen, Denmark: CGIAR Research Program on Climate Change, Agriculture and Food Security.

Buechler, S., Mekala, G.D., and Keraita, B. (2006). Wastewater use for urban and peri-urban agriculture. In van Veenhuizen, R. (ed.), *Cities Farming for the Future: Urban Agriculture for Green and Productive Cities.* The Philippines: RUAF Foundation/IDRC/IIRR, 241–272.

Burton, P., Lyons, J., Richards, C., Amati, M., Rose, N., Desfours, L., Pires, V., and Barclay, R. (2013). *Urban Food Security, Urban Resilience and Climate Change.* Gold Coast, Australia: National Climate Change Adaptation Research Facility.

Chambers, R., and Conway, G. (1988). *Sustainable Rural Livelihoods: Practical Concepts for the 21st Century.* IDS Discussion Paper 296, Brighton: IDS.

Chen, M., Xu, G., Wu, S., and Zheng, S. (2010). High-temperature hazards and prevention measurements for asphalt pavement. In: Proceedings of the International Conference on Mechanic Automation and Control Engineering (MACE), Wuhan, China, 26–28 June. Institute of Electrical and Electronics Engineers, New York, NY, pp. 1341–1344, doi:10.1109/ MACE.2010.5536275.

Cissé, O., Gueye, N. F. D., and Sy, M. (2005). Institutional and legal aspects of urban agriculture in French-speaking West Africa: from marginalization to legitimization. *Environment and Urbanization, 17*(2), 143–154.

Cofie, O., van Veenhuizen, R., and Dreschel, P. (2003). Contribution of Urban and Peri-urban Agriculture to Food Security in Sub Saharan Africa. IWMI-ETC paper to be presented at the Africa session of 3rd WWF, Kyoto, March 17. Retrieved from www.alnap.org/resource/7816

Cole, D.C., Lee-Smith, D., and Nasinyama, G.W. (eds.) (2008). *Healthy City Harvests: Generating Evidence to Guide Policy on Urban Agriculture.* Lima, Peru: CIP/Urban Harvest and Makerere University Press.

Crozier C.R., Polizzotto M., and Bradley L. (2012). *Minimizing Risks of Soil Contaminants in Urban Gardens.* Department of Horticultural Science, North Carolina State University. Published by North Carolina Cooperative Extension Service. Raleigh, NC. Retrieved from http://content.ces.ncsu.edu/20684.pdf.

DAI. (2006). Urban Agriculture Program for HIV Affected Children and Women (UAPHAW), Ethiopia Annual Progress Report. Addis Ababa, Ethiopia: Development Alternative Incorporated.

Danso, G., Drechsel, P., Wiafe-Antw, D.T., and Gyiele, L. (2002). Comparison of farm income and trade-offs of major urban, peri-urban and rural farming systems around Kumasi, Ghana. *Urban Agriculture Magazine, 7*, 5–6.

De Bon, H., Parrot, L., and Moustier, P. (2010). Sustainable urban agriculture in developing countries. A review. *Agronomy for Sustainable Development,* Springer Verlag (Germany), *30*(1), 21–32.

de Zeeuw H., and Dubbeling, M. (2008). *Cities, Food and Agriculture: Challenges and the Way Forward.* Leusden, The Netherlands: RUAF Foundation International Network of Resource Centres on Urban Agriculture and Food Security.

de Zeeuw, H., van Veenhuizen, R., and Dubbeling, M. (2011). The role of urban agriculture in building resilient cities in developing countries. *Journal of Agricultural Science, 149* (Supplement S1), 153–163.

Deelstra, T., and Girardet, H. (2000). *Urban Agriculture and Sustainable Cities* (Thematic Paper 2). Retrieved from http://citeseerx.ist.psu.edu/viewdoc/download?doi=10.1.1.168.4991&rep=rep1&type=pdf

Urban and peri-urban agriculture

Drechsel, P., and Kunze, D. (eds.) (2001). *Waste Composting for Urban and Peri-Urban Agriculture: Closing the Rural-Urban Nutrient Cycle in Sub-Saharan Africa.* Wallingford, UK.: IWMI/FAO/CABI, 229 pages.

Drechsel, P., Graefe, S., Sonou, M., and Cofie, O. (2006). *Informal Irrigation in Urban West Africa: An Overview.* Colombo, Sri Lanka: International Water Management Institute.

Drechsel, P., Keraita, B., Amoah, P., Raschid-Sally, L., Bahri, A., and Abaidoo, R.C. (2008). Reducing health risks from wastewater use in urban and peri-urban sub-Saharan Africa: applying the 2006 WHO guidelines. *Water Science & Technology, 57*(9), 1461–1466.

Drechsel, P., Cofie, O., and Danso, G. (2010, April). Closing the rural-urban food and nutrient loops in West Africa: a reality check. *Urban Agriculture Magazine, 23,* 8–10.

Dubbeling, M., and Merzthal, G. (2006). Sustaining urban agriculture requires the involvement of multiple stakeholders. In: van Veenuizen, R. (ed.) *Cities Farming for the Future, Urban Agriculture for Green and Productive Cities.* RUAF Foundation, IDRC and IIRR, pp. 19–52.

Dubbeling, M., Caton Campbell, M., Hoekstra, F., and van Veenhuizen, R. (2009). Editorial: building resilient cities, *Urban Agric. Mag.,* 3–11.

Egbuna, N.E. (2009). *Urban Agriculture: A Strategy for Poverty Reduction in Nigeria.* CBN.

EUKN. (2008). Sustainable Family-Farm Production, Bogota. European Urban Knowledge Network (January 30, 2008). Retrieved from www.eukn.org/E_library/Economy_Knowledge_Employment/ Economy_Knowledge_Employment/Sustainable_Family_farm_Production_Bogot%C3%A1

FAO. (1997). *Agriculture Food and Nutrition for Africa: A Resource Book for Teachers of Agriculture.* Rome, Italy: Food and Agriculture Organization of the United Nations.

FAO. (2008). *The State of Food Insecurity in the World 2008: High Food Prices and Food Security – Threats and Opportunities.* Rome, Italy: Food and Agriculture Organization of the United Nations.

FAO and ETC. (2010). Final Report – Urban and Peri-urban Agriculture on the Policy Agenda. FAO/ ETC joint Electronic Conference August 21–September 30, 2000. Retrieved from www.fao.org/ docrep/meeting/003/x6091e.htm

FAO, IFAD, and WFP. (2013). *The State of Food Insecurity in the World: The Multiple Dimensions of Food Security.* Rome, Italy: Food and Agriculture Organization of the United Nations.

Fischer, J., Lindenmayer, D., and Manning, A. (2006). Biodiversity, ecosystem function, and resilience: ten guiding principles for commodity production landscapes. *Frontiers in the Ecology and Environment, 4*(2), 80–86.

Foeken, D.W.J., and Owuor, S.O. (2008). Farming as a livelihood source for the urban poor of Nakuru, Kenya. *Geoforum, 39,* 1978–1990.

GTZ/GFA. (1999). Utilization of organic waste in (peri-) urban centres. Supraregional Sectoral Project. GTZ, Eschborn/Frankfurt. (incl. software programme)

Hamilton, A.J., Burry, K., Mok, H.-F., Barker, F., Grove, J.R., Williamson, V.G. (2013). Give peas a chance? Urban agriculture in developing countries. A review. *Agron Sustain Dev,* 45–73. doi: 10.1007/ s13593-013-0155-8

Heinegg, A., Maragos, P., Mason, E., Rabinowicz, J., Straccini, G., and Walsh, H. (2000). *Soil Contamination and Urban Agriculture: A Practical Guide to Soil Contamination Issues for Individuals and Groups.* Quebec, Canada: McGill University, McGill School of Environment.

Hodges, R., Buzby, J.C., and Bennett, B. (2011). Postharvest losses and waste in developed and less developed countries: opportunities to improve resource use. *The Journal of Agricultural Science, 149,* 37–45.

Holmer, R., and Drescher, A. (2006). Empowering Urban Poor Communities through Integrated Vegetable Production in Allotment Gardens: The Case of Cagayan de Oro City, Philippines. Proceedings of the FFTC-PCARRD International Workshop on Urban/Peri-Urban Agriculture in the Asian and Pacific Region, Tagaytay City, Philippines, May 22–26, pp. 20–40.

Hoornweg, D., and Munro-Faure, P. (2008). Urban Agriculture for Sustainable Poverty Alleviation and Food Security (Position paper). Final Draft October 3, 2008 (p. 8). The World Bank and FAO.

House, W., Ikiara, G., and McCormick, D. (1993). Urban self employment in Kenya. Panacea or viable strategy? *World Development, 21*(7) 1205–1223. Retrieved from www.ruaf.org/sites.default/files/chap 1 in October 2009.

Ingram, J., Ericksen, P., and Liverman, D. (eds.) (2010). *Food Security and Global Environmental Change.* London and Washington DC: Earthscan.

IPCC. (2014). Climate Change 2014: Impacts, Adaptation, and Vulnerability. Part A: Global and Sectoral Aspects. Contribution of Working Group II to the Fifth Assessment Report of the Intergovernmental Panel on Climate Change [Field, C.B., Barros, V.R., Dokken, D.J., Mach, K.J., Mastrandrea, M.D.,

Bilir, T.E., Chatterjee, M., Ebi, K.L., Estrada, Y.O., Genova, R.C., Girma, B., Kissel, E.S., Levy, A.N., MacCracken, S., Mastrandrea, P.R., and White, L.L (eds.)]. Cambridge, UK and New York, USA: Cambridge University Press, 1132 pp.

IPCC SREX. (2012). Special Report of Working Groups I and II of the Intergovernmental Panel on Climate Change (IPCC). Cambridge University Press, Cambridge, UK and New York, USA, pp. 339–392.

Jansma, J.E., Sukkel, W., Stilma, E.S.C., and Visser, A.J. (2012). The impact of local food production on food miles, fossil energy and greenhouse gas emission: the case of the Dutch city of Almere. In: Viljoen, A., and Wiskerke, J.S.C. (eds.) *Sustainable Food Planning: Evolving Theory and Practice.* Wageningen, The Netherlands: Wageningen Academic Publishers, pp. 307–321.

James, S.J., and James, C. (2010). The food cold-chain and climate change. *Food Research International,* *43*(7), 1944–1956.

Jones, P.G., and Thornton, P.K. (2003). The potential impacts of climate change on maize production in Africa and Latin America in 2055. *Global Environmental Change, 13*(1), 51–59.

Kaplan, R. (1973). Some psychological benefits of gardening. *Environment and Behavior, 5*(2), 145–162.

Karanja, N.F., Yeudall, M.N., Mbugua, S., Prain, G., Cole, D., Webb, A., Levy, J., Gore, C.D., and Sellen, D. (2010). Strengthening capacity for sustainable livelihoods and food security through urban agriculture among HIV and AIDS affected households in Nakuru, Kenya. *International Journal of Agricultural Sustainability, 8*(1 and 2), 40–53.

Keraita, B., and Drechsel, P. (2004). Agricultural use of untreated urban wastewater in Ghana. *Wastewater Use in Irrigated Agriculture: Confronting the Livelihood and Environmental Realities.* Wallingford: IWMI-IDRC-CABI, pp. 101–112.

Kovats, R.S., Edwards, S.J., Hajat, S., Armstrong, B.G., Ebi, K.L., and Menne, B. (2004). The effect of temperature on food poisoning: a time-series analysis of salmonellosis in ten European countries. *Epidemiology and Infection, 132*(3), 443–453.

Lee-Smith, D. (2010). Cities feeding people: an update on urban agriculture in equatorial Africa. *Environment and Urbanization, 22*(2), 483–499.

Lee-Smith, D., and Prain, G. (2006). Urban Agriculture and Health. FOCUS, 13 (Brief 13 of 16). Washington, DC: International Food Policy Research Institute.

Liverman, D., and Kapadia, K. (2010). Food systems and global environment: an overview. In: Ingram, J., Ericksen, P., and Liverman, D. (eds.) *Food Security and Global Environmental Change.* London, UK: Earthscan Ltd, pp. 3–24.

Lobell, D.B., and Field, C.B. (2007). Global scale climate–crop yield relationships and the impacts of recent warming. *Environmental Research Letters, 2*(1), 014002.

Lobell, D.B., Burke, M.B., Tebaldi, C., Mastrandrea, M.D., Falcon, W.P., and Naylor, R.L. (2008). Prioritizing climate change adaptation needs for food security in 2030. *Science, 319*(5863), 607–610.

Lobell, D.B., Schlenker, W., and Costa-Roberts, J. (2011). Climate trends and global crop production since 1980. *Science, 333*(6042), 616–620.

Lock, K., and de Zeeuw, H. (2001). *Health Risks Associated with Urban Agriculture (Annotated Bibliography on Urban Agriculture, ETC-RUAF).* Wageningen, The Netherlands: CTA.

Lwasa, S., Tenywa, M., Majaliwa Mwanjalolo, G.J., Prain, G., and Sengendo, H. (2009). Enhancing adaptation of poor urban dwellers to the effects of climate variability and change. 3 (Vol. 6, p. 332002). Presented at the IOP Conference Series: Earth and Environmental Science.

Lwasa, S., Mugagga, F., Wahab, B., Simon, D., Connors, J., and Griffith, C. (2014). Urban and peri-urban agriculture and forestry: transcending poverty alleviation to climate change mitigation and adaptation. *Urban Climate, 7,* pp. 92–106.

McGranahan, G., Satterthwaite, D., and Tacoli, C. (2004). Rural–urban change, boundary problems and environmental burdens. Human Settlements Working Paper Series Rural-Urban Interactions and Livelihood Strategies No. 10. London: IIED.

Makokha, A.O., Magoha, H.S., Wekesa, J.M., and Nakajugo, A. (2008). Environmental lead pollution and food safety around Kampala City in Uganda. *J. Applied Biosci., 12,* 642–649.

Mbabazi, J., Kwetegyeka, J., Ntale, M., and Wasswa, J. (2010). Ineffectiveness of Nakivubo wetland in filtering out heavy metals from untreated Kampala urban effluent prior to discharge into Lake Victoria, Uganda. *African Journal of Agricultural Research, 5*(4), 3431–3439.

Mkwambisi, D.D., Fraser, E.D.G., and Dougill, A.J. (2011). Urban agriculture and poverty reduction: evaluating how food production in cities contributes to food security, employment and income in Malawi. *Journal of International Development, 23*(2), 181–203.

Mlozi, M.R.S., Lupala, A., Chenyambuga, S.W., Liwenga. E., and Msogoya, T. (2014). *Building Urban Resilience: Assessing Urban and Peri-urban Agriculture in Dar es Salaam, Tanzania*. [Padgham, J. and J. Jabbour (eds.)]. Nairobi, Kenya: United Nations Environment Programme (UNEP).

Nabulo, G., Oryem-Origa, H., and Diamond, M. (2006). Assessment of lead, cadmium, and zinc contamination of roadside soils, surface films, and vegetables in Kampala City, Uganda. *Environmental Research*, *101*(1), 42–52.

Nabulo, G., Young, S.D., and Black, C.R. (2010). Assessing risk to human health from tropical leafy vegetables grown on contaminated urban soils. *Science of the Total Environment*, *408*, 5338–5351.

Nair, P.K.R., Kumar, B.M., and Nair, V. (2009). Agroforestry as a strategy for carbon sequestration. *Journal of Plant Nutrition and Soil Science*, *172*, 10–23.

Nelson, D.R., Adger, W.N., and Brown, K. (2007). Adaptation to environmental change: contributions of a resilience framework. *Annual Review of Environment and Resources*, *32*, 395–419.

Nolasco, C.L. (2009). A dimensão ecológica da agricultura urbana no município de Juiz de Fora/MG. (Master dissertation). Graduate Programme in Ecology/PGECOL, Universidade Federal de Juiz de Fora/UFJF. Juiz de Fora, Brazil.

OECD. (2012). *OECD Environmental Outline to 2050: The Consequences of Inaction*. OECD Publishing.

Pelling, M., and High, C. (2005). Understanding adaptation: what can social capital offer assessments of adaptive capacity? *Global Environmental Change*, *15*(4), 308–319.

Porter, J.R., Xie, L., Challinor, A.J., Cochrane, K., Howden, S.M., Iqbal, M.M., Lobell, D.B., and Travasso, M.I. (2014). Food security and food production systems. In: Climate Change 2014: Impacts, Adaptation, and Vulnerability. Part A: Global and Sectoral Aspects. Contribution of Working Group II to the Fifth Assessment Report of the Intergovernmental Panel on Climate [Field, C.B., Barros, V.R., Dokken, D.J., Mach, K.J., Mastrandrea, M.D., Bilir, T.E., Chatterjee, M., Ebi, K.L., Estrada, Y.O., Genova, R.C., Girma, B., Kissel, E.S., Levy, A.N., MacCracken, S., Mastrandrea, P.R., and White, L.L (eds.)]. Cambridge, UK and New York, USA: Cambridge University Press, pp. 485–533.

Prain, G., Karanja, N., and Lee-Smith, D. (2010). *African Urban Harvest: Agriculture in the Cities of Cameroon, Kenya and Uganda*. Springer, New York and IDRC Ottawa.

RUAF. (2014). *Urban Agriculture as a Climate Change and Disaster Risk Reduction Strategy*. Urban Agriculture Magazine. (No. 27, March).

Ruel, J.L., and Garrett, J.L. (2003). Features of Urban Food and Nutrition Security and Considerations for Successful Urban Programming. Paper prepared for the FAO technical workshop on "Globalization of food systems: impacts on food security and nutrition" (8–10 October 2003, Rome, Italy).

Schertenleib, R., Forster, D., and Belevi, D. (2004). An integrated approach to environmental sanitation and urban agriculture. *ActaHorticulturae*, *643*, 223–226.

Schmidt, S. (2012). Getting the policy right: urban agriculture in Dar es Salaam, Tanzania. *International Development Planning Review*, *34*(2), 129–145.

Seto, K.C., Fragkias, M., Güneralp, B., and Reilly, M.K. (2011). A meta-analysis of global urban land expansion. *PLoS ONE*, *6*(8), e23777. doi:10.1371/journal.pone.0023777

Sharma, R.K., Agrawal, M., and Marshall, F. (2007). Heavy metal contamination of soil and vegetables in suburban areas of Varanasi, India. *Ecotoxicology and Environment Safety*, doi:10.1016/j.ecoenv.2005.11.007.

Shoemaker, C.A. (ed.) (2002). *Interaction by Design: Bringing People and Plants Together for Health and Well-Being*, Proceedings of the Sixth International People Plant Symposium, Glencoe, IL, 2000. Ames, IA: Iowa State University Press.

Simon, D., McGregor, D., and Thompson, D. (2006). Contemporary perspectives on the peri-urban zones of cities in development areas. In: McGregor, D., Simon, D., and Thompson, D. (eds.) *Peri-Urban Interface: Approaches to Sustainable Natural and Human Resource Use*. London, UK: Earthscan Publications Ltd, pp. 3–17.

Sridhar, M.K.C. and Adeoye, G.O. 2003. Organo-mineral fertilizers from urban wastes: *Developments in Nigeria. The Nigerian Field*, *68*: 91–111.

Smit, J., Nasr, J., and Ratta, A. (2001). *Urban Agriculture: Food Jobs and Sustainable Cities*. The Urban Agriculture Network, Inc. A non-profit, 501 (c)(3) organization with the financial support and permission of the United Nations Development Programme (UNDP).

Smith, P., Bustamante, M., Ahammad, H., Clark, H., Dong, H., Elsiddig, E.A., Haberl, H., Harper, R., House, J., Jafari, M., Masera, O., Mbow, C., Ravindranath, N.H., Rice, C.W., Robledo Abad, C., Romanovskaya, A., Sperling, F., and Tubiello, F. (2014). Agriculture, Forestry and Other Land Use

(AFOLU). In: Climate Change 2014: Mitigation of Climate Change. Contribution of Working Group III to the Fifth Assessment Report of the Intergovernmental Panel on Climate Change [Edenhofer, O., Pichs-Madruga, R., Sokona, Y., Farahani, E., Kadner, S., Seyboth, K., Adler, A., Baum, I., Brunner, S., Eickemeier, P., Kriemann, B., Savolainen, J., Schlömer, S., von Stechow, C., Zwickel, T., and Minx, J.C. (eds.)]. Cambridge, UK and New York, USA: Cambridge University Press.

Stasinos, S., and Zabetakis, I. (2013). The uptake of nickel and chromium from irrigation water by potatoes, carrots and onions. *Ecotoxicology and Environmental Safety*, *91*, 122–128.

Stolze, M. (2010). Climate Change Policy Perspectives. Presentation at the UNCTAD/ENSEARCH International Conference on Climate Change, Agriculture and Related Trade Standards. Kuala Lumpur, 1–2 November.

Swalheim, S., and Dodman, D. (2008). *Building Resilience: How the Urban Poor Can Drive Climate Adaptation*. London: IIED.

Thebo, A.L., Drechsel, P., and Lambin, E.F. (2014). Global assessment of urban and peri-urban agriculture: irrigated and rainfed croplands. *Environmental Research Letters*, *9*, 114002.

Tidball, K.G., and Krasny, M. (2007). From risk to resilience: what role for community greening and civic ecology in cities? In: Wals, A. (ed.) *Social Learning Towards a more Sustainable World*. Wageningen, The Netherlands: Wageningen Academic Publishers, pp. 149–164.

UNCTAD. (2011). Assuring Food Security in Developing Countries under the Challenges of Climate Change: Key Trade and Development Issues of a Fundamental Transformation of Agriculture. United Nations Conference on Trade and Development.

UNDESA. (2014). World Urbanization Prospects: The 2014 Revision, Highlights. United Nations, Department of Economic and Social Affairs, Population Division (ST/ESA/SER.A/352).

UNFPA. (2007). *State of the World Population: Unleashing the Potential of Urban Growth*. New York: United Nations Population Fund.

University of Colombo. (2013). Environmental Impacts of Home Gardens in Kesbewa, Sri Lanka: Assessment of Food Miles and Reduction of Greenhouse Gas Emissions in KUC area, Sri Lanka. Report prepared for RUAF Foundation and CDKN.

van Rooijen, D.J., Biggs, T.W., Smout, I., and Drechsel, P. (2010). Urban growth, wastewater production and use in irrigated agriculture: a comparative study of Accra, Addis Ababa and Hyderabad. *Irrigation & Drainage Systems*, *24*(1/2), 53–64.

van Veenhuizen, R. (2012). Urban and Peri-urban Agriculture and Forestry (UPAF): an important strategy to building resilient cities? The role of urban agriculture in building resilient cities. Webinar ICLEI, 18 October 2012.

van Veenhuizen, R., and Danso, G. (2007). Profitability and sustainability of urban and peri-urban agriculture. FAO Agricultural Management, Marketing and Finance Occasional Paper No 19. Rome: FAO.

Von Braun, J. (2009). Food security and global environmental change: emerging challenges. *Environmental Science & Policy*, 12, 373–377.

Wahab, B., Sridhar, M.K., and Ayorinde, A.A. (2010). Improving food security through environmental management in Ibadan. *Urban Agriculture Magazine*, 23-Urban nutrient management, 25–26.

World Bank. (2007). Global Economic Prospects 2007: Managing the Next Wave of Globalization. Washington, DC: The World Bank.

Zezza, A. and Tasciotti, L. (2010). Urban agriculture, poverty, and food security: empirical evidence from a sample of developing countries. *Food Policy*, *35*(4), 265–273.

29

INTEGRATING BIODIVERSITY AND ECOSYSTEM SERVICES INTO URBAN PLANNING AND CONSERVATION

Robert McDonald

For the first time in the history of humanity, more than 50 percent of us now live in urban areas (UNPD, 2011). The next few decades will be the fastest period of urban growth: by 2050, more homes will be built than were built over centuries of urban development in Europe (McDonald, 2008). Rapidly urbanizing regions like Asia and Africa will add billions more people to their cities. In a sense, Asia and Africa are catching up with already urbanized societies in the United States and Europe, which already have a substantial proportion of their total population living in cities. However, even in the United States and Europe, urban population growth is continuing in some urban areas, driven by overall population growth or shifts in population among cities.

Urban planners, natural resource managers and government officials are increasingly asked to start planning for this urban world (McDonald, 2008). There are many issues that such planning must consider, but one of the most common concerns is increasing urban sustainability and resilience. This chapter first presents a broad overview of the three main goals of urban conservationists: conservation in cities, by cities and for cities. The remainder of the chapter focuses in on 'conservation for cities,' discussing how urban planning can be used to protect or restore the natural habitat that provides benefits to those living in cities. It discusses why markets cannot be expected to provide most ecosystem services to cities, which justifies government intervention. The chapter then presents one possible framework for integrating ecosystem services into urban planning processes, with a particular focus on comprehensive urban plans. Finally, the chapter shows why the geography of ecosystem service supply and demand suggests particular places that are optimal locations for conservation action to protect or restore ecosystem service provision, drawing links to some classic theories in urban economics.

Planning for an urban world

This section outlines the three different urban sustainability goals that urban planners are commonly asked to address: conservation in, by and for cities.

Robert McDonald

Conservation in cities

Often planners are asked to mitigate the direct impact of cities on biodiversity due to urban growth and land cover change. Urban growth is a significant global driver of land use conversion and deforestation. Urban areas occupy only around 3 percent of the Earth's land surface (McGranahan et al., 2006), although the actual number varies significantly depending on the definition of urban and the spatial grain of analysis (Schneider et al., 2009; Seto et al., 2010). However, cities are often located in riparian areas or along coastlines (places of disproportionately high biodiversity), so they have an even larger impact on global biodiversity than their area would suggest. For instance, McDonald et al. (2008) show that the majority of terrestrial ecoregions (comprising 62 percent of the Earth's land surface) are currently less than 1 percent urbanized and will experience little change through 2030. However, around 10 percent of terrestrial vertebrates are in ecoregions that are heavily impacted by urbanization, although these ecoregions only represent 0.3 percent of the Earth's land surface. More detailed modeling by Seto et al. (2012) shows that while less than 1 percent of all biodiversity hotspot areas (Mittermeier et al., 2004; Myers et al., 2000) were urbanized c. 2000, there are certain biodiversity hotspots such as the Mediterranean Basin, Atlantic Forest and California Floristic Province that have been heavily impacted by urbanization. Indeed, nearly all the urban land in Southeast Asia (27,000 km^2) is located in biodiversity hotspots (Güneralp and Seto, 2013).

Ecologists and natural resource managers are also asked to think about how to minimize the effect of fragmentation on the viability of rare elements of biodiversity. Urban development necessarily increases the amount of remnant natural habitat that is near a habitat/non-habitat edge (Murcia, 1995). This systematically alters conditions near the edge, affecting ecosystem structure and function (Fagan et al., 1999). For instance, at forest/non-forest edges, temperature is increased during the growing season due to greater solar insolation. Roads create a particular type of edge, with peculiar ecological effects (Forman, 2000), such as altering when and how bird species sing (Rheindt, 2003). Moreover, biotic interactions may change near edges; birds' nests, for instance, are more likely to be parasitized by cowbirds when they are near an edge (Lloyd et al., 2005). The conservation planning literature has often discussed how to protect biodiversity in areas of urban growth (e.g. Groves, 2003). The general strategy is to keep urban growth out of areas of high biodiversity and, wherever possible, keep houses clustered at high-density to minimize the total area of natural habitat impacted by urban growth.

Fundamentally, the task of conservation planning is an optimization problem: select the patches for conservation that achieve some conservation objective (e.g. minimize the loss of species) for the minimal cost (Sarkar et al., 2006). The first general principle identified in the conservation planning literature is the idea of complementarity between patches selected for conservation: the species list of one patch selected should not be entirely identical to the species list of another patch selected, but instead should complement it so that a large number of species are protected in total (Rodrigues and Gaston, 2002). Smarter analyses will incorporate another principle of conservation planning—vulnerability. Different parcels have different chances of being converted to another land use (perhaps through urban development) and, as a result, lose biodiversity. The effectiveness of conservation action is thus measured against the expected loss of biodiversity without conservation action, which can be defined as the vulnerability (probability of habitat conversion) multiplied by the biodiversity loss if conversion occurs (McDonald, 2009). Finally, analyses should consider parcel cost. Different parcels of natural habitat cost different amounts, both because they vary in area, but also because the per area cost can vary considerably across a planning landscape.

Conservation by cities

Overall, urban form of a metropolitan region determines per capita resource use by affecting how people live and structures their activities. For many resources, there are economies of scale to density: more dense settlements use less resources per capita. These economies of scale as well as arguments about the quality of life in more dense, walkable communities (Jacobs, 1961), lead to the 'compact city' (Dantzig, 1974) and 'smart growth' (APA, 2002) movements. These argue that a more compact urban form should be strived for as cities develop. For instance, more dense cities have substantially less per capita transportation sector energy use and hence less greenhouse gas (GHG) emissions (Kenworthy and Laube, 2000).

The same kind of relationship is true for many other resources. For instance, there are economies of scale in energy use for buildings. More dense cities tend to use less energy for heating, cooling and lighting than less dense cities, primarily because it is more energy efficient to heat and cool a large shared building rather than having to heat and cool multiple buildings. A typical study in Toronto compared suburban single-family dwellings to apartment buildings in the urban core. The suburban, low-density dwellings use around 80 percent more energy for building operations per capita than the apartment buildings, increasing GHG emissions by the same factor. The same study estimates that the energy used to create the building materials in the suburban home is 52 percent greater per capita than in the apartment buildings (Norman et al., 2006).

Similar arguments apply for water resources. In developed economies, the big difference in water use between apartment buildings and single-family residential homes is driven by the decreased need for water to irrigate lawns and landscaping. The same trend is true when comparing different neighborhoods of single-family residential homes with different lot sizes. Homes on larger lots have more lawn and landscape to irrigate and hence have higher per capita use of water. One study in Portland, Oregon, finds that for single-family houses in low-density neighborhoods (< 4 houses/acre), a unit increase in density (i.e. the addition of one extra house per acre) reduces water consumption by more than 30,000 liters per household per year (Chang et al., 2010).

These kinds of resource use efficiencies show why consideration of density in new cities in low- and middle-income cities is so important. For instance, if Chinese cities end up at the same average density as cities in high-income Asian countries such as Japan or South Korea then they will use significantly less energy for transportation and building operations than if they end up at the same average density as cities in Europe or the even less dense cities in the United States. Avoiding increased consumption of energy with more dense urban living will translate into significant avoided GHG emissions over the next decades (McDonald, 2008).

Conservation for cities

This chapter focuses primarily on the maintenance and creation of critical natural areas that provide important benefits to those living in cities. The patches of natural habitat that provide benefits to those in cities are increasingly called 'green infrastructure,' and planners are asked to incorporate them into urban planning processes. The term originated in the United States in the greenways movement and at first had a strong focus on preserving biodiversity and landscape connectivity. The term has broadened over time, as planners recognize an increasing variety of benefits from nature. In particular, green infrastructure often now refers to constructed wetlands and other natural habitats that help cities reduce the water runoff going into their stormwater drainage system. In this chapter, the term green infrastructure is used in its broadest possible sense, for any piece of nature that provides important benefits to those in a city.

The fundamental urban problem with ecosystem services

The central idea of 'conservation for cities' is that urban planners can use the natural world to meet some of the needs and desires of those in cities. While many desires are satisfied entirely with human technology, there are certain things that most urbanites would consider part of the good life that necessarily involve nature. All these things that cities demand from the natural world can be thought of, more positively, as ecosystem services or the benefits nature provides urban dwellers (see also Grimm et al., Chapter 14 in this volume, for this discussion). This chapter uses Boyd and Banzhaf's (2006, p. 8) definition, which defines ecosystem services as "the components of nature, directly enjoyed, consumed or used to yield human well-being." Note the emphasis on nature having direct value to human well-being.

Many different ecosystem services are important to urbanites' well-being. A short list of ecosystem services most relevant to cities is shown in Table 29.1, divided into the categories used by the Millennium Ecosystem Assessment (MEA, 2003). One category of ecosystem

Table 29.1 Most important ecosystem services for cities, classified according to the scheme of the MEA (2003)—also listed is the category of economic good they represent and the spatial scale at which they operate

Ecosystem service	Economic category	Spatial scale
Provisioning services:		
Agriculture	Private good	Regional to global
Water (quantity)	Private good	Upstream source watershed (100s km)
Cultural services:		
Aesthetic benefits	Public or common good	Area of daily travel by urbanites (10s km)
Recreation and tourism	Public or common good	Area of daily travel by urbanites (10s km)
Physical health	Public or common good	Area of daily travel by urbanites (10s km)
Mental health	Public or common good	Area of daily travel by urbanites (10s km)
Spiritual value	Public or common good	Varies—often local, but can be up to global
Biodiversity	Public or common good	Varies—global for existence value, local for direct interaction
Regulating services:		
Drinking water protection (water quality)	Public good	Upstream source watershed (100s km)
Stormwater mitigation	Public good	Downstream stormwater system (100s km)
Mitigating flood risk	Public good	Downstream flood-prone areas (100s km)
Coastal protection	Public good	Coastal zone (100s km)
Air purification (particulates, ozone)	Public good	Regional airshed (100s km)
Shade and heatwave mitigation	Public good	Varies with solar angle (< 100 m)
Supporting services:		
Soil formation	Not directly used	Varies—agriculture production zone
Nutrient cycling	Not directly used	Varies—agriculture production zone

services used by the MEA is provisioning services or the products people obtain from ecosystems such as food, fuel and fiber. Within this category, one of the most important for cities is the provision of fresh water in sufficient quantity. Municipalities supply water to their residents, who need water for drinking, sanitation, cleaning and other uses. In general, peripheral rural areas provide many provisioning services, which are then transported to the city center.

Another category of ecosystem services recognized by the MEA is cultural services. For cities, important ecosystem services in this category include the aesthetic benefits of natural areas as well as the recreational and health benefits that access to parks can provide. The category of regulating services refers to the benefits people obtain from the regulation of ecosystem function. In cities, some of the key services that wetland, riparian and other natural habitats provide are in mitigating flood risk along streams or coastlines. The natural world plays an important climate regulation role, and natural habitat may help reduce air pollution and regulate air quality within acceptable limits. A final category of ecosystem services is supporting services. In the terminology of Boyd and Banzhaf (2006), these are true ecosystem functions. Supporting ecosystem functions are rarely explicit goals of urban green infrastructure projects, although occasionally they are mentioned as ancillary benefits of a city's actions.

Urban economics, ecosystem services and the need for government involvement

Why are so many other ecosystem services less successfully provided by the free market to urban residents? Economists call the dysfunctional markets for most ecosystem services an example of market failure, which occurs systematically for certain types of goods and services. One important distinction is whether a good is rival or non-rival. A rival good is one where one person's use of the good prevents others from using it. Non-rival goods, by contrast, are not used up, and one person's use does not prevent another's use. Another important distinction is whether a good is excludable or non-excludable. An excludable good is one where it is possible to control who has access to the good (e.g. excluding all those who have not paid a fee). Non-excludable goods, on the other hand, are ones where it is not feasible to control access to the good (Kolstad, 2000).

Private goods—the agricultural produce urbanites buy at the supermarket, for instance— are rival and excludable. One cannot take home milk from the store unless one pays (excludable) and one person's purchase of the milk prevents others from purchasing it (rival). Other important natural resources for cities, like wood for timber and meat from ranching, function as private goods. However, many very important goods are 'public goods,' defined as non-rival and non-excludable. Urbanites' enjoyment of the aesthetic beauty of a park does not prevent others from enjoying it (non-rival), and most public parks are open to all those who wish to enter it (non-excludable). Another important category of goods is 'common goods,' those that are rival but non-excludable. Fishing in the open ocean is the classic example, since anyone can fish (non-excludable) but any fish that are caught and eaten are not there for others to use (rival).

Both economic theory and practical experience suggest that private goods are well provided to cities and society at large. Firms have financial incentives to bring private goods to cities, and the amount of private goods supplied is brought in line with the quantity of the good demanded via changes in price. In contrast, public goods and common goods are generally underprovided in cities, precisely because they are non-excludable. No firm could make money off the provision of these goods, since free riders (users who haven't paid) would just use the good for free.

Most ecosystem services that cities depend on are public or common goods (Table 29.1) and therefore free market actors will not adequately maintain natural habitats or the ecosystem services provided. For instance, urban parks for recreation will tend not to be provided by private land developers, at least at the provision level society would prefer. In contrast, agriculture and several other provisioning services create private goods and thus tend to be well provided by private markets. Indeed, only three out of 36 ecosystem services assessed by the MEA have been maintained or increased over time, and all three are private goods. For the remaining ecosystem services, there is no functioning market, so society has to find other ways to ensure those needs are met.

If actors in the private market have little incentive to consider many ecosystem services in their decision-making, then governments or other social organizations are justified in stepping in to ensure provision, either directly through policy or indirectly by giving firms an incentive to consider ecosystem services in their decisions. The solution to market failure is collective action to promote the public good. Urban planning and zoning is one of the key places where ecosystem services provision can be ensured. There are also many laws that try to promote the public good, usually for particular aspects of the environment. Apart from such policy solutions, there are also market-based mechanisms that fix market failure. Governments can create the rules and regulations to enable such mechanisms to work.

Hanley et al. (2013) list three kinds of market-based mechanisms that either correct some externality or adequately provide for public and common goods. First, cities could assign property rights to the public or common good, effectively making it a private good. For instance, some cities with previously free on-street parking have charged fees for spots dedicated to particular homeowners or businesses. A market for parking permits, whether legal or illegal, soon develops, and the price of permits on the market reflects the true value of a parking spot on that street. Second, cities could charge a fee or tax for the damage an externality causes to society. If the tax is equal to the true social damage of the externality (a Pigovian tax), then market actors will fully incorporate the externality into their economic decision-making. For instance, some cities like London have experimented with setting a price (a congestion charge) for each car that enters the city center during high traffic periods. Third, a city could require anyone creating a negative externality to obtain a permit for creating that externality. There are a finite number of permits in the system, which cumulatively allow for a socially optimal level of the negative externality. For instance, Singapore previously suffered from severe road congestion and moved to limit the number of automobiles on the road. Now, every car on the road is required to have a Certificate of Entitlement. There are a finite number of certificates, which may be purchased in an auction from the government.

A conservation framework for cities

As discussed above, cities have strong empirical and theoretical reasons to assume certain kinds of ecosystem services (those that are public or common goods) will be undersupplied by the free market, and they are justified in intervening to ensure the adequate provision of these ecosystem services to their citizens. The question that follows is how should a city go about systematically evaluating the possibilities for green infrastructure to satisfy its citizens' needs? The steps in this framework are derived from the rational planning model commonly used in urban planning (Berke et al., 2006). Many cities naturally turn to some planning framework like this one because the rational planning model is an ideal that many cities try to achieve. A related framework is presented in the Manual for Cities (TEEB, 2011). The Economics of Ecosystems and Biodiversity (TEEB) framework is focused on assessing the

Biodiversity and ecosystems for urban planning

value of ecosystem services to their residents, whereas the framework below includes planning, implementation and monitoring of ecosystem services projects.

In all the steps in this framework, it is important to keep in mind that green infrastructure cannot and should not be considered separate from gray infrastructure. While the two are sometimes substitutes for one another, more often they are complementary. The flood protection value of natural wetlands is only valuable because there are people and property that need protection from floods. So cities need to align their green infrastructure planning with their gray infrastructure planning. Six steps are presented below, although different cities will have to modify these steps to fit their unique political, socio-economic and ecological context.

Take inventory: what ecosystem services matter?

In this first step, cities begin by broadly considering the full suite of ecosystem services and determining those that matter most. Which ones matter depends on the problem or policy issue to be addressed. A city that evaluates ecosystem services in the context of climate change adaptation planning will define the problem one way: what actions, using natural infrastructure or gray infrastructure, should the city take to increase resilience to climate change (Prasad et al., 2009)? Some cities may assess their resiliency to multiple and different shocks and thus plan for ecosystem services to make the overall city more resilient (cf. Rockefeller Foundation and ARUP, 2014). Some cities may consider ecosystem services as part of their comprehensive planning process. Comprehensive plans, also known as master or general plans, aim to coordinate more specific decisions around zoning, transportation, parks and the other myriad things for which a city has to plan (Berke et al., 2006). Getting clarity on the key problem or policy issue to be addressed is essential and will shape the actions taken at every other stage in the framework.

A full consideration of the ecosystem services that are important in a city will be crucial, at a minimum, for identifying important co-benefits that should be included as part of the planning process. The goal of this phase is to quickly get from a large list of potentially important ecosystem services (see Table 29.1) to a short list of which ecosystem services really matter for a particular city and will be further evaluated in the planning process.

What critical natural systems provide those services?

The next step is determining which patches of natural habitat provide one or more of the important ecosystem services. For cities where the restoration of degraded natural habitat or the creation of novel patches of natural area is a possibility, the planning process must also consider where within the city restoration or creation would be most appropriate. For instance, in the source watershed of Sao Paulo, Brazil, scientists first mapped where reforestation was appropriate and then quantified its benefits to drinking water quality (Natural Capital Project, 2013). If possible, it is helpful to create quantitative estimates of the ecosystem services benefits provided, either in physical units (e.g. tons of sediment not eroded due to vegetation) or in monetary units. Not all habitat patches are equally important, and having some way to rank their importance allows for a transparent, defensible way to choose which habitat patches to try to protect or to restore (Groves, 2003). If advanced modeling efforts are not possible to rank patches, then sometimes expert opinion can help provide a semi-quantitative ranking of patches. Note that critical natural systems can also be located outside a city's municipal boundaries, sometimes far away, as discussed in the section on optimal location of conservation later in the chapter.

Identify threats to critical natural systems

In this step, the goal is to identify the threats that may reduce or destroy the ability of critical natural areas to provide ecosystem services (cf. Copeland et al., 2007). For situations in which the restoration of degraded natural infrastructure or the creation of novel patches of green infrastructure is a possibility, the planning process must consider spatially where restoration or creation would be most appropriate. This step is key because green infrastructure planning aims to increase ecosystem services provision relative to a baseline, status quo scenario of no action (McDonald, 2009). If there is little threat to an existing natural area, then efforts to protect it have little impact on levels of future ecosystem services provision. Conversely, if a piece of critical habitat is very likely to be lost under the baseline, status quo scenario, then conservation action significantly increases future ecosystem services provision above the baseline. The time frame over which threats are evaluated varies depending on the planning effort, with comprehensive or sustainability plans typically looking over several decades.

Identify opportunities to mitigate those threats

Cities next need to identify the opportunities or strategies to mitigate the threats to critical natural systems. Land protection is one obvious strategy (Brooks et al., 2004; Hoekstra et al., 2005; McDonald and Boucher, 2010), but there are many others. For instance, incentives to provide natural habitat and ecosystem services on private land can be another cost-effective strategy to mitigate threats or restore habitat (e.g. Goldman et al., 2008). Some natural habitat patches may also be easier to work in because of the low transaction costs of working there such as land already under city ownership. For situations where restoration or creation of new green infrastructure is a possibility, specific opportunities need to be defined. The outcome of this stage is a finite, well-defined set of proposed natural infrastructure options that seem worthy of further evaluation.

Develop a plan

In this phase, the city evaluates the various potential options and selects the best one. Sometimes this is done using formal cost-benefit analyses. To fully evaluate the return on investment (ROI) of a strategy, an analysis must synthesize information on the economic value of the ecosystem services provided, the threat to those services under the baseline scenario and the costs of implementing the strategy (Murdoch et al., 2007; Wilson et al., 2006). Specific techniques to estimate ROI for particular ecosystem services vary depending on the ecosystem service (McDonald, 2015). Often, however, a particular opportunity or strategy just makes the most sense to urban leaders and is selected without a formal cost-benefit strategy. After selecting the best opportunity or opportunities, cities have to develop plans to implement the strategy. In many ways this is standard business planning that is familiar to most organizational bureaucracies. However, some creativity is often required to figure out how to best correct the market failure that most ecosystem services represent.

Implementation and monitoring

Finally, the city moves to implementation. Strong leadership by municipal officials is essential, since many successful strategies to protect ecosystem services require staff to work across multiple departments in a city and to do new jobs that may be quite different from what they have done traditionally. After implementation, monitoring the green infrastructure program is a crucial and often ignored step, and helps ensure that the green infrastructure created is supplying the promised ecosystem services benefits. Moreover, information from monitoring programs can help refine management of existing natural infrastructure over time—the so-called adaptive management feedback loop (Aber et al., 2000; Allan and Stankey, 2009). If the city decides to further expand its portfolio of natural infrastructure, monitoring information can also be very helpful to make this new investment more efficient.

Optimal ecosystem service provision theory for cities

As cities move through the planning framework outlined above, they will naturally have to ask quantitative questions about where they should take conservation action to maintain or enhance ecosystem services. In this last section, a simple theory is presented for how to choose where to create or protect green infrastructure. This theory is most likely helpful during steps two and three of the above urban framework. The central insight of this theory is that both supply and demand matter for determining where to protect ecosystem services and can be modeled in a way analogous with classic urban economic theories.

Natural habitats must be within a certain distance from people for their ecosystem functions to be useful as an ecosystem service (McDonald, 2009). One common mistake made by urban planners is to focus on areas of ecosystem function and then to treat such zones as simple overlays in planning decisions. This approach misses a very important spatial dimension of ecosystem services, which is the importance of proximity between natural habitat and beneficiary. The transportability of different ecosystem services varies widely (see Table 29.1), which affects how close nature must be to benefit urbanites (McDonald, 2009). Some services operate over the scale of meters, like the shade from the street trees outside. Others, like access to parks for day-to-day recreation, operate over the scale of tens of kilometers. Water provision operates within watersheds, which have a unique element of upstream/downstream directionality: actions upstream affect water quantity and quality downstream, but actions downstream do not affect things upstream. Similarly, air quality in a region's 'airshed' depends on wind patterns, which define an upwind/downwind directionality. As can be seen, the transportability of an ecosystem service is not a simple function of Euclidean distance, but is determined by the physics of the ecosystem service in question.

Finally, some ecosystem services are global in scale. For instance, a ton of carbon sequestered anywhere on the Earth prevents a rise in global atmospheric GHG concentrations, since the Earth's atmosphere is relatively well mixed. Another interesting example is the option value of biodiversity (someday one might want to see or otherwise use that biodiversity) or the existence value of biodiversity (one might feel good just knowing that biodiversity is out there). These are global values, and their intensity does not decline with increasing distance between habitat and person.

Cities generally have a dense core with high population density and then lower population density as one heads into suburbs or exurbs. Even when cities have multiple cores, there is still a general gradient from dense urban to low-density rural. By definition, ecosystem services benefit people. Most people live and work in cities, so cities are centers of ecosystem services demand (McDonald, 2009). Natural habitat often provides greater ecosystem services when it is closer to the dense urban core than if it is located in a remote rural area. Thus, all else being equal, an urban planner or conservation practitioner should create or protect green infrastructure close to where people live and work. For instance, they might strive to protect patches of natural habitat in the urban core, as key areas of ecosystem services provision.

Robert McDonald

Box 29.1 Calculating optimal conservation action

To find the optimal place to preserve or protect habitat for ecosystem service provision, one needs to consider both the costs and benefits of conservation action in a particular place (see Figure 29.1). This is mathematically identical to bid rent theory for cities, which describes why different types of firms, farms or households are willing to pay (bid) up to a certain amount to locate at a certain distance from the city center (Alonso, 1964). Consider a factory that produces goods that it must transport to the city center for sale, with the cost of transport depending on the distance to the city center (km) and the transportability of the good ($/km). The factory's net profit, π, is:

$$\pi = PQ - T(m, Q) - R(m)$$

where m is the distance from the city center. This equation calculates the gross profit, defined as the quantity Q of goods they make times their market price P, minus their costs: transportation costs T, a function of m and Q, and the rent and the other costs of production R, a function of m. T decreases as you approach the city center, but at the expense of an increase in R. For any firm, there is a zone in which it is profitable to operate; if one is too close, the rent is too high, and if one is too far, the transport costs are too high (Richardson, 2013).

An urban planner has to make a similar calculation (McDonald, 2009). For an urban planner seeking to site a green infrastructure project, Q is the quantity of ecosystem services consumed, and P is their societal value. R represents the costs of maintaining or restoring natural habitat at a given distance m from the city center. T represents how distance between the natural habitat and people limit the quantity of ecosystem services consumed. Particular ecosystem services have different degrees of transportability, which affects the size of the term T.

However, there is a steep decrease in land prices or rent as one moves from the city center out to the suburbs and exurbs. For instance, house prices in San Francisco are more than triple that of suburban towns 50km from the city center (McDonald, 2009). This gradient in land prices reflects the intensity of the competition for land for different uses. Conserving or restoring natural habitat in the city center has a high opportunity cost; because the land could be used for many other purposes, it is very expensive. Conservation action in the center city is expensive for other reasons as well. Natural areas in an urban context are more expensive to maintain than natural areas in rural settings.

For a particular ecosystem service, there is a zone in which it is profitable to site green infrastructure. In the city center, the costs of doing projects may be too high, and in rural areas, there may be few ecosystem services provided (i.e. transport costs are too high). In between, there is a zone where conservation is worthwhile. If the city has a single dense core, these zones of profitability form concentric circles around the core. Since different ecosystem services have different transportability, the zones of optimal protection are at different locations for different services (see Figure 29.1).

A thorough planning process must consider what ecosystem services the municipal government wants to provide and to whom, and what it would cost to work in different locations. All of these considerations would affect where the infrastructure should work. Additionally, planners must consider what would happen to ecosystem services provision over

Biodiversity and ecosystems for urban planning

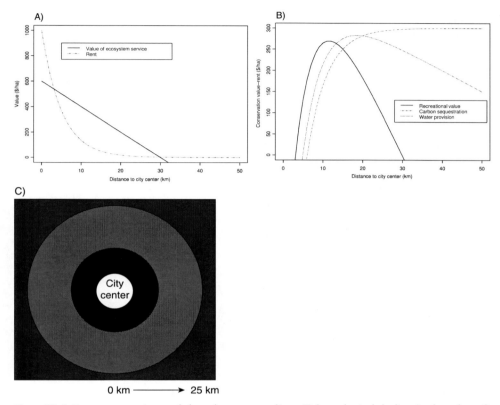

Figure 29.1 Ecosystem services and the urban rent gradient: A) hypothetical decline in the value of an ecosystem service that has some transport cost with distance to city center, compared with an urban rent gradient; B) the net value of a conservation action to preserve an ecosystem service as a function of distance from city center, for three hypothetical ecosystem services; C) a view of an idealized urban region, showing the optimal ecosystem service to preserve at that distance from a city (in the city center, rent is too expensive to justify any conservation action; as one proceeds farther from the city center, recreational value, water provision, and then carbon sequestration become the optimal investments)

Source: Adapted from McDonald (2009).

time without the conservation intervention. The expected gain of the conservation action is the increase in ecosystem services provision over the baseline case scenario of no action. There is a chance that even without conservation action, the natural area will still exist in the future and provide ecosystem services. On the other hand, the natural area could be degraded or destroyed in the future, stopping provision. For evaluating the ROI of conservation actions, an estimate of this probability of degradation in the baseline case is essential.

This conceptual framework gives insight into where ecosystem services projects should be located. Let's say a mayor is interested in increasing carbon sequestration by forests, as part of an overall plan to mitigate the city's contribution of carbon dioxide to the atmosphere. Since carbon sequestration operates on a global scale, the projects the city funds could be located anywhere. The opportunity costs of land and the cost of project management are cheaper farther away from the city center, so the city is always better off conducting the project in a rural area. For just this reason, most commercial sequestration projects trying to generate credits for one of the existing markets (e.g. Certified Emissions Reductions under the UN Framework Convention on Climate Change) are in rural areas. Some cities, of course, have programs like New York City's Plant a Million Trees program that plant trees within the urban extent. But these programs are focused on a whole suite of co-benefits, some quite local, and cannot be justified solely for their carbon sequestration value.

For urban parks to be a place of recreation for urban residents, they have to be located within a reasonable distance from where people live or work. Within this serviceshed, an optimal program would try to maximize the ROI of the action; where ROI is ecosystem services benefits of conservation action (creating the park) divided by the cost of the action (paying to create and maintain the park). This logic tends to lead to creation of urban parks for recreation at a moderate distance from urban centers and where the costs of land are lower than the city center. Finally, consider the benefits that street trees provide to aesthetics or to mitigate the urban heat island effect. These are very local benefits, which necessarily lead to conservation actions close to where people work or live. Working in a city center necessarily implies a high cost of conservation action per unit area. Cities often try to avoid this by creatively reusing vacant land or brownfields, where opportunity costs can be lower.

Key messages

- Three different urban sustainability goals that urban planners are commonly asked to address are conservation in cities (protecting biodiversity in urban areas), conservation by cities (reducing cities' resource use) and conservation for cities (the use of green infrastructure to provide ecosystem services to urbanites).
- Most ecosystem services are public or common goods and so will not be adequately provided by the free market, which is why public sector intervention to build green infrastructure or incentivize its creation is so fundamental.
- Cities planning to create green infrastructure generally go through six steps: taking inventory of which ecosystem services matter; designating critical natural areas for ecosystem services provision; identifying threats to those areas; identifying strategies to mitigate those threats; business planning; and implementation and monitoring.
- The different spatial scale of different ecosystem services implies different optimal places for conservation action to occur, with more local ecosystem services necessarily provided closer to the city center than more long-range ecosystem services.

Key research questions

- If there are tradeoffs between biodiversity protection, reducing natural resource use and the provision of green infrastructure, which facets of urban sustainability are most important?
- What are the most efficient policy mechanisms to give value to most ecosystem services, correcting the current state of market failure?
- How can cities best organize their planning and departments to consider green infrastructure as a unified whole rather than piecemeal?
- When areas of ecosystem services provision lie outside the municipal boundary, how can efficient cross-boundary governance arrangements be negotiated?
- What are institutional structures that can most overcome the collective action problems inherent in many green infrastructure creation projects with many beneficiaries?

References

Aber, J., Christensen, N., Fernandez, J., Franklin, J.F., Hidinger, L., Hunter, M., . . . van Miegroct, H. (2000). Applying ecological principles to management of the U.S. National Forest. *Issues in Ecology*, *6*, 1–20.

Allan, C., and Stankey, G. (2009). *Adaptive environmental management: A practioner's guide*. Kindle Store: Amazon Digital Services.

Alonso, W. (1964). *Location and land use: Toward a general theory of land rent*. Cambridge, MA: Harvard University Press.

APA. (2002). *Growing smart legislative guidebook: Model statutes for planning and the management of change*. Chicago, IL: American Planning Association.

Berke, P., Godshalk, D., and Kaiser, E. (2006). *Urban land use planning* (5th ed.). Champaign, IL: University of Illinois Press.

Boyd, J., and Banzhaf, S. (2006). What are ecosystem services? The need for standardized accounting units *RFF DP 06-02*. Washington, DC: Resources for the Future.

Brooks, T., Bakarr, M., Boucher, T., da Fonseca, G.A.B., Hilton-Taylor, C., Hoekstra, J., . . . Stuart, S. (2004). Coverage provided by the global protected-area system: is it enough? *Bioscience*, *54*(12), 1081–1091.

Chang, H., Parandvash, G.H., and Shandas, V. (2010). Spatial variations of single-family residential water consumption in Portland, Oregon. *Urban Geography*, *31*(7), 953–972.

Copeland, H., Ward, J., and Kiesecker, J. (2007). Assessing tradeoffs in biodiversity, vulnerability and cost when prioritizing conservation sites. *Journal of Conservation Planning*, *3*, 1–16.

Dantzig, G. (1974). *Compact city: Plan for a liveable urban environment*. New York, NY: W.H. Freeman & Co. Limited.

Fagan, W.E., Cantrell, R.S., and Cosner, C. (1999). How habitat edges change species interactions. *American Naturalist*, *153*(2), 165–182.

Forman, R. (2000). Estimate of the area affected ecologically by the road system in the United States. *Conservation Biology*, *14*, 31–35.

Goldman, R., Tallis, P., Kareiva, P., and Daily, G. (2008). Field evidence that ecosystem service projects support biodiversity and diversify options. *Proceedings of the National Academy of Sciences*, *105*(27), 9445–9448.

Groves, C. (2003). *Drafting a conservation blueprint: A practitioner's guide to planning for biodiversity*. Washington, DC: Island Press.

Güneralp, B., and Seto, K.C. (2013). Futures of global urban expansion: Uncertainties and implications for biodiversity conservation. *Environmental Research Letters*, *8*, 014025.

Hanley, N., Shogren, J., and White, B. (2013). *Introduction to environmental economics*. Oxford: Oxford University Press.

Hoekstra, J.M., Boucher, T.M., Ricketts, T.H., and Roberts, C. (2005). Confronting a biome crisis: Global disparities of habitat loss and protection. *Ecology Letters*, *8*(1), 23–29.

Jacobs, J. (1961). *The death and life of great American cities: The failure of town planning*. New York: Random House.

Kenworthy, J., and Laube, F. (2000). *An international sourcebook of automobile dependence in cities 1960–1990*. Boulder, CO: University of Colorado Press.

Kolstad, C.D. (2000). *Environmental economics*. New York: Oxford University Press.

Lloyd, P., Martin, T.E., Redmond, R.L., Langner, U., and Hart, M.M. (2005). Linking demographic effects of habitat fragmentation across landscapes to continental source-sink dynamics. *Ecological Applications*, *15*(5), 1504–1514.

McDonald, R.I. (2008). Global urbanization: Can ecologists identify a sustainable way forward? *Frontiers in Ecology and the Environment*, *6*(2), 99–104.

McDonald, R.I. (2009). Ecosystem service demand and supply along the urban-to-rural gradient. *Journal of Conservation Planning*, *5*, 1–14.

McDonald, R. I. 2015. *Conservation for cities: How to plan and build natural infrastructure*. Washington, DC: Island Press.

McDonald, R.I., and Boucher, T. (2010). Global development and the future of the protected area strategy. *Biological Conservation*, *144*, 383–392.

McDonald, R.I., Kareiva, P., and Forman, R. (2008). The implications of urban growth for global protected areas and biodiversity conservation. *Biological Conservation*, *141*, 1695–1703.

McGranahan, G., Marcotullio, P., Bai, X., Balk, D., Braga, T., Douglas, I., ... Zlotnik, H. (2006). Urban systems. In R. Hassan, R. Scholes and N. Ash (Eds.), *Ecosystems and human well-being: Current state and trends* (Vol. 1). Washington, DC: Island Press, pp. 795–825.

MEA. (2003). *Ecosystems and human well-being: A framework for assessment.* Washington, DC: Island Press.

Mittermeier, R.A., Robles-Gil, P., Hoffmann, M., Pilgrim, J.D., Brooks, T.B., Mittermeier, C.G., Lamoreux, J.L., and Fonseca, G.A.B. (2004). *Hotspots revisited: Earth's biologically richest and most endangered ecoregions.* Mexico City: CEMEX.

Murcia, C. (1995). Edge effects in fragmented forests – implications for conservation. *Trends in Ecology & Evolution, 10*(2), 58–62.

Murdoch, W., Polasky, S., Wilson, K.A., Possingham, H.P., Kareiva, P., and Shaw, R. (2007). Maximizing return on investment in conservation. *Biological Conservation, 139*, 375–388.

Myers, N., Mittermeier, R.A., Mittermeier, C.G., da Fonseca, G.A.B., and Kent, J. (2000). Biodiversity hotspots for conservation priorities. *Nature, 403*(6772), 853–858.

Natural Capital Project. (2013). Application of Invest's Sedimentation Retention model for restoration benefits forecast at Cantareira Water Supply System. Palo Alto, CA: Natural Capital Project, Stanford University. Online at http://ncp-dev.stanford.edu/~dataportal/natcap/NatCap%20Publications/Invest_Sediment_Retention_Cantareira.pdf.

Norman, J., MacLean, H.L., and Kennedy, C.A. (2006). Comparing high and low residential density: Life-cycle analysis of energy use and greenhouse gas emissions. *Journal of Urban Planning and Development, 132*(1), 10–21.

Prasad, N., Ranghieri, F., Shah, F., Trohanis, Z., Kessler, E., and Sinha, R. (2009). *Climate resilient cities: A primer on reducing vulnerabilities to disasters.* Washington, DC: The World Bank.

Rheindt, F.E. (2003). The impact of roads on birds: Does song frequency play a role in determining susceptibility to noise pollution? *Journal Fur Ornithologie, 144*(3), 295–306.

Richardson, H.W. (2013). *The new urban economics: And alternatives.* New York: Routledge.

Rockefeller Foundation, and ARUP. (2014). *City resilience framework.* New York, NY: Rockefeller Foundation.

Rodrigues, A.S.L., and Gaston, K.J. (2002). Optimisation in reserve selection procedures – why not? *Biological Conservation, 107*(1), 123–129.

Sarkar, S., Pressey, R.L., Faith, D.P., Margules, C.R., Fuller, T., Stoms, D.M., ... Andelman, S. (2006). Biodiversity conservation planning tools: Present status and challenges for the future. *Annual Review of Environment and Resources, 31*, 123–159.

Schneider, A., Friedl, M.A., and Potere, D. (2009). A new map of global urban extent from MODIS satellite data. *Environment Res. Lett., 4*, 0044003.

Seto, K.C., Sanchez-Rodriguez, R., and Fragkias, M. (2010). The new geography of contemporary urbanization and the environment. *Annual Review of Environment and Resources, 35*, 167–194.

Seto, K.C., Güneralp, B., and Hutyra, L.R. (2012). Global forecasts of urban expansion to 2030 and direct impacts on biodiversity and carbon pools. *Proceedings of the National Academy of Sciences of the United States of America, 109*(40), 16083–16088.

TEEB. (2011). *TEEB manual for cities: Ecosystem services in urban management.* The Economics of Ecosystems and Biodiversity.

UNPD. (2011). *World urbanization prospects: The 2011 revision.* New York: United Nations Population Division.

Wilson, K.A., McBride, M.F., Bode, M., and Possingham, H.P. (2006). Prioritizing global conservation efforts. *Nature, 440*(7082), 337–340.

30

THE POTENTIAL OF THE GREEN ECONOMY AND URBAN GREENING FOR ADDRESSING URBAN ENVIRONMENTAL CHANGE

David Simon

Although it has only relatively recently become a buzzword in the lexicon of environmental discourse internationally, the origin of the 'green economy' concept dates back to the 1980s, when sustainable development was popularized in the wake of the 1987 report on *Our Common Future* by the World Commission on Environment and Development (WCED, 1987). Possibly the first detailed effort to operationalize the concept was a report for the UK government's then Department of the Environment (Pearce et al., 1989). This environmental economic analysis examined different mechanisms for valuing the environment quantitatively in order to ensure that it was appropriately valued and included in economic decision-making as a set of resources rather than being exploited and polluted as something without specific value. Even environmental damage was to be valued. This report received wide acclaim, not least because the neoclassical economic approach underpinning the work offered market-based solutions and was thus more ideologically and politically acceptable. Despite the book's title, its focus was explicitly on sustainable development; no definition of 'green economy' was offered and the term does not even appear in the index.

Since then, the green economy has become shorthand for economic growth or development that is environmentally sympathetic, increases the efficiency of resource use and minimizes waste—in part through recognizing that most forms of waste represent actual or potential resources for other processes, which should be valued accordingly. The term has recently gained renewed currency in the context of attempts to mediate the effects and adapt to the realities of increasing climate/environmental change (EC). The Stern Review Report (Stern, 2007) was instrumental in drawing attention to the future economic savings to society by prompt action rather than delaying mitigation and adaptation. This landmark study contributed to the adoption in 2009 of the green economic approach by the G20 group of industrialized countries and to the work of the UN High-Level Panel on Global Sustainability (UNGSP, 2012). The latter, in turn, fed into the campaign within the UN System to secure a specific Sustainable Development Goal (SDG) on urbanization within the set of SDGs to succeed the Millennium Development Goals (MDGs) from 2016. Its contents include targets for urban greening and urban economic sustainability. More broadly, growing attention has been invested in economic greening in the context of urbanization and urban areas, and this chapter assesses the potential of such interventions.

Differentiating the 'green economy'

As with sustainable development itself, the 'green economy' is invoked in diverse ways and for widely different purposes. These range from self-interested or cynical 'greenwash', used to project a progressive and environmentally sensitive image of a company's activities, which may be polluting at one extreme, through a spectrum of incremental or reformist approaches to environmental mediation and enhancement, to holistic alternative approaches to human development and ecocentrism at the other extreme. Some interpretations also include elements of environmental justice or fairness, while in others these are separate issues. Bina (2013) provides a nuanced assessment of some of the tensions and compromises inherent in different approaches.

In view of this diversity of use, it is helpful to distinguish broadly between 'weak' and 'strong' green economic initiatives. The former category comprises interventions which are essentially positive and incremental or reformist in nature. In other words, their implementation does not require substantial changes to current production processes and behavior patterns. Examples include the recycling of domestic and commercial waste rather than sending it to landfill, or fitting movement-sensitive passive infrared (PIR) sensors to lighting that does not need to burn continuously. By contrast, 'strong' initiatives do involve more substantive changes to the status quo and make a commensurately larger contribution, although they often incur greater investment costs and may meet resistance from those whose vested interests might be challenged or who fear change for other reasons. A landmark example is the development of cleaner, more resource-efficient production technologies, and designing urban areas or entire new eco-cities to foster compact living and lower per capita emissions through integrated land use that minimizes commuting and single-purpose journeys.

The overarching objective of green schemes is to achieve sustainability by decoupling economic growth and human development from the longstanding linear or geometric increases in resource and energy consumption that have traditionally accompanied them. This involves increasing the efficiency of resource use (especially energy derived from fossil fuels), so that carbon emissions are reduced or eliminated (the so-called zero carbon approach).

The urgency of greening urban economies

As detailed in Part I of this Handbook, the global diversity of urban forms and trajectories limits the scope for generalizations. The importance of urban areas in the context of EC derives from their concentration of human populations, industries, motor vehicles and energy consumption, which together account for up to 80 percent of worldwide commercial energy consumption and the majority of greenhouse gas (GHG) emissions. The spectacular damage caused by extreme events like Hurricanes Katrina, Sandy and Haiyan in recent years tends to focus international media attention on the vulnerability of low-lying parts of coastal cities—and those who live or work in them—to the effects of EC, and thus the need for mitigation and adaptive coping strategies. The importance of cities as major sources of GHGs and thus as motors of EC as well as the challenges of mitigating those emissions and transforming urban areas to promote sustainability are, however, generally under-reported.

Evidence from different continents shows that overall urban energy consumption is a complex function of urban form and structure, industrial composition and technologies, residential layouts and densities, and the nature of transport systems (modal mix, extent of public versus private transport use and pricing strategies), with per capita consumption, emissions and vehicle kilometers travelled generally lower in denser, more compact urban segments and cities (Anderson et al., 1996; Kenworthy and Laube, 1999; Mindali et al., 2004; Bin and Dowlatabadi, 2005; Norman et al., 2006; Chen et al., 2008; Vance and Hedel, 2008; National Research Council, 2009; Hankey and Marshall, 2010). Comparative studies of emissions inventories have shown different relationships between per capita emissions in urban areas—and their respective national averages—depending on the energy sources and availability of scale economies of power generation and supply; the nature of urban economic production (sectoral mix and whether purely for domestic or both domestic and export markets) and its technological mix; residential and employment densities; average incomes and population densities; and on the extent of electric/electronic appliance and motorized transport use (Dodman, 2009; Ibrahim et al., 2012; Chavez and Ramaswami, 2013). One key way to reduce emissions is through investment in low carbon infrastructure, but the nature, extent and likely efficacy also vary according to individual city characteristics, which can nevertheless be categorized in relation to the above criteria (Kennedy et al., 2014). These are all pertinent considerations in planning for greater urban sustainability and implementing green economic policies and projects.

As a result of global economic restructuring since the 1980s, many traditional urban industrial heartlands and emissions hotspots in Organization for Economic Cooperation and Development (OECD) countries have deindustrialized and their post-industrial living environments have been transformed, with greatly reduced environmental pollution, except for motor vehicle emissions and PM_{10} micro particles. Some deindustrialized areas remain partly derelict or vacant, representing new opportunities for brownfield redevelopment in accordance with green economic principles (see below). By contrast, most of today's emissions hotspots are major, and often fast-growing industrial conurbations in low- and middle-income countries, where the infrastructure, services, governance institutions, resources and political will to address EC are often inadequate.

This is hardly surprising since, under such resource constraints, 'ordinary' and immediate economic and human development priorities are highly contested and often subject to clientelism or corrupt allocation. The basic problem is that until very recently EC was perceived as being a distant threat in the sense of impacts being felt mainly in several decades' time. In addition, the scale of predicted impacts would be well beyond the fiscal and technical capacity of poorer countries and cities to cope, even in an optimistic scenario. There is, furthermore, a widespread perception that any investment in mitigation or adaptation actions, however modest, would represent a diversion of resources away from development efforts; in other words, that there is a direct and intractable conflict between them. This represents a particular case of the long-held view that there are inevitable development–environment conflicts, in terms of which environmental conservation is held to be essentially a middle class and elite concern of little relevance to the poor majority. For the latter, employment and earning an income constitute the top priority, even at the expense of environmental pollution or degradation. Such issues are also pertinent in poor and deprived communities in wealthy countries—perhaps most extremely in inner-city 'slums' and the US 'rustbelt'—where 'social welfare' and 'urban regeneration' represent the discursive equivalent of 'development'. Yet, evidence is growing that locally appropriate and well-targeted interventions can avoid such conflicts, promoting mitigation and/or adaptation as well as enhancing immediate development or social needs (e.g. Rabinovitch, 1992; Schipper and Pelling, 2006; Parnell et al., 2007; Saunders, 2008; Dodman and Mitlin, 2011; Simon, 2011; Gasper et al., 2013; Satterthwaite and Dodman, 2013; Sánchez-Rodríguez, Chapter 24 in this volume).

A more specific variant of the development–environment conflict argument relating to green urban economic initiatives centers on the issue of jobs and employment. Traditional employers in energy-intensive and polluting industries seeking to defend their vested interests on the one hand, and (neo-)populist political, trade union and non-governmental leaders and community groups articulating the imperative of maximizing 'jobs now' ahead of tackling environmental problems on the other, often combine to oppose green economic agendas. Their concerns are that shutting dirty industries or forcing them to implement costly emissions reduction, pollution abatement or other environmental measures will raise costs and undermine the viability of their operations, jeopardizing the national and local tax revenues derived from them and threatening the jobs of their employees.

There is certainly some validity to this concern in respect of the aging heavy industrial plant that characterizes many iron and steel mills and petrochemical complexes in cities as diverse as Detroit and Milwaukee, Dresden and Lodz, Cape Town, Durban, Mombasa, Kolkota and Cubatão, for instance. Many older coal-fired power stations burn low-quality coal with high sulphur content. This constitutes a serious source of air pollution. China and India are distinctive in having expanded their coal-fired generating capacity massively in recent years, relying heavily on high-sulphur coal. The question must be asked, however, why most old industrial plants have been neglected for so long, rather than being modernized and made more energy efficient (and hence cleaner) as in other parts of the world, especially while those industries were highly profitable. Each case would need examining on its merits in relation to possible closure, modernization, upgrading and/or the implementation of abatement measures. Making successful transitions in large mineral-energy-industrial complexes will present difficult long-term challenges because of national and perhaps even regional strategic considerations. However, sometimes sustained social action by residents' groups living in the shadow of such plants and suffering from the pollution, as occurred in Cubatão (Brazil) and South Durban (South Africa), can be instrumental in promoting change (e.g. De Mello Lemos, 1998; Barnett and Scott, 2007; Scott and Barnett, 2009).

An important issue often overlooked by opponents of economic greening is that undertaking such remediation or new construction itself creates employment through the manufacture, installation and subsequent operation and maintenance of the retrofitted equipment and new plant. Technological change permitting a switch to lower carbon production also creates new jobs, as do new green industries such as the renewable energy sector and new materials, new eco-friendly building construction, factory and home insulation and other forms of intervention (see below). In relation particularly to poor and under-served urban areas or segments, additional employment along with enhanced service delivery and efficiency gains as well as improved environmental quality and health benefits can easily be obtained through the repair, rehabilitation, expansion and upgrading of defective or inadequate infrastructure to a standard better able to withstand expected EC. Examples range from roads and pedestrian walkways to water reticulation and wastewater drainage and sewerage systems, effluent and sewage treatment plants, and refuse collection and recycling (including the composting of organic waste). Particularly in the context of rapid urban growth, such interventions in the existing parts of cities as well as greener construction of new zones can offset or reduce the overall impact and cost of urbanization (Güneralp and Seto, 2012). The balance between aggregate employment losses and gains will differ by locality, but in many cases may be positive overall. To this must be added the health and environmental benefits to employees and the public.

This is the context in which efforts at both 'weak' and 'strong' urban economic greening should be understood and promoted. There is a broad spectrum of measures available at different scales and offering diverse benefits. While some require major strategic engagement and capital investment, others are sufficiently modest in scope, scale and cost to be implemented and/or maintained by individual households and community or neighborhood groups (Castán Broto and Bulkeley, 2013; Simon, 2013). The remainder of this chapter assesses several noteworthy categories of green economic intervention. Coverage is necessarily selective; some other topics, such as integrated transport and retrofitting existing buildings for energy efficiency, are referred to within these categories and are also addressed fully elsewhere in this volume.

The co-benefits of urban green economic investments

Renewable energy

Renewable energy exemplifies green economic investments that integrate rural, peri-urban and urban areas, sometimes literally, in terms of the generation and supply of electricity from on- and offshore wind farms, hydropower dams and geothermal vents, and sometimes figuratively in the sense that certain technologies like micro-wind turbines and solar panels are appropriate for installation on individual homes and commercial premises regardless of their location. Most biofuels, the final major renewable category, are used to power engines in urban areas. Many countries now have targets for renewable energy generation. In the UK, for instance, it is 20 percent of final energy (i.e. the forms of energy used by consumers) by 2020, in line with the European Union baseline target. It is unclear whether this will be achieved, in part because of changing subsidy policies in relation to the high uptake rate and an increasingly pro-nuclear policy as the government's principal route to meeting its energy emissions reduction targets. Some of the UK's largest wind farms lie just offshore, beyond river estuaries serving cities such as London and Liverpool, while the strong tidal bore in the Severn Estuary makes tidal power a potential future energy source for cities like Bristol, Newport and Cardiff.

Aggregate electricity demand is rising universally, through a combination of population growth, urbanization, increasing incomes and increased use of electric appliances (from computers to air conditioning units). Consequently, however ambitious and determinedly pursued a country's renewable energy target, this will comprise only a proportion of the total energy generation portfolio. The most dramatic turnaround in policy is that of China, which is pursuing an aggressive increase in generation by all methods, from coal through hydro-electric to nuclear, wind and solar. Its renewable target is 15 percent of primary generation by 2020. It has become the world's largest manufacturer of turbines and solar panels, originally principally for export, but now increasingly for domestic use too, and providing considerable employment as a result. Even before the latest surge, worldwide employment in renewables was estimated at three million in 2009, principally in China (1.12 million), Brazil (500,000, particularly in biofuels), the US (406,000), Germany (278,000, more than double the 2004 figure) and Spain (81,000). The total by 2030 is projected to exceed 20 million, 60 percent of them in biofuels and a third in solar—which provides 7–11 jobs per megawatt of capacity, although this is falling in line with the price of photovoltaic (PV) panels (Renner et al., 2008; ILO and EU, 2011: 6–7). Unfortunately, no urban–rural breakdown is available.

In rural areas and in parts of urban areas that are not served by the electricity grid or where power supplies are intermittent, private diesel generators have until very recently been the principal method of securing electricity supplies at the household level. However, solar

Box 30.1 Demonstrable co-benefits of PV installations projects: Nairobi and Cape Town

Two African examples illustrate the multiple co-benefits of urban PV schemes. The Kibera Community Youth Programme in Nairobi initiated a simple solar PV assembly project, providing employment in the city's largest shantytown and stimulating interest among groups in neighboring countries (UNEP, 2011a; Simon et al., 2011: 21). In the Kuyasa low-income housing improvement scheme in a high-density area in metropolitan Cape Town, solar water heating and roof insulation were retrofitted in 2,300 poorly built post-apartheid houses under South Africa's first urban Clean Development Mechanism (CDM) project. This provided on-the-job training and employment, while earning 2.82 tonnes of carbon credit per house annually. Besides reducing expenditure on heating fuel (and hence reducing emissions), the improved insulation also had the important co-benefit of a substantial decline in bronchial and related illness among residents, especially during winter, thus reducing household expenditure on medical bills and medicines, so that meager incomes can be used for other priorities (Meyer and Odeku, 2009; Simon et al., 2011: 21).

PV panels and micro-turbines are rapidly changing that. In most climates, PV cells represent a good investment, especially in view of the rapid decline in their purchase price over the last few years. They also provide security of supply for part of or all domestic needs, reducing or removing the previous disadvantage of not being on the grid, and perhaps conveying some advantages to that status. Because such indirect benefits, however real and substantive, were not the explicit reason for investing in the technology, they are known as co-benefits.

As with all interventions, however, comparative research in East and Southeast Asia shows the importance of the local appropriateness of the package of beneficiary control and senses of ownership of the project, precise technology, financing mechanism and affordability versus payback period of PV installation projects. Environmental concerns are seldom high or even relevant among poorer people's priorities, although they do benefit substantially if the projects are successful (Sovacool and D'Agostino, 2012). Most household-level actions are 'weak' measures but at larger scales they may well be 'stronger'.

Recycling, reuse and emissions mitigation

Progress with urban waste reduction and recycling has been made worldwide, albeit to very different extents and using diverse strategies. Many OECD countries have been only modestly successful in waste reduction, mainly through weak measures such as charging for plastic carry bags, reducing the volume and weight of some packaging and applying a deposit to returnable drinking bottles and cans to encourage return. EU targets have also promoted phased increases in recycling as a proportion of waste by imposing landfill taxes and penalties for failure to meet successive targets. Voluntary recycling schemes have gradually been replaced by mandatory municipal door-to-door collections and improved recycling facilities for bulky and specialized items. Diverse technologies are used to sort mixed household waste, ranging from magnets and other electro-mechanical devices to sieving and labor-intensive manual sorting on conveyor belts. Less easily recycled material may be incinerated, usually to produce electricity and avoid landfill. Waste reduction and recycling levels are now commonly around 50 percent. In Copenhagen, Denmark, around one-third of waste is recycled and the rest

Box 30.2 Community-based waste recycling: Cairo and Curitiba

Some 15 million people are estimated to be involved in the collection and processing of urban waste globally; 80,000 of them in Bangladesh alone (UN-HABITAT, 2013: 100). One of the most extensive and best-known such systems is operated in Cairo by the Coptic Christian minority Zabaleen community, which, over the last century, has developed a sophisticated and widespread door-to-door collection, sorting and recycling operation centered on their Al Mokattam facility. They account for an estimated 60 percent of the city's solid waste (Fahmi and Sutton, 2010; SDI, 2012). Nevertheless, in recent years, they have come under pressure from the city authorities, who are attempting to replace them with conventional, more 'modern' and privatized systems that currently handle some 30 percent of the waste. Displacing the Zabaleen to the urban periphery as envisaged would add to unemployment, raise costs and quite possibly reduce the efficiency and effectiveness of refuse collection. A novel system for exchanging recyclable materials for fresh fruit and vegetables, known as Cambio Verde (Green Exchange), has operated in the Brazilian city of Curitiba for over 20 years (UNEP, 2011b).

incinerated for energy generation. The city is a leader in district heating systems that provide integrated heating for local areas with high efficiency that saves around one million tonnes of CO_2 emissions (Simon et al., 2011: 22–23). Old landfill sites are being tapped for methane as a resource while also mitigating GHG emissions. Partly funded by the CDM, eThekwini Municipality in Durban, South Africa, now operates three methane capture sites—possibly the first in Africa—within its Integrated Waste Management program (Simon, 2013: 211). Stable old landfill sites are often being rehabilitated for use as brownfield developments, especially as sports fields and leisure facilities.

In lower income cities, regular door-to-door refuse collections by local authorities and/or private firms are geographically restricted, often to older, established areas occupied largely by the elites and middle classes, and the degree of recycling is often low. In shantytowns, 'slums' and other informal areas, services are generally inadequate and irregular, with the result that piles of uncollected refuse often arise on open land or roadways, representing obstructions and health hazards. Neighborhood upgrading schemes commonly include improved collection, but some community groups have organized themselves to address the problem and earn revenue through the sale of recyclable materials.

Labor-intensive composting of organic solid waste is also practiced in various cities worldwide, from San Francisco in the US to Dhaka in Bangladesh, either on a modest scale in particular areas or at a municipal or decentralized sub-municipal level, greatly reducing waste disposal costs and producing a revenue stream from the sale of compost that offsets part of the collection and processing costs. Some NGOs, community groups or entrepreneurial groups of residents undertake more modest such schemes, selling the compost to private homeowners for their gardens, to seedling nurseries and urban and peri-urban agriculturalists. Since much of this employment is semi-formal or informal, no reliable estimates exist.

Environmental rehabilitation, green planting and ecosystem services

Most modern urban areas are poorly adapted to sustainability requirements and the implications of EC, being generally designed against rather than with nature. This relates both to resource and waste flows into and out of towns and cities, and to the hazards and potential risks of current design. In contrast to many pre-colonial indigenous traditions, where urban structures, individual buildings and building materials were often appropriate to local conditions, contemporary Western-derived 'international' architecture, building materials and urban design are seldom adapted to different climatic, socio-economic and cultural conditions in tropical and subtropical zones. They are import- and energy-intensive, and when rolled out en masse for burgeoning urban populations including those unable to afford air conditioning or fans, for instance, can become oppressive. In urban areas where EC is leading to increased prevailing temperatures and reduced and more variable rainfall, heat island effects will become more acute, as will problems associated with wind, wind-blown sand and securing water and food supplies.

Contemporary cities everywhere increasingly comprise predominantly 'hard' infrastructure and surfaces, where bricks, steel, aluminium, glass, concrete and tarmac cover ever larger areas and are impervious to water. Hence, rainfall runoff has to be channelled into effective drainage systems, but these are already proving increasingly inadequate in areas experiencing increasing volumes and intensity of rainfall under EC, leading to increased flood risk. In coastal areas where intense rainfall events coincide with spring tides or storm surges, the effects can be particularly severe. Examples of this include one-off extreme events like the hurricanes mentioned above and severe individual storms, or more prolonged series of events such as the unprecedented winter storms and rainfall in the UK during the winter of 2013/14, which damaged and destroyed sea defenses and also inundated low-lying farmland, and inland towns and villages when saturated ground could not absorb additional rainfall and rivers burst their banks.

Addressing this challenge requires a multifaceted approach. At one level, design standards for drainage and water storage are increasingly being re-thought, although the cost of expanding capacity is high. At another level, some building materials and construction approaches are being modified, for instance using permeable paving bricks and slabs, so that water percolates through and infiltrates the soil rather than running off. This has a dual benefit—it moistens the subsoil and can help maintain the water table to nourish vegetation, and it also reduces the volume and intensity of runoff. These are now compulsory in British urban areas. Similarly, rainwater harvested from roofs, carparks and other large impervious surfaces can be stored on site and provide significant supplies for watering vegetation in gardens, road margins, parks and other recreational areas, cultivation of urban crops (see below) and, importantly, also for non-drinking domestic purposes such as flushing toilets, and washing bodies, dishes, laundry and motor vehicles. Underlying this is a necessary shift in thinking from seeing urban runoff as an increasingly expensive waste product to be disposed of, to being a valuable resource that should be stored and used for domestic and commercial tasks, and to maintain ecosystem services not requiring potable quality water. This is already being done by some individual households, often facilitated as part of local schemes, thereby helping to reduce the demand for expensively purified water, especially during dry periods. In slums and shantytowns lacking piped water supplies to each home, rainwater harvesting can have a particularly positive impact on overall access to water by poor households. To date, there has been little systematic application of rainwater harvesting by local authorities, although localized greywater storage and use on sports fields and parklands is becoming more accepted, sometimes providing an alternative to the use of borehole water and thereby reducing pressure on the water table (see Simon et al., 2011; UNEP 2011a; Elmqvist et al., 2013 for examples worldwide).

Complementing such efforts should be active greening of urban areas through additional planting of vegetation to provide tree cover for shade (preferably using indigenous species that are more resilient and economical in water demand). The consequent increased total biomass would act as local carbon sinks and help to mitigate heat island effects through shade and maintaining humidity. Turning wasteland or brownfield sites into parks, sports fields and other recreational areas can play a notable role, with added co-benefits of improving leisure facilities, thereby improving people's health. Manhattan Island in New York City provides illustrative examples in the Hudson River Park and the High Line Park, which are rehabilitated areas of derelict waterfront land and a disused overhead subway track respectively. Individual households can and do play significant roles, through maximizing the biomass on their land, turning balconies and flat roofs on multi-story buildings into rooftop gardens, or painting roofs silver to reflect sunlight in order to help cool the buildings and reduce the need for fans and air conditioners. Again, Manhattan represents a particularly good example. Encouraging urban and peri-urban agriculture and forestry at both individual household and larger commercial scales can also play a role in greening, with sometimes significant contributions to urban food supply and security, while providing considerable labor-intensive employment, especially but not exclusively for the urban poor (Lwasa et al., 2014).

In many urban situations, it has been recognized that 'hard' engineering interventions such as river canalization and seawall construction, undertaken in the past to mitigate or prevent flooding in vulnerable locations, often displace the flooding downstream or along the coast. As peak river flows and the severity and frequency of flooding increase due to the combined effects of urban expansion—not least through construction in floodplains without adequate protection—and the effects of EC, riverine and coastal flood defenses are proving inadequate, often being damaged or destroyed, while less well protected areas nearby suffer intensely (Rosenzweig et al., 2011; Roberts et al., 2012).

At the same time, there has been a growing realization that riverbank and coastal zones still protected by relatively intact reed beds and mangroves, and perennial or seasonal adjacent wetlands and meadows still functioning as natural floodplains, cope better and suffer less damage during flood events. Consequently, there are now many active programs of wetland, riparian and shoreline conservation and rehabilitation, generally using labor-intensive methods and often including environmental education for schoolchildren and youth (Simon et al., 2011). From South Africa to Taiwan, some urban river canalization is even being reversed in favor of vegetated banks (e.g. Chou, 2012), while restrictions on floodplain construction are being firmed up and enforcement improved. Although some flood defenses are being reinforced where deemed appropriate, along parts of the middle stretches of the River Thames in southeast England, for instance, recent flood relief schemes have included controlled water diversions into the artificial Jubilee River plus floodplains and overflow reservoirs, with wildlife habitat and leisure enhancements (www.thamesweb.co.uk/floodrelief/relief_bckgrnd. html), although these proved inadequate to cope with the unprecedented floods of early 2014. Construction of Dorney Lake rowing centre on the outskirts of Windsor, which hosted the rowing and kayaking events of the London Olympics in 2012, was enabled by one such scheme and illustrates the co-benefits of imaginative projects that turn problems into resources and opportunities.

All the initiatives outlined here are increasingly linked to the enhancement of urban biodiversity and an appreciation of the value of ecosystem services provided by conserved, restored and newly created natural areas and even household-scale urban greening (Elmqvist et al., 2013; Elmqvist et al., Chapter 10 in this volume). One distinctive variant is the Durban Metropolitan Open Space System (D'MOSS), a linked network of specially protected land and water-based areas, from parks and sports fields to rehabilitated riverbank and wetland areas spanning the eThekwini municipality and purposely including low- and high-income communities. This so-called 'green infrastructure' provides enhanced ecosystem services including climate resilience via the mechanisms described above, while increasing access to leisure areas, with associated public health benefits and hopefully reduced antisocial behavior by those with previously inadequate access (Roberts et al., 2012; Simon, 2013; Simon et al., 2011; Trundle and McEvoy, Chapter 19 in this volume).

This example also highlights the importance of integration at two levels. First, the network of green infrastructural nodes and corridors enables the whole to become more than the sum of its parts, both in terms of biodiversity enhancement and enabling movement of species rather than having them confined to often small individual locations. Second, D'MOSS itself forms part of a citywide approach to post-apartheid integration and development, which includes special attention to historically disadvantaged groups and the areas in which they live as well as to sustainability and climate adaptation/resilience (Roberts et al., 2012; Simon, 2013). Such principles of integration, equity and environmental justice promotion should be widely adopted and, wherever possible, linked to appropriate local, national or global targets like the MDGs or their successor SDGs in order to highlight the synergies and co-benefits between meeting development, social and EC objectives. This could also demonstrate that portfolios of complementary 'weak' interventions can have overall stronger impact.

Building elite climate-smart sustainability at the city scale: eco-city experiments

Recent years have seen the launch of several ambitious initiatives to build experimental new cities or communities designed on supposedly environmentally sustainable and low carbon principles. In contrast to the predominantly micro- or at best meso-scale interventions outlined above, such initiatives are claimed to point the way towards more sustainable future urbanism at scale and to characterize what is being done within existing urban areas. They are therefore being promoted aggressively, partly also in order to establish the environmental credentials of the city administrations or national governments behind the experiments, all of which have hitherto had reputations for unsustainable urbanization in their respective drives for modernity and international prominence.

The best-known 'eco-cities' or 'eco-districts'—all still in the design or construction stages—include Lavasa between Mumbai and Pune in India, Dongtan Eco-city on an island outside Shanghai, Masdar City in Abu Dhabi and the waterfront Songdo International Business District in Incheon (South Korea) (Datta, 2012; Chang and Sheppard, 2013; Cugurullo, 2013; Danish Architecture Centre, 2013; Meinhold, 2009; Kohn Pedersen Fox Associates, 2013; Shwayri, 2013). Although diverse in inspiration, specific objectives and design, they have all been commissioned from international architectural and urban design consultancy firms originating in Europe and North America and conveying ultra-modern and increasingly globalized 'Western' urban idioms and using the latest materials and low carbon technologies. Construction of Dongtan—the first design to be commissioned and the most ambitious in terms of having a design population of half a million living in energy self-sufficiency through a combination of energy-neutral buildings and entirely renewable on-site energy generation using biofuels, wind turbines and solar panels—has been delayed, so it has now been overtaken by the other two. The first phase of Songdo was completed already in 2009; when finished, the District as a whole should house 70,000 residents, augmented by up to 300,000 daily commuters living and working in solar-passive buildings with green roofs and benefiting from integrated, ultra-efficient public transport and extensive open space.

While it is naturally premature to judge how these 'strong' experiments will turn out and to what extent they offer genuinely sustainable and climate-resilient models, the designs and objectives demonstrate a clear orientation towards hi-tech modernity, as befits the 'international' ambitions of their sponsors and developers. As such, they are intended to house populations of professionals, living in essentially Western-style housing as nuclear families or young singles, many of them working for the kinds of transnational business, personal financial and other services firms that are likely to want and be able to afford such premises and housing. Conspicuously absent are other, more locally derived or hybrid visions of living and working in extended families; of non-Western housing styles and living arrangements; opportunities to practice more traditional, 'informal' or other livelihood activities that might be appropriate to poorer people; or facilities and services suited to elderly or other people with diverse needs. Indigenous architectural or urban design principles, which have rich histories in all three countries and were generally culturally and environmentally appropriate, also appear to find no resonance in the designs. As such, it is likely that these will turn out to be elite projects (Datta, 2012; Chang and Sheppard, 2013; Cugurullo, 2013; Shwayri, 2013). Locally based firms more attuned to the diversity and complexity of local conditions and to indigenous traditions that can be incorporated into forward-looking hybrid designs will perhaps be better positioned to produce more socially and economically inclusive urban designs appropriate to the needs and aspirations of the urban majority and the principles of which might prove replicable to diverse conditions elsewhere. The extent to which such designs prove scalable within the respective socio-cultural realms also remains to be seen. However, such initiatives are also likely to generate longer-term and more sustainable, locally appropriate construction industries as enduring elements of green urban economies.

Conclusions

Increasing attention to urban greening and sustainability around the world is a welcome development, although the diversity of interpretation and practice is great. Some claims have been shown to be largely cosmetic cover for continuing with inappropriate business as usual. Positive interventions within existing urban areas range in scale and initiator from individual households to neighborhood and community groups, private corporate projects, local authority schemes and public–private partnerships. There are no quick fixes or silver bullets, but portfolios of activities can have a substantial impact in addressing current problems and enhancing the liveability of towns and cities facing EC.

Importantly, this chapter has shown that the claimed conflicts between meeting 'ordinary' short-term social welfare or development needs and mitigating and/or adapting to EC for longer-term sustainability have been exaggerated. Carefully targeted and locally appropriate interventions often address both issues as well as have other co-benefits. This should be a game changer in favor of accelerated action in tackling social deprivation and promoting post-industrial inner-city regeneration in OECD countries or meeting basic needs and upgrading impoverished shantytowns in poor countries. Illustrative of how such challenges can be met strategically is UN-HABITAT's City Prosperity Index comprising five components (each made up of several sub-indices), of which only two are traditional economic indicators (productivity and infrastructure development); the other three reflect a more holistic and sustainable approach, namely quality of life, environmental sustainability, and equity and social inclusion (UN-HABITAT, 2013: 16–25).

Supposed eco-cities and other bold new vision statements of low carbon, energy-neutral urbanism are under construction in several countries, but may turn out to be elite projects of less value to the urban poor who comprise the majority of those most vulnerable and least resilient to EC. Nevertheless, the power of such ultra-modern visions seems irresistible to political leaderships, not least in countries and cities seeking global leadership roles. The Indian government of Narendra Modi, elected in 2014, is a case in point (Datta 2014, 2015). This is a dynamic topic worthy of critical research.

Green or environmentally friendly products are increasingly available, while sustainable/low-energy construction materials and processes are becoming more widely accepted, with new industry standards emerging. Social and economic acceptability in different regions remains under-researched, however. Similarly, truly green and sustainable economics still requires a profound shift from short-term convenience and profit towards longer-term economic and socio-environmental accounting. Indicative of such thinking that still awaits implementation is a shift from considering 'greenness' in the use of products towards a commodity life approach that accounts for the energy and resources used in the manufacture and transport of the product to the point of end use, and then towards its dismantling and recycling once no longer useful. One good example would be the production and distribution of PV panels and other energy generation equipment. Similarly, assessments of the relative efficiency and other merits of the various forms of renewable energy generation should be extended to include conventional energy generation and the subsidy regimes supporting each of them in different countries.

Key conclusions

- The green economy is no silver bullet and does involve trade-offs, but has substantial potential in urban contexts worldwide.
- Well-targeted interventions can provide economic opportunities, address climate change mitigation and adaptation as well as producing health or lifestyle benefits.
- Many modest or incremental aspects of urban greening and ecosystem services are readily being adopted, but the step change across thresholds required to generate more structural transformative change remains elusive.
- Eco-cities, smart cities and other invocations of bold new district or city scale sustainable urbanism are likely to become elite projects with resource demands and impacts that defy sustainability, especially in challenging environments.

Important research questions

- How do we achieve the required shift from short-termism to long-term economic thinking and planning required for urban sustainability?
- How do we calculate and implement product life cycle energy and carbon accounting?
- To what extent can current models of eco- and smart cities actually prove sustainable or what alternative and scalable conceptualizations are required for relevance in contexts of poverty and pressing development needs?
- How can we identify intervention points to promote transformative adaptation and greening?

References

Anderson, W. P., Kanaroglou, P. S. and Miller, E. J. (1996) Urban form, energy and the environment: A review of issues, evidence and policy, *Urban Studies*, vol. 33, no. 1, pp. 7–35.
Barnett, C. and Scott, D. (2007) Spaces of opposition: Activism and deliberation in post-apartheid environmental politics, *Environment and Planning A*, vol. 39, no. 11, pp. 2612–2631.

The green economy and urban greening

Bin, S. and Dowlatabadi, H. (2005) Consumer lifestyle approach to US energy use and the related CO_2 emissions, *Energy Policy*, vol. 33, no. 2, pp. 197–208.

Bina, O. (2013) The green economy and sustainable development: An uneasy balance, *Environment and Planning C: Government and Policy*, vol. 31, pp. 1023–1047.

Castán Broto, V. and Bulkeley, H. (2013) Maintaining climate change experiments: Urban political ecology and the everyday reconfiguration of urban infrastructure, *International Journal of Urban and Regional Research*, vol. 37, no. 6, pp. 1934–1948.

Chang, I-C. C. and Sheppard, E. (2013) China's eco-cities as variegated urban sustainability: Dongtan eco-city and Chongmin Eco-Island, *Journal of Urban Technology*, vol. 20, no. 1, pp. 57–75.

Chavez, A. and Ramaswami, A. (2013) Articulating an infrastructure supply-chain greenhouse gas (GHG) emissions footprint for cities: Mathematical relationships and policy relevance, *Energy Policy*, vol. 54, pp. 376–384. http://dx.doi.org/10.1016/j.enpol.2012.10.037.

Chen, H., Jia, B. and Lau, S. S. Y. (2008) Sustainable urban form for Chinese compact cities: Challenges of a rapid urbanized economy, *Habitat International*, vol. 32, no. 1, pp. 28–40.

Chou, R.-J. (2012) The problem of watercourse redevelopment: Disseminating new knowledge about flood risk perception in Taiwan's densely populated, typhoon-affected urban areas, *International Development Planning Review*, vol. 34, no. 3, pp. 241–267.

Cugurullo, F. (2013) How to build a sandcastle: An analysis of the genesis and development of Masdar city, *Journal of Urban Technology*, vol. 20, no. 1, pp. 23–37.

Danish Architecture Centre (2013) Dongtan: The world's first large-scale eco-city?, *Sustainable Cities*, available www.dac.dk/en/dac-cities/sustainable-cities-2/all-cases/energy/dongtan-the-worlds-first-large-scale-eco-city/?bbredirect=true, accessed June 9, 2013.

Datta, A. (2012) India's ecocity? Environment, urbanisation, and mobility in the making of Lavasa, *Environment and Planning C: Government and Policy*, vol. 30, pp. 982–996.

Datta, A. (2014) India's smart city craze: Big, green and doomed from the start? *The Guardian* 17 April. Available www.theguardian.com/cities/2014/apr/17/india-smart-city-dholera-flood-farmers-investors, accessed April 19, 2015.

Datta, A. (2015) New urban utopias of postcolonial India: Entrepreneurial urbanization and the smart city in Gujarat, *Dialogues in Human Geography* (in press).

De Mello Lemos, M. C. (1998) The politics of pollution control in Brazil: State actors and social movements cleaning up Cubatão, *World Development*, vol. 26, no. 1, pp. 75–87.

Dodman, D. (2009) Blaming cities for climate change? An analysis of urban greenhouse gas emissions inventories, *Environment and Urbanization*, vol. 21, no. 1, pp. 185–201.

Dodman, D. and Mitlin, D. (2011) Challenges for community-based adaptation: Discovering the potential for transformation, *Journal of International Development*, vol. 25, no. 5, pp. 640–659.

Elmqvist, T., Fragkias, M., Goodness, J., Güneralp, B., Marcotullio, P. J., McDonald, R. I., Parnell, S., Sendstad, M., Schewenius, M., Seto, K. C. and Wilkinson, C. (eds) (2013) *Urbanization, Biodiversity and Ecosystem Services: Challenges and Opportunities: A Global Assessment*, Dordrecht, Heidelberg, New York and London: Springer Open.

Fahmi, W. and Sutton, K. (2010) Cairo's contested garbage: Sustainable solid waste management and the Zabaleen's right to the city, *Sustainability*, vol. 2, no. 6, pp. 1765–1783.

Gasper, D., Portocarrero, A. V. and St. Clair, A. L. (2013) The framing of climate change and development: A comparative analysis of the Human Development Report 2007/8 and the World Development Report 2010, *Global Environmental Change*, vol. 23, no. 1, pp. 28–39.

Güneralp, B. and Seto, K. C. (2012) Can gains in efficiency offset the resource demands and CO_2 emissions from constructing and operating the built environment?, *Applied Geography*, vol. 32, no. 1, pp. 40–50.

Hankey, S. and Marshall, J. D. (2010) Impacts of urban form on future US passenger-vehicle greenhouse gas emissions, *Energy Policy*, vol. 38, no. 9, pp. 4880–4887.

Ibrahim, N., Sugar, I., Hoornweg, D. and Kennedy, C. (2012) Greenhouse gas emissions from cities: Comparison of international inventory frameworks, *Local Environment*, vol. 17, no. 2, pp. 223–241.

ILO and EU (2011) *Skills and Occupational Needs in Renwable Energy, 2011*, Geneva: International Labour Office.

Kennedy, C., Ibrahim, N. and Hoornweg, D. (2014) Low-carbon infrastructure strategies for cities, *Nature Climate Change*, vol. 4, pp. 343–346. DOI: 10.1038/NCLIMATE2160.

Kenworthy, J. and Laube, F. (1999) A global review of energy use in urban transport systems and its implications for urban transport and land-use policy, *Transportation Quarterly*, vol. 53, no. 4, pp. 23–48.

Kohn Pedersen Fox Associates (2013) Projects: "Songdo IBD", available www.kpf.com/project. asp?T=6&ID=9, accessed June 9, 2013.

Lwasa, S., Mugagga, F., Wahab, B., Simon, D., Connors, J. and Griffith, C. (2014) Urban and peri-urban agriculture and forestry: Transcending poverty alleviation to climate change mitigation and adaptation, *Urban Climate*, vol. 7, special issue on 'Urban Adaptation to EC: Governance, policy and planning', pp. 92–106. http://dx.doi.org/10.1016/j.uclim.2013.10.007.

Meinhold, B. (2009, September 4) Songdo IBD: South Korea's new eco-city. Retrieved from: http://inhabitat.com/songdo-ibd-south-koreas-new-eco-city/, accessed June 9, 2013.

Meyer, E. L. and Odeku, K. O. (2009) Climate change, energy, and sustainable development in South Africa: developing the African continent at the crossroads, *Sustainable Development Law & Policy*, vol. 9, no. 2, pp. 49–53, 74–5. The value of carbon credits was provided by Carl Wesselink in a presentation at the ICLEI Local Climate Solutions for Africa 2011 conference, Cape Town, March 1, 2011.

Mindali, O., Raveh, A. and Salomon, I. (2004) Urban density and energy consumption: A new look at old statistics, *Transportation Research Part A: Policy and Practice*, vol. 38, no. 2, pp. 143–162.

National Research Council (2009) *Driving and the Built Environment: The Effects of Compact Development on Motorized Travel, Energy Use, and CO2 Emissions*, Washington, DC: The National Academy of Sciences.

Norman, J., MacLean, H. and Kennedy, C. A. (2006) Comparing high and low residential density: A life-cycle analysis of energy use and greenhouse gas emissions, *Journal of Urban Planning and Development*, vol. 132, no. 1, pp. 10–21.

Parnell, S., Simon, D. and Vogel, C. (2007) Global environmental change: Conceptualising the growing challenges for cities in poor countries, *Area*, vol. 39, no. 3, pp. 357–369.

Pearce, D., Markandya, A. and Barbier, E. (1989) *Blueprint for a Green Economy*, London: Earthscan.

Rabinovitch, J. (1992) Curitiba: Towards sustainable urban development, *Environment and Urbanization*, vol. 4, no. 2, pp. 62–73.

Renner, M., Sweeney, S. and Kubit, J. (2008) *Green Jobs: Towards Decent Work in a Sustainable, Low-Carbon World*, Nairobi: United Nations Environment Programme.

Roberts, D., Boon, R., Diederichs, N. et al. (2012) Exploring ecosystem-based adaptation in Durban, South Africa: 'Learning by doing' at the local government coal face, *Environment and Urbanization*, vol. 24, no.1, pp. 1–29.

Rosenzweig, C., Solecki, W. D., Hammer, S. A. and Mehrotra, S. (eds) (2011). *Climate Change and Cities: First Assessment Report of the Urban Climate Change Research Network*, Cambridge: Cambridge University Press.

Satterthwaite, D. and Dodman, D. (eds) (2013) Special issue: Towards resilience and transformation for cities, *Environment and Urbanization*, vol. 25, no. 2, pp. 291–463.

Saunders, C. (2008) The stop climate chaos coalition: Climate change as a development issue, *Third World Quarterly*, vol. 29, no. 8, pp. 1509–1526.

Schipper, L. and Pelling, M. (2006) Disaster risk, climate change and international development: Scope for, and challenges to, integration, *Disasters*, vol. 30, no. 1, pp. 19–38.

Scott, D. and Barnett, C. (2009) Something in the air: Civic science and contentious environmental politics in post-apartheid South Africa, *Geoforum*, vol. 40, no. 3, pp. 373–382.

SDI (2012) *The Zabaleen of Cairo*, www.sdinet.org/blog/2012/03/14/zabaleen-cairo/, accessed February 8, 2014.

Shwayri, S. T. (2013) A model Korean ubiquitous eco-city: The politics of making Songdo, *Journal of Urban Technology*, vol. 2, no. 1, pp. 39–55.

Simon, D. (2011) Reconciling development with the challenges of climate change: Business as usual or a new paradigm?, in D. J. Kjosavik and P. Vedeld (eds), *The Political Economy of Environment and Development in a Globalised World: Exploring the Frontiers; Essays in honour of Nadarajah Shanmugaratnam*, Oslo: Tapir, Oslo, and Colombo: Social Scientists' Association of Sri Lanka, pp. 195–217.

Simon, D. (2013) Climate and environmental change and the potential for greening African cities, *Local Economy*, vol. 28, no. 2, pp. 203–217.

Simon, D., Fragkias, M., Leichenko, R., Sánchez-Rodríguez, R., Seto, K. C. and Solecki, W. (2011) *The Green Economy and the Prosperity of Cities*, Background Paper for the UN-HABITAT *State of World Cities Report 2012/3*, available at www.unhabitat.org/downloads/docs/GreenEconomyCity Prosperity5Aug11.pdf.

Sovacool, B. K. and D'Agostino, A. L. (2012) A comparative analysis of solar home system programmes in China, Laos, Mongolia and Papua New Guinea, *Progress in Development Studies*, vol. 12, no. 4, pp. 315–335.

The green economy and urban greening

Stern, N. (2007) *The Economics of Climate Change: The Stern Review*, Cambridge: Cambridge University Press.

UNEP (2011a) *Green Jobs: Towards Decent Work in a Sustainable, Low-Carbon World,* Nairobi: United Nations Environment Programme. See also: http://ecoswitch.com/articles/kibera-community-youth-programme-how-photovoltaics-are-helping-africa/.

UNEP (2011b) *Towards a Green Economy: Pathways to Sustainable Development and Poverty Eradication*, Nairobi: United Nations Environment Programme, Green Economy Initiative.

UN-HABITAT (2013) *State of the World's Cities 2012/2013: Prosperity of Cities*, New York: Routledge for UN-HABITAT.

UNGSP (2012) *Resilient People, Resilient Planet: A Future Worth Choosing. Report of the UN Secretary-General's High-Level Panel on Global Sustainability*, New York: United Nations. Available online at www.un.org/gsp/sites/default/files/attachments/GSP_Report_web_final.pdf, accessed August 13, 2012.

Vance, C. and Hedel, R. (2008) On the link between urban form and automobile use: Evidence from German survey data, *Land Economics*, vol. 84, no. 1, pp. 51–65.

WCED (World Commission on Environment and Development) (1987) *Our Common Future*, New York: Oxford University Press.

31

POSITIVE EXTERNALITIES IN THE URBAN BOUNDARY

The case of industrial symbiosis

Marian Chertow, Junming Zhu and Valerie Moye

Urban expansion and resulting land cover change increase environmental stressors and competition for physical resources at the local, regional and global levels. Too often, though, attention is placed on the costs rather than the benefits of agglomeration, on negative externalities rather than positive ones. This chapter examines urban industrial activities by inquiring whether and how the materials, energy and water that enter and pass through a local economy can be harnessed and used more than once. Such investigations fall under the general research area of 'industrial symbiosis,' a sub-field embedded in the intellectual tradition of industrial ecology (Graedel and Allenby 2010). Industrial symbiosis engages traditionally separate industries in a collective approach to competitive advantage involving physical resource sharing of materials, energy, water and/or by-products where both proximity and collaboration play key roles (Chertow 2000). Investigating industrial symbiosis at the urban scale asks whether and how industrial operations can be examined collectively rather than as a series of isolated facilities to reveal previously unrealized benefits. Successfully reusing resources locally makes the revealed benefits available across the resource chain from local to global.

This chapter argues that industrial symbiosis has multi-scale benefits and begins with an overview of its relationship to resource use and agglomeration economies. The second section describes in more depth how industrial symbiosis 'fits' into cities by examining where industrial clusters are likely to be sited with respect to urban centers based on six specific examples, and how to reap and extend the benefits of symbiosis. The third section discusses some of the lessons learned when it comes to quantifying resource inputs and outputs on a collective basis and draws on some of the authors' work in Karnataka, India and also at the city and state level in the US regarding boundaries and benefits. It also proposes a preliminary five-step model to calculate industrial symbiosis potential based on maximum feasible levels of resource exchange in a specific urban area. A short conclusion summarizes the findings of the chapter and areas for further research.

Industrial symbiosis: an introduction

The urban perspective is especially significant when thinking about resource flows. Famously, cities concentrate resources just as they concentrate people. Current research, especially in industrializing countries such as India and China, find that wealth is also concentrated in cities rather than in rural areas. Urban dwellers demand not only more resources such as water and energy, but also more extensive resources, reaching globally for imported goods that enhance the quality of life of these wealthier urbanites. Such consumption, in turn, affects water and energy use and also waste generation. It is important to recognize that when resource inputs are used more productively and waste is reduced, this can free up capacity at landfills, energy generation facilities and wastewater treatment plants for use in other parts of the community. If enough volume is gained, this, in turn, can postpone public infrastructure investments further into to the future, which preserves capital for cities as well as environmental capacity. Since urban areas are large users of resources, city residents and businesses find triple benefits of symbiosis: new revenues and employment from repurposing wastes, reduced costs of disposal or emissions and the marginal gain of infrastructure capacity returned to the community.

Local to global benefits

Many basic resources key to industrial symbiosis such as water and energy often must be supplied from beyond the urban boundary. Regarding energy, for example, power plants are found in only 7 percent of US counties (EPA 2007). While cities generally range from 10–100 miles, freight travels 600 miles on average, and food travels 1,200 miles (Ramaswami et al. 2012; Weber and Matthews 2008). The supply chains that feed cities are global in character and are getting longer. Thus, when industrial symbiosis makes possible the substitution of local by-products for imported energy and materials, there are significant associated savings in transportation and logistics. Urban consumption for infrastructure leads to very high demand for energy and pollution-intensive goods produced elsewhere, such as steel, cement and fuel (see Güneralp, Chapter 6 in this volume, for a detailed discussion of resource use and construction within the built environment). Indeed, the Intergovernmental Panel on Climate Change (IPCC) 5th Assessment Report results conclude that urban areas account for 71–76 percent of CO_2 emissions from global final energy use (Seto et al., 2014).

Because global trade not only moves resources and products, but also waste, local waste has become a global issue, too. The iconic case is the discard of consumer electronics. These products—televisions, cellphones, personal computers, etc.—are used globally and more intensively in industrialized countries. When they reach the end of life, however, they are transported to a few cities and towns, mainly in China and also in Nigeria, Kenya and India to be repaired, recovered or recycled by lower cost laborers. The recycling sector in these countries is generally under laxer environmental regulation and characterized by informality, such that local public health and environment are not duly considered (Williams et al. 2008). With respect to climate change, local landfills contribute 9 percent of methane emissions globally and waste from cities makes up most of that quantity (Vergara and Tchobanoglous 2012).

Understanding industrial symbiosis resource sharing

A clear way to visualize the positive effects of industrial symbiosis among actors is to think of the most familiar form of these resource exchanges—those that occur across two entities such as when one firm engages with another in cogeneration, supplying wastewater or material by-product reuse. When resource sharing grows from two organizations to more, network benefits can occur, increasing the number of exchange opportunities more rapidly than the number of exchange actors (see Figure 31.1).

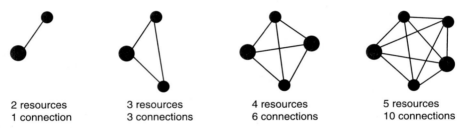

Figure 31.1 The growth of an industrial symbiosis network

Source: Laybourn and Morrissey (2009).

In practice, there are three principal resource-sharing activities that constitute industrial symbiosis. While there is no single definition or universal declaration that these three are always considered, they have been discussed in different ways by academics and practitioners. They are:

a) By-product exchanges—involves the material 'waste' from one organization becoming feedstock (or another form of productive reuse) for a second organization.
b) Utility/infrastructure sharing—involves shared access/management of utilities (electricity, water, wastewater and some others such as waste-to-energy power plants).
c) Service sharing—involves shared common services with clear environmental savings provided by a third party (e.g. recycling, landscaping, pipeline distribution).

Industrial symbiosis as a source of agglomeration benefits

For over 100 years scholars have been aware of three primary advantages accruing to businesses that spontaneously co-locate. Krugman (1991) summarizes these benefits as: the presence of a large and concentrated pool of firms and skilled workers (labor pooling); the availability of industry specific inputs at lower costs resulting from supplier economies of scale (input sharing); and the opportunities for information exchange essential to the innovation process (knowledge spillovers). These benefits of agglomeration, particularly within the same industry, are often referred to as *localization economies*.

The advantages that accrue to businesses spontaneously co-locating near different and unrelated industries are known as *urbanization economies*, and they help to define the vitality of urban diversity and innovation (Hoover 1937; Jacobs 1969). Additional urbanization economies include labor pooling mentioned above with a higher division of labor (Glaeser 1998; Quigley 1998), and also increased access to municipal services, public utilities, and transportation and communication systems (Duranton and Puga 2004; Mukkala 2004; Parr 2002). It is important to note that urban concentration can also bring familiar negative externalities such as increased costs of living and commuting, health-related costs, many types of pollution, congestion of local amenities and infrastructure, increased crime and other social problems (Glaeser 1998; Hanson 2001; Henderson 1994; Quigley, 1998).

Where in this spectrum does industrial symbiosis lie? Until recently, urbanization economies such as shared utilities and transportation, which embed environmental savings, were not explicitly recognized for their environmental benefits (Chertow et al. 2008). As discussed, the cooperative approach underlying industrial symbiosis can enhance conservation of water, energy and materials, substitute local 'wastes' for global resources and create new

Positive externalities in the urban boundary

Box 31.1   Examples of industrial symbiosis, from single to multi-resource exchanges

The examples below involving water, energy and materials, singularly and collectively indicate how bi-lateral relationships can expand to become networked industrial symbiosis in cities.

Water—single transfer: In the water constrained City of Guayama, Puerto Rico (population 45,000), power plant operators recognized that by using treated wastewater rather than fresh water for cooling, they would save the company $1.2 million per year (Chertow and Lombardi 2005). With the exchange in place, the area wastewater treatment plant, rather than discharge four million gallons per day of treated wastewater into the Caribbean Sea, reroutes it to the power plant in Guayama for use as cooling water.

Water—multiple transfers: An illustration of the shared water concept as a broader scale network occurs across the portside City of Kalundborg, Denmark (population 49,000). There, a network of firms including a pharmaceutical plant, an oil refinery and a power plant achieve private savings from water reuse by converting their individual water systems into a common resource pool (Jacobsen 2006). This shared activity has reduced overall system consumption by two million cubic meters/year of groundwater (530 million gallons) and one million cubic meters/year of surface water (265 million gallons).

Water and energy: In the mineral processing area of Kwinana, Australia in greater Perth, water is constrained and needs to be cycled and reused particularly for energy generation. By 2008, 15 resource-sharing projects involving 25 different water flows and 11 energy flows were established. They supply electricity generation, cogeneration, iron production, alumina production, pigment production and oil refining facilities with energy in the form of electricity, steam and fuel gas, and, in turn, receive treated and untreated wastewater. Collectively, more than two million cubic meters of wastewater had been recycled and reused annually among these firms (van Beers 2008).

Water, energy and materials: In Honolulu, Hawaii, the only coal-fired power plant on the Island of Oahu generates both 180 MW of electricity for the grid and also supplies steam to a nearby oil refinery. This cogeneration is facilitated by the purchase of 40 million gallons per year of boiler make-up water. The water is generated from recycled reverse osmosis water produced for very high quality industrial users in a publicly owned facility by the Honolulu Board of Water Supply. Multiple material by-products, common to urban areas that would ordinarily have been landfilled, instead supply high BTU local substitutes for globally sourced coal including used activated carbon from water filtration processes, recycled motor oil and tire shreds prepared for combustion by a tire reclaimer. For a detailed review of costs, lifecycle costs and quantities see Chertow and Miyata (2011) and Eckelman and Chertow (2013).

---

revenue streams through by-product reuse of steam, wastewater and discards. Some of the typical diseconomies of agglomeration such as high concentrations of pollution, may be ameliorated by the environmentally beneficial outcomes of industrial symbiosis. Thus, even while participating firms achieve private economic benefits, there are also public benefits created in the form of positive environmental externalities that accrue to the regional public to the extent that environmental harms are reduced or avoided and resources are conserved for community purposes (Chertow and Ehrenfeld 2012).

Table 31.1 Environmentally related benefits of industrial symbiosis in the UK for the period 2005–2012 as reported by Internatonal Synergies for the National Industrial Symbiosis Programme

Metrics	In year benefits★	Lifetime impact (max five years)
Landfill diversion	9 million tonnes	45 million tonnes
CO_2 reduction	8 million tonnes	39 million tonnes
Virgin material savings	12 million tonnes	58 million tonnes
Hazardous waste eliminated	0.4 million tonnes	2 million tonnes
Water savings	14 million tonnes	71 million tonnes
Cost savings	€ 243 million	€ 1.21 billion
Additional sales	€ 234 million	€ 1.71 billion
Jobs	10,000 +	???
Private investment	€ 374 million	???

Notes: € 40 million investment since 2005. ★ All outputs independently verified. Rate Euro £ 1 = € 1.18.

The National Industrial Symbiosis Programme (NISP) in the UK, begun in 2003, was the first country-wide industrial symbiosis network tasked with bringing together disparate industrial actors and developing inter-firm agreements and exchanges particularly in older industrial cities such as Birmingham and Manchester. Funded by a landfill tax, the money raised supported the initiation and growth of efforts to reduce waste and create and sustain employment. Table 31.1 shows reported results of NISP over seven years in categories that reflect local, regional and global environmental benefits of industrial symbiosis. Landfill diversion of materials is a local benefit, water saved is primarily a regional benefit and CO_2 reductions are a global benefit. Private investment of 374 million Euros and 10,000 jobs created or retained shows the intertwining of public and private benefits of industrial symbiosis.

The where and what of industrial symbiosis in urban areas

Linkages between industrial symbiosis and urban systems

Cities and urbanization processes open up channels for industrial symbiosis because they are characterized by high volumes of resource consumption, material accumulation and waste generation. In fact, half a century ago, Jane Jacobs (1969) observed that cities were waste-yielding mines and argued that as cities grew, even larger and more diverse mines would appear and support increased remanufacturing. She cites examples familiar to industrial symbiosis such as the production of garden compost and recovery of silver from restaurant garbage in Brooklyn, New York; a power plant capturing and converting sulfur dioxide emissions into sulfuric acid in Johnstown, Pennsylvania; and a cluster of at least one thousand remanufacturers of scrapped automobile parts centered in Chicago.

A modern example, mentioned earlier, and especially applicable for urban environments, is cogeneration. Power plants that simultaneously generate electricity and useful thermal energy for heating and sometimes cooling are known as cogeneration plants or combined heat and power (CHP) generators. A common use of previously wasted heat is for district heating, especially in Northern European countries. The location of CHP generators is usually close to population centers because agglomeration of residents and/or industrial facilities makes the efficient reuse of waste heat possible. In the US, CHP generator sites closely align with areas of higher population density.

The famous Spittelau Waste Incineration Plant in inner-city Vienna, Austria, not only features the benefit of cogeneration, but also helps the city manage its waste in a sustainable way. Armed with two waste-to-steam boilers, one generator, four heat exchangers and waste gas purification facilities, the plant receives 263,000 cubic meters of waste, generates 364,000 MWh of electricity and heats 190,000 homes and 4,200 public buildings in one year (Best Practices Hub Vienna 2002). It also became a tourist attraction owing to the intriguing design by a Viennese artist.

In addition to cogeneration's linkage with urban waste management and district heating, there are numerous projects for replacing fresh water with treated water from municipal sewage treatment plants in cooling processes, and collecting fly ash for cement production or other reuse. Sludge from water treatment plants, along with other organic waste produced inside cities, subject to environmental regulations, can be sent as fertilizers to nearby farms that supply food products for the cities. These resources—local waste converted to energy; steam for district heating; treated water substituted for fresh water; local fly ash in place of globally produced cement; and sludge used as fertilizer—all substitute local by-products for regional and international purchasing and shipping of fossil fuels and many other high volume goods.

While industrial symbiosis can be nurtured by the concentration of resources drawn to urban systems, it can, in turn, help mitigate urban environmental problems related to resource use and externalities. Japan initiated an Eco-Town program and provided generous investment subsidies for local recycling projects in 26 cities (van Berkel et al. 2009a). Several key projects in the City of Kawasaki alone diverted at least 565,000 tons of waste from incineration or landfill and created estimated benefits of 130 million USD annually (van Berkel et al. 2009b). In addition to previously mentioned benefits, for example, collecting and using heat that was previously wasted can reduce heat loss into the environment, moderate some urban heat island effects and reduce the residential and commercial energy demand for cooling. In particular, lower energy demand can help avoid electricity outages by saving large quantities of water in seasons when both water and energy demands are high. While energy and water security are important everywhere, this has become a vexing issue in rapidly growing economies such as China and India, where increasing electricity demand has been outpacing the growth of energy supply. Conventional energy efficiency measures also initiate this positive feedback loop in urban systems, but implementing them usually requires more extensive capital investment and technology adoption. When industrial symbiosis can be applied in combination with energy and water efficiency measures this can lead to extra air and water pollution mitigation by encouraging the collection of waste for reuse and remanufacturing.

Configurations of industrial symbiosis examples around urban areas

Examples used thus far indicate that industrial symbiosis is not necessarily an urban center phenomenon, but occurs frequently in smaller cities or on the peripheries of larger cities. The relative absence of large cities suggests the need for additional study resulting in further examination of over 20 well-documented cases of industrial symbiosis. The findings presented below illustrate six representative patterns of the two main spatial relationships identified—either symbiosis configured within local urban centers or peripheral to central metropolitan areas.

The first three cases describe symbiotic clusters within urban centers: a small city in Denmark; a medium city in Canada; and a large city in South Korea:

1   *Symbiosis in a small city: Kalundborg, Denmark.* The best-known case of industrial symbiosis is found in Kalundborg, Denmark, and has been developing and evolving for over 40 years (Ehrenfeld and Gertler 1997; Symbiosis Institute 2013). The small city and major companies involved in the symbiosis—a cogeneration plant, an oil refinery, a wallboard maker and a pharmaceutical company—are all located fairly closely together, circling around the seaport (see Figure 31.2). The short physical distance enables them to exchange energy and water, heat and sludge. The short mental distance among company and city managers illustrates the importance of a strong social network in the creation and development of symbiosis. It also shows how locally generated products substitute for international ones in global markets: whereas virgin gypsum used to be imported from Spain to provide raw material to the gypsum board manufacturer; today, synthetic gypsum is the by-product of the air pollution control system of the large local cogeneration plant. The plant has been powered by coal purchased on commodity markets, but for a short period used a fossil fuel from Venezuela thought to reduce CO_2 use, and today has substituted some local biomass for globally traded coal in one boiler.

2   *Symbiosis at the edge of a medium city: Halifax, Canada.* Across the Atlantic Ocean from Denmark is an eco-industrial development, the Burnside Industrial Park, in the former city of Dartmouth, Nova Scotia, Canada. Burnside is the largest industrial park in Atlantic Canada and hosts 1,300 businesses, most of which are small (Cote and Crawford 2003). An Eco-Efficiency Center, led by public and private partners including a university and

Figure 31.2 Industrial facilities in Kalundborg, Denmark are located around the local harbor
Source: Kalundborg Symbiosis (n.d.).

Positive externalities in the urban boundary

an electricity utility and supported by government agencies, has encouraged and engaged businesses in Burnside to cooperate in exchanging and reusing materials and waste, which greatly conserved industrial use of fresh water and the community's landfill resource (Cote and Crawford 2003). With respect to the urban configuration, most of Dartmouth has been dissolved to become part of metropolitan Halifax, the urban core of the newly formed Halifax Regional Municipality, which consists of several former cities and is the largest population center of Atlantic Canada with roughly 300,000 residents. The industrial park is located in a suburban community contiguous to the urban core.

3   *Symbiosis in a large city: Ulsan, South Korea.* Industrial symbiosis is also recognized and facilitated in the city of Ulsan, South Korea, with a population of one million. Ulsan, located in the south, is the most industrial part of the country and hosts large global companies importing raw materials from around the world such as Hyundai and SK Energy. The companies are clustered in industrial complexes around the city. Of all the urban configurations studied here, Ulsan comes the closest to supporting industrial symbiosis in the heart of the city. Initially, symbiotic exchanges happened spontaneously between industries, spurred on by stringent environmental regulations and associated economic benefits (Park et al. 2008). Later, the national Eco-Industrial Park (EIP) policy and local facilitation by the Ulsan branch of a federal industrial assistance organization, KICOX, have encouraged more development of symbioses in conventional industrial complexes (Behera et al. 2012).

The next three examples describe industrial symbiosis in three urbanized areas that are peripheral to central metropolitan areas: a coterminous city and county in the US; an industrial area in Western Australia; and a mega-city in China:

4   *Campbell Industrial Park, Honolulu, US.* Honolulu, Hawaii is officially known as the City and County of Honolulu, a consolidated city–county located in the State of Hawaii, coterminous with the Island of Oahu, with a population of almost one million. Only the county is incorporated, meaning that Honolulu includes both the state capital found within the largest urban agglomeration on the eastern part of the Island, and also the outlying districts of Oahu. Campbell Industrial Park is in Kapolei, an industrial community 20 miles west of the populous urban center. The industrial area is the only heavy industry site and the largest industrial park in Hawaii. Industrial symbiosis has been found in the park anchored by a cogeneration plant, which exchanges material, water or energy with eight other firms. Some of the firms are outside the park boundaries, but are also involved in symbiosis. In this example, both the industrial cluster and the urban core are geographically part of the same entity, but the distance between them illustrates an example of industrial symbiosis predominantly in the periphery (Chertow and Miyata 2011).

5   *Kwinana, Perth, Australia.* Similar in pattern to the Campbell Industrial Park, the Kwinana industrial area is an industrial suburb about 25 miles south of Perth, which is the capital and largest city of the State of Western Australia with a population of 1.9 million. Established in the 1950s, the industrial area features the largest and most diverse coastal strip in Western Australia. Located in a state with a rich endowment of natural resources, but with constraints on water supply, the area is dominated by heavy process industries such as alumina, nickel and oil refining, cement, and chemical processing, and also hosts utility operations for power generation, cogeneration and air separation. In total, over 100 resource exchanges have been identified in this suburb that is still closely connected to the city of Perth (van Beers et al. 2007).

Marian Chertow et al.

6. *Tianjin Economic-Technological Development Area (TEDA), Tianjin, China.* TEDA is a built-up area of China that has been dedicated to attracting foreign and domestic industrial investment since 1984 including over 70 Fortune 500 companies. It is part of the Tianjin metropolitan area with a total population of 13 million, located close to Tianjin's port and is 30 miles away from Tianjin downtown. In this case, TEDA is technically part of Tianjin but the industrial area has already grown so much that it has become the central business district of the later established Binhai New Area, which is a special zone of Tianjin for economic development and reform, and itself has a current population of over two million. Major industries in TEDA include electronics, machinery and automobiles, biotechnology and pharmaceuticals, and food and beverage. TEDA is one of the first national demonstration EIPs in China; many by-product exchanges and instances of utility sharing have been identified, with more than half formed between firms in TEDA and those outside. The local administrative agencies are crucial in TEDA's evolution as an EIP since they provide economic incentives, technology support and information (Shi et al. 2010).

These cases are characterized by different geographical locations, size of the urban systems and extent of symbioses. With the exception of Ulsan, however, they are not in the urban center nor are they in remote rural areas, but are found mainly in smaller cities or satellite cities. Kalundborg, Halifax and Ulsan are small- or medium-sized cities up to one million in population, where industrial symbiosis occurs in nearby communities adjacent to downtown. Kapolei, Kwinana and Binhai New Area are all zones in larger metropolitan areas, but removed from the urban cores, intentionally dedicated to industrial development areas. Although in the case of TEDA, the industrial cluster has become the center of its own satellite city, all of the locations feature close access to both highways and seaports to reach out to global markets.

Such a locational pattern of industrial clusters where symbiosis has been found traces back to the aforementioned discussion about urbanization and localization economies. Some researchers have suggested that large metropolitan areas usually have an industrial mix that is both diverse and constantly changing with products replaced quickly; whereas, specialized industries for standardized items tend to locate in smaller cities (Henderson 1997; Duranton and Puga 2000). A typical example is publishing and apparel design industries in large metropolitan areas such as New York or Tokyo, and the manufacturing of paper and textiles in other smaller cities. From a product life cycle perspective, Duranton and Puga (2001) further suggest that new firms tend to locate in large, diverse cities for urbanization economies, and move to specialized cities for localization economies if they survive and become mature.

To date, the localized industries in small and satellite cities appear to provide more opportunities for industrial symbiosis, as with the six examples mentioned above, and most other cases reported in the literature. Several reasons are suggested here. First, specialized, standardized production from localization economies has been associated with large, stable volumes of by-products and waste for symbiotic exchanges. Otherwise, fluctuations in the quantity or quality of by-products and wastes may prevent firms from reusing them. Second, relatively smaller city size means lower transportation costs for material exchanges, and lower construction costs for utility sharing, which would otherwise make symbiosis less viable economically. Third, smaller city size and proximity also foster mutual trust, recognition, knowledge exchange and shared norms, which tend to decrease transaction costs and therefore are increasingly considered important for industrial symbiosis development (Baas and Boons 2004; Chertow and Ehrenfeld 2012).

Particularly in East Asia, industrial relocation, usually to designated areas outside urban centers housing large-scale industrial complexes, has been pursued intentionally as an urban environmental management strategy (Bai 2002). Policies to build EIPs have followed and are encouraging potential symbiosis in these dedicated industrial areas. In general, Western countries are less directive than East Asian ones about where any individual company may wish to locate, so concentration and diffusion is more often left to the happenstance of markets.

Fostering industrial symbiosis in an urban context

The composition of socio-economic activities within an urban boundary changes over time and varies across cities. Urban systems consume global resources and both drive and are affected by global environmental change. While private benefits and localization economies have already provided impetus for the development of industrial symbiosis, there is room for expansion beyond traditional geographic and administrative boundaries of cities and urban centers. The resource efficiency and environmental benefits of industrial symbiosis increase when based on the functional flows of materials, energy and water that do not usually stop at the city limits. Thus, it is reasonable to define the urban boundary to include key functional areas (Brown and Holmes 1971) such as waste disposal sites, airports, power generation and other destinations that are most closely tied with local activities.

Such a comprehensive consideration of industrial symbiosis systems means that policy promotion for industrial symbiosis should not only be implemented at city and lower levels, but also be endorsed by higher level government to fulfill symbiosis beyond city boundaries. Even greater symbiotic potential can be harvested by tracing materials, energy and water flows and recognizing functional interdependency in and around urban systems. Shi and colleagues (2010) find that symbiotic exchanges for energy and water are mainly within China's TEDA, described above, and have average distances of 2.9 and 3.5 kilometers respectively, but those for materials are mainly between firms inside and outside TEDA and have an average distance of 28.2 kilometers. Two other cases of industrial symbiosis—the UK's National Industrial Symbiosis Programme and the waste recycling network in the Styria Region of Austria—are both on geographical scales much larger than single cities. Conventional regulations on the management of specific wastes and the treatment of certain pollution may need to be made flexible with regard to detailed treatment approaches and also become more performance-oriented to reduce legal barriers for recycling and reuse.

Policy-making can also help facilitate industrial symbiosis by internalizing public externalities and creating incentives. Negative environmental externalities can be internalized by removing inefficient subsidies and making prices of energy and pollution emissions reflect real economic and social costs. Although there have been reform efforts in several countries, the global cost of fossil fuel subsidies reached $544 billion in 2012, while subsidization for renewables was $101 billion (IEA 2013). Particularly, urban systems usually have higher pollution levels and shortage of landfills, which associate with higher marginal social costs for pollution emissions and waste disposal. These costs, if reflected back accurately to producers of wastes and pollution, can help them reduce pollution and waste in the first place and divert waste from landfills through symbiotic exchanges.

Early adopters of symbiotic practices absorb many of the costs of initiating this type of enterprise that may well benefit later adopters and society. Just as in the case of technological change, a firm that develops a new type of symbiosis bears transaction costs such as 1) searching for the waste that can be used as a feedstock or for industries that can use its waste; 2) negotiating with the other party; and 3) experimenting with and adjusting the quality and quantity of the exchanged waste. Other participants, if not involved in the development, at least generate information about the existence, characteristics and successfulness of the symbiosis. Urban systems may further increase positive externalities by facilitating the flow of knowledge and information resulting from agglomeration. The existence of positive externalities in the development and adoption of symbiosis indicates an undersupply of symbiosis. Therefore, public support is reasonable for those developers and early adopters, for example, through technical assistance, information platforms and adoption subsidies.

The lessons of scale and externalities play out differently in emerging versus developed economies. In the following section, however, lessons from symbiosis in and around the city of Mysore in India are discussed among other examples to describe a preliminary means of assessing industrial symbiosis potential if a full palette of resource exchanges could be realized.

The potential of industrial symbiosis as an urban intervention

There is a tremendous opportunity for industrial symbiosis to become established as an achievable intervention in support of sustainable urban development; highlighted here are lessons learned from India. As one of the world's largest emerging economies and most rapidly urbanizing countries (Mitra et al., Chapter 5 in this volume), India has experienced swift growth in both the urban and industrial sectors. A current rank order of India's leading industries by value of annual output reveals that five of the top six industrial sectors, from food processing to cement, are continuous process industries with large capacities for industrial symbiosis (Chertow 2010). The growth of cities across India requires the addition of 700–900 million square feet of floor space and 400 kilometers of metro rail and subways (McKinsey and Co. 2010). Thus, there is a continuing, massive flow of energy and materials entering India's growing cities being used and then distributed in India and globally. How can industries contribute to—rather than detract from—public health and quality of life for a burgeoning urban population? What is the potential to reduce urban resource consumption through industrial symbiosis and related activities of the industrial sector?

Moving from measuring industrial symbiosis in a defined industrial park, as described in several of the examples in the previous section, to the scale of a much larger, dynamic and fuzzy-edged urban area, poses a difficult challenge, but presents an interesting case study for industrial symbiosis as an urban intervention that can occur over a shorter time period rather than awaiting long-term urban planning and infrastructure replacement. Mysore, India, a city of one million in the southern state of Karnataka, is a good candidate for such further study based on its diversity of industries as well as data accessibility from local government authorities. Pharmaceuticals, textiles, software development and food are the major exportable products from the large-scale industries in and around Mysore.

The remainder of this section describes a five-step conceptual model to determine industrial symbiosis potential of an urban area. The five steps identified in Figure 31.3 are discussed below and are intended to stimulate discussion and highlight lessons learned about industrial symbiosis research in urban areas. There have been very few studies to date that attempt to quantify industrial symbiosis potential and its associated environmental benefits across broad geographic scales. This is due in part to the difficulty in accessing proprietary data from the vast number of industrial firms that exist across a city or region, and in part because little effort has gone into creating this data amalgamation. The lack of standardized reporting practices and metrics complicates data collection further.

The Mysore case study is based on current work by the authors related to a research project on Low Carbon Cities in India, China and the US, funded by the US National Science

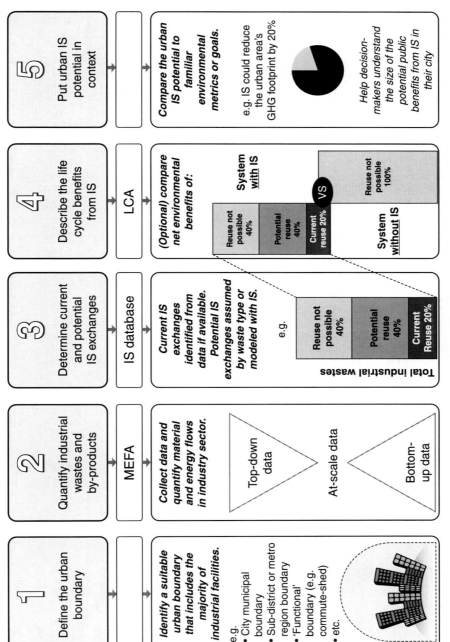

Figure 31.3 A conceptual model for determining industrial symbiosis potential in urban areas

Foundation, that explicitly examines industrial symbiosis as an urban intervention (PIRE n.d.). In the study, 'industrial symbiosis potential' of an urban area is defined as the sum of the wastes and by-products from all the industrial facilities in the urban area that could reasonably serve as resource inputs to other processes. This measured potential can offer a means of informing a city of what maximum opportunity is possible and how much is feasible.

Step 1. Define the urban boundary

In any material and energy flow analysis (MEFA), researchers must define a system boundary to establish which components are located within the system versus those that are located outside the system. Urban MEFA and environmental footprint analyses of cities typically use the city administrative boundary as the system boundary. However, as is discussed above, industrial symbiosis often occurs where manufacturing facilities are located: on the periphery of urban areas or along the transportation corridors of metro regions. Therefore, researchers interested in studying urban industrial symbiosis will likely need to expand their study boundary spatially to include outlying industrial areas that are functionally connected to a nearby urban core. Figure 31.4 shows the location of three industrial areas in relation to the municipal boundary (hatched area). In India, the next largest administrative unit beyond the level of the municipality is known as the taluk. The Mysore Taluk was chosen as the study boundary, in order to capture most of Mysore's industrial activity. The black points represent the locations of the first 22 facilities that were surveyed, deliberately choosing some at different locations throughout the city.

Step 2. Quantify industrial wastes and by-products

Once an urban boundary is defined, researchers compile a list of firms and other relevant organizations to be examined further to acquire data on industrial material use, reuse and disposal. Mysore Taluk is home to 1,028 operational firms. The Karnataka State Pollution Control Board records indicate whether each firm is located within urban or rural parts of the taluk as well as which ones are located within the Mysore Municipal Corporation

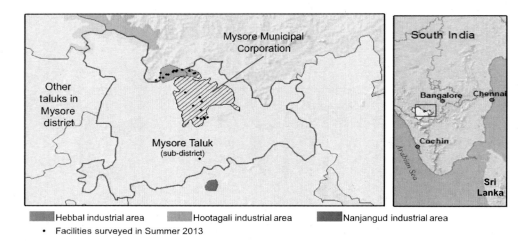

Figure 31.4 Geographic distribution of industrial parks around Mysore city boundary

Positive externalities in the urban boundary

Table 31.2 Number of firms in various industries located in rural or urban areas of Mysore

Geographic distribution of industry types in Mysore Taluk, India

Industry type	Rural	Urban		Total
		Inside city boundary	*Outside city boundary*	
Agriculture, forestry, fishing and hunting	4	1	8	13
Chemical manufacturing	10	19	32	61
Computer and electronic/electrical equipment manufacturing	29	22	75	126
Construction	8	3	24	35
Food manufacturing	70	28	35	133
Furniture and related product manufacturing			7	7
Nonmetallic mineral product manufacturing	49	9	22	80
Paper manufacturing	4	2	10	16
Petroleum and coal product manufacturing		4	5	9
Plastic and rubber product manufacturing	6	35	47	88
Primary metal manufacturing	3	12	29	44
Textile mills and textile product mills	6	10	15	31
Transportation equipment manufacturing	3	5	14	22
Waste industries (waste collection, treatment/disposal, remediation and other management services)		5	7	12
Wood product manufacturing	3	20	8	31
Other industries	41	171	108	320
Total	**236**	**346**	**446**	**1,028**

Table 31.2 shows the breakdown of industry types and where facilities from each type are located. Darkened cells highlight where each type of facility most often occurs. For instance, mining operations and food manufacturing are most often found in rural parts of the taluk. Industries such as chemical manufacturing, textile mills and primary metal manufacturing are most often located in the urban areas just outside the municipal boundary. The 'other' industries, primarily service-oriented operations such as hospitals, restaurants and hotels, occur most often within the municipal boundary. The list of the 1,028 facilities was used to estimate wastes and by-products.

Step 3. Determine current and potential industrial symbiosis exchanges

The ideal way to measure existing industrial symbiosis exchanges at the urban scale is to have at-scale data on waste/by-products—which equates to data for every individual facility. The industrial symbiosis potential of an urban area is comprised of all *current* waste exchanges plus the sum of all *possible but unrealized* waste exchanges across facilities in the urban area. An earlier study conducted in Pennsylvania indicates that potential waste/by-product reuse is approximately four times greater than the currently reported reused rate (Eckelman and Chertow 2009).

For Mysore, an initial survey of 22 facilities shows the destination of 1,441 tons of industrial wastes and by-products leaving the plants. Over half goes to recycling, 12 percent is used in symbiotic exchanges and 6 percent is reused off site for a total recovery rate of 70 percent. Several current symbiotic linkages were found including the use of spent coffee grounds from a food manufacturing facility as a biomass fuel for a tire manufacturer; use of sewage treatment sludge for fertilizer; and shared water services provided to several facilities by the Karnataka Industrial Areas Development Board.

Step 4. Describe the life cycle benefits from industrial symbiosis

The reuse and exchange of industrial wastes and by-products decreases the amount of virgin materials entering and the amount of wastes leaving the urban area's industrial system. Previous studies in the US have demonstrated net positive environmental benefits from avoided resource extraction and waste disposal at the city and state level (Eckelman and Chertow 2009; Chertow and Miyata 2011; Eckelman and Chertow 2013). A life cycle assessment can also estimate environmental benefits across a variety of upstream and downstream impact categories including primary energy use, greenhouse gas emissions, acidification, eutrophication and land use change.

Step 5. Put urban industrial symbiosis potential in context

To make an analysis of industrial symbiosis potential relevant to policy-makers, economic development organizations and citizens, it is helpful to describe the potential in the context of other more recognized environmental initiatives as a way of offering context. In the Pennsylvania study cited above, the state-wide primary energy savings of 13 PJ of primary energy from reducing waste by reusing industrial materials does not by itself make an impression on the public. This amount of energy saving, however, was found to be greater than the amount of energy generated by the entire state's renewable energy sector in 2009. Putting industrial symbiosis potential in context makes it easier for all parties to understand the public and private benefits of material and energy reuse.

Conclusion

The past two decades have witnessed the advance of industrial symbiosis both as a practice for sustainability and economic competitiveness and as a growing research field. This chapter demonstrates that there is a great deal of opportunity for industrial symbiosis in urban areas, which, in turn, extends to the global system especially when local resources can be substituted for global ones and local emissions such as methane from landfills are reduced at the source.

Additional findings of this chapter reveal that:

- The adoption of industrial symbiosis is associated with multi-scale benefits at the level of the firm, the urban community and globally by expanding existing environmental capacity through resource reuse and sharing. Not only do private financial and environmental benefits accrue to firms, but also public benefits accrue to the urban community and beyond. Just as increasing global environmental impact can negatively affect cities, so too can benefits created locally ameliorate at least some portion of global impact.
- Urban agglomeration of production, consumption and infrastructure provides opportunities for the development of industrial symbiosis. These opportunities are found most

commonly in local urban centers of smaller cities or on the peripheries of central metropolitan areas.

- Measuring industrial symbiosis potential offers a way to assess the total quantity of resources that could be recovered in an urban area through inter-organizational resource sharing. A conceptual model of how to do so is offered that, while data intensive, would enable a community to set targets and maximize resource value as part of a planning process.

Future research is important to continuing to advance industrial symbiosis as a positive approach to urban management as discussed below.

- Private benefits to participating firms and public benefits to the broader community have not been measured extensively so are not widely known. If the level of benefits follows the promising examples of what is currently available, then further research would likely find that industrial symbiosis is under supplied, which would provide additional motivation to public and private actors to continue to pursue industrial symbiosis.
- A means of identifying and recording potential symbiotic relationships between and among economic sectors is needed including information on price, quantity and environmental impacts for different urban contexts.
- Expanding the system boundary to study resource flows occurring across the urban and peri-urban landscape, in addition to those located within designated industrial clusters, may reveal new opportunities for industrial symbiosis and means to respond to global impacts by examining this broader scale.

References

Baas, L. W., and Boons, F. A. (2004). An industrial ecology project in practice: exploring the boundaries of decision-making levels in regional industrial systems. *Journal of Cleaner Production*, 12(8–10), 1073–1085. doi: 10.1016/j.jclepro.2004.02.005

Bai, X. (2002). Industrial relocation in Asia: a sound environmental management strategy? *Environment: Science and Policy for Sustainable Development*, 44(5), 8–21.

Behera, S. K., Kim, J.-H., Lee, S.-Y., Suh, S., and Park, H.-S. (2012). Evolution of 'designed' industrial symbiosis networks in the Ulsan Eco-industrial Park: 'research and development into business' as the enabling framework. *Journal of Cleaner Production*, 29, 103–112.

Best Practices Hub Vienna (2002). The Spittelau Waste Incineration Plant. www.bestpractices.at/main.php?page=vienna/best_practices/environment/waste_incineration&lang=en. Accessed December 2013.

Brown, L. A., and Holmes, J. (1971). The delimitation of functional regions, nodal regions, and hierarchies by functional distance approaches. *Journal of Regional Science*, 11, 57–72.

Chertow, M. (2000). Industrial symbiosis: literature and taxonomy. *Annual Review of Energy and Environment*, 25, 313–337.

Chertow, M. (2010). Conserving resources: the case for industrial symbiosis in India. *Energy Manager: A Quarterly Magazine of the Society of Energy Engineers and Managers*, (Oct–Dec), 12–17.

Chertow, M., and Lombardi, D. R. (2005). Quantifying economic and environmental benefits of co-located firms. *Environmental Science and Technology*, 39(17), 6535–6541.

Chertow, M., and Miyata, Y. (2011). Assessing collective firm behavior: comparing industrial symbiosis with possible alternatives for individual companies in Oahu, HI. *Business Strategy and the Environment*, 20(4), 266–280.

Chertow, M., and Ehrenfeld, J. (2012). Organizing self-organizing systems: toward a theory of industrial symbiosis. *Journal of Industrial Ecology*, 16(1), 13–27.

Chertow, M., Ashton, W., and Espinosa, J. (2008). Industrial symbiosis in Puerto Rico: environmentally related agglomeration economies. *Regional Studies*, 42(10), 1299–1312.

Cote, R., and Crawford, P. (2003). A case study in eco-industrial development. In E. Cohen-Rosenthal and J. Musnikow (eds.), *Eco-industrial Strategies: Unleashing Synergy between Economic Development and the Environment* (pp. 322–329). Sheffield, US: Greenleaf Publishing.

Duranton, G., and Puga, D. (2000). Diversity and specialisation in cities: why, where and when does it matter? *Urban Studies*, 37(3), 533–555.

Duranton, G., and Puga, D. (2001). Nursery cities: urban diversity, process innovation, and the life cycle of products. *American Economic Review*, 1454–1477.

Duranton, G., and Puga, D. (2004). Micro-foundations of urban agglomeration economies. In V. Henderson and J.-F. Thisse (Eds.), *Handbook of Regional and Urban Economics* (Vol. 4, pp. 2063–2117). Amsterdam, The Netherlands and Kidlington, Oxford, UK: Elsevier.

Eckelman, M. J., and Chertow, M. (2009). Quantifying life cycle environmental benefits from the reuse of industrial materials in Pennsylvania. *Environmental Science & Technology*, 43(7), 2550–2556.

Eckelman, M. J., and Chertow, M. (2013). Life cycle energy and environmental benefits of a US industrial symbiosis. *The International Journal of Life Cycle Assessment*, 18(8), 1524–1532.

Ehrenfeld, J., and Gertler, N. (1997). Industrial ecology in practice: the evolution of interdependence at Kalundborg. *Journal of Industrial Ecology*, 1(1), 67–73.

EPA (U.S. Environmental Protection Agency). 2007. eGRID: Emissions and generation resource integrated database. Updated 5 February 2009. www.epa.gov/cleanenergy/energy-resources/egrid/index.html#download. Accessed December 2014.

Glaeser, E. L. (1998). Are cities dying? *Journal of Economic Perspectives*, 12, 139–160.

Graedel, T. and Allenby, B. (2010). *Industrial Ecology and Sustainable Engineering*. Upper Saddle River, NJ, USA: Prentice Hall.

Hanson, G. H. (2001). Scale economies and the geographic concentration of industry. *Journal of Economic Geography*, 1(3), 255–276.

Henderson, I. V. (1994). Where does an industry locate? *Journal of Urban Economics*, 35(1), 83–104.

Henderson, V. (1997). Medium size cities. *Regional Science and Urban Economics*, 27(6), 583–612.

Hoover, E. (1937). *Location Theory and the Shoe and Leather Industries*. Cambridge, MA: Harvard University Press.

IEA (International Energy Agency). (2013). World Energy Outlook.

Jacobs, J. (1969). *The Economies of Cities*. New York: Random House.

Jacobsen, N. B. (2006). Industrial symbiosis in Kalundborg, Denmark: a quantitative assessment of economic and environmental aspects. *Journal of Industrial Ecology*, 10, 239–255.

Kalundborg Symbiosis (n.d.). Local area. Retrieved from: www.symbiosis.dk/en/lokalomraadet

Krugman, P. R. (1991). *Geography and Trade*. Cambridge, MA: MIT Press.

Laybourn, P., and Morrissey, M. (2009). *National Industrial Symbiosis Programme: The Pathway to a Low Carbon Sustainable Economy*. Birmingham, UK: International Synergies Ltd.

McKinsey & Company. (2010). Comparing urbanization in China and India. McKinsey Quarterly July 2010. www.mckinsey.com/insights/urbanization/comparing_urbanization_in_china_and_india

Mukkala, K. (2004). Agglomeration economies in the Finnish manufacturing sector. *Applied Economics*, 36(21), 2419–2427.

Park, H.-S., Rene, E. R., Choi, S.-M., and Chiu, A. S. F. (2008). Strategies for sustainable development of industrial park in Ulsan, South Korea – From spontaneous evolution to systematic expansion of industrial symbiosis. *Journal of Environmental Management*, 87(1), 1–13. doi: 10.1016/j.jenvman.2006.12.045

Parr, J. B. (2002). Missing elements in the analysis of agglomeration economies. *International Regional Science Review*, 25(2), 151–168.

PIRE (n.d.). US National Science Foundation Partnership for International Research and Education: Developing Low-Carbon Cities in the US, China, and India Through Integration Across Engineering, Environmental Sciences, Social Sciences, and Public Health. https://sites.google.com/a/umn.edu/pire-sustainable-cities/. Accessed December 2014.

Quigley, J. M. (1998). Urban diversity and economic growth. *The Journal of Economic Perspectives*, 12(2), 127–138.

Ramaswami, A., Weible, C., Main, D., Heikkila, T., Siddiki, S., Duvall, A., Pattison, A., and Bernard, M. (2012). Social ecological infrastructure systems framework for interdisciplinary study of sustainable city systems. *Journal of Industrial Ecology*, 16(6), 801–813.

Seto, K. C., Dhakal, S., Bigio, A., Blanco, H., Delgado, G. C., Dewar, D., Huang, L., Inaba, A., Kansal, A., Lwasa, S., McMahon, J. E., Müller, D. B., Murakami, J., Nagendra, H., and Ramaswami, A. (2014). Human settlements, infrastructure and spatial planning. In: *Climate Change 2014: Mitigation of Climate Change. Contribution of Working Group III to the Fifth Assessment Report of the Intergovernmental Panel on Climate Change* [Edenhofer, O., R. Pichs-Madruga, Y. Sokona, E. Farahani, S. Kadner, K. Seyboth, A. Adler, I. Baum, S. Brunner, P. Eickemeier, B. Kriemann, J. Savolainen, S. Schlömer, C. von Stechow, T. Zwickel and J.C. Minx (eds.)]. Cambridge, United Kingdom and New York, NY, USA: Cambridge University Press.

Shi, H., Chertow, M., and Song, Y. (2010). Developing country experience with eco-industrial parks: a case study of the Tianjin Economic-Technological Development Area in China. *Journal of Cleaner Production*, 18(3), 191–199. doi: 10.1016/j.jclepro.2009.10.002

Symbiosis Institute (2013). *Kalundborg Symbiosis Institute* [Online]. Available: www.symbiosis.dk/en/system. Accessed November 2013.

van Beers, D. (2008). Capturing Regional Synergies in the Kwinana Industrial Area: 2008 Status Report. Perth, WA: Centre for Sustainable Resource Processing.

van Beers, D., Bossilkov, A., Corder, G., and van Berkel, R. (2007). Industrial symbiosis in the Australian minerals industry: the cases of Kwinana and Gladstone. *Journal of Industrial Ecology*, 11(1), 55–72.

van Berkel, R., Fujita, T., Hashimoto, S., and Geng, Y. (2009a). Industrial and urban symbiosis in Japan: analysis of the eco-town program 1997–2006. *Journal of Environmental Management*, 90(3), 1544–1556. doi: 10.1016/j.jenvman.2008.11.010

van Berkel, R., Fujita, T., Hashimoto, S., and Fujii, M. (2009b). Quantitative assessment of urban and industrial symbiosis in Kawasaki, Japan. *Environmental Science and Technology*, 43(5), 1271–1281.

Vergara, S. E., and Tchobanoglous, G. (2012). Municipal solid waste and the environment: a global perspective. *Annual Review of Environment and Resources*, 37, 277–309.

Weber, C. L., and Matthews, H. S. (2008). Food-miles and the relative climate impacts of food choices in the United States. *Environmental Science & Technology*, 42(10), 3508–3513.

Williams, E., Kahhat R., Allenby, B., Kavazanjian, E., Kim, J., and Xu, M. (2008). Environmental, social, and economic implications of global reuse and recycling of personal computers. *Environmental Science & Technology*, 42(17), 6446–6454.

32

RESILIENT URBAN INFRASTRUCTURE FOR ADAPTING TO EXTREME ENVIRONMENTAL DISRUPTIONS

Rae Zimmerman

Critical infrastructure has become a centerpiece of public policy, planning and management. Some of this attention has been driven by catastrophic impacts of natural hazards and other extreme events on infrastructure, its users and those in its path, especially in urban settings. The World Bank (2013, p. 6) notes that infrastructure sectors account for 31 percent of the physical damage from hydrometeorologic disasters (based on 72 events between 1972 and 2013), and 17 percent of economic loss globally. The U.S. Global Change Research Program identifies infrastructure sectors as key targets of climate change impacts (Melillo et al., 2014). Concepts of resilience have emerged in many disciplines—physical, engineering, social, environmental and others (Gay and Sinha, 2013, p. 332; Vale, 2014, p. 192)—many of which address resilient infrastructure either directly or indirectly. Although threats to infrastructure and social environments are numerous and varied, a common thread exists that supports a unified approach to resilient infrastructure.

In this chapter, resilience as an infrastructure concept is presented first along with its importance and benefits. In the second section, definitions and dimensions of resilient infrastructure are highlighted as they have emerged across disciplines. In the third section, the common measure 'recovery' is introduced and examples of its application are presented across multiple infrastructures and types of system disturbances. In the fourth section, the role and associated challenges of green infrastructure in supporting resilient infrastructure are discussed as well as the role of infrastructure adaptation in response to many different threats simultaneously. Finally, lessons learned (fifth section) and key messages, and suggestions for future directions (sixth section) are presented that capture some of the subtleties of resilient infrastructure, especially within the framework of disaster management.[1]

The resilience concept in the context of urban infrastructure

Importance of resilience for infrastructure

The importance and visibility of resilient infrastructure has a long history of government attention globally. The World Economic Forum (WEF) (2014, pp. 8–9) underscores resilience as an essential part of responding to disasters and extending the 'life of assets' component of its Operations and Maintenance 'O&M Best Practices/Critical Success' strategy. In the US, government initiatives date at least from the 1990s, with attention to the relationship between infrastructure condition and resilience occurring much earlier. US national policy is reflected in the proclamations of the President of the United States that declared December 2012 and November 2013 critical infrastructure security and resilience months (U.S. White House, 2012; U.S. White House, 2013b).[2] The urban setting has been afforded considerable attention in the resilience literature, given the concentration of people, activities, and use of and reliance upon infrastructure in urban areas. Given the diversity of cities and the fact that disturbances can vary for different parts of a city, Vale (2014, pp. 195–196) cautions that the concept of resilience should be adaptable to such variation and that no one single meaning of resilience and its associated term of recovery be ascribed to the urban setting. However, common approaches are apparent in the resilience-oriented restoration of infrastructure in urban areas.

Infrastructure resilience has also emerged as an important element in understanding and coping with disasters (see Young, Chapter 16 in this volume for a discussion of extreme events and their impacts on urban systems). The National Academies call attention to resilient infrastructure in its report on disaster resilience identifying "indicators of the ability of critical infrastructure to recover rapidly from impacts" (National Academies, 2012, p. 4). The New York State (NYS) 2100 Commission (2013) focuses on resilient infrastructure following Hurricane Sandy in October 2012, as a basis for strengthening New York State against future catastrophes. This was followed by the Hurricane Sandy Rebuilding Task Force infrastructure resilience (2013, June, p. 49) guidelines as a basis for infrastructure investments. Seven guidelines were put forth, each with detailed criteria. The guidelines briefly are: (1) a comprehensive analysis for infrastructure investments, (2) transparency and inclusiveness in decision processes for project selection, investment and other issues as well as "measures that will advance the engagement of vulnerable and overburdened populations" (Hurricane Sandy Rebuilding Task Force, 2013, June, p. 51), (3) "regional resilience," (4) "long-term efficacy and fiscal sustainability," (5) "environmentally sustainable and innovative solutions," (6) "targeted financial incentives," and (7) "adherence to resilience performance standards (Hurricane Sandy Rebuilding Task Force, 2013, June, pp. 49–54)." The U.S. Department of Homeland Security (DHS) charged the Rand Corporation with an evaluation of how federal agencies were implementing the guidelines with respect to rebuilding following Hurricane Sandy and also whether the guidelines would be applicable to the allocation of infrastructure-related funds by the federal government in non-recovery situations (Finucane et al., 2014). Briefly, some of the findings are that pre-existing institutions developed after past disasters as plans, programs and other devices supported or eased the efforts following Sandy. Second, from an urban or community perspective, those areas that had frameworks for risk in place had a better chance of using the Sandy supplemental funds for long-term needs (Finucane et al., 2014, p. xi). Difficulties encountered were identified and translated into lessons learned and recommendations. These included, among others, the ability to adapt, apply or balance the guidelines for specific situations and use metrics also tailored to particular problems (Finucane et al., 2014, pp. 27–34).

Benefits of and need for resilient infrastructure for cities

It is commonly recognized that urban population is increasing throughout the world. Heilig, for example, identifies global trends that show urban population exceeded rural population by 2010 and is projected to more than double by 2100 (2012, slide 12). In the US, urban population is also anticipated to continue growing and exceed rural population growth (U.S. EPA, 2013b, p. 6). The United Nations Environmental Programme (UNEP) (2013, p. 26) cites figures for the percentage of the world population living in cities as being over 50 percent in 2007 and is expected to rise to almost 70 percent by 2050. Moreover, according to UNEP (2013, pp. 14–15), cities produce 80 percent of the world's GDP and consume three-quarters of the flows of global energy and materials. The fundamental issue is whether infrastructure services are keeping pace with and supporting these populations' needs and activities (Gay and Sinha, 2013, p. 331; UNEP, 2013, p. 17). Analyses of global infrastructure show an uneven picture of this gap.

Benefits of resilience for infrastructure are defined directly, in terms of monetary benefits, and indirectly, in terms of improvements in the service and quality of life for its users. Promoting resilient infrastructure and linking it to phases of disaster management that predate a catastrophe has been cited as being financially beneficial. There are many important non-monetary benefits, which are not easily quantifiable, but some financial benefits have been identified that provide an important context for pre-event resilient infrastructure. US and international organizations estimate that the total identified costs of storms and other emergencies are often in the billions of dollars for lives lost, business losses and property damage. Infrastructure costs, either directly or indirectly, are a growing part of those costs (World Bank, 2010; 2013).

Contributing to these costs, disabled infrastructure results in well-known delays in evacuation and recovery time for social and economic well-being, although the direct costs to infrastructure are not easily separated out. The World Bank (2013, p. v) indicates that the cost to restore the built environment including infrastructure can be 50 percent more post-disaster. A National Institute of Building Sciences (NIBS) Multi-Hazard Mitigation Council (2005, p. iii) study shows that, "On average, a dollar spent by FEMA on hazard mitigation (actions to reduce disaster losses) provides the nation about $4 in future benefits." Adaptation is also estimated to be a small share (HM Government, 2011, p. 43; citing the World Bank, 2010, p. 43). The World Bank (2010, p. 43) estimates that adapting infrastructure to climate change costs about 1–2 percent of infrastructure costs. Healy and Malhotra estimate that, "a $1 increase in preparedness spending resulted in approximately a $7.37 decrease in disaster damage" (Healy and Malhotra, 2009, p. 396; Zimmerman, 2014b, pp. 141–142), but infrastructure is not singled out. An important component of assessing the benefits of resilience, particularly for infrastructure, is the concept of the value of human life (Viscusi, 2003) and the value of investments in infrastructure in general for society (Zimmerman, 2014b). Thus, connection to estimates for lives saved from infrastructure improvements needs to be made.

Social significance and imperative of resilience

Infrastructure and social resilience are increasingly connected (Godschalk, 2003, p.137, citing Beatley, 1998). Tierney (2014) identifies the importance of social resilience in general in the context of disasters. Vale (2014) argues that resilience must aim at achieving social justice. Internationally, UNEP (2013) presents cases and models to incorporate the needs of those in poverty into the design of future infrastructure. The connection between resilient infrastructure and social resilience, enabling individuals, communities, societies and their institutions is a common theme emphasized by O'Rourke (2007), Verner and Petit (2013, p. 2) and others, as a necessary condition for the resilience of physical systems such as infrastructure.

Following Hurricane Sandy, the social dimensions of resilience in general as well as for infrastructure in support of social systems were particularly emphasized. The NYS 2100

Resilient urban infrastructure

Commission (2013, p. 7) defines resiliency specifically in social terms: "Building back better demands a focus on increased resilience: the ability of individuals, organizations, systems, and communities to bounce back more strongly from stresses and shocks." The NYS 2100 Commission (2013, p. 7) acknowledges the need for a broadly conceived concept of resilience that combines physical, social and environmental elements:

> Hard infrastructure improvements must be complemented by soft infrastructure and other resilience measures, for example, improving our institutional coordination, public communication, and rapid decision-making abilities will make us better able to recover from the catastrophic effects of natural disasters.

Along these lines, the Rockefeller Foundation's 100 Resilient Cities program defines resilience as:

> the capacity of individuals, communities and systems to survive, adapt and grow in the face of changes, even catastrophic incidents . . . in other words, building resilience is about making people, communities and systems better prepared to withstand catastrophic events—natural, climate change-driven, and man-made—and able to bounce back more quickly and sometimes even emerge stronger from those shocks and stresses.
>
> *(Rodin, 2013)*

Resilient infrastructure defined

The origin of the resiliency concept is identified by Brand (2009, p. 606) as arising during the 1960s and 1970s in the field of ecology, soon spreading to other science and social science disciplines. However, Alexander (2013) traces its origins back centuries earlier. Tierney (2014, Chapter 7) reviews the many different definitions of the resiliency concept as a basis for its applicability to communities in disasters. Schultz et al. (2012, p. ii, 2) note that the term resilience is ambiguous, referring either to change in a system, property of systems or goals that systems should attain (citing Klein et al., 2003), is dependent on contexts and objectives, and can change with boundary conditions. According to some, resilience is also dynamic in that it portrays a process and practice to move toward a desired state (Vale, 2014, pp. 191–192). Comfort (2012) and others define it as a process by which organizations and political systems are responsible for overseeing resilience and are themselves described in terms of resilience. Comfort et al. (2011, p. 19) apply this to the organizational structures that emerged to deal with recovery following the Haiti earthquake.

The concept of resilience in general shares many properties with infrastructure resilience. In the infrastructure setting, it is applied at multiple scales—individual infrastructure components and processes, within specific types of infrastructure, e.g. electric power (Reed et al., 2010), across entire infrastructure systems (U.S. White House, 2013a) and for numerous users or those affected ranging from individuals to whole communities, cities (Chang et al., 2014; Godschalk, 2003; Rodin, 2013; Vale, 2014, p. 191; Schulz et al., 2012), states and nations. From a physical infrastructure perspective, resilience applies to components, facilities, systems, and could even apply to the financial support to meet performance goals. Roe and Schulman (2012, p. 115) link the organizational and operational contexts for critical infrastructure at a more detailed level, noting that: "resilience for any critical infrastructure is the ability of its control room (dispatch center or operations unit) operators to bounce back or absorb a shock in the management of the critical service in question." Furthermore, they link infrastructure resilience directly to disaster management stages, dividing resilience into the post-impact (recovery) and pre-impact (precursor resilience) stages where operators make adjustments to maintain service outside of the view of the customers (Roe and Schulman, 2012, pp. 123–124).

While the resilience concept emerged later in the context of engineered systems in the risk literature (Haimes, 2009), the concept of resilient infrastructure became firmly situated in public policy during the early part of the 21st century. The term 'resilience' as it appeared directly or indirectly for infrastructure usually relates to some context, condition or action that has the potential to affect infrastructure. Some of the common definitions can be summarized as: (1) returning to some original state or 'bouncing back' (Vale, 2014, p. 198, citing Vale and Campanella, 2005; Paton and Johnston, 2006, p. 7), for example, in the context of electric power infrastructure (Reed et al., 2009, p. 174) or to a pre-disturbance state (National Research Council 2013, p. 152); (2) attaining a new, usually more desirable state (NYS 2100 Commission, 2013, pp. 1, 7; Côté and Darling, 2010) or 'bouncing forward' (see for example, Borenstein, 2014; Vale, 2014, p. 198); (3) being resistant to change or perturbation in the first place from some event (Mileti, 1999, pp. 32–33; Godschalk, 2003); and (4) as Litman (2006, p. 13) indicates at the systems level, avoiding catastrophic failure while variable, unanticipated conditions occur. The National Academies summarize all of these possibilities in its definition of resilience in connection with disasters in the following way: "resilience is the ability to prepare and plan for, absorb, recover from, and more successfully adapt to adverse events" (National Academies, 2012, p. 1).

The concept of resilience is defined in disciplines other than engineering and physical systems such as social and ecological systems that have relevance to infrastructure, especially given the breadth of adaptive strategies that complement conventional physical infrastructure. Definitions of resilience in different disciplines are compared by Folke (2006, p. 259) who points out that engineering, ecological and social-ecological resilience tend to emphasize respectively 'return time' and 'efficiency,' 'buffer capacity' and the ability to 'withstand shock' and 'maintain function' and development. The application of engineering resilience to infrastructure has included the incorporation of more adaptation planning and future orientation, reflecting the use of the term 'resilience' in the ecological and social sciences. Others have compared the way that resilience is defined across these three domains (Schultz et al., 2012, pp. 2–6; Wang and Blackmore, 2009).

Regulatory policy in the US has incorporated and reflects infrastructure resilience concepts in homeland security rules and regulations that provide models for other parts of the world. PPD-21, the US policy directive for infrastructure security and resilience, defines resilience as: "the ability to prepare for and adapt to changing conditions and withstand and recover rapidly from disruptions." Resilience as defined in federal infrastructure policy includes the ability to "withstand and recover from deliberate attacks, accidents, or naturally occurring threats or incidents" (U.S. White House, 2013a).

Definitions of engineering resilience related to infrastructure, like other applications of resilience, contain a number of components that are useful in operationalizing the resilience concept: a *perturbation*, disruption, disturbance or some event that changes some predefined state of the system, a baseline *state* against which the change is measured (not necessarily the original state) and a *measure* of the overall process of attaining resilience.

The *perturbation* or source of a change is usually unintended and adversely affects performance, use or stability of the infrastructure. Perturbation or disturbance has been used to characterize change and a variety of different forms of recovery or accommodation in ecosystems (Holling, 1973). Perturbations may not originate external to a given system: Gallopín (2006, p. 295) indicates that perturbations, while usually considered to occur from outside of a system, may also be internally driven. The National Research Council (NRC) (2013), in the context of climate change, uses intensity, duration and frequency as properties and consequences of disturbances. The timing of changes in a system and the relationship of these changes to the timing of the disturbance has also been addressed, for example, in terms of the concept of abruptness (National Research Council, 2013, pp. 26–28). Others have argued that a disturbance is not necessarily a precursor for the movement of a system toward another more resilient state, and that more gradual changes can be a framework for moving toward resilience (Vale, 2014, p. 192). In this chapter, natural hazards are used as the disruption to illustrate resilient infrastructure, but perturbations have many other origins such as intentional acts related to security breaches, cyber attacks, terrorist attacks, intentional acts of vandalism as well as accidents.

Definitions differ about the *state* against which resilience is measured. As indicated earlier, some definitions identify an earlier state or condition, or the ability to 'bounce back' after a major disturbance (Reed et al., 2009). Schultz et al. (2012, pp. ii, 1, 3) specifically define the state for engineering resilience as maintaining the performance of function during or following a disaster, and in some contexts this state is an equilibrium state (Schultz et al., 2012, p. 4). Others discuss an equilibrium point, as a desired or more preferred state (Côté and Darling, 2010). Still others point to a pre-disturbance state or one to which social systems can adapt (National Research Council, 2013, p. 67), while some identify a better future state. The latter concept has become the more commonly used. Folke (2006, p. 260) defines resilience as persisting as changes occur and as the ability to change to more desirable states. Farmani and Butler (2013, p. 1500), adopting Folke's definition, use the term for water supply distribution as having the ability to withstand adverse changes and achieve some future state or 'future proofing.' These also reflect the PPD-21 concept of the ability to adapt to changing conditions, rather than returning to the status quo. Culture and public perception play an important part in determining the reference state. Regulatory constraints and financing conditions can also predetermine how the reference state is defined.

Finally, *measuring resiliency* with respect to a baseline state after a disturbance is of central interest, but perhaps the most difficult to accomplish. Many have recognized the difficulty of actually measuring engineering resilience, since it involves knowledge of previous state conditions of the systems and linkages between a post-disaster and pre-disaster state, which is often not available (National Research Council, 2013, p. 67). Farmani and Butler (2013) use network theory as the basis for a measurement system for water systems, and Hart (2013) uses it as a basis for guiding infrastructure resiliency design, exemplifying an extensive tradition of using network theory to characterize infrastructure performance (Zimmerman, 2012b). The U.S. DHS has developed a Regional Resiliency Assessment Program that incorporates a number of risk-based tools that include recovery times (Verner and Petit, 2013, pp. 2–3; U.S. DHS, 2014b). Verner and Petit (2013, p. 2) underscore the need for measurement methodologies to integrate interconnections among different domains—social, environmental and technical—in infrastructure resilience measurement systems, and review a number of different measurement methods. A common measure of the infrastructure resiliency component of reverting infrastructure to another state is the speed or rate of recovery, which will be covered in detail below. Schultz et al. (2012, p. 5), for example, suggest this measure is the restoration rate for performance levels prior to a disturbance; there are other authors who reference different system states for the return.

Resilience has been discussed in relation to other similar concepts, such as recovery, restoration, robustness, sustainability, adaptation, adaptive capacity and vulnerability, which have important relationships to the resilience concept. Côté and Darling (2010) introduce a subtlety in the concept in the context of ecosystems that is potentially applicable to infrastructure systems, i.e. resiliency actually has two parts: resistance and recovery. Farmani and Butler (2013, p. 1496) note the similarity between robustness and resilience, defining robustness as "the persistence of a system's behavior under perturbations or conditions of uncertainty," which is closer to some of the definitions of resilience that maintain the status quo. Citing Gallopín (2006) and Anderies et al. (2004), Farmani and Butler (2013, p. 1496) further identify a robust system as being "able to perform its designated function (desirable state) without any modifications under change." Adaptation or adaptive capacity is another concept related to resiliency, which is the ability of a system to accommodate changes in system states or the environment (Paton and Johnston, 2006; Adger, 2006; Gallopín, 2006). Adaptation is often related to resilience through the concept of vulnerability, which is an essential component of the analysis of risk. According to Adger (2006, p. 270), vulnerability consists of "exposure and sensitivity to perturbations or external stresses, and the capacity to adapt." Gallopín (2006, p. 294) identifies relationships among resiliency and vulnerability and adaptive capacity, and particularly notes that both vulnerability and resiliency are usually defined relative to specific system disturbances and will vary according to the type of disturbance. Finally, resilience and sustainability are considered linked (Côté and Darling, 2010, p. 1). These links are critical for moving in the direction of improving resilient infrastructure.

Below is a further discussion of the operationalization of resilient infrastructure and more detailed dimensions of resilience in the context of recovery as a unit of measurement including capacity, flexibility, the ability to rebound and interconnected factors (Rodin, 2013; NYS 2100 Commission, 2013, pp. 24–25).

Recovery rates as a resiliency metric

Recovery rates defined and factors that influence the rates

A common way of measuring resiliency following catastrophic events is via recovery rates or restoration time, though caveats and limitations to the measure are important considerations in interpreting such rates. Stages in disaster management (mitigation, preparedness, response and rehabilitation) are an important context for defining recovery and its relationship to resilience, and the relationship between recovery and resilience can vary across different stages in the disaster management cycle (Gilbert, 2010, pp. 19–30, U.S. DHS, 2013). For example, precursors or pre-conditions that affect the strength of infrastructure can occur early in the disaster management cycle. These can later affect the ability of infrastructure to withstand impacts or to recover, hence its resilience.

The NRC (2013; Scheffer et al., 2012; van Nes and Scheffer, 2007) defines the recovery change rate in terms of changes in the state of a system or context (the 'regime'). Recovery depends on what recovers, for whom, how and when, and is often situation- and condition-specific (Vale, 2014). Moreover, in light of relationships among resilience in these different domains, it is important to understand that recovery may occur at different rates in different systems. The slow rate of recovery of social systems following a disaster is often magnified by the rate at which basic infrastructure services are restored. For example, following the Haiti earthquake, the United Nations notes that social restoration lagged substantially, as water and sanitation services lagged (United Nations Office for the Coordination of Humanitarian Affairs, 2014, p. 9). Furthermore, recovery over time does not necessarily occur linearly; it is subject to inflection points and the influence of single points of failure (Verner and Petit, 2013, p. 2). Infrastructure recovery can vary widely geographically due to variations in and exposure to initial perturbations. Recovery is often defined very specifically and separately for individual system components, since the vulnerability of each component can vary. Though the uncertainty in the duration of recovery is a major factor in managing outages, there are some similarities across different events.

As is the case for resilience of which recovery is a part, physical recovery of infrastructure is connected with social recovery. The US government defines recovery in the context of both social and physical systems including infrastructure as:

> those capabilities necessary to assist communities affected by an incident to recover, including but, not limited to, rebuilding infrastructure systems; providing adequate interim and long-term housing for survivors; restoring health, social and community services; promoting economic development; and restoring natural and cultural resources.
>
> *(U.S. White House, 2011, p. 6)*

Characteristics of overall resilience—capacity, flexibility, ongoing institutional adaptability to change and acknowledgement of interdependencies (NYS 2100 Commission, 2013, p. 25)—are important aspects of and foundations for the definition of recovery rates and these factors are addressed below as they shape recovery.

Capacity. Capacity is an important dimension of the balance between supply and demand, and includes social, ecological and environmental impacts. It is an old concept and appeared in the form of 'carrying capacity' in the mid-20th century, and is often referred to in the context of footprints (Wackernagel and Rees, 1996). System condition affects capacity, which, in turn, is influenced by age, poor construction and initial deficiencies in design or design flaws. Many pre-conditions or precursors influence recovery capacity. How severely infrastructure is disabled when an event occurs is often a function of initial condition and protective actions.

Flexibility. The ability to adapt to changing conditions enables infrastructure and its users to avoid or reduce impacts, for example, by shifting to alternative ways of providing infrastructure services or the resources needed for it to function, and using alternative locations or behavior in the short-term or long-term. Many infrastructure systems already have this capacity which has been invoked in emergencies. Zimmerman et al. (2014), for example, evaluate multi-modal connections between bus and rail transit in NYC, which potentially enables users to switch modes when one of the two systems is disabled.

Institutional adaptability. The National Academies (2012, p. 17) indicates that resilient infrastructure can refer simultaneously to physical facilities, the social systems that use and depend on them and the institutions that manage them. In any geographical area, resilience for any infrastructure type is the responsibility of organizations and jurisdictions at many different scales. Linkages among different infrastructures increase the complexity of those organizational and jurisdictional interactions. Communication is a key factor related to institutional characteristics affecting resilience and recovery, with effective communication producing less infrastructure damage and hence faster recovery. The National Transportation Safety Board's cases of catastrophic failures of bridges, rail and pipelines often point to communication between emergency responders, infrastructure managers and operators and between infrastructure managers and the public as key factors in recovery. These communication pathways are identified, for example, by Schultz et al. (2012, p. 30). Legal, regulatory and other institutional constraints influence recovery rates both positively and negatively. Exemptions, permit waivers and adaptation of various environmental laws during emergencies decrease recovery time (summarized by Zimmerman, 2012b). After the September 11, 2001 World Trade Center attacks, simplifying or waiving vessel docking permits in Manhattan allowed rapid transport of debris, and Fresh Kills landfill was reopened for debris disposal (Zimmerman, 2003, p. 259). Transportation vehicle fuel requirements under the Clean Air Act to reduce air pollution have been waived in times of disaster-related fuel shortages (U.S. EPA, 2011a, 2011b and summarized in Zimmerman, 2012b, p. 205). Similar waivers have occurred following major weather events. Priority setting is an institutional factor strongly influencing recovery rates. In the December 2010 NYC snowstorm, prioritizing streets for plowing affected how quickly services dependent on streets would recover. Supply chain management, especially for resources upon which infrastructure depends, also reflects a complex set of institutions key to recovery rate. Most importantly, the ability of institutions to retain 'slack' resources to bring to emergencies reduces infrastructure recovery time (Zimmerman, 2014c p. 29, citing Cyert and March).

Interdependencies. Infrastructure interconnections, defined by Rinaldi et al. (2001), play a key role as resilience factors. Interdependencies promote resiliency in normal times, but can produce catastrophic domino effects if some of the interconnections are destroyed in ways that unpredictably and adversely affect other connections. Zimmerman and Restrepo (2006) find that after the 2003 blackout in NYC, transit and transportation signals took over one and two times as long to come back online as it took for power restoration; in Cleveland and Detroit, water systems took over two and three times as long. These delays are typically due to the need for testing and readjustment of system components such as switches and signals in NYC or complex pump operations for water systems. Transportation, for example, is dependent on water management and electric power infrastructure. Water infrastructure can often obstruct transportation systems during flooding events.

Following Hurricane Sandy, transportation systems operated by the New York Metropolitan Transportation Authority (MTA) recovered in a shorter time period than expected due to the rapid removal of water and power restoration (Zimmerman, 2014c; Kaufman et al., 2012) and extensive measures are underway to harden the system for future storms (Mooney et al., 2013). The main constraint to system restoration after the mandatory pre-storm system shutdown of transit systems in NYC was the restoration of electric power, a key dependency. Water delivery systems were likewise constrained by electric power, with the lack of power to pump water to upper floors of high-rise buildings creating a major impairment. When a disabling event occurs, infrastructure interacts with many other support systems, and its recovery is also influenced by numerous social and environmental conditions. Environmental interactions include the build-up of debris, which can impede restoration of infrastructure and its services. Debris removal can take many months to accomplish, e.g. after the September 11, 2001 attacks (Zimmerman, 2003). The debris problem has been illustrated by many historic disasters (Zimmerman, 2012b, p. 195). One dramatic lesson involves interdependencies between infrastructure design and social and ecological factors. Cases of the removal of dams and highways due to design issues are a notable example, where improving the environment or responding to other social needs becomes paramount. For dams, American Rivers (McClain and Kober, 2013) reports the removal of 50–65 dams annually. Over a thousand dams (1,057) were removed through 2012, and between 1999–2012, about half of the total dams were removed. American Rivers identifies reasons for removal as ecological restoration, environmental resources improvement (water quality and water flow), obsolescence and public safety.

Recovery time measurements and factors influencing them are illustrated below for electric power and transportation following storm events.

Resilient urban infrastructure

Electric power restoration and recovery

Power failures are an important measure of the resiliency of electric power infrastructure systems. Weather is a key threat and source of perturbation to electric power infrastructure, since it can produce wind and water damage to that infrastructure. The ability of energy infrastructure to be restored reflects both the magnitude of weather impacts and the capacity of electrical systems to respond. Numerous measures usually apply separately to different energy system components: rigs, platforms, pipelines, port facilities, refineries and gas delivery systems. For example, refinery restoration duration during Hurricane Sandy ranged from three days for two refineries to 29 days for the Bayway refinery in Linden, reflecting its greater vulnerability given its low-lying location, the damage it sustained and the repair and maintenance required (U.S. DOE, 2013, pp. 14–15). Production activity supplements and relates to component measures as another dimension of impact and recovery, which has its own set of measures. For example, oil is measured in barrels per day and gas in cubic feet per day. For crude oil, following Hurricane Irene, the U.S. Department of Energy (DOE) reported that inputs fell 31 percent overall and recovered in about two to three weeks; following Hurricane Sandy, the decline in inputs was 28 percent and restoration did not occur for about a month, primarily attributed to the length of the Bayway refinery outage (U.S. DOE, 2013, p. 15).

Restoration duration for electric power can vary considerably by location for the same event. For example, the U.S. DOE (2009, pp. 5–6) points out that within two weeks following Hurricane Katrina's landfall, power was restored in three states, but Louisiana still had 40 percent of users without power due to flooding from the breaching of Lake Ponchartrain's levees. Gilbert (2010, p. 22) summarizes literature on lifeline recovery following the Kobe and Northridge earthquakes, noting considerable variability by type of lifeline and earthquake. In both earthquakes, electric power took the least amount of time to recover. Following Hurricane Sandy, Zimmerman (2014c) reports the recovery of most electric power within a few weeks in NYC, corroborating the findings of other studies; however, considerable variability by borough was found in electric power recovery.

The experience across numerous types of electric power systems has shown that regardless of season, outages have been increasing in number and duration. From 1965–2009, McLinn (2009) notes an increasing trend for 128 major power outages internationally. Simonoff et al. (2007, p. 552) analyzed outage data throughout the US ranging from the early 1990s through the early 2000s and find that between 1990 and 2004, regardless of season, outage incident rate increases by about 14 percent annually for the US, and compared to other seasons the rate for the summer was estimated about 75–135 percent higher. They also find that outage duration increased after mid-2000 (Simonoff et al., 2007, p. 561). It is apparent from outage duration that increased severity of storms during the first decade or so of the 21st century coincides with longer outage durations. Table 32.1 provides illustrations for the duration for some major outages in the 20th and 21st centuries in the US These figures are only for restoration of service to customers (not infrastructure) identified by the US.-Canada Power System Outage Task Force (2004) for the 2003 blackout and a few subsequent major outages from U.S. DOE reports. These few examples corroborate statistical analyses that show increasing outage frequency and duration (e.g. Simonoff et al., 2007).

Nuclear power plants portray a somewhat different recovery picture, measured in terms of reactor restoration times as shown in Table 32.2. It appears that plants that pre-emptively shut down took relatively longer to restore than those able to operate at reduced capacity, though more rigorous evaluation is needed to confirm this.

Rae Zimmerman

Table 32.1 Estimated duration of selected electric power outages and initial restoration time for selected storm events, US, 1965–2013

Year of event	# Customers affected at peak (in millions)	Duration (max) in days (% restored is in parenthesis if not 100%)	
1965–2003 (U.S.-Canada Power System Outage Task Force, 2004)			
November 9, 1965*	30.0	0.5	
July 13, 1977	9.0	1.1	
December 22, 1981	5.0	not available	
July 2, 1996	2.0	0.11	
July 3, 1996	7.5	0.4	
June 25, 1998	0.2	0.8	
July 22, 2003**	0.3	14.0	(McNeil et al., 2003)
August 14, 2003	50.0	4.0	
September 18, 2003	6.5	10.0 (75–90%)	(U.S. DOE, 2003a, b)
Selected major outages and outage durations (2005–2013) (U.S. DOE 2009, 2013; U.S. DOE various years; U.S. Department of Commerce, NOAA, 2013)			
August 30, 2005 (Hurricane Katrina)	2.7	24.0 (91%)	
September 25, 2005 (Hurricane Rita)	1.5	27.0 (97%)	
October 24, 2005 (Hurricane Wilma)	3.5	15.0 (97%)	
September 2, 2008 (Hurricane Gustav)***	1.3	10.0 (93%)	
September 14, 2008 (Hurricane Ike)	3.9	18.0	
August 27, 2011 (Hurricane Irene)	6.7	6.0	(U.S. DOE, 2013; U.S. Department of Commerce, NOAA, 2013)
October 29, 2012 (Hurricane Sandy)	8.7	13.0	(U.S. DOE, 2013; U.S. Department of Commerce, NOAA, 2013)
December 6, 2013 (Winter Storm)	0.3	1.8 (56%)	

Notes:
*New York City only.
**Pre-2004 events weather-related or initiated combined with subsequent system failures.
***Louisiana only.

The average recovery times presented in Tables 32.1 and 32.2 do not reflect the very long infrastructure recovery process that continues after initial service restoration. Ongoing repairs and replacements to fix individual components of the damaged infrastructure can take months or even years. For example, effects of salt water and water damage may impact underground and surface infrastructure components, and wind may impact ground infrastructure such as poles. After Hurricane Sandy, more time was needed after initial power restoration to ensure the safety of equipment and individual components before restoring power (NYS Moreland

Resilient urban infrastructure

Table 32.2 Estimated duration of selected nuclear power plant restoration times for Hurricanes Irene and Sandy, northeastern US

Hurricane Irene (2011)	Duration (in days) from impact to restoration	Hurricane Sandy (2012)	Duration (in days) from impact to restoration
Calvert Cliffs 2★	7	Nine Mile 1★	12
Oyster Creek★	4	Salem 1★	6
Brunswick 2	3	Susquehanna 2	6
Millstone 3	3	Indian Point 3★	4
Brunswick 1	2	Millstone 3	4
Millstone 2	2	Limerick 2	3
Limerick 1	2	Limerick 1	1
Limerick 2	2	Vermont Yankee	1

Note: An asterisk (★) indicates plants that were pre-emptively shutdown prior to the storm. Durations are compiled and computed from U.S. DOE (2013, p. 13, Table 5). Events are grouped by duration from largest to smallest, then alphabetically within a duration category if duration is the same for different facilities.

Commission, 2013, p. 59). The interface between the social system—the users of electric power—and electric power's physical infrastructure systems inevitably results in a longer recovery for social systems than the physical systems; that is, social resilience is typically not as strong as physical resilience. Many utilities are now introducing diverse measures to promote resiliency including operational measures and structural measures to elevate, seal and otherwise protect equipment (see for example Consolidated Edison Company of New York and Orange and Rockland Utilities, 2013).

Another more subtle measure of recovery, beyond the direct restoration of particular facilities and services, involves the adoption of policies and practices that address fundamental problems that undermine resilience. The recovery of nuclear power internationally, for example, is reflected in the extent to which countries throughout the world reinstituted nuclear power following the Tohoku earthquake and failure of the Fukushima nuclear power complex in Japan in March 2011. At the time of that disaster, countries were in various stages of adoption and reliance upon nuclear power, with France relying the most with almost three-quarters of its energy supplied by nuclear power in 2014 (Nuclear Energy Institute, 2014). Some countries such as Germany proposed immediate and future plans to close their nuclear facilities in a move toward greater market share of renewable sources of energy (Talbot, 2012). Other countries such as the US stopped licensing new units and increased safety standards. Still, others vacillated. Nuclear power plants were initially closed in Japan in September 2013 pending rigorous review. By October 2014, after a back and forth discussion, Japan approved two nuclear plants to restart production (Iwata, 2014), solidifying the country's reliance on nuclear power (Tabuchi, 2012). The impact of these policy discussions was felt most heavily in cities in which many of these plants were located.

Transportation recovery: rail transit

Many government-issued 'after-action' reports contain specific time periods during which transportation infrastructure recovered from disturbances. These recovery times are highly

Box 32.1   Rail transit recovery following the August 8, 2007 storm

On August 8, 2007, the New York area experienced an intense storm event that commenced at 6 a.m. during the start of the morning rush hour. The storm had region-wide estimated rainfall levels of 1.4 to 3.5 inches within two hours, which occurred alongside the high tide. It was unanticipated—no pre-emptive shutdown of city rail or bus transit occurred—but forced shutdowns occurred on 19 subway lines, all of Metro North Service (shutdown at 7:20 a.m.) and Long Island Railroad (LIRR) service on the Port Washington Line (shutdown at 6:46 a.m.) (MTA, 2007, pp. 2–3). The recovery times for various New York area rail transit systems can be estimated from the intervals given by MTA (2007), which are:

- New York City Transit (NYCT) (7 a.m.–3 p.m. 'midday'): 8 hrs
- Long Island Railroad (LIRR) (6:46 a.m.–12:55 p.m.): 6 hrs 9 min
- Metro-North Railroad (MNR) (7:20 a.m.–8:58 a.m.) 1 hr 38 min

variable depending on the extent of disruption and access to the disabled facilities. Zimmerman (2012b) reviews the ability of transportation infrastructure to recover in a selected number of disasters, and Zimmerman and Simonoff (2009) find that the recovery of rail transit after the September 11, 2001 attacks varied by system, but generally recovered within a few days or weeks after the attacks. Zimmerman (2014c) analyzes the recovery of individual subway lines and subway ridership in NYC following Hurricane Sandy using public transit data, and finds that 80 percent of the lines fully recovered within a couple of weeks and ridership within that period was down about 15 percent from the same period one year earlier (Zimmerman, 2014c). Thus, time to restore service immediately is a common measure of short-term recovery in transit systems. Nevertheless, there are many underlying factors that influence recovery that are a critical part of the recovery measure.

The way forward for resiliency with green infrastructure

The emphasis of resilient infrastructure definitions on pre-emptively invoking protective measures has resulted in numerous innovations across stages of infrastructure development and operation including: siting, design, construction, operation and maintenance. One such innovation is 'green infrastructure.' Green infrastructure is defined by the U.S. EPA as using water and natural areas to improve the environment:

> Green infrastructure uses vegetation, soils, and natural processes to manage water and create healthier urban environments. At the scale of a city or county, green infrastructure refers to the patchwork of natural areas that provides habitat, flood protection, cleaner air, and cleaner water.
>
> *(U.S. EPA, 2014a)*

Implicit in the concept is improvements in transportation, energy and water infrastructure by means of water, land and environmental management including vegetation controls, and promoting resiliency with these strategies (U.S. EPA, 2013a, U.S. EPA, 2014b). Green infrastructures reflect some of the latest infrastructure innovations and have been introduced along with traditional recovery measures for the effective rebounding of infrastructure systems after storms (Zimmerman, 2014c, pp. 24–29). The combined effects of climate change and extreme weather events on infrastructure have resulted in a convergence of green infrastructure and adaptation measures to reduce the consequences of such events. Three aspects of green infrastructure to support resilient infrastructure are noteworthy: (1) its relationship to ecological resilience and how it has borrowed from and been supported by that concept; (2) the ability of green infrastructure to provide an alternative to spatially concentrated facilities; and (3) the institutional buy-in to green infrastructure at all levels of government.

Green infrastructure and ecological resilience

Paralleling the growing interest in green infrastructure is renewed attention to ecological resilience (Pickett et al., 2014; Holling, 1973) and its integration with and support of green infrastructure concepts. As green infrastructure and the general use of natural systems move more to the forefront of infrastructure strategies, this will reinforce the importance of the concept of ecological resilience (Pickett et al., 2014), which is one of the areas where the resilience concept originated. Ecosystems are increasingly considered a key infrastructure for water management, especially for urban areas (U.S. EPA, 2010; U.S. EPA, 2013a), which has benefits not only for water infrastructure, but also for the protection of transportation and energy infrastructure. Concerns over climate change have reinforced interest in ecosystem resilience (Côté and Darling, 2010). However, an implicit orientation toward ecosystem resilience arose much earlier, relying on network and systems concepts to unify and integrate many disparate approaches to studying ecosystems (Forman et al., 2003) that are transferrable to green infrastructure especially in urban areas. Holling (1973), for example, develops new theories to detail ecosystem characteristics related to ecological resilience. Brand (2009, p. 606) defines ecological resilience as the ability to maintain a particular state while resisting disturbance, which not only refers to the concept of a state to be maintained, but also resistance to disturbance in the first place.

Green infrastructure as providing spatially distributed services

The scope of resilient infrastructure has expanded to encompass the movement toward green infrastructure which is diversified, decentralized and can become decoupled in contrast to more vulnerable spatially centralized or concentrated systems characteristic of many types of traditional infrastructure (Zimmerman, 2014c; Zimmerman, 2012a; Perrow, 2007). The expectation is that green infrastructures will increase resiliency of traditional infrastructure by expanding capacity through flexibility afforded by spatial decentralization, and thus decrease recovery time after disruption by reducing the initial extent of destruction. This strategy has achieved worldwide attention such as the need for combined civil infrastructure planning and land use planning that Riegel (2014) emphasizes for Germany. Many alternative designs for integrating infrastructure into land use decisions have been addressed for European cities (Beatley, 2012). After Hurricane Sandy, green infrastructure was emphasized at citywide levels, e.g. by the New York City (2013) plaNYC and for more specialized infrastructure analyses (New York City Environmental Protection, 2013; National Research Council, 2014). Many legal and regulatory changes are often needed to adopt green infrastructure strategies that were recognized prior to and following Hurricane Sandy (NYS, 2013; Parker, 2013a, p. 119; Parker, 2013b, pp. 136–137).

Some types of green infrastructure may not necessarily be resilient to extreme destruction while others are. The typhoon that hit the Philippines in 2013, for example, destroyed geothermal facilities (green infrastructure energy systems); the Philippines is second largest in geothermal energy production or capacity in the world (Bradsher, 2013). Nevertheless, green infrastructure came to the rescue after the Japanese earthquake in March 2011 when electric vehicles less dependent on gas infrastructure could traverse the rubble (Belson, 2011). As Bunkley (2011) indicates, batteries in electric vehicles were more durable or flexible sources of energy (Zimmerman, 2012b, p. 3). Advances in solar technology and financing mechanisms to support them are enabling storage capacity to increase and energy supply to be decoupled at the building level in disasters (Cardwell, 2013) as well as at the level of transportation vehicles. Meanwhile, studies of the relatively greater resilience of areas with protective vegetation and landforms in storm events have been reported in connection with Hurricane Sandy (Richard Stockton College of New Jersey Coastal Research Center, 2013). Similarly, in NYS, the Moreland Commission Report (NYS, 2013) according to Parker's review (2013b, p. 137) emphasizes two aspects of green infrastructure—increasing the capacity of wetlands to improve wetland restoration capability and expanding the extent of wetlands as a protective measure. The Hurricane Sandy Rebuilding Task Force (2013, pp. 73–79) heavily emphasizes the use of green infrastructure for resiliency along with the need for investments in those strategies.

Governmental support of green infrastructure

The green infrastructure concept is supported at many governmental levels, especially following extreme weather events (Hurricane Sandy Rebuilding Task Force, 2013). Such governmental support is critical for the success of green infrastructures and such institutional arrangements are discussed in more detail in the sub-section below: 'Institutional factors'. At the federal level, the U.S. EPA 2013 Strategy's focus on water systems underscores resiliency in its 'multi-benefit approach' for communities in the face of climate change. It notes that "green infrastructure offers an approach to increase resiliency and adaptability" (U.S. EPA, 2013b, p. 2). One of the objectives links green infrastructure to disaster planning for recovery and rebuilding, acknowledging its place at large scales "as a component of community resiliency and disaster relief" (U.S. EPA, 2013b, p. 3). The U.S. EPA Greening America's Capitals program (U.S. EPA, 2014b) promotes the intersection of green, resilient water infrastructure with transportation systems, acknowledging their interdependencies. The U.S. EPA, the Department of Housing and Urban Development (HUD) and the Department of Transportation (DOT) provide support to develop green infrastructure strategies, acting as models for cities. Eighteen cities received awards from 2010–2013 by the U.S. EPA, many in support of green infrastructure. Green infrastructure innovations in the US are described in a number of the U.S. EPA's green infrastructure cases in the US (U.S. EPA, 2010) and for European cities by Beatley (2012). Many of the elements adopted in neighborhoods or city sub-areas target improvements in green infrastructure in the form of stormwater runoff, drainage and flood protection. This often includes modifying street paving materials to increase permeability. Public transit connections, access and corridors for alternative transportation modes such as pedestrian and bike transportation are also included. Green infrastructure approaches reviewed for transportation by Zimmerman (2014c, pp. 25–26; Zimmerman, 2012b, p. 135) encompass water retention and removal. These efforts intersect with climate change initiatives for cities through programs from the U.S. Conference of Mayors (USCOM) program (www.usmayors.org/climateprotection/ClimateChange.asp),

ICLEI (Local Governments for Sustainability: www.icleiusa.org/) and C40 Cities (www. c40cities.org/), and Vale (2014) identifies many of the resilient city initiatives that have emerged out of these efforts.

Likewise, the U.S. Departments of Energy and Transportation promote green, resilient infrastructure within their domains that filter down to regional and local jurisdictions. The U.S. DOT, Federal Transit Administration (FTA) (2011) presents a wide range of adaptation measures to reduce impacts of climate change on transit. Meyer (2008) presents a range of design, construction and siting options for transportation infrastructure in the context of climate change adaptation. State-level initiatives have been mentioned earlier such as those of NYS's Moreland Commission (2013) following Hurricane Sandy.

Resilient infrastructure revisited: lessons learned

Threats to infrastructure and exposures of its users continue with each new event, yet a pattern of adaptation across disaster management stages is emerging that has the potential for increasing resilience prior to such events. Though these adaptations are often tailored to individual infrastructures and types of threats, a few broad lessons common to different infrastructures are emerging that reflect the dimensions of resilience cited earlier. Two areas in which such lessons emerge are: (1) diversification and flexibility of infrastructure services; and (2) institutional mechanisms to support resilient infrastructure that cut across policy, management and finance.

Diversification and flexibility

Diversifying infrastructure technologies and locations are critical to building in flexibility for resilience, regardless of the stage in disaster management. Many key infrastructure facilities and usage areas are often highly concentrated geographically, as described earlier (Zimmerman, 2014a; Zimmerman, 2012a; Perrow, 2007). Infrastructure represents a constellation of facilities, and supply chains and bottlenecks can act as limiting factors to the ability of these systems to reduce impacts and restore quickly. The disabled condition of the Bayway refinery after Hurricane Sandy, the largest serving the New York region, described in more detail earlier, contributed to delays in restoring energy. Henry Hub is an area where most pipelines converge and thus is a potential source of vulnerability for oil and gas transmission. Concentration and the density of urban areas, however, enhance some types of restoration. For example, after September 11, 2001, NYC used about 30 miles of cables to connect different parts of the electric grid to Lower Manhattan (Zimmerman, 2003, p. 252). This strategy could only be possible where lines and substations were located nearby. Also, decoupling infrastructure (e.g. solar power) retains urban density but allows for flexibility, and utilities are now incorporating this into new designs. Technologically, approaches to infrastructure design are emphasizing multiple ways for increasing resilience through diversification. For example, following Hurricane Sandy, Con Edison (Consolidated Edison Company of New York and Orange and Rockland Utilities, 2013) identifies a wide range of technologies to protect its infrastructure. Such portfolio approaches promote diversity and provide alternatives in emergencies. The NYS 2100 Commission (2013, p. 7) summarizes the diversification concept noting that: "Resilience means creating diversity and redundancy in our systems and rewiring their interconnections, which enables their functioning even when individual parts fail."

Institutional factors

Institutional factors need to address the longer-term recovery of social systems and their reliance upon physical systems, as both systems depend on one another. Infrastructure recovery requires robust social systems for a workforce and user support. Social systems require resilient infrastructure in order to restore pre-existing services. These institutional factors pertain to policy development, management and financial and other related resources.

Policy

Whether policy changes are aimed at infrastructure resilience is often reflected in how conditions change following disruptions. The nature of policy changes worldwide pertaining to the reliance on nuclear power following Fukushima was presented above in the section 'Electric power restoration and recovery.' Within four years after the Gulf oil blowout, offshore oil drilling in the Gulf appeared to be returning (Gilbert et al., 2014), which could be reversed due to cutbacks in response to the 2014 drop in oil prices (Reed, 2014). Other accidents such as bridge collapses and structural failures are important lessons for how infrastructure failures shape infrastructure policy and practices (Ratay, 2010).

Management

The work of Comfort cited above clearly links resilience to organization and management processes, which connects to the specifics of infrastructure resilience. The ability to create and maintain resource networks before, during and after a perturbation can substantially contribute to resilience. One key example is the use of 'mutual assistance' agreements. The ability to restore electricity in Florida after Hurricane Wilma, for example, was largely attributed to "the help of 18,000 workers from 33 states and Canada" (U.S. DOE, 2009, p. 7). Likewise, the ability to remove debris to expedite response and recovery in Lower Manhattan after the September 11, 2001 attacks under budget and timeframe constraints was attributed to access to construction management (described above in the section 'Recovery rates defined and factors that influence the rates'). In times of drought, transfers among water systems are possible and have occurred. For example, in the 1965 drought, provisions were made to transfer water between New York and New Jersey over the George Washington Bridge; although they were not actually used (New Jersey Department of Environmental Protection, 2007, pp. 1–5). Very widespread disruptions, however, create limits to these resource networks if everyone is tapping into the same resources.

Finance

Financial investments go hand in hand with infrastructure resilience (Hurricane Sandy Rebuilding Task Force, 2013), and finance becomes more critical in disasters. Gilbert (2010) summarizes literature and arguments that link financing of disaster recovery to both negative and positive outcomes. Gilbert (2010, p. 24) argues that following disasters, areas receiving funds can be ultimately worse off. Compressed spending for infrastructure might have been more effective if spread out over a longer time period, and the redistribution of funding following a disaster can deplete other areas that are in need, but not receiving funds. Nevertheless, certain industries benefit from recovery finance post-disaster such as the construction industry.

The timing of finance is particularly critical for recovery not only for infrastructure but also for the social and economic systems that depend on infrastructure. After Hurricane Sandy, people whose homes were destroyed recovered only slowly, often facing a long process of restoring damaged homes, obtaining resources to do so in light of often formidable and uncertain insurance rates and conditions for the receipt of disaster assistance, and often while dealing with temporary or permanent relocation. Those with fewer personal resources experienced the worst hardships (Parker, 2013a, pp. 114, 116 and 119). Meanwhile, large influxes of funds for infrastructure as well as community rebuilding followed the hurricane (see, for example, U.S. Housing and Urban Development, 2013).

As cities move more toward the incorporation of green infrastructure to promote resilience, financial mechanisms and resources will move more prominently to the forefront of vital institutional mechanisms to support such innovation (Hurricane Sandy Rebuilding Task Force, 2013). Worldwide, Merk et al. (2012) argue that cities are key promoters of green infrastructure, and thus investment must occur in, by and through cities. Urban areas, they argue, can provide financing by redesigning existing financial instruments and developing new ones. In OECD countries, Merk et al. (2012, p. 12) point out that in 2010 the investment in transportation, housing and environmental protection amounted to $59 billion (USD), with transportation accounting for the greatest share. Spending requirements for green infrastructure can be even greater than for traditional infrastructures when operation and maintenance expenditures are taken into account, though costs may decline over time as the adoption of the technologies becomes more widespread. The popularity of green infrastructure is illustrated by the ten urban areas that have spent 10–45 percent of their total expenditures on 'sectors with green potential' and even the other categories incorporate other types of infrastructure such as parks, electric power, waste management and roadways (Merk et al., 2012, pp. 15–16).

Summary and conclusion

A number of insights emerge in connection with infrastructure resiliency:

- The concept of resiliency, in general, and its application to infrastructure has various meanings and according to numerous studies, typically pertains to returning to a previous state, returning to a new, more desirable state, and resisting change in the first place.
- Resilience for infrastructure has a long history across many disciplines, and takes many different forms within and among urban areas; that is, the form of resilience can vary by time, place and social context.
- The connection between infrastructure resilience and social resilience has begun to emerge in the context of recent disasters and the continued attention to this linkage is critical.
- Although there are many measures of infrastructure resilience, recovery is a commonly used and useful measure when related factors such as investment and its equitable distribution, initial infrastructure condition and environmental characteristics are taken into account.
- Policies that cities develop or that are developed by states and nations that affect cities are emerging that promote resilience with respect to the form that infrastructure takes and the timing of innovations relative to disaster management stages.
- A systematic approach to the level and rate of financing to support resilient infrastructure alternatives is needed, especially for green infrastructure and renewable resource options.

Rae Zimmerman

- Future research is needed to identify and organize which innovations for infrastructure resiliency occur under what circumstances to promote consistent use across threats and hazards, to support economies of scale and avoid conflicts.

Acknowledgment

This material is based in part upon work supported by the National Science Foundation under Grant Number 1316335, titled "RAPID/Collaborative Research: Collection of Perishable Hurricane Sandy Data on Weather-Related Damage to Urban Power and Transit Infrastructure," R. Zimmerman, Principal Investigator at NYU (see references to Zimmerman, 2014c). The grant is a collaboration with the U. of Washington (lead) and Louisiana State University. Portions of this material are also based upon work supported by the National Science Foundation under NSF Grant Number 1441140. Any opinions, findings, and conclusions or recommendations expressed in this material are those of the author and do not necessarily reflect the views of the National Science Foundation.

Notes

1 Infrastructure sectors covered here emphasize seven of the 16 sectors defined by the U.S. Department of Homeland Security (DHS) (U.S DHS, 2014a) and Presidential Policy Directive (PPD)-21 (U.S. White House, 2013a) as critical infrastructures and analogous infrastructure categories used in other countries (Clemente, 2013, pp. 13–15). These infrastructures are: dams (U.S. DHS, 2010b), energy (U.S. DHS, 2010c), nuclear (U.S. DHS, 2010e), transportation (U.S. DHS, 2010f), water and wastewater (U.S. DHS, 2010g) and, in addition, communications (U.S. DHS, 2010a) and information technology (U.S. DHS, 2010d) sectors that relate to the resilience of other sectors.
2 The 2012 and 2013 proclamations included the concept of resilience, expanding on earlier proclamations from 2009, 2010 and 2011, which designated December as "critical infrastructure protection month" (Zimmerman, 2012b, p. 235). Presidential Policy Directive (PPD)-21 'Critical Infrastructure Security and Resilience' emphasizes the importance of resilience as a fundamental aspect of infrastructure security (U.S. White House, 2013a). PPD-21 calls for the development of a National Critical Infrastructure Security and Resilience (NCISR) Research and Development (R&D) Plan by the U.S. DHS. The 2013 revision of the National Infrastructure Protection Plan (NIPP) also emphasizes resilience and has as its subtitle, 'Partnering for Critical Infrastructure Security and Resilience' (U.S. DHS, 2013).

References

Adger, W. N. (2006). Vulnerability. *Global Environmental Change, 16(3)*, 268–281. doi: 10.1016/j.gloenvcha.2006.02.006

Alexander, D. E. (2013). Resilience and disaster risk reduction: An etymological journey. *Natural Hazards and Earth System Sciences, 13*, 2707–2716.

Anderies, J. M., Janssen, M. A., and Ostrom, E. (2004). *A framework to analyze the robustness of social-ecological system from an institutional perspective.* Available at www.ecologyandsociety.org/vol9/iss1/art18/

Beatley, T. (1998). The vision of sustainable communities. In Burby, R. J. (Ed.), *Cooperating with nature: Confronting natural hazards with land use planning for sustainable communities* (pp. 233–262). Washington, DC: Joseph Henry/National Academy Press.

Beatley, T. (2012). *Green cities of Europe: Global lessons on green urbanism.* Washington, DC: Island Press.

Belson, K. (2011, May 7). After disaster hit Japan, electric cars stepped up. *The New York Times.* Available at www.nytimes.com/2011/05/08/automobiles/08JAPAN.html?pagewanted=all&_r=0

Borenstein, D. (2014, February 24). Bouncing forward: Why "resilience" is important and needs a definition, the Wilson Center. Available at www.wilsoncenter.org/article/bouncing-forward-why-%E2%80%9Cresilience%E2%80%9D-important-and-needs-definition

Resilient urban infrastructure

Bradsher, K. (2013, November 21). Font of natural energy in the Philippines, crippled by nature. *The New York Times*. Available at www.nytimes.com/2013/11/22/world/asia/font-of-natural-energy-crippled-by-storms-natural-energy.html?hp

Brand, F. (2009). Critical natural capital revisited: Ecological resilience and sustainable development. *Ecological Economics*, *68(3)*, 605–612. doi: 10.1016/j.ecolecon.2008.09.013

Bunkley, N. (2011, December 21). Tsunami reveals durability of Nissan's Leaf. *The New York Times*. Available at www.nytimes.com/2011/12/22/business/tsunami-reveals-durability-of-nissans-leaf.html?scp=1&sq–isson%20Leaf&st=cse&_r=0

Cardwell, D. (2013, December 04). SolarCity to use batteries from tesla for energy storage. *The New York Times*. Available at www.nytimes.com/2013/12/05/business/energy-environment/solarcity-to-use-batteries-from-tesla-for-energy-storage.html?ref=business

Chang, S. E., McDaniels, T., Fox, J., Dhariwal, R., and Longstaff, H. (2014). Toward disaster-resilient cities: Characterizing resilience of infrastructure systems with expert judgments. *Risk Analysis*, *34(3)*, 416–434. doi: 10.1111/risa.12133

Clemente, D. (2013, February). Cyber security and global interdependence: What is critical? London, UK: Chatham House. Available at www.chathamhouse.org/sites/files/chathamhouse/public/Research/International%20Security/0213pr_cyber.pdf

Comfort, L. K. (2012). Designing disaster resilience and public policy: Comparative perspectives. *Journal of Comparative Policy Analysis: Research and Practice*, *14(2)*, 109–113. doi: 10.1080/13876988.2012.664709

Comfort, L. K., Siciliano, M. D., and Okada, A. (2011). Resilience, entropy, and efficiency in crisis management: The January 12, 2010, Haiti earthquake. *Risk, Hazards & Crisis in Public Policy*, *2(3)*, 1–25. doi: 10.2202/1944-4079.1089

Consolidated Edison Company of New York and Orange and Rockland Utilities (2013, June 20). *Post Sandy enhancement plan*. New York, NY: Consolidated Edison Company. Available at www.coned.com/publicissues/PDF/post_sandy_enhancement_plan.pdf

Côté, I. M., and Darling, E. S. (2010). Rethinking ecosystem resilience in the face of climate change. *PLoS Biology*, *8(7)*, E1000438. doi: 10.1371/journal.pbio.1000438

Cyert, R. M. and March, J.O. (1963) *A behavioral theory of the firm*. Englewood Cliffs, N.J.: Prentice-Hall, Inc.

Farmani, R., and Butler, D. (2013). Towards more resilient and adaptable water distribution systems under future demand uncertainty. *Water Science & Technology: Water Supply*, *13(6)*, 1495–1506. doi: 10.2166/ws.2013.161

Finucane, M. L., Clancy, N., Willis, H. H., and Knopman, D. (2014). The Hurricane Sandy Rebuilding Task Force's Infrastructure Resilience Guidelines, Santa Monica, CA: Rand. Available at www.rand.org/content/dam/rand/pubs/research_reports/RR800/RR841/RAND_RR841.pdf

Folke, C. (2006). Resilience: The emergence of a perspective for social–ecological systems analyses. *Global Environmental Change*, *16(3)*, 253–267. doi: 10.1016/j.gloenvcha.2006.04.002

Forman, R. T., Sperling D., Bissonette, J. A., Clevenger, A. P., Cutshall, C. D., Dale, V. H., Fahrig, L., France, R., Goldman, C. R., Heanue, K., Jones, J. A., Swanson, F. J., Turrentine, T., and Winter, T. C. (2003). *Road ecology: Science and solutions*. Washington, DC: Island Press.

Gallopín, G. C. (2006). Linkages between vulnerability, resilience, and adaptive capacity. *Global Environmental Change*, *16(3)*, 293–303. doi: 10.1016/j.gloenvcha.2006.02.004

Gay, L. F., and Sinha, S. K. (2013). Resilience of civil infrastructure systems: Literature review for improved asset management. *Int. J. Critical Infrastructures*, *9(4)*, 330–350.

Gilbert, D., Harder, A., and Scheck, J. (2014, November 21). Oil boom returns to Gulf after Deepwater Horizon Disaster. *The Wall Street Journal*. Available at http://online.wsj.com/articles/oil-rigs-return-to-gulf-after-deepwater-horizon-disaster-1416599464

Gilbert, S. W. (2010). *Disaster resilience: A guide to the literature*. Gaithersburg, MD: National Institute of Standards and Technology (NIST). Available at www.nist.gov/manuscript-publication-search.cfm?pub_id=906887

Godschalk, D. R. (2003). Urban hazard mitigation: Creating resilient cities. *Natural Hazards Review*, *4(3)*, 136–143. doi: 10.1061/(ASCE)1527-6988(2003)4:3(136)

Haimes, Y. Y. (2009). On the definition of resilience in systems. *Risk Analysis*, *29(4)*, 498–501. doi: 10.1111/j.1539-6924.2009.01216.x

Hart, S. D. (2013, December). A practical guide for designing resilient civil infrastructures. *The CIP Report*, *12(6)*, 12–14.

507

Healy, A., and Malhotra, N. (2009). Myopic voters and natural disaster policy. *American Political Science Review*, *103(3)*, 387–406. doi: 10.1017/S0003055409990104

Heilig, G. K. (2012, June 7). *World urbanization prospects: The 2011 revision*. Washington, DC: Live performance in Center for Strategic and International Studies (CSIS). Available at http://esa.un.org/unpd/wpp/ppt/CSIS/WUP_2011_CSIS_4.pdf

HM Government (2011). *Climate resilient infrastructure: Preparing for a changing climate*. Secretary of State for Environment, Food and Rural Affairs. Norwich, UK: The Stationery Office (TSO). Available at www.gov.uk/government/uploads/system/uploads/attachment_data/file/69269/climate-resilient-infrastructure-full.pdf

Holling, C. S. (1973). Resilience and stability of ecological systems. *Annual Review of Ecology and Systematics*, *4*, 1–23.

Hurricane Sandy Rebuilding Task Force (2013). Hurricane Sandy Rebuilding Strategy. Washington, DC: US Department of Housing and Urban Development. Available at http://portal.hud.gov/hudportal/documents/huddoc?id=hsrebuildingstrategy.pdf

Iwata, M. (2014, November 7) Japan nuclear reactors approved for restart. *The Wall Street Journal*. Available at http://online.wsj.com/articles/japan-nuclear-reactors-approved-for-restart-1415356593

Kaufman, S., Qing, C., Levenson, N., and Hanson, M. (2012). Transportation during and after Hurricane Sandy. New York, NY: NYU Wagner Rudin Center for Transportation. Available at http://wagner.nyu.edu/files/rudincenter/sandytransportation.pdf

Klein, R. J. T., Nichols, R. J., and Thomalla, F. (2003). Resilience to natural hazards: How useful is this concept? *Environmental Hazards*, *5*, 35–45.

Litman, T. (2006). Lessons from Katrina and Rita: What major disasters can teach transportation planners, Victoria, BC, Canada: Victoria Transport Policy Institute.

McClain, S., and Kober, A. (2013, March 12). *American rivers – Dams removed from 1999–2007*. Washington, DC: American Rivers. Available at www.americanrivers.org/assets/pdfs/dam-removal-docs/dams-removed-1998-to-2012.pdf

McLinn, J. (2009). Major power outages in the U.S., and around the world. *IEEE Reliability Society 2009 Annual Technology Report*. Available at http://rs.ieee.org/images/files/Publications/2009/2009-06.pdf

McNeil, S. et al. (2003, September 18). Presentation at annual meeting of the National Weather Association, *The Commercial Appeal* newspaper. Available at www.spc.noaa.gov/misc/AbtDerechos/casepages/kc1982mem2003pwrpage.htm#memphis.

Melillo, J. M., Richmond, T. C., and Yohe, G. W. Eds. (2014). Climate change impacts in the United States: The third national climate assessment. U.S. Global Change Research Program, 841 pp. doi:10.7930/J0Z31WJ2.

Merk, O., Saussier, S., Staropoli, C., Slack, E., and Kim, J.-H. (2012). Financing green urban infrastructure, OECD Regional Development Working Papers 2012/10. Geneva, Switzerland: OECD Publishing. Available at www.oecd.org/gov/regional-policy/WP_Financing_Green_Urban_Infrastructure.pdf

Meyer, M. D. (2008). Design standards for U.S. transportation infrastructure: The implications of climate change, 1–27. Available at http://onlinepubs.trb.org/onlinepubs/sr/sr290Meyer.pdf

Mileti, D. S. (1999). *Disasters by design: A reassessment of natural hazards in the United States*. Washington, DC: Joseph Henry Press.

Mooney, K. A., Cabrera, A., and Naik, M. (2013). MTA-NYCT's hurricane planning and infrastructure mitigation. ASCE Metropolitan Section Seminar: Impact of Sandy's Storm Surge on NY/NJ Infrastructure, April 8–9, 2013. Brooklyn, NY.

MTA (Metropolitan Transportation Authority) (2007, September). *August 8, 2007 storm report*. New York, NY: Metropolitan Transportation Authority, 1–115. Available at http://web.mta.info/mta/pdf/storm_report_2007.pdf

National Academies (2012). *Disaster resilience: A national imperative*. Washington, DC: The National Academies Press, Committee on Increasing National Resilience to Hazards and Disasters, Committee on Science, Engineering, and Public Policy.

National Institute of Building Sciences (NIBS), Multi-Hazard Mitigation Council (2005). Natural Hazard Mitigation Saves. Washington, DC: National Institute of Building Sciences. Available at http://c.ymcdn.com/sites/www.nibs.org/resource/resmgr/MMC/hms_vol1.pdf

National Research Council (2013). *Abrupt impacts of climate change. Anticipating surprises*. Washington, DC: National Academies Press.

National Research Council (2014). *An ecosystem services approach to assessing the impacts of the Deepwater Horizon oil spill in the Gulf of Mexico.* Washington, DC: National Academies Press.

New Jersey Department of Environmental Protection (2007). Interconnection Study Mitigation of Water Supply Emergencies, Trenton, NJ: NJ DEP. Available at www.nj.gov/dep/watersupply/pdf/interconnect-report.pdf

New York City (2013, June). *PlaNYC a stronger, more resilient New York.* Available at http://s-media.nyc.gov/agencies/sirr/SIRR_singles_Lo_res.pdf

New York City Environmental Protection (2013, October). NYC wastewater resilience plan. Climate risk assessment and adaptation study. Available at www.nyc.gov/html/dep/pdf/climate/climate-plan-single-page.pdf

Nuclear Energy Institute (2014, April). *World nuclear generation and capacity.* Washington, DC: Nuclear Energy Institute. Available at www.nei.org/Knowledge-Center/Nuclear-Statistics/World-Statistics/World-Nuclear-Generation-and-Capacity

NYS (New York State) 2100 Commission (2013). *Recommendations to improve the strength and resilience of the empire state's infrastructure.* Available at www.rockefellerfoundation.org/uploads/files/7c012997-176f-4e80-bf9c-b473ae9bbbf3.pdf

NYS (New York State) Moreland Commission on Utility Storm Preparation and Response (2013, June 22). *Final report.* Albany, NY. Available at http://utilitystormmanagement.moreland.ny.gov/

O'Rourke, T. D. (2007). Critical infrastructure, interdependencies, and resilience. *The Bridge, 37 (1),* 22–29.

Paton, D., and Johnston, D. (2006). *Disaster resilience: An integrated approach.* Springfield, IL: Charles C. Thomas.

Parker, J. L. (2013a, August). Extreme weather and its consequences: Adaptation and resilience are needed to address a changing world Part 1. *Environmental Law in New York, 24(8),* 111–120. Available at http://papers.ssrn.com/sol3/papers.cfm?abstract_id=2362460

Parker, J. L. (2013b, September). Extreme weather and its consequences: Adaptation and resilience are needed to address a changing world Part 2. *Environmental Law in New York, 24(9),* 133–142. Available at http://papers.ssrn.com/sol3/papers.cfm?abstract_id=2362502

Perrow, C. (2007). *The next catastrophe.* Princeton, NJ, USA: Princeton University Press.

Pickett, S. T., Mcgrath, B., Cadenasso, M., and Felson, A. J. (2014). Ecological resilience and resilient cities. *Building Research & Information, 42(2),* 143–157. doi: 10.1080/09613218.2014.850600

Ratay, R. T. (2010, December). Changes in codes, standards and practices following structural failures, part 1, *Bridges Structure* magazine, 16–19. Available at www.structuremag.org/wp-content/uploads/2014/08/C-CS-Ratay-Dec101.pdf

Reed, D., Kapur, K., and Christie, R. (2009, June). Methodology for assessing the resilience of networked infrastructure. *IEEE Systems Journal, 3(2),* 174–180. doi: 10.1109/JSYST.2009.2017396

Reed, D. A., Powell, M. D., and Westerman, J. M. (2010). Energy supply system performance for Hurricane Katrina. *Journal of Energy Engineering, 136(4),* 95–102. doi:10.1061/(ASCE)EY.1943-7897.0000028

Reed, S. (2014, December 10). BP to cut jobs as price of oil falls. *The New York Times.* Available at: www.nytimes.com/2014/12/11/business/bp-to-cut-jobs-as-price-of-oil-falls.html

Richard Stockton College of New Jersey Coastal Research Center (2013). Hurricane Sandy: Beach-dune performance assessment of New Jersey beach profile network sites. Port Republic, NJ: Coastal Research Center. Available at http://intraweb.stockton.edu/eyos/page.cfm

Riegel, C. (2014). Infrastructure resilience through regional spatial planning – prospects of a new legal principle in Germany. *Int. J. of Critical Infrastructures, 10(1),* 17–29.

Rinaldi, S., Peerenboom, J., and Kelly, T. (2001). Identifying, understanding, and analyzing critical infrastructure interdependencies. *IEEE Control Systems Magazine, 21(6),* 11–25. doi: 10.1109/37.969131

Rodin, J. (2013, July 1). The city resilient. The Rockefeller Foundation web site. New York, NY: The Rockefeller Foundation. Available at http://100resilientcities.rockefellerfoundation.org/blog/entry/the-city-resilient

Roe, E., and Schulman, P. R. (2012). Toward a comparative framework for measuring resilience in critical infrastructure systems. *Journal of Comparative Policy Analysis: Research and Practice, 14(2),* 114–125. doi: 10.1080/13876988.2012.664687

Scheffer, M., Carpenter, S. R., Lenton, T. M., Bascompte, J., Brock, W., Dakos, V., van de Koppel, J., van de Leemput, I. A., Levin, S. A., van Nes, E. H., Pascual, M., and Vandermeer, J. (2012). Anticipating critical transitions. *Science, 338(6105),* 344–348. doi: 10.1126/science.1225244

Schultz, M. T., McKay, S. I., and Hales, L. Z. (2012, August). The quantification and evolution of resilience in integrated coastal systems. Washington, D.C.: U.S. Army Corps of Engineers, ERDC TR-12-7.

Simonoff, J. S., Restrepo, C. E., and Zimmerman, R. (2007). Risk-management and risk-analysis-based decision tools for attacks on electric power. *Risk Analysis, 27(3)*, 547–570. doi: 10.1111/j.1539-6924.2007.00905.x

Tabuchi, H. (2012, September 19). Japan, under pressure, backs off goal to phase out nuclear power by 2040. *The New York Times*. Available at www.nytimes.com/2012/09/20/world/asia/japan-backs-off-of-goal-to-phase-out-nuclear-power-by-2040.html

Talbot, D. (2012, July/August). The Great German Energy Experiment. *Technology Review*. Available at www.technologyreview.com/featured-story/428145/the-great-german-energy-experiment/

Tierney, K. (2014). *The social roots of risk. Producing disasters, promoting resilience*. Stanford, CA: Stanford U. Press.

UNEP (United Nations Environmental Programme) (2013). City-level decoupling: Urban resource flows and the governance of infrastructure transitions. A Report of the Working Group on Cities of the International Resource Panel. Swilling, M., Robinson, B., Marvin, S., and Hodson, M. Paris, France.

United Nations Office for the Coordination of Humanitarian Affairs (2014). Haiti humanitarian action plan 2014. Available at https://docs.unocha.org/sites/dms/CAP/HAP_2014_Haiti.pdf

U.S.-Canada Power System Outage Task Force (2004, April). *Final report on the August 14, 2003 blackout in the United States and Canada: Causes and recommendations*. Available at http://energy.gov/sites/prod/files/oeprod/DocumentsandMedia/BlackoutFinal-Web.pdf

U.S. Department of Commerce, NOAA, NWS (2013, May 15). *Service assessment Hurricane/post-tropical cyclone Sandy*, October 22–29, 2012. Available at www.nws.noaa.gov/os/assessments/pdfs/Sandy13.pdf

U.S. DOE (Department of Energy) (2003a). Hurricane Isabel. A summary of energy impacts and OEA's response. Office of Energy Assurance. Available at www.oe.netl.doe.gov/docs/isabel/hurrisabel_report_100703.pdf

U.S. DOE (Department of Energy) (2003b, September 28). Hurricane Isabel situation report. Available at www.oe.netl.doe.gov/docs/isabel/hurrisabel_sitrep_092803_0700.pdf

U.S. DOE (Department of Energy) (2009). *Comparing the impacts of the 2005 and 2008 Hurricanes on U. S. energy infrastructure*. Available at www.oe.netl.doe.gov/docs/HurricaneComp0508r2.pdf

U.S. DOE (Department of Energy) (2013). *Comparing the impacts of northeast hurricanes on energy infrastructure*. Available at http://energy.gov/sites/prod/files/2013/04/f0/Northeast%20Storm%20Comparison_FINAL_041513c.pdf

U.S. DOE (Department of Energy) (various years). Situation reports. Available at www.oe.netl.doe.gov/print/emergency_sit_rep.aspx

U.S. DHS (2010a). Communications sector-specific plan: An annex to the national infrastructure protection plan. Washington, DC: U.S. DHS. Available at www.dhs.gov/xlibrary/assets/nipp-ssp-communications-2010.pdf

U.S. DHS (2010b). Dams sector-specific plan: An annex to the national infrastructure protection plan. Washington, DC: U.S. DHS. Available at www.dhs.gov/xlibrary/assets/nipp-ssp-dams-2010.pdf

U.S. DHS (2010c). Energy sector-specific plan: An annex to the national infrastructure protection plan. Washington, DC: U.S. DHS. Available at www.dhs.gov/xlibrary/assets/nipp-ssp-energy-2010.pdf

U.S. DHS (2010d). Information technology sector-specific plan: An annex to the national infrastructure protection plan. Washington, DC: U.S. DHS. Available at www.dhs.gov/sites/default/files/publications/IT%20Sector%20Specific%20Plan%202010.pdf

U.S. DHS (2010e). Nuclear reactors, materials, and waste sector-specific plan: An annex to the national infrastructure protection plan. Washington, DC: U.S. DHS. Available at www.dhs.gov/xlibrary/assets/nipp-ssp-nuclear-2010.pdf

U.S. DHS (2010f). Transportation systems sector-specific plan: An annex to the national infrastructure protection plan. Washington, DC: U.S. DHS. Available at www.dhs.gov/xlibrary/assets/nipp-ssp-transportation-systems-2010.pdf

U.S. DHS (2010g). Water and wastewater systems sector-specific plan: An annex to the national infrastructure protection plan. Washington, DC: U.S. DHS. Available at www.dhs.gov/xlibrary/assets/nipp-ssp-water-2010.pdf

U.S. DHS (2013). National infrastructure protection plan. Partnering for critical infrastructure security and resilience. Washington, DC: U.S. DHS. Available at www.dhs.gov/sites/default/files/publications/

Resilient urban infrastructure

NIPP%202013_Partnering%20for%20Critical%20Infrastructure%20Security%20and%20Resilience_508_0.pdf

U.S. DHS (2014a, June 12 [last updated]). Critical infrastructure sectors. Available at www.dhs.gov/critical-infrastructure-sectors

U.S. DHS (2014b, October 8) Regional resiliency assessment program. Available at www.dhs.gov/regional-resiliency-assessment-program

U.S. DOT (Department of Transportation) Federal Transit Administration (2011, August). Flooded bus barns and buckled rails: Public transportation and climate change adaptation. Available at www.fta.dot.gov/documents/FTA_0001_-Flooded_Bus_Barns_and_Buckled_Rails.pdf

U.S. EPA (Environmental Protection Agency) (2010, August). Green infrastructure case studies: Municipal policies for managing stormwater with green infrastructure. Washington, DC: U.S. Environmental Protection Agency. Available at http://nepis.epa.gov/Exe/ZyPDF.cgi/P100FTEM.PDF?Dockey=P100FTEM.PDF

U.S. EPA (Environmental Protection Agency) (2011a, September). Civil enforcement – fuel waivers. Available at www.epa.gov/compliance/civil/caa/fuelwaivers/

U.S. EPA (Environmental Protection Agency) (2011b, September 12). News Brief (HQ): EPA approves emergency fuel waiver for Pennsylvania. Washington, DC: U.S. EPA.

U.S. EPA (Environmental Protection Agency) (2013a, October). Green infrastructure strategic agenda. Available at http://water.epa.gov/infrastructure/greeninfrastructure/upload/2013_GI_FINAL_Agenda_101713.pdf

U.S. EPA (Environmental Protection Agency) (2013b). *Our built and natural environments: A technical review of the interactions among land use, transportation, and environmental quality*. Second edition. Washington, DC: EPA. Available at www.epa.gov/dced/pdf/b-and-n/b-and-n-EPA-231K13001.pdf

U.S. EPA (Environmental Protection Agency) (2014a, October 27 [updated]). What is green infrastructure? Green Infrastructure web site. Available at http://water.epa.gov/infrastructure/greeninfrastructure/gi_what.cfm

U.S. EPA (Environmental Protection Agency) (2014b, October 1). Greening America's capitals program. Available at www.epa.gov/smartgrowth/greencapitals.htm

U.S. Housing and Urban Development (HUD) (2013). HUD announces additional $5.1 billion in recovery funds for communities impacted by Hurricane Sandy. Available at http://portal.hud.gov/hudportal/HUD?src=/press/press_releases_media_advisories/2013/HUDNo.13-153

U.S. White House (2011, March 30). Presidential policy directive – PPD-8. National preparedness. Available at www.dhs.gov/xlibrary/assets/presidential-policy-directive-8-national-preparedness.pdf

U.S. White House (2012, November 30). Presidential proclamation – Critical infrastructure protection and resilience month, 2012. Available at www.whitehouse.gov/the-press-office/2012/11/30/presidential-proclamation-critical-infrastructure-protection-and-resilie

U.S. White House (2013a, February 12). Presidential policy directive – critical infrastructure security and resilience presidential policy directive. Available at https://www.whitehouse.gov/the-press-office/2013/02/12/presidential-policy-directive-critical-infrastructure-security-and-resil

U.S. White House (2013b, October 31). Presidential proclamation – Critical infrastructure security and resilience month, 2013. Available at www.whitehouse.gov/the-press-office/2013/10/31/presidential-proclamation-critical-infrastructure-security-and-resilience

van Nes, E. H., and Scheffer, M. (2007). Slow recovery from perturbations as a generic indicator of a nearby catastrophic shift. *The American Naturalist, 169(6)*, 738–747. doi: 10.1086/516845

Vale, L. J. (2014). The politics of resilient cities: Whose resilience and whose city? *Building Research & Information, 42(2)*, 191–201. doi: 10.1080/09613218.2014.850602

Vale, L. J., and Campanella, T. J. (eds.) (2005). *The resilient city: How modern cities recover from disaster*. New York, NY: Oxford University Press.

Verner, D., and Petit, F. (2013). Resilience assessment tools for critical infrastructure systems. *The CIP Report, 12(6)*, 2–5.

Viscusi, W. K. (2003). *The value of life: Estimates with risks by occupation and industry*. Cambridge, MA: Harvard University Press. Available at www.law.harvard.edu/programs/olin_center/papers/pdf/422.pdf

Wackernagel, M., and Rees, W. (1996). *Our ecological footprint*. Gabriola Island, B.C. and Stony Creek, CT: New Society Publishers.

Wang, C., and Blackmore, J. M. (2009). Resilience concepts for water resource systems. *Journal of Water Resources Planning and Management, 135(6)*, 528–536. doi: 10.1061/(ASCE)0733-9496(2009)135:6(528)

World Bank (2010). *The costs of adapting to climate change for infrastructure*. Washington, DC: The World Bank Group. Available at http://siteresources.worldbank.org/EXTCC/Resources/EACC-june2010.pdf

World Bank (2013). *Building resilience*. Washington, DC: The World Bank Group. Available at www.worldbank.org/content/dam/Worldbank/document/SDN/Full_Report_Building_Resilience_Integrating_Climate_Disaster_Risk_Development.pdf

World Economic Forum (WEF) (2014, April). *Strategic infrastructure steps to operate and maintain infrastructure efficiently and effectively*. Geneva, Switzerland: WEF. Available at www3.weforum.org/docs/WEF_IU_StrategicInfrastructureSteps_Report_2014.pdf

Zimmerman, R. (2003). Public infrastructure service flexibility for response and recovery in the September 11th, 2001 attacks at the World Trade Center. In *Beyond September 11th: An account of post-disaster research*. Boulder, CO: University of Colorado: Natural Hazards Research and Applications Information Center, Public Entity Risk Institute, and Institute for Civil Infrastructure Systems, pp. 241–268. Available at www.colorado.edu/hazards/publications/sp/sp39/sept11book_ch9_zimmerman.pdf

Zimmerman, R. (2012a). Critical infrastructure and interdependency revisited, Chapter 20 in *The McGraw-Hill homeland security handbook*, 2nd edition, edited by D. G. Kamien. New York, NY: The McGraw-Hill Companies, Inc. pp. 437–460.

Zimmerman, R. (2012b). *Transport, the environment and security: Making the connection*. Cheltenham, UK and Northampton, MA: Publishing, Ltd.

Zimmerman, R. (2014a). Network attributes of critical infrastructure, vulnerability, and consequence assessment, Chapter 372 in *Safety, reliability, risk and life-cycle performance of structures and infrastructures*, edited by G. Deodatis, B. R. Ellingwood, and D. M. Frangopol, London, UK: Taylor & Francis Group, CRC Press, pp. 2777–2782.

Zimmerman, R. (2014b). Strategies and considerations for investing in sustainable city infrastructure, Chapter 7 in *The Elgar companion to sustainable cities: Strategies, methods and outlook*, edited by D. A. Mazmanian, and H. Blanco, Cheltenham, UK: Edward Elgar Publishing, Ltd., pp. 133–153.

Zimmerman, R. (2014c, August). Planning restoration of vital infrastructure services following hurricane sandy: Lessons learned for energy and transportation. *J. Extreme Events*, *1(2)*. Available at www.worldscientific.com/doi/pdf/10.1142/S2345737614500043

Zimmerman, R., and Restrepo, C. E. (2006). The next step: Quantifying infrastructure interdependencies to improve security. *International Journal of Critical Infrastructures*, *2(2/3)*, 215–230. doi: 10.1504/IJCIS.2006.009439

Zimmerman, R., and Simonoff, J. S. (2009). Transportation density and opportunities for expediting recovery to promote security. *Journal of Applied Security Research*, *4(1–2)*, 48–59.

Zimmerman, R., Restrepo, C. E., Sellers, J., Amirapu, A., & Pearson, T. R. (2014). Promoting transportation flexibility in extreme events through multi-modal connectivity. U.S. Department of Transportation Region II Urban Transportation Research Center, New York, NY: NYU-Wagner, Final report. Available at: www.utrc2.org/sites/default/files/pubs/Final-NYU-Extreme-Events-Research-Report.pdf

33

SOFT AND HARD INFRASTRUCTURE CO-PRODUCTION AND LOCK-IN

The challenges for a post-carbon city

Stephanie Pincetl

Moving to post-carbon cities involves more than changing fuel sources. It requires deep and complex changes to city systems, and to the architecture of global economic activity. The economic primacy of city systems that has been enabled by a once in 30 million year energy source is also invisibly a result of the development of hard and soft infrastructures of socio-technical systems. Soft infrastructures are the codes, conventions, rules and policies that both guide the specifications of hard infrastructures (road widths and surface depths of materials, pipe sizes for sewage sanitation, specifications on pipe thickness and valves for gas pipelines) and their context as in city zoning policies that specify land uses, transportation plans that allocate roads and types of roads (and width) and their locations, or policies that decouple energy generation from energy distribution, as in the case of electricity in many industrialized countries. This relationship between hard and soft infrastructures is ubiquitous—there can be no hard infrastructure without a soft one. Airports cannot function without rules about air traffic. The soft infrastructures of air traffic rules are determined by the International Air Transport Association, made up of members from around the world who participate in working groups to develop codes and conventions including the hard infrastructures of airports such as landing strip widths and lengths. Global economic growth and the concentration of peoples in cities are predicated on the ability to exploit fossil energy. Its malleability, transportability, plasticity and enormous energetic content have enabled humans to transform the globe and increase economic activity to a level never before seen.

Hard and soft infrastructures operate in a chicken and egg relationship and the emergence of one engenders the other. However, they create lock-in. Once the port or airport facility has been constructed, once the fossil fuel electric generating plant has been built and the rules created to guide their construction developed, there is an immensely expensive infrastructure in place that has it adherents, both on the regulatory side and the infrastructure use side. The next iteration of the system—whether the rail route to distribute goods, or the grid to distribute the power—exhibits path dependency. New technologies or new rules disrupt that system and are often opposed. To move to a post-carbon future, both the hard and soft infrastructures, the social and the technical, must change, and they must do so in synch. The urban metabolism methodology presented in this chapter begins to unpack the carbon-fueled, infrastructure-enabled flow of resources into cities—how they are used and the waste flows produced—coupled with socio-economic and regulatory analysis. There are potentially other approaches to developing analyses helpful for a transition to a post-carbon future, but the goal is to enable the generation of other and additive approaches.

Fossil energy and the emergence of modern cities

The harnessing of fossil fuels has enabled the enormous shift of human populations to cities. As mentioned above, carbon energy and its concentrated, transportable, malleable and transformable properties and qualities have enabled humans and their societies to achieve levels of economic prosperity and consumption never before experienced (McNeill 2000, Smil 2008), although this prosperity and consumption is not equally shared around the world. That is, while fossil fuels have enabled the rise of a global middle class through inexpensive energy for the harvesting and processing of materials into products, and their broad distribution, this has not alleviated poverty. Paradoxically, the poorest of the world seem the most likely to experience the impact of carbon-induced climate change: sea level rise, drought, flooding, air pollution, degradation of availability of food, etc.

As we enter into the 21st century, the consequences of this affluence and the burning of carbon energy is an undercutting of the successes of the modern age. Climate change is ushering in climate events that destroy infrastructure and challenge the ability of governments and other organizations to respond (see the Zimmerman, Chapter 32 in this volume, for a discussion of urban infrastructure and extreme environmental disturbances)—the soft infrastructures of regulation, codes and conventions have not been developed to address these new issues. Contamination by synthetic and organic chemicals of groundwater in some places is proving to be irremediable. Water supplies for business-as-usual are at times insufficient, and the list goes on. The mutually constituted socio-technical infrastructures that make complex urban systems function are now in need of revisiting in order to reduce their environmental impacts and to respond to changing political, economic and cultural needs. Socio-technical infrastructures can also be described, more simply, as soft and hard infrastructures.

Over the past two centuries, soft infrastructures have developed with the ever more powerful ability to transform the Earth with fossil energy. These soft infrastructures are complex and operate at many scales, from banking systems that may support fracking investments for either natural gas or oil, to international trade agreements, to more prosaic rules about renewable energy portfolios. To transition away from a fossil fuel driven world, new post-carbon, post-modernist socio-technical infrastructures are needed. This transformation is most certainly intertwined with the future of the capitalist system and the system's intrinsic dependence on growth. How the world's economic system evolves is beyond this discussion, but will circumscribe potential outcomes of a shift from current socio-technical systems to a different way of organizing our cities, modes of production and hence relations with nature.

The modernist city

The key to making emerging industrial cities and their growing populations successful is adequate and appropriate infrastructure for economic growth, public health and efficiency of transactions of all kinds. This requires abundant resources and powerful energetic fuels for the creation of the hard infrastructures that are now in place in most industrialized cities. New knowledge, the result of scientific research, created standardized interworked technical infrastructures: sewers, water systems, railroad tracks, roads, electrical and telephone wires (Graham and Marvin 2001). At the same time, for the hard infrastructures to be built, operated and maintained, they need the development of soft infrastructures of governance including urban planning processes and regulations, the science of accounting and urban finance, the organizational structures of urban management from sanitation, parks, roads, building and safety departments to police departments and more.

Carbon infrastructures

Carbon fuels have facilitated the construction of cities and are fundamental to their structure. As fossil energy use grew, the exploration, extraction and treatment of the various fossil energies across the world became embedded in a complex global system of markets, companies and infrastructures that have enormous power and consequence. Rules, such as the oil depletion allowance or trade agreements that regulate the global transport of these fuels and their distribution, may seem distant to the question of urban infrastructure, but they are deeply important to current practices. They have evolved (produced each other) over a century along with the fossil fuel distribution infrastructure, an infrastructure that required immense financial and technical knowledge. This infrastructure constitutes the status quo that has inertia and created carbon lock-in (Unruh 2000) and path dependencies. Moreover, as Unruh (2002) points out, over time these systems become preferences, expectations and routines, and embedded in governmental institutions. These lock-ins arise because large technological systems, like electricity generation, distribution and end use, natural gas pipelines, petroleum refining and distribution, are complex systems of technologies embedded in powerful social contexts of public and private institutions—there are positive feedbacks among technological infrastructures and the organizations and institutions that create, diffuse and employ them (Unruh 2000).

Such lock-in inhibits policy action, even in the face of known global climate risk and the existence of alternatives, suggesting that the global energy transition could be prolonged and potentially difficult. It could require a fundamental shift in the global economy. Overcoming lock-in has been little explored (Unruh 2002). How it would be financed, what it would encompass, and how it could change urban morphology and daily life cannot really be predicted, but the transformations, over time, away from carbon infrastructures are likely to be substantial. The next sections explore potential approaches to identify soft and hard infrastructure lock-in and the challenges lock-in creates for moving toward a post-carbon city. This includes methods to better understand and quantify the flows of carbon and resources that enter urban systems and exit in the form of waste, and the use of this understanding to identify leverage points to prioritize decarbonization. Knowing the actual flows, tracing their cradle to grave lifecycle and measuring their impacts are all significant first steps. It enables prioritizing low- to post-carbon interventions, targeting the rules that support those flows and uses of the flows to enable shifts away from carbon-intensive coupled socio-technical systems. Alternatives that are less dependent on carbon-centralized infrastructures and some of the difficult tradeoffs that those may pose are explored.

Urban metabolism

Urban metabolism (UM) is the quantification of flows into cities and flows out of cities as waste. UM provides an instructive macro-view of the magnitude of urban resource dependencies and the waste that is generated (see Figure 33.1).

Stephanie Pincetl

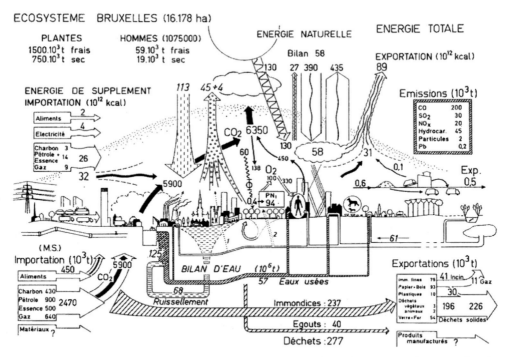

Figure 33.1 The urban metabolism of Brussels, Belgium, in the early 1970s; early comprehensive example of an urban metabolism analysis

Source: Duvigneaud and Denayeyer-De Smet (1977).

Kennedy et al. (2007) summarize research conducted on several flows for these cities: water supply and wastewater energy inputs, contaminant emissions and solid waste disposal for one year for an average US city in 1995; Brussels in the 1970s; Sydney in 1970; Hong Kong in 1971; Vienna in the 1990s; London in 2000; and Cape Town in 2000 (Kennedy et al. 2007, p. 48). Each of these cities has a different profile due to their history (pre-carbon hard infrastructural cores, like in Brussels, Vienna and London), geographical location (requiring more heating fuel, for example), wealth and regulatory regimes (soft infrastructures). In the case of Cape Town, apartheid was an important soft infrastructure in the uneven deployment of 'modernist infrastructure', advantaging white neighborhoods for water, sewage, electricity services and even roads and transportation systems. And yet, key to the contemporary population densities of each of these cities and the rapid acceleration of urbanization worldwide has been the availability of fossil fuels (Marcotullio et al. 2014). Each place has a different configuration of soft and hard infrastructures: European countries have complied with the soft infrastructures of the European Union in the contemporary period; Hong Kong's astonishing growth was a result of being a British protectorate, a center of financial trading with an exceptional port and successful manufacturing. Now under Chinese rule, the soft infrastructures of management are increasingly directed by China, changing both the regulatory regime and priorities in hard infrastructure investments. This change may lead to changes in energy sources, materials flows into the city for manufacturing and the handling of waste flows. Thus, understanding the coupled relationship between energy and material flows, and the soft infrastructures of management can yield insights into the interactions among socio-technical spheres.

Developing the urban metabolism accounting for cities quantifies their fossil fuel dependencies—the direct use of fossil fuels (gasoline, diesel and natural gas), and indirect, the fossil energy embodied in materials like concrete, steel or water. Such a first cut is an initial step in uncovering the explicit relationship between urbanization and fossil fuel consumption—the way in which hard and soft infrastructures co-produce the urban landscape. A deeper look at the distribution and patterns of consumption may then show spatial patterns across different people and economic sectors, building types and industries from which priorities for the reduction (through changes in the soft infrastructure of rules, codes and pricing) of the use of those fuels can be developed. Additional soft infrastructures might be legislation requiring a percentage of energy being generated by renewables; hard infrastructures could be the installation of wind machines or solar panels. Regardless, without a good sense of how much fossil energy is being used, it is difficult to determine how much is needed for it to be replaced. More granular analysis, by neighborhood, industry or sector (transportation), is even more useful, as it can help develop priorities for alternatives. For example, replacing household electricity use generated by coal with solar may be easier than powering the transportation system using solar energy. The hard infrastructures of each of these power-using sectors are likely different—as is the soft infrastructures of management of the sectors. Understanding these differences and how they inter-operate opens up opportunities for the potential shift away from fossil fuels, and the creation of alternative pathways—unlocking current path dependencies.

Mapping energy, water and resource flows

A step in unpacking this complex world of co-evolving and co-producing soft and hard infrastructures is to establish baselines of the flows onto geographical space. Such data is becoming more available, though often tends to be at quite high aggregations, or modeled data. Certainly such data is useful, and mapped onto basic maps using geographic information systems (GIS) coupled with land uses, hard infrastructures like water supplies, sewage treatment, rivers, electricity generating plants, solid waste facilities and others can begin to show not only their location, but concentrations of energy and materials use and flows in the city. What such mapping provides is insight into the places and infrastructures that are the most fossil fuel intensive and their spatial characteristics. Mapping and data gathering can establish the baselines of use over time, and a check on what is changing, where and why. If the long-term goal is to reduce the flows, increase local resilience and sustainability, without such actual analysis, it is more difficult to do so. Simultaneously constructing the soft infrastructures of management, regulation and oversight helps to further explain the patterns seen. One could call this doing a 'thick description' of a city's metabolism, or 'thick mapping'. It is a way to construct relationships that may not be obvious otherwise, producing new configurations of knowledge. Thick maps can emphasize context and meaning and are both denotative—showing factually the where and the when, and connotative, implying meaningful relations among the visual components (Presner et al. 2014).

For urban regions to develop strategies to reduce flows, information about how flows are distributed across space, across time and among uses is a first step for developing denotative (the where and when) and connotative (the meaningful relationships) analyses and knowledge. Thick mapping of data can also establish the baselines of use over time, and a check on what is changing, where and why. If the long-term goal is to reduce flows from far away systems

Stephanie Pincetl

to increase local resilience and sustainability, such analysis provides greater legibility of these complex systems. Simultaneously constructing the soft infrastructures of management, regulation and oversight, and developing visual configurations of knowledge, helps to further explain the patterns seen. This could entail mapping the territories of water delivery companies and how they are regulated—private, public or even non-regulated. Contrasting these regulatory regimes with parts of a city and respective water entities from which they receive water as well as the source or origin of those waters may provide great insights into the water/energy nexus, or where investments have or have not taken place. In the age of Google, where mapping and visual content from satellites, street shots and fly-overs are increasingly common, understanding complex urban systems through mapping tools can provide new insights. Collecting the data is, of course, only the first step; linking it to the regulatory—or soft infrastructure—is necessary in order to understand how both together create path dependencies.

As Presner et al. (2014) describe it, such thick maps are part of a blended trend, the interdisciplinary, interactive connecting of quantification of flows into the urban environment with how they are used, by whom and where, to do what, and the soft infrastructures of regulation, codes or conventions. Thick maps can elucidate the intricacies of law and regulation, shaped over time, and the logic and necessities of the moment though the tracing of hard infrastructure, its type, condition, connection to fossil energy and layout. This can help the understanding of how infrastructure can be the result of lock-in, the interacting, reinforcing relationship between hard and soft infrastructures. Existing systems have sunken investments, the operators of which have concerns about alternatives on their market and ability to recoup costs.

Intricacies of law and regulation, shaped over time, and the logic and necessities of the moment, can make infrastructure obdurate to change; they create lock-in. For example, successful water recycling could undermine existing water suppliers' economic viability since a competing water supply could undercut their market and ability to recoup sunken asset costs. Yet, wastewater is increasingly considered a significant resource. The hard and soft infrastructure system is a reflection of the issues and technologies of the early industrial city. Sewage was a serious health problem with centralized sewage treatment an enormous and expensive improvement; bacteriology was necessary and sewage sanitation expertise had to be developed (Melosi 2000). Cities needed to be convinced it was important to build the infrastructures and funding had to be found. Pipes had to be laid throughout the city, pumps placed in necessary places, and big swaths of land devoted to centralized sewage treatment plants. Fossil fuel energy was expended and made the system possible. Water quality regulations were developed to enforce this system. Now, under conditions of possible water scarcity, this system is an obstacle—it diverts water away from point of use, treats it as waste and the system has been both created and reinforced by rules that ensure water suppliers are fiscally viable and the city is free of disease. Of course, in developing cities, this is less of an issue, as centralized systems may not have been developed, but other obstacles may prevail: squatters may have built settlements ahead of infrastructure, making infrastructure difficult to lay down. The informal soft infrastructure of settlements and squatting due to poverty and lack of access to land and credit create hard infrastructures of housing and road, making areas hard to serve.

While the challenge is daunting, understanding how complex urban systems bring in, regulate and use resources in a detailed manner is an important step toward addressing infrastructure and infrastructure changes. Examining how the soft infrastructures reinforce hard infrastructures and vice versa is a first step in unraveling the system.

Embedded infrastructure

Embedded infrastructure accounting applies the quantification approach of urban metabolism retrospectively to account for the lifecycle cost of existing infrastructure and its impacts (Fraser and Chester 2013, Chester et al. 2012). One could think of this as accounting for the accumulated carbon in the atmosphere over time of city building and the pollution generated. Each building, street, pipe and wire required the excavation of Earth resources, their transformation into usable materials, their transport and installation. All along that supply chain, energy and water were required, pollution and waste were created and human health was impacted. Materials flow and lifecycle analyses can quantify this and reveal the fossil fuel energy that was necessary to create the existing city. To effectuate this, analysis requires importing models and analysis from the engineering professions. For example, there are US specification-based pavement lifecycle tools for environmental and economic effects (Horvath 2003) that can be joined with knowledge about how many street miles exist in a city (Fraser and Chester 2013). This enables calculating the embedded energy and materials that are in roadway infrastructure presently in US cities. Coupling this with rules that reinforce or structure transportation choices provides a window into the soft and hard infrastructures of this important aspect of a city's UM. Greenhouse gases (GHGs) and other impacts of the built environment can be assessed as well. Reyna and Chester (2013) pioneered this method and find that approximately 70 percent of embedded building energy and GHG emissions occurred between 1930 and 1980 in Los Angeles. They find that since 1980, Los Angeles has been deploying new building infrastructure at a slowing pace and is, instead, focused on using and upgrading existing building infrastructure.

Waste

Cities generate a great deal of waste. This primarily takes the form of air and water emissions, now often regulated by governments for their adverse environmental impacts. Smothered in air pollution and confronted with severely contaminated water, countries like China are now beginning to adopt US-like air and water quality standards (as has the EU), and to quantify air and water pollution. These too are part of an urban metabolism—the waste flows out—that can be associated with the different metabolic activities of the urban region. Pinpointing their sources and quantifying them is important in developing not only pollution control strategies, but perhaps also different manufacturing techniques and technologies to substitute for polluting materials and energy sources. Admittedly this has been a goal for a couple of decades now, known under the rubric of green chemistry. But to reduce energy use and pollution will require enormous new discoveries and changes. Soft infrastructures of finance, of competitive pressures, of scientific research and diffusion are important to take into consideration in this realm, as complex path dependencies and lock-ins make innovation risky and hard such that manufacturing processes, while perhaps improving efficiencies, may not make substitutions.

An example of successful substitution is in the area of circuit board-making—a citrus-based product substituted for chlorofluorocarbons (CFCs) (Bailey 1992, Missick 1993). The shift was in part due to the increasing national and international regulations (soft infrastructures) of CFCs due to their impact on the ozone layer. Today, those CFCs are also known as strong GHGs. Substitution in the circuit board industry was relatively easy, not requiring new manufacturing process changes—there was no strong lock-in. It also had the positive benefit of improving wastewater quality. In contrast, substituting CFCs in automobile air conditioning systems has been far more difficult; hydrofluorocarbons (HFCs) are less harmful to the ozone layer, but are significant GHGs. Selling cars without air conditioning would likely be uncompetitive due to the lack of an easy alternative substance. There is currently a lock-in between new car sales and air conditioning, at least for the affluent.

Toward post-carbon cities

Limiting the discussion to cities, recognizing that they are imbricated in the larger global economy and networks of power and infrastructure, it is clear that a transition to a post-carbon infrastructure is the challenge of the urban 21st century. Cities will differ widely depending on their climate zone, available resources (both physical and fiscal), existing urban morphology, political cultures and more. At the same time, the modernist city typology (and ideal) has infused nearly all urbanization processes around the world including the construction of centralized, fossil fuel dependent infrastructures—coal fired power plants in India and China, massive dam projects requiring huge fossil fuel-intensive infrastructures, large new port projects, and much more to facilitate urbanization and the consumption of materials. These investments are likely to constrain the future development of another kind, in part due to their high capital investment and creation of assets. Change is thus limited by decisions based on current knowledge, systems of power and investments, technical knowledge and ideas of modernity. A shift to another system is sometimes even unimaginable. However, just like the 20th century built the existing hard and soft infrastructures to address the issues of the time and created modern cities, techniques, technologies, rules and codes, more sustainable, resilient post-carbon cities will require the development of alternative hard and soft infrastructures.

As this discussion has suggested, understanding urban flows and a city's metabolism can be a point of departure for assessing urban carbon dependencies and, taking local specificities into account, developing new urban pathways. It is important to note that if the impacts of global climate change are taken seriously and policies and politics are evolved to reduce the dependence on carbon fuels, the post-carbon city is likely to look, feel and work very differently than a carbon-based city. If one of the goals is to reduce flows from far away places, brought into cities by using fossil energy, a city's own ability to source energy and other flows from its own hinterlands will likely be predicated on its morphology. Distributed infrastructure and decentralized systems have physical space requirements, thus their implementation is, at least partly, linked to urban form. Obviously, it would be desirable to also change global transportation fuels; this will no doubt be an aspect of a worldwide shift away from fossil fuel energy. What that might look like is hard to predict, but what happens in cities will be part of that change.

Distributed infrastructure

Distributed systems are ones that are located near the source or need; the scope and scale of distributed systems will be context specific. Distributed solar generation, for example, is a prime example of such an infrastructure and it has certain requirements to succeed. All solar systems are extensive to capture sunlight and need to have the proper exposure, so it may make less sense to install them in very cold, very cloudy climates. Even in cities that have the right climatic conditions, there are likely to be land use and building-level challenges that will emerge. To power the majority of a city's needs locally would require many, if not most, roofs to be covered with solar panels. Buildings' roofs would need to be built or retrofitted to accommodate those panels. Roof pitches would have to be reoriented to maximize solar exposure. Cities would look very different as a result. Moreover, new soft infrastructures regulating the 'ownership' of the panels and their management would be necessary, possibly entailing new business models for energy provision. Battery storage facilities would have to be sited. In already densely built cities, this could be difficult unless they could be placed underground—but under whose property? Both the panels and the battery storage units could be considered an eyesore or hazard. New zoning codes would have to be developed, or eminent domain may need to be used.

Distributed sanitation is yet another distributed infrastructure aimed to decrease reliance on both imported water and distant, large-scale sanitation plants. This involves small-scale sewage treatment plants—perhaps at the neighborhood level—that then integrate their pipes into local water distribution infrastructure. Some engineers are advocating this approach, especially in big cities where sewage treatment plant locations make the distribution of recycled water back into the urban system highly energy intensive and costly due to the laying of a vast new pipe infrastructure. But what will be the land use requirements of these new plants and neighborhood acceptance? As noted previously, new soft infrastructures of property rights and land use rules may need to evolve for this new distributed infrastructure. In many places, recycled water is not considered potable by law, so the soft rules of water quality will need to change. Depending on the regime of private property protections, this soft infrastructure could make change either more difficult or easier thus raising questions of public accountability and democratic process.

Other distributed water management approaches to increase local water supplies include increasing stormwater infiltration using permeable surfaces or creating infiltration opportunities in the urban fabric including de-paving and bioswales along streets. In California, distributed home-scale greywater systems have now been allowed by the state Department of Public Health. Both approaches, one to increase groundwater recharge, the other for domestic garden irrigation, are land intensive.

Decentralized systems

This approach is a close cousin to distributed systems, and the concept is being pioneered by the city of Sydney (Kinesis Consortium 2012). Fundamental to the plan to reduce dependency on carbon fuel inputs is an integrated network approach described as 'Green Transformers' that is based on trigeneration: the simultaneous production of electricity and the exploitation of waste heat from the generation process to supply heating and hot water needs, and used for cooling via a heat-driven chiller (p. 8) (see Chertow et al., Chapter 31 in this volume, for a more thorough discussion of urban industrial symbiosis). While the technical aspects of the plan are beyond this discussion, the reason it is decentralized is because it involves using a 250km radius around the city to source energy, rather than energy generation at the point of use. The policy aim (250km radius), a soft infrastructure of an official goal, is driving new hard infrastructures such as the cogeneration plants. This includes the use of agricultural waste to create renewable gas that will substitute directly for non-renewable natural gas, and the co-locating of cogeneration and cogeneration engines within the urban environment that makes it possible to collect waste heat and redistribute it to heat water and to cool buildings. The city is also planning to deploy decentralized wastewater treatment plants in the urban fabric for non-potable use, such as irrigating parks.

To target these ambitious programs, granular energy and water use analysis is being used, building-by-building, neighborhood-by-neighborhood. In this instance, the 250km hinterland radius means that, for Sidney, it can maintain the dense downtown areas, as the flows generated in this radius are sufficient to supply its needs. The density actually offers economies of scale relative to infrastructure investment. For less dense cities, distributed systems might work better. Thus a city's morphology determines which approaches are suitable to what land uses.

Any of these changes away from centralized systems dependent on fossil fuels will require vast new investments in cities and the co-evolution of hard and soft infrastructures. Choices will be culturally, economically and politically contingent. In places where there is a history and legal tradition of the protection of private property rights, a shift to distributed systems may require the modification of some of these rights—enabling solar systems on roofs or water infiltration on private property. In other places, such as in China, with its rapid and forced urbanization program, high densities may preclude distributed systems, rendering decentralized systems insufficient. The government is aiming to have 70 percent of its population urban by 2035 in 20,000–50,000 new high-rise apartment buildings. If this new urbanization is not being placed where there are sufficient local water, solar or other resources, it is hard to know how cities will be supplied without large, carbon-intensive infrastructures.

Next steps

Cities must and will change. The modernist city was built to respond to multiple interwoven crises of the early 20th century. No doubt we will rebuild current cities to respond to emerging issues of the 21st century. What is suggested here is a way to clarify the relationship between the hard infrastructures of urban morphology and the soft infrastructures of management of that morphology that operate at multiple scales. These are not two independent systems. Together, they create lock-in that constrains future options.

While change most likely does not happen in an orderly and methodical manner, providing coherent and coupled analysis using thick mapping of a city's metabolic flows, lifecycle analysis and other analyses of the flows over time and type can potentially infuse dialogue about change in a useful manner. Thinking through the implications of post-carbon infrastructure and the impacts on how cities look, function and the inherent tradeoffs is something that has not been widely discussed, other than in a utopian context.

New scientific approaches to cities are emerging that can provide better insights into how to understand the impacts of current cities including their historical legacy. Urban metabolism studies, urban ecology and engineering of distributed systems are beginning to provide analysis that can inform a transition to post-carbon cities.

In summary

- Cities are accreted over time, as the reflection of co-produced and interacting soft and hard infrastructures, and will differ in places and contexts.
- Disassembling their fossil fuel dependence and how the socio-technical infrastructures reinforce path dependencies can begin using urban metabolism analysis and 'thick mapping'.
- Post-carbon cities are likely to need distributed and decentralized energy and water systems, but to do so urbanization patterns must take these alternatives into account.

Key research questions

- What is the relationship between urban form and decarbonization of cities?
- Can decentralized and distributed alternative systems provide sufficient energy and resources to rapidly urbanizing areas? Are there significant geographical differences among different urbanizing regions of the world in this regard?
- How do hard and soft infrastructures create specific lock-ins in specific places?

Challenges for a post-carbon city

References

Bailey. E. (1992). Ozone layer gets lemon scent: environment: a Hughes worker creates a citrus formula for cleaning circuit boards, eliminating a use of CFCs. *Los Angeles Times* http://articles.latimes.com/1992-01-26/news/mn-1456_1_ozone-layer (accessed September 3, 2014).

Chester, M., Pincetl, S., and B. Allenby. (2012). Avoiding unintended tradeoffs by integrating life-cycle impact assessment with urban metabolism. *Current Opinion in Environmental Sustainability* 4: 451–457.

Duvigneaud, P., and Denayeyer-De Smet, S. (1977). L'Ecosystéme Urbs, in: *L'Ecosystème Urbain Bruxellois* (P. Duvigneaud, and P. Kestemont, eds.), Traveaux de la Section Belge du Programme Biologique International, Bruxelles, pp. 581–597.

Fraser, A., and Chester, M. (2013). Life-cycle greenhouse gas emissions and costs of the deployment of the Los Angeles roadway network. Arizona State University Digital Repository Report No. SSEBE-CESEM-2013-WPS-001. http://repository.asu.edu/items/16574.

Graham, S., and Marvin, S. (2001). *Splintering Urbanism, Networked Infrastructures, Technological Mobilities and the Urban Condition.* London: Routledge.

Horvath, A. (2003). *Parking Infrastructure: Energy, Emissions, and Automobile Life Cycle Environmental Accounting.* University of California Transportation Center Technical Report no. 683.

Kennedy, C., Cudihy, J., and Engel-Yan, J. (2007). The changing metabolism of cities. *Journal of Industrial Ecology* 11: 43–59.

Kinesis Consortium (2012). City of Sidney Decentralized Energy Master Plan, Trigeneration 2010–2030, prepared for the City of Sidney.

McNeill, J. R. (2000). *Something New Under the Sum. An Environmental History of the Twentieth-Century World.* New York: W.W. Norton & Company.

Marcotullio, P. J., Hughes, S., Sarzynski, A., Pincetl, S., Pena, L. S., Romero-Lankao, P., Seto, K. C., and Runfola, D. (2014). Urbanization and the carbon-cycle: Contributions from Social Science. *Earth's Future,* forthcoming. DOI: 10.1002/2014EF000257

Melosi, M. V. (2000). *The Sanitary City: Urban Infrastructure in America from Colonial Times to the Present.* Baltimore, MD: Johns Hopkins University Press.

Missick, P. (1993). *Health and Safety Impacts of Citrus-Based Terpenes in Printed Circuit Board Cleaning.* College of Engineering, University of Massachusetts Lowell: The Toxics Use Reduction Institute. June. www.turi.org/content/download/3762/46108/file/techreport6.pdf (accessed 9/3/14).

Presner T., Shepard, D., and Kawano, Y. (2014). *HyperCities, Thick Mapping in the Digital Humanities.* Cambridge, MA: Harvard University Press.

Reyna, J., and Chester, M. (2013). *Life-Cycle Embedded Infrastructure Energy Use and GHG Emissions in Los Angeles.* Arizona State University Digital Repository Report no. SSEBE-CESEM-2013-WPS-004 http://repository.asu.edu/items/18211.

Smil, V. (2008). *Energy in Nature and Society: General Energetics of Complex Systems.* Cambridge, MA: The MIT Press.

Unruh, G. C. (2000). Understanding carbon lock-in. *Energy Policy* 28: 817–830.

Unruh, G. C. (2002). Escaping carbon lock-in. *Energy Policy* 30: 317–325.

PART IV

Urbanization, global change and sustainability

Critical emerging integrative research

Part IV of the Handbook focuses on the needs for future research and action drawing upon nearly ten years of urbanization and global environmental change research while offering contributions from academics, practitioners and city officials alike. Parts I through III have clearly demonstrated that humanity is now living in an era of profound environmental change characterized by increasing complexity and uncertainty, but also great opportunity. A key lesson moving forward is that we must be flexible and continue to rethink how we conceptualize problems and our approaches to solving them. For example, 'adaptation', 'transition' and 'transformation' approaches are used to understand social change, governance and decision-making against the backdrop of urbanization and environmental changes, however, might they need to be re-configured and their assumptions re-assessed in order to create new integrative frameworks that are more solutions-oriented and better equipped to tackle the challenges we face today (Redman, Chapter 34)? Traditional governance structures are also increasingly being re-evaluated as the agency of cities increases and the intricate web of urban actors, formal and informal, continues to emerge as key players in tackling issues of global environmental change. The case study of Durban, South Africa is a prime example of how city-level action can influence and be influenced by the larger national and inter-national conversations and actions surrounding global environmental change issues (Roberts, Chapter 35), biodiversity and climate change, in particular.

Ultimately, urbanization is without a doubt a driver of societal transformation, but more exploration into how societies can effectively transition towards more livable and desirable low carbon futures is critical now in our immediate future. This includes addressing the challenges, needs and ways forward for climate change mitigation, in rapidly urbanizing developing cities, in particular, with high potential to be influenced along more sustainable and equitable urban trajectories informed by the 'mistakes' of their industrialized neighbors (Aylett et al., Chapter 36). How do we ensure that the multiplicity of perspectives and forms of knowledge are included in the low carbon transition and urban sustainability discourses? These will be critical components to address and understand if cities are to effectively create robust strategies for mitigating as well as adapting to the effects of urban GHG emissions that, as research indicates, will be as diverse as the cities themselves.

34
ADAPTATION, TRANSFORMATION AND TRANSITION
Approaches to the sustainability challenge

Charles L. Redman

It is widely agreed that we have entered an era of increasing complexity, unprecedented levels of interaction and a poor record of solving fundamental problems or predicting the future. Yet, we continue to use the tools and conceptual frameworks of the past to address the challenges of the future. Albert Einstein is often attributed with having said, "One cannot solve problems thinking in the same manner that created them." This is particularly true in the face of global climate change and the continuing process of rapid urbanization and integration of systems of cities across the globe.

During the past several decades, we have experienced a shift from the centralized government-based, nation-state toward more liberalized, market-based and decentralized decision-making structures in high-income societies. This has resulted in multi-scalar and increasingly diffuse policy-making structures and processes often focusing power in major cities and networks of cities (Loorbach, 2010). While this has promoted the growth of multi-national corporations as foci of economic activity and sometimes even of policy and governance itself, in low- and middle-income countries where the traditions of stability are weaker, it often results in instability and security breakdowns. As a result of these often divergent processes, there seems to be increasing acceptance of the position that top-down steering by government and the liberal free market approach are both outmoded as effective management mechanisms to address contemporary sustainability challenges by themselves, but at the same time, it is impractical to govern societal change without them (Loorbach, 2010). We need to take a new look at the underlying dynamics associated with governance, decision-making and social change and assess if they need to be reconfigured in new and forward looking solutions frameworks.

Adaptation, transformation and transition each represent a somewhat different approach to social change (as well as ecological, technological and other change), each with their own associated methodologies to address sustainability challenges. What are the implications for change of employing each of these approaches and their relevant, but often unvoiced, assumptions? Planning that would lead future cities on a pathway toward greater sustainability would benefit from input and discourses that allow the greatest range of potential outcomes. Cities are dominated by enormous investments in hard-to-change infrastructure and associated institutions that favor a business-as-usual approach. What are the specific factors that constrain and managers in seeking solutions that involve only modest refinements in the current system to induce changes they hope will lead to a sustainable pathway? How might we encourage a more open, inclusive framing of these challenges and a broader discourse on potential solutions including the possibility of radical change? It is argued here that facilitating more comprehensive consideration of future options starts with understanding the basic frameworks for change and how the vocabulary and underlying concepts condition potential outcomes.

Adaptation

The dictionary defines *adapt* (n.d.) as: to make fit, such as for a specific or new use or situation, often by modification. *Adapt* implies a modification according to changing circumstances, as distinguished from *adjust* that suggests bringing into a close and exact correspondence, and from *accommodate* that suggests yielding or compromising in order to affect a correspondence. Within a resilience theory framework, adaptation has been defined as the decision-making process and the set of actions undertaken to maintain the capacity to deal with current or future predicted change (Nelson et al., 2007). In other words, it involves efforts to refine a system to adjust to potential future shocks, stresses or other changing conditions in a way to maintain essential system functioning in a structure similar to the existing order.

Another perspective focuses on adaptability as capturing the capacity of a social-ecological system (SES) to learn, combine experience and knowledge, adjust its responses to changing external drivers and internal processes, and continue developing within the current stability domain or basin of attractions (Berkes et al., 2003; Folke et al., 2010). The usually unstated implication is that necessary changes to the system will be modest and incremental. Hence, adaptive strategies are often relatively conservative. Under the pressure of changing conditions, they work to maintain or return the system to the previous order or to one similar to it. Because humans are integral to SESs and their actions often dominate system futures, the adaptability of a system is sometimes associated with the capacity of the actors in the system to manage for resilience.

Adaptive strategies can often be very specific and localized, addressing an identified potential threat, the associated vulnerability and adjusting the system to successfully respond to the threat. Building levees to contain a river during floods may be effective up to specified limits, but it also influences subsequent land use in ways that make the system more dependent on further levee construction. A broader, less-specific perspective on taking an adaptive strategy is one that attempts to enhance the 'adaptive capacity' of the system in order to weather a wide range of potential shocks and stresses (Chapin et al., 2009; Nelson et al., 2007; Brown and Westaway, 2011). Enhancing a system's adaptive capacity is often cited as a major objective of resilience theorists in that it offers a strategic pathway to their central objective of maintaining system functioning (Gunderson and Holling, 2002). Adaptation is a favored approach in the current political arena in that it works to maintain elements of the established order and to address what are perceived of as likely near-term problems. Leaders and most citizens are more comfortable with adaptation in that it appears less radical than transformation, where one must work with the uncertainty of the outcomes and associated costs of major system restructuring.

Transformation

To *transform* (n.d.) is to change in composition or structure; to change the outward form or appearance; to change in character or condition. *Transformation* (n.d.) is an act, process or

Adaptation, transformation and transition

Table 34.1 System dynamics under adaptation, transition and transformation

Adaptation	Transition	Transformation
Incremental change	Major change, outcome predicted	Major, potentially fundamental change
Respond to shock	Action designed to achieve objective	Action in anticipation of major stresses
Maintain previous order	Transition to desired order	Create new order, open ended
Build adaptive capacity	Develop dynamics that result in desired system	Reorder system dynamics
Emergent properties guide trajectory	Trajectory and outcome predetermined	Build agency, leadership, change agents

instance of transforming. Interestingly, the mathematical definition of *transform* (n.d.) is the operation of changing (as by rotation or mapping) one configuration or expression into another in accordance with a mathematical rule. In the sustainability and resilience literature, transformation implies a more pervasive and radical reorganization of the SES than is intended by adaptive actions (see Table 34.1 for a comparison of concepts).

Nelson et al. (2007) define it as a fundamental alteration of the nature of a system once the current ecological, social or economic conditions become untenable or are undesirable. Walker et al. (2006) similarly see transformability as the capacity to create a fundamentally new system when the existing system is untenable. Hence, the transformation of systems dynamics may be required/desired in situations where a major threat is envisioned such as severe climate change, or where a significant malfunction in the system is recognized such as the 'poverty trap' that so frequently happens as large numbers of rural to urban migrants move into the periphery of low- and middle-income country cities (Marshall et al., 2010; Kates et al., 2012). The anticipated threat is often in the form of an enduring and increasing stress that is perceived as nearing a threshold or tipping point that could have dire implications for the system. Hence, the strategy is to take actions that would allow the system to move to and maintain a new set of dynamics that will operate within specified new and desirable values over the long-term.

Scholars have observed three phases in transformations of an SES: preparing the system for change; navigating the transformation by making use of a crisis as a window of opportunity for change; and, finally, building resilience of the new system regime (Olsson et al., 2004; Folke et al., 2010). Both a resilience approach to transformation and a transition management approach emphasize the early facilitation of transformative experiments at small scales and allow cross-learning and new initiatives to emerge. At that point, the two diverge in approach: whereas the transition model then determines the new goal and adopts a particular process for reaching it, a resilience approach would allow the new identity of the SES to emerge through interactions within and across scales of the system itself (Folke et al., 2010).

SESs can sometimes get trapped in very resilient, but undesirable regimes in which adaptation is not a sufficient option (Walker et al., 2006). To the extent that details of future states are left open and enhancing the general robustness of the system is the goal, transformation actions can be a part of resilience approaches, especially in cases where passing through the adaptive cycles results in regime shifts (Folke et al., 2010). Transformations are fundamental changes in SESs that result in different control variables defining the state of the system, shifts in the system's productive regime and often changes in scales of critical feedbacks (Chapin et al., 2009, pp. 328–329). Transformations involve taking substantial risks. They may have unintended consequences and may push the system over a feared tipping point. Moreover, they often are seen as potentially very expensive and have a greater uncertainty of outcomes than most adaptive actions. Hence, transformations may at first seem politically unpalatable and therefore benefit from participatory activities with stakeholders in order to build a broad consensus on the nature of the threat and to envision possible desirable outcomes. Unpalatable or not, many sustainability advocates suggest that fundamental transformations are necessary in order to put society on a pathway toward sustainability and urge their colleagues and students to take leadership roles in conceptualizing and implementing potentially transformative approaches (Hopwood et al., 2005; Leach et al., 2010; Wiek et al., 2011).

Transition

Transition (n.d.) is defined as a passage (movement, development or evolution) from one state, stage, subject, place or style to another. The key distinction of *transition* from *transformation* is that the outcome of the transition is known or defined in advance. A transition approach takes a more prescriptive stance toward governance and is explicitly normative in so far as it predefines desired outcomes. Hence, to the extent that transformative strategies are aimed at specific system outcomes, the approach has much in common with transition management—an approach that has gained strong support from many sustainability scientists and is the way the concept is used here. Transition management is a governance approach that is being developed with theoretical insights from complex system theory as well as on-going experiments and practical experience (Rotmans et al., 2001; Loorbach, 2010). The primary objectives of this type of strategy are persistent problems where purely analytical approaches will not suffice and participatory co-production of knowledge and desired outcomes is essential. The solutions are prescriptive to the extent that they propose specific trajectories of change resulting from case appropriate interventions. Steering of societal change is seen as a reflexive process of searching, learning and experimenting (Loorbach, 2010, p. 166). These efforts benefit from the presence of a variety of enabling factors such as effective leaders who act as change agents, knowledge of the implications of probable outcomes, and the financial and institutional resources to enact what may be radical changes.

Transformational adaptation

The high potential cost and uncertainty of substantial changes involved in transformation and transition planning have led some to suggest that the approach actually being taken in many cases should be thought of as transformational adaptation (Kates et al., 2012). As suggested above, adaptive responses are common, rely on familiar responses and often are applied incrementally. Differing from incremental adaptations, there are at least three classes of adaptation that Kates et al. (2012, p. 7156) describe as transformational: those that are adopted at a much larger scale or intensity; those that are truly new to a particular region or resource system; and those that transform places and shift locations. The difference is not always clear-cut. For example, are extensive seawall or levee constructions incremental or transformational? Kates et al. (2012) conceive these as transformational if they fundamentally change coastal land uses, but probably are not transformational if they primarily protect existing land uses. In resilience theory, one would view the adaptations as transformational if they acted to create

Adaptation, transformation and transition

new stability domains for development of a new stability landscape and if they propel the system across thresholds into a new development trajectory (Folke et al., 2010).

Assumptions and outcomes

The concepts and associated approaches of adaptation, transformation and transition are used in a variety of ways by researchers, practitioners and citizens. Within that range this chapter has provided a series of definitions that reflect, in part, how they are used in the literature and practice related to urban change. The intention is to focus on how the use of one or another of these concepts to organize an approach to decision-making and more broadly to governance implies some assumptions about how SESs operate as well as other assumptions about how researchers and practitioners can understand and predict the behavior of those systems. The 'smart cities' approach has gained widespread acceptance for its potential to address many of the major challenges facing cities—traffic congestion, pollution and resource use—by enhancing efficiency of current systems (Harrison et al., 2010; Calabrese et al., 2011). This would involve dramatically expanded efforts at monitoring, modeling expectable patterns, and then controlling the real-time flow of people and materials so as to avoid bottlenecks and more efficiently guide flows within the city.

As cities get larger, denser and more dependent on increasingly complex interactions, the incorporation of up-to-date sensor and information technologies seems logical and compelling to some. Clearly, there are potential benefits from this approach, but it must be recognized that it is based on the acceptance of the current system and building adaptations that directly address the perceived problems. The potential risk in taking this adaptation approach is that by enhancing a dependence on efficiency, one is also building into the urban system an increased vulnerability to rare, but expectable, shocks. An additional, indirect consequence of implementing smart cities approaches is that to make the flow of people more efficient, one must know where individuals are at all times—where they are going and possibly to control how they get there. The collection of this type of data could be benign, but the potential invasion of privacy and other misuses are quite disturbing. A more transformative approach would consider working to alter the underlying drivers of urban challenges. One might seek to diminish some of the root causes of increasing movement of goods and people by encouraging alterations in residence patterns to reduce travel to work and other activities, to transition to more distributed energy generating systems and possibly even expand urban agriculture to enhance local food security (see Pincetl, Chapter 33, and Griffith et al., Chapter 28, in this volume for more detailed discussions on these topics).

A broadly pursued objective of urban sustainability does not require us to accept one or another of these positions in advance, but to ensure that our discourse and framing does not preclude the full range of alternatives when designing and implementing plans for resilient, sustainable cities to flourish under future environmental and social conditions. In this chapter it is argued that the following potential elements of a conceptual framework and their associated assumptions about how systems can be understood and changed for the better implicitly directs research priorities and favors certain types of interventions over others:

- Adaptation is a relatively predictable low-risk approach that attempts to adjust the system so that it can weather shocks, without having to make major structural changes.
- Transformation is a more open-ended approach where major structural changes are deemed necessary to weather the expectable shocks and stresses, but the systemic outcome is not certain in advance.

- Transition implies an approach where participants suggest the desired outcome of potentially major structural changes and seek to develop interventions that would achieve these ends.

This chapter argues that it could be productive to distinguish between the approaches that pursue primarily resilience versus sustainability, yet employ in some manner all three approaches described above. It is suggested that the primary focus of resilience is on the process of building adaptive capacity (i.e. adaptation), whereas the primary focus of sustainability is on defining and strategically guiding urban systems toward sustainable outcomes (i.e. emphasizing transformation or transition) (Redman, 2014). This is not to deny that the practitioners pursuing either resilience or sustainability do not consider all three possible approaches in seeking to attain flexible, self-emergent and resilient system dynamics that at the same time exhibit sustainability characteristics such as net-zero carbon, social inclusiveness and protection of biodiversity. However, in most real world settings such as the contemporary drive toward smart cities, there are tradeoffs that require prioritization of objectives and choices of the strategies to reach these objectives. Moreover, different participants in these projects may hold different views on their abilities to predict or even to guide the trajectory of a system. The objective here is not to promote one of these approaches to the exclusion of the others, but to urge scientists and practitioners to recognize that there are alternate, acceptable, but not identical ways that we frame our challenges and potential solutions. The conundrum we face is that urban systems are extraordinarily complex, often exhibiting non-linear dynamics and are highly contextual, often being guided by charismatic individuals. They are also largely shaped by hard-to-predict, emergent properties. At the same time the goal of both those that adhere to a sustainability or resilience approach is to envision a sustainable, just world for the future and to design practical, risk minimizing strategies to reach that outcome. It is not argued here that there is a singular pathway and set of concepts to employ in analyzing the problem and designing solutions. Rather, it is suggested that multiple approaches may have a better chance of successful outcomes, especially if they are formulated with a consistency between their basic assumptions about system behavior and knowledge production as well as the strategies that are proposed to direct social change.

Future research directions that will lead to innovations in practice include developing:

- reflexive approaches for scientists and practitioners to be more aware of how the framing of the situation can influence the perception of possible alternative interventions;
- better ways to recognize whether intervention strategies focus on improved system dynamics or a specific range of sustainable outcomes; and,
- a methodology that would allow more effective employment of historical and cross-cultural case studies to provide a basis for initial evaluation of alternative interventions for modifying system dynamics or outcomes.

References

Adapt. n.d. In *Merriam-Webster online*. Retrieved from www.merriam-webster.com/dictionary/adapt

Berkes, F., J. Colding, and C. Folke. 2003. *Navigating Social–Ecological Systems: Building Resilience for Complexity and Change.* Cambridge University Press, Cambridge, UK.

Brown, K. and E. Westaway. 2011. Agency, capacity, and resilience to environmental change: lessons from human development, well-being, and disasters. *Annual Review of Environment and Resources.* 36(14):1–22.

Calabrese, F., M. Colonna, P. Lovisolo, D. Parata, and C. Ratti. 2011. Real-time urban monitoring using cell phones: a case study in Rome. *Intelligent Transportation Systems, IEEE Transaction on.* 12(1):141–151.

Chapin, F. III, G. Kofinas, and C. Folke. 2009. *Principles of Ecosystem Stewardship.* Springer, New York.

Folke, C., S. R. Carpenter, B. Walker, M. Scheffer, T. Chapin, and J. Rockstrom. 2010. Resilience thinking: integrating resilience, adaptability and transformability. *Ecology and Society.* 15(4):20ff.

Gunderson, L. H. and C. S. Holling. 2002. *Panarchy: Understanding Transformations in Human and Natural Systems.* Island Press, Washington, DC.

Harrison, C., B. Eckman, R. Hamilton, Hartswick, J. Kalagnanam, J. Paraszczak, and P. Williams. 2010. Foundations for smarter cities. *IBM Journal of Research and Development.* 54(4):1–16.

Hopwood, B., M. Mellor, and G. O'Brien. 2005. Sustainable development: mapping different approaches. *Sustainable Development.* 13:38–52.

Kates, R., W. Travis, and T. Wilbanks. 2012. Transformational adaptation when incremental adaptations to climate change are insufficient. *PNAS.* 109:7156–7161.

Leach, M., I. Scoones, and A. Stirling. 2010. *Dynamic Sustainabilities: Technology, Environment, Social Justice.* Earthscan, London.

Loorbach, D. 2010. Transition management for sustainable development: a prescriptive, complexity-based governance framework. *Governance: An International Journal of Policy, Administration, and Institutions.* 23(1):161–183.

Marshall, F., S. Stagl, and S. Thapa. 2010. Understanding peri-urban sustainability: the role of the resilience approach. *STEPS Working Paper* 38, STEPS Centre, Brighton, pp. 1–38.

Nelson, D. R., W. N. Adger, and K. Brown. 2007. Adaptation to environmental change: contributions of a resilience framework. *Annual Reviews Environment Resource.* 32:395–419.

Olsson, P., C. Folke, and F. Berkes. (2004). Adaptive co-management for building social-ecological resilience. *Environmental Management.* 34:75–90.

Redman, C. L. 2014. Should sustainability and resilience be combined or remain distinct pursuits? *Ecology and Society.* 19(2): 37 http://dx.doi.org/10.5751/ES-06390-190237

Rotmans, J., R. Kemp, and M. van Asselt. 2001. More evolution that revolution: transition management in public policy. *Foresight.* 3(1):15–31.

Transform. n.d. In *Merriam-Webster online.* Retrieved from www.merriam-webster.com/dictionary/transform

Transformation. n.d. In *Merriam-Webster online.* Retrieved from www.merriam-webster.com/dictionary/transformation

Transition. n.d. In *Merriam-Webster online.* Retrieved from www.merriam-webster.com/dictionary/transition

Walker, B., L. Gunderson, A. Kinzig, C. Folke, S. Carpenter, and L. Schultz. 2006. A handful of heuristics and some propositions for understanding resilience in social-ecological systems. *Ecology and Society.* 13(1):13.

Wiek, A., L. Withycombe, and C. Redman. 2011. Key competencies in sustainability: a reference framework for academic program development. *Sustainability Science.* 6:203–218.

35

CITY ACTION FOR GLOBAL ENVIRONMENTAL CHANGE
Assessment and case study of Durban, South Africa

Debra Roberts

A boundary-laden, urban future

Cities are amongst the oldest and most enduring forms of governance, and urbanization is one of the key defining global trends of the 21st century. On this 'planet of cities' (Angel et al., 2011) over half the world's population is already urban (UN-Habitat, 2013), and it is anticipated that up to two-thirds will live in urban areas by 2050 (UNDESA/Population Division, 2011). This is the most rapid period of urbanization in the world's history, and as a result it is projected that annual urban infrastructure investments will rise from current figures of 10 trillion USD to more than 20 trillion USD globally by 2025, with cities in the Global South being the focus of this investment (Dobbs et al., 2012). According to urbanist Richard Florida, "We will spend more and build more in the next century than we have in all of human history before this . . ." (Biello, 2013, p.1). The urban share of global investments and earnings will grow concomitantly during this period and reinforce the urban dominance of the global economy. Already, only 600 cities produce 60 percent of global GDP (McKinsey Global Institute, 2011).

From a global environmental perspective, this wave of urban-focused growth and development offers a unique, large-scale opportunity to leap-frog the unsustainable patterns of the past and put in place transformative development pathways focused on sustainability and equity that can assist in circumventing a 'difficult to avoid' collapse of socio-ecological systems (Ahmed, 2014, p.2) or a 'perfect storm' of critical environmental resource shortages (Beddington, n.d., p.1). This leap-frogging is critical as cities are currently home to extreme deprivation and environmental degradation despite their wealth creating capacity and economies of scale. One billion people live in urban slums globally (UN-Habitat, 2010, p.30), and the expansion of urban areas is set to result in a "considerable loss of habitats in key biodiversity hotspots" (Seto et al., 2012, p.1). A new global development path will, however, require the transformation of current political and socio-economic systems focused on GDP, profit and consumerism to ones that prioritize well-being, ecological integrity, sufficiency and equity (Costanza et al., 2014). Cities are central to this transformative process. There has thus been a global call for an urban sustainable development goal (SDG)[1] to be included

Case study: Durban, South Africa

within the commitments of the global post-2015 development agenda, in order to harness and maximize existing urban momentum and accelerate progress towards sustainable development (Urban SDG, n.d.). The current draft of the SDGs includes a draft goal focused on making "cities and human settlements inclusive, safe, resilient and sustainable" (OWG, 2014, p.5). The challenge will be to ensure that this goal is carried through into the final SDG agreement by Parties in 2015.

Whilst cities are important drivers of global change, urban areas do not exist in a vacuum. Stable, functioning, intact Earth systems are equally important in supporting the growing global network of cities and are "a prerequisite for a thriving global society" (Griggs et al., 2013, p.305). Recent research suggests that these key planetary systems may have been compromised by human activities and that there might be key life support systems (such as climate change, biodiversity loss and the nitrogen cycle) where the concept of planetary boundaries have already been exceeded (Rockström et al., 2009). While there are critiques of the concept of planetary boundaries (e.g. Nordhaus et al., 2012) the idea of thresholds still prompts the need to think about urban development within the context of a 'full world' (Costanza et al., 2000); in other words, a recognition that infinite urban growth on a finite world is simply not possible.

Given this background, the aim of this chapter is to reflect on relevant milestones in the evolution of the international climate change and biodiversity governance regimes and to use the case study of Durban (South Africa) to examine the extent to which these (and related national-level responses) have influenced city-level action and contributed to local-level action that addresses global environmental change (GEC).

The role of local authorities

Two of the three stressed planetary systems identified by Rockström et al. (2009)—biodiversity and climate—have received extensive global- and local-level policy attention. For both, the United Nations Conference on Environment and Development (the Earth Summit) held in Rio de Janeiro in 1992 was a critical moment and saw the opening for signature of both the United Nations Framework Convention on Climate Change (UNFCCC) and the Convention on Biodiversity (CBD) (see Figure 35.1). Agenda 21, a comprehensive sustainable development action plan intended to facilitate multi-sectoral action across all spheres of governance and to target all aspects of human environmental impact, was also approved at the Rio Earth Summit. Acknowledging that sustainable development at this scale could not be achieved by national governments alone, nine sectors of society (or 'major groups') were recognized through which citizens could organize and participate in international sustainable development efforts. These include business and industry; children and youth; farmers; indigenous peoples; local authorities; non-governmental organizations (NGOs); scientific and technological communities; women; workers and trade unions.

Local authorities constitute a particularly important stakeholder group within the 'major groups', as they are "the level of governance closest to the people" (United Nations, 1992, p.285). They are the planners and managers of the world's cities and construct, operate and maintain economic, social and environmental infrastructure, oversee land use, develop local environmental and sustainability frameworks and assist in the implementation of national and sub-national laws and policies (see Shi et al., Chapter 23 in this volume, for further discussion on urban climate adaptation planning and local governance capacity). At Rio+20 in 2012, the efforts and progress of local and sub-national stakeholders in meeting these sustainable development expectations were recognized, and the need to continue to promote

Debra Roberts

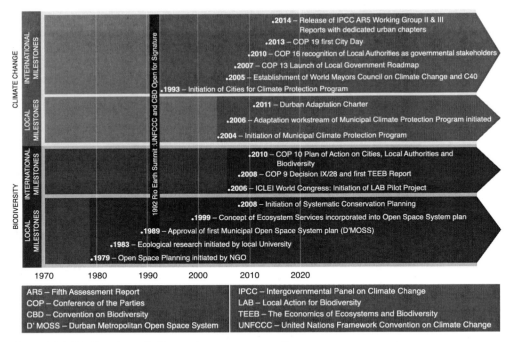

Figure 35.1 Biodiversity and climate change response milestones

"an integrated approach to planning and building sustainable cities and urban settlements, including through supporting local authorities", was re-emphasized (United Nations, 2012, para.135). Because of their strategic position, the capacity and willingness of local authorities to act often determines the extent to which the goals captured in multilateral environmental agreements (such as the CBD and UNFCCC) shift from policy-level aspiration to on-the-ground implementation (Revi et al., 2014a). A significant critique of the 'major groups' approach, however, is that by including local authority stakeholders as one of the nine NGOs, they remain functionally remote from their primary linkages to other spheres of government, and as a result their implementation potential has not been maximized (Otto-Zimmerman, 2011).

International climate change governance and related local government mobilization

In terms of the international climate change regime, local authorities (hereafter referred to as local governments) are amongst the earliest adopters of climate action. Eight months after the Rio Earth Summit, local governments began committing themselves to ICLEI's[2] Cities for Climate Protection (CCP) program (see Figure 35.1), creating one of the first transnational governance networks to address GEC (Betsill and Bulkeley, 2004). By comparison, it took the United Nations 13 years to agree to national-level commitments through the Kyoto Protocol. Given that neither the UNFCCC nor the related Kyoto Protocol reference the role of cities and their governing institutions, local governments began self-mobilizing through a coalition led by ICLEI and launched a Local Government Climate Roadmap in

Case study: Durban, South Africa

2007 in Bali at COP13-CMP3 (responding to the Bali Action Plan of national governments). The aim was to contribute to a strong and ambitious post-2012 global climate regime, which recognizes, engages and empowers local and sub-national governments to act. ICLEI has also played a key role in the establishment of an informal 'Friends of Cities' group in 2013 to facilitate a more structured dialogue and collaboration between UNFCCC Parties committed to urban issues and to increase support for local and sub-national governments' activities that contribute to national efforts to raise the level of ambition in the pre-2020 period.

An important step forward in acknowledging the central role of local government in addressing climate change was taken at COP16-CMP6[3] of the UNFCCC in Cancun in 2010 when local government was recognized as a governmental stakeholder[4] for the first time. Just prior to COP16-CMP6, local governments had demonstrated their ongoing and increasing commitment in this regard through the adoption and signing of the mitigation-focused Mexico City Pact at the World Mayors Summit on Climate. They also agreed to the carbon*n* Cities Climate Registry as a global reporting mechanism for measurable, reportable and verifiable local climate action by local governments. COP19-CMP9 in Warsaw in 2013 saw the first Cities Day hosted by a COP Presidency, and a number of urban references were included in the COP decisions and conclusions. This was largely the result of the growing recognition amongst Parties that the climate negotiations cannot remain 'urban blind' given the high proportion of greenhouse gas (GHG) emissions generated by city-based activities and the vulnerability of urban areas to the impacts of climate change (Revi et al., 2014b). Key amongst the urban-focused decisions was the need to facilitate the sharing of city- and sub-national level climate change mitigation and adaptation experiences and best practices to promote information exchange and cooperation[5] amongst the Parties. To initiate this, an urban forum was convened during the ADP (Ad Hoc Working Group on the Durban Platform for Enhanced Action) session[6] held in conjunction with the 40th session of the subsidiary bodies of the Convention in Bonn in 2014.

This meeting followed the release of the Working Group II and III reports of the Fifth Assessment report by the Intergovernmental Panel on Climate Change (IPCC), which included the first ever chapters focused on urban adaptation and mitigation, respectively. The IPCC reports provide the clearest evidence yet of the need for a clear climate change response mandate to be assigned to cities and for local government to be empowered and resourced in fulfilling this role. Within the UNFCCC system itself, there is also the potential for a greater urban focus in the Nairobi Work Program (NWP).[7] This flows from COP19 decisions that identify human settlements as an issue to be considered and the need for the NWP to better inform adaptation planning and actions at the regional, national and sub-national levels.[8] There have also been discussions within some Party delegations of the need for National Adaptation Plans (NAPs) and associated Local Adaptation Plans (LAPs) to be included as a commitment for all Parties in the much anticipated 2015 global climate agreement at COP-21 in Paris. Such a requirement would focus global attention on urban adaptation issues, but its success would depend on the ability to increase the amount of available global adaptation finance and to ensure that local governments have direct access to such funds. Until such financial and institutional support is in place, systemic, widespread and measureable urban change in support of climate change adaptation is unlikely.

Further examples of international self-mobilization by local government include the establishment of the World Mayors Council on Climate Change (WMCCC), which is an alliance of local government leaders advocating for the enhanced engagement of local governments as governmental stakeholders in the multilateral climate change and global sustainability negotiations. Founded in December 2005 soon after the Kyoto Protocol entered into force by the Mayor of Kyoto, Yorikane Masumoto, the WMCCC has also convened the Mayors' Adaptation Forum (MAF) (in collaboration with the City of Bonn and ICLEI). The latter acts as the leadership segment of ICLEI's annual Resilient Cities Congress, which focuses on climate change adaptation and climate resilient development. The MAF also provides an important venue for reporting on the implementation of the Durban Adaptation Charter (discussed below). The C40 (2012) provides an additional forum for local government action and is a network of megacities aimed at reducing GHG emissions and climate risks and impacts, both locally and globally. Initiated in 2005 by the former Mayor of London, Ken Livingstone, the first meeting of 18 megacities led to an agreement to reduce GHG emissions through *inter alia* procurement policies and alliances to accelerate the uptake of climate-friendly technologies. This marked the beginning of the C40 Climate Leadership Group and a partnership with the Clinton Climate Initiative (CCI) further enhanced emissions reductions efforts at the city level. The fact that cities continue to demonstrate their ability and willingness to act faster than nation states makes them necessary and critical partners in any 2015 global climate agreement.

This inclusion is particularly necessary given that local governments are not only mobilizing trans-nationally, but also at national and regional levels. In the US, for example, a growing number of mayors have signed the US Conference of Mayors' Climate Protection Agreement in order to advance the goals of the Kyoto Protocol through local leadership and action. Following the commencement of the Kyoto Protocol, the Agreement was launched by Mayor of Seattle, Greg Nickels, and commits cities to strive to meet or beat the Kyoto Protocol targets; urges their state governments and the federal government to enact policies and programs to meet or beat the GHG emission reduction target suggested for the US in the Kyoto Protocol; and urges the US Congress to pass GHG reduction legislation, which would establish a national emission trading system (United States Conference of Mayors, n.d.). Similarly, in Europe, the Covenant of Mayors is a movement of local and regional governments that voluntarily commit to increasing energy efficiency and use of renewable energy sources. The Covenant was launched by the European Commission after the 2008 adoption of the European Union Climate and Energy Package in order to endorse and support the efforts of local governments in implementing sustainable energy policies (EDGAR, n.d.). Covenant signatories aim to meet and exceed the European Union 20 percent CO_2 reduction objective by 2020, and there are recent moves towards the development of an adaptation equivalent of the Covenant.

Outside of the formal UNFCCC processes, other United Nations agencies such as UN-Habitat and the United Nations Environmental Program (UNEP) have also been working to encourage cities to address climate change. For example, UN-Habitat's Cities and Climate Change Initiative (CCCI), launched in 2008, seeks to enhance the adaptive capacity and mitigation activities of small- to medium-sized cities in low- to middle-income countries in innovative ways that are pro-poor and supported by appropriate tools (Bulkeley and Tuts, 2013). UNEP similarly launched a Campaign on Cities and Climate Change to engage cities in the global climate debate and assist them in reducing GHG emissions through awareness raising material, workshops, training and tools, and involving cities in international climate change meetings (UNEP, n.d.). Most recently, the UN Secretary-General appointed former New York Mayor, Michael Bloomberg, as special envoy for cities and climate change to assist in "consultations with mayors and related key stakeholders, in order to raise political will and mobilize action among cities" (Nichols, 2014, p.1). This is a significant signal of the increasing importance accorded to the role of cities in addressing global climate change.

Case study: Durban, South Africa

The Durban case study: climate change adaptation

None of the international level GEC interventions are of any value, however, unless they translate into local action. In this regard, the city of Durban (South Africa) provides a useful case study, as it is a recognized leader in both the climate change adaptation and biodiversity fields and provides the opportunity to examine what factors have influenced local-level action over time (see Figure 35.1). The duality of South Africa's cities, with their mix of developmental characteristics, also means that many of the lessons learned in Durban will have application in Africa and beyond. A recent paper by Roberts and O'Donoghue (2013) provides a detailed analysis of the Durban experience in addressing climate change adaptation and is repeated and updated here.

The first climate action taken in Durban was the city's participation in ICLEI's Cities for Climate Protection Program in the early 2000s, but this ended once the international funding was exhausted. The key catalyst in mainstreaming climate change action at the local government level was the participation of the Head of the (then) Environmental Management Department (later renamed the Environmental Planning and Climate Protection Department) in an advanced international environmental management program in 2004. This provided an opportunity for an in-depth engagement with the science of climate change and highlighted the significance of climate change impacts for the biodiversity planning being conducted by the city. This was a critical factor in the subsequent initiation of the Municipal Climate Protection Program, creating a local level activist for change and underscoring the importance of capacity-building and champion-building in institutionalizing complex environmental issues such as climate change, at the local government level. The development of the program since 2004 has been phased and opportunistic in nature as a result of several factors: the newness of the field, the lack of suitable precedents to guide action and the limited interest, leadership, institutional support and resources available for climate change work in the city. A phased approach has, however, ensured that the experience and knowledge gained through early interventions have been used to shape and refine subsequent actions and thinking.

Following an initial climate change impact analysis undertaken at the start of the Municipal Climate Protection Program, the work plan was extended through the development of an adaptation work stream in 2006. This is made up of several separate components:

- municipal adaptation (i.e. adaptation activities linked to the key line functions of local government);
- community-based adaptation (i.e. adaptation activities focused on improving the adaptive capacity of local communities);
- a strong ecosystem-based component present in both the municipal and community interventions;
- a series of urban management interventions that address specific climate change challenges (e.g. the urban heat island, increased storm water runoff, water conservation and sea level rise); and,
- several actions taken to mainstream climate protection action and develop locally appropriate tools to facilitate uptake.

The focus in all components has been on the initiation of projects that represent a 'no regrets' approach and are beneficial under a range of climate change scenarios. The majority of this work was initially funded using the municipal biodiversity budget, with the first dedicated climate change funding being received in the 2010–2011 financial year. Various sources of international funding have been used to supplement municipal resources as and when they have become available. Development of the Municipal Climate Protection Program has also necessitated the identification and capacitation of sectoral champions to act as change agents to secure the support of other departments and to help re-cast climate change as a development issue. This has helped minimize the marginalization associated with the program being initiated by an environmental department.

Similarly, lobbying has been required at the national and international level to create an appropriate and supportive external governance context for the Municipal Climate Protection Program. A unique opportunity to address the low priority assigned to adaptation within the broader climate change debate emerged with South Africa's hosting of the United Nation's Framework Convention on Climate Change COP17-CMP7 in 2011. As the host city, Durban worked with other members of a local government partnership to organize and host an adaptation-focused international local government convention at COP17-CMP7. The key outcome of this convention was the Durban Adaptation Charter (DAC), signed by 107 mayors representing more than 950 local governments worldwide, with the majority of signatories being from the Global South.[9] Post-COP17-CMP7, Durban has continued to work with members of the original local government partnership, as well as a group of new international partners to ensure the effective implementation of the DAC by focusing on the development of hub cities capable of driving context-relevant adaptation action at a sub-national and regional level. By thinking locally, Durban has been able to act globally in helping to influence the debate and significance assigned to urban adaptation. This has, however, come at a significant cost as limited human resources have meant that the international focus has compromised local adaptation work, which has had to take a back seat for the last three years. This situation is slowly changing as more climate staff members have recently been appointed, but it does indicate the difficult choices facing proactive local governments in addressing GEC.

Because of its leadership role, Durban is increasingly drawn into or is being profiled in international processes. For example, one of the large-scale reforestation projects that constitute the city's Community Ecosystem Based Adaptation (CEBA) initiative was acknowledged as a Momentum for Change Project at the UNFCCC's COP17-CMP7, while the head of the Environmental Planning and Climate Protection Department was selected as a lead author for Chapter 8 (Urban Areas) of Working Group II's contribution to the IPCC's Fifth Assessment Report. At a political level, as a result of hosting the local government convention at COP17-CMP7, the Mayor of Durban, James Nxumalo, has also emerged as a global champion for climate change adaptation. He has been elected as a member of ICLEI's Regional Executive Committee for Africa, where he is responsible for the Adaptation and Disaster Risk Reduction Portfolio. He is also the African representative on ICLEI's Global Executive Committee and was recently elected (2012) as one of three vice-presidents holding the Resilience, Climate Adaptation and DAC portfolio. In addition, in 2012, eThekwini Municipality appointed a new municipal manager who has expressed a desire to pursue a clear and centralized environmental and climate change agenda that builds on the legacy of COP17-CMP7. The fact that climate change is easily linked to the local development agenda has been a key factor in facilitating the emergence of this high level support, but whether this can be sustained in the face of mounting and often difficult to control development pressures is debatable.

A further confounding factor is that local government in South Africa has not been assigned any specific climate change response mandate. The National Climate Change Response Strategy released in 2011 simply acknowledges the significant role to be played by local government in addressing climate change (Government of the Republic of South Africa, 2011), but gives no further guidance on what that role should be in legislative or policy terms. The closest link with national government is currently through Durban's provision of technical support to the national UNFCCC negotiating team in the arena of adaptation and urban and local government issues. Durban has, however, proceeded to develop its own Climate Change Strategy (addressing both adaptation and mitigation) and is beginning to focus on regional engagement with neighboring municipalities in order to address climate change adaptation in a more integrated and synergistic way. This approach is based on lessons learned from a city-to-city partnership with Fort Lauderdale and Broward County in the US (initiated as part of the operationalization of the DAC) where a similar regional compact has been in place for over five years.

International biodiversity governance and related local government mobilization

Climate change is currently the global environmental *cause celebre*, and as a result, biodiversity issues are often overshadowed and insufficiently prioritized, despite the fact that biodiversity is a critical element in ensuring local level sustainability and enhancing adaptive capacity (Roberts et al., 2012). As such, cities are key in addressing the global biodiversity crisis— particularly as biodiversity thresholds are likely to be local and regional, rather than global in nature (Nordhaus et al., 2012). Mobilization by local governments around biodiversity issues is, however, relatively recent (see Figure 35.1). At the 2006 ICLEI World Congress in Cape Town (South Africa), the cities of Cape Town and Durban proposed the establishment of an international pilot project aimed at encouraging a commitment to biodiversity planning and management at the local level. Both of these cities are located in global biodiversity hotspots and have long histories of urban open space planning. Their cooperation and extensive experience in the field were key factors in boosting the credibility of the proposal and in ensuring that sufficient experience was available to technically and administratively support the resulting three-year Local Action for Biodiversity (LAB) pilot project. The institutional support provided by the International Union for Conservation of Nature (IUCN)'s Countdown 2010 campaign was also a significant factor in raising the international profile of the initiative. The LAB project involved 21 pioneer cities (who also self-funded the program) reporting on the state of local level biodiversity; politically supporting the protection and management of that biodiversity through the signing of the Durban Commitment (LAB, n.d.); developing action plans; and beginning project-based implementation of the plan. The success of this pilot project resulted in biodiversity being incorporated into ICLEI's international work program as a strategic theme and has provided the impetus for the establishment of the Biodiversity Cities Centre in the ICLEI Africa office in Cape Town. This incorporation of a biodiversity element into ICLEI's strategic work plan was regarded by the participating cities as a critical step in building international support for the protection of urban biodiversity and affirms the idea that such trans-national networks are favored because of the political support they offer local action (Betsill and Bulkeley, 2004). It also demonstrated local government's ability to proactively reach out to the international sphere, rather than just waiting passively for the downward migration of policy directives.

At the same time as the LAB project was being initiated, a meeting was convened in 2007 in Curitiba (Brazil) by Mayor Richa of Curitiba and the Secretariat of the CBD, which led to a proposal for a Global Partnership on Cities and Biodiversity. This was to support cities in the sustainable planning and management of their biodiversity resources and the implementation of practices that would support national, regional and international biodiversity strategies and plans (Mader and Hillel, n.d.). This close alignment between local and international biodiversity agendas contrasts strongly with the climate regime where the two work streams are distinct and largely separate. The formal launch of the Global Partnership on Cities and Biodiversity took place at the IUCN World Conservation Congress in Barcelona in 2008, and in the same year, the Parties discussed the role of local authorities in the implementation of the CBD at COP9 in Bonn (Germany). This resulted in the adoption of Decision IX/28 (CBD, n.d.(a)), which encourages Parties to recognize the role of cities in national strategies and plans and invites them to support and assist cities in implementing the Convention at the local level. The Global Partnership and a number of CBD Parties proposed supporting this decision with a broader plan of action, culminating in the adoption of the 2011–2020 CBD Plan of Action on Cities, Local Authorities and Biodiversity (CBD, n.d.(b)) at CBD COP10 in Nagoya (Japan) in 2010. This Plan of Action has resulted in at least one national government (Nepal) including a local government framework for biodiversity planning in its National Biodiversity Strategy and Action Plan and there is a view that more will follow (A. Mader, personal communication, February 27, 2014). The significant progress made in incorporating local government issues into the international biodiversity regime is now being used as motivation for similar progress to be made in the climate regime.

The Durban case study: biodiversity

The biodiversity and climate change adaptation functions in Durban have been established at very different times in the city's history. Biodiversity planning has been ongoing (under the banner of urban open space planning) since 1979 (see Figure 35.1). During this time, it has been championed by a succession of different organizations using different methodologies. Unlike climate change, where the drivers of local-level action have been exogenous, the catalyst for early biodiversity action in Durban was endogenous. Initially the primary focus was on the protection of conservation-worthy areas identified by a local NGO. By the mid-1980s this had evolved—as a result of ecological research undertaken by a local university—into a concern for the creation of ecologically viable networks focused on minimum critical areas and corridor widths. By the late 1990s local government had taken full control of the biodiversity planning process and introduced the concept of ecosystem services[10] as a key design informant. This marked the emergence of a more holistic understanding of the relationship between natural areas and sustainable urban development and human well-being and was informed by the emergence of new tools such as resource economics. The key drivers of this change were, first, the democratization of South African society and, second, the global prioritization of sustainable development following the 1992 Rio Earth Summit. The emergence of post-1994 political leadership, which prioritized poverty, economic development and meeting basic needs over less tangible ecological and conservation needs was a particularly important consideration in this paradigm shift. The use of resource economics was useful in this context, as it helped challenge prevailing perceptions of the open space system as an elitist resource focused on plant and animal requirements, and linked it directly to the ability to meet basic human needs and ensure long-term sustainability. Interestingly this use of resource economics substantially pre-dates (by about a decade) the subsequent international reporting on the Economics of Ecosystems and Biodiversity (TEEB) (beginning in 2008) and highlights the ability of local governments to assimilate and operationalize new ideas more rapidly than their national and international equivalents.

Despite its initial success, the political impact of the ecosystem services concept has weakened over time and has contributed to a broader experience base which suggests that new and

Case study: Durban, South Africa

transformative environmental ideas generally have a limited political shelf life in Durban (despite the fact that they might remain scientifically valid). As a result, there is a need to consistently 'refresh' the motivation for action. In response to biodiversity coming under increasing pressure from development aimed at addressing the escalating poverty and unemployment,[11] there has thus been a need to develop new scientifically based decision-making processes to support biodiversity protection. In this regard, the emergence of new tools such as fine-scale systematic conservation planning have facilitated a move away from reliance on expert opinion to a computer-based analysis of representability and irreplaceability, which provides a more objective and replicable foundation for biodiversity planning. The endorsement of systematic conservation planning by provincial and national government as part of the voluntary, but legally mandated preparation of bioregional plans under the National Environmental Management: Biodiversity Act (No. 10 of 2004), combined with local initiatives that link biodiversity management and restoration to job creation and poverty alleviation, currently provides the best motivation for sustaining the progress made in institutionalizing and mainstreaming biodiversity planning in Durban in an era of short-term, politically focused decision-making.

Given that local-level mobilization and the use of new tools (such as resource economics) significantly pre-dates any equivalent international mobilization, local government in Durban has only been minimally influenced by the international biodiversity regime. Durban's self-motivated involvement in the LAB pilot project was linked to the need to create a broader global consensus around the importance of urban biodiversity that could be used to influence the views of local leadership. This has not transpired, and even with the emergence of the CBD's Plan of Action on Cities, Local Authorities and Biodiversity, very little of the international debate has produced action or resources that intersect with local needs. National policy and legal requirements have, however, been more influential. The identification of threatened ecosystems and species; provisions for the proclamation of nature reserves and protection of forests; and the regulation of invasive alien species, all provide guidance for local-level planning. A significant challenge, however, is the fact that the constitutional mandate for biodiversity planning and protection in South Africa is assigned to national and provincial spheres of government with no obvious equivalent responsibility for local government. In Durban this resulted in a legal challenge from a land owner aggrieved by local government's attempts to protect biodiversity through targeted town planning interventions. The outcome of this high court challenge was the confirmation of the existence of a constitutional mandate for local government to undertake local-level biodiversity planning and has created a major governance breakthrough for local government country-wide. Despite this, the issue is not completely resolved, as this decision could still be challenged in a higher court. The need for an uncontested legal mandate for local-level biodiversity planning and management is as vital as the requirement for political will, appropriate technical and scientific capacity, institutional coordination, and accessible and reliable funding. Until the mandate issue can be addressed it is likely that local-level action will proceed largely independently of national frameworks, operating in the 'grey' institutional spaces created by passionate and innovative local champions whose energy and drive replace the traditional institutional platforms created by funded, legal mandates, and whose power is derived from local expertise and well-articulated moral positions (Bulkeley, 2005).

Future developments

The experience gained in institutionalizing biodiversity and climate change planning and implementation in Durban has made it possible to begin exploring the opportunities for integrated city-wide and city-led interventions to address GEC. Locally driven initiatives include contributing to the establishment of a multi-stakeholder partnership to address the role of ecological infrastructure in increasing water security and adaptive capacity in a large river catchment. At a global level, Durban's participation in the Rockefeller Foundation's recently established 100 Resilient Cities Centennial Challenge[12] will result in the development of the city's first integrated Resilience Strategy, building off the work already done in the climate change, biodiversity and water sectors. This opportunity to explore what transformation and resilience might mean locally has the potential to have a more direct impact on Durban than either the CBD or UNFCCC, simply because it has the capacity to address and resource local-level needs directly, cutting out the usual middlemen of national governments and multilateral agencies who have their own interpretations of how local actors should respond to international challenges. This of course assumes that this normative role will not, in turn, be assumed by the philanthropic funders. That local governments are seeking to subvert top-down control is perhaps not surprising in a globalized world where trans-national forces are increasingly circumventing the nation state. As sociologist Daniel Bell observes, "The nation-state is becoming too small for the big problems of life and too big for the small problems of life" (Biello, 2013, p.1).

Concluding messages

- In the past it has been assumed that the challenges posed by GEC should be addressed through global environmental governance structures such as the UNFCCC and CBD. The multi-sectoral biodiversity and climate change adaptation action and advocacy undertaken by cities such as Durban, however, raise the question of whether new forms of GEC governance are on the rise.
- Responding to the fact that their self-leadership, strategic importance and commitment to action have been underutilized by international and national government institutions, cities like Durban are tackling GEC through experimentation and by using available tools such as control over local use planning and budgets to protect biodiversity and increase adaptive capacity. This is occurring even under circumstances where political will is low or marginal, or pre-dependent on a clear link being made with the local development agenda. This represents a potential new form of international power as cities utilize this local action to identify others of a similar mindset and begin to forge trans-national networks through organizations such as ICLEI (e.g. CCP, 100 Resilient Cities) or through the initiative of champion cities (e.g. C40, LAB and the DAC) using these horizontal governance networks to develop and support more comprehensive, equitable and sustainable approaches to urban development (Bulkeley and Betsill, 2013).
- These independent local government policy entrepreneurs also challenge the notion that global environmental governance is a cascading process in which agreements are struck by nation states at the international level and then "passed down to be implemented through domestic processes within those states" (Betsill and Bulkeley, 2004 p.473). As cities and their local governments increasingly seek the freedom to deploy more decentralized models of environmental governance, there is a need to recalibrate international and national processes to accommodate and support such local-level action. Such 'rescaling' (Bulkeley, 2005, p.893) of prevailing governance systems will, however, challenge existing power bases and will require a transformation in mindset and a recognition that cities represent a new and critically important sphere of authority within environmental

Case study: Durban, South Africa

governance structures (Bulkeley, 2005) that should be supported rather than constrained by higher-level governance (Klein et al., 2014).

Key research questions to explore in the future

- How does one develop detailed city case studies that identify how transformative and integrative processes capable of addressing GEC emerge at the local level?
- How does one realistically and cost-effectively measure and assess the efficacy of city-level action in addressing GEC?
- How does one create a means of knowledge co-production that ensures that practitioners' perspectives are adequately captured and influence international GEC discourse?

Notes

1 One of the main outcomes of the 2012 Rio+20 Conference was the launching of a process to develop a set of SDGs which will build upon and replace the millennium development goals (MDGs). An inter-governmental Open Working Group was established by the United Nations to develop a proposal for these new goals.
2 ICLEI-Local Governments for Sustainability is the focal point for the Local Governments and Municipal Authorities (LGMA) constituency in the inter-governmental climate negotiations.
3 The abbreviation COP-CMP refers to a Conference of the Parties also serving as a meeting of the Parties to the UNFCCC.
4 1/CP16 on "Outcome of the Work of the Ad Hoc Working Group on Long-term Cooperative Action under the Convention (AWG-LCA)," paragraph 7 (UNFCCC, 2010).
5 1/CP19, paragraph 5b (UNFCCC, 2013a).
6 Conclusions proposed by the Co-Chairs as adopted by the ADP, paragraph 4d (UNFCCC, 2013b).
7 The objective of NWP is to improve the understanding and assessment of impacts, vulnerability and adaptation to climate change and make informed decisions on practical adaptation actions and measures to respond to climate change (UNFCCC, n.d.).
8 Decision 17/CP19 NWP on impacts, vulnerability and adaptation to climate change, paragraph 5. (UNFCCC, 2013c).
9 Currently the Charter has over 1,000 signatories.
10 They had taken control of the direct and indirect contributions of ecosystems to human well-being. The concept "ecosystem goods and services" is synonymous with ecosystem services (TEEB, n.d.).
11 Of the eight metropolitan areas in South Africa, Durban has the highest percentage of people living in poverty (EThekwini Municipality, 2013).
12 These cities will receive technical support and resources for developing and implementing plans for urban resilience over the next three years.

References

Ahmed, N. (2014). NASA-funded study: industrial civilisation headed for 'irreversible collapse'?. *The Guardian*. Retrieved from www.theguardian.com/environment/earth-insight/2014/mar/14/nasa-civilisation-irreversible-collapse-study-scientists.

Angel, S., with Parent, J., Civco, D.L. and Blei, A.M. (2011). *Making Room for a Planet of Cities*. Cambridge MA, USA: Lincoln Institute of land Policy.

Beddington, J. (n.d.). Food, energy, water and the climate: a perfect storm of global events. Retrieved from http://webarchive.nationalarchives.gov.uk/20121212135622/http://www.bis.gov.uk/assets/goscience/docs/p/perfect-storm-paper.pdf.

Betsill, M. and Bulkeley, H. (2004). Transnational networks and global environmental governance: the cities for climate protection program. *International Studies Quarterly*, 48, 471–493.

Biello, D. (2013). Can cities solve climate change?. *Scientific American*. Retrieved from www.scientificamerican.com/article.cfm?id=cities-as-solutions-to-climate-change.

Bulkeley, H. (2005). Reconfiguring environmental governance: towards a politics of scales and networks. *Political Geography, 24,* 875–902.

Bulkeley, H. and Betsill, M.M. (2013). Revisiting the urban politics of climate change. *Environmental Politics, 22:1,* 136–154.

Bulkeley, H. and Tuts, R. (2013). Understanding urban vulnerability, adaptation and resilience in the context of climate change. *Local Environment, 18(6),* 646–662.

C40 (2012). Mayors of the world's largest cities demonstrate progress in greenhouse gas reductions and launch two new initiatives. Retrieved from http://c40.org/c40blog/mayors-of-the-world%E2%80%99s-largest-cities-demonstrate-progress-in-greenhouse-gas-reductions-and-launch-two-new-initiatives.

CBD (Convention on Biological Diversity) (n.d.(a)). COP 9 Decision IX/28: Promoting engagement of cities and local authorities. Retrieved from www.cbd.int/decision/cop/default.shtml?id= 11671.

CBD (Convention on Biological Diversity) (n.d.(b)). Decision adopted by the Conference of the Parties to the Convention on Biological Diversity at its Tenth Meeting X/22. Plan of Action on Subnational Governments, Cities and Other Local Authorities for Biodiversity. Retrieved from www.cbd.int/doc/decisions/cop-10/cop-10-dec-22-en.doc.

Costanza, R., Daly, H., Folke, C., Hawken, P., Holling, C.S., McMichael, A.J., Pimentel, D. and Rapport, D. (2000). Managing our environmental portfolio. *BioScience, 50:2,* 149–155.

Costanza, R., Kubiszewski, I., Giovannini, E., Lovins, H., McGlade, J., Pickett, K.E., Ragnarsdóttir, K.V., Roberts, D., De Vogli, R. and Wilkinson, R. (2014). Time to leave GDP behind. Comment in *Nature, 505,* 283–285.

Dobbs, R., Remes, J., Manyika, J., Roxburgh, C., Smit, S. and Schaer F. (2012). *Urban World: Cities and the Rise of the Consuming Class.* McKinsey Global Institute. Retrieved from www.mckinsey.com/insights/urbanization/urban_world_cities_and_the_rise_of_the_consuming_class.

EDGAR (Emission Database for Global Atmospheric Research) (n.d.). Covenant of Mayors. Retrieved from http://edgar.jrc.ec.europa.eu/com/index.php.

EThekwini Municipality (2013). *Economic and Job Creation Strategy 2013.* Durban, South Africa: Economic Development and Investment Promotion Unit.

Government of the Republic of South Africa (2011). National Climate Change Response White Paper. Pretoria, South Africa: Government Printer.

Griggs, D., Stafford-Smith, M., Gaffney, O., Rockström, J., Öhman, M.C., Shyamsundar, P., Steffen, W., Glaser, G., Kanie, N. and Noble, I. (2013). Sustainable development goals for people and planet. *Nature, 495,* 305–307. DOI:10.1038/495305a.

Klein, R.J.T., Midgley, G.F., Preston, B.L., Alam, M., Berkhout, F.G.H., Dow, K. and Shaw, M.R. (2014). Chapter 16. Adaptation opportunities, constraints, and limits. In: Climate Change 2014: Impacts, Adaptation, and Vulnerability. Part A: Global and Sectoral Aspects. Contribution of Working Group II to the Fifth Assessment Report of the Intergovernmental Panel on Climate Change [Field, C.B., V.R. Barros, D.J. Dokken, K.J. Mach, M.D. Mastrandrea, T.E. Bilir, M. Chatterjee, K.L. Ebi, Y.O. Estrada, R.C. Genova, B. Girma, E.S. Kissel, A.N. Levy, S. MacCracken, P.R. Mastrandrea, and L.L. White (eds.)]. Cambridge University Press, Cambridge, UK and New York, NY, USA, pp. 899–943.

LAB (Local Action for Biodiversity) (n.d.). The Durban Commitment: local governments for biodiversity. Retrieved from http://archive.iclei.org/fileadmin/template/project_templates/localaction biodiversity/user_upload/LAB_Files/Resources_webpage/DC_English_August_2011.pdf.

McKinsey Global Institute (2011). *Urban World: Mapping the Economic Power of Cities.* McKinsey Global Institute. Retrieved from www.mckinsey.com/insights/urbanization/urban_world.

Mader, A. and Hillel, O. (n.d.). *Local Governments: Key Implementers of the Three Rio Conventions.* Biodiversity and Climate Change. Issue Paper No. 9. UNEP. Retrieved from www.cbd.int/doc/publications/unep-cbd-issue-papers/unep-cbd-issue-papers-09-en.pdf.

Nichols, M. (2014). U.N. appoints former NYC Mayor Bloomberg cities, climate change envoy. *Reuters.* Retrieved from www.reuters.com/article/2014/01/31/us-climate-un-bloomberg-idUSBREA0U 02Q20140131.

Nordhaus, E., Shellenberger, M. and Blomquist, L. (2012). *The Planetary Boundary Hypothesis: A review of the Evidence.* Oakland, California, USA: The Breakthrough Institute.

Otto-Zimmerman K. (2011). Global environmental governance: the role of local governments. *Sustainable Development Insights, March 7,* 1–8. Retrieved from http://local2012.iclei.org/fileadmin/files/

Case study: Durban, South Africa

Pardee_Center_SD_insight_GEG_and_the_role_of_Local_Governments_by_Konrad_Otto-Zimmermann.pdf.

OWG (Open Working Group) (2014). Untitled report. Retrieved from http://sustainabledevelopment.un.org/content/documents/4518SDGs_FINAL_Proposal%20of%20OWG_19%20July%20at%201320hrsver3.pdf.

Revi, A., Satterthwaite, D., Aragón-Durand, A., Corfee-Morlot, J., Kiunsi, R.B.R., Pelling, M., Roberts, D., Solecki, W., Gajjar, S.P. and Sverdlik, A. (2014a). Towards transformative adaptation in cities: the IPCC's Fifth Assessment. *Environment and Urbanization, 26(1)*, 11–28.

Revi, A., Satterthwaite, D.E., Aragón-Durand, F., Corfee-Morlot, J., Kiunsi, R.B.R., Pelling, M., Roberts, D.C. and Solecki, W. (2014b). Chapter 8. Urban Areas. In: Climate Change 2014: Impacts, Adaptation, and Vulnerability. Part A: Global and Sectoral Aspects. Contribution of Working Group II to the Fifth Assessment Report of the Intergovernmental Panel on Climate Change [Field, C.B., V.R. Barros, D.J. Dokken, K.J. Mach, M.D. Mastrandrea, T.E. Bilir, M. Chatterjee, K.L. Ebi, Y.O. Estrada, R.C. Genova, B. Girma, E.S. Kissel, A.N. Levy, S. MacCracken, P.R. Mastrandrea, and L.L. White (eds.)]. Cambridge University Press, Cambridge, UK and New York, NY, USA, pp. 535–612.

Roberts, D. and O'Donoghue, S. (2013). Urban environmental challenges and climate change action in Durban, South Africa. *Environment and Urbanization, 25(2)*, 299–319.

Roberts, D., Boon, R., Diederichs, N., Douwes, E., Govender, N., McInnes, A., Mclean, C., O'Donoghue, S. and Spires, M. (2012). Exploring ecosystem-based adaptation in Durban, South Africa: 'learning-by-doing' at the local government coal face. *Environment and Urbanization, 24(1)*, 167–195.

Rockström, J., Steffen, W., Noone, K., Persson, Å., Chapin, F.S. III, Lambin, E., Lenton, T.M., Scheffer, M., Folke, C., Schellnhuber, H., Nykvist, B., De Wit, C.A., Hughes, T., van der Leeuw, S., Rodhe, H., Sörlin, S., Snyder, P.K., Costanza, R., Svedin, U., Falkenmark, M., Karlberg, L., Corell, R.W., Fabry, V.J., Hansen, J., Walker, B., Liverman, D., Richardson, K., Crutzen, P. and Foley, J. (2009). Planetary boundaries: exploring the safe operating space for humanity. *Ecology and Society, 14(2)*, 32. Retrieved from www.ecologyandsociety.org/vol14/iss2/art32/.

Seto, K.C., Güneralp, B. and Hutyrac, L.R. (2012). Global forecasts of urban expansion to 2030 and direct impacts on biodiversity and carbon pools. *PNAS October 2 (109)*, 40, 16083–16088.

TEEB (The Economics of Ecosystems and Biodiversity) (n.d.). Glossary of terms. Retrieved from www.teebweb.org/resources/glossary-of-terms.

UNDESA (United Nations Department of Economic and Social Affairs)/Population Division (2011). *World Urbanization Prospects: The 2011 Revision.* Retrieved from http://esa.un.org/unup/Documentation/highlights.html.

UNEP (United Nations Environment Program) (n.d.). Cities and Climate Change. Retrieved from www.unep.org/urban_environment/issues/climate_change.asp.

UNFCCC (United Nations Framework Convention on Climate Change) (2010). Report of the Conference of the Parties on its Sixteenth Session, held in Cancun from 29 November to 10 December 2010. Retrieved from http://unfccc.int/resource/docs/2010/cop16/eng/07a01.pdf.

UNFCCC (United Nations Framework Convention on Climate Change) (2013a). Decision 1/CP.19 Further Advancing the Durban Platform. Retrieved from http://unfccc.int/resource/docs/2013/cop19/eng/10a01.pdf#page=3.

UNFCCC (United Nations Framework Convention on Climate Change) (2013b). Conclusions Proposed by the Co-Chairs as Adopted by the ADP. Retrieved from https://unfccc.int/files/bodies/awg/application/pdf/adp_conclusions_as_adopted.pdf.

UNFCCC (United Nations Framework Convention on Climate Change) (2013c). Decision 17/CP.19 Nairobi Work Programme on Impacts, Vulnerability and Adaptation to Climate Change. Retrieved from http://unfccc.int/resource/docs/2013/cop19/eng/10a02.pdf#page=4.

UNFCCC (United Nations Framework Convention on Climate Change) (n.d.). Nairobi Work Programme on Impacts, Vulnerability and Adaptation to Climate Change (NWP) – Understanding Vulnerability, Fostering Adaptation. Retrieved from http://unfccc.int/adaptation/workstreams/nairobi_work_programme/items/3633.php.

UN-Habitat (2010). *State of the World's Cities 2010/2011: Bridging the Urban Divide.* London: Earthscan. Retrieved from www.unhabitat.org/pmss/listItemDetails.aspx?publicationID=2917.

UN-Habitat (2013). *State of the World's Cities 2012/2013: Prosperity of Cities.* New York: Routledge. Retrieved from www.unhabitat.org/pmss/listItemDetails.aspx?publicationID=3387.

United Nations (1992). *Agenda 21.* Retrieved from https://docs.google.com/gview?url=http://sustainabledevelopment.un.org/content/documents/Agenda21.pdf&embedded=true.

United Nations (2012). *The Future We Want*. Rio+20 United Nations Conference on Sustainable Development. Rio de Janeiro, Brazil. Retrieved from www.uncsd2012.org/content/documents/727The%20Future%20We%20Want%2019%20June%201230pm.pdf.

United States Conference of Mayors (n.d.). U.S. Conference of Mayors Climate Protection Agreement Retrieved from www.usmayors.org/climateprotection/agreement.htm.

Urban SDG (n.d.). Campaign for an Urban Sustainable Development Goal. Retrieved from http://urbansdg.org/.

36

CLIMATE CHANGE MITIGATION IN RAPIDLY DEVELOPING CITIES

Alexander Aylett, Boyd Dionysius Joeman,
Benoit Lefevre, Andrés Luque-Ayala, Atiqur Rahman,
Debra Roberts, Sarah Ward and Mark Watkins

According to the Intergovernmental Panel on Climate Change Fifth Assessment Report (IPCC AR5), urban areas are responsible for up to 76 percent of global energy use and generate about three-quarters of carbon emissions (Seto et al., 2014). Cities worldwide are faced with the challenge of reducing emissions while also maintaining economic growth. Because there are such large variations in per capita urban greenhouse gas (GHG) emissions— from a low of 0.5 tCO_2/capita to more than 190 tCO_2/capita—strategies to reduce urban GHG emissions will need to vary significantly. Indeed, recent research suggests that cities must diversify strategies to most effectively mitigate urban GHG emissions (Creutzig et al., 2015). Cities that have mature infrastructure including built-out road networks and established transit patterns, face different challenges for mitigation than rapidly growing cities in low- and middle-income countries, where infrastructure is not well developed, and transportation systems are not locked-in. However, rapidly growing cities also may be challenged by limited financial and governance capital, thus creating what has been called the 'governance paradox', where the greatest opportunities for GHG emissions reduction may be in places where institutional capacities are lowest (Seto et al., 2014).

The cities built today will have lasting effects on GHG emissions over the next century. The energy required to mine and process the materials used to construct and operate the built environment determines emissions profiles now and also for decades to come. In developing countries, urbanization is taking place at lower levels of economic development, but is large in scale. For most of these cities, climate change *adaptation*, not mitigation has greater urgency. This is because per capita GHG emissions are still relatively low, but there are a number of exceptions such as cities with large manufacturing bases that have high per capita emissions. Several large Chinese cities such as Shanghai and Tianjin have per capita GHG emissions of 14 and 13 tCO_2eq/capita, which are levels similar to per capita emissions of some Western cities such as Los Angeles (~ 14 tCO_2eq/capita) and Toronto (~ 13 tCO_2eq/capita) (Sugar et al., 2012).

A significant challenge for climate change mitigation is not the level of emissions today, but rather what rapid and global scale urbanization means for future energy demand and associated emissions. In India, the growth in motorization and passenger transport is expected to result in a four-fold increase in transport emissions between 2000 and 2020 (Singh, 2006). Similarly, the continued growth of the transportation sector in China is expected to result in a five-time increase in transportation emissions between 2007 and 2050 (Ou et al., 2010). Because future urban population growth is expected to take place primarily in developing countries, there is a great opportunity to shape urban futures and hence urban energy, consumption patterns and ultimately emissions.

This chapter is a compilation of thought pieces from multiple urban practitioners and scholars who were asked to consider a set of questions related to the challenges, needs and ways forward for climate change mitigation with particular emphasis on developing cities. The narratives are the authors' responses that have been woven together to offer a multiplicity of perspectives, ranging from generalizations applicable to growing cities, to insights from specific case studies. Together, they represent diverse and often contrasting viewpoints— which in some cases are not shared by all authors involved. However, these perspectives reflect the complexity of the challenges and opportunities faced by rapidly developing cities, as experienced by those implementing and devising new ways of embedding climate mitigation in city planning in the Global South. The contributors include academics, planners and government officials with expertise in researching, communicating and implementing urban low carbon development efforts and wider-community (e.g. industry, public) engagement of these issues. The content herein is an attempt to contextualize the issues and supplement the information produced in the IPCC AR5 Working Group III, Chapter 12, 'Human Settlements, Infrastructure and Spatial Planning' (Seto et al., 2014).

Why is mitigation important for rapidly developing cities in particular?

The cities we build today will be those we live in over the next century. In developing countries, urbanization is taking place at a faster rate, so acting now is urgent and a long-term commitment. The poor and vulnerable members of society, who suffer the most during weather and climate-related extreme events, should be given top priority. An important question is: *how can we mitigate climate change while dealing with the impacts that threaten all countries, but developing countries and cities in particular?* Developing countries are likely to bear the brunt of 75 percent of the costs of damages resulting from climate change impacts (World Bank, 2010), and most developing cities are already facing a number of challenges, e.g. how to provide basic minimum services like safe drinking water, sanitation, electricity, housing and connectivity to millions living in often very harsh conditions. As climate change adds a layer of urgency and complexity to these existing problems, mitigation becomes a significant need.

Rather than seeing a reduction in GHG emissions as a burden that diminishes the development potentials of cities, mitigation can be a goal supporting cities in achieving a development agenda around prosperity and sustainable growth, improved living conditions and poverty reduction. Not 'developing despite', but 'developing by' reducing GHG emissions can become the attitude of every city planner and policy-maker. In the case of India, for example, most communities and institutions have a history of coping with uncertainty and extreme events. However, the current pressure on resources, high population densities and ongoing rapid economic, technological and slower social changes imply that a mix of institutional, market- and community-led mitigation and adaptation interventions will be necessary if future losses are to be within 'tolerable limits'. Mitigation in rapidly developing cities offers vital opportunities to increase the adaptive capacity of vulnerable cities through co-benefits such as natural resource protection, watershed management and job creation, while at the same time allowing them to avoid the developmental mistakes of the Global North by moving directly to, for example, efficient public transport and renewable energy.

We must decide today on which path we want to build our cities: with dependence on outdated decisions and technologies that were made without the awareness of climate change we have today, or in recognition of these new environmental realities to pursue low carbon pathways of development, creating cities and urban societies that are sustainable. Although many governments still hesitate at this crossroads, the emergence of new paradigms aligning development and sustainability around the concept of green growth shows possibilities and interest to walk along the low carbon pathway.

Cities are also labs of innovation, and can bring a fresh perspective to the United Nations Framework Convention on Climate Change (UNFCCC) negotiations. Cities around the world have come up with creative solutions to solving climate and sustainability challenges that are unique in their local context, but city-typical at their core. Increasingly, cities are creating new and joining established networks of learning, so that solutions and challenges in one city can be applied or learned in another city. In this way, cities around the world, regardless of size and location, are becoming role models that foster mutual knowledge and technology transfer. They can become test-beds around the possibility of win–win solutions. By doing so, cities can become important protagonists in the climate arena, providing confidence to national governments to embrace a low carbon development pathway.

What are the challenges or opportunities with respect to mitigation in rapidly developing cities?

Cities have trouble reducing emissions when it comes to areas like residential and commercial energy use as well as emissions related to transportation or local industry. In many cities, over 90 percent of urban emissions come from these four sectors. If we aren't making progress in those areas, then we aren't making progress. Unlike changing the light bulbs in city hall, reducing emissions across the city as a whole requires strong partnerships with local businesses and communities. Some cities have nurtured these partnerships, and they are paying dividends in terms of an increased ability to implement ambitious climate change strategies. Currently, community groups and civil society organizations are helping to contribute to their cities' climate change efforts. Local industries and businesses, in comparison, are often missing in action. Only a very small minority of cities report that the private sector is actively engaged in helping to design or implement their climate change strategies (Aylett, 2014). Internally, economic development units within municipal bureaucracies are also among the least engaged with the issue. It is not that the private sector holds the key to dealing with climate change, but as drivers of innovation, pools of capital and areas for emissions reduction, they are clearly important partners, who, in many cases remain absent. The objective in reducing even a fraction of GHG emissions cannot be achieved without the strong acceptance, commitment and involvement of all stakeholders. Political will, commitment, funding as well as public involvement are all critical.

The case of Malaysia provides an example of this; the importance of having strong partnerships and communication across stakeholder groups has been identified by the Iskandar Regional Development Authority (IRDA; www.irda.com.my) for influencing the successful mitigation efforts of the 'Low Carbon Society Blueprint for Iskandar Malaysia' (LCSBPIM). Since its global launching in 2012, the IRDA, a statutory body tasked with driving sustainability by working with multiple stakeholders, has embarked on a GHG emissions reduction trajectory by aiming to reduce emissions by 40 percent by 2025 (base year 2005). The IRDA duly notes the important role the governance structure plays in the implementation and acceptance of the LCSBPIM. In the case of Iskandar, Malaysia, where there are at least three levels of government (federal, state and local), policy coordination and acceptance are critical. A major challenge is continuing acceptance and ownership of the LCSBPIM by all (e.g. stakeholders, NGOs, business and commercial interests and the public) in order to ensure their awareness of and involvement in the implementation of the LCSBPIM. At the same time, it is important that the public sees visible evidence of the low carbon society. Terms such as 'climate change mitigation', 'adaptation', 'carbon credits' and 'sustainable developments' are very hard to grasp for most people. In this sense, monitoring the progress of implemented Low Carbon Society programs together with communication is important in order to ensure that they actually show: (a) visible physical changes that people can relate to and enjoy; (b) changes that have an impact on their lives (e.g. new bicycle lanes, footpaths, new park); and (c) GHG-reduction percentages. This serves as a reminder and lesson for new projects that if such efforts are made to help keep the public regularly informed and involved, they are more likely to remain interested and engaged.

Moreover, in the Global South, more fundamental challenges exist for successful climate change mitigation. These relate to the structural configuration of cities and their ongoing transformation: a deeply unequal social structure, which coexists alongside a steep increase in consumption by its growing middle classes. At one end of the spectrum, higher social mobility in the context of current economic development models means that the rapidly growing middle classes are constituted through significant increases in material consumption, inevitably resulting in an increase in GHG emissions. At the other, low carbon strategies often fail to provide benefits to the low-income population, failing to transform the pre-existing structural conditions of inequality, something which at best takes the low carbon objective away from local policy priorities and at worst furthers such conditions of inequality. These twin challenges highlight the extent to which the relationship between mitigation and development is both complex and contested. To bridge this, mitigation approaches need to engage with a broader discussion around development modes. If development is seen as a highly political process of establishing collective priorities around shared futures, then climate change mitigation needs to be imagined beyond the simple reach of GHG emissions accounting, and fully engage with a debate around the nature and practice of development.

Key questions required to structure such a debate relate to how certain development modes imply both carbon-intensive infrastructures and social practices, and how to imagine alternative development modes based on low carbon infrastructures. Climate change offers an opportunity for re-thinking urban infrastructure—as the materiality of development—in a way that contributes both to low carbon futures and fairer societies. Reviewing the politics of low carbon interventions means recognizing that these are not apolitical or post-political processes (Swyngedouw, 2010), as by their very nature they are likely to affect how specific sectors of society access key resources.

Are current approaches and planning strategies effective for reducing GHG emissions? What's working or not working?

It is still too early to tell if planning approaches and strategies are working, as most of the interventions are project-based with very little up-scaling evident. We will only truly be able to assess the impact once the carbon cities climate registry has amassed sufficient trending data to allow a more objective assessment of impact. However, there are indications of movements in the right direction, which can be considered 'successes' in their own right as well as valuable lessons learned that can be shared across municipalities in their efforts to plan and implement mitigation strategies. For example, in Cape Town, having real institutional change within the city is integral to having an effective strategy. Establishing a strong foundation and having strong and consistent champions means that energy and climate change have a fairly high level of recognition in the city. Institutional shifts within political and administrative levels towards prioritizing these issues as well as continuing efforts with respect to energy inventories and reporting, energy futures modeling, sea level rise research and water scarcity modeling are all very important for establishing the overall picture and for cross-city comparisons.

Also significant is developing links and networks with experts outside of the city such as local and global universities, other cities and being an active part of the global climate change conversation and climate change politics by participating in relevant global events. All of these initiatives require a strategy to identify the most crucial issues and what is going to place these issues on the appropriate stage to make them visible. In the case of Cape Town, the city has established an 'energy and climate change' committee and a city team that incorporates energy and climate change work as a dedicated part of their job, instead of simply as an add-on. The energy and climate change committee and team create platforms for working across departments, which is of course fundamental to such a cross-cutting area of work. This is precisely the core challenge with urban responses to climate change—that it is not an isolated issue. Responses must weave climate-smart principles throughout the fabric of our cities; however, that's not how cities do things. Climate change responses tend to be much more isolated, like the new kid at school playing off at the margins of how cities get their 'real work' done; or at least they were—something significant has changed.

Cities around the world report that they are now mainstreaming climate change as an issue into the plans that govern the overall direction of their development. For example, 75 percent of ICLEI-member cities recently surveyed as part of the MIT-ICLEI Urban Climate Change Governance Survey (UCGS) (Aylett, 2014, 2015) report that in addition to dedicated climate change plans, they are building climate change into their long-range and sectoral plans. Climate change responses, in other words, are becoming part of a new DNA for how we approach urbanization. This could be the key to unlocking the kind of ambitious cross-sectoral policies if we are really going to tackle urban emissions. It is as much about the environment as it is about other key issues like transportation, economic development or health. That means that if we want effective responses to climate change, they have to be cross-cutting; however, this is not without significant challenges. Take, for example, the ASI principle; the worldwide recognized message, which is quite clear: 'Avoid–Shift–Improve' (ASI), but difficult to implement. This approach assumes that urban spatial organizations determine a large part of urban energy consumption and GHG emissions, and are mainly influenced by transport–land-use interaction. To 'avoid' travel means to reduce the number and length of trips, but without limiting accessibility. Also, the need for travel can be reduced, e.g. with new communication technologies for teleconferences. Avoiding goes far beyond the transport sector, integrating cross-sectorial solutions; alignment and coordination between policies developed by various administrations (urban development, fiscal balance of taxes and subsidies, finance, energy, transport, etc.) is not an easy task. Institutional reforms and strong political willingness are needed.

To 'shift' trips refers to the goal to use more environmentally friendly modes of transport, which in most cases means more trips by public transport, bicycle and foot, and less car driving. The overall goal is to ensure accessibility of facilities, services and goods, but in a preferable low carbon manner. However, given earmarked budget, municipal financial constraints and the risk perception from private investors, the paradigm shift requires policy reform and political leadership. To close the financial gap between needs and investments to support the adoption of a low carbon pathway, we need to shift and increase the current financial flows. Mobilizing private finance must play a more significant role. This could happen if appropriate combinations of financial instrument and public policy are in place. To 'improve' the urban transport system also means to improve the technologies we use in our daily trips. Improving fuel quality, increasing fuel efficiency, reducing energy consumption of vehicles, making bicycle lanes safer or providing real-time information at bus stations—there is a huge range of things we can do to improve technologies used to travel; some are low-hanging fruit, many are win–win solutions, but others require improved accessibility and further development and deployment. Another challenge of this approach, as with all other ones, is to adjust promising solutions to the local context in which they are applied in order to make them successful solutions.

However, focusing solely on technology overlooks how successful mitigation strategies rely on deep engagements with social and political realities. Framing climate change mitigation primarily as a technological problem is a common misconception within the world of low carbon transitions. This framing also fails to recognize the decisively social and political nature of the very techniques used for the promotion of low carbon technologies. New conceptual tools are required to bridge the social and technological dimensions involved in the making of low carbon societies. For example, recent climate change research drawing on conceptual frameworks from critical social theory and post-structuralism is enabling a dialogue between the multiple techno-material and socio-political factors at play (Bulkeley et al., 2015). Drawing on Foucault's governmentality (2009), low carbon transitions can be understood through an engagement with the material or discursive, social or political techniques that are being used for the promotion of low carbon futures—rather than only focusing on the techno-material configuration of technologies (Bulkeley et al., 2014).

In an application of this approach, comparative studies examining the adoption of low carbon energy systems in Brazilian and Indian cities showcase this process as a function of two sets of governmental techniques: techniques of calculation (such as local energy baselines, renewable energy master plans and GHG emissions reduction targets) and techniques of standardization (including energy efficiency certificates, codes of professional practice and quality standards). Both sets of techniques bring about new ways of constituting energy practices in the city, and through this, new opportunities for implementing a low carbon future (Luque-Ayala, 2014). Re-imagining techniques for governing a low carbon transition brings forward a new set of priorities and opportunities. For example, in the Brazilian and Indian case described above, involving the municipality alongside other urban stakeholders in large scale energy planning encouraged a new set of (material, social and economic) circulations. In the promotion of these new (and low carbon) circulations, techniques of standardization (e.g. the standardization of solar hot water systems for social housing, as in São Paulo, Brazil) appear to have a key advantage over techniques of calculation (e.g. local energy baselines, as in Thane, in the Mumbai Metropolitan Region, India): by using a relational approach that firmly considers material qualities, standardization accounts for cross-sectorial interaction in the search for low carbon systems. This means opening up possibilities for strategies that integrate resource flows and infrastructures (such as energy and water, two key resources involved in local energy strategies based on the use of solar hot water systems) and deliver lower GHGs as part of this integration. Calculative techniques on their own, in contrast, appeared to restrict the intervention to a single resource domain (e.g. energy), failing to account for the multiple material relationships involved in and required for establishing low carbon urban systems (Luque-Ayala, 2014).

Figure 36.1 Installation of solar hot water systems on the roof of a luxury residential tower in Thane, Mumbai Metropolitan Region (India)

Photo credit: Andrés Luque-Ayala

What is needed for a shift towards comprehensive and cross-sectoral mitigation strategies?

The urban mitigation chapter, 'Human Settlements, Infrastructure, and Spatial Planning' of Working Group III of the IPCC AR5 concludes that thousands of cities are undertaking climate action plans, but that their aggregate impact on urban emissions is uncertain. This is in part because the majority of cities that have developed climate action plans have not conducted GHG inventories to understand the key sources of their emissions or to offer a benchmark to assess future emissions reductions. The chapter also finds that the majority of city climate action plans focus on improving energy efficiency of individual sectors such as transport and buildings, but few plans consider land use planning strategies and cross-sectoral measures.

The reality is that this is true; cities are laboring under the weight of siloed bureaucracies designed to respond to the problems of another century. It's also generally difficult to measure the overall impact of city actions, considering limited knowledge about those actions as well as multiplier and rebound effects. A first step to get closer to the goal of aggregated data availability is the development of tools that allow us to evaluate potentials and track impacts of action and policies. For this purpose, organizations such as the World Resources Institute (WRI), ICLEI and the C40 jointly developed the Global Protocol for Community-Scale GHG-Emissions (GPC). By harmonizing the measurement and reporting of GHG

Figure 36.2 Solar hot water systems displaying the PROCEL Seal for Energy Savings in São Paulo (Brazil)

Photo credit: Andrés Luque-Ayala
Note: The PROCEL Seal is an energy efficiency standard that forms part of the Brazilian Labelling Programme.

emissions at the city-level, the GPC aims at identifying major emissions sources and their reduction potential. Those sources are not limited to energy generation and consumption, but can be found throughout the whole life cycle of a product, for instance. WRI is also developing a standard to track impacts of action and policies, and another one to measure progress toward goals.

Those tools can really change the game and allow better recognition and support—financial and capacity building—to local actions. The large adoption of GPC and other such standards will allow for the creation of consistent, credible information on the state of climate policy and mitigation goals and the identification of important information gaps in national climate policies. Taken together, this work improves the implementation and design of national level mitigation policies, drives countries to adopt more ambitious GHG reduction goals, and promotes more transparency and accuracy in the GHG-emissions tracking. That said, we should progressively have a better view on the road toward Paris, COP21. Many municipal and subnational authorities presented their actions and objectives at the UN climate summit in NYC (September 2014). Cities are active in the UNFCCC process and are gaining momentum, with different countries supporting vertical integration strategies. More information on potential and gaps, more support and more ambition could come out of this process.

Another means to enhance the work on land use planning and cross-sectorial solutions is the aforementioned ASI approach, which serves as a valuable guide in the field of urban transport planning. For instance, by avoiding the functional segregation into residential and commercial areas and instead striving to a functional mixed-use development of cities, people can reach work places, supermarkets and health facilities in shorter distances. Hence, the length of trips is reduced and one multipurpose trip can replace multiple single trips, altogether avoiding travel and saving GHG emissions. Obviously cities tackle the 'lower hanging fruit' first. Energy efficiency (and diversifying energy supply) is most often the easiest for cities and has the quickest returns (savings, jobs). While doing these 'easier' activities, they must constantly be laying the groundwork to tackle these bigger, longer-term issues such as spatial form/land use change and public transport. However, the only cities in the world where public transport is profitable are located in Asia. South African cities, for example, can only dream of such density and integration. Massive investments in infrastructure development, public transport subsidies and concerted land use change programs are required and not easy. Responding effectively to climate change, therefore, challenges traditional structures because the issue simply doesn't fit neatly into any one institutional unit. Coordinating transportation and land use planning is one of the most effective things that a city can do to reduce its emissions. However, there are many cities where those issues are managed by different agencies with only limited collaboration across institutional lines. The same can be said for links between climate policies and areas like health and economic development, for example.

According to the U.S. Geological Survey (USGS), cities that are managing to mainstream climate planning across multiple agencies are using a combination of formal and informal measures. Tactics for building internal networks between departments dominate the strategies that are identified as most effective. Among these are more formal interventions, such as creating interdepartmental climate change or sustainability-focused working groups, or hiring or designating staff within local government agencies to work with the core climate planning team. But, even more effective are informal interventions based on person-to-person exchanges and trust—specifically creating informal channels of communication between those responsible for climate planning and staff within other agencies, building personal contacts and trust between the climate planning team and staff elsewhere in the municipality, and reaching beyond areas of municipal control, coordinating effective cross-sectoral partnerships. This is an opportunity to reach the bulk of urban emissions, but action at this scale is also the kind of action that will have a more profound positive impact on local economies and quality of life.

What further research or action is needed with respect to mitigation in order to successfully develop future low carbon cities?

The challenge over the next decade is to turn the curve, otherwise staying below suggested temperature increases becomes very unfeasible. Acting now is a matter of urgency and a long-term commitment. The IPCC AR5 (2014) shows that it is possible to stay below two degrees Celsius and outlines the way to do so. We know what needs to be done—this is not in question. What is needed is to build a pipe of bankable sustainable projects, programs and policies; we need research, diffusion and action on both sides—demand and supply—from all sciences. More research needs to be conducted at the local level, particularly with regard to making the economic case for interventions and laying out the policy, financial and legal requirements. Research must keep pace as new data and information (such as land use developments) emerge. We are beginning to see with global governmental bodies such as the UN, IPCC and global networks like C40 and ICLEI, that cities are occupying a much more important place and voice internationally. At the annual COPs there is now a day devoted to cities and the IPCC AR5 Working Group II report includes a dedicated urban chapter (Revi et al., 2014). This is a significant change to recognizing the role of cities and the greater drive and flexibility cities have to make urban development more sustainable—national governments are severely constrained by global posturing and politics—cities are not.

Scientists have outlined roadmaps, plans, policies and identified where there is still work to be done. The IPCC AR5, as one example, gives the timing required to avoid the worst impacts, outlines the sectoral pathways to decarbonize and identifies the key governance areas of work including the importance of cities and urban infrastructure and overall governance. It shows how smart, efficient urban design is critical to avoid locking in carbon-intensive infrastructure for decades. It is now the policy-makers that must take up these roadmaps, adapt them to their countries and implement them. The good news is that the actions, as shown by the IPCC AR5 Working Group III, can address many of the issues developing countries are facing today—air pollution, job creation, energy security and energy services. However, there is no clear pathway for the cities in the Global South to leap frog from their current state into a low carbon and climate resilient future. Until this story can be articulated in a way that is comprehensive and feasible given current resources, major change is going to be slow.

Access to the right kinds of funding is still a big obstacle—this is something which global governmental bodies such as the UN could facilitate for cities. The Iskandar Regional Development Authority (IRDA), for example, which is charged with developing and implementing mitigation strategies, continues to look for funds from outside sources and establish collaborations and partnerships with local, national as well as international partners. In general, the complexity of accessing and the amount of climate finance has been disappointing. Often funds are not available for what is needed; one can get money for processes, but not for capital for implemention, or for staff to do the work. Another challenge concerns getting industries interested and involved in projects to reduce GHG emissions. More detailed reporting and assessments (e.g. the 'Climate smart cities: The economics of low carbon cities' report from the IRDA and Centre for Low Carbon Futures at Leeds University, UK) are needed that make a compelling economic case for climate action at the city scale, including evidence based on investment requirements, returns and carbon savings of a wide range of low carbon measures, which can be used to underpin and strengthen applications for investment from a range of sources.

The low carbon transition can also be an opportunity to advance progressive and democratic approaches, through the recognition of issues of urban equity and social justice and through the mobilization of climate interventions toward solving pre-existing urban challenges and structural inequalities (Luque et al., 2013). But achieving this commendable objective is not a straight-forward matter. For example, how could solar energy systems or low carbon transport play a role in shaping, transforming or preserving the subject's position (as rich or poor, subjugated or empowered) in society? Can renewable and low carbon technologies become instruments for democratization and social justice? And, if so, how? How can these initiatives create new forms of exclusion or inclusion? In the cities of the Global South, characterized by overlapping environmental and social priorities, advancing a low carbon agenda with a limited understanding of its socio-political implications can result in greater inequality, the intensification of forms of social exclusion and/or regressive approaches toward both resources and the environment (Bulkeley et al., 2013).

An understanding of the low carbon transition as primarily a social and political endeavor should lie at the heart of any future research agenda around climate mitigation. It is through such explicit recognition of the social and the political that both climate change researchers and practitioners can meaningfully contribute to a much needed debate around a different set of collective futures and a new set of priorities guiding a low carbon world. Advancing this research agenda requires revisiting classic debates around urban politics and social justice (e.g. Harvey, 1996; Harvey, 2009; Rawls, 1999), through specific re-interpretations in the context of low carbon urbanism. For example, drawing on Harvey (2009), the interface between low carbon urbanism, democratization and social justice could be considered from various perspectives at different levels of complexity: first, as a matter of distribution and thus equity; second, in relation to a response to contemporary urban challenges, where social justice is problematized beyond issues of distribution through an engagement with issues of production, efficiency, capital and the configuration of consumption; and finally, by considering the role of social movements and protest in the configuration of the city's low carbon development vision and tactics, where social justice is recognized not as eternal justice and morality, but rather as contingent to social processes and to the multiple viewpoints and re-interpretations that emerge from resistance (Luque-Ayala, 2014).

References

Aylett, A. (2014). *Progress and Challenges in the Urban Governance of Climate Change: Results of a Global Survey*. Cambridge, MA: MIT.

Aylett, A. (2015). "Relational Agency and the Local Governance of Climate Change: International Trends and an American Exemplar." In *The Urban Climate Challenge: Re-thinking the Role of Cities in the Global Climate Regime*. Eds. Craig Johnson, Noah Toly, and Heike Schroeder. New York, NY: Routledge, pp. 156–179.

Bulkeley, H., Carmin, J., Castan Broto, V., et al. (2013). Climate justice and global cities: Mapping the emerging discourses. *Global Environmental Change* 23: 914–925.

Bulkeley, H., Castán Broto, V. and Edwards, G.A. (2015). *An Urban Politics of Climate Change Experimentation and the Governing of Socio-Technical Transitions*. New York: Routledge.

Bulkeley, H., Luque-Ayala, A. and Silver, J. (2014). Housing and the (re) configuration of energy provision in Cape Town and São Paulo: Making space for a progressive urban climate politics? *Political Geography* 40: 25–34.

Creutzig, F., Baiocchi, G., Bierkandt, R., Pichler, P.P. and Seto, K.C. (2015). Global typology of urban energy use and potentials for an urbanization mitigtion wedge. *Proceedings of the National Academy of Sciences of the United States of America* 112(20): 6283–6288.

Foucault, M. (2009). *Security, Territory, Population: Lectures at the College de France 1977–1978*. New York: Palgrave Macmillan.

Harvey, D. (1996). *Justice, Nature and the Geography of Difference*. Oxford: Blackwell.

Harvey, D. (2009). *Social Justice and the City*. Athens (GA): University of Georgia Press.

IPCC. (2014). *Climate Change 2014: Mitigation of Climate Change. Contribution of Working Group III to the Fifth Assessment Report of the Intergovernmental Panel on Climate Change*. Eds. O. Edenhofer, R. Pichs-Madruga, Y. Sokona, E. Farahani, S. Kadner, K. Seyboth, A. Adler, I. Baum, S. Brunner, P. Eickemeier, B. Kriemann, J. Savolainen, S. Schlömer, C. von Stechow, T. Zwickel and J.C. Minx. Cambridge, UK and New York, NY, USA: Cambridge University Press.

Luque-Ayala, A. (2014). Reconfiguring the city in the global south: Rationalities, techniques and subjectivities in the local governance of energy. *Geography*. Durham: Durham University.

Luque, A., Edwards, G.A. and Lalande, C. (2013). The local governance of climate change: New tools to respond to old limitations in Esmeraldas, Ecuador. *Local Environment* 18(6): 738–751.

Ou, X., Zhang, X. and Chang, S. (2010). Scenario analysis on alternative fuel/vehicle for China's future road transport: Life-cycle energy demand and GHG emissions, *Energy Policy* 38(8): 3943–3956.

Rawls, J. (1999). *A Theory of Justice*. Cambridge, MA: Harvard University Press.

Revi, A., Satterthwaite, D.E., Aragón-Durand, F., Corfee-Morlot, J., Kiunsi, R.B.R., Pelling, M., Roberts, D.C. and Solecki, W. (2014). "Urban Areas." In *Climate Change 2014: Impacts, Adaptation, and Vulnerability. Part A: Global and Sectoral Aspects. Contribution of Working Group II to the Fifth Assessment Report of the Intergovernmental Panel on Climate Change*. Eds. C.B. Field, V.R. Barros, D.J. Dokken, K.J. Mach, M.D. Mastrandrea, T.E. Bilir, M. Chatterjee, K.L. Ebi, Y.O. Estrada, R.C. Genova, B. Girma,

E.S. Kissel, A.N. Levy, S. MacCracken, P.R. Mastrandrea and L.L. White. Cambridge, UK and New York, NY, USA: Cambridge University Press, pp. 535–612.

Seto, K.C., Dhakal, S., Bigio, A., Blanco, H., Delgado, G.C., Dewar, D., Huang, L., Inaba, A., Kansal, A., Lwasa, S., McMahon, J.E., Müller, D.B., Murakami, J., Nagendra, H. and Ramaswami, A. (2014). Human Settlements, Infrastructure and Spatial Planning. In *Climate Change 2014: Mitigation of Climate Change. Contribution of Working Group III to the Fifth Assessment Report of the Intergovernmental Panel on Climate Change.* Eds. O. Edenhofer, R. Pichs-Madruga, Y. Sokona, E. Farahani, S. Kadner, K. Seyboth, A. Adler, I. Baum, S. Brunner, P. Eickemeier, B. Kriemann, J. Savolainen, S. Schlömer, C. von Stechow, T. Zwickel and J.C. Minx. Cambridge, UK and New York, NY, USA: Cambridge University Press, pp.

Singh, S.K. (2006). Future mobility in India: Implications for energy demand and CO2emission. *Transport Policy* 13(5): 398–412.

Sugar, L., Kennedy, C. and Leman, E. (2012). Greenhouse gas emissions from Chinese cities. *Journal of Industrial Ecology* 16(4): 552–563.

Swyngedouw, E. (2010). Apocalypse forever? Post-political populism and the spectre of climate change. *Theory, Culture & Society* 27(2–3): 213–232.

World Bank (2010). Cities and Climate Change: An Urgent Agenda (Volume 10, December). Washington, D.C.: The International Bank for Reconstruction and Development/The World Bank. http:// siteresources.worldbank.org/INTUWM/Resources/340232-1205330656272/CitiesandClimate Change.pdf

CONCLUSION

The road ahead for urbanization and sustainability research

Karen C. Seto and William D. Solecki

This Handbook attempts to synthesize and assess a rapidly growing literature on urbanization and global environmental change. With chapter contributions from a wide range of researchers and practitioners, the volume aims to provide perspective and insight into the diverse body of knowledge.

What the chapters collectively show

Together, the chapters present five important themes. First, it is clear that the interaction between urbanization and global environmental change is quite diverse, complex and dynamic. Embedded in these interactions are the actions of individuals, organizations and institutions within a weave of defined formal (e.g. laws, regulations) and informal practices and arrangements (e.g. norms, behaviors). The pace and scale of these interactions are rapidly evolving and the understanding of their impact has accelerated.

Second, the chapters also clearly illustrate that the connections between the process of urbanization and conditions of global environmental change occur at multiple spatial scales. While the environmental impacts of urbanization have been perceived as a local phenomenon (e.g. air pollution, waste management), the research presented in these chapters defines a new perspective on the effects of urban change that has evolved over the last decade: urbanization results in a number of impacts that manifest at the global scale. These global scale impacts can be viewed in one of two ways: that urbanization occurs in multiple places around the world simultaneously and hence the local scale impacts are occurring globally. For example, the transformation of farmland, forest and other land systems to new urban land has implications for the local hydrology, air quality and agricultural capacity. A second view of the global impacts of urbanization is that aggregated globally, urbanization can change regional and global biophysical dynamics. One example of this is the dominance of urban areas in global energy use and greenhouse gas (GHG) emissions. As urbanization continues to unfold through this century, the chapters collectively suggest that we can expect the urban signal on Earth system processes to become clearer.

A third theme of the volume points to the importance of collective and coordinated action across multiple urban areas. Because urbanization is occurring in thousands of places around the world simultaneously, the impact of any single urban area on global processes will become relatively smaller over time. Consequently, the actions taken at any single urban place will have little effect on the whole, making it increasingly important for urban areas to engage in collective and coordinated action. One example of this where we see tremendous promise is in the area of climate change mitigation. The reduction in GHG emissions for any single urban area will have small impacts lowering global urban GHG emissions. However, if thousands of cities simultaneously reduce their carbon footprint, the global implications are likely to be profound. For example, the Mayor's Compact agreed to by 228 cities in 2014 representing almost 500 million urban residents has committed to 13 gigatons of carbon emissions reduction by 2050.

A fourth theme within the chapters is that the mass of new knowledge and information has the potential to make a significant impact on policy. In many cases, advances in research are already being incorporated into public policy debates and decision-making. The connection between new climate risk information and the actions taken by governments is particularly strong and illustrative of the emerging science-policy linkage being defined. This feedback to governance has been established in many urban sectors and infrastructure systems such as water and energy supply, public health and transportation. Social media and other crowd sourcing activities have become important sources of data and information about everyday practices (e.g. commuting patterns, water use) as well as during moments of extreme events (e.g. evacuation planning and disaster response). Much of this activity increasingly is taking place under the context of co-production of knowledge in which learning is taking place through scientific, decision-maker and stakeholder communities acting together.

A fifth and absolutely key theme presented in the chapters is that the window of opportunity is small and closing as the urbanization process over the next few decades will lock-in behavior, environmental impacts and ultimately the sustainability trajectory for the long-term. Especially in the case of long-lived infrastructure and buildings, the technology and materials used today will define energy profiles for the coming decades. Significant opportunities, however, do exist to shape urbanization pathways, especially for places where infrastructures and urban form have not been established. Prospects are especially strong in places in sub-Saharan Africa and India, where urbanization levels are still relatively low (less than 50 percent), but where the rate of urban population growth is high. We must keep in mind that the majority of the urban areas of tomorrow have not been built.

Over the next 15 years, we can expect an area of about 10,000 hectares (100 sq. km) becoming urban every day. This will result in the largest investment in and construction of the built environment in human history, and massive demand for materials, energy, land and water. Estimates suggest that 25–30 trillion USD will be invested in urban infrastructure worldwide between 2010 and 2030; 100 billion USD per year in China alone. Often called the 'skeleton' of a city, the built environment establishes the fundamental character of an urban area. This infrastructure will form the center point of urban development, from street layouts and other transport routes to sewer and power lines, with lasting effects on urban form and urban life throughout this century and beyond. Once established, basic urban structure is difficult to change and forge social and economic relations. As Winston Churchill once remarked, "We shape our buildings, and afterwards our buildings shape us." If the initial conditions of urbanization do not enable or foster environmental and human well-being, the options for transforming an established urban area towards more sustainable conditions may be severely limited. Thus, initial decisions regarding urban infrastructure are critical for sustainability.

For example, New York City today illustrates the power of legacy effects on current sustainability. The City achieves some measures of sustainability and low per capita GHG emissions by the fact it has a well-established and extensive rail and bus transit system combined with urban form and highly mixed land uses that enable walking and the use of public transport. Mixed medium- and high-density land uses over small areas allow for housing and places of employment to be in close proximity, and create highly accessible places and neighborhoods. For instance, the contemporary street layout of the Wall Street area, the financial district in New York City, can trace its origins to the mid-1600s. In fact, Broadway, the boulevard which bisects part of the area, was originally a Native American trail. The configuration of the streets in this part of Manhattan is unlike the rectangular grid patterns typical in other parts of New York such as the Upper West and Upper East Sides.

Of course, urban form can be transformed such as in the case of contemporary Beijing. The twisting lanes, haphazard street layout and irregular intersections which were characteristic of Beijing during the Ming Dynasty (1368–1644) up through the early 1980s, have nearly all disappeared from the city due to the massive urban redevelopment that has occurred over the past 30 years. With the exception of a few places, namely parks and along main thoroughfares, much of the built environment in Beijing has been redesigned to create relatively uniform and standardized landscapes of industrial and residential compounds. There are many other examples around the world where large-scale urban, often transport, infrastructure has been re-envisioned and re-designed (e.g. the Central Artery/Tunnel Project in Boston, also known as the 'Big Dig'; the Cheonggyecheon Stream in Seoul; the Madrid RIO project in Madrid). These redevelopment efforts reshape the physical structure and ultimately the character of a city, but not without significant financial costs. A clear takeaway lesson is that infrastructure has a long life and that it significantly shapes a city's character for decades if not longer. Hence, the effect of past decision-making is both profound and long-term.

Needs and directions for the science of urbanization and global change

While it is clear that worldwide urbanization has global scale impacts, there are large uncertainties around the specifics of this relationship. We have hundreds of case studies around the world embedded in the chapters of this volume on the localized environmental impacts of urbanization, such as on land use, biodiversity, hydrology and regional climatology. We also know the aggregate impacts of urbanization such as the global water use, GHG emissions and energy use. However, we have very little understanding of the nature of the relationship between *how* urban areas are developed or the *processes* within urban areas and the resulting environmental outcomes. What are the benefits and tradeoffs associated with one type of urban development versus another? How and why does urbanization affect energy use and GHG emissions and the carbon cycle? How do politics and policies affect urban demand for energy? And most fundamentally, what pathways of urbanization—across different geographies and contexts—can lead to more sustainable outcomes? To address these questions, we need more information and better knowledge about fundamental relationships between urbanization in all settings and environmental change across multiple spatial scales.

Despite a diverse group and large number of scholars studying cities, research on urbanization has just begun to examine the intersection between the process of urbanization and other environmental systems (e.g. such as ecological systems, water supply). Moreover and most fundamentally, research in this area is largely *atheoretical*, empirically based and biased towards case studies typically involving European or North American cities, with only a few studies that span the full range of urban places or experiences (see Box C.1). The limitation of a large and representative sample of urban areas worldwide is in part due to the paucity of available data. Data gaps are widest in low- and middle-income country cities, and areas

Karen C. Seto and William D. Solecki

Box C.1   What we need from an urban science: not only planning, but also operations

Colin Harrison
IBM Distinguished Engineer Emeritus

The study of urbanization centers on the long-term evolution of cities, as population influx leads to increased demands for housing, industry, transportation, services, resources and so forth. These demands then become the input to planning exercises that apply known practices to meet them. However, we still have little scientific understanding of these practices and their potential for innovation. For example, urbanization needs a foundation that enables the rational comparison of per capita energy consumption across various methods of planning, construction, and operation of the built environment and the public and private services that it supports. The ability to perform such comparisons is still very much lacking. The 2014 Urbanization and Global Environmental Change Synthesis Conference revealed that we do not even have conclusive evidence for the widely held hypothesis that larger cities are more energy efficient than small cities. The sustainability programs adopted by cities around the world have led to a confusion of vague, unquantified and inconsistent metrics that defy the development of the comparisons required to put urbanization on a solid foundation. The urbanization community urgently needs to develop consistent scientific methods for evaluating such practices.

Furthermore, these current approaches are largely based on incremental improvements to traditional practices. Some of these, for example, providing better thermal insulation for buildings, have high potential, but others that attempt to modify complex systems produce only marginal results. There are trends underway that point to a second industrial revolution, to the replacement of scheduled services by on-demand or pre-emptive services, and to the decomposition of centralized, top-down services that will bring much greater changes in resource consumption and in how cities are planned and operated. A simple estimate shows that replacing private cars by on-demand, self-driving cars would reduce the number of vehicles on the streets by 75 percent with a corresponding reduction in emissions and in the need for car parking facilities. Hence, urbanization needs to consider the role that social and technical innovation can bring to cities.

---

experiencing rapid urbanization—which are currently the countries with emerging economies or low incomes. Even in high-income countries, significant data gaps are present. In some cases, shrinking government budgets further curtail the capacity for systematic data gathering. Throughout the world, the need for more urban data collection systems is loud and clear.

These pressures on structured and established data collection such as by government agencies, come at a time when there has been a dramatic growth in urban-related data. Much of these 'big data' come not from government or private sources (e.g. utility companies), but from social media, the internet and other user-contributed sources. It is possible that data from mobile phones, credit cards and social media can transform our understanding of urbanization and global environmental change. However, without a larger theoretical frame with which to examine these relationships and dynamics, there is a danger that big data will only provide additional sample points without a deeper systematic understanding. Critically, without a theoretical frame, each additional case study becomes simply that: a case study.

The road ahead for urbanization research

Moving away from opportunistic, *ad hoc* case studies to a more systematic assessment would allow for the development of axioms, postulates and theories that help explain fundamental relationships between urbanization and global change.

For example, given that urban areas contribute to about 75 percent of global GHG emissions, one task of the climate change mitigation research community is to identify mitigation potential of cities around the world. Much of what we know about urban emissions is either derived from top-down data such as disaggregated national data or not comparable across cities due to differences in collection methods, boundary definitions and variables of interest. Currently, our understanding of climate change mitigation potentials of cities is derived from climate action plans for thousands of individual cities. While not very meaningful by itself, the combined effect of these mitigation activities for the global set of cities can be significant. Yet, we need to ensure that consistent datasets are developed that can be examined as a large statistical sample of urban areas. This requires systematic sampling from the entire population of urban areas and not just a few. Importantly, we need better theory to ground these relationships. Two key concepts for the emerging science of urban systems are that of urban scaling and cities as social reactors. Together, they have the potential to provide the empirical evidence to test hypotheses about urban systems as well as provide a new framework to examine the system of cities—a new kind of complex system (see Box C.2).

Box C.2   Key concepts for an urban science: urban scaling and cities as social reactors

Luis Bettencourt

Sante Fe Institute for Complexity

Urban scaling is the empirical observation that most average properties of cities are predictable non-linear functions of city size. Many urban quantities are observed to obey scaling 'laws' of the form $Y(N) = Y_0 N^b$, where N is the city's population, Y_0 is a pre-factor independent of population (but dependent on time, level of national development, etc.) and b is a scaling exponent. Much of the interest in urban scaling results from the observation that the exponents b are general to all cities, regardless of level of socio-economic development, the nature of local culture or the quality of urban infrastructure, thus justifying the term 'law'. Socio-economic quantities such as the size of a city's economy (GDP), its labor productivity (total wages), its innovation rate (patents) or its level of conflict (number of violent crimes) are all characterized by $b > 1^a$ 1.10–1.20. This per capita increase in socio-economic rates with city size explains why larger cities are more expensive and also more violent than smaller towns: it is known as *increasing returns to scale*. The total area of urban infrastructure and the land area surface of a city are characterized by $b < 1^a$ 5/6 and $b < 1^a$ 2/3, respectively. Because this behavior implies that people in larger cities occupy, on average, less space per capita, these effects are known as *economies of scale*. Urban scaling can be explained by general models of cities resulting from the interplay between social and spatial networks. According to these models, cities realize social *network effects* mediated by built space driven by the general (net) benefits of social interactions minus transportation costs. Because urban scaling sets up a general expectation for urban metrics characterizing each given its size, it can be used to construct scale-independent measures of urban performance and to rank cities across an entire urban system. Urban scaling is intimately related to the concept of *agglomeration effects* in urban economics.

Cities are social reactors is the idea that the principal function of cities is to promote and maintain faster rates of human social interactions. As a result, people in larger cities can come in contact with more people over the same amount of time and traveling the same distance. These effects have been measured by sociologists using traditional survey methods and more recently, on larger scales, through data from cell phone networks. Although the effect is observed to hold on average, there is substantial individual variation from person to person. According to urban scaling theory, the net attractive nature of social interactions between people is at the basis for this effect, which shares some qualitative properties with the dynamics of stars and other systems with negative specific heat. Such systems display the common feature that, as the number of their elements increases, the system speeds up thus accelerating the rate of interactions between people, masses, etc. This means that rates of interaction and their products increase faster than the size of the system, making cities more productive per capita and stars brighter per unit mass as their size increases. This effect is known as *increasing returns to scale* in economics. The acceleration of social interactions with city size also lies at the root of several interesting effects relating to the diversity, size and productivity of cities. Higher rates of social interactions allow, in general, for greater division of labor and knowledge, as each person or organization can depend on a larger and more diverse number of others for their immediate needs. This interdependence, in turn, tends to lead to the faster growth of knowledge through deeper and more specialized learning and to its recombination through more connected socio-economic networks. In this way, the idea of *cities as social reactors* emphasizes the role of urban areas as collective learning systems, which may help explain the advantages of urbanized societies in the face of the fact that real wages (earnings minus costs) stay flat with city size. This idea also stresses the importance of social and spatial inclusion of marginal populations in each city as a means to maximize urban network effects and their socio-economic potential for the development of each citizen and of the system in general.

Opportunities and challenges for a new science

A new science that defines urbanization as a local and global process and that connects urbanization to global environmental change will have a set of requirements. The science must be able to describe and provide a pathway to explain the connections between the various scales (from the household to the city and to the global) at which decisions and activities take place that affect the conditions of urbanization. The science also must ontologically and epistemologically present urbanization as a system process driven by flows (e.g. speed, volume and direction of materials, resources and ideas), stocks (including the concentration and pattern of distribution of materials, resources and ideas) and feedbacks and response mechanisms (e.g. patterns of interactions, transitions and system-level tipping points). And, perhaps most profoundly, an urbanization science needs to explain urbanization in all settings (e.g. low-, middle- and high-income cities; rapidly growing and slowly growing) and conceive it as a continual process both in established cities and new and developing cities. Like all sciences, an urbanization science must be able to develop and test hypotheses and make predictions about unknown locations or the future. It will involve experimentation— often in the form of taking existing urban places or processes as living labs. It must be able to generate principles and properties about urban systems.

To achieve this level of success, a new science of urbanization will face a set of challenges, but also be buoyed by a number of opportunities. Here we discuss a few of them, with the

The road ahead for urbanization research

aim of being explicitly provocative and in the hope that the challenges can be overcome and the opportunities realized.

Challenge #1: Use of old tropes, theorems and analytical lenses

Irrespective of the urban theory or disciplinary frame that one espouses, it is very likely that it has limited applicability to 21st century urban phenomenon. Classic urban economists, from Isard to Mills, focused on a single city. Similarly, urbanists such as Burgess, Alonso, Hall and Howard all examined one city at a time. Even the revered Jane Jacobs focused her attention on intra-city issues. Yet at a time when there is a weekly increase in the global urban population by 1.3 million, there is urgency to understand not one city at a time, but all cities, and to examine variation across geographies, development levels and institutional contexts. While there is much to be learned by applying established frameworks, and recognizing that not all paradigms are irrelevant, much more needs to be done to develop new theories to explain the current wave of urbanization. The development of new theories must not be limited to orthodox concepts or conventional wisdom.

Challenge #2: Ignoring (or, not explicitly accounting for) informality, political economies and inequality

Much of what we have learned about urbanization in Europe and North America assumes that formal institutions and modes of governance (including rational planning with market economics) shape urban outcomes, and that the observations and data collection reflect key urbanization processes underway. However, in many parts of the world, urbanization today is taking place in extralegal and non-acknowledged settings. Many who are affected by or affecting urbanization are excluded from the formal urban decision-making process. These people (e.g. migrant laborers or floating populations), places (e.g. informal settlements) and conditions (e.g. the informal economy) often are not captured in official statistics or the conceptual frameworks about urbanization (see Box C.3). Contemporary urbanization by definition is part formal and part informal, and shaped by local and global political economies that result in growing inequalities within and between cities. Given the difficulties in obtaining data on these informal processes, it will be both a challenge and a necessity to acknowledge and include them in our studies of urbanization science.

Challenge #3: Ossified academe that is resistant to change

Academe is not only slow to change; it is resistant to change. Traditional urban studies and urban planning departments in general do not consider global change. Global change departments give marginal—if any—consideration to urban areas or human processes. There are many programs, centers and schools around the world that are trying to embrace a new urban academic agenda. Some of these changes involve initiatives that take existing programs and merge them; others are making new faculty cluster hires. While these developments are promising for a new era of urban research, the merging of existing departments and a few new faculty hires will not change the culture of academia around urban research, the evaluation and promotion process or peer-reviewed research journals. We need more integrated programs and studies that bring together disparate perspectives on urbanization science. Once we have a single urban program

Box C.3 The future of the science of cities, urbanism and urbanization: the places left behind

Colin Harrison

IBM Distinguished Engineer Emeritus

At the other end of the pipelines of urbanization are the small towns and villages left behind by those choosing an urban life. While this is most obvious in Asia and Africa, it affects also the developed countries as old industries decline and workers must seek new opportunities. The loss of population through urbanization creates many problems for these places. First, those leaving are predominantly the younger and better educated, who could have become local entrepreneurs, and so the future for economic growth is diminished. Second, the older population remaining will increasingly require social and medical services for which both skills and money will be harder to find. Third, the loss of population leaves 'hollowed out' places that would benefit from consolidation or re-densification to bring the population closer to public and private services and to reduce future dependency on driving. In extreme cases, isolated communities become unsustainable and the remaining population must be relocated to a nearby town. A related issue is the decline in value of residential and commercial property, often the primary assets of the community, which creates a further decline in equality. Fourth, the fixed costs of operating and maintaining infrastructure and services must now be borne by a smaller tax base. A related issue is the effect of under-utilization of infrastructure; for example, water systems with reduced flows need to be flushed to prevent them from silting up, which is an additional waste of water. Fifth, the decline of these communities represents a loss of diversity in the cultural heritage.

There is much study of the economics of growing or at least stable settlements, but little on these problems of decline. Yet, even when urbanization is well advanced, a significant fraction of the population will remain in these places left behind. The burden of dealing with these problems will ultimately fall on regional or national governments and these bodies should consider these additional costs of urbanization.

that includes faculty from a range of disciplines including but not limited to economists, engineers, sociologists, hydrologists, ecologists, architects, engineers, historians, geographers, planners and designers, then we will know that real change in academe has been achieved.

Although these challenges are significant, they are not insurmountable. Moreover, the current excitement, enthusiasm and attention on urbanization and cities represent a redefining moment for urbanization science. Here is what we consider the key opportunities for the development of a new science.

Opportunity #1: Urbanization as a sustainability challenge and a sustainable development opportunity

In 1992, the Secretary General of the UN Conference on Environment and Development, Maurice Strong, stated, "the battle to ensure that our planet remains a hospitable and sustainable home for the human species will be won or lost in the major urban areas." That prescient statement is even truer today, with mounting scientific evidence pointing to

The road ahead for urbanization research

potential sustainability tipping points driven by urbanization: planetary urbanization is accelerating evolutionary changes in many species; the embodied GHG emissions to construct the infrastructure for tomorrow's cities *exceed* the emissions targets to limit climate change to 2 degrees Celsius relative to pre-industrial levels; energy efficiency gains are being overshadowed by the magnitude of the increase in energy demand due to urbanization and industrialization. While urbanization at the planetary scale has significant environmental challenges, for many countries and local regions, urbanization is an opportunity to improve sanitation, provide access to modern energy and clean water, and increase education and health services, especially to women and children. Urbanization is increasingly seen as a means to lift millions out of poverty through employment and education opportunities. Nowhere are the challenges and opportunities of urbanization-led sustainable development more evident than in China, where the country is pursuing a landmark Urbanization Plan (see Box C.4). Thus, there is an enormous opportunity to align the development agenda, which is largely focused on human well-being, with the environmental sustainability and especially, climate change agenda. This puts urban research at the fore of local, national and international policy, and ultimately science, agendas.

Box C.4 Urbanization and environmental change in China: challenges and opportunities

Weiqi Zhou and Zhiyun Ouyang
State Key Laboratory of Urban and Regional Ecology, Research Center for Eco-Environmental Sciences, Chinese Academy of Sciences

A brief history of urbanization in China

China has been an agricultural country for thousands of years. Rural residents, 90 percent of the entire population, constituted the solid majority at the establishment of the People's Republic of China. For a long time, the proportion of urban population in China has been lower than the global average. However, starting from the late 1970s, with the 'open-door' policies and economic reform, the growth of China's cities, and the transformation of its population from predominantly rural to mainly city-dwelling, took off. Now, more than half of the Chinese population lives in cities.

It is certain that this trend will continue. In the recently released National New-type Urbanization Plan, China sets the target to raise its urban population proportion by 1 percent each year, to reach 60 percent by 2020. That is, every year, there will be approximately 18 million people, the size of the total population in the Los Angeles Metropolitan area, added into established or new cities.

This rapid, magnificent and virtually uncontrolled urbanization process in China has dramatically changed, and will continue to change, the local, regional and even global environment. Such changes greatly alter ecological functions and processes, with significant consequences for ecosystem services and human well-being. In fact, many Chinese cities are facing tremendous challenges, particularly environmental pollution and ecological degradation—smog, water shortage and pollution, urban heat islands, to name just a few.

A look forward: challenges and opportunities of future urbanization

Continued urbanization is likely to lead to more serious ecological and environmental problems. However, urbanization is considered to be critical to deliver a more ecologically sustainable and resource-efficient world because the per person environmental impact of city dwellers is generally lower than people in the countryside in developed countries. Under huge population pressure, urbanization may also be the ultimate solution for China to enhance sustainability (Bai et al. 2014). Here, we want to bring attention to the following three key issues that may constitute great challenges to future Chinese cities, yet all are potential opportunities to enhance urban sustainability across the country:

1   *The revision of the household registration system, also known as the Hukou system.* The Hukou system registers each citizen as either rural or urban, and people are only entitled to full social benefits and infrastructure in the specific place in which they are registered. China has 250 million migrant workers who now work and live in cities, but only have access to social infrastructures (i.e. education, health care and social security) as rural residents. In order to encourage urbanization outside the largest megacites, China revised the household registration system to open up small and medium cities to rural migrants in July, 2014. We expect a pulse in migration to these cities as a result. Are these small-medium cities ready for more residents? Such a pulse in urban population, if it happens, will be challenging for any city both socially and ecologically. At the same time, however, the migrations of rural people to cities would greatly relieve the pressure on local environment, and increase the resource efficiency for both the urban and rural areas. How this new policy influences China's future urbanization, and what the social and ecological consequences are, will be not only interesting, but very likely profound.

2   *Focusing urbanization in a number of urban agglomerations.* The National New-type Urbanization Plan explicitly focuses China's desire for urbanization in a number of urban agglomerations. A few of those designated urban agglomerations with a relatively long development history such as the Yangtze River Delta, Beijing-Tianjin-Hebei (Jingjinji), and Pearl River Delta, already have a network of cities. But many of the planned urban agglomerations still only have a few scattered cities. In the relatively well-developed urban agglomerations, urbanization has been one of the predominant forces driving local and regional environment and ecosystem change. The relatively high concentrations of cities and population already have resulted in deteriorated regional air and water quality. For example, PM2.5 is significantly higher in urban agglomerations compared to other regions (Figure C.1). With the expected population increase in these urban agglomerations, these regions may face more and more environmental and ecological challenges. Meanwhile, many of the environmental issues caused by urbanization, taking air pollution again as an example, can only be effectively addressed through coordinated regional development. Therefore, focusing on urban agglomerations, with better regional planning, provides an enormous opportunity to address regional environment issues.

3   *Within-city dynamics provide both challenges and opportunities for improving cities.* Chinese cities are more than simply expanding outward, but they also have a high degree of within-city dynamics. Even within the very well developed urban cores of cities such as Beijing and Shanghai, a significant proportion of the lands could be dramatically changed (Figure C.2). For example, within the fifth-ring road in Beijing, a very well developed urban area, there was an increase of

new greenspace of approximately 60 km² from 2005 to 2009. Meanwhile, 29 km² greenspace was converted to buildings and pavement. Rapid within-city changes bring great social and ecological challenges. For example, redevelopment of the urban villages – old settlements engulfed by city expansion – may lead to dramatic changes in social network and life styles of the dwellers. However, these changes also provide opportunities to introduce green technologies and practices in new constructions, and to build better neighborhoods that facilitate social interactions.

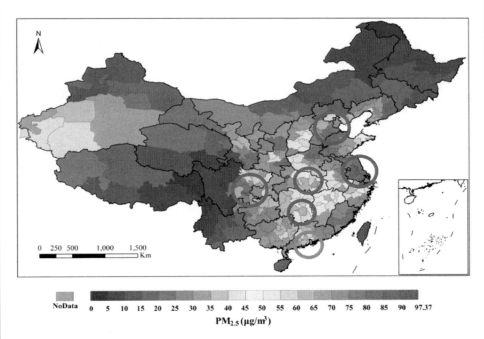

Figure C.1 PM$_{2.5}$ is significantly higher in urban agglomerations compared to other regions in China

Adapted from Han et al. (2014).
A detailed, colored version of this figure can be found on the book's website, www.routledge.com/9780415732260

References

Bai, X., Shi, P., and Liu, Y. 2014. Realizing China's urban dream. *Nature* 509:158–160.
Han, L., Zhou, W., Li, W., and Li, L. 2014. Impact of urbanization level on urban air quality: A case of fine particles (PM2.5) in Chinese cities. *Environmental Pollution* 194:163–170.
Qian, Y., Zhou, W., Li, W., and Han, L. 2015. Understanding the dynamic of greenspace in the urbanized area of Beijing based on high resolution satellite images. *Urban Forestry and Urban Greening* 14:39–47.

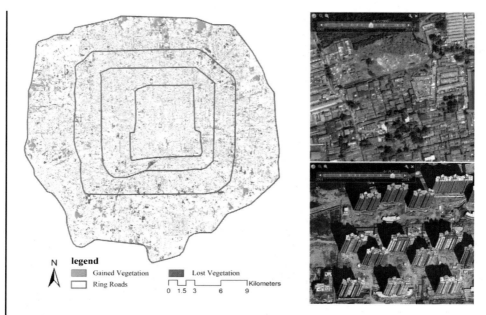

Figure C.2 High degree of within-city dynamics in the well-developed urban cores of Beijing

Note: Left panel: changes in green cover during 2005 and 2010 within the 5th ring road of Beijing (adapted from Qian et al. 2015); right panel: an urban village (an image collected on August 2, 2008) was converted into high-rise apartments (an image collected on September 15, 2012) (images from Google Earth).

Opportunity #2: New tools and new data

Just as the microscope changed science because it provided a new spatial scale of analysis, the current data revolution will transform everything we know about the *where*, *who* and *what*. While some have called this the 'big data' era, in the urban context, the data revolution is more than just about the increase in volume and sources of data. Rather, the type of data that are now available for research is changing *who* and *what* are observed and recorded, their periodicity and temporality as well as their routes and patterns (see Box C.5). Because we are changing *who* and *what* gets counted, the data revolution is profoundly reshaping *what matters*. It used to be that we fitted our data to match our analytical tools. Now we have so much new data that the tools have to be refitted to match the data. New ways of thinking and integrating data and perspectives are giving rise to new areas of research. A word of caution about the data revolution: Not everything that matters can be counted, and not everything that can be counted matters.

The road ahead for urbanization research

Box C.5   Satellite-derived data on global urban human settlements

Martino Pesaresi and Thomas Kemper

European Commission, Joint Research Centre, Institute for Protection and Security of the Citizens, Global Security and Crisis Management Unit, Italy

As it is increasingly clear that urban settlements are significant drivers of global change, policy-makers and scientists at all spatial scales require accurate and globally consistent data for evidence-driven decision-making to test hypotheses, develop theories and concepts, monitor and understand trends, and explore alternative scenarios of development, urbanization and the future.

Since the 1960s, space and airborne remote sensing technologies have contributed to creating globally consistent data over time that describe the physical characteristics of the global atmosphere, ocean and land systems. Census surveys and archeological records have been used to report on the size and geographic location of populations and economic activities for more than 5,000 years. Similarly, cadastral and geometrical land surveys have been essential elements in developing an understanding of human environments and human settlements since the beginning of recorded history.

Today, the majority of human beings spend a dominant part of their lifetime in urban environments. From a material and practical point of view, the urban environment includes built-up structures and their soft and hard infrastructures including buildings, roads, walkways, squares and urban green space. In its entirety, the aforementioned can be described as the basic, physical and material elements of the human settlements.

An understanding of the physical characteristics of human settlements globally, especially urban settlements, is absolutely critical for a large number of local and global issues, including climate change, biodiversity, crisis management and disaster risk reduction, housing and urban development, and poverty reduction. However, despite their importance and a long history of advances in survey techniques, basic information about human settlements globally remains unknown. For example, how many square/cubic meters of built-up space, in both horizontal and vertical dimensions, have been built in the last 40 years? What are the spatial and temporal patterns of urban development? Who lives where and do they have access to basic infrastructure services? What are the material living conditions and population densities of different human settlements around the world? "We may believe we know the single branch of the tree on which we sit, but we ignore the tree and the forest within which it grows" (GEO GHS Working Group, 2014).

If collected by census survey techniques and if made available, these data are very local, while large inconsistencies and large data gaps exist at the global level inherited from different national standards, nomenclature and resource availability for census data collection (United Nations Statistics Division, 2012). Similar issues are related to fine-scale land surveys and cadastral data: they are very expensive to be produced, collected and managed. From the global surface perspective, they are a rarity rather than the normal. Large-scale land surveys using standard remote sensing technologies can map the global land mass at low costs, but only using low spatial resolution (kilometric scale) sensors in input, thus introducing large bias in the description of human settlements.

The density, heterogeneity and dynamics of human settlements and their interactions with the environment are fundamental pieces of information we need to have in order to understand how to keep within the regenerative capacities and limits of our planet. However, the current picture of the human footprint is incomplete.

The majority of small and medium-sized settlements, critical for accounting and understanding the impact of people on the globe, remain largely invisible. The big dots may be visible, but not the all-important connections between them. And, the truly vulnerable such as those dwelling in refugee camps, informal settlements and slums are effectively missing from our global understanding of urban settlements.

(GEO GHS Working Group, 2014)

The Global Human Settlement project (GLOB-HS) supported by the European Commission, Joint Research Centre has the objective to design and test new technologies to generate global fine-scale representations of the physical characteristics of human settlements. In particular, the GLOB-HS project is focused on innovative automatic image information extraction processes, using metric and decametric scale satellite data input (Pesaresi et al., 2013). The target information collected by the GLOB-HS project is the built-up structure or 'building', aggregated in built-up areas and then settlements according to explicit spatial composition laws. They are the primary sign and empirical evidences of our human presence on Earth's surface that are observable by current remote sensors. As opposed to conventional remote sensing practices based on concepts of land cover 'urban' or 'impervious surface', the GLOB-HS approach is a continuous quantitative measure centered around the presence of buildings and their spatial patterns, making the information gathering independent from any abstract definition of rural versus urban (Pesaresi and Ehrlich, 2009). The GLOB-HS project assumes an inclusive concept of 'building', including temporary structures observable in refugee camps and internally displaced people camps, and poor structures of informal settlements and slums. From the GLOB-HS methodological perspective, automatic information gathering processes are the necessary conditions for sustainable global detailed surveys, but also for the reproducibility and public control of the information, thus contributing to the objectivity and evidence-based support to the decision processes. In 2015, the GLOB-HS will produce the first global map of the evolution of human settlements using Landsat data for the years 1975, 1990, 2000 and 2014 with a spatial resolution ranging from 15 to 75 meters.

References

GEO GHS Working Group. (2014). *Statement for a Global Human Settlement Partnership.* Retrieved from GEO Group of Earth Observation: www.earthobservations.org/ghs.php

Pesaresi, M. and Ehrlich, D. (2009). Chapter 3 – A methodology to quantify built-up structures from optical VHR imagery. In P. Gamba and M. Herold (Eds.), *Global Mapping of Human Settlement Experiences, Datasets, and Prospects* (pp. 27–58). Boca Raton, FL: CRC Press.

Pesaresi, M., Huadong, G., Blaes, X., Ehrlich, D., Ferri, S., Gueguen, L., . . . Syrris, V. (2013, Vol 6). A global human settlement layer from optical HR/VHR RS data: concept and first results. *IEEE Journal of Selected Topics in Applied Earth Observations and Remote Sensing*, 2102–2131.

United Nations Statistics Division. (2012). *Housing.* Retrieved from http://unstats.un.org/unsd/demographic/sconcerns/housing/default.htm

Opportunity #3: New modes of creating and sharing knowledge

New forms of data and analyses are flattening the world of knowledge creation. Consequently, we are in an era where the co-production of knowledge is increasingly the norm, challenging the conventional model where knowledge flows one way from scientists and researchers to stakeholders and decision-makers. Rather than the 'science to policy interface,' we now have new concepts about 'co-designed research' that involves active engagement and collaboration among scientists and stakeholders. These new modes of creating knowledge will require including new perspectives into research as well as dismantling traditional models of the scientific process. Importantly, for urbanization science, these new modes of creating and sharing knowledge are likely to produce more actionable knowledge and accelerate the uptake of science into decision-making. It will fundamentally change science-policy interfaces and hopefully lead to more science-informed urban futures.

Final thoughts

Embedded in the connections between urbanization and global environmental change are the promise and prospect of more sustainable pathways of urbanization. Since the beginnings of human civilization, people have clustered in groups, communities and other forms of settlements. Over time, human settlements have become increasingly socially organized, physically structured, economically specialized and urban. Human settlements have helped meet our needs for security, livelihood and social exchange and provided the mechanisms to realize these needs via customs, norms and laws. Beyond these fundamental needs, cities are centers for innovation and creativity, arts and culture, theaters and museums, music and dance. Cities also have become associated with conspicuous consumption, waste and pollution.

The history of cities and urbanization in many ways is the history of the human experience. It is profound that the connection between people and their cities has become evermore critical not only for the reproduction of the everyday practice, as expressed in urbanized areas throughout the world, but also now for the global community and the world's biophysical systems. The chapters in this volume speak directly to these ongoing tensions and how they are captured by the interactions between urbanization and global environmental change. The chapters go beyond simple description and speak of urbanization *in* global environmental change and global changes *in* urbanization. It is through this dialogue that key questions of sustainability may be realized and effectively answered.

INDEX

adaptation 411–12; climate adaptation planning 98, 336–47, 350–9, 427; climate justice 331; Durban (South Africa) 537–41; ecosystem-based 204, 206, 209; gender 312; high-income cities 396–8; resilient infrastructure 490–4; spatial planning 368–73; transformation and transition 528–30; urban greening 455, 457; urban/peri-urban agriculture 433; vulnerability/hazard 215–17, 223–5; water/hydrology 268, 270–1

adaptive capacity 206, 216–22, 314, 333, 336, 339–47, 357–8, 365, 421, 424–6, 434, 493–4, 528–32, 538–44, 551

Africa: adaptation 340–3, 350, 539–40; agriculture 302, 423–9, 434; biodiversity 541–4; construction 81; diet 29–32, 37, 298, 422; ecosystem services 147–8; employment 297; energy consumption 108–11; extreme events 237–8; heat storage 175; mitigation 413–14; renewable energy 460–1; sanitation 297; urbanization 51–4, 140–2, 329, 406, 441, 562–8; urban vulnerability 221, 340, 351; water 255, 263–8, 296, 430

agglomeration 9–21, 52, 56, 110, 157–8, 472–5, 570

agricultural land loss 43–4, 69–71

agricultural production, environmental impacts 29–31, 45–7, 433

air pollution 81–2, 182, 215–16, 220, 458, 496, 519

aridity 99–101, 144, 191, 231, 286

Asia: adaptation 340–6; conservation 147–8; construction 81–2; consumption 127; density 443; diet 29–32, 34, 37, 297, 422; economy 21, 109, 112; employment 297; energy use 83–5, 99, 108–9, 184, 443; extreme events 232–7, 338–40; GHG emissions 109–12; industry 443; informal settlements 54, 351; renewable energy 443;

Southeast 21, 32, 85, 129, 232, 240, 263, 406, 460; transportation 386, 443, 557; urban agriculture 426–9; urbanization 42, 51–4, 60, 67, 129, 221, 329, 350, 406, 441–2, 568; vulnerability 217, 221, 232, 240; water 99, 184, 240, 263, 429

Bangalore (India) 66

Beijing-Tianjin-Hebei 44, 478

biodiversity 27, 139–49, 203, 423, 441–52, 463, 534–6, 541–4

Boston 191, 269

buildings: climate 169–71, 182–3; density 57, 443; ecosystems 205; energy use and emissions 115–17; high income cities 380–1, 385–8; posts-carbon 520–2; resource use 77–87; spatial planning 367; urban form 564; urban greening 285, 462–4

built environments 3–4, 28, 31, 35–8, 77–87, 107, 115–18, 158, 206, 224, 229, 283, 327, 352, 355, 365–8, 380, 471, 490, 519, 549, 562–4

C40 377, 397, 503, 538, 556, 558

Cape Town 460

Caracas 132

carbon cycle 107, 180–1, 190–3

carbon sequestration 181, 367, 427–8, 451–2

Chennai (India) 65–7

China: agricultural land 42–8; built environment 562; energy use/emissions 110, 112, 115, 328, 550; extreme climate events 229, 241; food consumption 29, 34, 36; future urbanization 569–71; industrial symbiosis 470, 478–80; post carbon cities 516, 519–20, 522; resource use 79–82, 84; spatial planning 371; suburbanization 128–9, 133; urban density 51, 52, 57; urban expansion 140, 145–6; urban greening 458–9;

576

Index

urban precipitation 160–1; water 31, 255, 257, 265, 267

climate 3, 8, 45, 50, 74, 77, 84, 95, 103, 113–16, 144, 152–3, 158, 162–3, 169–70, 172–84, 189–90, 196, 201, 213–14, 220, 229–31, 235–7, 255, 266, 277, 279, 282–3, 312, 314, 340, 342, 371, 395–6, 423, 427, 431, 445, 460, 520, 535; adaptation planning 98, 336–47, 350–9, 427; governance 354, 536; hydroclimate 152, 158, 161; justice 331–3; microclimates 152–3, 204–5, 264, 427; models 154, 163, 232, 267, 269–71, 421, 429; policy and planning 4, 20, 148, 310–20, 330, 337, 342, 345, 354, 407, 409, 412–15, 434, 535, 543, 550, 555–8, 565; resilience 310, 321, 325, 338, 346, 412–25, 434, 464, 538, 558; risks and hazards 216, 221, 230, 238–41, 271, 284, 311, 314, 328, 354, 357–8, 368, 408, 411, 414, 424, 429, 515, 538, 550, 562; scholarship and research 153–4, 313, 434, 559; sensitivity 93, 99, 102; uncertainties 99, 102, 268, 422

climate change 3–4, 10–12, 18–23, 27–36, 69, 74, 83, 93–85, 100–2, 140, 144–8, 152, 158, 162–3, 169, 180–4, 201–4, 213–25, 229–31, 240–1, 247–8, 262, 266–72, 276, 282–9, 296–9, 306, 310–26, 328–32, 336–41, 345, 350–9, 364–73, 377–98, 406–15, 421–2, 447–55, 471, 488–93, 501–3, 514, 525, 535–44, 549–59, 562–73

climate change adaptation 206, 223–5, 337–9, 357, 368–9, 433, 537–42, 549

climate change mitigation 20, 325, 366–7, 371–2, 377–97, 406–15, 427, 537, 549–59, 565

climate variability 229–31, 236, 357, 369, 435

co-benefits 277, 286, 316–17, 320, 329, 367, 370, 383, 395, 409, 411, 424, 428, 452, 459–65, 520, 527

conservation 44, 83, 441–54; biodiversity 145–7, 203, 207, 368–9, 457; industrial symbiosis 472, 542–3; water 95–8, 539

consumption: energy 19–20, 57, 78, 84, 106–17, 126, 179, 190–1, 194, 253, 257, 315–16, 321, 328, 366–7, 372, 377, 383, 386–8, 392, 413, 443, 456–7, 517, 553–6, 564; food 27–38, 194, 297–301, 425, 431; patterns 7, 10–11, 15, 22, 60–1, 77–82, 86, 125–35, 147, 196, 232, 276, 284, 296, 312, 315–19, 366, 407–10, 422, 471–4, 480–4, 514, 520, 550–2, 559, 575; water 94–102, 126, 247, 249–50, 254–8, 270, 430, 443, 473

COP21 6, 556

co-production 513–24, 530, 562, 575

cultivated land 31, 42–5

Dallas 250

Delhi 67

density 12, 54–8, 85, 98, 113, 116, 128, 221, 269, 281, 287, 367, 382–3, 392–3, 443, 503, 521

developing cities 51–2, 160, 518, 549–59

developing country cities 299, 549–50

Dhaka (Bangladesh) 82

diet 28–30, 36, 258, 293–5, 304, 422, 428

disaster risk reduction 223, 311, 337, 352, 357, 540, 573 *see also* extreme events (disasters)

drought 95, 98–100, 229–30, 236–41, 254, 266–7, 283, 340, 408, 504

Durban (South Africa) 463, 537–41

economic growth 9–23, 64–7, 69, 111–12, 221–2, 382–3, 413–15, 455–6, 513–14

ecosystem services 144–8, 203–10, 276, 279, 282, 288, 329, 372, 414, 421, 432–3, 441–52, 461–4, 542

embodied energy 79–81, 82–5

energy: consumption 19–20, 57, 78, 84, 106–17, 126, 179, 190–1, 194, 253, 257, 315–16, 321, 328, 366–7, 372, 377, 383, 386–8, 392, 413, 443, 456–7, 517, 553–6, 564; demand 223, 311, 337, 352, 357, 540, 573; infrastructure 84–7, 106, 109, 115–18, 205, 229, 231, 234, 237, 311, 318, 378–84, 395, 413, 497, 555, 562; use 18–19, 77, 82–6, 106–19, 175, 257, 365–7, 382, 388, 427–8, 459, 515

energy-water nexus 98–9

environmental justice 119, 148, 216, 327–33, 464

Europe: adaptation 338–43, 353; conservation 83, 147; consumption 128–34; density 443; diet 32, 34, 424; economy 413; ecosystems 204; energy use 108, 115, 384, 387, 459, 475, 538; extreme events 95, 235, 237–9, 278, 328; gender 312, 315–17; GHG emissions 110, 366, 389, 391, 394; heat storage 175; housing stock 381; infrastructure 501–2, 516; nutrient cycling 195; precipitation 158–61; recycling 330; transportation 390; urban nature agriculture 426–9; urbanization 42, 52–8, 78, 95, 140, 169, 441, 464, 563, 567; water 96, 250, 267–8

exposure 3, 134, 152–3, 171, 206, 214–16, 220–3, 231–3, 238–40, 267, 278–9, 287, 313–14, 337, 424, 427, 430, 494–5, 503, 520

extreme events (disasters) 3, 27, 32, 162, 201, 208, 221, 224, 229–31, 234–41, 254, 266–7, 272, 315, 338–40, 357, 368–9, 424, 427, 456, 462, 488–96, 500–5, 550–1, 562; cyclones 230, 234; drought 230; flooding 208, 224, 240, 424; heatwaves 208, 230; landslides 233–2343; precipitation 230–3; storm surges 224 *see also* disaster risk reduction

flooding: ecosystem services 207–9, 445; extreme events 230–5, 237–40; food forest 36; gender and equity 314–15; impermeable surfaces 144; power loss 497; urban and peri-urban agriculture 423–4, 427, 430; urban precipitation 152–3, 158; vulnerability and risk 216–21, 463; water quality 264–7; waterscape urbanism 329 *see also* hydrology; water

Index

flux and cycling (energy, nutrient) 114, 188–97, 266 and 272, 432–5, 444

food: consumption 27–38, 298–301, 425, 431; infrastructure 27, 30–5, 202, 298, 304–5, 421; markets 249, 298–9; prices 293–305, 423; security 35–6, 46–7, 421–4, 429–31

gender 27–8, 201, 214–19, 310–21, 331, 338

global change 4, 93, 158, 208–10, 268, 325, 525, 563–5, 567, 573

global environmental change 1–6, 7–8, 10, 12, 18, 20–3, 28, 50, 93, 101, 125, 133, 190, 196, 201–2, 208–10, 218, 271–2, 310, 325, 331, 364, 424, 429, 434, 479, 525, 534–6, 539–40, 544–5, 561, 564–6, 575

globalization 125–34

governance 67, 96, 98, 118–19, 145–8, 223–4, 336–47, 354, 368, 515, 530–1, 535–43, 549, 552–3, 567

greenhouse gas emissions 30, 77–81, 106–19, 181, 282, 408

green infrastructure 83, 202, 277, 280–9, 329, 365, 443–52, 500–5

greening: economy 22, 413, 456–9, 466; emissions 389, 412–14; health 202, 276–89, 383; infrastructure 502; irrigation 170; urban greening 427, 455–6, 462–6

habitat 10, 126, 139–49, 203–7, 219, 231, 288, 329, 431, 441–50, 463, 500, 534

Halifax 477

health: biodiversity 7; ecosystem 232, 328, 333; healthcare 294–6, 570; watershed 96 *see also* public health

high-income cities 377–98

hotspots 52, 139–48, 189–90, 195, 254, 258, 442, 541

housing: agricultural land 45; economic ties 111; fast-growing cities 350–1; high-income cities 381–3; justice 328; low-density 98; low-income cities 412; spatial configuration 54, 56, 563; spatial planning 366–7, 371; suburban 125–35; timber 82; urban greening 285–8, 460, 465, 502; vulnerability 216, 220

hunger 295–7, 299–303, 422, 431

Hyderabad 65–6

hydrology 7, 93, 144, 152–3, 160, 180, 219, 233, 235–9, 252, 258, 262–9, 272, 427, 561–3 *see also* flooding; water

ICLEI 338–9, 377, 397, 408–9, 503, 536–41, 553, 556, 557

impervious surfaces 50, 55–6, 162, 182, 190, 230, 264–5, 270, 367, 462

industrial symbiosis 470–85, 521, 542–3

inequality: gender 314–17, 321; income 20–1, 133–5, 215, 229–33, 238–41, 314–17, 341, 552, 558, 567

informal economy 302, 318, 567

infrastructure: adaptation 337, 346, 351–2, 365–9, 396, 411, 488; blue 206; energy 84–7, 106, 109, 115–18, 205, 229, 231, 234, 237, 311, 318, 378–84, 395, 413, 497, 555, 562; food 27, 30–5, 202, 298, 304–5, 421; gray 206–9, 278, 287, 329, 447; green 83, 202–9, 276–7, 281–9, 326, 329, 365, 443–52, 464, 488, 500–5; mitigation 377–8, 383–4, 389, 392, 395, 397–8, 415, 457, 569; natural 447–9, 535, 544; observational 161; resilient 488–506; soft and hard 513–22, 573; transportation 65, 69, 116, 128, 205, 229, 233, 311, 318, 378–80, 386, 389–90, 392, 395, 397, 407, 424, 496, 500, 503, 562–3; urban 3–4, 9–10, 14, 17–18, 27, 31–3, 45–7, 52–4, 57, 67–9, 73, 77–8, 81–7, 93–5, 106, 109, 115–18, 127–33, 158–60, 169, 175, 183, 203–9, 214–21, 230, 237–8, 241, 279, 284, 299, 303, 314, 318, 325–8, 336, 350, 356, 365–8, 371, 378, 380–1, 392–3, 398, 407, 412–13, 457–8, 462–5, 471–2, 480, 484, 488–506, 514–15, 527, 534, 549–52, 555–8, 562–5, 568–70, 573; vulnerability 152, 158, 172, 201, 209, 219, 221–3, 229, 233–5, 408, 427; water 47, 93–5, 98, 101–2, 116, 127, 205, 209, 220, 223, 247, 252–8, 262–6, 272, 299, 311, 318, 338, 496, 500, 544, 555, 562

IPBES 146, 203

IPCC Fifth Assessment Report (AR5) 366, 538, 549, 555, 558

irrigation 29–31, 98–9, 179–80, 248, 255, 428, 430–1, 433

Jakarta 129, 132

Kenya 31, 32, 300, 426–7, 429, 471

knowledge spillovers 11, 472

Kolkata 64–72, 67, 71–4, 233

Kuala Lumpur 129

land cover change 42–6, 64–6, 69–71, 140, 190–2, 470

landslides 220–1, 234–6, 238, 240, 427

land use: adaptation planning 337, 342, 352, 412, 528, 530; biodiversity 145–8; conflicts 133; conservation 442; energy consumption 20; food consumption/production 37, 43–7, 302, 422, 422–3, 432–3; GHG/C/N 188–91, 194–5, 383, 397, 408, 553; green infrastructure 287, 501; hazards/urban vulnerability 152, 216, 224, 233, 235–6; India 64–6, 69–71, 73; infrastructure 517, 520–1; livelihood 456; mixed 366, 383, 563; physical urban growth 18, 219–20; resource use/energy consumption 84, 86–7, 117; spatial planning 364–76, 391–3, 411, 413, 555, 557–8;

578

Index

suburban 134, 175; urban dynamics 4, 50–63; water management/quality 98, 262, 264–5, 267–70; zoning 513
LEED certification 84, 387, 396
Lifecycle 116, 346, 473, 515, 519
lifecycle analysis (LCA) 86
livelihood 2, 47–8, 50–1, 61, 69, 74, 82, 147, 201, 207, 216, 224–5, 229, 303, 306, 314, 318, 325, 337, 351–2, 356, 408, 421, 424, 425–9, 434, 465, 575
lock-in 85, 116, 380–3, 513–22, 562
Los Angeles 569
low carbon transition 525, 554, 558–9
low-income cities 248, 297, 332, 406–15, 424

Manchester 279
Material and energy flow analysis (MEFA) 482
Material Flow Analysis (MFA) 86
metabolism 176–8, 191, 514–22
metropolitan 2, 15–21, 44, 67, 84–7, 108, 117, 132–5, 152, 202, 251, 277, 286, 297, 407, 412, 443, 476–8, 477, 485; Bangalore 66; Beijing-Tianjin-Hebei 44, 478; Boston 191, 269; Cape Town 460; Caracas 132; Chennai 65–7; Dallas 250; Delhi 67; Dhaka 82; Durban 463; GDP 17–21; GHG emissions 18–20; Halifax 477; Hyderabad 65–6; Jakarta 129, 132; Kolkata 64–72; Kuala Lumpur 129; Los Angeles 569; Manchester 279; Mexico City 80, 129; Miami 133, 250; Minneapolis-Saint Paul 116; Monterrey 80; Mumbai 66–7, 554–5; New York 478, 496; Pune 66; Rio de Janeiro 236; Sao Paulo 231, 236–8; Tokyo 388, 478; water use 238, 250–4, 269
Metropolitan Meteorological Experiment 153
Mexico City 80, 129
Miami 133, 250
microclimates 152–3, 204–5, 264
Middle East: Arab Spring 332; energy use 108; GHG emissions 110; water supply 255
migration 10–12, 22, 42, 45–8, 52, 66–8, 73–4, 133, 140, 241, 256, 306, 310, 541, 570
Minneapolis-Saint Paul 116
Monterrey 80
Mumbai 66–7, 554–5

Nationally Appropriate Mitigation Actions (NAMAs) 406–7, 409–11
New York 478, 496
Nigeria 160, 299, 424, 427–8, 471
nitrogen 79, 180, 193–5, 263, 265, 268–9, 428
nitrogen cycle 193–5, 276, 535
nutrient gradient 189–90

pathways 2, 3, 6, 7, 112, 139, 153, 162, 201–2, 210, 237, 265, 271, 293, 306, 328, 422, 495, 517; adaptation 336–9, 397, 406, 414, 424–5, 433;

urbanization 213, 217, 406, 520, 534, 551, 558, 562–3, 575
peri-urban 44, 47, 58, 129, 147, 277–8, 284–5, 302, 372, 407, 414, 421–34, 459, 461, 463
planning: adaptation 331, 336, 337–48, 350–9, 369, 412, 447, 492, 530, 535, 537; business 448, 452; climate change 4, 20, 148, 313, 325, 354, 407, 409, 415, 543, 550, 557; conservation and biodiversity 83, 142, 147, 442, 448, 539, 541–3; emergency 234, 239, 241, 328, 502, 562; energy 554; land use 20, 37, 84, 98, 145–6, 216, 364, 369, 372, 383, 397, 501, 555–7; resilience 208; spatial 237, 325, 364–73, 382, 391–3, 398, 409–13; transportation 20, 317, 371, 383, 389, 410, 557; urban-rural 47–8; water 254, 258 *see also* urban planning
population density 18–20, 51–2, 57, 113, 223, 238, 248–55, 367, 383, 449, 475
population growth 29, 51–2, 66–8, 73–4, 93, 101–2, 126–7, 129, 145, 262–3, 287–8, 379–81, 441
power laws 22
precipitation 29, 50, 95, 152–63, 171, 214–15, 231–2, 253, 263, 268–70
public health 21, 29, 33, 36, 47, 50–2, 74, 82, 96, 144, 158, 161, 179, 196, 202, 207, 215–16, 220, 229–30, 236–40, 276–89, 294–6, 301, 305, 310, 315–17, 320, 328–30, 356, 370, 383, 422, 426, 430–5, 444–5, 458–66, 471–2, 480, 495, 500, 514, 518–21, 553, 557, 562, 569 *see also* health
Pune (India) 66

race 60, 311, 329
rapidly growing cities 69, 82, 112, 350–9, 549
resilience: adaptation planning 338, 351, 358, 527–30; ecosystem services 204, 208–10, 447; extreme weather events 153, 372; flooding 329; food systems 32–3; gender-based 317–18; infrastructure 488–505; urban 215–18; urban and peri-urban agriculture 421–34; urban greening 280, 463–4
Rio de Janeiro 236
risks: agricultural lands 44; climate change adaptation 336–43, 352–4, 357–8, 368–71; ecosystems 445, 462; ecosystem services 207, 209; extreme events 229-241 213–25; food supply chains 32–3, 299; infrastructure 493–4; low-income cities 408–11; precipitation 152–8; urban and peri-urban agriculture 423–7, 429–30; urban greening and health 277–84; vulnerability 213–25; water supply 258
rural-urban gradient 45, 47, 57–60
rural-urban migration 45–6, 66–8, 71–2, 126–7, 256

Sao Paulo 231, 236–8
scaling 15, 22, 109–10, 565–6

Index

sea level rise 208, 224, 229–32, 231–2, 328, 346, 408, 424, 553

Shanghai 112, 129–30, 132–3, 232–3, 241, 265, 549, 570

South America: extreme events 235–8; urbanization 52, 238; water 255

spatial planning 52–4, 364–73, 391–3, 409–14

species richness 139–44

sprawl 18, 53, 56, 113, 129, 265, 366–7, 372, 393, 412

suburbs 95, 125–35, 170, 173–5, 264–5, 285–6, 443, 449–50, 477

supply chains 31–5, 206, 250, 424, 496 *see also* teleconnections

sustainability: adaptation and transformation 527–32; agriculture 46–7, 69, 83–4; conservation 441, 448, 452; disaster vulnerability 221, 225; eco-cities 464–6; economics 455–7, 489; energy 108; environmental justice 327–33, 339, 341; food system 36–7, 303; global 2–6, 525, 534–7; green growth 551–2; nutrient 188, 197; resilience 493–4; suburbanization 125, 134; urban and peri-urban agriculture 425, 429, 432, 434; urban greening 461, 464; water 248, 257, 517–18

TEEB 203, 446, 542

teleconnections 3, 12, 22–3, 50, 60, 112, 333 *see also* supply chains

Tokyo 388, 478

transformation 2–3, 10, 14, 21, 42–5, 58, 66, 68, 98, 106, 125, 127, 129–33, 139, 189, 197, 205, 217, 219, 284, 320, 326–7, 333, 370, 383, 386, 391, 397, 409, 412, 414, 426, 429, 433, 514–15, 519, 525, 527–32, 534, 544, 552, 561, 569

transition 2–5, 7, 15, 27–9, 34–7, 42, 47, 58, 64, 66–7, 73, 74, 77, 101, 130–1, 152, 219, 221, 230, 254, 298, 326, 327–8, 332–3, 382, 384, 386, 397, 409, 413, 422, 458, 514–15, 520, 522, 525, 527–32, 554, 558–9, 566

transportation: ecosystems 204–5, 450; emissions 20–1, 108–9, 113, 192, 378, 383, 496; energy use 443; expansive physical urban growth 18; food systems 27, 31–2, 46, 69, 112, 298, 424; fossil fuels 407, 409–13; infrastructure 380, 386–98; nitrogen 193–4; planning 20, 317, 371, 383, 389, 410, 557; population density 57, 60, 128, 130; post carbon-city 516–20, 553, 557, 564–5; rail 499–503, 513; sea level rise 231, 234; spatial planning 364–7, 371–2; urban boundary 471–2, 482–3, 496

UNFCCC 317, 535–8, 540–1, 551, 556

UN Habitat 6, 20–1, 51–2, 54, 127, 134, 217, 221–3, 231, 238–9, 253, 296, 300, 310–11, 338, 346, 350–2, 406–8, 413, 426, 434, 461, 465, 534, 538

urban agriculture 35–6, 301–5, 330, 411–12, 421–35, 463

urban and peri-urban agriculture (UPA) 35–6, 421–35

urban canyon 171–2

urban energy exchanges 172–9

urban expansion 18, 31, 42–4, 51, 64–5, 69–74, 84, 140, 145–7, 190, 367, 391, 463

urban form 3–4, 20, 50–1, 84–7, 115–17, 145; carbon and GHG emissions 190–1, 241, 318, 366–7, 372, 391–2, 520, 522; climate 152, 154; compact 443, 456–7, 562–3; ecosystem services 372, 377–8, 380, 382; radiation 173; transportation 386, 411–12; urban greening 277–8; water 272

urban greening 202, 276–89, 383, 427–8, 455–66

urban heat island 84, 95, 153, 181–2, 220, 278, 286, 427, 475

urbanity 50

urbanization 1–6, 7–8, 9–23, 27–37, 42–8, 50–61, 64–74, 77–87, 93, 106–19, 125–35, 139–49, 152–5, 158–62, 169–84, 188–97, 201–2, 206–9, 213–25, 229, 233, 237–8, 247–58, 262–72, 277–8, 294–9, 305, 310–21, 325–32, 364, 367, 377, 406, 412, 414, 421–34, 442, 455–9, 464, 472–8, 516–17, 520–2, 525, 527, 534, 549–53, 561–75

urbanization science 1, 10, 22–3, 564–8, 575

urban land teleconnections 22, 50

urban planning 4, 37, 65–8, 72–4, 101, 118, 129, 152–4, 201, 210, 241, 284–8, 304, 311, 318, 320, 352–5, 364, 371–2, 394, 412–13, 415, 422, 431–3, 441–3, 446–52, 457, 466, 480, 485, 501, 521, 527, 536, 541, 543, 550, 553, 564, 567, 570; ecosystem services 144–5, 325; equity 332, 351–5; food systems 28, 37; industrial symbiosis 480; infrastructure 83; smart growth 57, 288; spatial planning 366, 371, 431–3; water 158, 220, 254, 318 *see also* planning

urban systems 2–5, 7, 102, 112, 176, 194, 201–2, 214, 248, 325, 336, 355, 377, 398, 409, 474–80, 489, 514–15, 518, 521, 531–2, 555, 565–6

vulnerability: climate adaptation planning 337–8, 340–2, 368–70; climate change impacts 271, 537; climate mitigation 407–8; disaster risk reduction 357–8; extreme events 230–6; flooding and precipitation 144, 153–4, 158; food 297–8; gender 312–15, 318; green infrastructure 286–7; infrastructure 209, 493–5; urban agriculture 421, 425–7; urban vulnerability and risk 213–25

waste 34–5, 109, 116, 196, 238, 250, 257, 301, 313, 330, 378–80, 391–2

wastewater: agricultural land 47; discharge 238; energy consumption 99; high-income cities

Index

391–2; industrial symbiosis 471–3; post-carbon city 247–61, 516, 518, 519, 521; treatment plants and management 195, 204–5, 218, 231, 247–8, 250–1; urban and peri-urban agriculture 428, 430, 433; urban greening 458; water availability/supply 253–8; water quality 262–3

water: consumption 126, 247, 249–50, 254–8, 270, 430, 443, 473; demand 93–103; infrastructure 47, 93–5, 98, 101–2, 116, 127, 205, 209, 220, 223, 247, 252–8, 262–6, 272, 299, 311, 318, 338, 496, 500, 544, 555, 562; quality 262–72; scarcity 31, 47, 248–56; supply 93, 247–58 *see also* flooding; hydrology

zoning: adaptation strategies 352; conservation 446–7; distributed infrastructure 521; flooding 209; residential 134; smart growth 393; soft and hard infrastructure 513; spatial planning 265–9, 371–2; urban and peri-urban agriculture 432–3; urban sprawl 18; vulnerability and risk 223–4; water resources 270

Taylor & Francis eBooks

Helping you to choose the right eBooks for your Library

Add Routledge titles to your library's digital collection today. Taylor and Francis ebooks contains over 50,000 titles in the Humanities, Social Sciences, Behavioural Sciences, Built Environment and Law.

Choose from a range of subject packages or create your own!

Benefits for you
- Free MARC records
- COUNTER-compliant usage statistics
- Flexible purchase and pricing options
- All titles DRM-free.

Benefits for your user
- Off-site, anytime access via Athens or referring URL
- Print or copy pages or chapters
- Full content search
- Bookmark, highlight and annotate text
- Access to thousands of pages of quality research at the click of a button.

REQUEST YOUR FREE INSTITUTIONAL TRIAL TODAY

Free Trials Available
We offer free trials to qualifying academic, corporate and government customers.

eCollections – Choose from over 30 subject eCollections, including:

Archaeology	Language Learning
Architecture	Law
Asian Studies	Literature
Business & Management	Media & Communication
Classical Studies	Middle East Studies
Construction	Music
Creative & Media Arts	Philosophy
Criminology & Criminal Justice	Planning
Economics	Politics
Education	Psychology & Mental Health
Energy	Religion
Engineering	Security
English Language & Linguistics	Social Work
Environment & Sustainability	Sociology
Geography	Sport
Health Studies	Theatre & Performance
History	Tourism, Hospitality & Events

For more information, pricing enquiries or to order a free trial, please contact your local sales team:
www.tandfebooks.com/page/sales

 Routledge Taylor & Francis Group | The home of Routledge books | **www.tandfebooks.com**